D1422980

The
Biology and Taxonomy
of the
Solanaceae

Reports of Linnean Symposia

Speciation in Tropical Environments (Lowe-McConnell) — *Biological Journal of the Linnean Society* Vol. *1* 1969 pp. 1–246

New Research in Plant Anatomy (Robson, Cutler & Gregory) — Supplement 1 to the *Botanical Journal of the Linnean Society* Vol. *63* 1970

Early Mammals (Kermack & Kermack) — Supplement 1 to the *Zoological Journal of the Linnean Society* Vol. *50* 1971

The Biology and Chemistry of the Umbelliferae (Heywood) — Supplement 1 to the *Botanical Journal of the Linnean Society* Vol. *64* 1971

Behavioural Aspects of Parasite Transmission (Canning & Wright) — Supplement 1 to the *Zoological Journal of the Linnean Society* Vol. *51* 1972

The Phylogeny and Classification of the Ferns (Jermy, Crabbe & Thomas) — Supplement 1 to the *Botanical Journal of the Linnean Society* Vol. *67* 1973

Interrelationships of Fishes (Greenwood, Miles & Patterson) — Supplement 1 to the *Zoological Journal of the Linnean Society* Vol. *53* 1973

The Biology of the Male Gamete (Duckett & Racey) — Supplement 1 to the *Biological Journal of the Linnean Society* Vol. *7* 1975

Continued as the Linnean Society Symposium Series

No. 1 — **The Evolutionary Significance of the Exine** (Ferguson & Muller) (1976)

No. 2 — **Tropical Trees. Variation, Breeding and Conservation** (Burley & Styles) (1976)

No. 3 — **Morphology and Biology of Reptiles** (Bellairs & Cox) (1976)

No. 4 — **Problems in Vertebrate Evolution** (Andrews, Miles & Walker) (1977)

No. 5 — **Ecological Effects of Pesticides** (Perring & Mellanby) (1977)

No. 6 — **The Pollination of Flowers by Insects** (Richards) (1978) (*Botanical Society of the British Isles Conference Report, No. 16*)

No. 7 — this volume (1979)

Also published

Botanical Journal of the Linnean Society, Vol. *73*, Nos 1–3 July/Sept./Oct. 1976

Zoological Journal of the Linnean Society, Vol. *63*, Nos 1 & 2 May/June 1978

The Biology of Bracken (Perring & Gardiner)

Sea Spiders (Pycnogonida) (Fry)

Linnean Society Symposium Series   Number 7

The
Biology and Taxonomy
of the
Solanaceae

Editors

J. G. Hawkes
R. N. Lester   A. D. Skelding

Department of Plant Biology,
University of Birmingham

Published for the Linnean Society of London by Academic Press

ACADEMIC PRESS INC. (LONDON) LIMITED
24/28 Oval Road
London NW1 7DX
(Registered Office)
(Registered number 5985 14)

US edition published by
ACADEMIC PRESS INC.
111 Fifth Avenue
New York
New York 10003

Printed in Great Britain by
Henry Ling Ltd., The Dorset Press, Dorchester, Dorset

Foreword

Symposia are expensive events in time, money and effort. Their value can be greatly enhanced by the publication of well edited, attractively produced symposium volumes.

Here we have such a volume, reporting papers given at meetings in Birmingham University, on the economically important plant family, Solanaceae.

The symposium was jointly sponsored by the Linnean Society of London and the Department of Plant Biology at Birmingham University.

Those who took part have an excellent record of the meetings, to which they can refer to refresh their memories. Those who were unable to attend may benefit by reading about the wide range of studies currently of importance in the family. Our thanks are due to the organisers and editors, J. G. Hawkes, R. N. Lester and A. D. Skelding.

The value and impact of such meetings and their consequent publications will be judged by the stimulation they provide for productive research. If the resolutions proposed by the Conference are carried out, I am certain that we shall all benefit materially.

DAVID F. CUTLER

Botanical Secretary
Linnean Society of London

Symposium on

The Biology and Taxonomy of the Solanaceae

held at Birmingham University, 13–17 July 1976

Session I   *Family and generic classification*
                Chairman: J. G. Hawkes

Session II   *Floristics and phytogeography*
                Chairman: W. G. D'Arcy

Session III   *Biosystematics of sections of Solanum*
                Chairman: D. E. Symon

Session IV   *Floral biology and cytogenetics*
                Chairman: C. B. Heiser

Session V   *Evolution, haploidy and polyploidy*
                Chairman: J. G. Th. Hermsen

Session VI   *Incompatibility and genome relationships*
                Chairman: I. Manton

Session VII   *Chemotaxonomy and tropane alkaloids*
                Chairman: J. G. Vaughan

Session VIII   *Steroid alkaloids*
                Chairman: T. Swain

Session IX   *Biosystematics of genera*
                Chairman: A. T. Hunziker

Session X   *Morphology, fine structure and physiology*
                Chairman: J. L. Gentry

Session XI   *Contributed ten-minute papers*
                Chairman: W. H. Eshbaugh

Session XII   *Biosystematics and numerical taxonomy of domesticated species*
                Chairman: C. M. Rick

Session XIII   *Ethnobotany*
                Chairman: B. Pickersgill

Session XIV   *Future collaboration. Newsletter. Resolutions*
                Chairman: J. G. Hawkes

Contributors

ANDERSON, G. J. *Biological Sciences Group, University of Connecticut, Storrs, 06268, U.S.A.* (p. 549)

AVERETT, J. E. *Department of Biology, University of Missouri—St Louis, St Louis, Missouri 63121, U.S.A.* (p. 493)

BERRY, M. *School of Pharmacy, Liverpool Polytechnic, Byrom St., Liverpool, L3 3AF, U.K.* (p. 505)

BESSIS, J. *Laboratoire de Biologie Cellulaire, Faculté des Sciences, Université de Dijon, 21000—Dijon, France* (p. 321)

BRADLEY, V. *Department of Chemistry, Monash University, Victoria, Australia, 3168* (p. 203)

BURTON, D. L. *Department of Plant Sciences, Indiana University, Bloomington, Indiana 47401, U.S.A.* (p. 513)

CHILD, A. *Ryburn, Middle St., Nafferton, Driffield, East Yorkshire, U.K.* (p. 345)

COLLINS, D. J. *Department of Chemistry, Monash University, Victoria, Australia, 3168* (p. 203)

COUTTS, R. H. A. *Department of Botany & Plant Technology, Imperial College of Science & Technology, London, SW7 2AZ, U.K.* (p. 371)

D'ARCY, W. G. *Missouri Botanic Garden, 2345, Tower Grove Avenue, St Louis, Missouri 63110, U.S.A.* (p. 3)

DAVIES, M. E. *Department of Plant Biology, University of Birmingham, P.O. Box 363, Edgbaston, Birmingham, B15 2TT, U.K.* (p. 231)

DEB, D. B. *Indian Botanic Garden, Botanical Survey of India, Howrah 711103, India* (p. 87)

DRYSDALE, R. B. *Department of Microbiology, University of Birmingham, P.O. Box 363, Edgbaston, Birmingham, B15 2TT, U.K.* (p. 237)

DUMAS DE VAULX, R. *Station d'Amelioration des Plantes Maraichères, I.N.R.A., Centre de Recherches Agronomiques d'Avignon, 84140 Montfavet—Avignon, France* (p. 455)

EASTWOOD, F. W. *Department of Chemistry, Monash University, Victoria, Australia, 3168* (p. 203)

EDMONDS, J. M. *Botany School, Downing St., Cambridge, CB2 3EA, U.K.* (p. 52)

ESHBAUGH, W. H. *Department of Botany, Miami University, Oxford, Ohio 45056, U.S.A.* (p. 701)

EVANS, W. C. *Department of Pharmacy, University of Nottingham, Nottingham, U.K.* (p. 241)

FORD, J. E. *Department of Pathology, Scarborough General Hospital, Scarborough, North Yorkshire, U.K.* (p. 237)

GBILE, Z. O. *Forestry Research Institute of Nigeria, P.M.B. 5054, Ibadan, Nigeria* (pp. 113, 335)

GENTRY, J. L. *Bebb Herbarium, Department of Botany—Microbiology, University of Oklahoma, 770 Van Vleet Oval, Norman, Oklahoma 73019, U.S.A.* (p. 327)

GROUT, B. W. W. *Department of Biology, Faculty of Science, North East Polytechnic, Romford Road, London, E15 4LZ, U.K.* (p. 377)

GRUN, P. *Department of Biology, 202 Buckhout Laboratory, Pennsylvania State University, University Park, Pennsylvania 16802, U.S.A.* (p. 655)

GUTTMAN, S. I. *Department of Zoology, Miami University, Oxford, Ohio 45056, U.S.A.* (p. 701)

GUYOT, M. *Laboratoire de Biologie Cellulaire, Faculté des Sciences, Université de Dijon, 21000—Dijon, France* (p. 321)

HAEGI, L. *Waite Agricultural Research Institute, University of Adelaide, P.M.B.1., Glen Osmond, South Australia 5064, Australia* (p. 121)

HAMMOND, H. D. *Department of Biological Science, State University College at Brockport, Brockport, New York 14420, U.S.A.* (p. 357)

HARBORNE, J. B. *Department of Botany, University of Reading, Reading, RG6 2AS, U.K.* (p. 257)

HAWKES, J. G. *Department of Plant Biology, University of Birmingham, P.O. Box 363, Edgbaston, Birmingham, B15 2TT, U.K.* (p. 637)

HEISER, C. B. *Department of Biology, Indiana University, Bloomington, Indiana 47401, U.S.A.* (pp. 513, 679)

HENSHAW, G. G. *Department of Plant Biology, University of Birmingham, P.O. Box 363, Edgbaston, Birmingham, B15 2TT, U.K.* (p. 377)

HEPPER, F. N. *The Herbarium, Royal Botanic Gardens, Kew, Richmond, Surrey, TW9 3AB, U.K.* (p. 131)

HERMSEN, J. G. Th. *Institute for Agricultural Plant Breeding (I.v.P.), Agricultural University, Lawickse Allee 166, Wageningen, The Netherlands* (pp. 445, 647)

HOGENBOOM, N. G. *Institute for Horticultural Plant Breeding, Mansholtlaan 15, Wageningen, The Netherlands* (p. 435)

HUNZIKER, A. T. *Museo Botánico, Universidad Nacional de Córdoba, Casilla de Correo 495, 5000 Córdoba, Argentina* (p. 49)

IRVINE, M. C. *Department of Chemistry, Monash University, Victoria, Australia, 3168* (p. 203)

JACKSON, B. P. *School of Pharmacy, Sunderland Polytechnic, Galen Building, Green Terrace, Sunderland, SR1 3SD, U.K.* (p. 505)

KHAN, R. *Botany Department, Aligarh Muslim University, Aligarh, India* (p. 629)

LESTER, R. N. *Department of Plant Biology, University of Birmingham, P.O. Box 363, Edgbaston, Birmingham, B15 2TT, U.K.* (pp. 285, 615)

McCANCE, D. J. *Microbiology Department, Guy's Hospital Medical School, London Bridge, London, SE1, U.K.* (p. 237)

McLEOD, M. J. *Department of Zoology, Miami University, Oxford, Ohio 45056, U.S.A.* (p. 701)

McNEILL, J. *Biosystematics Research Institute, Agriculture Canada, Ottawa, Ontario, Canada* (p. 679)

MÁTHÉ, I., JR. *Research Institute for Botany, Hungarian Academy of Sciences, Vácrátót, Hungary, H-2163* (p. 211)

MÁTHÉ, I., SR. *Research Institute for Botany, Hungarian Academy of Sciences, Vácrátót, Hungary, H-2163* (p. 211)

MEHRA, K. L. *National Bureau of Plant Genetic Resources, I.A.R.I. Campus, New Delhi, India* (p. 161)

MILLER, L. M. F. *Glaxo Research Ltd, Sefton Park, Stoke Poges, Bucks., SL2 4DZ, U.K.* (p. 231)

NEE, M. *Botany Department, University of Wisconsin, Madison, Wisconsin 53706, U.S.A.* (p. 569)

OMIDIJI, M. O. *Institute of Agricultural Research and Training, University of Ife, P.M.B. 5029, Ibadan, Nigeria* (p. 599)

PANDEY, K. K. *Genetics Unit, Grassland Division, D.S.I.R., Palmerston North, New Zealand* (p. 421)

PARMENTIER, F. *Laboratorium voor Algemene Scheikunde, Rijksuniversitair Centrum Antwerpen, Groenenborgerlaan 171, 2020, Antwerpen, Belgium* (p. 269)

PEARCE, K. G. *Department of Botany, University of Malaya, Pantai Valley, Kuala Lumpur 22–11, Malaysia* (p. 615)

PETERSON, N. *Department of Prehistory & Anthropology, The Australian National University, Box 4 P.O. Canberra A.C.T. 2600, Australia* (p. 171)

PICKERSGILL, B. *Department of Agricultural Botany, University of Reading, Reading, RG6 2AS, U.K.* (p. 679)

PLOWMAN, T. *Botanical Museum, Harvard University, Oxford St., Cambridge, Massachusetts 02138, U.S.A.* (p. 475)

POCHARD, E. *Station d'Amelioration des Plantes Maraichères, I.N.R.A., Centre de Recherches Agronomiques d'Avignon, 84140 Montfavet—Avignon, France* (p. 455)

QUAGLIOTTI, L. *Institute of Plant Breeding and Seed Production, Via Giuria 15, 10126 Turin, Italy* (p. 399)

RAMANNA, M. S. *Institute for Agricultural Plant Breeding (I.v.P.), Agricultural University, Lawicksee Allee 166, Wageningen, The Netherlands* (p. 647)

RAO, N. N. *Ithanager, Tenali-522 201, Andhra Pradesh, India* (p. 605)

REID, W. W. *Chemistry Department, Queen Elizabeth College, Campden Hill, London, W8 7AH, U.K.* (p. 273)

RICK, C. M. *Department of Vegetable Crops, University of California, Davis, California 95616, U.S.A.* (p. 667)

RODDICK, J. G. *Department of Biological Sciences, University of Exeter, Exeter, EX4 4QG, U.K.* (p. 223)

ROE, K. E. *Life Sciences Library, Pennsylvania State University, University Park, Pennsylvania 16802, U.S.A.* (p. 563)

SAWICKA, E. *Potato Research Institute, Mlochów, Poland* (p. 445)

SCHILLING, E. E. *Department of Plant Sciences, Indiana University, Bloomington, Indiana 47401, U.S.A.* (p. 513)

SCHREIBER, K. *Institute of Plant Biochemistry, Academy of Sciences of the G.D.R. 402, Halle (Saale), Weinberg, German Democratic Republic* (p. 193)

SCHULTES, R. E. *Botanical Museum, Harvard University, Oxford St., Cambridge, Massachusetts 02138, U.S.A.* (p. 137)

SEITHE, A. *Mommenpesch 2, D—4150 Krefeld—29 (Hüls), West Germany* (p. 307)

SOWUNMI, M. A. *Department of Archaeology, University of Ibadan, Ibadan, Nigeria* (p. 335)

STEGEMANN, H. *Institut für Biochemie, Biologischen Bundesanstalt, D—3300 Braunschweig, Messeweg 11, West Germany* (p. 279)

SWAIN, T. *Department of Biology, Biological Science Center, Boston University, Boston, Massachusetts 02215, U.S.A.* (p. 257)

SWAN, J. M. *Department of Chemistry, Monash University, Victoria, Australia, 3168* (p. 203)

SYMON, D. E. *Waite Agricultural Research Institute, Department of Agronomy, University of Adelaide, P.M.B.1., Glen Osmond, South Australia, 5064, Australia* (pp. 125, 203, 385)

WESTCOTT, R. J. *Department of Plant Biology, University of Birmingham, P.O. Box 363, Edgbaston, Birmingham, B15 2TT, U.K.* (p. 377)

WHALEN, M. D. *L. H. Bailey Hortorium, 467 Mann Library, Cornell University, Ithaca, New York 14853, U.S.A.* (p. 581)

Preface

This volume is based on an international symposium on the Biology and Taxonomy of the Solanaceae which was organized by the staff of the Department of Plant Biology at the University of Birmingham, U.K., and which took place from 13th to 17th July 1976. It was jointly sponsored by the Linnean Society of London and the Department of Plant Biology of the University of Birmingham.

The chapters presented in this volume follow fairly closely the papers read at the Symposium, though quite a number of them have been re-written and most have been updated where necessary. The volume also includes the Resolutions agreed to during the last session of the Symposium.

Symposia and edited volumes on families of flowering plants have become something of a tradition in recent years. At their best they can be stimulating and innovative. At worst they could become dull and repetitive, following a tradition without the stimulus of new ideas and clearly-felt needs. We sincerely trust that the present volume will not be viewed in that light, as an eclectic mixture of unrelated facts, since our aims before and during the Symposium, as well as when editing this book, have been to try and draw together speakers from a wide range of disciplines, all of whom were working on the Solanaceae. We hoped that the interactions and cross-fertilization of ideas arising from the conference and the informal contacts resulting from the mixing of workers from very diverse disciplines would help to promote a better understanding of the Solanaceae and its biochemical, taxonomic, cytogenetical and evolutionary relationships.

Of the really large families of flowering plants the Solanaceae is perhaps one that is least well understood. It contains some of the most benign plants on the one hand (potatoes, tomatoes, peppers) and some of the most sinister on the other (tobacco, mandrake, henbane). Yet the alkaloids of these and most other genera are proving of considerable value to mankind and could be even more valuable to us as our level of understanding of them grows.

There is, however, a danger that taxonomists, geneticists, cytologists, biochemists and others in clearly defined disciplines will miss the wider implications of their work through ignorance of other studies on solanaceous taxa. We all tend to specialize in a small group of plants or in a narrowly restricted discipline, and the Symposium, just as our Solanaceae Newsletter, was aimed to promote understanding across subject boundaries and to discuss techniques and results arising from their use.

We have re-arranged many of the contributions to the conference in what we hope to be something of a logical progression. To begin with, in the first section on Taxonomy and Floristics, we have presented a general systematic overview of the family (D'Arcy) and its representation in the main world continental masses (Hunziker—South America; Deb—India; Haegi and Symon—Australia; and Gbile—part of Africa). There are of course certain gaps, but the major distribution areas and patterns of variation are covered, with a paper on nomen-clature (Hepper) dealing with this difficult but essential matter.

In the second section we introduce the subject of folk uses and ethnobotanical aspects of Solanaceae, again, as far as possible to cover the main continental masses (Schultes—New World; Mehra—India; Petersen—Australia). Much more needs to be learned in this field, and it is hoped that these contributions will stimulate others to fill the gaps in our knowledge.

From folk and aboriginal uses we pass in sections three and four to the scientific analyses of the chemical components of Solanaceae. This is of course a vast field which two sections of a symposium volume cannot hope to cover adequately. Nevertheless, we believe its value will be clearly manifest, in showing the tremen-dous actual and potential biochemical importance to mankind of this family.

Sections five and six deal with the structure and development of some groups of plants in Solanaceae, again, admittedly from only a few aspects. Whilst listening to the conference papers and editing the chapters we have realized only too clearly just how much we do *not* know in these fields. However, something is better than nothing, and the distinguished contributions will, we hope, serve to stimulate fresh work.

The seventh section on floral biology, incompatibility and haploidy might perhaps have been entitled "cytogenetics", or "reproductive biology". Together with sections two to six, this section attempts to present "disciplinary" or "problem-oriented" themes, and could logically have led on to several sections on plant breeding. Rightly or wrongly, we decided when organizing the conference not to venture into the plant breeding field. We know only too well what an interesting but overwhelming body of research exists on the breeding of potatoes, tomatoes, peppers, egg-plants and several other crops. So the present section is intended as an introduction to cytogenetics and plant breeding, based on the reproductive biology of certain cultigens.

Finally, in sections eight and nine we attempt to illustrate the well-known proverb "the proof of the pudding is in the eating"! How can we use our knowledge described in the previous sections to promote a real understanding of the Solanaceae on an experimental and evolutionary basis. We have therefore tried to group chapters on genera and sections of the family in the eighth section and chapters on the domesticated taxa in the ninth, illustrating the use of at least some of the techniques, themes and disciplines described in the previous ones. These sections are thus syntheses of knowledge and techniques, applied to certain groups, and helping to provide a deeper understanding of them.

Finally, in section ten we have set out the resolutions resulting from our conference, stemming not only from the formal session on this subject but much more from the informal ones by day and by night when we and all our fellow "solanogues" discussed the future of our work.

We should not like to end this preface without paying tribute to our former colleague, David Langley, who acted so cheerfully and efficiently as Conference

Secretary. We should also like to record our sincere thanks to many other colleagues who helped with the conference, especially David Radley, Sarah Marsh and others who grew and continue to grow Solanaceae material for us in our research gardens. Last but not least, our gratitude goes to Sue Davies and Barbara Hamilton for excellent typing and other secretarial help in preparing the manuscripts for publication.

We should also like to thank the British Council, the Commonwealth Foundation and the Royal Society for financial support.

J. G. HAWKES
R. N. LESTER
A. D. SKELDING

Department of Plant Biology
University of Birmingham

Contents

CONTENTS

I. Taxonomy and floristics

1. The classification of the Solanaceae

W. G. D'ARCY

Missouri Botanical Garden, St Louis, Missouri, U.S.A.

The development of thought on the classification of the Solanaceae is traced from classical times to the present day, with particular reference to the boundaries of the family, its relationships to others such as Nolanaceae and Scrophulariaceae, and the general arrangement of tribes and genera.

Phytogeographical data indicate the strong concentration of the family in South America and the tendency to endemism of tribes and genera in distinct areas throughout the world. The value of experimental studies, particularly biochemical ones, in helping to elucidate or strengthen classificatory systems is emphasized.

A detailed taxonomic listing of genera, together with bibliographical references and a list of Floras which include substantial accounts of the family, are included as appendixes.

CONTENTS

EARLY SOLANACEAE: TO 1789

One of the first references to the Solanaceae is in the *Dioscorides Codex* of A.D. 815 where plants of *Solanum nigrum*, *Physalis alkekengi*, and a species of *Mandragora* are illustrated. The *Physalis* was labelled "*Physallis*", but the *Solanum* went under the name *Strychnos*, and the *Mandragora* was not labelled. These were plants of pharmacological repute which reappeared in a number of medieval herbals. By the time of Caspar Bauhin's *Pinax* (1623) the Solanaceae appear as a group including *Solanum* (now under that name), *Atropa*, *Physalis*, and also *Mirabilis* and *Paris* which are placed in other families today. In 1700, Tournefort recognized seven genera, placing those with soft fruit into a distinct group. He recognized both *Lycopersicon* and *Melongena* as distinct from *Solanum*. Feuillée (1725) also recognized *Lycopersicon*, and he noted *Cestrum* (as *Hediunda*).

Linnaeus leaned heavily on the Pinax for generic concepts, and in his *Species Plantarum* (1753) and *Genera Plantarum* (1754) he recognized two groups of what we now call Solanaceae. In the first group, within his Pentandria Monogyna, were *Datura*, *Hyoscyamus*, *Nicotiana*, *Mandragora*, *Atropa*, *Physalis*, *Solanum*, and *Capsicum*. Slightly separate, but still within the Pentandria Monogyna, were *Brunfelsia*, *Cestrum*, and *Lycium*. From his arrangement of genera, he evidently believed these genera to relate closely to *Verbascum*, *Strychnos*, *Chironia*, *Cordia*, and *Chrysophyllum*. Although we do not include these genera in the Solanaceae today, most of them are placed in families that we believe are of close affinity.

The second group of Linnaean Solanaceae were *Browallia* and *Schwenckia*, which he placed somewhat distantly in his Didynamia Angiosperma. Just why Linnaeus separated *Brunfelsia* from *Browallia* in this way is not clear, for they both have four stamens and are now considered to be closely related. The group of genera from Linnaeus' Pentandria Monogyna have come down to us today as as undisputed core-assembly of the Solanaceae, but it was almost a century later before *Schwenckia* and *Browallia* were recognized as members of the family. Linnaeus had material of *Nolana* but did not describe it, leaving it for his son (1762) to incorporate this unusual solanaceous group into the scientific literature. Nine of the 13 genera of Solanaceae recognized by Linnaeus were native or widely cultivated in Europe at that time. An important departure from previous and contemporary practice was Linnaeus' grouping of *Lycopersicon* and *Melongena* under *Solanum*.

Contemporary with Linnaeus and also a student of the Solanaceae was Philip Miller, whose *Gardeners' Dictionary* was published in eight editions of differing content. He recognized the genus *Lycopersicon* to include both potatoes and tomatoes and to be distinct from *Solanum*. A similar proposal emerged about a century and a half later when Börner (1912) proposed the genus *Solanopsis*.

Other workers of the Linnaean era contributed to the understanding of the Solanaceae. Nikolaus J. Jacquin described a number of new species but did not present a new arrangement. One of his genera, *Aquartia*, is today placed in *Solanum*, but his *Scopolia* is still accepted. Jean Baptiste Aublet (1775) described *Witheringia* and *Bassovia*, but his generic concepts were poorly understood, and it is only in recent years that workers have incorporated the nomenclatural consequences of these early names. During this period and slightly later, genera were proposed by Conrad Moench, Friedrich Medikus, John Hill, Duhamel du Monceau, and others, but their names were not validly published.

In 1789 Antoine Laurent de Jussieu published his *Genera Plantarum* where genera were grouped into what we today call families. The Order Solaneae was the eighth in his eighth class, Plantae Dicotyledones Monopetalae, and it is from this arrangement that the Solanaceae as a family dates for nomenclatural purposes. In the family he recognized three groups: those with capsular fruits, those with baccate fruits, and those which were related to the Solanaceae but perhaps not properly part of it. The first of these groups included *Verbascum* (now Scrophulariaceae), and the third group included *Bontia* (Myoporaceae) and *Crescentia* (Bignoniaceae). In all, Jussieu placed 15 now-recognized genera in the Solanaceae, and he followed Linnaeus in placing *Lycopersicon* under *Solanum*. Jussieu also recognized *Browallia* and *Schwenckia* in his Scrophulareae and *Nolana* in his Boragineae.

From the time of Jussieu's *Genera Plantarum*, many botanists have tried to create arrangements within the family to reflect generic and intergeneric relationships. Before considering these, we will move ahead about a century to consider the two frameworks of Solanaceae classification most in use today.

THE SOLANACEAE IN BENTHAM & HOOKER AND IN ENGLER & PRANTL

Two works of the second half of the 19th century still dominate the classification systems of vascular plants today. These are Bentham & Hooker's *Genera Plantarum* prepared in England, and Engler & Prantl's *Die Natürlichen Pflanzenfamilien* prepared in Germany. In the first, the Solanaceae was written by George Bentham and published in 1876, and in the second, the Solanaceae was written by R. von Wettstein and published in 1895. Each listed and numbered all the genera considered to comprise the family, and each proposed a tribal structure. Each treatment provided a short morphological and geographic description of each genus. Bentham recognized 67 genera and Wettstein recognized 76 genera including six genera of doubtful placement. Although the two systems agree in general, there are differences in the placement of many genera. Perhaps the most conspicuous difference is the treatment of *Nolana*. Bentham gave it tribal status under the Convolvulaceae while Wettstein gave it family status next to the Solanaceae. Although a number of genera have been proposed since, there has been reluctance on the part of the taxonomic community to recognize innovations since these important works. Botanists have felt that adjustments are necessary to these systems, but they seem to be waiting for a new revision of the whole family before accepting piecemeal changes.

Wettstein's treatment has received much greater following than that of Bentham although there are those who feel that Bentham's arrangement of genera is more natural (e.g. Hegnauer, 1973). Wettstein's treatment came later and was part of the Engler and Prantl arrangement of families which became almost universal in herbaria and texts outside the British Empire. Engler and Prantl's opus ostensibly incorporated the evolutionary theory expounded by Charles Darwin in 1859, and it was well illustrated.

DUNAL AND DE CANDOLLE

Between the *Genera Plantarum* of Jussieu and the *Pflanzenfamilien* treatment of Wettstein, many botanists made important contributions to the understanding of the Solanaceae, but a central figure was Michel Félix Dunal who worked in collaboration with Augustin Pyramus de Candolle.

Dunal was born in Montpellier, France, in 1789, and he prepared his doctorate under the tutelage of de Candolle who was then director of the Institut de Botanique and botanical garden at Montpellier. The thesis was published in 1813 comprising a treatise on medicinal and economic uses of *Solanum* and a description of morphology, a taxonomic revision of all the 235 then known species, and 26 illustrations of species treated. The illustrations are line drawings prepared by Node-Veran, a noted botanical illustrator of Montpellier.

The political climate of Europe in the early 19th century was turbulent, and soon after Dunal's thesis was published, José Mariano Mociño arrived in Montpellier seeking asylum from political events in Spain. About the same time, de Candolle made preparation to return to his native Geneva as he feared the arrival in Montpellier of Napoleon whom he had displeased some years previously in Paris. Perhaps in the hope that he would be appointed director in de Candolle's stead, Dunal began a revision of his thesis. A summary of this was published in 1816 as the *Solanorum Generumque Affinium Synopsis*. This summary covered 320 species, many of them new, and while the descriptions are fragmentary, they are sufficient to validate the names used. Many of the new species of the *Synopsis* cite plates. These are line drawings by Node-Veran, never published, which are inserted in the general collections at the Institut de Botanique at Montpellier. The drawings appear to have been prepared from specimens now at the Laboratoire de Phanerogamie, Paris, but in the folders in Paris there are often several specimens labelled with the same name, superficially similar but sometimes representing different species. For typification of these names it is necessary to compare the plates at Montpellier with the plants in Paris. In the *Synopsis* there are many other new species based on material of Jean B.A.P.M. de Lamarck and on material of Alexander von Humboldt and Aimé Jacques Bonpland. Lamarck subsequently published his *Encyclopédie Méthodique* (1793, 1804, 1813) and Supplements, elaborating slightly on Dunal's descriptions. Humboldt & Bonpland subsequently published their *Nova Genera* (1818) with ample descriptions and some plates, but the promised second edition of Dunal's thesis never appeared. When I visited Montpellier in 1972, M. Granel turned up what we take to be the second edition manuscript at the Bibliothéque de l'Academie des Sciences in Montpellier. This consists of a copy of the *Synopsis* with numerous fragmentary marginal notes by Dunal.

When Mociño came to seek refuge with de Candolle in 1817 he brought with him a large tome of colored plates done in Mexico by Echeverría, La Cerda, and other artists. This was the product of the royal expedition of Charles III to the New World which had lasted from 1787 to 1803. When de Candolle soon left for Geneva, Mociño asked that he take the drawings with him, but after a short time, Mociño had a change of heart and requested his illustrations back. De Candolle mobilized local artistic talent in Geneva to copy the work before returning it, and the product is still in Geneva, the *Icones Florae Mexicanae*. Tracings have been made of these copies and distributed to many herbaria, but they have never been published. Among the drawings in Geneva are a number of Solanaceae done by Node-Veran, indicating that copying had begun before de Candolle's departure from Montpellier, perhaps under the influence of Dunal. Sesse & Mociño's *Plantae Novae Hispaniae* and *Flora Mexicana* did not appear in print until the end of the century (1887–91, 1891–97), but several botanists published new species, some of them Solanaceae, based on the drawing-copies in Geneva. These authors never saw the original drawings nor the plants from which they were drawn, many of which are now at Madrid. Mociño's tome of drawings has disappeared.

Although Dunal was not appointed to succeed de Candolle, the two men apparently remained friends, for Dunal was asked to contribute the Solanaceae treatment to the *Prodromus*. This appeared in 1852, some years after the death of de Candolle in Geneva in 1841. Dunal visited Paris again in 1845 when he reviewed his earlier work. The treatment in the de Candolle *Prodromus* is the last published attempt to revise all the species and genera in the Solanaceae. It included 920 species of *Solanum* and revised 59 other genera. For many groups it is still the latest revision available. In the *Prodromus*, Dunal included *Nolana* in the Solanaceae. *Salpiglossis* and related genera were treated separately by George Bentham in the Scrophulariaceae (Bentham, 1846). The *Prodromus* is one of the most influential works to appear on the Solanaceae in the 19th century.

In the herbarium at Montpellier are numerous fragments of Solanaceae annotated by Dunal in the years around 1846. Either Dunal visited Geneva and took them as he was preparing the treatment for the *Prodromus*, or more likely, he was sent all the Geneva material and retained fragments for his later study. In any event, this raises the question of which specimens should be used to typify the species of Solanaceae in the *Prodromus*. To my knowledge, no one has tried matching the plants to verify that they represent the same species.

AFTER JUSSIEU

Soon after the recognition of the Solanaceae as a family by Jussieu in 1789, Adanson (1763) in France, and Necker (1790) in Germany published general systems which recognized some Solanaceae but made no innovations. Ruiz & Pavón (1794, 1798a, b) published important works on the flora of Peru and Chile which recognized new genera, e.g. *Fabiana*, *Juanulloa*, *Nierembergia*, *Saracha* and *Sessea* and many new species, but they followed Linnaean procedure and did not recognize a family Solanaceae. From these efforts on, a steady flow of major and minor works added new genera and species to the known inventory of Solanaceae, so that each treatment of the family as a whole incorporated new information on the diversity of the group but not always new insight into the relationships of its parts. Table 1.1 lists the major systems that have dealt with the family as a whole.

Table 1.1. Conspecti of the Solanaceae

Brown, R.	1810	Miers, J.	1846–49
H.B.K.	1818	Dunal, M. F.	1852
Nees von Esenbeck	c. 1837	Bentham, G.	1876
Don, G.	1838	Wettstein, R. von	1895
Endlicher, Meisner	1839–41	Baehni, C.	1946

Perhaps the first to propose a new classification of Jussieu's Solanaceae was Robert Brown (1810) who made two major groupings, a "Section" *Solaneae verae* with plicate corollas, stamen number equal to corolla lobes, and strongly curved embryo, versus an unnamed "Section" with non-plicate corollas, didynamous stamens and the embryo only slightly curved. Soon after, Humboldt, Bonpland & Kunth (1818) proposed a "Section" *Capsulares* and a "Section" *Baccatae*, and they divided *Solanum* into a number of groups using polynomials for the divisions. At about the same time, Roemer & Schultes (1819) moved *Nolana* into the Solanaceae. In this period, Pouchet (1829) elaborated on the uses and morphology of the genera known to him.

A major step in ordering the genera of the Solanaceae was presented by George Don, an Englishman, whose *Gardeners' Guide* (1838) was a general review of the plant kingdom. He recognized seven tribes and included 43 genera. Each tribe was given a reasonably ample diagnosis and the names are familiar to modern workers: Solaneae, Nicotianeae, Datureae, Francisceae (*Brunfelsia*), Anthocerceae, Nolaneae, and Cestrineae. Besides his tribal classification, Don presented infrastructure for the genus *Solanum* which embraced 412 species, and also for *Capsicum* (33 species), *Physalis* (34), *Lycium* (34), *Nicotiana* (40), and *Datura* (14). Don separated *Verbascum* from the Solanaceae into a family of its own. Although Dunal and Wettstein are much better known today, the treatment by Don was at least as important, providing a modern basis from which improvements to understanding the Solanaceae could evolve. This system was repeated almost unchanged by Walpers (1844) in his system.

Appearing at about the same time as Don's gardening system was Nees von Esenbeck's work on the German flora. Although the title page reads 1845, it appeared much earlier as Don referred to it in 1838. Nees recognized four tribes, but as he confined his treatment to the ten genera present in Germany, one cannot discern what he might have proposed for a classification of the family in its full amplitude. So far as it goes, Nees' system is similar but not identical to that of Don.

Another system published just after that of George Don was that of Endlicher (1839, 1841) and Meisner (1840) which proposed two "suborders", Curvembryae and Rectembryae, following the embryo distinctions noted earlier by Robert Brown. Under these suborders, or subfamilies, were ranked six tribes and 42 genera. This system differed little from that of George Don, but it assembled a tribe Hyoscyameae from elements Don had placed in the Nicotianeae and Solaneae, and it placed *Nolana* in its own family.

During the years between 1814 and 1840, Constantine S. Rafinesque proposed over 40 genera in the Solanaceae. Few of these are recognized today, and some cannot be definitely identified. An astute taxonomist, Rafinesque often published in obscure journals (see Merrill, 1949) and usually did not elaborate sufficiently on his concepts for ready acceptance. While many of his names have been formally rejected or placed in synonymy, it is always prudent to review Rafinesque's names before proposing new generic segregates.

In 1846 the German, Otto Sendtner, provided the Solanaceae for the *Flora Brasiliensis* of Martius. This work was similar in many ways to the *Flora Peruviana et Chilensis* by Ruiz & Pavon (1798–1802) a half-century earlier. It provided no infra-familial structure but did present a key to the 22 genera recognized. This work separated *Cestrum* and *Metternichia* into a distinct family.

John Miers was an engineer who lived in various parts of South America between 1819 and 1838. Upon his return to England he proceeded to publish his botanical finds which were principally in the Solanaceae. Miers monographed *Lycium* (1854) with 69 species. He published a series of papers in Hooker's *London Journal of Botany* and in the *Annals and Magazine of Natural History* which were reprinted in two editions as *Illustrations of South American Plants*, a two-volume work. Dates of issue of the various parts are somewhat confusing. Miers' publications ran from 1844 to 1871. His illustrations, which he did himself, are excellent, but his nomenclatural practices deviate from those of other botanists, and his names must be inspected with care. A number of taxa he described have not been collected since.

Miers (1848) separated the traditional Solanaceae into three families: the Sclerophylaceae (*Grabowskia*), the Nolanaceae (including *Dichondra*, Convolvulaceae), and the Solanaceae. He later (Miers, 1849) further split out much of the Solanaceae into the Atropaceae, including *Nicotiana*, *Datura* and a number of other genera. Within this Atropaceae he erected ten tribes. In this publication he erected eight tribes for the newly circumscribed Solanaceae. Miers' 1846 publication had placed the tribes in two "suborders", the Rectembryae and Curvembryae, but for his 1849 Atropaceae, Miers drew from both suborders.

By 1852 when Dunal published the Solanaceae for the de Candolle *Prodromus*, there was a large body of information to draw on as well his own long-standing studies. Besides the classifications and floristic sources noted above, there was a large number of species descriptions and a growing number of monographs and morphological studies. Fingerhuth (1832) had monographed *Capsicum* using a quite advanced format. Nees had monographed *Physalis* (1831). Several writers had monographed *Solanum*, adding new species each time. Both Miers (1845) and Sentdner (1845) had revised *Cyphomandra*, Bernhardi (1833) had revised *Datura*, and Lehmann (1818) had monographed *Nicotiana*. There were also general morphological surveys such as Gaertner's (1791) work on seeds and fruits.

Dunal included the Nolaneae, which by now had four genera, as a tribe in the Solanaceae. Within the tribe Solaneae he erected nine subtribes. He followed Miers to some degree, placing *Nierembergia* and *Fabiana* together in a separate tribe, in separating *Cestrum* and *Sessea*, and in placing *Sessea* and *Nicotiana* in a tribe together; but in most respects Dunal's scheme was new.

George Bentham's treatment of the Salpiglossidae in the Convolvulaceae of the *Prodromus* recognized two tribes, one for *Duboisia*, *Anthocercis* and *Schwenckia*, and the other for *Salpiglossis* and six other South American genera: *Leptoglossis*, *Browallia*, *Brunfelsia*, *Heteranthia*, *Salpiglossis* and *Schizanthus*.

It was during the second half of the 19th century that the conspecti of Bentham and Wettstein appeared. Many new species were discovered by botanists exploring new regions, but many of these descriptions did not appear in print until the early part of the 20th century. Von Mueller in Australia, Palmer in Mexico, Fendler in Panamá and Trinidad, Spruce, Rusby, Glazious, Poeppig, Ule, Triana, Bang, Buchtein and others in South America collected large numbers of Solanaceae specimens from areas of great plant diversity. A number of important morphological studies appeared, such as Cauvet's (1864) study of leaf arrangement and chemistry, Payer (1857) on floral development, Eichler's (1875) *Blüthendiagramme*, Chatin (1874) on embryology, Vesque (1885) on leaves, and Solereder (1898–99) on stem anatomy. These were all useful background for the great conspecti of Bentham and Wettstein. The only tribal arrangement between those of Dunal and Bentham was that of Milne Edwards (1864) who made some adjustments to Dunal's scheme.

In the first quarter of the 20th century, the study of the Solanaceae was dominated by the German, Georg Bitter, who between 1912 and 1927 revised *Solanum* in Africa making use of important new collections from German expeditions. Bitter erected a partial infrastructure for *Solanum* as a whole, and he dealt with a number of other genera in varying degree. He was also active in *Solanum* and other genera of South America. A full bibliography was published by Weber (1928), and an index to all species of his African work (Weber, 1931) and to his publications in *Fedde Repertorium* (1896) assist the reader in finding various elements of his work. Bitter was a correspondent of Hassler who established the genus *Lycianthes*. Shortly thereafter Bitter himself monographed *Lycianthes*, segregating some 150 species of *Solanum* into this genus. Although Bitter's original material was destroyed at Berlin during World War II, many duplicates are present in other herbaria.

STUDIES IN THE CURRENT PERIOD

Seithe (1962) published a monumental documented study of hair types and development in *Solanum* and other genera, and she extracted Bitter's classification from the literature and integrated it into a useable conspectus of sections. She added the new rank, "Chorus Subgenera".

The only attempt to arrange the whole of the Solanaceae in the present century was that of Charles Baehni (1946) of Geneva. Baehni studied prefloration of calyces and corollas and he erected a new system suggesting pathways of evolution between the major groups. Baehni worked with about 80 genera. After a study of wood, he excluded *Henoonia* from the Solanaceae, but he included *Goetzea*. He did not study *Nolana*. Baehni recognized five tribes and a number of subtribes. Baehni's system was followed by Pojarkova (1955) but has not drawn much other attention.

During the 20th century, a number of important generic revisions have appeared which are listed in the Appendix. Perhaps most noteworthy are Goodspeed's (1954) study of *Nicotiana*, and Hawkes' (1956, 1963; Hawkes & Hjerting, 1969) and Correll's (1952, 1962) revisions of the potatoes. Other revisions notable for the large numbers of species treated were Bitter (1919) on *Lycianthes*, and Francey (1935–36) on *Cestrum*.

During this period, as earlier, Floras contributed much to our knowledge of the family. Noteworthy for their size and importance were Reiche's (1910) *Flora of Chile*, and Wright (1904, 1906) on Africa. Two of the most recent floristic treatments of the family are Gentry (1974) and D'Arcy (1973) for Guatemala and Panamá. In spite of frontispiece dates, Gentry's treatment was published first. Although not strictly either a Revision or a Flora, Morton's (1976) treatment of *Solanum* in Argentina, published posthumously under the guidance of Lyman Smith and A. Hunziker, notes 132 species and is a landmark in the study of the family in South America. The Appendix notes the most important floristic works for study of the Solanaceae.

In the early 20th century, scientists became aware of chromosomes and their importance for systematic and breeding studies. Many chromosome counts have been published providing part of the basis for major realignment of the family. Jørgensen's work on *Solanum* ploidy levels (1928) was of great use in cytology in general. Blakeslee's (Avery *et al.*, 1959) work on *Datura* was of similar impact. Menzel (1951) studied *Physalis*. Gottschalk (1954, 1956, 1958), von Wangenheim *et al.* (1957), Baylis (1958, 1963), Roe (1966b), Randall and Symon (1975) and many others made important contributions to the cytology of the Solanaceae. Much of this is reviewed by Magoon *et al.* (1961). Hand in hand with cytology is genetics and plant breeding, and the reviews of this field for *Capsicum* by Lippert *et al.* (1966) and for *Solanum* by Howard (1970) are good examples of what can be derived from this sort of study.

Important morphological studies of various sorts have contributed to 20th century knowledge. Seithe's (1962) study of trichomes; Danert (1958, 1967) on inflorescences and (1969) on stone cell concretions; Gunn & Gaffney (1974) on seeds and Murray (1945) on floral vasculature are notable. Schürhoff (1925) studied embryology and Maheshwari (1950) reviewed the study of embryology with good attention to the Solanaceae. The number of studies appearing in this century of systematic import is large. There is now a rapidly growing literature on chemical features of these plants. Somewhat similar in intent, papers on serological comparison such as that of Tucker (1969; see also Hawkes & Tucker, 1968) provide useful insights into relations in the family.

REVISIONS AND ADDITIONS SINCE WETTSTEIN

Wettstein's 1895 treatment of the Solanaceae and Nolanaceae together included 80 genera with 1693 species. Bentham's 1873 treatment included 80 genera with nine of doubtful placement. A few of the genera recognized by Bentham were discarded or submerged by Wettstein; *Himeranthus*, *Leptoglossis*, *Reyesia*, and *Lycopersicon*. In the near-century since Wettstein's treatment, many novelties have been described, and revisions have altered the disposition of a number of genera. There are now about 3600 names in the genus *Solanum* and perhaps 1400 good species, while Wettstein noted only 900 species. The number of genera has not increased greatly, now about 85, but the number of species in the family now stands at nearly 3000, nearly double the number known to Wettstein. When *Solanum* is subtracted, the family includes about 1400 species, still almost double the number known to Wettstein. Appendix I lists all names used for genera in the Solanaceae with an indication of size, geography, subfamily and disposition.

Since Wettstein's synopsis, a number of earlier genera have been revived by various writers: *Brugmansia*, *Lycopersicon*, *Jaltomata*, *Deprea*, *Quincula* and *Witheringia;* and *Lycianthes* has been segregated from *Solanum*. Several new genera, almost all small and of local occurrence, and some of dubious status have been recognized by workers since Wettstein: *Archiphysalis*, *Atropanthe*, *Atrichodendron*, *Combera*, *Hunzikeria*, *Leptofeddea*, *Leucophysalis*, *Pantacantha*, *Parabouchetia*, *Pauia*, *Protoschwenkia*, *Rahowardiana*, *Sesseopsis*. Elsewhere in this volume (Chapter 2), Hunziker has proposed acceptance of several new and resurrected solanaceous genera. The genera *Goetzea*, *Espadaea*, *Coeloneurum* and *Henoonia*, small groups from the Greater Antilles, have been moved to the family Goetzeaceae which appears to have no close relationship to the Solanaceae.

Although Appendix I presents a clear-cut list of the genera now recognized in the Solanaceae, the problem of generic boundaries is far from resolved, and is an impediment to understanding of relationships. In almost every group there is at least some question of generic delimitation, and in some cases, the confusion is large. The genera to be recognized for the species now treated as *Dunalia*, *Acnistus*, and *Iochroma* are recondite and the line between *Saracha* and *Hebecladus* is unclear. Many of the species now placed in *Salpichroa* may belong in a genus of close relationship to *Nectouxia*. Haegi and Gentry comment elsewhere in this Volume (Chapters 5 and 25) on the problems of separating the four Australian endemic genera *Duboisia*, *Anthocercis*, *Anthotroche* and *Isandra* of von Mueller. Under study for some time by myself and by Gentry are the small salpiglossoid genera *Bouchetia*, *Leptoglossis*, *Reyesia*, etc. which are confused as to the taxa to be recognized and the names to be applied. Gentry's pollen analysis may clarify the situation. The physaloid genera are poorly defined: should *Quincula* be separated?; are *Deprea* and *Withania* distinct?; what is the relationship of *Scopolia* to *Chamaesaracha* and *Leucophysalis*? We cannot regard the present list of solanaceous genera as fixed, and there are likely to be name changes in long-established groups as new knowledge helps in clarifying generic limits. Much of this problem will be solved by studying plants occurring far away from the long-botanized areas of temperate Europe and North America.

With recent exclusions of a number of unrelated elements from the Solanaceae, the recognized genera appear to be firmly placed in the family. Only one important question comes to mind: *Schizanthus* was found to have different endosperm from other members of the family (Schürhoff, 1925), it appears to have different pollen (Gentry, this Volume, Chapter 25), and Tucker (1969) found no serological correspondence with other Solanaceous genera he studied. On the other hand, Souèges (1936) stated that, based on embryological considerations, *Schizanthus* belongs in the Solanaceae and not with the Scrophulariaceae or Verbenaceae.

PHYLOGENY AND DIVISIONS WITHIN THE FAMILY

In all, about ten schemes (Table 1.1) have been proposed for ordering the genera within the Solanaceae, and most of them differ in important ways. A recurring theme relates to the curvature of the embryo, which was recognized by Robert Brown in 1810. There is a reasonable division in the family between those genera with straight or slightly curved embryos and those with circinnately coiled embryos. Some useful terminology was illustrated by Gunn & Gaffney (1974), and most genera have been surveyed for this character at some time.

The genera with coiled embryos often have a number of other characters: complicated or primitive floral vasculature (Murray, 1945); accrescent calyces; actinomorphic corollas; five stamens inserted low in the corolla tube; small stigmas; baccate fruits; compressed seeds; an embryo more or less uniform in diameter; a chromosome base number 12, sometimes with euploidy but seldom with aneuploidy, and stenopalynous pollen. The genera with straight or slightly curved embryos often have a number of contrasting characters: reduced vasculature; calyces not accrescent; corollas often zygomorphic; four or fewer stamens inserted high in the corolla tube; elaborate stigmas; capsular fruits; prismatic seeds; cotyledons frequently wider than the rest of the embryo; chromosome base numbers various but seldom 12, and frequent aneuploidy; and eurypalynous pollen. There is also a dichotomy in the alkaloids present in the two groups (see Hegnauer, 1973). These two groups are best recognized as separate subfamilies, the Solanoideae and Cestroideae. They were recognized as such ("suborders") by Endlicher (1841) but his names Curvembryae and Rectembryae do not conform to the current *Code*.

Another recurring theme of Solanaceae classification is the close relationship of *Nolana*. No recent literature argues against this relationship, and many writers have pointed out similarities and some have included it in the Solanaceae. Johnston (1936) believed that *Nolana* and its relative *Alona* evolved from South American members of the Solanoideae. The main difference between these two genera and the rest of the Solanaceae is in the fruit (Johnston, 1936; Ferreyra, pers. comm.), but all other significant characters correspond with those of the Solanaceae. Di Fulvio (1969, 1971) noted similarities in embryology and floral morphology, and Danert (1958) and Figdor (1909) noted similar branching patterns. Because any full scale discussion of the Solanaceae must also consider *Nolana*, and because there are no other groups of plants known to be so closely related, it is appropriate to include *Nolana* and *Alona* within the Solanaceae as the subfamily Nolanoideae. This was done by many workers of the past century and again recently by Thorne (1968).

Hutchinson (1969) recently proposed a family Salpiglossidaceae as an intermediate between the Solanaceae and Scrophulariaceae. This family, which included a number of genera from the Solanaceae, is not supported by chemical and other data.

Table 1.2 notes the distribution of the tribes and subtribes of Wettstein, Bentham and Baehni into the three subfamilies of Solanaceae. Table 1.3 shows how the genera are disposed into the three subfamilies, and Hunziker (this Volume, Chapter 2) has placed the South American genera into tribes. Chemical data may assist in establishing tribal divisions, e.g. withanolide compounds are found only in *Acnistus*, *Nicandra*, *Jaborosa*, *Physalis* and *Withania*.

The Solanoideae includes many genera of undisputed close affinity, *Solanum*, *Physalis*, *Capsicum*, etc., and a number of genera which are not so evidently of

Table 1.2. Subfamilial structure of the Solanaceae

Subfamily	Wettstein (1895) Tribes & subtribes	Bentham (1876) Tribes	Baehni (1946) Tribes & subtribes
I. Solanoideae	Nicandreae	Solaneae	Solaneae
	Solaneae	Atropeae	—Solaninae
	—Lyciinae	Hyoscyamae	—Sarachinae
	—Hyoscyaminae		—Margaranthinae
	—Solaninae		—Physalidinae
	—Mandragorinae		—Iochrominae
	Daturae		Atropeae
			—Atropinae
			—Markeinae
			—Hyoscyaminae
			Nicotianeae
			—Datureae
			—Nicandrae
II. Cestroideae	Cestreae	Cestrineae	Nicotianeae
	—Cestrinae	Salpiglossidae	—Nicotianinae
	—Nicotianinae	(Scrophulariac.)	Anthocercideae
	Salpiglossidae		Salpiglossidae
III. Nolanoideae	Nolanaceae	Nolaneae	not considered
	(family)	(Convolvulac.)	
Excluded (as family)	Cestreae-Goetzeinae		

this alliance such as *Datura*, *Brugmansia*, and *Markea*. *Datura* and *Brugmansia* have prismatic seeds, and the stamens are inserted relatively high in the corolla tube, but a chromosome base number of 12, bent embryos, and some chemical data suggest an origin within Solanoideae. Pichenot (1956) was able to graft plants of *Solanum* onto *Datura* rootstocks. *Solandra*, *Markea*, *Rahowardiana*, *Metternichia*, and *Iochroma* all have somewhat tubular flowers but they have five stamens, actinomorphic corollas, and most have baccate fruits so they are

Table 1.3. Conspectus of the Solanaceae. (D'Arcy, 1975)

I. SOLANOIDEAE

1. Acnistus	18. Hyoscyamus	35. Pantacantha
2. Archiphysalis	19. Iochroma	36. Pauia
3. Athenaea	20. Jaborosa	37. Phrodus
4. Atrichodendron	21. Jaltomata	38. Physalis
5. Atropa	22. Latua	39. Physoclaina
6. Atropanthe	23. Leucophysalis	40. Quincula
7. Brugmansia	24. Lycianthes	41. Rahowardiana
8. Capsicum	25. Lycium	42. Salpichroa
9. Chamaesaracha	26. Lycopersicon	43. Saracha
10. Cyphomandra	27. Mandragora	44. Scopolia
11. Datura	28. Margaranthus	45. Solandra
12. Deprea	29. Markea	46. Solanum
13. Discopodium	30. Mellissia	47. Trechonaetes
14. Dunalia	31. Metternichia	48. Trianaea
15. Exodeconus	32. Nectouxia	49. Triguera
16. Grabowskia	33. Nothocestrum	50. Withania
17. Hebecladus	34. Oryctes	51. Witheringia

II. CESTROIDEAE

52. Anthocercis	62. Fabiana	73. Saccardophytum
53. Anthotroche	63. Heteranthia	74. Salpiglossis
54. Benthamiella	64. Isandra	75. Schizanthus
55. Brachyhelus	65. Leptofeddea	76. Schwenckia
56. Browallia	66. Leptoglossis	77. Sclerophylax
57. Brunfelsia	67. Melananthus	78. Sessea
58. Cestrum	68. Nicotiana	79. Sesseopsis
59. Combera	69. Parabouchetia	80. Streptosolen
60. Dittostigma	70. Petunia	81. Vestia
61. Duboisia	71. Reyesia	

III. NOLANOIDEAE

82. Alona
83. Nolana

tentatively placed in the Solanoideae. *Nicandra* has been given special status by some workers because of its five-locular ovary and its corolla aestivation, but in most features these plants are much like herbaceous species of *Solanum* or *Physalis*, and in any garden population, ovaries commonly range from three to five locular. Tucker (1969) found serological relationships compatible with placement in the Solanoideae. Alkaloid chemistry (see Hegnauer, 1973) tends to speak for a unity of this subfamily with hyoscyamin, hygrin and scopolamin of general occurrence.

The Cestroideae are more advanced in vasculature, chromosome number, stamen and ovule reduction, zygomorphy and other features. It appears that the two subfamilies have long undergone independent evolution. The Cestroideae are strongly concentrated in South America with few representatives in the Old World. Placement of the four Australian genera is uncertain, and they have chemical properties in common with both subfamilies (Barnard, 1952). They are placed in the Cestroideae following the tradition of Wettstein and Bentham, but with considerable reservation. The Salpiglossidae appear to have evolved from a *Cestrum–Petunia–Nicotiana* ancestor, perhaps in several lines, and their separation together into a tribe or other taxon is unwarranted.

THE SOLANACEAE AND OTHER FAMILIES

Bartling (1830) placed the Solanaceae in an alliance of families having sympetalous corollas, epipetalous stamens, and connate carpels. In the Englerian system this alliance, the Tubiflorae, comprised 22 families, including the Convolvulaceae, Scrophulariaceae, Boraginaceae, Verbenaceae, etc. More recently the Tubiflorae has been broken into smaller units. Six phylogenists, Bessey (1915), Cronquist (1968), Hutchinson (1959), Melchior (1964), Takhtajan (1969), Thorne (1968), have recently placed the Solanaceae in an order which includes the Solanaceae, Nolanaceae and Convolvulaceae. These families were placed next to a group which includes the Scrophulariaceae, Acanthaceae, Gesneriaceae and other families. A number of workers (Wettstein, 1892; Corner, 1976) have indicated belief that the Tubiflorae is an unnatural assemblage of families, but the recent divisions into smaller orders may not have meant improvement. On the other hand, Jay & Gonnet (1974) found chemical evidence supporting unity of the Tubiflorae.

Several workers consider the Solanaceae to be closely related to the Convolvulaceae, but this may be incorrect. Placentation, calyx structure, the origin of primary phloem (Mirande, 1922), and the differential possession of calcium oxalate and secretory laticifers argues against a close relationship. Corner (1976) argued against a close relationship on the basis of seed morphology, and Shah & Patel (1968) found that nodal anatomy in the Solanaceae is unlike that in *Ipomoea* (Convolvulaceae). On the other hand, Tucker (1969) found some correspondence between *Datura* and *Calystegia* (Convolvulaceae). Hegnauer (1973) supports an affinity on chemical grounds.

Hitchcock (1936) made a case for a conjunction of the Solanaceae with the Verbenaceae through a *Lycium–Grabowskia–Citharexylem* sequence. He claimed the last two genera differ mainly in embryo curvature. Some species of *Citharexylem* have raphids (Metcalfe & Chalk, 1950), a verbenaceous character, and some have no raphids but druses, a solanaceous character. Species of *Clerodendron* have alternate leaves as juveniles and the fruits are not unlike some of those in the Solanoideae. D'Arcy & Keating (1973) and Inamdar & Patel (1969) reported that members of the Solanaceae have unusual degenerate and solitary guard cells, and Inamdar (1969) reported these in the Verbenaceae as well.

A close relationship is often proposed between the Solanaceae and the Scrophulariaceae, and this subject was reviewed by Thieret (1967). Some workers, Bentham (1873), Wettstein (1891), Baillon (1887), Henslow (1893) and Hitchcock (1932) have suggested particular genera as transitional between the two families. Pennell (1935) and Robertson (1891) have discounted some of these possibilities, and Robyns (1930, 1932) showed that the ovary is oblique with respect to the floral axis in the Solanaceae but is straight in the Scrophulariaceae and some other members of the Tubiflorae. Besides differences in floral symmetry, the two families differ in alkaloid chemistry (Hegnauer, 1964) and position of the primary phloem (Mirande, 1922; Metcalfe & Chalk, 1950). Tucker (1969) found greater serological correspondence between the Solanaceae and the Scrophulariaceae than the Convolvulaceae. Carlquist (1975) found similarities in the parenchyma placement in woods of the Solanaceae and the Scrophulariaceae, Verbenaceae, Theophrastaceae and several other families. In discussing the placement of *Melananthus*, Solereder (1891) contrasted the Solanaceae and Scrophulariaceae as did Walters (1969) in discussing *Schizanthus*.

Species of *Ehretia* and *Solanum* are superficially similar, and there may be a close relationship between other members of the Boraginaceae and Solanaceae. Bicollateral bundles (internal phloem) have been found in *Heliotropium paniculatum* (Metcalfe & Chalk, 1950), and the nature of branching is sometimes similar in the two families (Danert, 1958). Many members of the Boraginaceae have accrescent calyces, and Eichler (1875) described *Echium vulgare* as having an oblique ovary orientation. Mirande (1922) suggested the Nolanoideae as intermediate between the Solanaceae and the Boraginaceae. But Chatin (1874) noted that massive endosperm occurs in the seeds of Solanaceae and Scrophulariaceae whereas it is much reduced in the Boraginaceae and Labiatae. In studying endosperm formation, Samuelsson (1913) noted similarity between the Solanaceae (except *Schizanthus*) and the Boraginaceae.

Robyns suggested a relationship between the Saxifragaceae and the Solanaceae based on floral symmetry, and Hutchinson (1969) proposed a similar relationship. Souèges (1937) suggested a relationship between the Solanaceae and *Amaranthus* in the Centrospermae based on embryogenesis, but this notion no longer attracts much support. Radlkofer (1889) discussed relationships between the Solanaceae, Theophrastaceae and Sapotaceae.

Although generally regarded as useful for systematics at higher levels, chemical data can suggest affinity with a number of diverse and unlikely families, e.g. Umbelliferae (coumarins), Crassulaceae (cf. Hegnauer, 1964: 584). In recent years there has been considerable publication of family arrangements which derive families of the Tubiflorae from dillenioid, rosoid, or other remote stocks, often with the assumption that the Compositae are intimately related to these Tubifloral families (as Asterales). Evidence for these opinions is left largely to the imagination.

It would seem that the Solanaceae is related to some, but perhaps not all families in the Tubiflorae. Melchior (1964) tabulated assumptions about primitive conditions within the Tubiflorae, and it would appear that members of the Solanaceae are more or less primitive compared with many other families in the order. Incidentally, application of these same assumptions shows that members of the Solanoideae are more primitive than the Cestroideae.

SOME NOTES ON PHYTOGEOGRAPHY

The Solanaceae embraces some 84 genera and almost 3000 species which occur on every vegetated continent in the world. The biggest genus, *Solanum*, has the greatest geographic amplitude, and members of sect. *Solanum* such as *Solanum americanum* Mill. have the greatest geographic and elevational amplitude in the family. However, only ten genera are native to both New and Old Worlds and only *Physalis*, *Solanum* and *Lycium* are widely distributed across both the Eastern and Western Hemispheres. With the exception of *Leucophysalis* and perhaps *Capsicum*, other genera such as *Datura*, *Schwenckia*, *Cestrum*, *Nicotiana*, and *Lycianthes*, occurring both in the Americas and outside are probably of recent occurrence in the Old World, since their Old World species are few and not greatly differentiated from New World relatives.

Thirteen genera are confined to the Old World, all of them in subfamily Solanoideae. Many of them range from the Mediterranean to the Himalayas or Japan, and only *Atrichodendron* occurs well south of the equator. In Africa, only *Discopodium* on tropical mountains and *Triguera* at the Pillars of Hercules show evolution at the generic level, although there are many species of *Solanum* and several of *Lycium*.

Two genera occur on oceanic islands, *Mellissia* (St. Helena) and *Nothocestrum* (Hawaiian Islands), and the factors leading to the dispersal and differentiation of these genera may be similar to the factors accounting for the differentiation of *Discopodium* and *Triguera*.

Australia has a rich development of *Solanum*, some species being of considerable morphological interest. While ovaries in the genus are usually two-locular, in Australia and Madagascar some species have four-locular ovaries. At the generic level Australia harbours four unusual endemic genera, *Duboisia*, *Anthocercis*, *Anthotroche* and *Isandra*, each with two or more species. Although placed in the Cestroideae, their immediate relatives on other continents are not yet identifiable.

North America has few endemic genera, only *Quincula*, *Oryctes*, *Chamaesaracha* and *Margaranthus* which ranges to the Antilles, but there is elaborate development of *Physalis*, and *Datura*. *Habrothamnus*, a section of *Cestrum* centred in Mexico, with inflated corollas, numerous ovules, longer anthers and calyx, the stamens inserted lower on the corolla tube, and a distinctively coloured endosperm, may have an ancestral relationship to the numerous species now found throughout tropical America. The South American genera *Salpiglossis*, *Leptoglossis*, *Petunia*, *Nierembergia*, and *Nicotiana* all have one or a few disjunct representatives in Mexico. These genera are centred in different parts of South America, and their disjunct species are native to different parts of North America. *Lycium* is also disjunct, with numerous species in both North and South America. No genera of Solanaceae are endemic to the Antilles or to northeastern South America, Venezuela, Surinam, or Brazil north of the Amazon, but the most primitive species of *Brunfelsia* occur in the Guyana uplands.

The centre of diversity in the Solanaceae at the generic level is in western and southern South America. About 59 genera are native to the continent, and of these 28 are endemic to South America and the Galapagos. Eighteen genera are restricted to temperate South America. Several genera are restricted to tropical Central America and to tropical America: perhaps more will be established when plants from these regions are better understood. The majority of genera of the Cestroideae and all of the Nolanoideae are present in South America.

Darrah (1939) reviewed the fossil record of the Solanaceae, and Couper (1954) noted a solanaceous-like spore from New Zealand, but these findings are in dispute and there may now be no accepted fossil record for the Solanaceae.

STUDIES STILL NEEDED

There are important gaps in our knowledge of the Solanaceae. Several large genera have never been revised, and many have no revisions for more than limited portions of their ranges. *Cyphomandra*, *Hyoscyamus* and *Mandragora* are examples. Many countries or regions are still not covered by useful Floras, and there is no good way to assess the solanaceous diversity in these regions. Extrapolating from neighbouring Floras is a poor makeshift. A large part of the knowledge of the Solanaceae comes from Floras which are generally held in low repute in scientific and academic communities. Floras are often seen only as a means of determining plants, so the published format often largely omits notes of a biological or comparative nature. These may in the long run be more valuable than the keys and nomenclature presentation that appears in print.

Morphological and chemical studies of a comparative nature are a rich source of useful data, but frequently the amplitude of species or genera is short of the ideal. Much of this sort of work is undertaken by scientists with no intrinsic interest in the Solanaceae but who find its plants useful subjects for manipulation. It is usually more satisfactory for such workers to use plants whose systematics are well understood and documented. Thus, increased systematics studies tend to encourage work on the group.

The chemical literature is becoming massive, and as it is usually published in minutia one item at a time, review papers are especially important to summarize new information. Papers by Sander (1963), Schreiber (1968), and Hegnauer (1973) are especially useful but must be updated as new information accumulates.

Pollination studies appear from time to time and some studies of an ethnobotanical nature have been undertaken, but they have only begun to elaborate the many ways in which man and other animals have affected the Solanaceae.

An important difficulty in assessing generic relationships in the family is the paucity of collections in many groups. In some cases, genera or critical species are extinct or nearly so. *Mellissia* is known from only two or three specimens at Kew, and *Oryctes* has been collected only seven times. All species of *Nothocestrum* are seriously endangered or extinct. The wild habitat for much of the *Habrothamnus* group of *Cestrum* has been converted to farms and industry. The type species of *Leucophysalis*, formerly recorded from eight U.S. States and four Canadian Provinces has been collected only in Michigan and Ontario since World War II; this in an area of considerable collecting activity. *Rahowardiana* is known from only one lowland cloud forest in Panamá that is now being cut off for agriculture. The Goetzeaceae are nearly extinct in the Antilles and some species can no longer be found (Liogier, pers. comm.). Other genera and many other species are similarly endangered in other parts of the world. It is important therefore, that studies of living plants, collections of seed and of herbarium and spirit material proceed rapidly while there is still evidence of much of the disappearing diversity of this essential plant family.

ACKNOWLEDGEMENT

Support from the Penrose Fund of the American Philosophical Society is gratefully acknowledged.

APPENDIX I

Generic index and bibliography

About 256 generic names have been used in the Solanaceae, many of them redundant or wrongly assigned to the family. Only about 90 genera are currently recognized. The following index includes all such names except for some obvious spelling variants. It also includes references to literature of systematic interest. All names are followed by author citation and date of publication. The list was prepared with assistance from Mrs Ellen Farr, Index Nominum Genericorum, Smithsonian Institution.

Names currently accepted are preceded by an indication of subfamily: *Solanoideae, **Cestroideae, ***Nolanoideae. Numbers of species and ranges are estimates by the present writer or are taken from Airy Shaw (1966). The first literature citations following recognized genera are those of general systematic value. Then follow more specialized references of morphological or miscellaneous nature (M:). In many cases Floras are the only important treatments of genera beyond their initial publication, and so Floras are sometimes cited here as well. A number of important older works have not been incorporated in the index, e.g. Wettstein, 1895; Bentham, 1873; Dunal, 1852, various papers by Miers, etc., as most genera published before these works are treated in them.

Names not currently accepted are usually followed by at least one bibliographic reference supporting the assignment indicated. In cases of questionable assignment, this is indicated. In a few cases, assignment to synonymy under particular genera is provided here for the first time. An identity sign ($\equiv$) is used where names are based on the same type species, while an equals sign ($=$) is used when the types are different. Generic names rejected through conservation procedures under the *International Code of Botanical Nomenclature* are listed in Stafleu (1972).

Several sorts of reference deal with the Solanaceae at a level above the genus, and they do not readily catalogue under the generic format. The following subject categories include papers of systematic importance. In addition, many papers on chemistry and pharmacology are not included here. The reader is referred to the chemical and pharmaceutical abstracting and review periodicals for literature on these subjects.

BIOLOGY: Baylis, 1968; Bell, 1959; Bowers, 1975; Cockerell & Porter, 1899; Free, 1975; Hardin *et al.*, 1972; Harris & Kuchs, 1902; Heiser, 1969a; Henslow, 1893; Knuth, 1906; Linsley, 1960, 1962; Linsley & Cazier, 1963; Michener, 1962; Nothmann, 1973; Rick & Bowman, 1961; Rusbridge, 1976; Symon (this Volume); Werth, 1956; Utech & Kawano, 1975; Vogel, 1958.

ETHNOBOTANY AND ECONOMIC BOTANY: Bristol, 1969; Davenport, 1970; Jenkins, 1948; Legge, 1974; Patino, 1963, 1964; Peterson (this Volume); Pickersgill, 1969a, b; Plowman *et al.*, 1971; Romero-Casteñeda, 1961a, b; Romero-Casteñeda & Schultes, 1962; Salaman, 1946, 1949; Schultes, 1956; Todd, 1882; Ugent, 1968; Vargas, 1936; Watson, 1927; Yarnell, 1959, 1965.

MORPHOLOGY AND MISCELLANY (for other titles the reader is referred to Metcalfe & Chalk, 1950): Arneson & Durbin, 1968; Baehni, 1946; Baillon, 1887; Chatin, 1874; Dahlgren, 1923, 1975a; Danert, 1958; Doulot, 1889; Guignard, 1902; Hegnauer, 1973; Inamdar & Patel, 1969; Kreusch, 1933; Levin, 1976; Martin, 1946; Murray, 1945; Namikawa, 1919; Robyns, 1930, 1932, 1936; Sandwith, 1936; Schreiber, 1954, 1968, 1974; Scott *et al.*, 1957; Seithe, 1962; Shah & Patel, 1968; Solereder, 1898–99; Souèges, 1907, 1922, 1937; Swift, 1962; Tucker, 1969; Wojciechowska, 1972.

PALYNOLOGY*: Armbruster & Oenike, 1929; Basak, 1967; Bell, 1959; Bellartz, 1956; Berg & Schmidt, 1861; Chaubal & Deodikar, 1963; DiFulvio, 1975; Erdtman, 1952, 1954, 1969; Erdtman *et al.*, 1963; Faegri & Iversen, 1950; Ferguson & Coolidge, 1932; Fritzsche, 1832; González, 1969; Griebel, 1930; Griffith & Henfrey, 1875; Heusser, 1971; Huang, 1968, 1972; Ikuse, 1956; Jain & Nanda, 1966; Khan & Rehman, 1966; Kolreuter, 1811; Kostoff, 1926; Mallik *et al.*, 1964; Martin, 1969; Martin & Drew, 1969; de Mendia, 1939; Mohl, 1934; Murry & Eshbaugh, 1971; Nair, 1965; Natarajan, 1957; Oetker, 1888; Poddubnaja-Arnoldi, 1936; Purkinje, 1830; Quiros, 1975; Raj & Suryakanta, 1970; Salgado-Labouriau, 1969; Sassen, 1964; Schnizlein, 1843–70; Schwanitz, 1967; Selling, 1947; Sharma, 1974; Ting-Su, 1949; Wang, 1960; Winkler, 1916; Zander, 1935, 1941, 1951.

WOOD HISTOLOGY: Burgerstein, 1908, 1912; Cozzo, 1946; D'Arcy, 1970; Descole & O'Donell, 1937; Greguss, 1945; Greiss, 1945; Kuhlmann, 1934; Metcalfe & Chalk, 1950; Pfeiffer, 1926; Record, 1933, 1934, 1936, 1939; Record & Hess, 1943; Williams, 1936.

*Listing prepared with the assistance of Johnnie L. Gentry.

*Acnistus Schott (1829). 50 Tropical America
   Sleumer, 1950 (included in Dunalia). Hunziker, 1960 (accepts Acnistus).
   M: Amshoff, 1957; Baker & Simmonds, 1953; Britton & Wilson, 1925; Gentry & Standley, 1974; Hitchcock, 1932; Macbride, 1962(as Dunalia); Millán, 1931; Smith & Downs, 1966; Rojas, 1974; Standley & Morton, 1938; Wiggins & Porter, 1971.

Alibrexia Miers (1845) = Nolana L.f. (Johnston, 1936).

Alicabon Raf. (1838) ≡ Withania Pauq.

Alkekengi Mill. (1754) ≡ Physalis L.

***Alona Lindl. (1844). 6 Chile.
   Johnston, 1936.
   F: Reiche, 1910.

Amatula Med. (1787) = Lycopersicon Mill.

Amphipleis Raf. (1837) = Nicotiana L. (Merrill, 1949).

Androcera Nutt. (1818) = Solanum L. (D'Arcy, 1973; Dunal, 1852; Whalen, 1976).

Anisodus Link ex Spreng. (1824) = Scopolia Jacq. (Wettstein, 1895); removed from Scopolia (Pascher, 1909).

**Anthocercis Labill. (1806). 20 Australia
   M: Baehni, 1944; Bentham, 1869; Black, 1952.

**Anthotroche Endl. (1839). 6 Australia

Antimion Raf. (1840) = Lycopersicon Mill. (Merrill, 1949).

Apemon Raf. (1837) = Datura L. (D'Arcy, 1973; Merrill, 1949).

Aplocarya Lindl. (1844) = Nolana L.f. (Johnston, 1936).

Aquartia Jacq. (1760) = Solanum L. (D'Arcy, 1973; Schulz, 1909–10).

*Archiphysalis Kuang (1966). 3 China, Japan, relates to Leucophysalis.

Armeniastrum Lemaire (1854) = Espadaea A. Rich. (Miers, 1871), Goetzeaceae, excluded from Solanaceae.

Artorhiza Raf. (1840) = Solanum L. (D'Arcy, 1973, 1974a; Merrill, 1949).

Ascleia Raf. (1838) = Hydrolea L., Hydrophyllaceae.

*Athenaea Sendt. (1846). 7 Brasil, conserved over Athenaea Adans. (Compositae) and Deprea Raf., but see D'Arcy, 1973.

*Atrichodendron Gagnep. (1950). 1 Indochina, relates to Lycium?

*Atropa L. (1753). 4 Mediterranean to Himalayas.
    Marzell, 1927.
    M: Bartorelli, 1922; Eisendrath, 1964; Feinstein & Slama, 1940; George, 1946; Hegelmaier, 1886; Souèges, 1920b; Wallis & Butterfield, 1939.

*Atropanthe Pascher (1909). 1 China, relates to Scopolia.

Aureliana Sendt. (1846), non Boehm. (1760). = Capsicum (Smith & Downs, 1966); = Athenaea (Hunziker, this Volume, Chapter 2).

Bargemontia Gaud. (1841) = Nolana L. (Johnston, 1936).

Bassovia Aublet (1775) = Solanum L. (D'Arcy, 1973, 1974a; Hunziker, 1969a).

Battata Hill (1765) = Solanum L. (D'Arcy, 1973).

Belenia Decaisne in Jacquem. (1844) = Physochlaina G. Don (Kuang & Lu, 1974; Wettstein, 1895).

Belladona Duham. (1755) ≡ Atropa L.

Belladona Mill. (1754) ≡ Atropa L.

Bellinia R. & S. (1819) ≡ Saracha R. & P.

**Benthamiella Speg. (1883). 10 Patagonia, relates to Nicotiana. Skottsberg, 1916; Soriano, 1948.

Blenococoës Raf. ex Jacks. (1893) = Nierembergia R. & P. (Merrill, 1949).

Blenocoes Raf. (1837) = Nicotiana L.

Bosleria A. Nels. (1905) = Solanum (D'Arcy, 1974b).

Bouchetia Dun. in DC. (1852) = Salpiglossis R. & P. (D'Arcy, in prep.).

Brachistus Miers (1849) = Witheringia L'Her. (D'Arcy, 1973; Gentry & Standley, 1974; Hunziker, 1969a).

**Brachyhelus (Benth.) O. Ktze. (1903). 4 Brasil, removed from Schwenckia.

**Browallia L. (1753). 2 tropical America, widely naturalized.
    D'Arcy, 1973
    M: Mohan, 1968

*Brugmansia Pers. (1805). 15 tropical America (see also Datura).
    De Wolf, 1956; Lagerheim, 1895; Lockwood, 1973; Safford, 1921.
    M: Bristol, 1969; Menninger, 1966; Rendle, 1921.

**Brunfelsia L. [Brunsfelsia] (1753). 25 tropical America.
    Plowman, 1973.
    M: Benoist, 1928; Hahmann, 1920.

Brunfelsiopsis (Urb.) O. Ktze. (1903) = Brunfelsia L. (D'Arcy, 1973; Plowman, 1973).

Cacabus Bernh. (1839) ≡ Exodeconus Raf.

Calibrachoa LaLlave & Lex. (1825) = Petunia Juss. (Fries, 1911).

Calydermos R. & P. (1799) ≡ Nicandra Adans. (Airy Shaw, 1966).

Cantalea Raf. (1838) ≡ Trozelia Raf. = Lycium L. (Merrill, 1949).

*Capsicum L. (1753). 10 tropical America.
    D'Arcy & Eshbaugh, 1974; Fingerhuth, 1832; Heiser & Smith, 1953;
    Hunziker, 1969a; Irish, 1898; Smith & Heiser, 1957.
    M: Ballard *et al.*, 1970; Davenport, 1970; Emboden, 1961; Eshbaugh, 1970,
    1971, 1974; Ferrari & Aillaud, 1971; Guillard, 1901; Heiser & Pickersgill,
    1969, 1975; Kostoff, 1926; Lippert *et al.*, 1966; Martin *et al.*, 1932;
    Munting, 1974; Ohta, 1962; Pickersgill, 1966, 1969a, b, 1971a, b;
    Raghuvanshi & Singh, 1972; Shopova, 1966; Smith & Heiser, 1951;
    Yaqub & Smith, 1971.

Ceratocaulos (Bernh.) Reichenb. (1837) = Datura L. (Fosberg, 1959; Safford, 1921).

Ceranthera Raf. (1819) = Solanum L. (D'Arcy, 1973, 1974a).

**Cestrum L. (1753). 175 tropical America, ?Old World, Australia
    Francey, 1935–36; Morton, 1936; Pittier, 1932; Schulz, 1909–10; Scolnik,
    1954.
    M: Ahmad, 1964a; Falcao & Alencastre, 1974; Montiero, 1968; Overland,
    1960; Shah & Suryamarayana, 1967.

Chaenesthes Miers (1845) = Iochroma Benth. (Wettstein, 1895).

Chaetochilus Vahl. (1804) = Schwenckia L. (Wettstein, 1895).

*Chamaesaracha (A. Gray) A. Gray (1876). 7 U.S. & Mexico.
    Averett, 1973; Rydberg, 1896.

Cleochroma Miers (1848) = Iochroma Benth. (Wettstein, 1895).

Cliocarpus Miers (1849) = Solanum L. (D'Arcy, 1973).

Codochonia Dun. (1852) = Acnistus Schott. (Hunziker, 1960).

Coeloneurum Radlk. (1890). Goetzeaceae, excluded from Solanaceae.

**Combera Sandw. (1936). 2 temperate South America, relates to Petunia?
    Ricardi, 1962.

*Cuatresia A. T. Hunz. (1977). See Hunziker, this Volume, Chapter 2.

Cyathostyles Schott ex Meisner (1840) = Cyphomandra Sendt. (D'Arcy, 1973).

Cyclostigma Phil. (1870) = Leptoglossis Benth.

Cyphanthera Miers (1853) = Anthocercis Labill. (Airy Shaw, 1966).

*Cyphomandra Sendt. (1845). 30 tropical America.
    M: Smith & Downs, 1966.

Dalea Mill. (1754) ≡ Browallia L.

Dartus Lour. (1790) = Maesa Forsk. Myrsinaceae (Airy Shaw, 1966).

*Datura L. (1753). 10 Mexico, Australia and India (see also Brugmansia).
 Avery, *et al.*, 1959; Bernhardi, 1833; De Wolf, 1956; Fosberg, 1959; Haegi, 1976; Safford, 1921.
 M: Barclay, 1959; Danert, 1973; Ewan, 1944; Gopal *et al.*, 1969; Rand, 1963; Satina & Blakeslee, 1940; Schultes, 1956; Timmerman, 1927.

*Deprea Raf. (1838). 2 tropical America.
 M: D'Arcy, 1974.

Diamonon Raf. (1837) = Solanum L. (D'Arcy, 1973; Schulz, 1909–10).

Dictyocalyx Hook. f. (1846) = Exodeconus Raf.

Dierbachia Spreng. (1824) ≡ Dunalia H.B.K. (Hunziker, 1960).

Diplukion Raf. (1838) = Iochroma Benth. (Stafleu, 1972).

*Discopodium Hochst. (1844). 1 tropical African mountains.

Diskion Raf. (1838) ≡ Saracha R. & P.

**Dittostigma Phil. (1870). 1 Argentina, relates to Nicotiana?

Dolia Lindl. (1844) = Nolana L. (Johnston, 1936).

Dolichosiphon Phil. (1873) = Jaborosa Juss. (Reiche, 1910).

Dorystigma Miers (1845) = Jaborosa Juss. (Wettstein, 1895).

**Duboisia R. Br. (1810). 3 Australia.
 Barnard, 1952.
 M: Beauvisage, 1897; Bentham, 1869; Griffin, 1965.

Duckeodendron Kuhlm. (1925). Duckeodendraceae (Kuhlmann, 1934); Apocynaceae (Record, 1933), excluded from Solanaceae.

Dulcamara Moench (1794) = Solanum L. (D'Arcy, 1973).

*Dunalia H.B.K. (1818). 7 or 30 western South America.
 Hunziker, 1960 (excludes Acnistus); Sleumer, 1950 (includes Acnistus).
 Macbride, 1962; Reiche, 1910.

Dyssochroma Miers (1849) = Markea L. C. Rich. (Cuatrecasas, 1958a), but see Hunziker, this Volume, Chapter 2; Solereder, 1898.

Eadesia F. v. Muell. (1858) = Anthocercis Labill. (Airy Shaw, 1966).

Ectozoma Miers (1849) = Juanulloa R. & P. (Wettstein, 1895); = Markea L. C. Rich. (D'Arcy, 1973).

Elisia Milano (1847) = Brugmansia Pers. (Rendle, 1921).

Entrecasteauxia Montr. (1860) = Duboisia R. Br. (Beauvisage, 1897; Heine, 1976).

Ephaiola Raf. (1838) = Acnistus Schott.

Eplateia Raf. (1838) ≡ Acnistus Schott.

Espadaea A. Rich. (1850). Goetzeaceae, excluded from Solanaceae (Miers, 1871).

Eucapnia Raf. (1837) = Nicotiana L. (Goodspeed, 1954).

Eutheta Standl. (1931) = Melasma Bergius, Scrophulariaceae (Gentry, 1973a).

Evoista Raf. (1838) = Lycium L. (Merrill, 1949).

*Exodeconus Raf. (1838). 6 South America, Galapagos.
 M: Bernhardi, 1839; Bitter, 1921c; Svenson, 1946.

**Fabiana R. & P. (1794). 25 temperate South America.
 Reiche, 1910.

Fontqueriella Rothm. (1940) = Triguera Cav. (Hansen & Hansen, 1973).

Franciscea Pohl (1826) = Brunfelsia L. (D'Arcy, 1973; Plowman, 1973).

Fregirardia Dun. ex Raff.-Del. (1849) = Cestrum (Hunziker, 1967).

Goetzea Wydl. (1830). Goetzeaceae, excluded from Solanaceae (Miers, 1871).

*Grabowskia Schlecht. (1832). 6 South America.

Gubleria Gaud. (1851) = Nolana L.f. (Johnston, 1932).

Habrothamnus Endl. (1839) = Cestrum L. (Francey, 1935–36).

*Hawkesiophyton A. T. Hunz. (1977). See Hunziker, this Volume.

*Hebecladus Miers (1845). 8 western South America, relates to Jaltomata.
    Bitter, 1921a, b, (*18:* 99–112; *19:* 265–270).
    M: Baker, 1870; Macbride, 1962 (treats with Saracha).

Henoonia Griseb. (1866). Goetzeaceae, excluded from Solanaceae (Baehni, 1943);
    Sapotaceae, excluded from Solanaceae (Kramer, 1939; Radlkofer, 1889; Record,
    1939).

Herschellia Bowd. ex Reichenb. (1837) = Physalis L. (Dunal, 1852).

**Heteranthia Nees & Mart. (1823). 1 Brasil.
    M: Metcalfe & Chalk, 1950; Solereder, 1915.

Huanaca Raf. (1838) = Dunalia (Hunziker, 1960; Sleumer, 1950).

**Hunzikeria D'Arcy (1976). 1 Texas, Mexico.

*Hyoscyamus L. (1753). 20 Mediterranean and Asia.
    Pojarkova, 1942.
    M: Marzell, 1927; Raghavan, 1976; Souèges, 1920b; Weinert, 1972.

*Iochroma Benth. (1845). 20 tropical South America.

**Isandra F. v. Muell. (1883). 2 Australia.
    M: Solereder, 1898.

*Jaborosa Juss. (1789). 20 South America.
    M: Cabrera, 1965; Hunziker, 1967; Muñoz, 1959; Philippi, 1857; Reiche,
    1910.

*Jaltomata Schlecht. (1838). 4 tropical and subtropical America, removed from
    Saracha R. & P.
    M: Bernhardi, 1839; Gentry, 1973b, 1974; Gentry & Standley, 1974;
    Jönsson, 1881; Macbride, 1962; Menzel, 1950; Morton, 1938.

Jasminoides Duham. (1755) ≡ Lycium L. (Hitchcock, 1932).

Johnsonia Necker (1790) = Lycium L. (Hitchcock, 1932).

*Juanulloa R. & P. (1794). 10 tropical America.
    Cuatrecasas, 1958b.
    M: Geremicca, 1902.

Kokabus Raf. (1838) = Hebecladus Miers (Stafleu, 1972).

Kukolis Raf. (1838) = Hebecladus Miers (Stafleu, 1972).

Langsdorfia Raf. (1837) = Nicotiana L. (Goodspeed, 1954).

Larnax Miers (1849) = Deprea Raf. (D'Arcy, 1973), but see Hunziker (this Volume).

*Latua Phil. (1858). 1 Chile.
    Plowman *et al.*, 1971.

Laureria Schlecht. (1834) = Juanulloa R. & P. (D'Arcy, 1973; Gentry & Standley,
    1974).

Lehmannia Spreng. (1817) = Nicotiana L. (Goodspeed, 1954).

Leloutrea Gaud. (1852) = Nolana L.f. (Johnston, 1936).

Leptofeddea Diels (1919) = Leptoglossis Benth.

Leptoglossis Benth. (1845). 1 Peru.
    M: Monachino, 1940.

Leptophragma Benth. ex Dun. (1852) = Petunia Juss. (Wettstein, 1895).

Leucanthea Scheele (1853) = Salpiglossis R. & P. (Hemsley, 1881–82).

***Leucophysalis** Rydb. (1896). 9 North America, eastern Asia, removed from Chamae-saracha, Physalis.
    Averett, 1971, and this Volume; Kuang & Lu, 1965.

Levana Raf. (1840) = Vestia Willd. (Merrill, 1949).

Lithophytum Brandeg. (1911). Verbenaceae, excluded from Solanaceae (D'Arcy & Keating, 1973).

Lomeria Raf. (1838) = Cestrum L. (Merrill, 1949).

Lonchestigma Dun. (1852) = Jaborosa Juss. (Reiche, 1910; Wettstein, 1895).

***Lycianthes** (Dun.) Hassl. (1917). 200 tropical America, eastern Asia, removed from Solanum.
    Bitter, 1920; Morton, 1944.
    M: D'Arcy, 1973; Hassler, 1917.

Lycioplesium Miers (1845) = Dunalia H.B.K. (Hunziker, 1960).

***Lycium** L. (1753). 100 widespread, warm temperate.
    Barkley, 1953; Dammer, 1913a; Feinbrun, 1968; Hitchcock, 1936; Miers, 1854; Pojarkova, 1955; Terracciano, 1890.
    M: Barkley, 1954; Ross, 1975; Weitz, 1921a, b; Wright, 1904.

Lycomela Fabr. (1763) ≡ Lycopersicon Mill.

***Lycopersicon** Mill. (1754). 8 western South America, Galapagos.
    Boswell *et al.*, 1933; Luckwill, 1943; Muller, 1940; Scasso & Millán, 1930.
    M: Butler, 1952; Byrne *et al.*, 1975; Clayberg *et al.*, 1966; Cooper, 1927; Ecole, 1974; Jenkins, 1948; Khush & Rick, 1966; Menzel, 1964; Muller, 1940; Ocana, 1972; Rick, 1974; Rick & Bowman, 1961; Rick & Khush, 1966; Rusbridge, 1976; Sander, 1963; Toma, 1966; Woodcock, 1935.

Lycopersicum Hill (1765) = Lycopersicon Mill.

***Mandragora** L. (1753). 6 Mediterranean to Himalayas.
    M: Bouquet, 1952.

***Margaranthus** Schlecht. (1838). 1 United States and Mexico, Antilles.
    M: Averett & Judd, 1977; Menzel, 1950; Rydberg, 1896.

***Markea** L. C. Rich. (1792). 18 tropical America.
    Cuatrecasas, 1958, 1959; D'Arcy, 1973.

Mathaea Vell. (1825) = Schwenckia L. (Wettstein, 1895).

Matthissonia Raddi (1820) = Schwenckia L. (Wettstein, 1895).

Melananthus Walp. (1850). 5 tropical America.
    Cabrera, 1952.
    M: Solereder, 1891; Suessenguth & Beyerle, 1938; Urban, 1922.

***Mellissia** Hook. f. (1867). 1 St. Helena.

Melongena Mill. (1754) = Solanum L. (D'Arcy, 1973, 1974a).

Merinthopodium Donn. Sm. (1897) = Markea L. C. Rich. (Cuatrecasas, 1958a; D'Arcy, 1973).

Methysticodendron R. E. Schultes (1955) = Brugmansia Pers. (D'Arcy, 1973).

*Metternichia Mikan (1823). 1 eastern Brazil.

Meyenia Schlecht. (1833) = Cestrum L. (Francey, 1935–36).

Microschwenkia Benth. ex Hemsl. (1882) = Melananthus Walp. (Wettstein, 1895; Gentry & Standley, 1974).

*Nectouxia H.B.K. (1818). 1 Mexico.

Neudorfia Adans. (1763) = Nolana L.f. (Johnston, 1936).

*Nicandra Adans. (1763). 1 Peru, widely naturalized.
> M: Dupuy, 1964; Fernándes, 1969; Fernald, 1950; Hogstad, 1912.

**Nicotiana L. (1753). 60 America, South Pacific, Australia, South West Africa.
> Burbidge, 1960; Comes, 1899; Goodspeed, 1954; Lehmann, 1818; Millán, 1928.
> M: Anastasia, 1920; Avery, 1933a, 1933b, 1934; Bentley & Wolf, 1945; Biraghi, 1929; East, 1935; Heiser, 1969a; Khan & Rehman, 1966; Merxmüller & Buttler, 1975; Mukherjee, 1974a, b; Souèges, 1920a; Splendore, 1906.

**Nierembergia R. & P. (1794). 35 America, especially Argentina.
> Millán, 1941.
> M: Di Fulvio, 1975a, b.

***Nolana L.f. (1762). 57 western South America.
> Bentham, 1845; Ferreyra, 1961; Johnston, 1936.
> M: Di Fulvio, 1969, 1971; Johnston, 1929; Mirande, 1922; Reiche, 1910; Saunders, 1935, 1936.

Normania Lowe (1872) = Solanum L. (D'Arcy, 1973, 1974a).

*Nothocestrum A. Gray (1862). 5 Hawaiian Islands.

Nycterium Vent. (1805) = Solanum L. (D'Arcy, 1973, 1974a).

Oplukion Raf. (1838) ≡ Lycium L.

Opsago Raf. (1838) = Withania Pauq. (Merrill, 1949).

Orinocoa Raf. (1838) = Deprea Raf. (Merrill, 1949).

*Oryctes Wats. (1871). 1 Nevada, California.
> M: Wiggins, 1951.

Osteocarpus Phil. (1884) = Alona L. (Johnston, 1936).

Otilix Raf. (1830) ≡ Lycianthes (Dun.) Hassl.

Ovaria Fabr. (1763) = Solanum L. (Airy Shaw, 1966).

Pachysolen Phil. (1895) = Nolana L.f. (Johnston, 1936; Reiche, 1910).

Pallavicinia De Not. (1847) = Cyphomandra (D'Arcy, 1973).

*Pantacantha Speg. (1902). 1 Patagonia, relates to Lycium L.?
> M: Soriano, 1948.

Panzeria Gmel. (1791) = Lycium L. (Hitchcock, 1932).

**Parabouchetia Baill. (1887). 1 South America.

Parascopolia Baill. (1888) = Lycianthes (Dun.) Hassl. (Stafleu, 1972).

Parmentiera Raf. (1840) = Solanum L. (D'Arcy, 1973; Merrill, 1949).

Parqui Adans. (1763) = Cestrum L. (D'Arcy, 1973).

*Pauia Deb & Dutta (1965). 1 Assam, relates to Capsicum, Lycianthes?

Pederlea Raf. (1838) = Acnistus Schott (Merrill, 1949).

Pentagonia Fabr. (1763) ≡ Nicandra Adans. (Airy Shaw, 1966).

Pentaphiltrum Reichenb. (1841) = Physalis L. (Wettstein, 1895).

Perieteris Raf. (1837) = Nicotiana L. (Goodspeed, 1954; Merrill, 1949).

Periloba Raf. (1838) = Nolana L.f. (Johnston, 1936).

Perizoma (Miers) Lindl. (1846) = Salpichroa Miers (Shinners, 1962).

Petagnia Raf. (1814) = Solanum L. (Merrill, 1949).

****Petunia** Juss. (1803). 40 America, especially Brasil.
>Fries, 1911.
>M: Biraghi, 1929; Cooper, 1946; Ferguson, 1927; Ferguson & Coolidge, 1932; Sassen, 1964; Souèges, 1936.

Pheliandra Werderm. (1940) = Solanum L. (Hunziker, 1967).

****Phrodus** Miers (1849). 4 Chile.
>M: Reiche, 1910.

Physaliastrum Makino (1914) = Leucophysalis (Averett, this Volume).
>Kuang & Lu, 1965.

***Physalis** L. (1753). 100 widespread, especially Mexico.
>Bernhardi, 1839; Gray, 1875; Menzel, 1951; Rydberg, 1896; Waterfall, 1958, 1967.
>M: Baldwin & Speese, 1951; Fernándes, 1969, 1970; Hara & Kurosawa, 1952; Legge, 1974; Menzel, 1950.

Physalodes Boehmer (1760) ≡ Nicandra Adans.

Physaloides Moench (1794) = Withania Pauq. (Airy Shaw, 1966).

***Physochlaina** G. Don (1838). 6 China.
>Kuang & Lu, 1974.

Pionandra Miers (1845) = Cyphomandra Sendt. (D'Arcy, 1973).

Plicula Raf. (1838) = Acnistus Schott (Hunziker, 1960).

Poecilochroma Miers ≡ Saracha R. & P.

Polydiclis (G. Don) Miers (1849) = Nicotiana L. (Goodspeed, 1954).

Portaea Tenore (1846) = Juanulloa R. & P. (D'Arcy, 1973).

Poortmannia Drake (1892) = Trianaea Planch. (Cuatrecasas, 1958; Solereder, 1898).

****Protoschwenkia** Solereder (1898). 1 Bolivia.

***Przewalskia** Maxim. (1882). 2 Central Asia.
>M: Pascher, 1910a, b.

Pseudocapsicum Med. (1789) = Solanum L. (D'Arcy, 1973).

Pseudodatura van Zijp (1920) = Brugmansia Pers. (D'Arcy, 1973).

Pteroglossis Miers (1850) = Reysia Clos.

Pukanthus Raf. (1838) = Grabowskia Schlecht. (Merrill, 1949).

Puneeria Stocks (1849) = Withania Pauq. (Airy Shaw, 1966),

***Quincula** Raf. (1832). 1 North America, removed from Physalis.
>M: Averett, this Volume; Menzel, 1950; Rydberg, 1896.

***Rahowardiana** D'Arcy (1973). 1 Panamá, relates to Markea?

Rayera Gaud. (1851) = Alona L. (Johnston, 1936).

Retzia Thunb. (1776). Retziaceae, excluded from Solanaceae (Goldblatt & Keating, 1976; Herbst, 1972).

**Reyesia Clos in Gay (1849) 4 Chile, Argentina.

Rhopalostigma Phil. (1860) = Phrodus Miers (Reiche, 1910).

Saccardophytum Speg. (1902) = Benthamiella (Soriano, 1948).

Sairanthus G. Don (1838) = Nicotiana L. (Goodspeed, 1954).

*Salpichroa Miers (1845). 17 South America.
> M: Benoist, 1938; Shinners, 1962; West, 1952.

**Salpiglossis R. & P. (1794). 18 temperate South America.
> Werdermann, 1928.
> M: Baillon, 1887; Bentham, 1846; Correll, 1970; Dale, 1937; Gay, 1849; Johnston, 1929.

*Saracha R. & P. (1794). 3 South America, includes Poecilochroma, excludes Jaltomata.
> Bernhardi, 1839; Bitter, 1921a, b; Gentry, 1973b; Macbride, 1962; Morton, 1944.

Sarcophysa Miers (1849) = Juanulloa R. & P. (Cuatrecasas, 1958a; D'Arcy, 1973), but see Hunziker, this Volume.

**Schizanthus R. & P. (1794). 8 Chile.
> Reiche, 1910; Sudzuki, 1969; Vilmorin, 1896; Walters, 1969.
> M: Souèges, 1936.

*Schultesianthus A. T. Hunz. (1977). See Hunziker, this Volume.

**Schwenckia L. (1764). 25 America, Africa.
> Bentham, 1869.
> M: D'Arcy, 1973; Heine, 1963b; Solereder, 1898.

Schwenkiopsis Damm. (1916). 1 Andes, relates to Schwenckia?

**Sclerophylax Miers (1848). 12 Argentina.
> Di Fulvio, 1961.

*Scopolia Jacq. (1764). 6 Mediterranean to Himalayas.
> Weinert, 1972.
> M: Marzell, 1927; Mills & Jackson, 1972; Warming, 1869.

Scubulon Raf. (1840) = Solanum or Lycopersicon (D'Arcy, 1973).

**Sessea R. & P. (1794). 5 Andes.
> M: Bitter, 1922a; Francey, 1933, 1934.

**Sesseopsis Hassl. (1917). 2 South America, relates to Sessea?
> M: Bitter, 1922b.

Sicklera Sendt. (1846) = Witheringia L'Her. (D'Arcy, 1973; Hunziker, 1969a).

Siphaulax Raf. (1837) = Nicotiana L. (Goodspeed, 1954).

Siphonema Raf. (1837) = Nierembergia R. & P. (Merrill, 1949).

Solanastrum Fabr. (1759) = Solanum L. (Airy Shaw, 1966).

*Solandra Sw. (1787). 8 tropical America.
> DeWolf, 1955; Martínez, 1966.
> M: Cuatrecasas, 1958a; D'Arcy, 1973; Solereder, 1898.

Solanocharis Bitt. (1918) = Solanum L. (Hunziker, 1967).

Solanopsis Börner (1912) = Solanum L. (D'Arcy, 1973).

*Solanum L. 1400 widespread.

Revisions	*Section or group*
Anderson, 1975	Basarthrum
Bartlett, 1909	Androceras
—— 1958	Archaesolanum
Baylis, 1963	Solanum
Bhaduri, 1951	Melongena
Correll, 1952	Petota
—— 1962	Anarrhichomenum, Basarthrum, Neolycopersicon, Petota
—— 1967	Basarthrum, Neolycopersicon, Petota
Danert, 1970	General System
D'Arcy, 1972	General System
Edmonds, 1972	Solanum
Fernald, 1900	Melongena (Torvum)
Gilli, 1970	General System
Hawkes, 1953, 1963	Petota
Hawkes & Hjerting, 1969	Petota
Heiser, 1955, 1963, 1969a, b	Solanum
—— 1968, 1971, 1972	Lasiocarpum
—— 1969a	Petota
Henderson, 1974	Solanum
Herasimenko, 1970	Archaesolanum
Morton, 1944	Leiodendron
Ochoa, 1962	Petota
Roe, 1967, 1972	Brevantherum
Seithe, 1962	General System

Morphology and miscellany

Ahmad, 1964b	Epidermis
Bhaduri, 1932	Embryology, *S. melongena*
Brücher, 1966, 1974a, b	Basarthrum, Petota
Danert, 1969	Stone cells (fruits)
Dnyansagar & Cooper, 1960	Seed, *S. phureja*
Economidou & Yannitsaros, 1975	*S. elaeagnifolium*
Gibson, 1974	Trichomes, Petota
Gottschalk, 1954, 1956, 1958	Cytology
Haeupler, 1974	sect. Solanum
Harris, 1903	Polygamous flowers
Heiser, 1964	*S. muricatum*
Holm, 1920	*S. carolinense*
Howard, 1970	*S. tuberosum*
Kessel, 1974	Inheritance
Jørgensen, 1928	Heteroploidy
Magoon *et al.*, 1961	Cytology
Magtang, 1936	*S. melongena*
Máthé & Máthé, 1973	*S. dulcamara*
Mohan & Singh, 1968	*S. torvum* seed
Morretes, 1969	Leaves
Raj & Suryakanta, 1970	*S. xanthocarpum* pollen
Randall & Symon, 1976	Chromosomes
Roe, 1966a, b	Juvenile forms
—— 1971	Trichome terminology
Sander, 1963	Chemistry
Schreiber, 1954, 1968, 1974	Alkaloids
Seithe, 1962	Trichomes

*Solanum—*continued*
Sharma, 1974	Pollen
Sinha, 1950	Cytology
Smith, 1931	*S. melongena*
Swaminathan & Howard, 1953	Cytology and genetics
Thiel, 1931	*S. melongena* axis
Ugent, 1970	Wild potato origins
Wangenheim *et al.*, 1957	Cytology
Zutshi, 1974	Cytology

Sorema Lindl. (1844) = Nolana L.f. (Johnston, 1936).

Sterrhymenia Griseb. (1874) = Sclerophylax Miers (Di Fulvio, 1961).

Stigmatococca Willd. (1827) = Ardisia L., Myrsinaceae, excluded from Solanaceae (Airy Shaw, 1966).

Stimenes Raf. (1837) = Nierembergia R. & P.

Stimomphis Raf. (1837) = Petunia Juss. (Fries, 1911).

Stimoryne Raf. (1837) = Petunia Juss. (Fries, 1911).

Stramonium Mill. (1754) ≡ Datura L.

**Streptosolen Miers (1850). 1 Peru.
        M: Macbride, 1962.

Streptostigma Regel (1853) = Exodeconus Raf.

Swartsia Gmel. (1791) = Solandra Sw. (Stafleu, 1972).

Tabacum Gilibert (1781) = Nicotiana L. (Goodspeed, 1954).

Tabacus Moench (1794) = Nicotiana L. (Goodspeed, 1954).

Teganium Schmeidel (1763) = Nolana L.f. (Johnston, 1936).

Teremis Raf. (1838) = Lycium L. (Hitchcock, 1932).

Thinogeton Benth. (1845) = Exodeconus Raf.

*Trechonaetes Miers (1845). 3 Chile and Argentina.
        Reiche, 1910.

*Trianaea Planch. & Linden (1853). 4 tropical America.
        M: Cuatrecasas, 1958a, 1959; Solereder, 1898.

*Triguera Cav. (1786). 1 western Mediterranean.
        Hansen & Hansen, 1973.

        M: Moore, 1972; Robyns, 1930.

Triliena Raf. (1838) = Acnistus Schott (Merrill, 1949).

Trozelia Raf. (1838) = Acnistus Schott (Hunziker, 1960).

Tubocapsicum Makino (1908) = Capsicum (Hunziker, 1958).

Tunaria O. Ktze. (1898) = Cantua Lam., Polemoniaceae, excluded from Solanaceae. (Mattfeld & Bitter, 1922).

Ulloa Pers. (1805) ≡ Juanulloa R. & P.

Ulticona Raf. (1838) = Hebecladus Miers (Stafleu, 1972).

Valerioa Standl. & Steyerm. (1938) = Peltanthera Benth., Loganiaceae, excluded from Solanaceae (Hunziker & Di Fulvio, 1957).

Valteta Raf. (1838) = Iochroma Benth. (Stafleu, 1972).

Vassobia Rusby (1907) = Witheringia L'Her. (D'Arcy, 1973), but see Hunziker, this Volume.

Velpeaulia Gaud. (1851) = Nolana L.f. (Johnston, 1936).

Vestia Willd. (1809). 1 Chile.
  Reiche, 1910

Waddingtonia Phil. (1860) = Nicotiana L. (Goodspeed, 1954).

Wadea Raf. (1838) = Cestrum L. (D'Arcy, 1973).

Walkeria Mill. ex Ehret (1764) = Nolana L.f. (Johnston, 1936).

*Withania Pauq. 10 Old World.
  M: Abraham, 1968; Fernandes, 1969.

*Witheringia L'Her. (1789). 20 from tropical America, removed from Capsicum, Brachistus, Bassovia, etc.
  Hunziker, 1969.
  M: D'Arcy, 1973; Gentry & Standley, 1974.

Zwingera Hofer (1762) = Nolana L.f. (Johnston, 1936).

APPENDIX II

Floras treating Solanaceae

Floras	Region
Amshoff, 1957	Cuba
Backer, 1965	Java
Baker & Simmonds, 1953	Trinidad & Tobago
Bitter, 1919b	Papua
Black, 1952	South Australia
Blatter & Mallard, 1954	India (cultivated)
Brezhnev, 1958	U.S.S.R. (cultivated)
Correll, 1970	Texas
Dammer, 1913b	Bolivia
D'Arcy, 1973	Panamá
D'Arcy, 1974a	Florida
Deb, 1961	Manipur
—— 1969	Bhutan
Fernald, 1950	N.E. North America
Gay, 1845–54	Chile
Grisebach, 1879	Argentina
Hayek & Markgraf, 1931	Balkans
Heine, 1963	W. tropical Africa
Heine et al., 1969	Southwest Africa
Jepson, 1943, 1951	California
Kuzeneva, 1966	Murmansk region
Lawrence, 1941, 1960a, b	U.S.A. (cultivated)

Floras	Region
Marchesi, 1965	Uruguay
Martínez, 1949	Argentina (cultivated)
Marzell, 1927	Central Europe
Millán, 1931	Argentina
Moore, 1972	Europe
Munz & Keck, 1959	California
Pojarkova, 1955	U.S.S.R.
Probst, 1949	Europe (adventives)
Reiche, 1910	Chile
Rojas, 1974	Venezuela
Santapau, 1948	Bombay
Schulz, 1909–10	Antilles
Seckt, 1929–30	Argentina
Small, 1933	S.E. North America
Smith & Downs, 1966	Sta. Catarina (Brasil)
Standley & Morton, 1938	Costa Rica
Troncoso & Torres, 1974	Chiloé Islands
Wiggins, 1951	W. North America
Wiggins & Porter, 1974	Ecuador (Galapagos)
Williams & Williams, 1951	Trinidad (cultivated)

REFERENCES

ABRAHAM, A., 1968. A chemotaxonomic study of *Withania somnifera* (L.) Dun. *Phytochemistry*, 7: 957–962.

ADANSON, M., 1763. *Familles des Plantes*, 2 vols. Oct. Paris.

AHMAD, K. J., 1964a. Cuticular studies with special reference to abnormal stomatal cells in *Cestrum*. *Journal of the Indian Botanical Society*, 43: 165–177.

AHMAD, K. J., 1964b. Epidermal studies in *Solanum*. *Lloydia*, 27(3): 243–250.

AIRY SHAW, H. K., 1966. In J. C. Willis (Ed.), *A Dictionary of Flowering Plants and Ferns*, 7th ed. Cambridge.

AMSHOFF, DRA. JANE G., 1957. Fam. 6. Solanaceae. In Hno. Leon, & Hno. Alain (Liogier), *Flora of Cuba*, 4: 345–386.

ANASTASIA, G. E., 1920. Le forme elementari della composizione dei vegetali, o l'origine della specie (Filogenesi delle Nicotianae, delle Primulaceae e delle Violae). *Bol. Tec. 4. R. 1st. Sci. Speriment. del Tabacco*: 43 pp. Scafati, Italy.

ANDERSON, G. J., 1975. The variation and evolution of selected species of *Solanum* section *Basarthrum*. *Brittonia*, 27: 209–222.

ARMBRUSTER, L. & OENIKE, G. 1929. Die Pollenformen als Mittel zur Honigherkunftsbestimmung. *Bücherei fur Bienenkunde*, 10. Neumünster i. Hilst. (n.v.).

ARNESON, P. A. & DURBIN, R. D., 1968. The sensitivity of fungi to a-tomatine. *Phytopathology*, 58: 536–537 (n.v.).

AUBLET, J. B. C. F., 1775. *Histoire des Plantes de la Guiane francaise*, 4 vols. Qu. London & Paris.

AVERETT, J. E., 1970. [1971] New combinations in the Solaneae (Solanaceae) and comments regarding the taxonomy of *Leucophysalis*. *Annals of Missouri Botanical Garden*, 57: 380–382.

AVERETT, J. E., 1973. Biosystematic study of *Chamaesaracha* (Solanaceae). *Rhodora*, 75: 325–365.

AVERETT, J. E. & JUDD, J. W., 1977. Flavonoid chemistry of *Margaranthus*. *Biochemical Systematics and Ecology*, 5: 279–280.

AVERY, A. G., SATINA, S. & RIETSEMA, J., 1959. Blakeslee: the genus *Datura*. *Chronica Botanica*, 20. New York: Ronald Press.

AVERY, G. S., Jr., 1933a. Structure and germination of tobacco seed and the developmental anatomy of the seedling plant. *American Journal of Botany*, 20: 309–327.

AVERY, G. S., Jr., 1933b. Structure and development of the tobacco leaf. *American Journal of Botany*, 20: 565–592.

AVERY, G. S., Jr., 1934. Structural responses to the practice of topping tobacco plants: a study of cell size, cell number, leaf size and veinage of leaves at different levels on the stalk. *Botanical Gazette*, 96: 314–329.

BACKER, C. A., 1965. Solanaceae. In C. A. Backer & R. C. Bakhuizen van den Brink, Jr. (Eds.), *Flora of Java*, 2: 464–483.

BAEHNI, C., 1943. *Henoonia*, type d'une famille nouvelle? *Boissiera*, 7: 346.

BAEHNI, C., 1944. Organogénie de la fleur chez l'*Anthocercis littorea* Labill. (Solanée). *Bulletin de la Société Botanique Suisse*, 54: 640.

BAEHNI, C., 1946. L'ouverture du bouton chez les fleurs de Solanées. *Candollea*, 10: 400–492.

BAILLON, H., 1887. Sur quelques types du groupe intermédiaire aux Solanacées et aux Scrophulariacées. *Bulletin de la Société Linnéenne de Paris*, 1(83): 660–663.

BAKER, J. G., 1870. *Hebecladus ventricosus*. In W. W. Saunders (Ed.), *Ref. Bot.*, 3: t. 208. (n.v.).

BAKER, R. E. D. & SIMMONDS, N. W., 1953. Solanaceae. In R. O. Williams (Ed.), *Flora of Trinidad and Tobago*, 2: 241–272.

BALDWIN, J. T. & SPEESE, B. M., 1951. Cytogeography of *Physalis* in West Africa. *Bulletin of the Torrey Botanical Club*, 78: 254–257.

BALLARD, R. E., McCLURE, J. W., ESHBAUGH, W. H. & WILSON, K. G., 1970. A chemosystematic study of selected taxa of *Capsicum*. *American Journal of Botany*, 57: 225–233.

BARCLAY, A. S., 1959. New considerations in an old genus: *Datura*. *Botanical Museum Leaflets*, 18: 245–272.

BARKLEY, F. A., 1953. *Lycium* in Argentina. *Lilloa*, 26: 177–238.

BARKLEY, F. A., 1954. Notes on *Lycium*. *Lloydia*, 17: 332–334.

BARNARD, C., 1952. The *Duboisias* of Australia. *Economic Botany*, 6: 3–17.

BARTLETT, H. H., 1909. *Solanum* (*Androceras*) of Mexico and the southern United States. *Contributions from the Gray Herbarium of Harvard University*, 36.

BARTLING, F. T., 1830. *Ordines Naturales Plantarum*. Gottingae.

BARTORELLI, I., 1922. Di un nuova carattere farmacognostico della *Belladonna* (*Atropa belladonna* L.). *Annali di Botanica*, 15: 273–275.

BASAK, R. K., 1967. The pollen grains of Solanaceae. *Bulletin of the Botanical Society of Bengal*, 21: 49–58.

BAUHIN, C., 1623. *Pinax Theatri Botanici. . .* Basel.

BAYLIS, G. T. S., 1958. A cytogenetical study of New Zealand forms of *Solanum nigrum* L., *S. nodiflorum* Jacq., and *S. gracile* Otto. *Transactions of the Royal Society of New Zealand*, 85(3): 379–385.

BAYLIS, G. T. S., 1963. A cytogenetical study of the *Solanum aviculare* species complex. *Australian Journal of Botany*, 11: 168–177

BAYLIS, G. T. S., 1968. Day length and flowering in the *Solanum aviculare* group. *New Zealand Journal of Botany, 6:* 221–225.

BEAUVISAGE, G. E. C., 1897. Deuxième note sur l'herbier du R. P. Montrousier. Le genre *Entrecasteauxia* Montr. *Annales de la Société Botanique de Lyon, 22:* 71–76.

BELL, C. R., 1959. Mineral nutrition and flower to flower pollen size variation. *American Journal of Botany,* 46(9): 621–624.

BELLARTZ, S., 1956. Das Pollenschlauchwachstum nach arteigener und artfremder Bestaubung einiger Solanaceen und die Inhaltsstoffe ihres Pollens und ihrer Griffel. *Planta, 47:* 588–612.

BENOIST, R., 1928. Une nouvelle espèce de *Brunfelsia* (Solanacées), plante magique des Indiens du Haut-Amazone. *Bulletin de la Société Botanique de France, 75:* 295.

BENOIST, R., 1938. Nouvelles espèces du genre Salpichron (Solanées). *Bulletin de la Société Botanique de France, 85:* 53–56; 408–410.

BENTHAM, G., 1845. *Botany of the Voyage of the Sulphur:* 141–143.

BENTHAM, G., 1846. Scrophulariaceae. In A. P. De Candolle (Ed.), *Prodromus Systematis Naturalis Regni Vegetabilis, 10:* 186–679.

BENTHAM, G., 1869. [1868] Solanaceae, *4:* 442–470. Scrophulariaceae, *4:* 471–523 (Duboisia, Anthocercis). *Flora Australiensis.* London.

BENTHAM, G. & HOOKER, J. D., 1873a. [1876] Family (ordo) 113. Convolvulaceae. In, *Genera Plantarum,* 2(1): 865–881.

BENTHAM, G. & HOOKER, J. D., 1873b. [1876] Family (ordo) 114. Solanaceae. In, *Genera Plantarum,* 2(1): 882–913.

BENTLEY, N. J. & WOLF, F. A., 1945. Glandular leaf hairs of oriental tobacco. *Bulletin of the Torrey Botanical Club, 72:* 345–360.

BERG, O. C. & SCHMIDT, C. F., 1861. *Darstellung und Beschreibung der offizinellen Pflanzen.* Leipzig. (2nd ed., A. Meyer & K. Schumann (Eds), 1893–1902.)

BERNHARDI, J. J., 1833. Über die Arten der Gattung *Datura. Linnaea, 8:* (litteratur-Bericht): 115–144. Reprint from Trommsdorf, *Neues Journal für Pharmacie,* 26(1): 118–158 (n.v.).

BERNHARDI, J. J., 1839. Über *Saracha* und *Physalis. Linnaea, 13:* 357–362.

BESSEY, C. A., 1915. The phylogenetic taxonomy of flowering plants. *Annals of Missouri Botanical Garden, 2:* 109–164.

BHADURI, P. N., 1932. The development of the ovule and embryo sac in *Solanum melongena* L. *Journal of the Indian Botanical Society, 11:* 202–224.

BHADURI, P. N., 1951. Inter-relationship of non-tuberiferous species of *Solanum* with some consideration on the origin of brinjal (*S. melongena* L.). *Indian Journal of Genetics and Plant Breeding, 11:* 75–82.

BIRAGHI, D. A., 1929. Impollinazioni tra *Nicotiana rustica* var. *brasila* e *Petunia* sp. e loro effetti. *Annali di Botanica, 18:* 216–222.

BITTER, G., 1919a. [1920] Die Gattung *Lycianthes. Abhandlungen Naturwissenschaftlichen Verein zu Bremen,* 24(2): 292–520.

BITTER, G., 1919b. Die papuasischen Arten von *Solanum. Botanische Jahrbücher für Systematik, Pflanzengeschichte und Pflanzengeographie, 55:* 59–113.

BITTER, G., 1921a. Zur Gliederung der Gattung *Saracha* und zur Kenntnis einiger ihrer bemerkenswerten Arten. *Feddes Repertorium Specierum Novarum Regni Vegetabilis, 17:* 338–346.

BITTER, G., 1921b. Eine verkannte *Hebecladus*-Art und ihre Bedeutung für die Stellung der Gattung in der Tribus der Solaneae. *Feddes Repertorium Specierum Novarum Regni Vegetabilis, 17:* 246–251.

BITTER, G., 1921c. Zur Gattung *Cacabus* Bernh. *Feddes Repertorium Specierum Novarum Regni Vegetabilis, 17:* 242–245.

BITTER, G., 1922a. Georg Bitter, Zur Gattung *Sessea. Feddes Repertorium Specierum Novarum Regni Vegetabilis, 18:* 199–225.

BITTER, G., 1922b. Georg Bitter, *Sesseopsis vestioides* (Schlecht.) Bitt. nov., comb. *Feddes Repertorium Specierum Novarum Regni Vegetabilis, 18:* 225–227.

BITTER, G., 1923. Namenverzeichnis (index to Solana Africana). *Feddes Repertorium Specierum Novarum Regni Vegetabilis Beih., 16:* 312–320.

BLACK, J. M., 1952. [1957] *Flora of South Australia,* Part III, 2nd ed., *Suppl.,* 1965: 270–278. Adelaide.

BLAKESLEE, A. F., AVERY, A. G., SATINA, S. & RIETSAMA, J., 1959. *The Genus* Datura. *Chronica Botanica, 20:* 289 pp. New York: Ronald Press.

BLATTER, E. & MALLARD, W. S., 1954. *Some Beautiful Indian Trees,* 2nd ed. rev. by Wm. T. Stearn. Bombay: Bombay Nat. Hist. Soc.

BÖRNER, C., 1912 [1913] *Solanum* L. und *Solanopsis* gen. nov. *Abhandlungen hrsg. vom Naturwissenschaftlichen Verein zu Bremen, 21:* 282.

BOSWELL, V. R., PEARSON, O. L. H., WORK, P., BROWN, H. D., MacGILLIVRAY, J. H. and others, 1933. Descriptions of types of principal American varieties of tomatoes. *U.S.D.A. Miscellaneous Publication,* No. 160: 23 pp.

BOUQUET, J., 1952. La *Mandragora* en Afrique du Nord. *Bulletin de la Société des Sciences Naturelles Tunisia, 5:* 29–44.

BOWERS, K. A. W., 1975. The pollination ecology of *Solanum rostratum* (Solanaceae). *American Journal of Botany, 62:* 633–638.

BREZHNEV, D. D., 1958. Solanaceae. In P. M. Zhukovsky (Ed.), *Flora of Cultivated Plants, 22:* 531 pp. Leningrad.

BRISTOL, M. L., 1966. Notes on the species of tree *Daturas*. *Botanical Museum Leaflets, 21:* 229–347.

BRISTOL, M. L., 1969. Tree *Datura* drugs of Colombian Sibundoy. *Botanical Museum Leaflets, 22*(5): 165–227.

BRITTON, N. L. & WILSON, P., 1925. Solanaceae. In, *Botany of Porto Rico and the Virgin Islands. Scientific Survey of Porto Rico and the Virgin Islands, 6:* 162–179; 561–562.

BROWN, R., 1810. *Prodromus Florae Novae Hollandiae et Insulae Van-Diemen*. London.

BRÜCHER, H., 1966. *Solanum caripense* H.B.K. (subsect. Basarthrum) in Venezuela. *Feddes Repertorium Specierum Novarum Regni Vegetabilis, 73:* 216–221.

BRÜCHER, H., 1974a. Über Art-Begriff und Art-Bildung bei *Solanum* (sect. Tuberarium). *Beiträge zur Biologie der Pflanzen, 50:* 393–429.

BRÜCHER, H., 1974b. Die Section Tuberarium des Genus *Solanum* in Paraguay. Zur Wiederauffindung der Originalstandorte der Wildkartoffeln *S. chacoense* Bitter und *S. malmeanum* Bitter. *Berichte der Deutschen Botanischen Gesellschaft, 87*(3): 405–420. [July 1975].

BURBIDGE, N. T., 1960. The Australian species of *Nicotiana* L. (Solanaceae). *Australian Journal of Botany, 8:* 342–380.

BURGERSTEIN, A., 1908. Anatomische Untersuchungen samoanischer Hölzer. *Denkschriften der Akademie der Wissenschaften Wien (Math.-nat. Kl.), 84:* 455–458.

BURGERSTEIN, A., 1912. Anatomische Untersuchungen argentinischer Hölzer des K. K. naturhistorischen Hofmuseums in Wien. *Annalen des K. K. Naturhistorischen Museums, Wien, 26:* 1–36.

BUTLER, L., 1952. The linkage map of the tomato. *Journal of Heredity, 43:* 25–35.

BYRNE, J. M., COLLINS, K. A., CASHAU, P. F. & AUNG, L. H., 1975. Adventitious root development from the seedling hypocotyl of *Lycopersicon esculentum*. *American Journal of Botany, 62*(7): 731–737.

CABRERA, A. L., 1952. El genero '*Melananthus*' (Solanaceae) en la republica Argentina. *Boletin de la Sociedad Argentina de Botanica, 4:* 192–194.

CABRERA, A. L., 1965. *Flora de la Provincia de Buenos Aires*. Buenos Aires, Solanaceae: *5:* 190–250.

CARLQUIST, S., 1975. *Ecological Strategies of Xylem Evolution:* 299 pp. Univ. Calif. Press.

CAUVET, D., 1864. *Des Solanées:* 152 pp. Strasbourg.

CHATIN, J., 1874. Études sur le développement de l'ovule et de la graine dans les Scrophulariaées, les Solanacées, les Borraginées et les Liabiées. *Annales des Sciences Naturelles, 5 Ser Botanique, 19:* 5–107, figs.

CHAUBAL, P. D. & DEODIKAR, G. B., 1963. Pollen grains of poisonous plants—1: Poisonous pollen in honey samples from western Chats (India). *Grana Palynologica, 4:* 393–397.

CLAYBERG, C. D., BUTLER, L., KERR, E. A., RICK, C. M. & ROBINSON, R. W., 1966. Third list of known genes in the tomato. *Journal of Heredity, 57:* 189–196.

COCKERELL, T. D. A. & PORTER, W., 1899. Observations on bees, with descriptions of new genera and species. *Annals and Magazine of Natural History (Ser. 7), 4:* 403–421.

COMES, O., 1899. *Monographie du Genre* Nicotiana *Comprenent le Classement Botanique des Tabacs Industriels*. Naples.

COOPER, D. C., 1927. Anatomy and development of tomato flower. *Botanical Gazette, 83:* 399.

COOPER, D. C., 1946. Double fertilization in *Petunia*. *American Journal of Botany, 33:* 54–57, f. 1–10.

CORNER, E. J. H., 1976. *The Seeds of Dicotyledons*, 2 vols. Cambridge.

CORRELL, D. S., 1952. Section Tuberarium of the genus *Solanum* of North America and Central America. *U.S.D.A. Agricultural Monograph*, No. 11.

CORRELL, D. S., 1962. *The Potato and its Wild Relatives*. Texas: Renner.

CORRELL, D. S., 1967. Solanaceae (parte). In Correll (J. F. Macbride), *Flora of Peru. Field Mus. Bot., 8:* V-B (2).

CORRELL, D. S., 1970. Solanaceae (Fam. 159). In D. S. Correll & M. C. Johnston (Eds), *Manual of the Vascular Plants of Texas. Solanaceae:* 1386–1408. Texas: Renner.

COUPER, R. A., 1954. *Triorites spinosus* Couper. In, *Plant Microfossils from New Zealand*, No. 1. *Transactions of the Royal Society of New Zealand, 81*(4): 479–483.

COZZO, D., 1946. Los géneros de fanerógamas Argentinas con radios leñosos altos en su leño secundario. *Revista Argentina de Agronomia (Buenos Aires), 13:* 207–210.

CRONQUIST, A., 1968. *The Evolution and Classification of Flowering Plants:* 396 p. New York: Riverside Press.

CUATRECASAS, J., 1958a. Notes on American Solanaceae. *Feddes Repertorium Specierum Novarum Regni Vegetabilis, 61:* 74–86.

CUATRECASAS, J., 1958b. The Colombian species of *Juanulloa*. *Brittonia, 10:* 146–150.

CUATRECASAS, J., 1959. New chiropterophilous Solanaceae from Colombia. *Journal of the Washington Academy of Sciences, 49:* 269–272.

DAHLGREN, K. V. O., 1923. Notes on the ab initio cellular endosperm. *Botaniska Notiser*, 1923: 1–24.

DAHLGREN, R., 1975a. A system of classification of the Angiosperms to be used to demonstrate the distribution of characters. *Botaniska Notiser, 128:* 119–147.

DAHLGREN, R., 1975b. The distribution of characters within an Angiosperm system I. Some embryological characters. *Botaniska Notiser, 128:* 181–197.

DALE, L. E., 1937. A series of multiple alleles especially affecting the corolla in *Salpiglossis*. *American Journal of Botany, 24:* 651–656.

DAMMER, U., 1913a. Solanaceae Africanae II. *Lycium. Botanische Jahrbücher für Systematik Pflanzengeschichte und Pflanzengeographie, 48:* 225–260 (n.v.).

DAMMER, U., 1913b. Solanaceae. In J. Perkins, Beiträge zur Flora von Bolivia. *Botanische Jahrbücher für Systematik Pflanzengeschichte und Pflanzengeographie, 49:* 215–217.

DANERT, S., 1958. Die Verzweigung der Solanaceen im reproduktiven Bereich. *Abhandlungen der Deutschen Akademie der Wissenschaften zu Berlin, 6:* 1–292.

DANERT, S., 1967. Die Verzweigung als infragenerisches Gruppenmerkmal in der Gattung *Solanum* L. *Kulturpflanze, 15:* 225–292.

DANERT, S., 1969. Über die Entwicklung der Steinzellkonkretionen in der Gattung *Solanum*. *Kulturpflanze, 17:* 299–311.

DANERT, S., 1970. Infragenerische Taxa der Gattung *Solanum* L. *Kulturpflanze, 18:* 253–297.

DANERT, S., 1973. About development of fruits and centric vascular bundles in genus *Datura* (Solanaceae). *Kulturpflanze, 21:* 119–193.

D'ARCY, W. G., 1970 [1971]. Solanaceae studies I. *Annals of Missouri Botancial Garden, 57:* 258–263.

D'ARCY, W. G., 1972. Solanaceae studies II: typification of subdivisions of *Solanum*. *Annals of Missouri Botanical Garden, 59:* 262–278.

D'ARCY, W. G., 1973 [3 July 74]. Solanaceae. In R. E. Woodson, Jr., R. W. Schery & collaborators, Flora of Panamá. *Annals of Missouri Botanical Garden, 60:* 573–780.

D'ARCY, W. G., 1974a. *Solanum* and its close relatives in Florida. *Annals of Missouri Botanical Garden, 61:* 819–867.

D'ARCY, W. G., 1974b. Reduction of *Bosleria* (Solanaceae). *Annals of Missouri Botanical Garden, 61:* 906.

D'ARCY, W. G., 1975. The Solanaceae: an overview. *Solanaceae Newsletter, 2:* 8–15.

D'ARCY, W. G. & ESHBAUGH, W. H., 1974. New World peppers (*Capsicum*-Solanaceae) North of Colombia: A Resumé. *Baileya, 19*(3): 93–105.

D'ARCY, W. G. & KEATING, R. C., 1973. The affinities of *Lithophytum*: a transfer from Solanaceae to Verbenaceae. *Brittonia, 25:* 213–225.

DARRAH, W. C., 1939. *Textbook of Paleobotany.* New York: Appleton-Century.

DARWIN, C. R., 1859. *On the Origin of Species by means of Natural Selection, or the Preservation of Favoured Races in the Struggle for Life:* 502 p. London.

DAVENPORT, W. A., Jr., 1970. Progress report on the domestication of *Capsicum* (chili peppers). *Proceedings of the Association of American Geographers, 2:* 46–47 (n.v.).

DEB, D. B., 1961. Solanaceae *in* Dicot. Plants of Manipur. *Bulletin of the Botanical Survey of India, 3:* 253–350.

DEB, D. B., 1969. Solanaceae. In, A contribution to the flora of Bhutan. *Bulletin of the Botanical Society of Bengal, 22:* 169–217.

DESCOLE, H. R. & O'DONELL, C. A., 1937. Estudios anatomicos en el leno de plantas tucumanas. *Lilloa, 1:* 75–93.

DeWOLF, G. P., Jr., 1955. Notes on the Cultivated Solanaceae 1. *Solandra. Baileya, 3:* 173–175.

DeWOLF, G. P., Jr., 1956. Notes on Cultivated Solanaceae 2. *Datura. Baileya, 4:* 12–23.

DI FULVIO, T. E., 1961. El Genero *Sclerophylax* (Solanaceae). *Kurtziana, 1:* 9–103.

DI FULVIO, T. E., 1969. Embriologia de *Nolana paradoxa* (Nolanaceae). *Kurtziana, 5:* 39–54.

DI FULVIO, T. E., 1971. Morfologia floral de *Nolana paradoxa* (Nolanaceae) con especial referencia a la organizacion del gineceo. *Kurtziana, 6:* 41–51.

DI FULVIO, T. E., 1975. [1976]a. Sobre el polen de *Nierembergia* (Solanaceae). *Kurtziana, 9:* 87–91.

DI FULVIO, T. E., 1975. [1976]b. Recuentos cromosomicos en *Nierembergia* (Solanaceae). *Kurtziana, 9:* 141–142.

DNYANSAGAR, V. R. & COOPER, D. C., 1960. Development of the seed of *Solanum phureja*. *American Journal of Botany, 47:* 176–186.

DON, G., 1838. *A General History of the Dichlamydeous Plants . . .*, 4 vols. Solanaceae, *4:* 397–488. London.

DOULOT, M. H., 1889. Recherches sur le periderme. *Annales des Sciences Naturelles* (Ser. 7), *10:* 325–395.

DUNAL, M. F., 1813. *Histoire Naturelle, Medicinale et Economique des Solanum.* Paris, Strasbourg, Montpellier.

DUNAL, M. F., 1816. *Solanorum Generumque Affinium Synopsis seu Solanorum Historiae.* Montpellier.

DUNAL, M. F., 1852. Solanaceae. In A. P. De Candolle (Ed.), *Prodromus Systematis Naturalis Regni Vegetabilis, 13*(1): 1–690.

DUPUY, P., 1964. Formation de fleurs anormales, reguliers, irregulieres ou zygomorphes par action de l'acido 2,4,-dichlorophenoxyacetique, chez le *Nicandra physaloides* L. *Compte Rendu Hebdomadaire des Séances de l'Academie des Sciences, 258:* 1307–1310 (n.v.).

EAST, E. M., 1935. Genetic reactions in *Nicotiana*. *Genetics, 20:* 403–451 (n.v.).

ECOLE, D., 1974. Sympodial functioning in two Solanaceae of genus *Lycopersicum*. *Revue de Cytologie et de Biologie Végétales, 37:* 127–160 (n.v.).

ECONOMIDOU, E. & YANNITSAROS, A., 1975. Recherches sur la flore adventice de Grèce. III. Morphologie, development et phenologie du *Solanum elaeagnifolium* Cav. *Candollea, 30:* 29–41.

EDMONDS, J., 1972. A synopsis of the taxonomy of *Solanum* sect. *Solanum* (Maurella) in South America. *Kew Bulletin, 27:* 95–114.

EDWARDS, A. MILNE, 1864. *De la Famille des Solanacées:* thesis, 137 p. Paris: E. Martinet.

EICHLER, A. W., 1875. *Blüthendiagramme Construirt und Erläutert*. Leipzig.

EISENDRATH, E. R., 1964. Illustrations of *Atropa belladonna* in the Sixteenth Century Herbals. *Medical and Biological Illustration, 14*(1): 27–35; (2): 96–104.

EMBODEN, W. A., Jr., 1961. A preliminary study of the crossing relationships of *Capsicum baccatum*. *Butler University Botanical Studies, 14:* 1–5.

ENDLICHER, S. L., 1839, 1841, *Genera Plantarum:* 662–699; 1403–1404. Suppl. II: 60–65. 1842. Vienna.

ERDTMAN, G., 1952. *Pollen Morphology and Plant Taxonomy—Angiosperms*. Stockholm: Almquist & Wiksell.

ERDTMAN, G., 1954. Pollen morphology and plant taxonomy. *Botaniska Notiser:* 65–81.

ERDTMAN, G., 1969. *Handbook of Palynology. Morphology–Taxonomy–Ecology. An Introduction to the Study of Pollen Grains and Spores:* 1–486. Copenhagen.

ERDTMAN, G., PRAGLOWSKI, J. & NILSSON, S., 1963. *An Introduction to a Scandinavian Pollen Flora*, II: 1–89, Uppsala.

ESHBAUGH, W. H., 1970. A biosystematic and evolutionary study of *Capsicum baccatum* (Solanaceae). *Brittonia, 22:* 21–43.

ESHBAUGH, W. H., 1971. A biosystematic and evolutionary study of *Capsicum pubescens* Ruiz & Pav. *Yearbook of the American Philosophical Society:* 315–316.

ESHBAUGH, W. H., 1974. Variation and evolution of *Capsicum pubescens*. *American Journal of Botany, 61:* 42.

EWAN, J., 1944. Taxonomic history of perennial southwestern *Datura meteloides*. *Rhodora, 46:* 317–323.

FAEGRI, K. & IVERSEN, J., 1950. *Textbook of Modern Pollen Analysis*. Copenhagen.

FALCÃO, W. F. de A. & ALENCASTRE, F. M. M. R., 1974. Anatomia e morphologia da especie *Cestrum laevigatum* Schlecht. (Solanaceae). *Brasil Florest., 5:* 65–72.

FEDDE, F., 1896. *Beitrage zur Vergleichenden Anatomie der Solanaceae*. Dissert. Schreiber, Breslau (n.v.).

FEINBRUN, N., 1968. The genus *Lycium* in the flora orientalis region. *Collectanea Botanica a Barcinonensi Botanico Instituto Edita, 7:* 359–379.

FEINSTEIN, B. S. & SLAMA, F. J., 1940. Differentiation of *Belladonna, Digitalis, Hyoscyamus* and *Stramonium* leaves. *Journal of the American Pharmacological Association, 29:* 370 (n.v.).

FERGUSON, M., 1927. A cytological and a genetical study of *Petunia* I. *Bulletin of the Torrey Botanical Club, 54:* 657–664.

FERGUSON, M. & COLLIDGE, E., 1932. A cytological and a genetical study of *Petunia* IV. Pollen grains and the method of studying them. *American Journal of Botany, 19:* 644–658.

FERNALD, M. L., 1900. A revision of the Mexican and Central American *Solanums* of the subsection Torvaria. *Proceedings of the American Academy, 35:* 557–562.

FERNALD, M. L., 1950. Solanaceae. In Gray, *Manual of Botany*, 8th ed.: 1251–1260.

FERNANDES, R. B., 1969. Os generos *Nicandra* Adans., *Physalis* L., e *Withania* Pauq. no Ultramar Portugues. *Garcia de Orta, 17*(3): 275–288.

FERNANDES, R. B., 1970. Sur l'identification d'une espèce de *Physalis* souspontanée au Portugal. *Boletim da Sociedade Broteriana* (Sér. II), *49:* 343–367.

FERRARI, J.-P. & AILLAUD, G., 1971 [1972]. Bibliographie du genre *Capsicum*. *Journal d'Agriculture Tropicale et de Botanique Applique, 18:* 385–480.

FERREYRA, R., 1961. Revision de las especies peruanas del genero *Nolana* (Nolanaceae). *Memorias del Museo de Historia Natural 'Javier Prado', 12,* 72 pp. Lima: Univ. de San Marcos.

FEUILLÉE, L. É., 1725. *Journal des Observations Physiques, Mathématiques et Botaniques*, 3 vols. Paris.

FIGDOR, W., 1909. *Die Erscheinung der Anisophyllie:* 174 p. Solanaceae: 85–86. Leipzig & Vienna.

FINGERHUTH, A., 1832. *Monographia Generis Capsici:* 32+ pp. Dusseldorf.

FOSBERG, F. R., 1959. Nomenclatural notes on *Datura* L. *Taxon, 8:* 52–57.

FRANCEY, P., 1933. Beitrag zur Kenntnis der Gattung *Sessea*. *Notizblatt des Botanischen Gartens und Museums zu Berlin, 11:* 879.

FRANCEY, P., 1934. Übersicht über die Gattung *Sessea*. *Notizblatt des Botanischen Gartens und Museums zu Berlin, 11:* 978.

FRANCEY, P., 1935–36. Monographie du genre *Cestrum* L. *Candollea, 6:* 46–398; 7: 1–132.

FREE, J. B., 1975. Pollination of *Capsicum frutescens* L., *Capsicum annuum* L., and *Solanum melongena* L. (Solanaceae) in Jamaica. *Tropical Agriculture (Trinidad), 52:* 353–357.

FRIES, R. E., 1911. Die Arten der Gattung *Petunia*. *Kongliga Svenska Vetenskapsakademiens Handlingar*, *46*(5): 1–72.

FRITZSCHE, J., 1832. *Beitrage zur Kenntniss des Pollen, 1:* 48 p. Berlin (n.v.).

GAERTNER, J., 1791. *De Fructibus et Seminibus Plantarum, 2:* 236–244.

GAGNEPAIN, F., 1950. *Atrichodendron* Gagnep., n.g. Solanacearum. *Notulae Systematicae, Paris, 14:* 29.

GAY, C., 1845–54. Flora Chilena. In, *Historia Física y Política de Chile*, 8 vols. Solanaceae, *5* (1849).

GENTRY, J. L., 1973a. Studies in Mexican and Central American Solanaceae. *Phytologia, 26:* 265–278.

GENTRY, J. L., 1973b. Restoration of the genus *Jaltomata* (Solanaceae). *Phytologia, 27:* 286–288.

GENTRY, J. L., 1974. The generic name *Saracha* Ruiz & Pavon (Solanaceae). *Fieldiana: Botany, 36:* 69–72.

GENTRY, J. L. & STANDLEY, P. C., 1974. Solanaceae. In, *Flora of Guatemala*, Part 10, Nos. 1 & 2: 151 pp.

GEORGE, E., 1946. The palisade ratio values of *Atropa belladonna* and *Atropa acuminata*. *Quarterly Journal of Pharmacy and Pharmacology, 19:* 144–154 (n.v.).

GEREMICCA, M., 1902. Note preliminari morfo-istologische su la *Juanulloa aurantiaca*. *Bull. Soc. Nat. Napoli ser. I, 15:* 61–76.

GIBSON, R. W., 1974. Aphid-trapping glandular hairs on hybrids of *Solanum tuberosum and Solanum berthaultii*. *Potato Research, 17:* 152–154.

GILLI, A., 1970. Bestimmungsschlüssel der Subgenera und Sektionen von *Solanum*. *Feddes Repertorium Specierum Novarum Regni Vegetabilis, 81:* 429–435.

GOLDBLATT, P. & KEATING, R. C., 1976. Chromosome cytology, pollen structure and relationship of *Retzia capensis*. *Annals of Missouri Botanical Garden, 63:* 321–325.

GONZALEZ, Q. L., 1969. Morfologia polinica: la flora del valle del Mezquital, Hidalgo. *Palaeoecologia (Mexico), 3:* 1–185 (n.v.).

GOODSPEED, T. H., 1954. The genus *Nicotiana*. *Chronica Botanica, 16*(1–6): 536.

GOPAL, SHARMA, K. & DUNN, D. B., 1969. Environmental modifications of leaf surface traits in *Datura stramonium*. *Canadian Journal of Botany, 47:* 1211–1216.

GOTTSCHALK, W., 1954. Die Chromosomenstruktur der Solanaceen und der Berücksichtigung Phylogenetischer Fragestellungen. *Chromosoma, 6:* 539–626.

GOTTSCHALK, W., 1956. Das Konjugationsverhalten partiell homologer Chromosomen. *Chromosoma, 7:* 708–725.

GOTTSCHALK, W., 1958. Über die Anwendung cytologischer Methoden für die Bearbeitung phylogenetischer Fragestellung bei den Solanaceen. Eine Erwiderung. *Zeitschrift für Pflanzenzüchtung, 39:* 47–70 (n.v.).

GRAY, A., 1875. Contributions to the Botany of North America. III. Synopsis of the North American species *Physalis*. *Proceedings of the American Academy of Arts and Science, 10:* 62.

GREGUSS, P., 1945. *Bestimmung der Mitteleuropaischen Laubholzer und Straucher auf Xylotomischer Grundlage:* 183 pp. Budapest (n.v.).

GREISS, E. A. M., 1945. On the rate of water conduction in the wood of certain trees in Egypt. *Bulletin of the Faculty of Science, Fouad I University, 25:* 35–54 (n.v.).

GRIEBEL, C., 1930. Zur Pollenanalyse des Honigs (I. Mitteilung). *Zeitschrift für Untersuchung Lebensmittel*, 59.

GRIFFIN, W. J., 1965. The alkaloids of *Duboisia leichhardtii*. *Australian Journal of Pharmacology, 36:* 128–131.

GRIFFITH, J. W. & HENFREY, A., 1875. *The Micrographic Dictionary: a Guide to the Examination and Investigation of the Structure and Nature of Microscopic Objects*, 3rd ed. London (n.v.).

GRISEBACH, A., 1879. *Symbolae ad Floram Argentinam*. Gottingen: Dietrich.

GUINARD, L., 1902. La double fécondation chez les Solanées. *Journal de Botanique, Paris, 16:* 145–167.

GUILLARD, F., 1901. *Les Piments des Solanées: Etude Historique botanique des Piments du Genre* Capsicum. Lens-le-Saunier, L. Declume: 124 p. (Kew Catalogue). (n.v.).

GUNN, C. R. & GAFFNEY, F. B., 1974. Seed characteristics of 42 economically important species of Solanaceae in the United States. *Technical Bulletin 1471. Agricultural Research Service, U.S.D.A.* Feb, 1974.

HAEGI, L., 1976. Taxonomic account of *Datura* L. (Solanaceae) in Australia. *Australian Journal of Botany. 24:* 415–435.

HAEUPLER, H., 1974. *Solanum nitidibaccatum* Bitter und *Solanum sarachoides* Sendtner em. Bitter zwei gut unterscheidbare Nachtschattenarten aus der Sektion *Solanum*. *Göttinger Flor. Rundbr., 8:* 98–105.

HAHMANN, C., 1920. Beiträge zur anatomischen Kenntnis der *Brunfelsia hopeana* Benth., im besonderen deren Wurzel, Radix Manaca. *Angewandte Botanik, 2:* 113–133, 179–191 (n.v.).

HANSEN, A. & HANSEN, C., 1973. *Verbascum osbeckii* L. and the *Triguera* species (Solanaceae). A revision. *Lagascalia, 3:* 183–193.

HARA, H. & KUROSAWA, S., 1952. *Physalis alkekengi* and its variation in East Asia. *Journal of Japanese Botany, 27:* 247–253 (English abstract).

HARDIN, J. W., DOERKSEN, G., HERNDON, D., HOBSON, M. & THOMAS, F., 1972. Pollination ecology and floral biology of four weedy genera in southern Oklahoma. *Southwestern Naturalist, 16:* 403–412.

HARRIS, J. A., 1903. Polygamy and certain floral abnormalities in *Solanum*. *Transactions of the Academy of Science of St. Louis, 13:* 185–202.

HARRIS, J. A. & KUCHS, O. M., 1902. Observations on the pollination of *Solanum rostratum* Dunal and *Cassia chamaecrista* L. *Kansas Science Bulletin, 1:* 15–41.

HASSLER, E., 1917a. Solanaceae Austro-Americanae imprimis Paraguarienses I. *Lycianthes* (Dun.) Hassler. *Annuaire du Conservatoire et Jardin Botanique Genéve, 20:* 173–183.

HAWKES, J. G., 1956. A revision of the tuberbearing *Solanums*. I. *Scottish Plant Breeding Station Record, 1956:* 37–109.

HAWKES, J. G., 1963. A revision of the tuberbearing *Solanums*. II. *Scottish Plant Breeding Station Record, 1963:* 71–181.

HAWKES, J. G. & HJERTING, J. P., 1969. *The Potatoes of Argentina, Brazil, Paraguay and Uruguay*. Oxford.

HAWKES, J. G. & SMITH, P., 1965. Continental drift and the age of angiosperm genera. *Nature, 207:* 48–51.

HAWKES, J. G. & TUCKER, W. G., 1968. Serological assessment of relationships in a flowering plant family (Solanaceae). In J. G. Hawkes (Ed.), *Systematics Association Special Volume* No. 2. *Chemotaxonomy and Serotaxonomy:* 77–78.

HAYEK, A. & MARKGRAF, Fr., 1931. Prodromus florae peninsulae Balcanicae. Solanaceae. *Repert. Spec. Nov. Beih., 30*(2): 95–105.

HEGELMAIER, F., 1886. Zur Entwicklung endospermatischer Gewebekörper. *Botanische Zeitung, 44.*

HEGNAUER, R., 1964. *Chemotaxonomie der Pflanzen III. Crassulaceae:* 572–584.

HEGNAUER, R., 1973a. *Chemotaxonomie der Pflanzen VI. Scrophulariaceae:* 343–386; nachträge 746–749, 787.

HEGNAUER, R., 1973b. *Chemotaxonomie der Pflanzen VI. Solanaceae:* 403–452; nachträge 750–754; 788–790.

HEINE, H., 1963a. Solanaceae. In F. N. Hepper (Ed.), *Flora of West Tropical Africa, 2*, 2nd ed.: 325–335.

HEINE, H., 1963b. On the correct spelling of the generic name *Schwenckia* D. van Royen ex L. (Solanaceae), with a note about Martin Wilhelm Schwenke. *Kew Bulletin, 16:* 465–469.

HEINE, H., 1976. Solanacées. In A. Aubreville & J. F. Leroy (Eds), *Flore de la Nouvelle Caledonie, 7:* 119–212.

HEINE, H., PODLECH, D. & ROESSLER, H., 1969. Solanacae. In H. Merxmüller (Ed.), *Prodromus einer Flora von Südwest-afrika*, (#124) *124:* 1–17.

HEISER, C. B., Jr., 1955. The *Solanum nigrum* complex in Costa Rica. *Ceiba, 4:* 293–299.

HEISER, C. B., 1963. *Solanum* (*Morella*) in Ecuador. Author's translation from the Spanish. *Ciencia y Naturaleza, 6:* 50–58.

HEISER, C. B., 1964. Origin and variability of the pepino (*Solanum muricatum*): a preliminary report. *Baileya, 12:* 151–158.

HEISER, C. B., 1968. Some Ecuadorian and Colombian *Solanums* with edible fruit. *Ciencia y Naturaleza 11:* 1–9.

HEISER, C. B., 1969a. *Nightshades, the Paradoxical Plants:* 200 p. San Francisco: Freeman.

HEISER, C. B., 1969b. *Solanum caripense* y el origen de *Solanum muricatum*. *Revista Politecnica (Ecuador), 1:* 1–7.

HEISER, C. B., 1971. Notes on *Solanum* (*Leptostemonum*) in Latin America. *Baileya, 18:* 59–65.

HEISER, C. B., 1972. The relationships of the naranjilla, *Solanum quitoense*. *Biotropica, 4:* 77–84.

HEISER, C. B. & PICKERSGILL, B., 1969. Names for the cultivated *Capsicum* species (Solanaceae). *Taxon, 18:* 277–283.

HEISER, C. B. & PICKERSGILL, B., 1975. Names for the Bird Peppers (*Capsicum*-Solanaceae). *Baileya, 19:* 151–156.

HEISER, C. B. & SMITH, P. G., 1953. The cultivated *Capsicum* peppers. *Economic Botany, 7:* 214–227.

HEMSLEY, W. B., 1881–82. *Biologia Centrali-Americana. Solanaceae, 2*. London.

HENDERSON, R. J. F., 1974. *Solanum nigrum* L. (Solanaceae) and related species in Australia. *Contributions of the Queensland Herbarium, 16:* 1–78.

HENSLOW, G., 1893. *The Origin of Floral Structures through Insects and other Agencies*. 349 p. London.

HERASIMENKO, I. I., 1965. New forms of *Solanum* L. subgenus *Archaesolanum*. *Byulleten' Glavnogo Botanicheskogo Sada, 59:* 71–73 (in Russian) (n.v.).

HERASIMENKO, I. I., 1970. Conspectus subgeneris *Archaesolanum* Bitt. ex Marz. generis *Solanum* L. *Nov. Syst. Pl. Vasc., 7:* 270–275 (in Russian) (n.v.).

HERBST, E. E., 1972. 'N morphologiese ondersoek van Retzia capensis Thunb. M.S. thesis, unpubl. Universiteit van Pretoria.

HEUSSER, C. J., 1971. *Pollen and spores of Chile:* 1–67. Tucson: Univ. Arizona Press (n.v.).

HITCHCOCK, C. L., 1932. A monographic study of the genus *Lycium* in the Western Hemisphere. *Annals of Missouri Botanical Garden, 19:* 179–374.

HOGSTAD, R., 1912. A morphological and chemical study of *Nicandra physaloides* (L.) Pers. *Journal of the American Pharmacology Association, 12:* 576–582 (n.v.).

HOGSTAD, R., 1920. *Solanum carolinense* L. *Merck's Rep., 19:* 249, see *Botanisches Zentralblatt, 114:* 606 (n.v.).

HOWARD, H. W., 1970. *Genetics of the Potato*, Solanum tuberosum: 126 pp. London: Logos Press.

HUANG, T., 1968. Pollen grains of Formosan Plants (4). *Taiwania, 14:* 254–261.

HUANG, T., 1972. *Pollen Flora of Taiwan:* 297 pp. National Taiwan Univ., Botany Dept. Press.

HUMBOLDT, F. H. A. von, BONPLAND, A. J. & KUNTH, C. S., 1818. *Nova Genera et Species Plantarum*, Sect. 3. In, *Voyage aux Régions Équinoctiales du Nouveau Continent,* 2: 372–375; *3:* 1–64. Paris.

HUNZIKER, A. T., 1958. Synopsis of the genus *Capsicum. 8 Congres International de Botanique:* 73–74.

HUNZIKER, A. T., 1960. Estudios sobre Solanaceae. II. Sinopsis taxonomica del genero *Dunalia* H.B.K. *Boletín de la Academia Nacional de Ciencias en Córdoba,* 41: 211–243.

HUNZIKER, A. T., 1967. Estudios sobre Solanaceae IV. Una especie nueva y dos notas criticas. *Kurtziana,* 4: 131–138.

HUNZIKER, A. T., 1969a. Estudios sobre Solanaceae V. Contribucion al conocimiento de *Capsicum* y generos afines (*Witheringia, Acnistus, Athenaea,* etc.), primera parte. *Kurtziana,* 5: 101–179.

HUNZIKER, A. T., 1969b. Estudios sobre Solanaceae VI. Contribucion al conocimiento de *Capsicum* y generos afines (*Witheringia, Acnistus, Athenaea,* etc.), secunda parte. *Kurtziana,* 5: 393–399.

HUNZIKER, A. T., 1971. Estudios sobre Solanaceae. VII. Contribucion al conocimiento de *Capsicum* y generos afines. (*Witheringia, Acnistus, Athenaea,* etc.), tercera parte. *Kurtziana,* 6: 241–259.

HUNZIKER, A. T., 1977. Estudios sobre Solanaceae VIII. Novedades varias. *Kurtziana, 10:* 7–50.

HUNZIKER, A. T. & DI FULVIO, E., 1957. Observaciones morfologicas sobre *Peltanthera* (Loganiaceae), con referencia a su posicion sistematical *Boletín de la Academia Nacional de Ciencias en Córdoba, 40:* 217–228.

HUTCHINSON, J., 1959. *The Families of Flowering Plants*, 2nd ed. Oxford.

HUTCHINSON, J., 1969. *Evolution and Phylogeny of Flowering Plants:* 717 p. London: Academic Press.

IKUSE, M., 1956. *Pollen Grains of Japan:* 1–304. Tokyo: Hirokawa.

INAMDAR, J. A., 1969. Epidermal structure and ontogeny of stomata in some Verbenaceae. *Annals of Botany* (*N.S.*), *33:* 55–66.

INAMDAR, J. A. & PATEL, R. C., 1969. Development of the stomata in some Solanaceae. *Flora, 158:* 462–472.

IRISH, H. C., 1898. A revision of the genus *Capsicum* with especial reference to garden varieties. *Report Missouri Botanical Garden, 9:* 53–110.

JAIN, R. K. & NANDA, S., 1966. Pollen morphology of some desert plants of Pilani, Rajasthan. *Palynologica Bulletin* (*Lucknow*), *II & III:* 56, 69 (n.v.).

JAY, M. & GONNET, J. F., 1974. Les flavonoïds de deux Lentibulariacées: *Pinguicula vulgaris* et *Utricularia vulgaris. Biochemical Systematics and Ecology, 2:* 47–51.

JENKINS, J. A., 1948. The origin of the cultivated tomato. *Economic Botany,* 2(4): 379–392.

JENSEN, S. R., NIELSON, B. J. & DAHLGREN, R., 1975. Iridoid compounds, their occurrence and systematic importance in the angiosperms. *Botaniska Notiser, 128:* 148–180.

JEPSON, W. L., 1925. Solanaceae. *A Manual of the Flowering Plants of California:* 886–895. 1951 reprint, Univ. of Calif. Press.

JEPSON, W. L., 1943. Solanaceae. In, *A Flora of California,* 3(II): 449–464.

JOHNSTON, I. M., 1929. Papers on the flora of northern Chile. *Contributions from the Gray Herbarium, 85:* 1–180. Solanaceae: 112–115; 159–161; 177–179. Nolanaceae: 103–112; 156–159.

JOHNSTON, I. M., 1936. A study of the Nolanaceae. *Contributions from the Gray Herbarium, 112:* 1–83.

JÖNSSON, B., 1881. Ytterligare bidrag till kännedomen om Angiospermernas embryosäckutveckling. *Botaniska Notiser, 6:* 169–187 (n.v.).

JØRGENSEN, C. A., 1928. The experimental formation of heteroploid plants in the genus *Solanum. Journal of Genetics, 19:* 133–210.

JUSSIEU, A. L. de, 1789. *Genera Plantarum*. Paris.

KESSEL, R., 1974. Inheritance of split stigma character in diploid *Solanum* species. *Potato Research, 17:* 227–233.

KHAN, H. A. & REHMAN, K., 1966. Pollen grains of some cultivated species of *Nicotiana. Science and Culture, 32:* 145–146.

KHUSH, G. S. & RICK, C. M., 1966. The origin, identification, and cytogenetic behavior of tomato monosomics. *Chromosoma, 18:* 407–420.

KNUTH, P. A., 1906. *Handbook of Flower Pollination.* (English Transl. J. R. A. Davis) Oxford.

KOELREUTER, J. G., 1811. Dissertationis de antherarum pulvere continuatio. Sectio quarta. De figura antherarum pulveris. *Mémoires de l'Académie Impériale des Sciences de St. Petersbourg, 3.*

KOSTOFF, D., 1926. Die Bildung der Pollenkorner bei einigen Varietaten von *Capsicum annuum. Jahrb. Univ. Sofia, 4:* 101–126 (n.v.).

KRAMER, P. R., 1939. The woods of *Billia, Cashalia, Henoonia,* and *Juliania. Tropical Woods, 58:* 3.

KREUSCH, W., 1933. Über Entwicklungsgeschichte und Vorkommen des Kalzium-oxalates in Solanaceen. *Botanisches Zentralblatt, 56:* 410–431 (n.v.).

KUANG, K.-Z., & LU, A.-M., 1965. Revisio Physaliastrorum Makino. *Acta Phytotax. Sin., 10:* 347–355.

KUANG, K.-Z. & LU, A.-M., 1974. De speciebus sinensibus generis Physochlainae. *Acta Phytotaxonomica Sinica*, *12*(4): 407–413.

KUHLMANN, J. G., 1934. Notas sôbra o género *Duckeodendron*. *Archivas do Instituto de Biologia Vegetal*, *1*: 35–37.

KUNTH, C. S., 1823. Solanaceae. In F. H. A. von Humboldt & A. J. Bonpland (Eds), *Synopsis Plantarum*, 2: 146–188.

KUZENEVA, O. I., 1966. Solanaceae. In A. I. Pojarkova (Ed.), *Flora Murmanskoi Oblasti*, 5: 93–96. (Flora of the Murmansk Region.)

LAGERHEIM, G., 1895. Monographie der ecuadorischen Arten der Gattung *Brugmansia* Pers. *Engl. Bot. Jahrb.*, *20*: 655, t. 11.

LAMARCK, J. B. A. P. M. DE, 1793. [1794] *Tab. Encyclopedique et methodique . . .*, *2*: 2–31.

LAMARCK, J. B. A. P. M. DE, 1804. [1794 fide Jussieu, *Annales du Musie d'Histoire Naturelle, Paris 3*: 121.] Morelle (*Solanum*). In, *Encyclopedie méthodique. Botanique*, *4*: 278–309. Paris.

LAMARCK, J. B. A. P. M. DE, 1813. Morelle (*Solanum*). In, *Encyclopedie Méthodique. Botanique* Suppl., *3*: 738–780.

LAWRENCE, G. H. M., 1941. Solanaceae, etc. In L. H. Bailey & E. Z. Bailey (Eds), *Hortus Second*, new ed. Macmillan.

LAWRENCE, G. H. M., 1960a. The cultivated species of *Solanum*. *Baileya*, *8*(1): 20–35.

LAWRENCE, G. H. M., 1960b. Notes on cultivated *Solanums*. *Baileya*, *8*(2): 75–76.

LEGGE, A. P., 1974. Notes on the history, cultivation and uses of *Physalis peruviana* L. *Journal of the Royal Horticultural Society*, *99*: 310–314.

LEHMANN, J. G. C., 1818. *Generis Nicotianarum Historia*: 52 pp. Hamburg.

LEVIN, D. A., 1976. Alkaloid-bearing plants: an ecogeographic perspective. *American Naturalist*, *110*: 261–284.

LINNAEUS, C., 1753. *Species Plantarum*, 2 vols. Stockholm.

LINNAEUS, C., 1754. *Genera Plantarum*, 5th ed. Stockholm.

LINNAEUS, C. von, Jr., 1762. *Decas Prima Plantarum Rariorum Horti Upsaliensis*. Stockholm.

LINSLEY, E. G., 1960. Observations on some matinal bees at flowers of *Cucurbita*, *Ipomoea* and *Datura* in desert areas of New Mexico and southeastern Arizona. *Journal of the New York Entomological Society*, *68*: 13–20.

LINSLEY, E. G., 1962. The colletid *Ptiloglossa arizonensis* Timberlake, a matinal pollinator of *Solanum*. *Pan-Pacific Entomologist*, *38*: 75–82.

LINSLEY, E. G. & CAZIER, M. A., 1963. Further observations on bees which take pollen from plants of the genus *Solanum*. *Pan-Pacific Entomologist*, *39*: 1–18.

LIPPERT, L. F., SMITH, P. G. & BERGH, B. O., 1966. Cytogenetics of the vegetable crops: garden pepper. *Botanical Gazette*, *136*: 20–26.

LOCKWOOD, T. E., 1973. Generic recognition of *Brugmansia*. *Botanical Museum Leaflets*, *23*(6): 273–283.

LUCKWILL, L. C., 1943. The genus *Lycopersicon*. *Aberdeen University Studies*, No. 120: 43 pp.

MACBRIDE, J. F., 1962. Solanaceae (parte). In, *Flora of Peru. Field Museum Bot.*, *8*, V-B (1).

MAGOON, M. L., RAMANUJAM, S. & COOPER, D. C., 1961. Cytogenetical studies in relation to the origin and differentiation of species in the genus *Solanum* L. *Caryologia*, *15*: 151–252.

MAGTANG, M. V., 1936. Floral biology and morphology of the egg plant. *Philippine Agriculturalist*, *25*: 30–53.

MAHESHWARI, P., 1950. *An Introduction to the Embryology of Angiosperms*. New York.

MALLIK, N. *et al.*, 1964. Pollen morphology of some Pakistani medicinal plants. *Pakistan Journal of Scientific and Industrial Research*, *7*(2): 130–136 (n.v.).

MARCHESI, E. H., 1965. Plantas nuevas o poco conocidas de la flora Uruguaya I. *Comunicaciones Botanicas del Museo de Histoiria Natural de Montevideo*, *44*: 1–10.

MARTIN, A. C., 1946. The comparative internal morphology of seeds. *American Midland Naturalist*, *36*: 513–660.

MARTIN, J. N., ERWIN, A. T. & LOUNSB, C. C., 1932. Nectaries of *Capsicum*. *Iowa State College Journal of Science*, *6*: 277–284.

MARTIN, P. S., 1969. Pollen analysis and the scanning electron microscope. *Proceedings of II Annual Scanning Electron Microscope Symposium, I. I. T. Research Institute*, Chicago, 1–14. (n.v.).

MARTIN, P. S. & DREW, M., 1969. Scanning electron photomicrographs of southwestern pollen grains. *Journal of the Arizona Academy of Science*, *5*: 147–176 (n.v.).

MARTÍNEZ, C. P., 1949. Los '*Solanum*' ornamentales cultivados en Argentina (abstract). *Lilloa*, *17*: 49.

MARTÍNEZ, M., 1966. Las *Solandras* de Mexico, con una especies nueva. *Anales del Instituto de Biologiá Universidad de México*, *37*: 97–106.

MARZELL, H., 1927. Fam. 116. Solanaceae. In G. Hegi, *Illustrierte Flora von Mittel-Europa*, *4*: 2548–2625.

MÁTHÉ, I. & MÁTHÉ, I., Jr., 1973. Data to European area of chemical taxa of *Solanum dulcamara* L. *Acta Botanica Academiae Scientiarum Hungaricae*, *19*: 441–451.

MATTFELD, J. & BITTER, G., 1922. Genus *Tunaria* O.K. e *Solanaceis* excludendum (= *Cantua pirifolia* Juss.). *Feddes Repertorium Specierum Novarum Regni Vegetabilis, 18:* 299–300.

MEISNER, C. F., 1840. Nolanaceae and Solanaceae. In, *Plantarum Vascularium Genera:* 275–278.

MELCHIOR, H., 1964. Solanaceae, *2:* 444–447. In A. Engler (Ed.), *Syllabus der Pflanzenfamilien,* 12th ed. Berlin.

MENDIA, C. de, 1939. *Subsidios Para o Estudo dos Granulos de Polen da Flora Melifera.* Thesis, Lisbon (n.v.).

MENNINGER, E. A., 1966. *Datura* species in Florida gardens. *American Horticultural Magazine, 45:* 375–387.

MENZEL, M. Y., 1950. Cytotaxonomic observations on some genera of the Solanae: *Margaranthus, Saracha,* and *Quincula. American Journal of Botany, 37:* 25–30.

MENZEL, M. Y., 1951. The cytotaxonomy and genetics of *Physalis. Proceedings of the American Philosophical Society, 95(2):* 132–183.

MENZEL, M. Y., 1964. Preferential chromosome pairing in allotetraploid *Lycopersicon esculentum—Solanum lycopersicoides. Genetics, 50:* 855–862.

MERRILL, E. D., 1949. *Index Rafinesquianus:* 296 pp. Jamaica Plain, Massachusetts.

MERXMÜLLER, H. & BUTTLER, K. P., 1975. *Nicotiana* in der Afrikanischen Namibein pflanzengeographisches und phylogenetisches Rätsel. *Mitteilungen Botanischen München, 12:* 91–104.

METCALFE, C. R. & CHALK, L., 1950. Fam. 197. Solanaceae. In, *Anatomy of the Dicotyledons, 2:* 965–978. Oxford.

MICHENER, C. D., 1962. An interesting method of pollen collecting by bees from flowers with tubular anthers. *Rev. Biol. Trop. 10:* 167–175.

MIERS, J., 1848–51; 1850–57. Notes on the Solanaceae. *Annals and Magazine of Natural History, (Ser. II), 3–6* (pages var.); *Hooker's London Journal of Botany, 4–7* (pages var.); *Hooker's Journal of Botany, 1:* 65–67; 192; 225–226.

MIERS, J., 1854. On the genus *Lycium. Annals and Magazine of Natural History (Ser. 2), 14:* 1–20; 131–141; 182–194; 336–346.

MIERS, J., 1871. On the genera *Goetzia* and *Espadea. Transactions of the Linnean Society of London, 27:* 187.

MILLÁN, A. R., 1928. Las especieas del genero *Nicotiana* de la flora Argentina. *Revista de la Facultad de Agronomia, Universidad de Buenos Aires, 6:* 169–240 (n.v.).

MILLÁN, A. R., 1931. Solanáceas argentinas. Clave para la determinación de los géneros. *Boletin de Ministerio de Agricultura Nacional, 30(1):* 1–21. Argentina.

MILLÁN, A. R., 1941. Revision de las especies del genero *Nierembergia* (Solanaceae). *Darwiniana, 5:* 487–549. 1941; *4:* 331. 1942.

MILLER, P., 1731–68. *The Gardeners Dictionary,* 8 eds. London.

MILLS, D. E. & JACKSON, B., 1972. *Scopolia lurida* Dunal: the structure of the leaves and stem. *Journal of Pharmacy and Pharmacology, 24:* 234–235 (n.v.).

MIRANDE, M., 1922. Sur l'origine morphologique du liber interne des Nolanacées et la position systematique de cette famille. *Compte Rendu Hebdomadaire des Séances de l'Academie des Sciences, 175:* 375–376.

MOHAN, K., 1968. Morphological studies in Solanaceae II. Morphology, development and structure of seed of *Browallia demissa* Linn. *Proceedings of the National Institute of Sciences of India, B34:* 142–148.

MOHAN, K. & SINGH, B., 1968. Morphological studies in the Solanaceae III—Structure and development of seed of *Solanum torvum* Sw. *Journal of the Indian Botanical Society, 48:* 338–345.

MOHL, H., 1934. *Beitrage zur Anatomie und Physiologie der Gewachse. Erstes Heft. Über den Bau und die Formen der Pollenkorner.* Bern (n.v.).

MONACHINO, J., 1940. *Leptoglossis.* In H. N. Moldenke (Ed.), *The Flora of Extra-tropical South America. Lilloa, 5:* 435–436.

MONTEIRO, H. DA C., FILHO, 1968. Plantae Vellozianae Criticae. *Revista da Faculdade de Ciencias, Universidade de Lisboa, (2 ser.), 15(2):* 269–274.

MOORE, D. M. (Ed.), 1972. Solanaceae. In T. G. Tutin *et al.* (Eds), *Flora Europaea, 3:* 193–201.

MORRETES, B. L. DE, 1969. Contribuioao ao estudio da anatomia das folhas de plantas de cerrado: iii. *Boletim da Faculdade de Filosofia, Ciências e Letras, 24:* 7–32.

MORTON, C. V., 1936. The genus *Cestrum* in Guatemala. *Journal of the Arnold Arboretum, 17:* 341–349.

MORTON, C. V., 1938. Notes on the genus *Saracha. Proceedings of the Biological Society of Washington, 51:* 75–78.

MORTON, C. V., 1944. *Some South American Species of* Solanum: 260 pp. Córdoba: Acad. Nac. Cienc.

MORTON, C. V., 1976. *A Revision of the Argentine Species of* Solanum. Córdoba.

MUKHERJEE, D., 1974a. Pharmacognostic study of some market varieties of tobacco leaves. *Bulletin of the Botanical Society of Bengal, 28:* 91–94.

MUKHERJEE, D., 1974b. Pharmacognostic study of two market varieties of tobacco leaves. *Bulletin of the Botanical Society of Bengal, 28:* 95–99.

MULLER, C. H., 1940. A revision of the genus *Lycopersicon. U.S.D.A. Miscellaneous Publication, No. 382:* 29 pp.

MUÑOZ-PIZARRO, CARLOS, 1959. Solanaceae. In, *Sinopsis de la Flora Chilena:* 188–191. Santiago.

MUNTING, A. J., 1974. Development of flower and fruit of *Capsicum annuum* L. *Acta Botanica Neerlandica, 23:* 415–432.

MUNZ, PHILIP, A. & KECK, D. B., 1959. Solanaceae (Fam. 62). In, *A California Flora:* 590–603. Berkeley: Univ. Calif. Press.

MURRAY, M. A., 1945. Carpellary and placental structure in the Solanaceae. *Botanical Gazette, 107:* 243–260.

MURRY, L. E. & ESHBAUGH, W. H., 1971. A palynological study of the Solaninae (Solanaceae). *Grana, 11:* 65–78.

NAIR, P. K. K., 1965. Pollen grains of Western Himalayan plants. *Asia Monographs, India, 1:* VIII + 1–102. (n.v.).

NAMIKAWA, I., 1919. Über das öffnen der antheren bei einigen solanaceen. *Botanical Magazine (Tokyo), 33:* 62–69.

NATARAJAN, A. T., 1957. Studies in the morphology of pollen grains. Tubiflorae. *Phyton, 8:* 21–42 (n.v.).

NECKER, N. J. de, 1970. *Elementa Botanica.* Neuwied.

NEES VON ESENBECK, C. G., 1831. Versuch einer Verstandigung über die Arten der Gattung *Physalis. Linnaea, 6:* 431–483.

NEES VON ESENBECK, T. F. L., *c.* 1845. Solanaceae. In, *Genera Plantarum Florae Germanicae, 1* (pages not numbered).

NOTHMANN, J., 1973. Morphogenetic effects of low temperature stress on flowers of egg plant, *Solanum melongena* L. *Israel Journal of Botany, 22:* 231–235.

OCAÑA, G. G., 1972. Production comercial de tomate a nivel del mar durante la estacion lluviosa en suelos contaminados por el agente causal de la marchitez bacteriana. *La Estrella de Panama, 28 July 1972:* 11, 10.

OCHOA, C. M., 1962. *Los* Solanum *Tuberiferos Silvestres del Peru (Secc. Tuberarium, Subsecc. Hyperbasarthrum):* 297 pp. Lima.

OETKER, A., 1888. *Zeigt der Pollen in den Unterabtheilungen der Pflanzen-Familien charakteristische Unterschiede?* Thesis, Freiberg i.B. Berlin.

OHTA, Y., 1962. *Genetical Studies in the Genus* Capsicum: 91 pp. Kihara Institute for Biological Research, Yokahama (n.v.).

OVERLAND, L., 1960. Endogenous rhythm in opening and odor of flowers of *Cestrum nocturnum. American Journal of Botany, 47:* 378–382.

PASCHER, A., 1909. *Atropanthe,* eine neue Gattung der Solanaceen. *Österreichische Botanische Zeitschrift, 59:* 329–331.

PASCHER, A., 1910a. Der Aufbau des Sprosses bei *Przewalskia tangutica* Maximowicz. *Flora, 100:* 295–304.

PASCHER, A., 1910b. Über Gitterkelche, einen neuen biologischen Kelchtypus der Nachtschattengewächse. *Flora, N.F., 1:* 273.

PATINO, V. M., 1963, 1964. *Plantas Cultivadas y Animales Domesticos en America Equinoccial.* Cali (Colombia). I. *Frutales:* 547. Lycopersicon, Cyphomandra, Physalis, Solanum (pp. 104–110). II. *Plantas alimenticias:* 364. Potatoes, peppers, tomatoes: 217–222; 195–196; 71–89.

PAYER, J.-B., 1857. *Traite d'Organogenie Comparée de la Fleur,* 2 vols. Solanaceae: 539–540. Nolanaceae: 599–600. Paris.

PENNELL, F. W., 1935. The Scrophulariaceae of eastern temperate North America. *Monographs, Academy of Natural Sciences of Philadelphia, 1:* 1–650.

PFEIFFER, H., 1926. Das abnorme Dickenwachstum. In Linsbauer, *Handbuch der Pflanzenanatomie, 9.* Berlin (n.v.).

PHILIPPI, R. A., 1857. Ueber *Jaborosa* Jussieu. *Botanische Zeitung, 15:* 719–726.

PICHENOT, M., 1956. Essai de greffage de *Solanum sisymbrifolium* Lamk. sur *Datura stramonium* L., *Acad. Sci.,* 17 Dec. 1956.

PICKERSGILL, B., 1966. *The Variability and Relationships of* Capsicum chinense *Jacq.* Ph.D. dissert. Indiana University, Bloomington, (n.v.).

PICKERSGILL, B., 1969a. The domestication of chili peppers. In P. Ucko & G. W. Dimbleby (Eds), *The Domestication and Exploitation of Plants and Animals:* 443–450. London: Duckworth.

PICKERSGILL, B., 1969b. The archaeological record of chili peppers (*Capsicum* sp.) and the sequence of plant domestication in Peru. *American Antiquity, 34:* 54–61 (n.v.).

PICKERSGILL, B., 1971a. Reply to Davenport (letter). *Professional Geographer, 32:* 169–170.

PICKERSGILL, B., 1971b. Relationships between weedy and cultivated forms in some species of chili peppers (genus *Capsicum*). *Evolution, 25:* 683–691.

PITTIER, H., 1932. Studies in Solanaceae: I. The species of *Cestrum* collected in Venezuela up to 1930. *Journal of the Washington Academy of Sciences, 22:* 25–37.

PLOWMAN, T., 1973. *The South American Species of* Brunfelsia *(Solanaceae).* Ph.D. thesis, Harvard University.

PLOWMAN, T., GYLLENHAAL, L. O. & LINDGREN, J. E., 1971. *Latua pubiflora*. Magic plant from southern Chile. *Botanical Museum Leaflets, 23*(2): 61–92.

PODDUBNAJA-ARNOLDI, V., 1926. Beobachtungen über die Keimung des Pollens einiger Pflanzen auf Künstlichem Nährboden. *Planta, 25:* 502–529 (n.v.).

POJARKOVA, A. I., 1942. Contribution to the systematics of the species of henbane related to *Hyoscyamus reticulatus* L. *Journal Botanique de l'URSS, 27:* 117–130.

POJARKOVA, A. I., 1955. Solanaceae. In B. K. Shishkin & E. G. Bobov (Eds), *Flora SSSR, 22:* 1–117.

POUCHET, A. F., 1829. *Histoire Naturelle et Médicale de la Famille des Solanées:* 187 pp. Rouen.

PROBST, R., 1949. *Wolladventiflora Mitteleuropas:* 193 pp. Solothrun.

PURKINJE, J. E., 1830. *De Cellulis Antherarum Fibrosis nec non de Granorum Pollinarium Formis: Commentatio Phytotomica.* Pressburg (n.v.).

QUIROS, C. F., 1975. Exine pattern of a hybrid between *Lycopersicon esculentum* and *Solanum pennellii*. *Journal of Heredity, 66:* 45–47.

RADLKOFER, L., 1889. Zur Klärung von Theophrasta und der Theophrasteen, unter Ubertragung dahin gerechneter Pflanzen zu den Sapotaceen und Solanaceen. *Sitzungberichte der Akademie der Wissenschaften in München, (Math.-Phys.), 19:* 221.

RAGHAVAN, V., 1976. Role of the generative cell in androgenesis in Henbane. *Science, 191:* 388–389.

RAGHUVANSHI, R. K. & SINGH, D., 1972. Epidermal studies in *Capsicum* L. *Journal of the Indian Botanical Society, 51:* 311–319.

RAJ, B. & SURYAKANTA, 1970. Palynological studies in some south Indian weeds. III. *Palynological Bulletin, 6:* 70–83.

RAND, G., 1963. Organographie de la capsule du *Datura stramonium* L. *Bulletin Société Botanique de France, 110:* 216–237.

RANDELL, B. R. & SYMON, D. E., 1976. Chromosome numbers in Australian *Solanum* species. *Australian Journal of Botany, 24:* 369–379.

RECORD, S. J., 1933. The woods of *Rhabdodendron* and *Duckeodendron*. *Tropical Woods, 33:* 6–10.

RECORD, S. J., 1934. *Identification of the Timbers of Temperate North America:* 196 pp. New York.

RECORD, S. J., 1936. Classification of various anatomical features of dicotyledonous woods. *Tropical Woods, 47:* 12–27.

RECORD, S. J., 1939. American woods of the family Sapotaceae. *Tropical Woods, 59:* 35.

RECORD, S. J. & HESS, R. W., 1943. *Timbers of the New World.* New Haven.

REICHE, C., 1910. Solanaceas. In, *Flora de Chile, 5:* 309–410.

REICHENBACH, H. T[G.]. L., 1828. *Conspectus Regni Vegetabilis*. Solanaceae: 125–126. Leipzig.

RENDLE, A. B., 1921. Elisia—an overlooked genus name. *Journal of Botany, 59:* 261–264.

RICARDI, M., 1962. El genero *Combera* (Solanaceae). *Gayana, 4:* 3–13.

RICK, C. M. & BOWMAN, R. I., 1961. Galapagos Tomatoes and Tortoises. *Evolution, 15:* 407–417.

RICK, C. M. & KUSH, G. S., 1966. Chromosome engineering. In Riley & Lewis (Ed.), *Lycopersicon Chromosome Manipulations and Plant Genetics.* London: Oliver & Boyd.

RICK, C. M., ZOBEL, R. W. & FORBES, J. F., 1974. Four Peroxidase loci in red-fruited tomato species: genetics and geographic distribution. *Proceedings of the National Academy of Sciences of the U.S.A., 71:* 835–839.

ROBERTSON, C., 1891. Flowers and insects, Asclepiadaceae to Scrophulariaceae. *Transactions of the Academy of Sciences of St. Louis, 5:* 569–598.

ROBYNS, W., 1930 [1931]. L'organisation florale des Solanacées zygomorphes. *Mémoires de l'Academie r. de Belgique. Classe des Sciences, 11*(8): 82 pp.

ROBYNS, W., 1932. L'étude detaillee des formes florales et son importance pour la systematique (Solanacées, Labiatées). *Bulletin de l'Association Française pour l'Avancement des Sciences,* 264–270.

ROBYNS, W., 1936. Sur des phénomènes de pleiomerie et de synanthie dans *Salpiglossis sinuata* Ruiz et Pav. *Bulletin de l'Academie r. de Belgique. Classe des Sciences (Ser. 5), 12:* 1080–1098.

ROE, K. E., 1966a. Juvenile forms in *Solanum mitlense* and *S. blodgettii* and their importance in taxonomy. *Sida, 2:* 381–385.

ROE, K., 1966b. Chromosome numbers in some Solanaceae. *Sida, 3*(3): 153–155.

ROE, K., 1967. A revision of *Solanum* sect. *Brevantherum* in North and Central America. *Brittonia, 19:* 353–373.

ROE, K., 1971. Terminology of hairs in the genus *Solanum*. *Taxon, 20:* 501–508.

ROE, K., 1972. A revision of *Solanum* section *Brevantherum* (Solanaceae). *Brittonia, 24:* 239–278.

ROEMER, J. J. & SCHULTES, J. A., 1819. *Caroli a Linné Equitis Systema Vegetabilium . . .* Ed. 15. 7 vols. Solanaceae, 4. Stuttgart.

ROJAS, C. E. BENITEZ DE, 1974. Los géneros de las Solanaceae de Venezuela. *Revista de la Facultad de Agronomia, Universidad Central de Venezuela, 7*(3): 25–108.

ROMERO-CASTAÑEDA, R., 1961a. *Frutas Silvestres de Colombia, 1.* Bogotá.

ROMERO-CASTAÑEDA, R., 1961b. El lulo, una fruta de importancia economica. *Agricultura Tropical*, *17*(4): 214.

ROMERO-CASTAÑEDA, R. & SCHULTES, R. E., 1962. Edible fruits of *Solanum* in Colombia. *Botanical Museum Leaflets*, *19*(10): 235–286.

ROSS, J. H., 1975. The typification of *Lycium inerme*. *Bothalia*, *11*: 491–493.

RUIZ, L. H. & PAVON, J. A., 1794. *Florae Peruvianae et Chilensis Prodromus*. Madrid.

RUIZ, L. H. & PAVON, J. A., 1798. *Systema Vegetabilium Florae Peruvianae et Chilensis*. Madrid.

RUIZ, L. H. & PAVON, J. A., 1798–1802. *Flora Peruviana et Chilensis*, 3 vols. Madrid.

RUSBRIDGE, J. W., 1976. The factors affecting the germination of tomato seeds. *Report of the Rugby School Natural History Society*, *108*: 14–17.

RYDBERG, P. A., 1896. The North American species of *Physalis* and related genera. *Memoirs of the Torrey Botanical Club*, *4*: 297–374.

SAFFORD, W. D., 1921. Synopsis of the genus *Datura*. *Journal of the Washington Academy of Science*, *11*: 173–182.

SALAMAN, R. N., 1946. The early European potato, its character and place of origin. *Journal of the Linnean Society (Botany)*, *55*: 185–190.

SALAMAN, R. N., 1949. *The History and Social Influence of the Potato*: 685 pp. Cambridge University Press.

SALGADO-LABOURIAU, M. L., FREIRE DE CARVALHO & CAVALCANTE, P. B., 1969. Pollen grains of plants of the 'Cerrado' XXI—Ebenaceae, Nyctaginaceae, Rhamnaceae and Solanaceae. *Boletim do Museu Paraense 'Emilio Goeldi' (N. S.)*, *32*: 1–14.

SAMUELSSON, G., 1913. Studien über die Entwicklungsgeschichte der Blüten einiger Bicornes Typen. *Svensk Botanisk Tidskrift*, *7*: 97.

SANDER, H., 1963. Zur Chemotaxonomie der Gattung *Solanum* (sensu ampl.). *Botanische Jahrbücher für Systematik, Pflanzengeschichte und Pflanzengeographie*, *82*: 404–428.

SANDWITH, N. Y., 1936. *Combera paradoxa* Sandwith. *Hooker's Icones Plantarum*, *3325*: 1–5.

SANTAPAU, H., 1948. Notes on the Solanaceae of Bombay. *Journal of the Bombay Natural History Society*, *47*: 652–662.

SASSEN, M. M. S., 1964. Fine structure of *Petunia* pollen grain and pollen tube. *Acta Botanica Neerlandica*, *13*: 175–181.

SATINA, S. & BLAKESLEE, A. F., 1940. Periclinal chimeras in *Datura stamonium* in relation to development of leaf and flower. *American Journal of Botany*, *28*: 862–871.

SAUNDERS, E. R., 1935. The history, origin, and characters of certain interspecific hybrids in *Nolana* and their relation to *Nolana paradoxa*. *Journal of Genetics*, *29*: 337–419.

SAUNDERS, E. R., 1936. On certain unique features of the gynoecium in Nolanaceae. *New Phytologist*, *35*: 423–430.

SCASSO, J. M. & MILLÁN, A. R., 1930. Ensayo de classificacion de variedades de tomate. *Boletin del Ministerio de Agricultura (Argentina)*, *29*: 267–296.

SCHNIZLEIN, A., 1843–70. *Iconographia Familiarum Naturalium Regni Vegetabilis*, *II*: 144, 148. Bonn.

SCHREIBER, K. VON, 1954. Die Glykoalkaloide der Solanaceen. *Chemische Technik 6*(?12): 648–657.

SCHREIBER, K. VON, 1968. Steroid alkaloids. The *Solanum* group. In R. H. F. Manske (Ed.), *The Alkaloids*, *X*: 1–192 New York: Academic Press.

SCHREIBER, K. VON, 1974. Synthesis and biogenesis of steroidal *Solanum* alkaloids. *Transactions of the Biochemical Society*, *2*: 1–25.

SCHULTES, R. E., 1956. A new plant source of narcotic drugs: *Methystichodendron amesianum*. *Bulletin of Narcotics*, *8*: 1–4.

SCHULZ, OTTO E., 1909–10. Solanacearum genera nonnulla. In, *Symbolae Antillanae*, *6*: 140–279. Leipzig.

SCHÜRHOFF, P. N., 1925. Die Haploidgeneration der Blütenpflanzen (siphonogamen Embryophyten). *Botanisch Jahrbücher für Systematik, Pflanzengeschichte und Pflanzengeographie*, *59*: 198–293.

SCHWANITZ, G., 1967. Untersuchungen zur postmeiotischen mikrosporogenese—II. Vergleichende Analyse der Pollenentwicklung sub-und emers bluehender Arten. *Pollen et Spores*, *9*(2): 183–209 (n.v.).

SCOLNIK, R., 1954. Sinopsis de las especies de *Cestrum* de Argentina, Chile y Uruguay. *Revista Argentina Agronomica*, *21*(1): 25–32.

SCOTT, W. E., MA, R. M., SCHAFFER, P. S. & FONTAINE, T. D., 1957. A survey of selected Solanaceae for alkaloids. *Journ. Amer. Pharm. Ass.*, *46*(5): 302–304.

SECKT, H., 1929–30. *Flora Cordobensis*: 632 p. Cordoba.

SEITHE, A., (née von Hoff), 1962. Die Haararten der Gattung *Solanum* L. und ihre taxomomische Verwertung. *Botanische Jahrbücher für Systematik Pflanzengeschichte und Pflanzengeographie*, *81*(3): 261–336.

SELLING, O. H., 1947. Studies in Hawaiian pollen statistics—part II. The pollens of the Hawaiian phanerogams. *Special Publications. Bernice Pauahi Bishop Museum*, *38*: 1–430 (n.v.).

SENDTNER, O., 1845. Monographia Cyphomandrae, novi Solanacearum generis. *Flora*, *28*: 161–176.

SENDTNER, O., 1846. Solanaceae, Cestrineae. In C. F. P. von Martius (Ed.), *Flora Brasiliensis*, *10*.

SESSE Y LACASTA, M. de & MOCIÑO, J. M., 1887–91. *Plantae Novae Hispaniae*. Mexico.

SESSE Y LACASTA, M. de & MOCIÑO, J. M., 1891–96. *Flora Mexicana*. Mexico.

SHAH, G. L. & SURYAMARAYANA, D., 1967. Floral variations in three species of *Cestrum* Linn. vis. *C diurnum* Linn., *C. elegans* Schlecht., and *C. nocturnum* Linn. *Journal of the Bombay Natural History Society*, *63:* 456–459.

SHAH, J. J. & PATEL, J. D., 1968. Vascular inter-relationships in some Solanaceae Members. *Biologia Plantarum*, *11:* 329–333.

SHARMA, B. D., 1974. Contributions to the palynotaxonomy of genus *Solanum* Linn. *Journal of Palynology*, *10:* 51–68.

SHINNERS, L. H., 1962. *Salpichroa originifolia* instead of *S. rhomboidea* (Solanaceae). *Leaflets in Western Botany*, *9:* 257–259.

SHOPOVA, M., 1966. Studies in the genus *Capsicum*. I. Species differentiation. *Chromosoma*, *19:* 340–348.

SINHA, N. P., 1950. On the somatic chromosomes of some non-tuberous *Solanum* species. *Current Science*, *11:* 348.

SKOTTSBERG, C. J. F., 1916. *Benthamiella* Speg. und *Saccardophytum* Speg. *Botanische Jahrbücher für Systematik Pflanzengeschichte und Pflanzengeographie*, *54:* 44–50.

SLEUMER, H., 1950. Estudios sobre el genero *Dunalia* H.B.K. *Lilloa*, *23:* 117–142.

SMALL, J. K., 1933. *Manual of the Southeastern Flora*. Solanaceae: 1105–1121. Univ. N. Carolina Press.

SMITH, L. B. & DOWNS, R. J., 1966. Solanaceas. In P. R. Reitz (Ed.), *Flora Ilustrada Catarinense*, Parte I.

SMITH, O., 1931. Characteristics associated with abortion and intersexual flowers in the eggplant. *Journal of Agricultural Research*, *43:* 83–94.

SMITH, P. G. & HEISER, C. B. Jr, 1951. Taxonomic and genetic studies on the cultivated peppers: *Capsicum annuum* L. and *C. frutescens* L. *American Journal of Botany*, *38:* 362–368.

SMITH, P. G. & HEISER, C. B. Jr, 1957. Taxonomy of *Capsicum sinense* Jacq. and the geographic distribution of the cultivated *Capsicum* species. *Bulletin of the Torrey Botanical Club*, *84:* 413–420.

SOLEREDER, H., 1891 [1892]. Über die Versetzung der Gattung *Melananthus* Walp. von den Phrymaceen zu den Solanaceen. *Bericht der Deutschen Botanischen Gesellschaft*, *9:* 65–91.

SOLEREDER, H., 1898. Zwei Beiträge zur Systematik der Solanaceen. *Bericht der Deutschen Botanischen Gesellschaft*, *16:* 242–260.

SOLEREDER, H., 1898–99. *Systematische Anatomie der Dicotyledonen*: 984 pp. Stuttgart.

SOLEREDER, H., 1915. Über die Versetzung der Gattung *Heteranthia* von den Scrophulariaceen zu den Solanaceen. *Beihefte zum Botanischen Zentralblatt*, *33*(2): 113–117.

SORIANO, A., 1948. El genero *Benthamiella* (Solanaceae). *Darwiniana*, *8:* 233–262.

SOUÈGES, R., 1907. *Développement et Structure du Tégument séminal chez les Solanacées*: 1–124 pp. Thesis. Fac. des Sci. de Paris, (n.v.).

SOUÈGES, R., 1920a. Embryogénie des Solanacées. Développement de l'embryon chez les *Nicotiana*. *Compte Rendu Hebdomadaire des Séances de l'Academie des Sciences*, *170:* 1125–1127 (n.v.).

SOUÈGES, R., 1920b. Embryogénie des Solanacées. Développement de l'embryon chez les *Hyoscyamus* et *Atropa*. *Compte Rendu Hebdomadaire des Séances de l'Academie des Sciences*, *170:* 1279–1281 (n.v.).

SOUÈGES, R., 1922. Recherches sur l'embryogénie des Solanacées. *Bulletin Société Botanique de France*, *69:* 163–178, 236–241, 352–365, 555–585 (n.v.).

SOUÈGES, R., 1936. Développement de l'embryon chez le *Schizanthus* et les *Petunias*. *Bulletin Société Botanique de France*, *83:* 570–577 (n.v.).

SOUÈGES, R., 1937. L'embryon chez les *Amaranthus*. Relations embryologiques entre les Solanacées et les Centrospermales. *Bulletin Société Botanique de France*, *84:* 242–255.

SPLENDORE, A., 1906. *Sinossi Descrittiva ed Iconografia dei Semi del Genere* Nicotiana. Portici.

STAFLEU, F. A., 1972. *International Code of Botanical Nomenclature*: 426 p. Utrecht.

STANDLEY, P. C. & MORTON, C. V., 1938. Solanaceae. In, Flora of Costa Rica. *Field Museum Botany*, *18:* 1035–1099.

SUDZUKI, F., 1969. Thesis, unpubl. Ingeniero Agronomo. cf. Walters 1969 (n.v.).

SUESSENGUTH, K. & BEYERLE, 1938. *Melananthus luetzelburgii* Suessenguth & Beyerle, nov. sp. *Feddes Repertorium Specierum Novarum Regni Vegetabilis*, *44:* 34.

SVENSON, H. K., 1946. Vegetation of Ecuador, Peru, and Galapagos Islands. *American Journal of Botany*, *33:* 481–485.

SWAMINATHAN, M. S. & HOWARD, H. W., 1953. The cytology and genetics of the potato (*Solanum tuberosum*) and related species. *Genetica*, *16:* 1–192.

SWIFT, L. H. 1962. *Venation Patterns in the Lamina of Solanaceous seedlings*. Doctoral Dissert. unpubl. Univ. Nebraska.

TAKHTAJAN, A., 1969. *Flowering Plants: Origin and Dispersal*. Washington: Smithsonian Institution Press.

TERRACCIANO, A., 1890. Contributo alla storia del genere *Lycium*. *Malpighia*, *4:* 472–540.

THIEL, A. F., 1931. Anatomy of the primary axis of *Solanum melongena*. *Botanical Gazette*, *92:* 407–419.

THIERET, J. W., 1967. Supraspecific classification in the Scrophulariaceae: a review. *Sida*, *3*(2): 87–106.

THORNE, R., 1968. Synopsis of a putatively phylogenetic classification of the flowering plants. *Aliso*, *6:* 57–66.

TIMMERMAN, H. A., 1927. *Stramonium* and other species of *Datura:* a comparative study of the structure of their leaves. *Pharmaceutical Journal and Pharmacist, 64* er. *4:* 735–746.

TING-SU, 1949. Illustration of pollen grains of some Chinese plants. *Botaniska Notiser, 4:* 277–282 (n.v.).

TODD, J. E., 1882. On the flowers of *Solanum rostratum* and *Cassia chamaecrista. American Naturalist, 16L:* 281–287.

TOMA, C., 1966. O anomalie aparte la *Lycopersicum esculentum* Mill. *Ann. Sti. Univ. Al. I. Cuza Iasi, Biol., 12:* 419–422 (n.v.).

TOURNEFORT, J. P. de, 1700. *Institutiones Rei Herbariae, Edito Altera,* 3 vols. Paris.

TRONCOSO, A. & TORRES, M. R., 1974. Estudio de la Vegetacion y Florula de la Isla de Quinchao (Chiloe). *Boletín Museo Nacional de Historia Natural, Chile, 33:* 65–107.

TUCKER, W. G., 1969. Serotaxonomy of the Solanaceae: a preliminary survey. *Annals of Botany, 33:* 1–23.

UGENT, D., 1968. The potato in Mexico: geography and primitive culture. *Economic Botany, 22:* 108–123.

UGENT, D., 1970. *Solanum raphanifolium,* a Peruvian wild potato species of hybrid origin. *Botanical Gazette, 131:* 225–233.

URBAN, I., 1922. Sertum Antillarum 14. *Feddes Repertorium Specierum Novarum Regni Vegetabilis, 18:* 23–24.

UTECH, F. H. & KAWANO, S., 1975. Spectral polymorphisms in angiosperm flowers determined by differential ultraviolet reflectance. *Botanical Magazine, 88:* 9–30.

VARGAS, C., CÉSAR, 1936. *El* Solanum tuberosum *a través del Desenvolvimiento de las Actividades Humanas:* 73 pp. Thesis, Univ. de Cuzco.

VESQUE, I., 1885. Caractères des principales familles gamopétales tirés de l'anatomie de la feuille. *Annales des Sciences Naturelles, (Sér. 7), 1:* 183–360. pl. 9–15.

VILMORIN, P. L. F. L. de, 1896. Gattung 845 *Schizanthus* Ruiz & Pav. Spaltblume Vilmorin's, *Blumen-gärtnerei:* 772–774. Berlin.

VOGEL, S., 1958. Fledermausblumen in Südamerika. *Österreichische Botanische Zeitschrift, 104:* 491–530.

WALLIS, T. E. & BUTTERFIELD, R., 1939. The flower of *Atropa belladonna* Linn. *Quarterly Journal of Pharmacy and Pharmacology, 12:* 511–533 (n.v.).

WALPERS, W. G., 1844. Solanaceae, Ordo CLII. In, *Repertorium Botanices Systematicae, III:* 5–126. Leipzig.

WALTERS, D., 1969. *A Revision of the Genus* Schizanthus *(Solanaceae).* Ph.D. Thesis, Indiana Univ.

WANG, F. H., 1960. *Pollen Grains of China:* 276 pp. (in Chinese) (n.v.).

WANGENHEIM, K.-H. F. VON, FRANDSEN, N. O. & ROSS, H., 1957. Ursache der Kreuzungsschwierig-keiten zwischen *Solanum tuberosum* L. und *S. acaule* Bitt. bzw. *S. stoloniferum* Schlechtd. et Bouché. *Zeitschrift für Pflanzenzüchtung, 34:* 7–48 (n.v.).

WARMING, E., 1869. Nogle bemaerkninger om *Scopolia atropoides* og Andre Solaneer (Danish). *Botanisk Tidsskrift, 3:* 39–66.

WEINERT, E., 1972. Zur Taxonomie und Chorologie der Gattung *Scopolia* Jacq. *Feddes Repertorium Specierum Novarum Regni Vegetabilis, 82:* 617–628.

WATERFALL, U. T., 1958. A taxonomic study of the genus *Physalis* in North America north of Mexico. *Rhodora, 60:* 107–114; 128–142; 152–173.

WATERFALL, U. T., 1967. *Physalis* in Mexico, Central America and the West Indies. *Rhodora, 69:* 82–120; 202–239; 319–329.

WATSON, J. A. S., 1927. *The Potato, its History, Varieties, Culture and Diseases:* 264 pp. Edinburgh & London: Oliver & Boyd.

WEBER, C. A., 1928. Georg Bitter. *Bericht der Deutschen Botanischen Gesellschaft, 46* (1 Generalvers Heft 1): 148–156.

WEBER, C. A., 1931. Gesamtverzeichnis von Band XI-XX. *Feddes Repertorium Specierum Novarum Regni Vegetabilis,* Index.

WEITZ, R., 1921a. Le lyciet (*Lycium vulgare* Dunal): recherches botaniques, chimiques et pharmacologiques. *Bulletin de Sciences Pharmacologiques, 28:* 503–508; 562–568 (n.v.).

WEITZ, R., 1921b. *Les* Lycium *Europeens et Exotique; Recherches Historiques, Botaniques, Chimiques et Pharmacologiques:* 202 pp. Vigot. Paris: Lons-le Saunier, L. Declume.

WERDERMANN, E., 1928. Übersicht über die in Chile vorkommenden Arten der Gattung *Salpiglossis* Ruiz et Pav. (Solanaceae). *Notizblatt des K. Botanischen Gartens und Museums zu Berlin, 10:* 472–475.

WERTH, E., 1956. *Bau und Leben der Blumen.* Stuttgart.

WEST, E., 1952. *Perizoma,* a potential weed pest. *Circular, University of Florida Agricultural Experiment Stations,* S-54.

WETTSTEIN, R. VON, 1891. Scrophulariaceae. In A. Engler & K. Prantl (Eds), *Die Natürlichen Pflanzen-familien, 4*(3b): 39–107.

WETTSTEIN, R. VON, 1892. Über die Systematik der Solanaceen. *Verhandlungen de Zoologisch-Botanischen Gesellschaft in Wien, 42* (Sitzber.): 29.

WETTSTEIN, R. VON, 1895. Solanaceae. In A. Engler & K. Prantl (Eds), *Die Natürlichen Pflanzenfamilien, 4*(3b): 4–38: Nachtrage 292–293.

WHALEN, M. D., 1976. New taxa of *Solanum* section Androceras from Mexico and adjacent United States. *Wrightia*, *5:* 228–239.

WIGGINS, I. L., 1951. Family 135. Solanaceae. In L. Abrams (Ed.), *Illustrated Flora of the Pacific States, 3:* 662–685. Stanford Univ. Press.

WIGGINS, I. L. & PORTER, D. M., 1971. *Flora of the Galapagos Islands:* 454–481.

WILLIAMS, Ll., 1936. *Woods of Northeastern Peru.* Chicago: Field Museum.

WILLIAMS, R. O. & WILLIAMS, R. O. Jr., 1951. *The Useful and Ornamental Plants of Trinidad and Tobago:* 335 pp. Port of Spain, Trinidad.

WINKLER, H., 1916. Ueber die experimentelle Erzeugung von Pflanzen mit abweichenden Chromosomenzahlen. *Zeitschrift für Botanik, 8* (n.v.).

WOJCIECHOWSKA, B., 1972. Studia systematyczne nad nasioname rodziny Solanaceae Pers. (Systematic studies on the seeds of the Solanaceae family. *Monographiae Botanicae, 36:* 151–178.

WOODCOCK, E. F., 1935. Vegetative anatomy of the tomato (*Lycopersicum esculentum* Mill.). I. Stem structure. *Papers from the Michigan Academy of Science, Arts and Letters, 21:* 215–222 (n.v.).

WRIGHT, C. H., 1904. Solanaceae (Order XCIV). In W. T. Thiselton-Dyer (Ed.), *Flora Capensis,* 4(2): 87–121.

WRIGHT, C. H., 1906. Solanaceae (Ordo XCI). In W. T. Thiselton-Dyer (Ed.), *Flora of Tropical Africa, IV,* Sect. 2, Pt. II.

YAQUB, C. M. & SMITH, P. G., 1971. Nature and inheritance of self-incompatibility in *Capsicum pubescens* and *C. cardenasii. Hilgardia, 40:* 459–470 (n.v.).

YARNELL, R. A., 1959. Evidence for prehistoric Pueblo use of *Datura. El Palacio, 66:* 176–178.

YARNELL, R. A., 1965. Implications of distinctive flora on Pueblo ruins. *American Anthropology, 67:* 662–674.

ZANDER, E., 1935, 1941, 1951. *Beitrage zur Herkunftsbestimmung bei Honig, I–V. Pollengestaltung und Herkunftsbestimmung bei Blutenhonig. I,* Berlin 1935; *III,* Leipzig 1941; *V,* Leipzig 1951 (n.v.).

ZUTSHI, U., 1974. Meiotic studies in some exotic non-tuberous species of *Solanum. Cytologia, 39:* 225–232.

2. South American Solanaceae: a synoptic survey

ARMANDO T. HUNZIKER

Museo Botánico, Universidad Nacional de Córdoba, Argentina

This chapter appraises and summarizes the available data on the taxomony of the South American Solanaceae, giving a key to the subfamilies and tribes. It provides an enumeration of the tribes and genera with notes on certain species, their distribution and their habitats. Centres of diversity, levels of endemism and distributional peculiarities are also discussed.

CONTENTS

INTRODUCTION

Although the Solanaceae is a family of world-wide importance its greatest concentration of diversity is to be found in South America. For this and other reasons most authorities believe that it originated in this area, spreading to other continents at a very early period, perhaps before the ancient continent of Gondwanaland became fragmented.

The very complexity of the family in South America has made it difficult to attempt synoptic studies, and a great deal of research is still needed.

The present chapter represents an attempt to synthesize information at the tribal and generic levels, and it is hoped that this will act as an incentive to other workers. Special care has been taken to focus attention on the problems of classification at tribal and generic levels. In this connection it is relevant to point out how important experimental data have been in elucidating taxonomic problems in this family. Further experimental work will therefore be welcome, especially in the field of chemotaxonomy. Nevertheless, I believe that the classical techniques of morphology (exomorphology, combined with anatomy, embryology, caryology, palynology, etc.) are still of primary importance and are far from being exhausted.

KEY TO THE SUBFAMILIES AND TRIBES

1. Seeds discoidal or more or less kidney-shaped, flat, compressed; embryo circinnate or curved . . . Subfamily I. SOLANOIDEAE
    2. Embryo with incumbent or sometimes oblique cotyledons; endosperm usually abundant.

49

    3. Flower buds with corolla lobes never overlapping (aestivation valvate, induplicate, plicate or conduplicate).

        4. Filaments inserted at the base or near the base of the anthers.

            5. Aestivation valvate, induplicate, or plicate, never conduplicate-contorted . . . . . **1.** *Solaneae*

            5′. Aestivation conduplicate-contorted . . **2.** *Datureae*

        4′. Filaments inserted generally near the centre of the anthers, on the dorsal face . . . . . **3.** *Jaboroseae*

    3′. Flower buds with overlapping corolla lobes (corolla aestivation imbricate, cochlear or quincuncial).

        4. Filaments inserted on the back of the anthers; thecae separated from each other for their lower third or even higher up; gynoecium 2-carpellary; all the inflorescence internodes very short and of the same length   **4.** *Lycieae*

        4′. Filaments inserted at the base of the anthers.

            5. Calyx and corolla separated by a long internode; filaments geniculate at point of insertion on the corolla . . . . **5.** *Nicandreae*

            5′. Calyx and corolla not separated by a long internode; filaments straight at point of insertion on the corolla . . . . **6.** *Solandreae*

  2′. Embryo generally with accumbent or sometimes oblique cotyledons; endosperm scanty; aestivation of the corolla quincuncial, cochlear or imbricate (valvate only in *Merinthopodium* and *Dyssochroma*) . . . . . . **7.** *Juanulloeae*

1′. Seeds prismatic, reniform, or subglobose, or of a different form but never discoidal-compressed; embryo straight or bent, but then with incumbent or oblique cotyledons . . . .
                        Subfamily II. CESTROIDEAE Schlecht.

  2. Pedicels articulated; corolla actinomorphic; shrubs or trees
                                      8. *Cestreae*

  2′. Pedicels lacking an articulation; corolla actinomorphic or zygomorphic; herbs or shrubs.

    3. Stigmas not umbonate; anthers with small not coalescing connectives.

        4. Prefloration contorted-conduplicate, imbricate-conduplicate, imbricate or cochlear; anthers generally with an ovate, elliptic or almost circular outline; both thecae generally of similar length (excepting some genera of *Salpiglossideae*) and united from the apex to their basal third or even less (sometimes free almost for half their length).

            5. Stamens 5, rarely 4, or even 2, usually inserted at the same level (when the androecium is dimerous, there are no traces of staminodes), the anthers of one

flower similar to each other; corolla actinomorphic
or slightly zygomorphic .   .   .   .   . **9.** *Nicotianeae*

5'. Stamens 4, or 2; when the androecium is tetra-
merous, it is usually didynamous (compared with
the lower ones, the upper stamens show anthers of
different size, or with uneven thecae—owing to the
partial abortion of one of the two, or monothecic—
owing to a total abortion of the other); when the
androecium is dimerous there are 3 staminodes;
corolla clearly zygomorphic .   .   . **12.** *Salpiglossideae*

4'. Prefloration valvate or induplicate, sometimes slightly
contorted; stamens 4 or 2; anthers with a linear outline;
both thecae usually of different length, and coalescing
from the apex to approximately their basal fourth;
corolla zygomorphic .   .   .   .   . **10.** *Schwenckieae*

3'. Stigmas umbonate; anthers coalescent by means of their
protruding connectives .   .   .   .   . **11.** *Parabouchetieae*

ENUMERATION AND DISCUSSION OF THE TRIBES AND GENERA
Tribe I. SOLANEAE
(Fig. 2.1)

The largest tribe in the family with 18 genera and *c*. 1250 species. The aestivation
is never imbricate nor contorted, and the filaments adhere to the base of the
anthers or near it. $x = 12$.

Special references: Bitter (1921), Gentry (1973), Hawkes & Hjerting (1969),
Heiser & Pickersgill (1969), Hunziker (1950, 1954, 1960, 1961a, b, 1967, 1969a, b,
1971, 1977), Miers (1849–1857), Morton (1938), Muller (1940), Sandwith (1938),
Waterfall (1958, 1967).*

1. *Solanum* L.

In this genus, as with *Cyphomandra* and *Lycopersicon*, the filaments are united
at the base, forming a ring adnate to the base of the corolla. The absence of a
nectariferous disk is a remarkable character of obvious relevance to the floral
biology. One of the largest genera of vascular plants, with over 1500 species all
around the world. The estimate of 1000/1100 species for South America (including
Lycianthes Hassler) is undoubtedly conservative. *S. tuberosum* L. (with two
subspecies and other related cultigens) evolved originally in the Peru-Bolivia area,
and is nowadays one of the most important food plants of the world. Besides this
there are about a dozen non-tuberiferous ornamentals cultivated as garden plants.

2. *Cyphomandra* Sendt.

A tropical genus inhabiting Central America, the Antilles and South America,
with the bulk of its *c*. 40 species in Brazil; around ten reach Bolivia and two or

---

* For the sake of brevity, the special references on tribes and their genera do not include books or articles
dealing with the family as a whole (of taxonomic, floristic, or morphological nature), such as those of Bentham
& Hooker (1876), Baehni (1946), Erdtman (1966), Fedde (1896), Foster (1958), Gentry & Standley (1974),
Hemsley (1882), Philippi (1860, 1891), Reiche (1910), Remy (1849), Ruiz & Pavón (1794, 1798–99), Sendtner
(1846), Smith & Downs (1966), Tucker (1969), Wettstein (1891b).

three northern Argentina. The delimitation between *Cyphomandra* and *Solanum* is unclear and badly needs revision. The well known "tomate de árbol", *Cyphomandra betacea* (Cav.) Sendt., is cultivated as a food crop in many parts of the world.

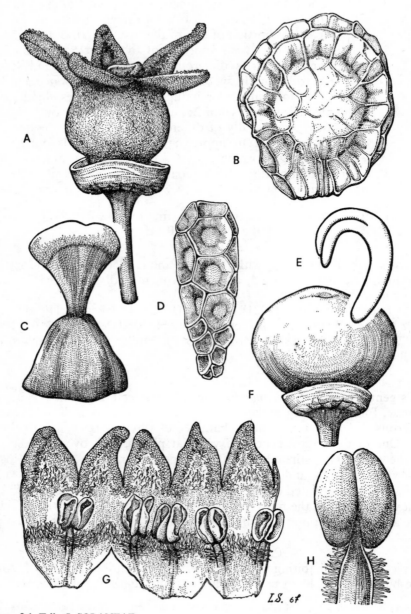

Figure 2.1. Tribe I. SOLANEAE.
Witheringia mortonii A. T. Hunz. (*Skutch* 2790). A, Flower (× 8); B, D, seed, front and lateral views (× 30); C, gynoecium (× 20); E, embryo (× 40); F, fruit (× 5); G, internal view of corolla (× 10); H, stamen, dorsal view (× 20).

3. *Lycopersicon* Miller

A small genus of six species inhabiting coastal western South America, from Colombia to northern Chile. Its stamens with elongated connectives are unique, and, judging from the importance of this floral whorl in the understanding of the taxonomy and evolution of the family as a whole, it seems quite right to accept the genus, at least until modern studies of comparative morphology are carried out. *L. esculentum* Miller is the traditional "tomato", a food plant much cultivated in many countries all over the world.

4. *Jaltomata* Schlecht.

I am unable to draw a line of separation between *Jaltomata* and *Hebecladus* Miers. Both groups share a number of common attributes such as: pedicels articulated, insertion of filaments on the central face of the anthers, filaments enlarged at the base, well developed disk, corolla lobes valvate, fruits with accrescent calyx and juicy mesocarp, etc. They differ exclusively in the corolla form which is tubular or campanulate-tubular in *Hebecladus*, rotate or campanulate-rotate in *Jaltomata*, but as there are intermediate forms it looks as if both could be considered sections of one genus. It is to be hoped then, that the studies currently in progress in the United States on these plants (D'Arcy, pers. comm.) will clarify the question. For the time being, I prefer to consider both groups as belonging to only one genus, with a total of *c.* 20–24 species, one in southern Arizona, a few in Mexico, Central America and the Antilles, and the great majority in western Andean South America (from the Cordillera de la Costa in Venezuela, to Mapiri in Bolivia, the main centre of concentration being between southern Colombia and Peru). Towards the east, the genus reaches the Amazonian territory at the border between Peru and Brazil: *J. repandidentata* (Dunal) A. T. Hunziker and *J. procumbens* (Cav.) Gentry seem to be the most common and widespread species, ranging from southwestern United States to Peru.

5. *Cuatresia* A. T. Hunz.

Cuatresia and *Jaltomata*, *sensu lato*, are the only genera in the tribe with the filaments inserted on the ventral face of the anthers. The absence of a pedicel articulation and the filaments not enlarged at their point of insertion distinguish *Cuatresia* from *Jaltomata*. The genus is composed of six species of shrubs or small trees, growing mainly in Central America and northern South America: *C. riparia* (H.B.K.) A. T. Hunz. Costa Rica, Colombia, etc.; *C. physocalycia* (Donn. Smith) A. T. Hunz. Guatemala; *C. plowmanii* A. T. Hunz. Colombia; etc. An undetermined species reaches Bolivia (Santa Cruz de la Sierra).

6. *Athenaea* Sendt.

A small Brazilian genus of nine species of woody plants, easily distinguished by their corolla lobes being valvate in the bud and longer than the rest of the mature corolla. With respect to the androecium, the small filaments are usually shorter or slightly longer than the anthers; the basal part of the filament is enlarged, and as happens in *Capsicum*, each one has two indentations, the whole structure being inserted on the basal part of the corolla tube. There are two groups of species:

Athenaea picta (Mart.) Sendt., *A. cuspidata* Witas., *A. pogogena* (Moricand) Sendt., *A. martiana* Sendt., and *A. anonacea* Sendt., belong to the first one, with strongly accrescent calyces. In the other group, on the contrary, the calyx does not enlarge during fructification; these species, for example *A. brasiliana* A. T. Hunz., have usually been assigned to a different genus: *Aureliana* Sendt. Since the accrescence of the calyx is the only apparent differential character between *Aureliana* and *Athenaea*, I do not think that at present the former entity can be accepted at the generic level. However, as the vasculature of the calyx seems different in both groups, it is possible that future studies may justify the separation between them. For this reason I postpone the transference of the epithets from *Aureliana* to *Athenaea*.

7. *Witheringia* L'Herit.

Generally tropical herbs, ranging from Mexico to Bolivia, usually between 900 m to 2000 m alt.; the main concentration of species lies in Central America and north-western South America. The polymorphic *W. solanacea* L'Herit. is the most widespread species: from southern Mexico to Central America, the Antilles, western America and southern Bolivia. In all, the genus (*sensu stricto*, i.e. excluding *Brachistus* Miers) comprises around 16 species, only five in South America.

8. *Larnax* Miers

One of the genera whose early recognition we owe to the clearsightedness of John Miers (1849), *Larnax*, was later submerged under *Athenaea* by Bentham and Hooker (1876: 890), and, since then forgotten. Tropical herbs, quite different from the woody species of *Athenaea*, on account of calyx and stamen peculiarities. Strictly South American, *Larnax* is native to Colombia, Ecuador, Peru and Bolivia, reaching the western border of the Hylea. Around ten species: *L. subtriflora* (R. et P.) Miers, *L. sachapapa* A. T. Hunz., *L. steyermarkii* A. T. Hunz., *L. hawkesii* A. T. Hunz., *L. peruviana* (Zahlbr.) A. T. Hunz., etc. Some botanists have wrongly referred some of these species to *Withania* Pauquy, an Old World genus with a substantially different androecium structure.

9. *Capsicum* L.

About 25 species (with some varieties) growing spontaneously in America, together with five that are cultivated all over the world. Floral and carpological characters nowadays permit clear circumscription of this genus, whose delimitation has been a puzzle until recently. *C. ciliatum* (H.B.K.) O.K. (Mexico, C. America and S. America to northern Peru) and *C. annuum* var. *glabriusculum* (Dunal) Heiser et Pickersgill (southern United States to northern Peru; also in the Antilles) are the species with the greatest geographical range.

The main concentration of species lies in eastern Brazil, Bolivia, Paraguay and northern Argentina where there are about 17 species (*C. baccatum* L., *C. dusenii* Bitt., *C. parvifolium* Sendt., *C. chacoense* A. T. Hunz., etc.). Central America has only one autochthonous species: *C. lanceolatum* (Greenm.) Morton et Standley, but in western South America there are five: *C. dimorphum* (Miers) O.K., *C. galapagoense* A. T. Hunz., etc.

The cultivated species are *C. annuum* L., *C. frutescens* L., *C. pubescens* R. et P., *C. chinense* Jacq. and *C. baccatum* L.

10. *Physalis* L.*

With very few representatives in the Old World, *Physalis* has its principal centre of concentration of species in Mexico (70 species), radiating towards the United States, Central America and the Caribbean. Of the total of 90/95 species of the genus, in South America there are only 12 tropical and subtropical species: *P. hygrophila* Mart., *P. peruviana* L., *P. margaranthoides* Rusby, *P. surinamensis* Miq., *P. neesiana* Sendt., *P. viscosa* L., etc. Some species are ancient garden plants like the European *P. alkekengi* L. grown mainly on account of its coloured inflated calyx.

12. *Vassobia* Rusby†

A clearly defined genus, both from the morphological and phytogeographical point of view. The androecium is much the same as in *Capsicum* and *Athenaea*, but the corolla and fruits are quite different. Four species of shrubs or small trees; in Sect. *Vassobia: V. dichotoma* (Rusby) Bitt. from southern Bolivia, and *V. breviflora* (Sendt.) A. T. Hunz., which is the most widespread species (southern Brazil, Paraguay, Uruguay, Bolivia and northern Argentina). The remaining pair of species belongs to sect. *Eriolarynx* A. T. Hunz.: *V. lorentzii* (Dammer) A. T. Hunz. in northwestern Argentina and *V. fasciculata* (Miers) A. T. Hunz., in southern Peru and Bolivia.

13. *Exodeconus* Rafin. [*Cacabus* Bernh.]

Thanks to a kind report of Dr W. G. D'Arcy, I was made aware that *Cacabus* Bernh. is antedated by the name *Exodeconus* of Rafinesque. Around eight to ten species with very marked ecological adaptation, mainly prostrate ephemerals of the desert pacific coast from Tarapacá northwards to Peru, southern Ecuador and the Galápagos Islands. *E. miersii* (Hook. *f.*) D'Arcy is endemic to these islands.‡

14. *Dunalia* H.B.K.

Very easily distinguished by its peculiar androecium, *Dunalia* grows in the Andes between 1600 to 3700 m alt., from Colombia to Argentina (31° lat. S). Section *Dunalia* has two species in Colombia and Ecuador: *D. solanacea* H.B.K. and *D. trianaei* Dammer; in section *Pauciflorae* Miers the remaining five species are allocated: *D. cyanea* Dunal (Peru, between 1700 to 1900 m), *D. spinosa*

---

* In 1961 I published *P. spruceana* sp. nov.; after 15 years of additional experience with the taxonomy of the family, I arrived at the conclusion that this curious species should be excluded from *Physalis* L. It combines the floral characters of *Capsicum* with the fruiting ones of *Chamaesaracha* A. Gray. I have discussed the subject with Dr J. Averett (Saint Louis, U.S.A.), and he thinks that it does not fit at all in *Chamaesaracha*. For the time being, until studies that Averett and myself are going to undertake on this annual of Amazonian Peru, I shall not name the taxon at the generic level, although I do count it as one of the valid genera of South America (genus No. 11).

† Genus No. 11 is referred to in the footnote attached to the heading of *Physalis*.

‡ The names of the other species belonging to *Exodeconus* have not yet been transferred by D'Arcy, and I refrain therefore to mention them.

(Meyen) Dammer in northern Chile, Bolivia and Peru, up to 3700 m alt., etc.; *D. brachyacantha* Miers reaches the hills of Córdoba in central Argentina from southern Peru and Bolivia.

15. *Acnistus* Schott

The valvate aestivation of the oblong corolla lobes, the absence of folds in the corolla tube, etc. characterize this small genus of two tropical woody species: *A. arborescens* (L.) Schlecht. (this is the more common species, growing in Mexico, Central America, W. Indies, and in South America: Colombia, Venezuela, Ecuador, Peru and Brazil) and *A. confertiflorus* Miers (Ecuador and northern Peru).

16. *Saracha* R. et P.*

Between 2700 and 4000 m, along the Andes from Venezuela to Bolivia, grow the three species of this curious genus: *S. punctata* R. et P. (Ecuador, Peru, Bolivia), *S. quitoensis* (Hook.) Miers (Venezuela, Colombia, Ecuador) and *S. ovata* (Miers) A. T. Hunz. (endemic to the Venezuelan "páramos"). Together with this peculiar geographic distribution, *Saracha* is unique on account of its uncommon carpological structure: *S. punctata* has baccate fruits, with 12 to 15 seeds and two to nine sclerotic bodies immersed in the mesocarp; the other two species, instead, show drupaceous fruits with two or three endocarpids, which remain embedded in the flesh; each of these endocarpic units enclose from 2 to 12 seeds.

17. *Iochroma* Bentham

Entirely South American, this genus embraces 15 woody species, many with ornamental value owing to their richly coloured corollas (orange-red, yellow, scarlet, bluish, etc.). There are three sections: Sect. *Iochroma* has nine species growing in humid north-western South America (Colombia to Peru); for example *I. gesnerioides* H.B.K., in Colombia and Ecuador, *I. calycinum* Benth. in Ecuador, *I. cornifolium* (H.B.K.) Miers in Ecuador and Peru, etc. The most southerly species is *I. australe* Griseb. in southern Bolivia and northwestern Argentina; to this section belongs also the Ecuadorean *I. cyaneum* (Lindl.) M. L. Green, not uncommonly cultivated in gardens for its blue flowers. Section *Spinosa* A. T. Hunz. (microphyllic spiny plants often with anthocyanin pigments in the flowers) comprises *I. horridum* (H.B.K.) A. T. Hunz., native of Peru, *I. cardenasianum* A. T. Hunz. (Bolivia), and a couple of undescribed Peruvian species. Finally in sect. *Lehmannia* A. T. Hunz., briefly characterized by its yellow flowers and spineless branches, there are two species: *I. lehmannii* Bitt. (Ecuador and Peru) and *I. ellipticum* (Hook. *f.*) A. T. Hunz. (endemic to the Galápagos Islands).

18. *Deprea* Rafin.

A genus rightly reinstated by D'Arcy in 1973 (p. 624). Concerning the doubts

---

* Authorities on nomenclature are of the opinion that this name does not clash with *Saraca* L., applied to a valid genus of the Fabaceae. Very reluctantly I accept their point of view, confessing, by the way, that I cannot understand it.

expressed by my eminent colleague about the distinction between *Deprea* and *Physalis*, I believe they are invalid, because the tubular corollas of *Deprea* are absolutely different to the rotate or campanulate-rotate ones of *Physalis*. With one species in Mexico and another in Costa Rica, the genus extends from Venezuela and Colombia to Bolivia, with an estimated total of five or six species: *D. orinocensis* (H.B.K.) Rafin., *D. sylvarum* (Standl. et Morton) A. T. Hunz., *D. cardenasiana* A. T. Hunz., etc.

Tribe II. DATUREAE Reichenb.
(Fig. 2.2)

The conduplicate-contorted aestivation of the corolla is, among others, the common attribute of the two genera recognized in this tribe: *Datura* and *Brugmansia*. $x = 12$.
Special references: Barclay (1959), Don (1837), Lagerheim (1895), Lockwood (1973, 1976), Reichenbach (1828), Schultes (1955).

19. *Datura* L.

Annual plants with inflorescences predominantly dichasial; flowers erect, with circumscissile calyx; ovary bicarpellate and tetralocular; capsules with seeds lacking prominent caruncle or not carunculate at all, up to 6 mm long and 4.5 mm wide. About ten species usually growing in open disturbed areas. *D. metel* L. is believed to be of Asiatic origin; *D. leichhardtii* F. Muell. has two varieties: one in Mexico and Guatemala, the other in Australia. The remaining species are American; their centre of distribution and diversity includes Mexico and the south-western United States; in South America are generally found *D. ferox* L. ("chamico", with poisonous seeds), *D. innoxia* Miller, and *D. metel* L.

20. *Brugmansia* Persoon

Shrubs or small trees with pendulous or inclined flowers; calyx not circum-scissile; ovary bicarpellate and bilocular; dry berries with large seeds (7 to 12 mm long and 5 to 8 mm wide) lacking a caruncle and usually with a thick corky episperm. Five species and natural hybrids living from sea level to around 3000 m alt., in the Andes of Colombia, Ecuador, Peru and possibly Bolivia; *B. arborea* (L.) Lagerheim and *B. suaveolens* (Willd.) Bercht. et Presl are cultivated as ornamentals, and others play an important role in ceremonial and initiation rites.

Tribe III. JABOROSEAE Miers
(Fig. 2.3)

With the corolla lobes never overlapping in the bud, the tribe is characterized by the insertion of the filaments on the dorsal face of the anthers, near the centre. It is composed of four American genera; *Nectouxia* is a monotypic endemic of Mexico; the three others are strictly South American. $x = 12$.
Special reference: Hunziker (1967).

Figure 2.2. Tribe II. DATUREAE.
Datura ferox L. (*Subils* 2200). A, I, Polar and lateral views of a flower bud showing corolla aestivation (×7.5 and ×3); B, habit (×0.5); C, E, cross sections of ovary at 2/3 and 1/3 of its height (×5.5); D, stigma (×7.5); F, G, ventral and dorsal views of a young anther before its dehiscence (×5.5); H, sector of the corolla, internal view (×1); J, L, N, seed, lateral and hilar views and cross section (×5.5); K, embryo (×3); M, longitudinal section of ovary (×4); O, lateral view of gynoecium (×7.5).

Figure 2.3. Tribe III. JABOROSEAE.
Trechonaetes sativa Miers (*A. T. Hunziker* 21670 and 20462). A, K, Lateral and internal views of flower (×2); B, cross section of a tricarpellate ovary (×7.5); C, stigma (×7.5); D, distal portion of a branch (×0.5); E, stamen showing the attachment of the filament on the back of the anther (×7.5); F, G, cross section of a fresh seed with details of the thickening of the tangential internal wall of the episperm cells (F ×25; G ×5); H, fruit (×1); I, embryo (×7.5); J, root system (×0.25); L, flowering node (×0.5); M, lateral view of a dried seed (×7.5).

21. *Jaborosa* Jussieu

Leaves pinnatifid. The ovaries with two to five carpels and slightly developed disks are distinctive features of this genus and *Trechonaetes*, in contrast to *Salpichroa*.

Twelve species from Patagonia northwards; in the east, *J. integrifolia* Lam. and *J. runcinata* Lam. inhabit Uruguay (Canelones, Río Negro) and north-eastern Argentina (Buenos Aires, Entre Ríos, etc.). The majority of the species grow in the dry central and western parts of Argentina: *J. bergii* Hieron., *J. leucotricha* (Speg.) A. T. Hunz., *J. odonelliana* A. T. Hunz., *J. oxipetala* Speg., etc.; *J. caulescens* Gill. et Hook. is an Andean element distributed in the largest area of the genus (Argentina: Mendoza; Chile; Bolivia; Peru: Puno, Tacna). In southern Chile (Ñuble) is found a very rare species: *J. volkmanni* (Phil.) Reich., but the southernmost one is *J. magellanica* (Griseb.) Dusen (Argentina: Santa Cruz and Tierra del Fuego).

22. *Trechonaetes* Miers

Extremely close to *Jaborosa*; it is not certain whether *Trechonaetes* should be considered a genus by itself, or a section of the preceding one. Campanulate-rotate corollas or almost rotate ones, and the stamens inserted at a lower level than in *Jaborosa*, are the only differences between the two. *T. laciniata* Miers (Chile: Cordillera de Linares; Argentina: Mendoza) and *T. sativa* Miers (northern Argentina to southern Bolivia) are the species known up to now and are strictly Andean in their distribution.

23. *Salpichroa* Miers

An assemblage of about 22 species characterized by entire leaves, a bi-carpellary ovary and a large disk. Two groups can be easily recognised. The first is represented by *S. origanifolia* (Lam.) Thell. with small urceolate corollas, growing in southern Bolivia, central and northern Argentina (Buenos Aires, Corrientes, Córdoba, Salta, Tucumán, etc.), Uruguay and southern Brazil (R.G. do Sul); sometimes cultivated, it has been collected as an adventive in Europe (England, Portugal, Italy, etc.), Africa (Cameroon, Transvaal, etc.), Australia, U.S.A., etc. The second group of species is essentially Andine: from Venezuela in the north (Trujillo) through Colombia, Ecuador, Peru and Bolivia, to northern Argentina (Jujuy to Catamarca); these are especially abundant in Peru, like *S. dependens* (Hook.) Miers, *S. longiflora* Benoist, *S. didierana* Jaubert, etc.

Tribe IV. LYCIEAE A. T. Hunz.
(Fig. 2.4)

Woody plants, sometimes halophytic. Flower buds with overlapping corolla lobes. Filaments inserted on the back of the anthers; thecae free from each other for their lower third or even higher up. Gynoecium 2-carpellary. Fruit baccate or drupaceous. $x = 12$.

Special references: Barkley (1953), Hitchcock (1932), Hunziker (1977).

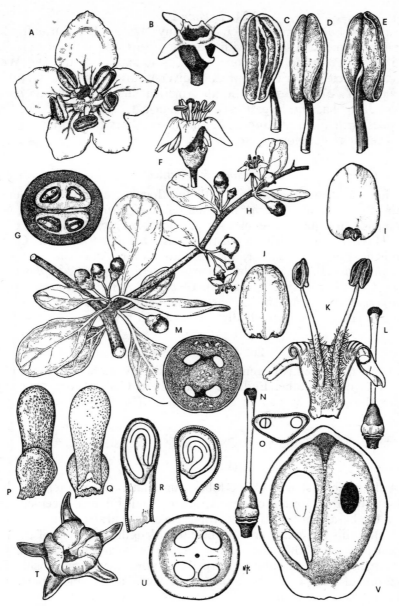

Figure 2.4. Tribe IV. LYCIEAE.
Grabowskia duplicata Arnott (*Anton* 21). A, Front view of the slightly zygomorphic flower (×4);
B, calyx (×4); C, ventral face of an anther after dehiscence (×12.5); D, E, ventral and dorsal faces
of a younger anther (×12.5); F, flower (×2); G, cross section of a fruit (two pyrenes, each one with
two biseminate locules) (×3); H, branchlet with fertile brachyblasts (×1); I, J, external and internal
faces of a pyrene showing its basal aperture (×3); K, internal view of a portion of the corolla (×4);
L, N, gynoecium, showing the basal nectary (×4); M, U, ovary, cross sections (M at the level where
each carpel is unilocular; U, further up where the ovary becomes tetralocular) (×12); O, cross
section of a seed (×9); P, Q, dorsal and ventral views of the pair of seeds of one locule (×6); R, S, the
two seeds of one locule in longitudinal section (×6); T, above view of a flower bud with its cochlear
aestivation (×4); V, ovary after fertilization, in longitudinal section, showing the two dissimilar
erect ovules of one of the four locules (×8).

24. *Lycium* L.

Ovary always bilocular, with few to many ovules. Fruit usually baccate, excepting a few American species with indurated pericarp (in this case each locule bears one seed). Cosmopolitan genus, with 35–40 species in the Old World; in America *c.* 40 species, with two centres of diversification: Arizona (*c.* 10 species) and Argentina (*c.* 27 species); in South America, in all, around 35 species. *L. americanum* Jacq. has the largest area (Venezuela to Argentina); but there are also endemics with reduced ranges: *L. vimineum* Miers (Argentina), *L. comberi* Hitchc. (Argentina: Patagonia), etc.

25. *Grabowskia* Schlecht.

Aestivation cochlear. Each carpel with two locules, each of the four locules with a pair of ovules (one superior and one inferior). Fruit drupaceous, with two pyrenes immersed in a juicy sweet mesocarp. About six species; one is *G. geniculata* (Fernald) Hitchc., with disjunct distribution: Mexico (Puebla), Galápagos, Bolivia, north-western Argentina; the rest are South American, for example: *G. boerhaaviaefolia* (L. *f.*) Schlecht.* (Peru to Argentina), *G. duplicata* Arnott (Paraguay, eastern Argentina, Uruguay), *G. obtusa* Arnott (western Argentina), *G. megalosperma* Speg. (Argentina: eastern Patagonia), and *G. glauca* Johnston (Chile: Antofagasta). The genus badly needs careful revision.

26. *Phrodus* Miers

Differs from the other two genera of the tribe by its actinomorphic corolla and the plicate corolla tube. Endemic to northern Chile (Provs. Atacama and Coquimbo): *P. bridgesii* Miers and *P. microphyllus* (Miers) Miers.

Tribe V. NICANDREAE Wettst.

(Fig. 2.5)

Herbs with hollow stems. Calyx lobes auriculate. Long internode between calyx and corolla. Prefloration of corolla quincuncial. Filaments geniculate when adhering to the basal portion of the corolla; thecae separated from each other at their lower end, this being a fourth of their total length; insertion of the filament strictly basal between both thecae. Gynoecium 5 or 4-carpelled. Fruit enveloped by the accrescent calyx; mature pericarp thin, rather dry, semitranslucent, with five concretions of stone cells of the size of the seeds, at its apical part. $x = 10$, 11. The only genus is *Nicandra*.
Special references: Bitter (1903), Kaniewski (1965).

27. *Nicandra* Adanson

Monotypic genus: *N. physalodes* (L.) Gaertn. Thought to be a Peruvian plant, it has long been cultivated as an ornamental and is nowadays found as a ruderal in tropical and subtropical areas all over the world, presumably as an escape.

---

* Most probably *G. geniculata* and *G. boerhaaviaefolia* are two names for one and the same species.

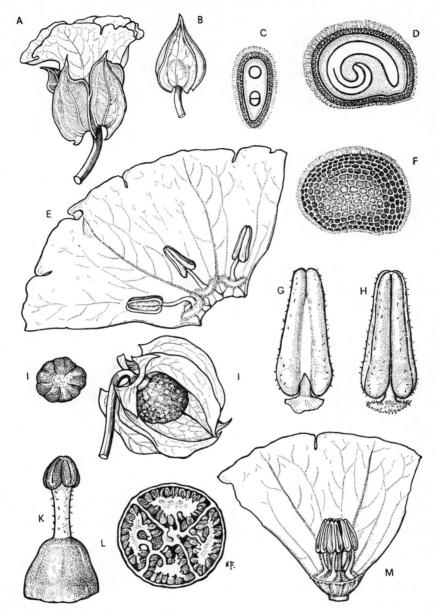

Figure 2.5. Tribe V. NICANDREAE.
Nicandra physalodes (L.) Gaertner (*A. T. Hunziker* 19626). A, Flower (×1.3); B, flower bud (×1.7); C, seed, cross section (×17); D, seed, longitudinal section (×17); E, portion of expanded corolla (×1.7); F, seed (×17); G, H, anther, dorsal and ventral views (×8.5); I, stigma top view, (×13.6); J, fruit, with one of the lobes of the accrescent calyx removed (×0.85); K, gynoecium (×6.8); L, fruit, cross section (×0.85); M, flower, with pedicel, calyx and most of the corolla removed showing the peculiar arrangement of the stamens around the gynoecium (×1.7).

Tribe VI. SOLANDREAE Miers
(Fig. 2.6)

Scandent shrubs or high-climbing lianas, frequently epiphytic. Corolla with overlapping corolla lobes in the bud. Pericarp juicy. Seeds with rather abundant endosperm; embryo curved with incumbent slightly oblique cotyledons. Two genera: *Solandra* and *Trianaea*; future studies may prove that they should be separated into different tribes or subtribes. Chromosome counts are known only for *Solandra:* $x = 12$.

Special references: Cuatrecasas (1958b, 1959), Drake del Castillo (1892), Hunziker (1977), Martínez (1967), Romero-Castañeda (1972), Solereder (1898).

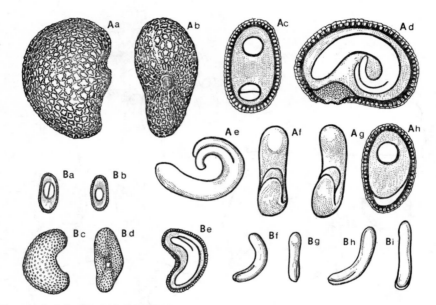

Figure 2.6. Tribe VI. SOLANDREAE.
A, *Solandra sp.* (*Matuda* 37316); B, *Trianaea sp.* (*Bristol* 884). Aa, Ab, Bc, Bd, Seeds, lateral and hilar views; Ac, Ah, cross sections of seeds at approximately 1/3 distance from each pole; Ad, Be, longitudinal sections of seeds; Ae–Ag, Bf–Bi, embryos; Ba, Bb, cross sections of seeds at approximately 1/4 distance from each pole (all ×7.5).

28. *Solandra* Swartz

Corolla large, 10 to 36 cm long, much exceeding the calyx. Androecium zygomorphic: filaments declinate, inserted on the base of the anthers, between the thecae; anthers 7 to 11 mm long. Ovary slightly sunken in the receptacle, 2-carpellate, 4-locular; style declinate. Seeds 4.2 to 5.2 mm long, 2.6 to 3.5 mm wide.

A tropical genus with about nine or ten species; *S. paraensis* Ducke grows in eastern Brazil (Pará) and French Guiana, having the largest flowers in the family (up to 36 cm long); *S. grandiflora* Swartz and *S. longiflora* Tuss. are native to the Antilles (Cuba, Jamaica, etc.). The range of the genus also includes Mexico,

Central America and Venezuela, at altitudes between 1500 to 2500 m. I am at a loss to understand *S. boliviana* Britton: its procumbent habit, the blue-purple corolla tube, etc. would point to its exclusion from the genus; nevertheless, the photograph of its type collection—the only specimen known heretofore—shows that, for the time being, it should not be removed from its present place in the system. Three or four species are planted as ornamentals (*S. grandiflora*, etc.).

29. *Trianaea* Planchon et Linden

Corolla small (*c.* 5 cm long), equal or slightly longer than the calyx; androecium actinomorphic, anthers ventrifixed, 11 to 20 cm long. Ovary 4–5-carpellate, 8–10-locular. Seeds kidney-shaped, *c.* 2.4 mm long, *c.* 1.6 mm broad, with less endosperm and more compressed than in *Solandra*.

A very distinct genus of four Andean species, extending from northern Colombia to Ecuador, i.e.: *T. neovisae* Romero (Colombia: Magdalena), *T. nobilis* Planchon et Linden (Ecuador), *T. speciosa* (Drake del Cast.) Soler. (Ecuador), and *T. spectabilis* Cuatrec. (Colombia: Cauca).

Tribe VII. JUANULLOEAE A. T. Hunz.
(Fig. 2.7)

A natural taxonomic assemblage of eight neotropical genera inhabiting the rain forests of Mexico, Central and South America; woody plants generally epiphytic on trees, and in some cases (four species of *Markea*) myrmecophilous. In *Merinthopodium* the corolla is valvate, but the remaining genera have a quincuncial, cochlear or imbricate arrangement of the corolla lobes.

The androecium is the floral whorl with the best characters for generic delimitation: *Dyssochroma*, *Markea* and *Schultesianthus* have basifixed anthers, but in the other genera the insertion of the filament is strictly dorsal. *Ectozoma* is unique because of its monadelphous stamens. All the genera show a 2-carpellary gynoecium.

The structure of the seed is remarkable: episperm with shallow cells, endosperm scanty, and embryo with oblique or almost always accumbent cotyledons. The only genus of the tribe growing outside South America is *Rahowardiana* D'Arcy, from Panamá. It is most probable that future information may necessitate the separation of this tribe as a third subfamily; for instance, nothing is known about the chromosome numbers of these plants.
Special references: Cuatrecasas (1958a), Hunziker (1977).

30. *Juanulloa* R. et P.

This genus has the largest area of distribution in the tribe; it comprises nine or ten species, native to southern Mexico to Bolivia, along the Andes, from 200 m or less to 2500 m alt. *J. parviflora* (Ducke) Cuatr. is the only species that reaches the Brazilian Amazonian basin, having flowers of only 20 mm long. Among the others, can be cited the following: *J. speciosa* (Miers) Dunal (Colombia: Quindío), *J. ochracea* Cuatr. (Colombia: Putumayo; Peru: Loreto), *J. ferruginea* Cuatr. (Colombia: Putumayo), *J. mexicana* (Schlecht.) Miers (Mexico, Central America, Colombia), *J. parasitica* R. et P. (Peru), *J. membranacea* Rusby (Bolivia), *J. pedunculata* Rusby (Bolivia), etc.

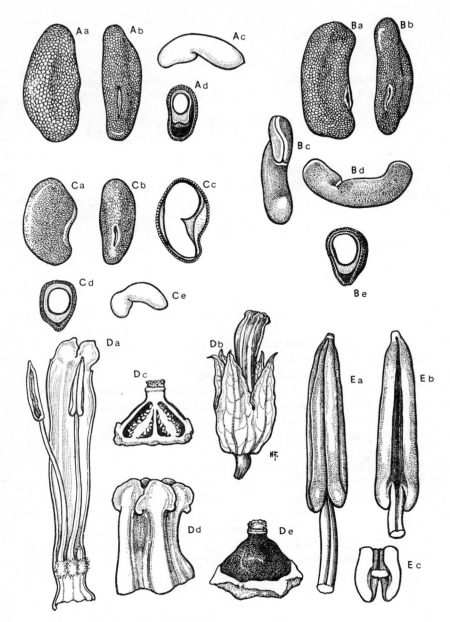

Figure 2.7. Tribe VII. JUANULLOEAE.
A, *Juanulloa sp.* (*Purpus* 2435); B, *Juanulloa sp.* (*Krukoff* 5733); C, *Juanulloa sp.* (*Cárdenas*, XI–1947); D, *Juanulloa sp.* (*Scolnik* et al. 588); E, *Juanulloa sargii* Donn. Smith (*Metcalf* et al. 30095). Aa–Ab, Ba–Bb, Ca–Cb, Seeds lateral and hilar views respectively (*c.* ×7.2); Ad, Be, Cd, cross sections of seeds, near the hilum (*c.* ×7.2); Cc, longitudinal section of seed, (*c.* ×7.2); Ac, Bc–Bd, Ce, embryos, (*c.* ×7.2); Da, sector of corolla, internal view (×1.7); Db, flower (×0.85); Dc, ovary, longitudinal section (*c.* ×5); Dd, apex of corolla showing aestivation (×3.6); De, ovary (*c.* ×5); Ea, Eb, anther, dorsal and ventral faces (*c.* ×4.2); Ec, base of anther, ventral face, the filament removed (*c.* ×4.2).

31. *Ectozoma* Miers

Monotypic genus: *E. pavonii* Miers (= *Salpichroma cuspidatum* Dunal) endemic to the rain forests of Ecuador (Prov. Pichincha and Guayas) at low altitudes (200–800 m alt.) and Peru (San Martín)*. Closely related to *Juanulloa*, it differs, among other attributes, in its monadelphous androecium in which the filaments are fused in a narrow basal ring that adheres to the corolla.

32. *Merinthopodium* Donnell Smith

The anthers, the seeds and the embryo are similar to those of *Juanulloa*, but the stamens are exserted instead of included, and the vernation of the corolla is valvate. A small genus with at least five species: *M. neuranthum* (Hemsley) Donnell Smith (Guatemala to Panamá), *M. leptesthemium* Blake (British Honduras), *M. campanulatum* Donnell Smith (Guatemala), *M. uniflorum* (Lundell) A. T. Hunz. (Chiapas) and *M. pendulum* (Cuatr.) A. T. Hunz. (Colombia: Norte de Santander). In Colombia (Cundinamarca) grows also *Markea vogelii* Cuatr., which belongs to *Merinthopodium* (it is known from a single and very poor specimen); the difference, if any, from *M. uniflorum* lies in the size of the anthers, which are slightly longer in the only flower of the type specimen; it looks as if in the future *Merinthopodium uniflorum* will be the valid name for the South American material from Cundinamarca.

33. *Markea* L. C. Richard

Anthers as in *Dyssochroma*, but the vernation is imbricate or cochlear, and the calyx usually has cuspidate laciniae with prominent nerves.

According to the delimitation of *Markea* followed here, this genus comprises at least the following eight species in South America: *M. camponoti* Ducke (Brazil: Pará and east of Amazonia; Venezuela; Guayana), *M. coccinea* L. C. Richard (Brazil: Amazonia; Guayana), *M. formicarum* Dammer (Brazil: western Amazonia), *M. longiflora* Miers (Trinidad†), *M. lopezii* Cuatrec. et A. T. Hunz. (Colombia), *M. porphyrobaphes* Sandw. (Guayana), *M. sessiliflora* Ducke (Brazil: Pará), and *M. reticulata* Steyerm. et Maguire (Venezuela); outside South America is *M. venosa* Standl. et Steyerm. (Costa Rica).

Four species from eastern South America are myrmecophilous (*M. camponoti*, *M. coccinea*, *M. formicarum* and *M. sessiliflora*), the ants belonging to the genera *Azteca* and *Camponotus*.

34. *Schultesianthus* A. T. Hunz.

This genus differs from *Markea* by its fleshy thick calyx (becoming coriaceous or even woody) with immersed nerves, by its declinate style and filaments, and its shorter anthers (4 to 8 mm long); furthermore, the arcuate embryo is very peculiar: the accumbent cotyledons are almost as broad as long. *S. leucanthus*

---

* The Peruvian material may belong to a different species.

† The holotype is the single collection heretofore known for this species; the dissection of the flower could prove that it belongs to *Schultesianthus*, but for the time being until better materials are collected, it is retained in *Markea* on account of its calyx.

(Donn. Smith) A. T. Hunz. ranges from Mexico (Chiapas) to Panamá, being the species with the largest geographical distribution; *S. megalandrus* (Dunal) A. T. Hunz. is restricted to Venezuela (Carabobo, Yaracuy, Sucre); the remaining, species are Colombian: *S. coriaceus* (O.K.) A. T. Hunz. (Antioquia, Cundina-marca, Boyacá, etc.), *S. odoriferus* (Cuatr.) A. T. Hunz. (Valle), and *S. suaveolens* (Standl.) A. T. Hunz. The genus reaches Ecuador and Peru, but the identity of the respective species is not clear.

35. *Dyssochroma* Miers

A small Brazilian genus akin to *Markea* and *Schultesianthus* on account of its basifixed anthers; nevertheless, its valvate prefloration together with its acutish and revolute corolla lobes, clearly separate it from both. *Dyssochroma* has two species in Brazil: *D. viridiflora* (Sims) Miers (Minas Gerais, Rio de Janeiro, Santos and S. Catarina), with exserted anthers, and *D. longipes* (Sendt.) Miers (its area overlaps in part with that of the preceding species, inhabiting S. Paulo, S. Catarina and Paraná) with larger flowers, included anthers and shorter corolla lobes. It is worth pointing out that up to now mature fruits with ripe seeds have never been collected.

36. *Hawkesiophyton* A. T. Hunz.

Together with *Rahowardiana*, *Hawkesiophyton* has anthers with confluent thecae, a feature that separates both genera from *Ectozoma*, *Juanulloa* and *Merinthopodium*, where the thecae are not confluent. Apart from the inflorescence and other secondary differential features, *Rahowardiana* has filaments longer than the anthers and with a peculiar thickening at their point of insertion on the corolla; furthermore the dehiscence of each theca is complete in *Rahowardiana*, but in *Hawkesiophyton* it does not reach the basal border. For the time being, three species are known, i.e.: *H. panamense* (Standley) A. T. Hunz. (Panamá; Colombia: Antioquia), *H. ulei* (Dammer) A. T. Hunz. (Brazil: Amazonas, Pará; Peru: Loreto, San Martín), and *H. klugii* A. T. Hunz. (Peru: Loreto).

Tribe VIII. CESTREAE Don
(Fig. 2.8)

Woody plants with flowers articulated on the peduncles. Calyx 5-dentate, never torn after anthesis. Aestivation inflexo-valvate slightly contorted; corolla with a short upper fold on the tube. Stamens 5, equal (a few species of *Cestrum* have one or two shorter filaments), free from each other, included (exserted only in *Vestia* and *Metternichia*); filaments longer than the anthers, generally with a short geniculate part at their proximal end (exceptionally this bending is located at the portion of the filament already adhering to the corolla), free portion many times shorter than the connate one, adhering to the back of the anthers; anthers small (0.5 to 2.5 mm long) almost circular in outline or somewhat oblong; thecae introrse, dehiscing independently, free from each other in the basal third or to a slightly higher level. Ovary on a prominent disk, surrounded by a cyathium (except *Metternichia* which lacks the cyathium). Fruit a capsule or a berry. Seeds never discoidal. Embryo straight or slightly curved at the apex of the cotyledons.

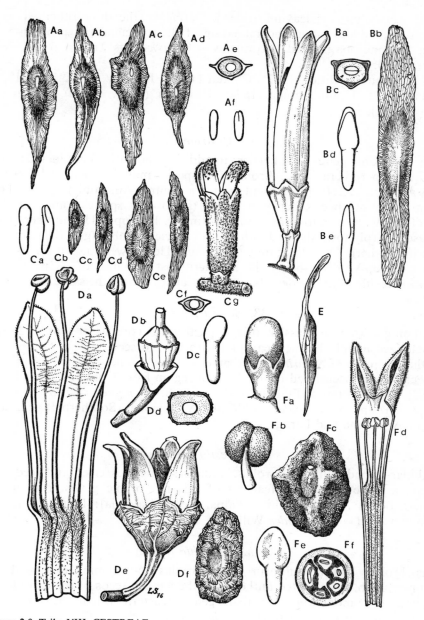

Figure 2.8. Tribe VIII. CESTREAE.
A, *Sessea dependens* R. et P. (*Rusby* 2625); B, *S. brasiliensis* Toledo (*Handro* 1126); C, *S. stipulata* R. et P. (*Scolnik* 547); D, *Vestia foetida* (R. et P.) Hoffm. (Corolla: *Hollermayer* 325; others: *A. T. Hunziker* 20152); E, *Sessea regnellii* Taubert (*Smith et Reitz* 12925); F, *Cestrum parqui* L'Herit. (*A. T. Hunziker* 23085). Aa–Ad, Bb, Cb–Ce, E, Seeds, hilar view (×5); Df, Fc, seeds, hilar view (×7); Ae, Bc, Cf, Dd, seeds, cross sections (×7); Af, Bd–Be, Ca, Dc, Fe, embryos (×7); Ba (×3·4), Cg (×3·4), De (×1·2), Fa (×1·2), fruits; Db, ovary partly enveloped by the persistent base of the corolla (×2·5); Da, Fd, portion of expanded corolla, internal view (×2·5); Fb, anther, dorsal view (×9·5); Ff, fruit, cross section (×7).

Four woody genera; leaving aside *Metternichia*, the three others seem to be ill-founded, and when our understanding of their peculiarities progresses in the future, they may be merged into one. $x = 8$ (no counts known for *Sessea* and *Metternichia*).

Special references: Bitter (1922), Chodat (1916), Francey (1933, 1934, 1935–36), Hassler (1917), Hoffmannsegg (1824), Hunziker (1967, 1969b), Mikan (1823), Scolnik (1954).

37. *Cestrum* L.

This is the only genus with berries; in addition the anthers are included and the seeds lack a peripheral wing. Around 250 species mainly in tropical America; in South America *c.* 150 spp. Absent from the central Amazonian basin, the main concentration of species being in Brazil (almost 50 species), and the Andean region from Venezuela to northern Argentina (*c.* 100 spp.). The number of species decreases sharply towards the south; the 18 species inhabiting Bolivia are reduced to ten in northern Argentina; only *C. parqui* L'Herit., well known for its toxicity, reaches Chile, Valdivia being the southernmost province of its range; on the Argentinian side of the Andes, this species reaches almost the same latitude (41°S).

38. *Sessea* R. et P.

Including *Sesseopsis* Hassler, this tropical genus consists of around 25 species, characterized mainly by their capsules with winged seeds; all in South America with the exception of *S. farinosa* (Urban et Ekman) Francey, which inhabits Haiti; 22 species are concentrated in the Andean region from Colombia to Bolivia, and the only Brazilian representatives of the genus are *S. regnellii* Taub. (M. Gerais to S. Catarina) and *S. vestioides* (Schlecht.) A. T. Hunz. (Southern Brazil, adjacent parts of Argentina and Paraguay).

39. *Vestia* Willd.

Tubular-infundibuliform corollas, anthers strongly exserted and seeds lacking a wing, are the prominent features of this monotypic genus from temperate southern South America (Chile: Valparaiso to Valdivia): *V. foetida* (R. et P.) Hoffmannsegg (= *V. lycioides* Willd.).

40. *Metternichia* Mikan

The broadly infundibuliform long corollas (50–60 mm) expanded into a prominent limb, the exserted anthers and the winged seeds of the capsules, easily set apart this monotypic Brazilian genus (eastern Brazil: Rio de Janeiro, Corcovado, Cabo Frio, etc.): *M. principis* Mik.

Tribe IX. NICOTIANEAE G. Don
(Fig. 2.9)

Nine genera of herbaceous or, less frequently, woody plants. Confluent thecae in the androecium are an exceptional feature; generally the stamens carry anthers with independent thecae; in *Benthamiella* a pair of species possess only two

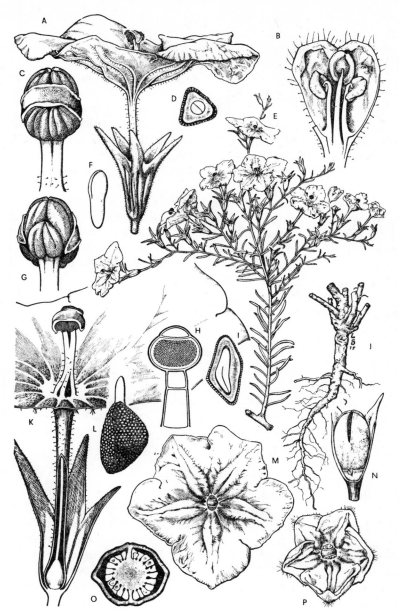

Figure 2.9. Tribe IX. NICOTIANEAE.
Nierembergia hippomanica Miers (*Di Fulvio* 356, 376). A, Flower (×2.2); B, flower bud, upper portion of a longitudinal section (note the ventrifixed and extrorse anthers) (×5.5); C, G, stigma embracing the anthers in a flower bud (×7.5); D, seed, cross section (×20); E, flowering branch (×0.5); F, embryo (×20); H, capitate hair of the corolla limb, near the insertion of the filaments (×225); I, seed, longitudinal section (×20); J, root and stem base (×0.5); K, flower, lateral view with most of the corolla limb, a portion of its tube and half of the calyx removed, showing the absence of a disk at the base of the ovary, the insertion of the stamens, etc. (×4); L, germinating seed (×20); M, flower, top view (×1·5); N, capsule, with part of the calyx excised, to show the persistent base of the corolla (×3); O, ovary, cross section (×15); P, expanding flower bud with cochlear aestivation (×2.2).

stamens, the rest of the genera with a pentamerous androecium, almost without exception.

From the phytogeographical point of view it is noteworthy for its almost exclusively southern distribution: only *Nicotiana* has some autochthonous species in the North American continent. No chromosome counts known for *Benthamiella*, *Combera* and *Pantacantha*; in *Nicotiana* $x = 7, 8, 9, 10, 11, 12$; in *Petunia* $x = 7, 9$; in *Latua* and *Fabiana* $x = 9$; in *Nierembergia* $x = 8$.
Special references: Burbidge (1960), Dammer (1906), Fries (1911), Goodspeed (1954), Hieronymus (1881), Merxmüller & Buttler (1975), Millán (1926, 1928, 1941), Morton (1944), Philippi (1870), Plowman *et al.* (1971), Ricardi (1962, 1966), Sandwith (1936, 1939), Solereder (1915), Soriano (1948), Spegazzini (1902).

41. *Nicotiana* L.

Around 67 species; with the exception of *N. africana* Merxm. (Africa: Namibia) and 20 species from Australia and islands of the South Pacific, the remaining 46 are American (37 occur in South America; eighteen of these are found in Argentina). A few are soft woody shrubs (e.g. *N. glauca* Graham), but the majority are herbs. *N. tabacum* L. is an amphiploid, apparently originating in northwestern Argentina; the other cultivated tobacco is the cultigen *N. rustica* L. *Dittostigma* Phil. is one of the many synonyms of this genus (Hieronymus, 1881: 58; Millán, 1931: 16).

42. *Latua* Phil.

Woody plants up to 10 m tall, spiny, with heteroblastic growth. Monotypic genus: *Latua pubiflora* (Griseb.) Baillon, endemic to southern Chile (between Valdivia and Chiloé), is a curious species with narcotic and toxic properties due to its content of atropine and scopolamine in stem, seeds and leaves. Its transference from the Solaneae to the Nicotianeae is based on the morphology of the flower and its chromosome number.

43. *Petunia* Juss.

Closely allied to *Nicotiana*, the genus *Petunia* differs mainly in its solitary flowers. In all it consists of around 40 species; *P. parviflora* Juss. has the largest range: North and Central America, Cuba and South America; the others are strictly South American from Minas Gerais and southern Bolivia to eastern Patagonia; the main concentration of species is found in southern Brazil, Paraguay, northeastern Argentina and Uruguay. *P. integrifolia* (Hook.) Schinz et Thell. (= *P. violacea* Lindl.) and *P. axillaris* (Lam.) B.S.P., together with hybrids of both, are cultivated as ornamentals.

44. *Fabiana* R. et P.

Small ericoid shrubs or chamaephytes, sometimes with brachyblasts. Around nine species in western South America from 17° (Peru: Tacna) to 50° lat. S. (Argentina: Santa Cruz) at altitudes varying from 200 m to 4500 m. Some species have a restricted distribution area: for example, the two with heteroblastic branching: *F. bryoides* Phil. (Chile: Atacama; Argentina: Catamarca, Jujuy, Salta) and *F. friesii* Dammer, both at high elevations in the "puna". *F. imbricata*

R. et P., sometimes grown for ornament, shows on the contrary a wide range, extending from Bolivia to southern Chile (Ñuble and Valdivia) and Argentina (Neuquén and Chubut).

45. *Combera* Sandwith

A curious small genus of Andean southwestern South America, from 1700 to 2300 m alt. The flowers are unique on account of the small hypanthium around the base of the ovary. Its two species are herbaceous: *C. paradoxa* Sandw. (Argentina: Neuquén; Chile: Ñuble) and *C. minima* Sandw. (Chile: Valdivia).

46. *Pantacantha* Speg.

A Patagonian monotypic endemic genus of woody plants with two striking features: calyx with spiny divisions and cristate seeds. *P. ameghinoi* Speg. (Argentina: southern Mendoza, Chubut).

47. *Benthamiella* Speg.

Patagonian chamaephytes, agreeing with the previous genus in having a pair of opposite bracteoles near the calyx. Its 16 species extend southwards of 45° lat. S. almost exclusively in Argentina (Chubut; Santa Cruz); only *B. azorella* (Skottsb.) Soriano is known to enter Chilean territory (Andes of Magallanes).

48. *Nierembergia* R. et P.

The hypocrateriform corolla, the five stamens inserted on or near the apex of its narrow tube and the absence of a nectariferous disk easily separates this genus from *Bouchetia*. Both genera share a cochlear aestivation, laminar stigma, and similar filaments. Around 23 species, some (e.g. *N. hippomanica* Miers) with poisonous properties; *N. angustifolia* H.B.K. (Mexico) is the only North American one. The genus fails to appear in Central America, and the bulk of the species grow in Argentina (21 species) reaching northern Patagonia (Rio Negro and Neuquén). *N. repens* R. et P., has the largest area, from Colombia, Ecuador, Peru and Chile to Uruguay and eastern Argentina.

49. *Bouchetia* Dunal in DC.

Small disjunct genus of two species, very close to *Nierembergia*. The South American entity is *B. anomala* (Miers) Britton et Rusby, a native of southern Brazil, Uruguay, and north-eastern Argentina; it is very similar to *B. erecta* Dun., inhabiting North America (U.S.A.: Texas; Mexico: Puebla, Hidalgo, etc.) and Guatemala, but a good number of peculiarities keep them apart easily.

Tribe X. SCHWENCKIEAE A. T. Hunz.
(Fig. 2.10)

Three tropical genera are placed in this tribe, clearly distinguished by the racemose inflorescences, the valvate-induplicate aestivation, the slightly capitate stigmata, the androecium with 2 or 4 fertile stamens, etc. It is unfortunate that no chromosome counts have been made for any of its members.
Special references: Carvalho (1966a, b, 1969a, b, 1971), Heine (1963), Hepper (1976), Hunziker (1977), Schumacher (1828), Solereder (1891, 1898).

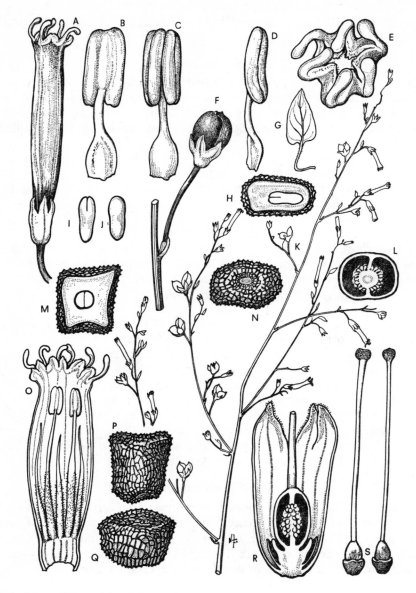

Figure 2.10. Tribe X. SCHWENCKIEAE.
Schwenckia americana L. (*A. T. Hunziker* 5478). A, Flower, lateral view (×5); B, C, D, stamen, ventral, dorsal and lateral views (×17); E, flower, top view (×17); F, fruit (×3.4); G, basal leaf (×0.4); H, seed, longitudinal section (×34); I, J, embryo (×25); K, upper part of flowering branch (×0.85); L, ovary, cross section (×17); M, N, seed, cross section and hilar view (×34); O, expanded corolla showing two stamens and three staminodes (×5); P, Q, seeds, lateral views (×34); R, flower with corolla removed, longitudinal section, with its prominent disc (×15); S, gynoecium, side views (×5).

50. *Schwenckia* L.

Between 25 to 30 species, usually herbaceous, with a bizarre corolla structure, and with ventrifixed anthers; peculiar to tropical and subtropical regions, from Central America and the Antilles to north-eastern Argentina. The commonest and most widespread species is *S. americana* L., which has been found also in Africa (in disturbed places in Guinea), as early as the beginning of the previous century.

51. *Protoschwenkia* Soler.

This genus is characterized by the absence of interlobular appendages in the corolla, by its dorsifixed anthers, and by two histological peculiarities: the stomata are surrounded by several normal epidermal cells, and the pith and cortical cells contain small, acicular or prismatic crystals of calcium oxalate (instead of drusae). A monotypic genus inhabiting the rain forests of Bolivia (Nor-Yungas, Coripata, Larecaja, etc.) and Brazil (Matto Grosso); the species is *P. mandonii* Soler.

52. *Melananthus* Walp.

A unique genus in the family on account of its uniovulate ovary with five tropical herbaceous species; *M. cubensis* Urban occupies the largest area (Cuba: Oriente and Camagüey; Brazil: Roraima and Minas Gerais); three appear to be endemics: *M. guatemalensis* (Benth.) Solereder (southeastern Guatemala), *M. fasciculatus* (Benth.) Solereder (Brazil: Guanabara), and *M. ulei* Carv. (Brazil: Piauí). *Melananthus multiflorus* Carv. marks out the southern limit of the genus, reaching Argentina (Corrientes: Isla de las Damas at Rio Paraná) from neighbouring Brazil (Rio Grande do Sul). *Melananthus* is native also in Colombia (Tolima), but the species is still undetermined.

Tribe XI. PARABOUCHETIEAE A. T. Hunz.
53. *Parabouchetia* Baillon

A monotypic tribe, very different from all the others at present known in the family, to a point that it is possible that in the future it may be removed from it. The androecium and gynoecium are really odd: the five stamens with very short filiform filaments are inserted not far from the base of the corolla tube, and have the anthers coalescent by their protruding and enlarged connectives, leaving a small central space occupied by the umbonate stigma; in addition, the few epitropic ovules hang from the placenta. It has intra-xylary phloem (T. E. Di Fulvio, pers. comm.). The only species is *P. brasiliensis* Baillon. The rarity of this herb is astounding. Collected in central Brazil by Burchell on 26 October 1828, the type specimen, the only one known heretofore, comes from a place between S. Bénto and Rio Cangálho in the state of Goiás (near the junction of parallel 12°S. and meridian 47°W.), towards the NNE of Brazilia.
Special references: Baillon (1887, 1888).

Tribe XII. SALPIGLOSSIDEAE Bentham
(Fig. 2.11)

The morphology of the androecium has paramount importance in the understanding of the genera of this tribe; the number of fertile and sterile stamens, the

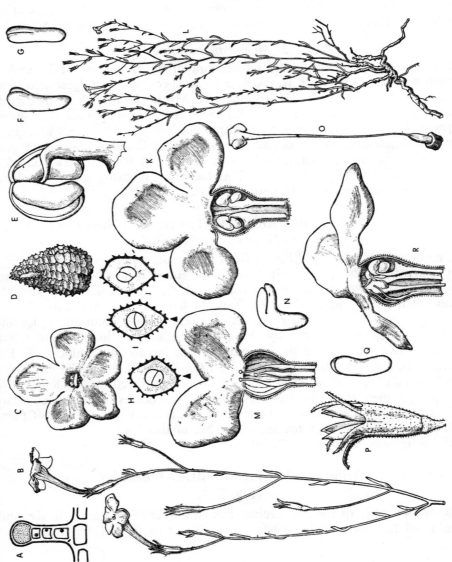

Figure 2.11.
Tribe XII. SALPIGLOSSIDEAE.

Leptoglossis linifolia (Miers) Griseb. (*Di Fulvio* 144; *A. T. Hunziker* 11359). A, Glandular corolla hair (outer surface) (× c. 185); B, flowering branch (× 1.5); C, flower, top view, (× 3); D, seed, hilar view (× 37); E, stamen, ventral face (× 15); F, G, N, Q, embryos (× 30); H-J, seeds, cross sections, the small triangle at the base indicating the position of the hilum (× 30); M, K, the two halves of the upper part of the corolla, one with three staminodes, the other with two stamens (× 6); O, gynoecium (× 3.7); P, fruit (× 3.7); R, upper part of the flower, vertical section, showing the arrangement of the stamens in relation to the stigma and the staminodes (× 6).

amount of abortion and the number of staminodes, the number of thecae, their size and the degree and level of adnation between them, the form of the filament and the way it connects with the anther, the pollen, etc. are all features which should be carefully investigated, together with those of the corolla, the gynoecium, and the disk.

Seven genera in South America: *Streptosolen*, *Browallia* and *Brunfelsia* in the north, the other four (*Salpiglossis*, *Leptoglossis*, *Reyesia* and *Schizanthus*) mainly in the western southern part. $x = 10$ (*Leptoglossis*), 10, 11 (*Schizanthus*), 11 (*Browallia*, *Salpiglossis*), 11, 12 (*Brunfelsia*).

Special references: Bentham (1835, 1845), Bureau (1863), Clos (1849a), Dahlgren (1922), Diels (1919), Hunziker (1977), Johnston (1929a, b), Juel (1911), Philippi (1870), Plowman (1973, 1974, 1976), Samuelsson (1913), Wawra (1866), Werdermann (1928).

54. *Salpiglossis* R. et P.

According to the limits here assigned to this genus (corolla 3.2 to 4.5 cm long; dithecic anthers with both thecae always independent) it consists of two Chilean species: *S. sinuata* R. et P. (grown in gardens on account of the corolla variations in markings and venation) and *S. spinescens* Clos.

55. *Leptoglossis* Benth.

Ephemerals, or herbaceous perennials, with anthers showing ventral insertion of the filaments. *L. schwenckioides* Benth., *L. lomana* (Diels) A. T. Hunz. and some others growing in Peru; in central, western and northern Argentina *L. linifolia* (Miers) Griseb. In total, around five species, very different from those of *Salpiglossis* or *Reyesia*.

56. *Reyesia* Clos

Considered a section of *Salpiglossis* by some authors, *Reyesia* is here assigned generic rank, at least for the time being, considering its peculiar androecium and its tiny flowers (corollas of 7 to 12 mm long). Small genus of four or five species: *R. chilensis* Gay (Chile: Antofagasta and Atacama) and *R. parviflora* (Phil.) A. T. Hunz. (Chile: Antofagasta to Coquimbo; western Argentina: San Juan to Mendoza), etc.

57. *Browallia* L.

A neotropical genus of annual herbaceous plants, extending from southern Mexico, Central America, and the Antilles to Bolivia. At present its taxonomy is obscure; at least two species are recognized, but its number may increase as soon as sound and critical studies are initiated: *B. americana* L., and *B. speciosa* Hook. Both have ornamental value, and therefore are cultivated in tropical and sub-tropical areas; usually they escape from cultivation and are common elements in disturbed land, rendering their original distribution hard to define.

58. *Streptosolen* Miers

A very close relative of *Browallia*, this is an easily distinguishable genus with suffruticose or shrubby habit, and a somewhat twisted corolla. Only one species:

S. jamesonii (Benth.) Miers; the colourful yellow-orange to orange-red corolla is mainly responsible for its cultivation as an ornamental. Grows in Ecuador (Loja, Azuay) and Peru (Cajamarca, Piura), from 1200 to 2300 m alt.; the type specimen was collected by Jameson "in Nova Granada inter Mivir et Naranjas", a place that I have not been able to locate; if it really grows in Colombia it appears to be a very rare species there.

59. *Schizanthus* R. et P.

Very easy to recognize by its deeply cleft corolla lobes, its leaves usually pinnatisect or pinnate, and its minute or almost obsolete stigma, this genus has been questioned as a member of the Solanaceae on the grounds of its serological relationship and its pollen; these arguments would point to a "scrophulariaceous" nature of the genus. Before any final decision is made, it is prudent to await additional and less fragmentary evidence properly correlated with other taxonomic data. For instance, it is worthwhile remembering that the nuclear endosperm of *Schizanthus* (Samuelsson, 1913: 141; Dahlgren, 1922: 81) is not an exception in Solanaceae, being a feature as well of *Capsicum*, *Lycium*, etc. (Davis, 1966: 246). But in the Scrophulariaceae, as a rule, the endosperm is cellular (Schnarf, 1931: 176); therefore, as regards the endosperm, there are no objections to maintaining *Schizanthus* in its present place in the system.

Between 15 to 20 annual herbs from Chile (Atacama to Llanquihue), one entering Argentina (*S. grahami* Gillies: Mendoza and Neuquén); this particular species, together with *S. retusus* Hooker and *S. pinnatus* R. et P., are cultivated as ornamentals.

60. *Brunfelsia* L.

The only woody genus of the tribe, with a total of 40 species in tropical America: one in Panamá, 20 in the Antilles, 20 in South America (Plowman, pers. comm.); *B. pilosa* Plowman reaches northeastern Argentina, marking the southern limit of the genus. It is an important genus for several cultivated ornamental species, as well as from the ethnobotanical point of view (hallucinogens).

HETERANTHIA, A GENUS OF UNCERTAIN SYSTEMATIC POSITION

Heteranthia Nees et Mart. (with the species *H. brasiliensis* Nees et Mart.), is another highly problematic Brazilian monotypic genus. Described in 1823 as a close relative of *Limosella* L., Bentham & Hooker (1876: 926) kept it rather reluctantly in the Scrophulariaceae, near *Leucophyllum* Humb. et Bonpl.; Wettstein (1891c: 107) followed an almost similar path, maintaining it in the same family, although at its very end, as a genus of uncertain systematic position. Finally, Solereder (1915) propounded its transference to the Solanaceae, near *Schwenckia* and *Browallia*, by reason of its internal phloem.

At this moment, I am not prepared to make a decision on this matter and I prefer to postpone it until new information, for example, of palynological or caryological nature, is at hand.

Special references: Nees von Esenbeck & Martius (1823), Solereder (1898), Wawra (1866).

SOME PHYTOGEOGRAPHICAL CONSIDERATIONS

Of the 60 genera of Solanaceae native to South America, 32 are restricted to its territory, whereas the remaining 28 occur also in some other land mass.

(a) *Endemics*. Western tropical Andean South America is the region that ranks first in the number of endemic genera, with the following ten genera confined to it: *Nicandra, Lycopersicon, Brugmansia, Streptosolen, Dunalia, Iochroma, Saracha, Protoschwenkia, Trianaea* and *Ectozoma*. In second place may be mentioned eastern tropical South America with only four endemic genera, i.e. *Metternichia, Athenaea, Parabouchetia* and *Dyssochroma*. Four other genera can be assigned to tropical and subtropical South America: *Vassobia, Leptoglossis, Salpichroa* and an undescribed genus (no. 11) discussed in Tribe *Solaneae*. In the desert-like coast of the Pacific from northern Peru to northern Chile grow *Exodeconus* and *Phrodus*, and in south-western humid South America the monotypic genera *Latua* and *Vestia* are native.

Strictly southern Andean genera are *Schizanthus, Salpiglossis* and *Reyesia*, which inhabit almost exclusively the western slopes of the Andes; on the eastern slopes and reaching Patagonia from southern Peru, grow *Jaborosa, Trechonaetes* and *Fabiana*. Finally, there are three others strictly Patagonian: *Combera, Pantacantha* and *Benthamiella*.

(b) *Widespread genera*. Apart from the three cosmopolitan and very well known genera *Solanum, Lycium* and *Physalis*, the only genus extending from South America to Africa and Australia is *Nicotiana**. Covering the whole continent of America are *Acnistus, Capsicum, Jaltomata, Datura, Cestrum* and *Brunfelsia*. Five genera grow in Central and South America: *Cuatresia, Larnax, Deprea, Markea* and *Hawkesiophyton*. The only entity common to the Antilles and South America is *Sessea*, but there are five which are found also in Central America: *Cyphomandra, Witheringia, Schwenckia, Melananthus* and *Browallia*. On the contrary, *Juanulloa, Merinthopodium* and *Schultesianthus* do not inhabit the Antilles, and their ranges extend without interruption from Mexico to South America. The case of *Solandra* is different: the majority of the species are natives of Mexico, Central America and the West Indies, and a few extend along north-eastern South America (Venezuela, Guianas, Brazil) reaching as far south as São Paulo. Finally there are four disjunct genera (North and South America): *Grabowskia, Petunia, Nierembergia* and a genus discussed in Tribe *Nicotianeae* (see genus no. 49).

THE SOLANACEAE AND RELATED FAMILIES

The following is an attempt to summarize in a key the main diagnostic features of the Solanaceae, which distinguish it from four closely related families. Goetzeaceae is native of the Antilles; the others are all South American.

1. Embryo straight or curved, with comparatively slender cotyledons: these are somewhat shorter or longer than the well developed radicle. Pollen grains without small spinules on the sexine.

---

* Mention should also be made of the curious disjunction (see p. 57) between the two varieties of *Datura leichhardtii* (Mexico, Guatemala and Australia).

2. Ovules erect, sometimes pendant but not from the upper part of
the ovary. Fruit a berry, capsule, dry berry, drupe, or schizo-
carp.
   3. Fruit never schizocarpic. Carpels 2, rarely 5 (or 4, or 3).
      4. Baccate or capsular fruits, exceptionally drupaceous
      (only in *Grabowskia*, and then with 2 bilocular pyrenes
      —in each locule with one or two seeds—immersed in a
      juicy mesocarp). Seeds discoidal, prismatic, reniform or
      subglobose. Embryo straight or curved
                       1. SOLANACEAE Jussieu
      4'. Drupaceous fruits: pyrene solitary, unilocular and
      uniseminate, immersed in a fibrous mesocarp. Seeds
      horseshoe shaped. Embryo vermiform, curved
                2. DUCKEODENDRACEAE Kuhlmann[1]
   3'. Fruit schizocarpic: at maturity it separates in 3 to many
   uni- or pluriseminate mericarps. Embryo curved. Carpels 5,
   or 10 to 25 (when 10 or more, in 2 verticils).
                    3. NOLANACEAE Dumortier[2]
2'. Ovules hanging from the upper part of the ovary. Embryo
straight or curved. Fruit a carcerule
                 4. SCLEROPHYLACACEAE Miers[3]
1'. Embryo straight with thick, fleshy cotyledons: these are far larger
than the tiny radicle. Pollen grains with spinuliferous exine
          5. GOETZEACEAE Miers corr. Airy Shaw[4]

All five families have internal phloem, a feature which has never been recorded
in the Scrophulariaceae.

Special references:
[1] Airy Shaw (1965), Kuhlmann (1925, 1930, 1934, 1947), Record (1933), Record
and Hess (1942).
[2] Clos (1849b), Di Fulvio (1969, 1971), Ferreyra (1961), Johnston (1936),
Wettstein (1891a).
[3] Di Fulvio (1961).
[4] Airy Shaw (1965), Baehni (1943), Erdtman (1966), Miers (1869), Radlkoffer
(1888, 1889), Richard (1838–42), Wydler (1830).

CONCLUSIONS

(1) According to this synopsis the Solanaceae in South America is a family of
60 genera with around 1800 species. For its arrangement the author has followed
the system of Wettstein (1891b) but introducing fundamental modifications in
order to include the information accumulated in the last 80 years, together with
concepts of his own in relation to the tribal arrangement and the position of many
genera.

(2) Taking into consideration characters of seeds, embryos and caryotypes the
family is subdivided into two subfamilies: *Solanoideae* and *Cestroideae*; these
embrace 12 tribes (seven in *Solanoideae*; five in *Cestroideae*), characterized mainly
by peculiarities of the aestivation, the androecium, and the gynoecium.

(3) The list of genera and the estimate of species for each subfamily and tribe is the following one:

I. Subfamily SOLANOIDEAE: 7 tribes, 36 genera, about 1400 spp.
       Tribe 1.   Solaneae: 18 genera, about 1250 spp.
       Tribe 2.   Datureae: 2 genera, about 8 spp.
       Tribe 3.   Jaboroseae: 3 genera, about 36 spp.
       Tribe 4.   Lycieae: 3 genera, about 43 spp.
       Tribe 5.   Nicandreae: 1 genus, 1 sp.
       Tribe 6.   Solandreae: 2 genera, about 14 spp.
       Tribe 7.   Juanulloeae: 7 genera, about 33 spp.

II. Subfamily CESTROIDEAE: 5 tribes, 24 genera, about 413 spp.
       Tribe 8.   Cestreae: 4 genera, about 176 spp.
       Tribe 9.   Nicotianeae: 9 genera, about 130 spp.
       Tribe 10. Schwenckieae: 3 genera, about 34 spp.
       Tribe 11. Parabouchetieae: 1 genus, 1 sp.
       Tribe 12. Salpiglossideae: 7 genera, about 72 spp.

(4) *Heteranthia* is a Brazilian monotypic genus of uncertain systematic position. Whether or not it should be maintained in the Solanaceae is a question for future research; the available information does not at present permit any sound decision on the matter.

(5) On the basis of characters of the embryos, ovules, pollen grains and fruits, the author believes that the following families should be accepted, in order to accommodate genera that, at some time or other, have been thought to belong to Solanaceae: Duckeodendraceae (*Duckeodendron* Kuhlmann), Nolanaceae (*Nolana* L., and *Alona* Lindl.), Sclerophylacaceae (*Sclerophylax* Miers), and Goetzeaceae (*Goetzea* Wydler, *Espadaea* A. Rich., *Henoonia* Griseb., and *Coeloneurum* Radlk.).

ACKNOWLEDGEMENTS

I should like to express my gratitude to those who helped me with this survey. Mrs L. Sánchez and Mrs N. Flury have prepared the illustrations, and Mr R. Münch did a lot of photographic work; I am also indebted to my colleagues Dr T. E. Di Fulvio, Dr A. E. Cocucci, Dr R. Subils, Mrs A. M. Anton and Dr L. Ariza, who assisted me in many ways, helping to solve morphological problems, making special observations, etc. My sincere thanks are tendered as well to Mrs E. G. de Mautino, who was always ready to search for bibliographical references, or to find herbarium specimens. Last but not least, I must thank Professor J. G. Hawkes for encouraging me to write this conspectus of the South American Solanaceae and for inviting me to participate in the symposium at which the paper was read. I should also like to express my gratitude to the Consejo Nacional de Investigaciones Científicas y Técnicas, Argentina, for its unfailing support.

REFERENCES

AIRY SHAW, H. K., 1965. Diagnoses of new families, new names, etc. for the seventh edition of WILLIS'S Dictionary. *Kew Bulletin, 18* (2): 249–273.

BAEHNI, Ch., 1943. *Henoonia*, type d'une famille nouvelle? *Boissiera, 7:* 346–358.
BAEHNI, Ch., 1946. L'ouverture du bouton chez les fleurs de Solanées. *Candollea, 10:* 399–492.

BAILLON, H. E., 1887. Sur quelques types du groupes intermediares aux Solanacées et aux Scrofulariacées. *Bulletin Mensual de la Société Linnéene de Paris, 83:* 660–663.

BAILLON, H. E., 1888. *Histoire des Plantes, 9:* 281–359. Paris.

BARCLAY, A. S., 1959. *Studies in the Genus* Datura *(Solanaceae). I. Taxonomy of Subgenus* Datura: 221 pp. Unpubl. thesis, Harvard University.

BARKLEY, F. A., 1953. *Lycium* in Argentina. *Lilloa, 26:* 177–238.

BENITEZ DE ROJAS, C. E., 1974. Los géneros de las Solanaceae de Venezuela. *Revista de la Facultad de Agronomia Universidad Central de Venezuela, 7* (3): 25–108.

BENTHAM, G., 1835. Revision of Scrophulariaceae; tribe *Salpiglossideae. Bot. Reg., 21.*

BENTHAM, G., 1845. *The Botany of the Voyage of H.M.S. Sulphur:* 1–194. *Leptoglossis* gen. nov.: 143. London.

BENTHAM, G., 1846. Scrophulariaceae. Subord. et Trib. I. *Salpiglossideae.* In A. P. De Candolle (Ed.), *Prodromus Systematis Naturalis Regni Vegetabilis, 10:* 190–203.

BENTHAM, G. & HOOKER, J. D., 1876. *Genera Plantarum, 2* (2): 882–980, 1244–1245.

BITTER, G., 1903. Die Rassen der *Nicandra physaloides. Beihefte zum Botanischen Zentralblatt, 14:* 145–176.

BITTER, G., 1921. Zur Gattung *Cacabus* Bernh. *Repertorium Novarum Specierum Regni Vegetabilis, 17:* 243–245.

BITTER, G., 1922. Zur Gattung *Sessea. Repertorium Novarum Specierum Regni Vegetabilis, 18:* 199–225.

BRONGNIART, A., 1843. *Enumération des Genres de Plantes Cultivés au Muséum d'Histoire Naturelle de Paris* . . . 1–136 [2ᵉ ed., 1850, 1–237]. Paris.

BURBIDGE, N. T., 1960. The Australian species of *Nicotiana* L. *(Solanaceae). Australian Journal of Botany, 8* (3): 342–380.

BUREAU, E., 1863. Etudes sur les genres *Reyesia* et *Monttea* et observations . . . etc. *Bulletin Société Botanique de France, 10:* 1–13.

CARVALHO, L. d'A. F. DE., 1966a. O gênero *Melananthus* no Brasil (Solanaceae). *Sellowia, 18* (18): 51–66.

CARVALHO, L. d'A. F. DE., 1966b. O gênero *Protoschwenckia* no Brasil (Solanaceae). *Sellowia, 18* (18): 67–72.

CARVALHO, L. d'A. F. DE., 1969a. Duas novas espécies de *Schwenckia* (Solanaceae). *Loefgrenia, 33:* 1–3.

CARVALHO, L. d'A. F. DE., 1969b. Novitates Schwenckiarum I. *Loefgrenia, 37:* 1–4.

CARVALHO, L. d'A. F. DE., 1971. Novitates Schwenckiarum II. (Solanaceae). *Rodriguesia, 26* (38): 247–249.

CHODAT, R., 1916. Géobotanique et étude critique de quelques Solanées paraguayennes. *Bulletin de la Société Botanique de Genève,* (Ser. 2), *8:* 142–160.

CLOS, D., 1849a. Bignoniáceas. In C. Gay, *Historia Física y Política de Chile, Botánica, 4:* 408–420. [p. 418: *Reyesia*]. Paris.

CLOS, D., 1849b. Nolanaceas, Escrofularineas. In C. Gay, *Historia Física y Política de Chile, Botánica, 5:* 100–188. Paris.

CUATRECASAS, J., 1958a. The Colombian species of *Juanulloa Brittonia, 10* (3): 146–150.

CUATRECASAS, J., 1958b. Notes on American Solanaceae. *Repertorium Novarum Specierum Regni Vegetabilis, 61* (1): 74–86.

CUATRECASAS, J., 1959. New chiropterophilous Solanaceae from Colombia. *Journal of the Washington Academy of Sciences, 49* (8): 269–272.

DAHLGREN, K. V. O., 1922. Die Embriologie der Loganiaceen-Gattung *Spigelia. Svensk Botanisk Tidskrift, 16:* 77–87.

DAMMER, U., 1906. Solanaceae americanae. *Botanische Jahrbücher für Systematik, Pflanzengeschichte und Pflanzengeographie, 37:* 167–171.

D'ARCY, W. G., 1973. Family 170. Solanaceae, Flora of Panamá. *Annals of Missouri Botanical Garden, 60* (3): 573–780.

DAVIS, G. L., 1966. *Systematic Embryology of the Angiosperms, I–VIII:* 1–528, Solanaceae: 246–247. New York: J. Wiley.

DIELS, L., 1919. *Leptofeddea* Diels, ein neue Gattung der Solanaceen aus Perú. *Repertorium Novarum Specierum Regni Vegetabilis, 16:* 193.

DI FULVIO, T. E., 1961. El género *Sclerophylax* (Solanaceae). Estudio anatómico, embriológico y cariológico con especial referencia a la taxonomía. *Kurtziana, 1:* 9–103.

DI FULVIO, T. E., 1969. Embriología de *Nolana paradoxa* (Nolanaceae). *Kurtziana, 5:* 39–54.

DI FULVIO, T. E., 1971. Morfología floral de *Nolana paradoxa* (Nolanaceae), con especial referencia a la organización del gineceo. *Kurtziana, 6:* 41–51.

DON, G., 1837. *A general system of gardening and botany, 4.* Solanaceae: 397–488. London.

DRAKE DEL CASTILLO, M., 1892. Note sur une plante nouvelle des Andes. *Bulletin de la Société Philomatique de Paris* (8e. Ser.), *4:* 128–129.

DUNAL, M. F., 1852. Solanaceae. In A. P. De Candolle (Ed.), *Prodromus Systematis Naturalis Regni Vegetabilis, 13* (1): 1–741. Paris.

ERDTMAN, G., 1966. *Pollen Morphology and Plant Taxonomy of Angiosperms:* i–xii, 1–553. Corrected reprint ed. 1952. New York & London: Hafner.

FEDDE, F., 1896. Beiträge zur vergleichenden Anatomie der Solanaceae: 1–48. *Arbeit aus dem Botanischen Gartens zu Universität Breslau.*

FERREYRA, R., 1961. Revisión de las especies peruanas del género *Nolana* (Nolanaceae). *Memorias del Museo de Historia Natural 'Javier Prado', 12:* 1–70.

FOSTER, R. C., 1958. A catalogue of ferns and flowering plants of Bolivia. *Contributions from the Gray Herbarium of Harvard University, 184:* 1–223.

FRANCEY, P., 1933. Beitrag zur Kenntnis der Gattung *Sessea. Notizblatt des Botanischen Gartens und Museums zu Berlin, 11:* 879–883.

FRANCEY, P., 1934. Übersicht über die Gattung *Sessea. Notizblatt des Botanischen Gartens und Museums zu Berlin, 11:* 978–990.

FRANCEY, P., 1935–36. Monographie du genre *Cestrum* L. *Candollea, 6:* 46–398 (1935), *7:* 1–132 (1936).

FRIES, R. A., 1911. Die Arten der Gattung *Petunia. Svenska Vetenskapsakademiens Handlingar* (III), *4* (5): 3–72.

GENTRY, J. L., 1973. Restoration of the Genus *Jaltomata* (Solanaceae). *Phytologia, 27* (4): 286–288.

GENTRY, J. L. & STANDLEY, P. C., 1974. Flora of Guatemala. Solanaceae. *Fieldiana: Botany, 24,* 10 (1–2): 1–151.

GOODSPEED, T. H., 1954. *The Genus Nicotiana:* i–xxi, 1–536. Waltham, Mass.: Chronica Botanica Co.

HASSLER, E., 1917. Solanaceae Austro-Americanae imprimis Paraguarienses. *Annuaire de Conservatoire et Jardin Botanique Genève, 20:* 173–189.

HAWKES, J. G. & HJERTING, J. P., 1969. *The Potatoes of Argentina, Brazil, Paraguay and Uruguay. A Biosystematic Study:* i–xxiii, 1–525. Oxford: Clarendon Press.

HEINE, H., 1963. On the correct spelling of the generic name *Schwenckia* D. van Royen ex L. (Solanaceae, with a note about Martin Wilhelm Schwencke). *Kew Bulletin, 16* (3): 465–469.

HEISER, C. B. & PICKERSGILL, B., 1969. Names for the cultivated *Capsicum* species (Solanaceae). *Taxon, 18:* 277–283.

HEMSLEY, W. B., 1882. *Biologia Centrali Americana (Botany), 2:* 404–438.

HEPPER, F. N., 1976. *The West African herbaria of Isert and Thonning.* Kew: The Bentham-Moxon Trust.

HIERONYMUS, G., 1881. Sertum sanjuaninum. *Boletín de la Academia Nacional de Ciencias en Córdoba, 4* (1): 1–73. Solanaceae: 57–61. Argentina.

HITCHCOCK, Ch. L., 1932. A monographic study of the genus *Lycium* of the Western Hemisphere. *Annals of Missouri Botanical Garden, 19:* 179–374.

HOFFMANNSEGG, I. C., 1824. *Verzeichniss der Pflanzenkulturen in den Gräflich Hoffmannseggischen Gärten zu Dresden und Rammenau:* 1–310. Dresden.

HUNZIKER, A. T., 1950. Estudios sobre Solanaceae. I. Sinopsis de las especies silvestres de *Capsicum* de Argentina y Paraguay. *Darwiniana, 9:* 225–247.

HUNZIKER, A. T., 1954. Synopsis of the genus *Capsicum. Compte Rendu des Séances Rapp. et Comn. Sect.* 3, 4, 5, 6: 73–74.

HUNZIKER, A. T., 1960. Estudios sobre Solanaceae II. Sinopsis taxonómica del género *Dunalia* H.B.K. *Boletín de la Academia Nacional de Ciencias en Córdoba, 41* (2): 211–244.

HUNZIKER, A. T., 1961a. Estudios sobre Solanaceae III. *Kurtziana, 1:* 207–216.

HUNZIKER, A. T., 1961b. Noticia sobre el cultivo de *Capsicum baccatum* (Solanaceae) en Argentina. *Kurtziana, 1:* 303.

HUNZIKER, A. T., 1967. Estudios sobre Solanaceae. IV. *Kurtziana, 4:* 131–138.

HUNZIKER, A. T., 1969a. Estudios sobre Solanaceae. V. *Kurtziana, 5:* 101–179.

HUNZIKER, A. T., 1969b. Estudios sobre Solanaceae. VI. *Kurtziana, 5:* 393–399.

HUNZIKER, A. T., 1971. Estudios sobre Solanaceae. VII. *Kurtziana, 6:* 241–259.

HUNZIKER, A. T., 1977. Estudios sobre Solanaceae. VIII. *Kurtziana, 10:* 7–50.

JOHNSTON, I. M., 1929a. Papers on the flora of northern Chile. Contributions from the *Gray Herbarium of Harvard University, 85:* 1–172.

JOHNSTON, I. M., 1929b. Some undescribed species from Perú. *Contributions from the Gray Herbarium of Harvard University, 85:* 172–180.

JOHNSTON, I. M., 1936. A study of the Nolanaceae. *Proceedings of the American Academy, 71:* 1–87.

JUEL, H. O., 1911. Om blommans byggnad hos *Browallia.* In, *Biol. Arb. Tilegnede Eug. Warming Poa Hans 70 aars . . . etc.:* 9–118. Copenhagen.

KANIEWSKI, K., 1965. Fruit histogenesis in *Nicandra physaloides* (L.) Gaertn. *Bulletin de l'Académie Polonaise des Sciences, 13* (9): 553–556.

KUHLMANN, J. G., 1925. Plantas novas . . . etc. *Archivos do Jardim de Botanico, 4:* 346–365. *Duckeodendron:* 362.

84 A. T. HUNZIKER

KUHLMANN, J. G., 1930. Contribuição para o conhecimiento de algunas novas especies da região amazonica ... etc. *Archivos do Jardim Botanico, 5:* 203–209. *Duckeodendron:* 209.

KUHLMANN, J. G., 1934. Notas sobre o gênero *Duckeodendron. Archivos do Instituto Biologia Vegetal, 1* (1): 35–37.

KUHLMANN, J. G., 1947. Duckeodendraceae Kuhlmann (Nova familia). *Archivos do Serviço Florestal do Brasil, 3:* 7–8.

LAGERHEIM, G., 1895. Monographie der ecuadorianischen Arten der Gattung *Brugmansia* Persoon. *Botanische Jahrbücher für Systematik, Pflanzengeschichte und Pflanzengeographie, 20:* 655–668.

LOCKWOOD, T. E., 1973. Generic recognition of *Brugmansia. Botanical Museum Leaflets, 23*(6): 273–281.

LOCKWOOD, T. E., 1976. Systematics of the genus *Brugmansia. Abstracts of International Symposium on Biology and Taxonomy of Solanaceae.* Abstracts of papers: 54–55. Birmingham University.

MARTÍNEZ, M., 1967. Las Solandras de México, con una especie nueva. *Anales del Instituto de Biologia, Universidad de México, 37:* 97–106.

MERXMÜLLER, H. & BUTTLER, K. P., 1975. *Nicotiana* in der afrikanischen Namib. Ein pflanzengeographisches und phylogenetisches Rätsel. *Mitteilungen aus der Botanischen Staatssammlung, 12:* 91–104.

MIERS, J., 1849–57. *Illustrations of South American plants, I:* i–iv, 1–183; 1850, *II:* i–iv, 1–150; Appendix 1–79; 1849–57. London: Baillière.

MIERS, J., 1869. On the genera *Goetzia* and *Espadea. Transactions of the Linnean Society of London, 27:* 187–195.

MIKAN, J. Ch., 1823. *Delectus Florae et Faunae Brasiliensis, 3.* Wien.

MILLÁN, A. R., 1926. Notas críticas sobre las Nicotianas de la Flora argentina. *Revista de la Facultad de Agronomia y Veterinaria, Universidad de Buenos Aires, 5* (2): 172–188.

MILLÁN, A. R., 1928. Las especies del género 'Nicotiana' de la Flora argentina. *Revista de la Facultad de Agronomia y Veterinaria, Universidad de Buenos Aires, 6* (2): 169–216.

MILLÁN, A. R., 1931. Solanáceas argentinas. Clave para la determinación de los géneros. *Boletín del Ministerio de Agricultura, 30* (1): 3–21. Buenos Aires.

MILLÁN, A. R., 1941. Revisión de las especies del género *Nierembergia (Solanaceae). Darwiniana, 5:* 487–547.

MORTON, C. V., 1938. Notes on the genus *Saracha. Proceedings of the Biological Society of Washington, 51:* 75–78.

MORTON, C. V., 1944. Notes on *Bouchetia. Contributions from the United States National Herbarium, 29* (1): 72–73.

MULLER, C. H., 1940. A revision of the genus *Lycopersicon. Miscellaneous Publications United States Department of Agriculture, 382:* 1–28.

NEES VON ESENBECK, C. G. & VON MARTIUS, K. F. P., 1823. Beitrag zur Flora Brasiliens von Maximilian ... etc. *Nov. Act. Phys.-Med. Nat. Curios., 11:* 3–88. *Heteranthia:* 41–43.

PHILIPPI, R. A., 1860. *Viaje al Desierto de Atacama:* i–iv, 1–236. Solanaceae, etc.: 214–220.

PHILIPPI, R. A., 1870. Sertum mendocinum alterum. *Anales de la Universidad de Chile, 36:* 159–212. *Dittostigma:* 194; *Cyclostigma:* 197.

PHILIPPI, R. A., 1891. Catalogus praevius plantarum in itinere ad Tarapaca ... etc. *Anales del Museo Nacional de Chile, 2a Ser. Bot., 1:* I–VIII, 1–96, Santiago. Solanaceae, etc.: 61–70.

PLOWMAN, T. C., 1973. Four new Brunfelsias from northwestern South America. *Botanical Museum Leaflets, 23* (6): 246–272.

PLOWMAN, T. C., 1974. Two new Brazilian species of *Brunfelsia. Botanical Museum Leaflets, 24* (2): 37–48.

PLOWMAN, T. C., 1976. Systematics and biogeography of *Brunfelsia. Abstracts of International Symposium on Biology and Taxonomy of the Solanaceae:* 40. Birmingham University.

PLOWMAN, T., GYLLENHAAL, L. O. & LINDGREN, J. E., 1971. *Latua pubiflora,* magic plant from southern Chile. *Botanical Museum Leaflets, 23* (2): 61–92.

RADLKOFER, L., 1888. Ueber die Versetzung der Gattung *Henoonia* von den Sapotaceen zu den Solanaceen. *Sitzungsberichte der Bayerischen Akademie der Wissenschaften zu München (Math.-Physik.), 18* (3): 405–421.

RADLKOFER, L., 1889. Zur Klärung von *Theophrasta* und der Theophrasteen, unter Übertragung dahin gerechneter Pflanzen zu den Sapotaceen und Solanaceen. *Sitzungsberichte der Bayerischen Akademie der Wissenschaften zu München (Math.-Physik.), 19:* 221–281.

RECORD, S. J., 1933. The woods of *Rhabdodendron* and *Duckeodendron. Tropical Woods, 33:* 6–10.

RECORD, S. J. & HESS, R. W., 1942. *Timbers of the New World:* i–xv, 1–640. New Haven: Yale Univ. Press.

REICHE, K., 1910. Solanáceas. *Flora de Chile, 5:* 309–410. Santiago.

REICHENBACH, H. Th. L., 1828. *Conspectus Regni Vegetabilis* ... Pars Prima ...; i–xiv, 1–294. Lipsiae.

REMY, J., 1849. Solaneas. In C. Gay, *Historia Física y Política de Chile, Botánica, 5:* 38–100. Paris, Santiago.

RICARDI, M., 1962. El género *Combera (Solanaceae). Gayana, 4:* 3–13.

RICARDI, M., 1966. Plantas interesantes o nuevas para Chile. *Gayana, 14:* 3–29.

RICHARD, A., 1838–42. *Espadaea.* In Ramón de la Sagra, *Hist. Phys. Pol. Nat. de l'île de Cuba, 9* (1) Bot. Pl. Vasc. Atlas t. 65. Paris.

ROMERO-CASTAÑEDA, R., 1972. Apuntes botánicos, IV. *Mutisia, 38:* 1–18.

RUIZ, H. & PAVON, J., 1794. *Flora Peruvianae et Chilensis Prodromus.* i–xxii, 1–154. Reprint 1965. Lehre: J. Cramer.

RUIZ, H. & PAVON, J., 1798–99. *Flora Peruvianae et Chilensis, 1:* 13 (1798); *2:* 27–47 (1799).

SAMUELSSON, G., 1913. Studien über die Entwicklungsgeschichte der Blüten einiger Bicornes-Typen. *Svensk Botanisk Tidskrift, 7:* 97–188.

SANDWITH, N. Y., 1936. *Combera paradoxa* Sandwith. *Hooker's Icones Plantarum,* tab. 3325.

SANDWITH, N. Y., 1938. The correct name of the "tree tomato". *Chronica Botanica, 4:* 225.

SANDWITH, N. Y., 1939. *Combera minima* Sandwith. *Hooker's Icones Plantarum,* tab. 3325.

SCHNARF, K., 1931. *Vergleichende Embryologie der Angiospermen:* 1–354. Solanaceae, Scrophulariaceae: 174–180. Berlin: Gebr. Borntraeger.

SCHULTES, R. E., 1955. A new narcotic genus from the Amazon slope of the Colombian Andes. *Botanical Museum Leaflets, 17* (1): 1–11.

SCHUMACHER, F. C., 1828. Beskrivelse af Guineiske planter. *Det Kongelige danske videnskabernes Selskabs Naturvidenskabelige og mathematiske Afhandlinger, 3:* 28–29. Kjöbenhavn.

SCOLNIK, R., 1954. Las especies de *Cestrum* de la Argentina, Chile y Uruguay. *Revista Fac. Cienc. Ex. Fis. Nat. Ser. Cienc. Nat., 3:* 1–104. Córdoba, Argentina.

SENDTNER, O., 1846. Solanaceae et *Cestrineae.* In C. F. P. von Martius (Ed.), *Flora Brasiliensis* (f. 6), *10:* 1–228. Vindobonae et Lipsiae.

SMITH, L. B. & DOWNS, R. J., 1966. Solanáceas. In P. R. Reitz (Ed.), *Flora Illustrada Catarinense,* 1–321. Itajai, Brasil.

SOLEREDER, H., 1891. Über die Versetzung der Gattung *Melananthus* Walp. von der Phrymaceen zu den Solanaceen. *Berichte der Deutschen Botanischen Gesellschaft, 9:* 65–85.

SOLEREDER, H., 1898. Zwei Beiträge zur Systematik der Solanaceen. *Berichte der Deutschen Botanischen Gesellschaft, 16:* 242–260.

SOLEREDER, H., 1915. Über die Versetzung der Gattung *Heteranthia* von de Scrophulariaceen zu den Solanaceen. *Beihefte zum Botanischen Zentralblatt, 22,* Abt. 2, Heft 1: 113–117.

SORIANO, A., 1948. El género *Benthamiella* (Solanaceae). *Darwiniana, 8* (2–3): 233–262.

SPEGAZZINI, C., 1902. Nova addenda ad floram patagonicam II. *Anales de la Sociedad Cientifica Argentina, 53:* 149–150 [reprint: 51, 52].

TUCKER, W. G., 1969. Serotaxonomy of the Solanaceae: a preliminary survey. *Annals of Botany (NS), 33* (129): 1–23.

WATERFALL, U. T., 1958. A taxonomic study of the genus *Physalis* in North America, north of Mexico. *Rhodora, 60:* 107–114, 128–142, 152–173.

WATERFALL, U. T., 1967. *Physalis* in Mexico, Central America and the West Indies. *Rhodora, 69:* 82–120, 203–239, 319–329.

WAWRA, H., 1866. *Botanische Ergebnisse. Heteranthia:* 82.

WERDERMANN, E., 1928. Übersicht über die in Chile vorkommenden Arten der Gattung *Salpiglossis* Ruiz et Pav. (Solanaceae). *Notizblatt des Botanischen Gartens und Museum zu Berlin, 10* (95): 472–475.

WETTSTEIN, R. VON, 1891a. *Nolanaceae.* In H. G. A. Engler & K. A. E. Prantl (Eds), *Die Natürliche Pflanzenfamilien, 4* (3b): 1–4. Berlin.

WETTSTEIN, R. VON, 1891b. Solanaceae. In H. G. A. Engler & K. A. E. Prantl (Eds), *Die Natürliche Pflanzenfamilien, 4* (3b): 4–38.

WETTSTEIN, R. VON, 1891c. Scrophulariaceae. In H. G. A. Engler & K. A. E. Prantl (Eds), *Die Natürliche Pflanzenfamilien, 4* (3b): 39–107.

WYDLER, H., 1830. Plantarum quarundam descriptiones. *Linnaea, 5:* 423–430.

3. Solanaceae in India

D. B. DEB

Botanical Survey of India, Calcutta

The Solanaceae in India are represented by 24 genera of which 10 are native. There are 108 species recorded, of which 33 appear to be native, 16 naturalized, 40 cultivated and 19 cultivated experimentally. The taxonomy and distribution of certain taxa are discussed, and their distribution is shown by means of maps based on well-authenticated records.

CONTENTS

INTRODUCTION

The solanaceous plants of India are cited in ancient Sanskrit literature. The generic name *Datura*, derived from the Bengali vernacular name, is mentioned as Dhustura or Dhattura in ancient Sanskrit literature, particularly in Charaka and Susruta (Mazumder, 1929) which were written before America was known to Europe. Brinjal is cited as Bartaku in the Amarkosha which was written in A.D. 1100. The earliest scientific literature describing Indian plants is credited to Garcia (1563) followed by Rheede (1678–1703), Plukenet (1695), Hermann (1687), Burman (1737) and Rumphius (1741–50), contributing to the pre-Linnean literature which formed the basis of "Species Plantarum" (Linnaeus, 1753).

Linnaeus (1753) treated 7 genera and 15 species from Malabar, Ceylon and Coromandel coasts. Roxburgh (1820–24, 1832) was the first to give an all India treatment of the family and described 9 genera and 28 species. Nees (1837) contributed to the knowledge of 8 genera and 47 species from Eastern India. Voigt (1845) treated 5 genera comprising 11 species cultivated in Calcutta and Serampore. Clarke (1883) described 14 genera and 42 species referable to the present day delimitation of India. Chatterjee (1939) estimated 14 genera and 58 species from the Indian subcontinent. He considered 16 species as endemic.

Indian materials have not been given any general treatment after Clarke (1883), though many important contributions have been made on these plants from other parts of the world. However, Prain (1903), Cooke (1905), Duthie (1911), Haines (1922), Gamble (1923), Kanjilal, Kanjilal, Das & De (1939), Santapau (1948), Deb (1961, 1972), Yamazaki (1966), Deb & Dutta (1965, 1974) Deb, Sengupta &

Malick (1969), Sengupta (1973) and others treated the family in their respective regional studies.

An attempt is made in the present paper to assess the Indian material in the light of recent researches on taxonomic delimitations and nomenclatural implications and also to trace the distribution of species on the basis of well authenticated occurrences.

The paper is based on field studies conducted by the author in connection with a floristic survey of regional Floras during the past two and a half decades, and materials extant in Indian herbaria. The taxonomic revision on which this work is based will appear elsewhere.

Occurrences of taxa on the basis of well authenticated herbarium records are plotted on distribution maps. The distribution outside India is taken mostly from the literature.

TAXONOMIC STUDY

Efforts to classify Indian material have resulted in significant changes in nomenclature and taxonomy of certain taxa, some of which are presented below.

Solanum lysimachioides Wall. in Roxburgh, *Flora Indica*, 2: 257.1824 (Type: *Wall. Cat.* 2609! CAL) was described on the basis of a plant grown in the Indian Botanic Garden, Calcutta, in 1823 from seeds collected from Nepal in 1821.

S. macrodon Wall. was collected from Sylhet, now in Bangladesh, in December 1824 by F. de Silva. Nees in *Transactions of the Linnean Society of London*, 17: 43.1837 distinguished it from *S. lysimachioides* in woody stem, longer and narrower leaves, many flowered peduncle and sulcate calyx. Dunal (1852) treated them as distinct species but Clarke (1883) reduced *S. lysimachioides* to a variety of *S. macrodon* for slender procumbent stem often rooting at the nodes and solitary pedicels. Bitter (1920) treated them as two distinct species under the genus *Lycianthes* Hassl. emend. Bitter.

Solanum biflorum Lour. *Flora Cochinchina*, 1: 129. 1790 was described on the basis of a gathering from Cochinchina. Several species subsequently described by Roxburgh (1832), Nees (1837), Dunal (1852) and Blume (1856) from various countries, namely *S. decemdentatum* Roxb. (Nepal), *S. decemfidum* Nees (China), *S. mollissimum* Bl. (Malaya), *S. zollingeri* Dun. (Java), *S. cauleryanum* Dun. (China) were reduced to synonymy of *S. biflorum* by Clarke in Hooker, *Flora of British India*, 4: 231. 1883, thereby recognizing its variability and enlarging its distribution. In consideration of the phenotypic plasticity, *S. lysimachioides* and *S. macrodon* are treated here as subspecies of *S. biflorum* under the genus *Lycianthes*, as follows:

L. *biflora* (Lour.) Bitt. subsp. *macrodon* (Wall. ex Nees) Deb, 1978.

L. *biflora* (Lour.) Bitt. subsp. *lysimachioides* (Wall.) Deb, 1978.

In consideration of the phenotypic plasticity of *S. laeve*, *S. crassipetalum*, *S. bigeminatum*, *S. subtruncatum* and *S. kaitisis*, these taxa are treated as conspecific under *Lycianthes*, as follows:

L. *laevis* (Dun.) Bitt. subsp. *crassipetala* (Wall.) Deb, 1978.

L. *laevis* (Dun.) Bitt. subsp. *bigeminata* (Nees) Deb, 1978.

L. *laevis* (Dun.) Bitt. subsp. *subtruncata* (Wall. ex Dun.) Deb, 1978.

L. *laevis* (Dun.) Bitt. subsp. *kaitisis* (Dun.) Deb, 1978.

L. *laevis* (Dun.) Bitt. subsp. *kaitisis* (Dun.) Deb var. *gouakai* (Dun.) Deb, 1978.

Solanum hovei Dun. in De Candolle, *Prodromus 13*(1): 311. 1852 (Type: Dholka, near the river Sabarmati in Gujarat, 1788, *A.P. Hove* s.n. in BM-photo! in herb. CAL) was not cited by Clarke (1883), or any other investigator. Sengupta (1964) determined two gatherings as this. While describing this species Dunal (1852) considered its affinity with *S. carolinense* L., a tropical American species, but differing in floral characters in which respect it was treated as similar to *S. indicum*. Sengupta (1964) distinguished it from *S. indicum* by its scant pubescence and

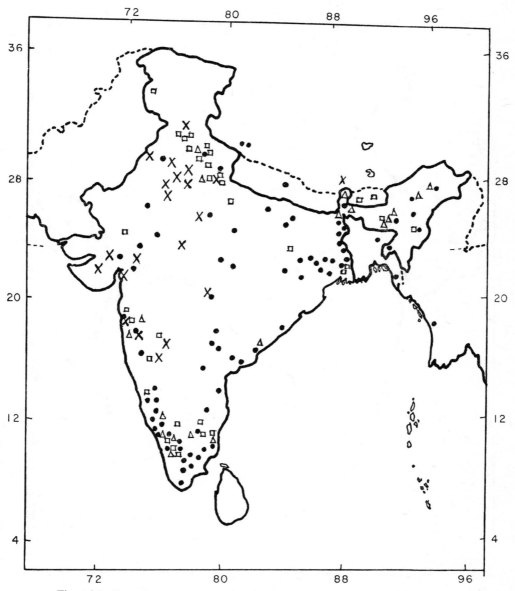

Figure 3.1. *Datura* L.: ×, *D. inoxia;* ●, *D. metel;* □, *D. stramonium;* △, *D. suaveolens.*

unarmed nature of the inflorescence with recurved fruiting pedicels. On the
other hand, *S. indicum* L. is a very widely distributed taxon growing in varying
ecological situations. It shows phenotypic plasticity, which created occasions
for describing new taxa from time to time, which were only to be suppressed
subsequently. Distinguishing characters of *S. hovei* cited by Sengupta (1964)
are evident on a large number of gatherings in CAL, BSIS, BSI, BLAT, the like
of which must be found in other herbaria also. A close look at them shows that

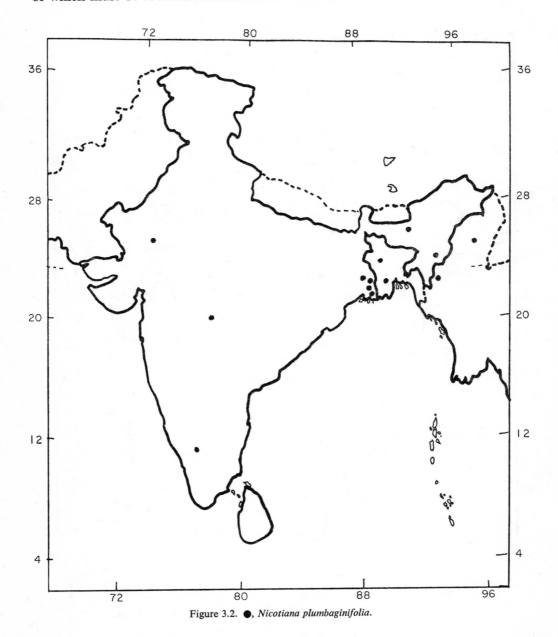

Figure 3.2. ●, *Nicotiana plumbaginifolia*.

they are continuous, linking them with *S. indicum*. Hence *S. hovei* Dun. is relegated to a synonym of *S. indicum*.

Solanum esuriale Lindl., native to Australia, was reported from Poona by Vartak (1957). Raghavan & Wadhwa (1970) subsequently showed that Vartak's report (1957) was based on misidentification of the specimens which represented mostly *S. elaeagnifolium* Cav. *S. mccanni* Sant. in *Journal of the Bombay Natural History Society*, *47*: 653. 1948 (Type: Khandala, 18.10.1943, *Santapau* 2972 & 2973 BLAT.) has been rightly reduced to a synonym of *S. surattense* by Venkatareddi in *Bulletin of the Botanical Survey of India*, *11*: 209. 1969.

S. melongena L. varies widely in habit, flower bearing and type of fruit. *S. incanum* has many characteristics in common with *S. melongena* and is distinguished from the latter mainly by the shape of the fruit, density of prickles and by the fruits being borne in clusters. *S. insanum* is distinguished mainly by solitary flowers, small rounded fruits and the nature of the petal lobes and petal tips. These taxa closely resemble each other and have been assigned various taxonomic statuses. *S. incanum* has been treated as a species by Linnaeus (1753), Dunal (1852), Cooke (1905) and Duthie (1911), whereas Nees (1837) and Kuntze (1891) treated it as a variety of *S. melongena*. Likewise Linnaeus (1753), Willdenow (1806) and Roxburgh (1832) treated *S. incanum* as a species, but Nees (1837), Voigt (1845) and Prain (1903) treated it as a variety. Clarke (1883) treated all these taxa as conspecific, under *S. melongena* but maintained *S. coagulans* as a species, though it is sometimes treated as a synonym of *S. incanum* (see Pearce & Lester, this Volume, Chapter 48).

Experimental hybrids between *S. incanum*, *S. insanum* and *S. melongena* in reciprocal combinations have been obtained by Rao *et al.* (1970; and Rao, this Volume, Chapter 47). Viswanathan (1975) reported on the occurrence of a natural hybrid between a cultivar of *S. melongena* and *S. melongena* var. *incanum* which produced different types of fruits ranging between those of the two parents. All these studies strongly indicate that *S. incanum* and *S. insanum* should be given varietal status under *S. melongena*.

The taxonomy of the *S. nigrum* complex has been the subject of studies in India, as in other parts of the world. (See also Heiser, Burton & Schilling and Edmonds, this Volume, Chapters 40 and 41.)

In the strict sense, *S. nigrum* breaks up into a number of taxa distinguishable on correlated morphological and cytological features; these are sometimes treated as species, when the name *S. nigrum* is restricted to a hexaploid taxon ($2n = 72$), while *S. nodiflorum* Jacq. and *S. villosum* Mill. are restricted to diploid ($2n = 24$) and tetraploid ($2n = 48$) taxa respectively. A somatic chromosome number of 72 is also reported for *S. nodiflorum* subsp. *nodiflorum*. (Gerasimenko & Reznikova, 1968; Henderson, 1974). Henderson (1974) noted putative hybrids between *S. nodiflorum* subsp. *nutans* and *S. gracilius* in the field, and spontaneous hybrids under glasshouse conditions between hexaploid *S. scabrum* and diploid *nodiflorum* subsp. *nodiflorum*. He observed a number of apparent hybrids between taxa of different ploidy levels within herbarium materials he studied, and treated *S. villosum*, *S. luteum* and *S. rubrum* as synonymous. Rao, Khan & Khan (1971), also observed that *S. luteum* and *S. villosum* represented one taxon. Bhaduri (1933) and Tandon & Rao (1966) distinguished three types of plants growing sympatrically in

Bengal and Delhi respectively. Henderson recognized ten species for the *S. nigrum* complex in Australia and gave a key to the species, in which he considered several characters which are not yet known to have any such taxonomic significance in this genus. Stebbins (1950) suggested that *S. nigrum* L. *sensu stricto* might have arisen through hybridization. Tandon & Rao (1966) confirmed it by producing a plant by crossing, which was morphologically identical to hexaploid *S. nigrum*. These plants occur extensively throughout India in varying habitats. In an attempt

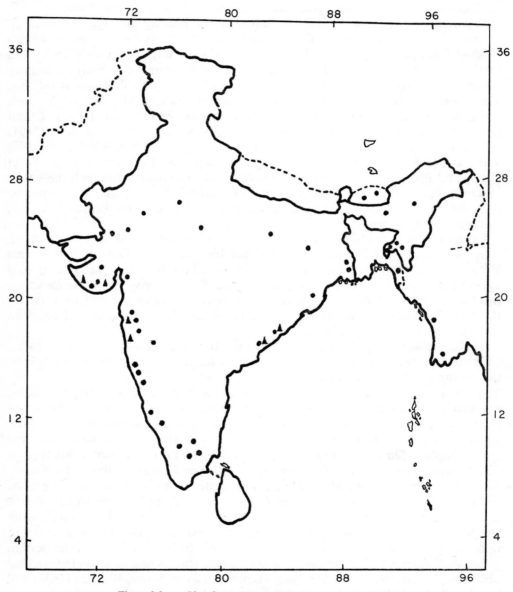

Figure 3.3. ▲, *Physalis virginiana* var. *sonorae*; ●, *P. minima*.

to sort out the specimens extant in CAL and BSIS in the light of recent studies, it came to light that intergrading characters are frequent and continuous. In consideration of this evidence *S. nigrum* L. is treated here in the broad sense.

S. purpureilineatum Sabnis & Bhatt. in *Bulletin of the Botanical Survey of India, 12*: 258. 1970 (Type: Gujarat, Hampheshwar near temple, Chholandepur Forest Div., 28.10.1971, *S & T* 582! CAL) was proposed for a rare weed in gardens and road sides in Baroda and in cultivated fields, river banks and waste places in some localities in north and central Gujarat. This taxon was compared with *S. burbankii* Bitt. in *Feddes Repertorium Specierum Novarum Regni Vegetabilis, 12*: 83. 1913, a species of uncertain provenance described from the U.S.A. Examination of the type (*S & T* 582 extant in CAL) and scrutiny of the description show that it agrees with the description of *S. villosum* Mill. except for the presence of sclerotic granules. Nevertheless, Henderson (1974) reports in *S. nodiflorum* absence of sclerotic granules in subsp. *nodiflorum* and their presence in subsp. *nutans*, thereby evidently indicating that the presence or absence of sclerotic granules may not be of specific significance. Thus, *S. purpureilineatum* is relegated to a synonym of *S. villosum*.

DISTRIBUTION AND STATUS OF THE TAXA

The Solanaceae are represented by 24 genera and 108 species in India. Ten genera and 34 species are native to the country, while 14 genera and 74 species are exotic.

Table 3.1. Status of the genera and species of Solanaceae in India

| Genus | No. of species in the world | Native | Exotic | | | Total |
			Naturalized	Cultivated	Experimental stage	
Atropa	4	1		1		2
Browallia	6			2		2
Brunfelsia	30			2	3	5
Capsicum	20			4		4
Cestrum	150			3	4	7
Cyphomandra	30			1		1
Datura	24		3	4		7
Hyoscyamus	20	2		1		3
Iochroma	25			1		1
Lycianthes	200	2			1	3
Lycium	80	3				3
Lycopersicon	7			1	5	6
Mandragora	6	1				1
Nicandra	1		1			1
Nicotiana	21		1	2	1	4
Pauia	1	1				1
Petunia	40			3		3
Physalis	100		3	3		6
Physochlaina	10	1				1
Scopolia	6	1				1
Solandra	10			1	1	2
Solanum	1700	19	8	10	4	41
Streptosolen	1			1		1
Withania	10	2				2
Total 24	2502	33	16	40	19	108

Genera native to India

Ten genera are native to India. *Pauia* Deb et Dutta (1965) is a monotypic genus recently described from Arunachal. *Pauia belladonna* Deb et Dutta is known only from the type collection, from Tirap District adjacent to Burma. It could not be collected again. Ripe fruits are not yet known. *Mandragora caulescens* C. B. Clarke grows abundantly in Sikkim and Darjeeling, between 3000 m and 3900 m in altitude, and extends to West China (Fig. 3.4). *Physochlaina*

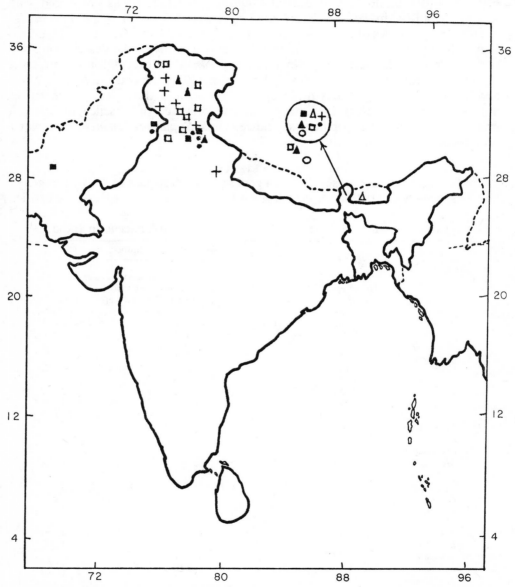

Figure 3.4. +, *Atropa acuminata;* ●, *Hyoscyamus niger;* ○, *H. pusillus;* △, *Mandragora caulescens;* □, *Physochlaina praealta;* ▲, *Scopolia anomala;* ■, *Withania coagulans.*

praealta (Walp.) Hook.f. occurs in Ladak (Kashmir) (Abrol & Chopra, 1962), Himachal Pradesh and the Punjab, extending to Tibet, between 2250 m and 4500 m in altitude (Fig. 3.4). *Scopolia anomala* (Link & Otto) Airy Shaw is more abundant in occurrence than any of the above taxa, and grows in Kashmir, Himachal Pradesh, Nepal, Sikkim, Bhutan and Tibet (Fig. 3.4).

Atropa acuminata Royle occurs in Barmula, Kinnaur, Simla and Nainital Districts (Fig. 3.4). *A. belladonna* L. is distributed from Southern Europe

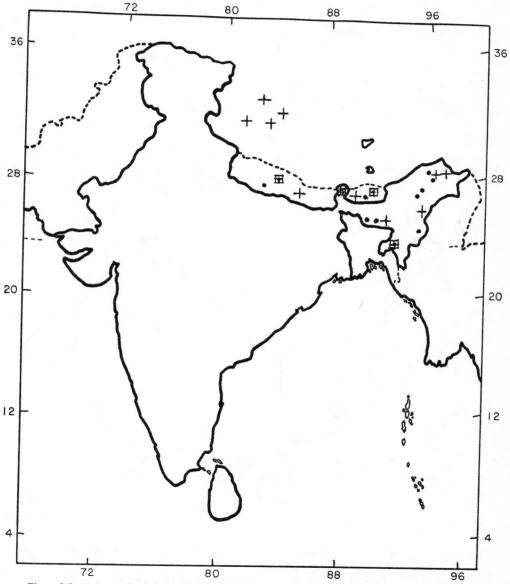

Figure 3.5. ⊞, *Lycianthes biflora;* ●, *L. biflora* subsp. *macrodon;* +, *L. biflora* subsp. *lysimachioides.*

to the Caucasus and North Iran, and is cultivated in Kashmir as a medicinal plant.

Hyoscyamus L. is represented by three species—two native and one cultivated. *H. niger* L. is distributed from Tibet to Kashmir and Pakistan, occurring between 1500 m and 2300 m in altitude (Fig. 3.4). *H. pusillus* L. is distributed from Egypt to Tibet, occurring in India at Ladak (Kashmir), at 3000 m (Fig. 3.4). *H. muticus* L. is cultivated in Kashmir as a medicinal plant.

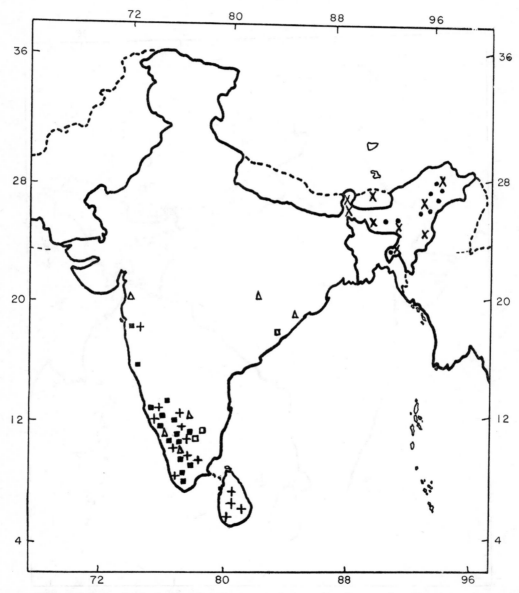

Figure 3.6. *Lycianthes laevis;* △, subsp. *laevis;* ×, subsp. *crassipetala;* ■, subsp. *bigeminata;* ●, subsp. *subtruncata;* +, subsp. *kaitisis;* □, subsp. *kaitisis* var. *gouakai.*

Lycianthes (Dun.) Hassl. emend. Bitter is represented by three species—two native and one introduced experimentally. *L. biflora* is widely distributed from the Eastern Himalayas and West China to Burma, Cochin China and Malaysia (Fig. 3.5). Subsp. *biflora* occurs in Nepal, Sikkim, Bhutan, Nagaland and Tripura. Subsp. *macrodon* occurs in Nepal, Sikkim, Bhutan, Arunachal, Manipur, Meghalaya and Bangladesh. Subsp. *lysimachioides* occurs in Nepal, Arunachal Nagaland, Mizoram and Meghalaya.

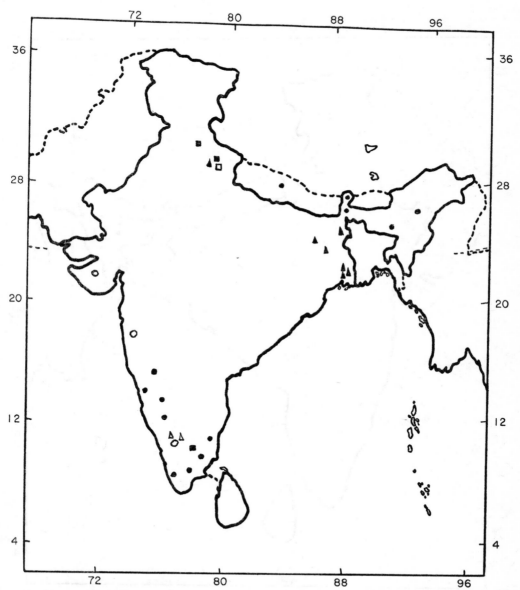

Figure 3.7. ●, *Solanum aculeatissimum;* ○, *S. elaeagnifolium;* ▲, *S. glaucophyllum;* △, *S. grandiflorum;* ■, *S. hispidum;* □, *S. pseudocapsicum.*

L. laevis is distributed in South and East India, Java and Ceylon. The sub-species occur as follows (Fig. 3.6). Subsp. *crassipetala:* East Nepal, Sikkim, Bhutan, Darjeeling, Arunachal, Assam, Khasi Hills, Mizoram, Manipur, Burma in subtropical and temperate climate. Subsp. *bigeminata:* Karnataka, Kerala, Andhra, Tamil Nadu and Ceylon. Subsp. *subtruncata:* Sikkim, Arunachal, Nagaland, Manipur, Assam, Khasi Hills, Mizoram, Tripura, Bangladesh, Bihar, Upper Burma. Subsp. *kaitisis* var. *kaitisis:* Maharashtra southwards to Ceylon.

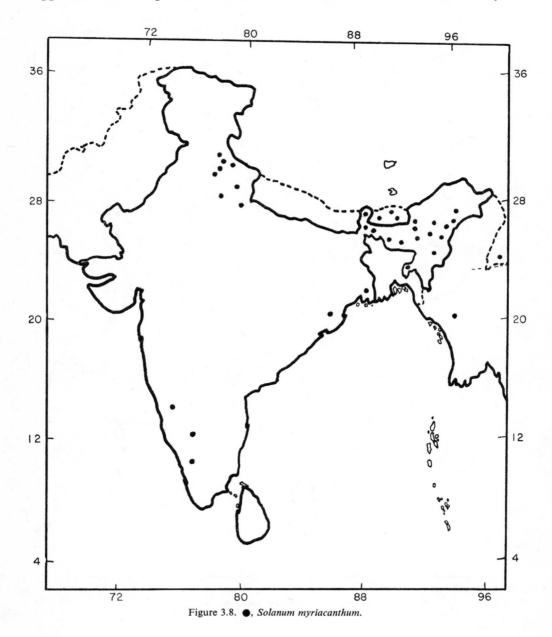

Figure 3.8. ●, *Solanum myriacanthum.*

Subsp. *kaitisis* var. *gouakai:* Pulney and Kodaikanal.

L. rantonetti (Carr. ex Lusc.) Bitt. is experimentally cultivated at Dehra Dun and Bombay.

Lycium L. is represented by three species. *L. europaeum* L. is distributed from the Mediterranean region to Central Asia, extending to Northwest and West India, in Uttar Pradesh, the Punjab, Gujarat and Maharashtra. *L. barbarum* L., known to be distributed in West Asia, occurs in Himachal Pradesh, Haryana, the

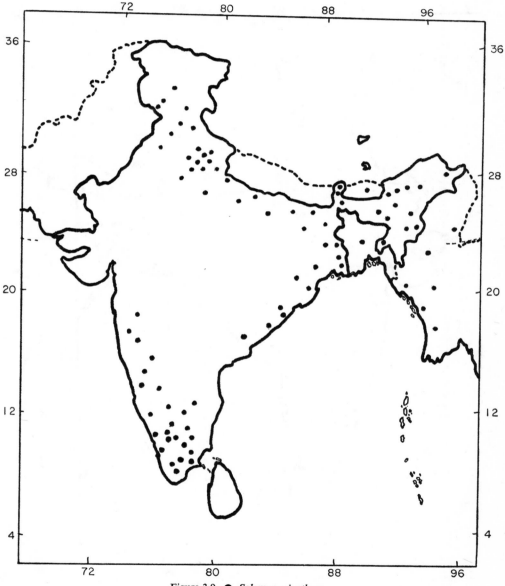

Figure 3.9. ●, *Solanum erianthum.*

Punjab, Rajasthan, Gujarat and Maharashtra. This species is very variable in
the arrangement, size and shape of the leaf, length of corolla and relative length
of corolla tube and lobes. It does not appear to be specifically distinct from the
former. *L. ruthenicum* Murr., distributed from Central Asia to Afghanistan and
Pakistan occurs in North Kashmir. This also, is not clearly distinguishable from
the other species. These three taxa deserve further study to understand their
taxonomic status.

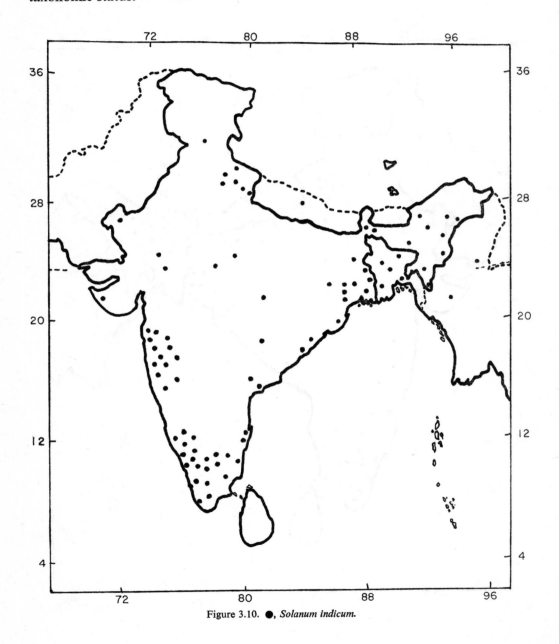

Figure 3.10. ●, *Solanum indicum.*

The genus *Solanum* L. is represented by 41 species—22 exotic and 19 native.

Species under experimental introduction are: *S. aviculare* Forst.f. in Kashmir (Dutta, 1963), *S. macrophyllum* Dun., *S. mammosum* L. and *S. marginatum* L.f. in the Nilgiris and *S. rostratum* Dun. at Uttar Kashi (Uttar Pradesh) are the alien species experimentally cultivated.

Species under cultivation are: *S. tuberosum* L., very extensively cultivated all over the country. *S. capsicoides* Mart., *S. clavatum* Rusby, *S. convolvulus* Sendt.,

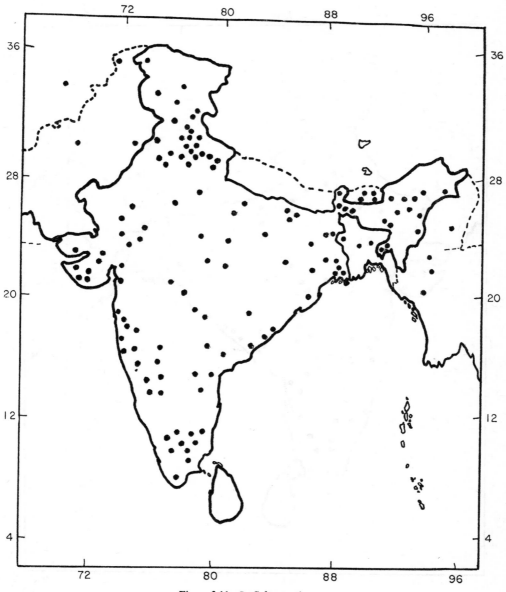

Figure 3.11. ●, *Solanum nigrum*.

S. jasminoides Pax, *S. macranthum* Dun., *S. seaforthianum* Andr. and *S. wendlandii* Hook.f. are naturalized and cultivated for ornamental purposes but are not found wild.

Species naturalized and running wild are as follows: *S. aculeatissimum* Jacq. was included in the Flora of India by Clarke (1883) on the basis of T. Thomson's collection from Singapore. Evidently it was not known from India at that time. Gamble (1923) reported it for India for the first time from Quilon. Bezbarua &

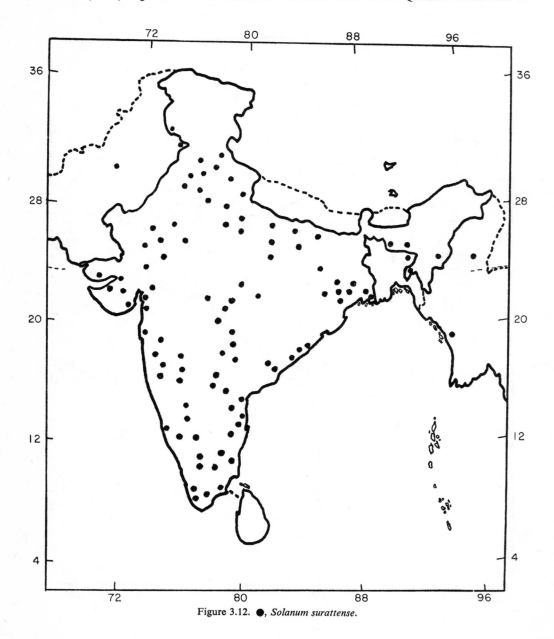

Figure 3.12. ●, *Solanum surattense*.

Bezbarua (1963) reported it from Jorhat. Within a short period it has spread rapidly in damp and shaded situations in Tamil Nadu, Karnataka, Kerala, Assam, Maghalaya, West Bengal, Sikkim and Nepal (Fig. 3.7). *S. elaeagnifolium* Cav. was collected from Valparai (Tamil Nadu) and grown in the Indian Botanic Garden, Calcutta during N. Wallich's time (Dunal, 1852). When and how it reached Valparai is not known. It is now a weed in cultivated fields, gardens and waste places in Tamil Nadu and Maharashtra (Fig. 3.7) and is spreading rapidly

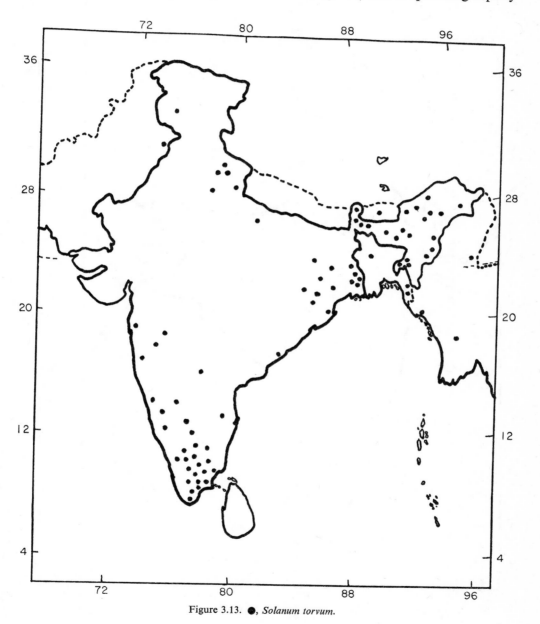

Figure 3.13. ●, *Solanum torvum.*

(Raghavan & Wadhwa, 1970). *S. hispidum* Pers., a Mexican species, is wild in
Dehra Dun, Simla and Shevaroy Hills (Rao, 1969) (Fig. 3.7). *S. pseudocapsicum*
L. is naturalized and wild at Dehra Dun (Fig. 3.7). *S. sisymbriifolium* Lamk., a
South American plant, is wild in the Nilgiris, Pulneys, Khasi Hills and North
Bengal. *S. glaucophyllum* Desf., a South American plant growing wild (Fig. 3.7)
in West Bengal, Bihar and Dehra Dun (Sengupta, 1969) was thought to be native
to India and described as *S. argenteum* Prain. *S. grandiflorum* Ruiz & Pavon, a

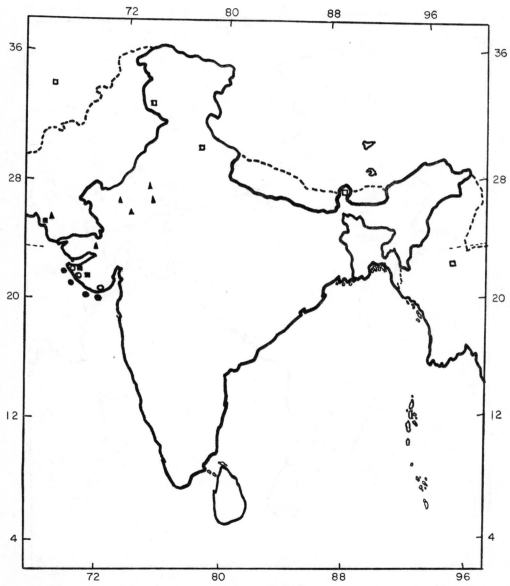

Figure 3.14. ▲, *Solanum albicaule;* ●, *S. arundo;* ○, *S. dubium;* □, *S. dulcamara;* ■, *S. gracilipes.*

Mexican plant, is wild in the Nilgiris (Fig. 3.7). It was initially thought to be a
South Indian species and named *S. wightii* Nees. *S. myriacanthum* Dun.* was
also thought of as native to India and was named *S. khasianum* C. B. Clarke
(Deb, Sengupta & Malick, 1969; Deb, 1975). This is now widely spread in the foot-
hills of the Himalayas, extending to Meghalaya and Assam, and also to the
Nilgiris, in the subtropical climatic zone (Fig. 3.8).

Species native to the country are: *S. erianthum* D. Don (Fig. 3.9), *S. indicum*

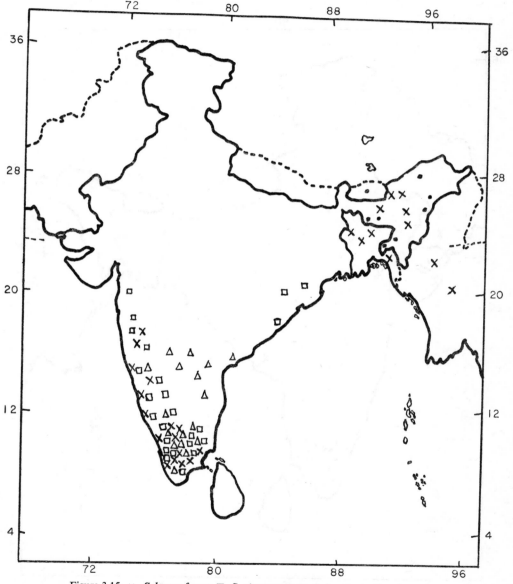

Figure 3.15. ×, *Solanum ferox*; □, *S. giganteum*; ●, *S. kurzii*; △, *S. pubescens*.

* also *Solanum viarum* Dun., see footnote on p. 231.—Eds.

L. (Fig. 3.10), *S. nigrum* L. (Fig. 3.11), *S. surattense* Burm.f. (Fig. 3.12), and *S. torvum* Sw. (Fig. 3.13). They are very common in India and are widely distributed in south east Asia. *S. nigrum, S. surattense* and *S. torvum* extend to Australia, tropical Africa and America. *S. dulcamara* L., distributed from Europe to west and central Asia, occurs in the Himalayas from Kashmir to Sikkim from 1200 m to 2400 m (Fig. 3.14). *S. albicaule* Kotschy ex Dun. extends from tropical Africa to the arid zone in Rajasthan and Cutch (Gujarat) (Fig. 3.14). *S. arundo* Mattei

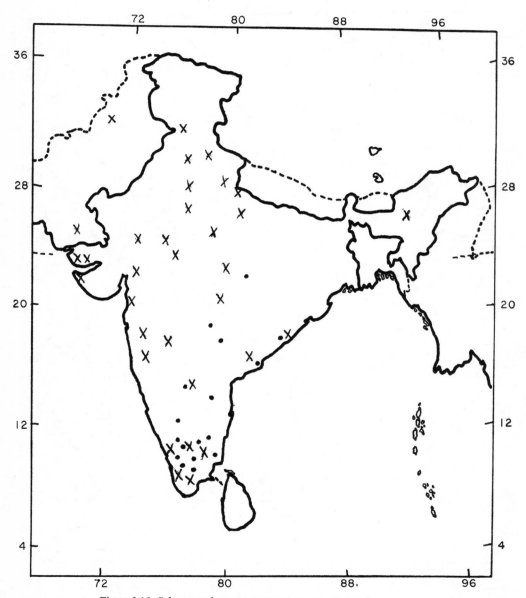

Figure 3.16. *Solanum melongena:* ×, var. *incanum;* ●, var. *insanum.*

(Sengupta, 1963) of east tropical Africa, extending from the southern part of Somalia to the drier parts of Kenya and Tanzania occurs in the coastal regions of the Kathiwar peninsula (Fig. 3.14). *S. dubium* Frassen (Rao & Safui, 1963) is distributed in the eastern coast of Africa and the western coast of India (Fig. 3.14). *S. trilobatum* L. inhabits the tidal swamps along the coasts. *S. gracilipes* Dcne extends from Pakistan to Okha and the Pari hills in Gujarat (Fig. 3.14). *S. giganteum* Jacq. and *S. pubescens* Willd. are distributed in south India (Fig.

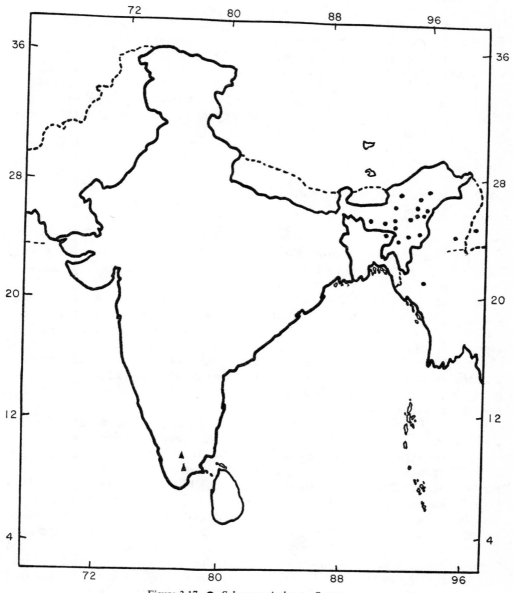

Figure 3.17. ●, *Solanum spirale;* ▲, *S. vagum.*

3.15) and Sri Lanka, the former extending to Africa. *S. barbisetum* Nees and *S. kurzii* Brace ex Prain grow from the foothills of the eastern Himalayas to Burma and Bangladesh (Fig. 3.15). *S. ferox* L. is distributed in south and east India from Assam to Burma and Japan (Fig. 3.15).

S. cornigerum Anders, known as wild brinjal and taxonomically nearer to and probably an old relative of *S. melongena*, is localized in upper Burma. It is not yet recorded even from the contiguous parts of India.

S. melongena L., as the type variety, is found only under cultivation, but the wild varieties *insanum* and *incanum* are widely distributed. The variety *incanum* is found throughout India, while *insanum* is restricted to some extent (Fig. 3.16). From a study of ancient records it appears to have been originally cultivated in this country whence it spread through Iran to Egypt and other North African countries, as well as to Turkey and the Balkans (Mazumder, 1929). The abundance, extensive distribution and phenotypic variability of the varieties *incanum* and *insanum* tend to support the hypothesis that *S. melongena* originated in India.

S. spirale is distributed from Assam and Meghalaya to Bangladesh and Burma, whereas *S. vagum* is localized in Tamil Nadu (Fig. 3.17).

Withania Pauquy is represented by two species. *W. coagulans* (Stocks) Dun. is distributed from Himachal Pradesh and the Punjab to Afghanistan (Fig. 3.4). *W. somnifera* is wild in dry subtropical regions in Gujarat and Rajasthan and is commonly cultivated in other parts of the country.

Exotic naturalized genera

Datura L., *Nicandra* Adans., *Nicotiana* L. and *Physalis* L. are exotic genera, some species of which have become naturalized, often covering large areas in varying habitats.

Datura L. is represented by seven species. *D. metel* L., *D. stramonium* L. and *D. inoxia* Mill. grow wild in waste places and abandoned cultivations (Fig. 3.1). *D. metel* is very common throughout the country in tropical and warm climates. Next in abundance is *D. stramonium*, distributed on the hills up to an altitude of about 2400 m. *D. inoxia* is scattered in north and west India in tropical climates. *D. arborea* L. is usually cultivated on hill stations and grows as an escape. *D. quercifolia* H.B.K., *D. sanguinea* Ruiz. & Pav. and *D. suaveolens* H.B. ex Willd. are infrequently cultivated in the garden.

Nicandra Adans., a monotypic genus native to the Andes, is naturalized in the subtropical regions of India and is very often a weed of cultivation.

Nicotiana L. is represented by four species. Tobacco was introduced in 1508 in the Deccan and took about 100 years to be carried to other parts of the country for large scale cultivation. It was exported to the Red Sea ports as early as 1619 (Watt, 1908). *N. tabacum* L. and *N. rustica* L. are cultivated for tobacco, the former very extensively in many parts of India and the latter to a very limited extent. It seems *N. rustica* was initially taken up for cultivation on a large scale but due to its very restricted temperature and moisture requirements, it has been replaced by *N. tabacum*. *N. alata* Link & Otto has been recently introduced into the gardens at Bombay and Kodaikanal. *N. plumbaginifolia* Viv. is naturalized and grows wild in many parts of India as a weed, particularly in sandy and damp places (Fig. 3.2). As this plant was not mentioned by Roxburgh (1820–24) it is assumed that it reached India after 1813.

Six species of *Physalis* L. are found in India. *P. alkekengi* L. is occasionally cultivated in gardens. *P. angulata* L. is more frequently cultivated. *P. virginiana* Mill. var. *sonorae* (Torr.) Waterfall (Syn. *P. longifolia* Nutt.) runs wild on the east and west coasts, and *P. minima* L. is a common weed throughout India (Fig. 3.3). *P. peruviana* L. is widely cultivated for the edible fruit. *P. ixocarpa* Brot. is recently introduced for the edible fruit.

Exotic genera under cultivation

Ten genera, namely, *Browallia* L., *Brunfelsia* L., *Capsicum* L., *Cestrum* L., *Cyphomandra* Mart. ex Sendtn., *Iochroma* Benth., *Lycopersicon* Mill., *Petunia* Juss., *Solandra* Sw. and *Streptosolen* Miers, represented by 33 species, occur only under cultivation.

Browallia americana L., *B. viscosa* H.B.K., *Brunfelsia americana* L., *B. hopeana* Benth., *Cestrum aurantiacum* Lindl., *C. diurnum* L., *C. nocturnum* L., *Petunia axillaris* Benth., *P. violacea* Lindl., *P. hybrida* Vilm. and *Solandra grandiflora* Sw. are commonly cultivated in gardens as ornamental plants. *Streptosolen jamesonii* Miers is cultivated in temperate gardens. *Iochroma lanceolata* Miers is occasionally cultivated in the garden. *Brunfelsia calycina* Benth., *B. latifolia* Benth., *B. nitida* Benth., *Cestrum dumetorum* Schl., *C. fasciculatum* Miers, *C. parqui* L'Herit., *C. purpureum* Standl. and *Streptosolen viridiflora* Sim. are casually or experimentally cultivated.

Capsicum was introduced into India at Goa in the middle of the 16th century from the West Indies under the name of Pernambuco Pepper (Garcia, 1563; Clusius, 1579) and was soon cultivated on a large scale and exported (Watt, 1908). Roxburgh (1832) described six species, while Clarke (1883) recognized only three. He treated *C. annuum* var. *annuum* as *C. frutescens*. In *C. minimum* he not only included *C. annuum* var. *glabriusculum* but also probably the one now recognized as *C. chinense* Jacq. Irish (1898) and Watt (1908) recognized two species, namely *C. annuum* and *C. frutescens*. Bailey (1923) accepted only one species and preferred the name *C. frutescens* to *C. annuum*. In recent years Smith & Heiser (1951, 1957), Heiser & Smith (1953), Filov (1956), Shinners (1956), Terpo (1966), Heiser & Pickersgill (1969, 1975), Eshbaugh (1968, 1970) and D'Arcy & Eshbaugh (1973) have contributed to the taxonomy and nomenclature of the cultivated species. In the light of these studies the genus *Capsicum* L. is represented by four species, under cultivation as an agricultural crop.

C. annuum L. var. *annuum* is extensively cultivated throughout India as the main source of Chilli of commerce. *C. frutescens* L. is cultivated for green Chilli, to a lesser extent, mostly in domestic compounds, particularly in tribal localities. *C. annuum* var. *glabriusculum* (Dun.) Heiser f. & Pickersgill and *C. baccatum* L. var. *pendulum* (Willd.) Heiser f. & Pickersgill are cultivated to a very limited extent, the former only in cold climates. *C. chinense* is hardly distinguishable from and probably is an extreme form of *C. frutescens*. It was originally described from India as *C. luteum* Lamarck (1793). According to Heiser, Eshbaugh & Pickersgill (1971), and Heiser (1975), *C. frutescens* and *C. chinense* form a complex of wild and cultivated peppers comparable to the range of forms included within *C. annuum* or *C. baccatum*.

Cyphomandra betacea (Lar.) Sendtn. is cultivated in hill stations for the edible tree tomato.

Lycopersicon esculentum L. was introduced into India in the last century. It is now cultivated extensively throughout the country for the edible fruit. *L. pimpinellifolium* Mill., *L. peruvianum* Mill., *L. hirsutum* H. & B., *L. pissisi* Phil. and *L. glandulosum* C. H. Muller are experimentally cultivated at the Indian Agricultural Research Institute, Delhi (Sastri, 1962).

ACKNOWLEDGEMENTS

The author records his gratitude to Prof. J. G. Hawkes, Head of the Department of Plant Biology, Birmingham University, for inviting him to read a paper in the Symposium. He is grateful to the authorities of the Royal Botanic Gardens, Kew and Edinburgh, British Museum (Natural History), London, Linnean Society, London and Museum National d'Histoire Naturelle, Paris for facilities to consult their respective herbaria; to the Secretary, Commonwealth Museum Association, London, for financing a part of his stay abroad, and to Prof. P. V. Bole of St. Xaviers' College, Bombay, Dr R. S. Rao of Central National Herbarium, Calcutta and Dr R. S. Raghavan of Western Circle, Poona, for facilities to consult their respective herbaria. Thanks are also due to Mr K. C. Sahni, Officer-in-charge, Systematic Botany Branch, Forest Research Institute, Dehra Dun and to Drs S. K. Jain, A. S. Rao and J. Joseph of the Botanical Survey of India for furnishing information on the distribution of certain taxa from their respective herbaria.

REFERENCES

ABROL, B. K. & CHOPRA, I. C., 1962. Some vegetable drug resources of Ladak (Little Tibet). *Current Science, 31*(8): 324–326.

BAILEY, L. H., 1923. *Capsicum. Gentes Herbarum, 1:* 128–129.

BEZBARUA, H. P. & BEZBARUA, B., 1963. *Solanum aculeatissimum* Jacq. A new record for northeastern India. *Journal of the Bombay Natural History Society, 60*(3): 759–761.

BURMAN, J., 1737. *Thesaurus Zeylanicus Exhibens Plantas in Insula Zeylans Nascentes; Interquas Plurimae Novae Species et Genera Inveniuntur.* Omnia iconibus illustrata et descripta. Amsterdam.

CHATTERJEE, D., 1939. Studies on the endemic flora of India and Burma. *Journal of the Asiatic Society of Bengal, 5*(1): 19–67.

CLARKE, C. B., 1883. Solanaceae. In J. D. Hooker (Ed.), *Flora of British India, 4:* 228–246. Kent: L. Reeve.

CLUSIUS, CAROL, 1579. *Armatum et Simplicium Aliquot Medicamentorum apud Indos Nascentium Historia.* Primum quiden Lusitanica lingua D. Indogos conscripta D. Garcia Ab Orto Proregis Indiae medica. Antwerp.

COOKE, Th., 1905. *The Flora of the Presidency of Bombay, 2:* 261–273. London: Taylor & Francis.

D'ARCY, W. G. & ESHBAUGH, W. H., 1973. The name for the common bird pepper. *Phytologia, 25:* 350.

DEB, D. B., 1961. Dicotyledonous plants of Manipur Territory. *Bulletin of the Botanical Survey of India, 3:* 253–350.

DEB, D. B., 1972. *The Flora of Tripura State.* Unpubl. MSS.

DEB, D. B., 1975. Conspecificity of *Solanum khasianum* Cl. & *S. myriacanthum* Dun. In J. G. Hawkes, (Ed.). *Solanaceae Newsletter,* No. 2: 33–34.

DEB, D. B. & DUTTA, R., 1974. Contribution to the Flora of Tirap Frontier Division. *Journal of the Bombay Natural History Society, 71*(2): 266–267.

DEB, D. B. & DUTTA, R., 1965. *Pauia* Deb et Dutta—A remarkable new genus and species from Tirap Frontier Division. *Indian Forester, 91*(6): 363–366.

DEB, D. B., SENGUPTA, G. & MALICK, K. C., 1969. A contribution to the Flora of Bhutan. *Bulletin of the Botanical Society of Bengal, 22*(2): 169–217.

DUNAL, M. F., 1852. In A. P. De Candolle (Ed.), *Prodromus Systematis Naturalis Regni Vegetabilis, 13*(1): 1–690.

DUTHIE, J. F., 1911. *Flora of the Upper Gangetic Plain and of the Adjacent Siwalik and Sub-Himalayan Tracts, 2:* 123–135. Calcutta: Govt. Press.

DUTTA, A. K., 1963. Introduction of *Solanum aviculare* Forst.f. in Kashmir. *Indian Journal of Pharmacy, 25*(5): 160–161.

ESHBAUGH, W. H., 1968. A nomenclatural note on the genus *Capsicum. Taxon, 17:* 51–52.

ESHBAUGH, W. H., 1970. A biosystematic and evolutionary study of *Capsicum baccatum* (Solanaceae). *Brittonia, 22:* 31–43.

FILOV, A. I., 1956. *Peppers and Eggplants.* Moskva, Gosudarstvennos Izdatal'-stvo Selskoziaist vennoi Literatura.

GAMBLE, J. S., 1923. *Flora of the Presidency of Madras:* 931–937. London: Secretary of State for India.

GARCIA AB ORTA, 1563. *Colloquios dos Simples, e Drogas e Cousas Mediçinais da India, e assi Dalguas Frutas. Achadas, etc.* Goa. English translation by Sir Clements Markham, 1913. London: Henry Southern.

HAINES, H. H., 1922. *The Botany of Bihar and Orissa:* 606–617. London: Adlard & Sons & West Newman.

HEISER, C. B., 1975. Names for the Bird Peppers (*Capsicum*—Solanaceae). *Baileya, 19*(4): 151–156.

HEISER, C. B., ESHBAUGH, W. H. & PICKERSGILL, B., 1971. Letter to Editor. *Professional Geographer, 23*(2): 169–170.

HEISER, C. B. & PICKERSGILL, B., 1969. Names for the cultivated *Capsicum* species (Solanaceae). *Taxon, 18:* 277–283.

HERMANN, PAUL, 1687. *Horti Academici Lugduno-batavi Catalogus.* Leiden.

IRISH, H. C., 1898. A revision of the genus *Capsicum* with special reference to garden varieties. *Report. Missouri Botanical Garden, 9:* 53–110.

KANJILAL, U. N., KANJILAL, P. C., DAS, A. & DE, R. N., 1939. *Flora of Assam, 3:* 363–375. Shillong: Govt. of Assam.

LAMARCK, J. B. A. P. M. de, 1793. *Tableau Encyclopedique et Methodique des Trois Regnes de la Nature, 5*(1). Botanique. Paris.

LINNAEUS, C., 1753. *Species Plantarum.* Stockholm.

MAZUMDER, G. P., 1929. Vanaspathi: *Plants and Plant Life as in Indian Treaties and Traditions.* Calcutta University Press.

NEES VON ESENBECK, FREDERIK, 1837. Monograph of the East Indian Solanaceae. *Transactions of the Linnean Society of London, 17:* 37–82.

PLUKENET, L., 1695. Opera omnia botanica. *Phytographia, 2–5.*

PRAIN, D., 1903. *Bengal Plants, 2:* 742–753.

RAGHAVAN, R. S. & WADHWA, B. M., 1970. *Solanum elaeagnifolium* Cav.—an addition to Indian Flora. *Maharashtra Vidyan Mandir Patrika, 5*(1, 2): 22–24.

RAO, A. V. N., 1969. *Solanum hispidum* Pers.—A new record for South India. *Bulletin of the Botanical Survey of India: 11:* 197–198.

RAO, T. A. & SAFUI, B., 1963. Distribution of some rare plants along Saurashtra coast and neighbouring Islands. *Proceedings of the Indian Academy of Sciences, 58B*(6): 362–366.

RHEEDE, TOT DRAKESTEIN, H. A. VAN, 1678–1703. Hortus indicus malabaricus, *Continens Regni Malabarci apud Indos Celeberrimi omnis Generis Plantas Rariores, Latinis, Malabaricis, Arabicis et Bramanum Characteribus Expressas.* 12 vols. Amsterdam.

ROXBURGH, W., 1820–24. In William Carey (Ed.), *Flora Indica: or Descriptions of Indian Plants,* 2 vols. to which are added descriptions of plants by N. Wallich. Serampore.

ROXBURGH, W., 1832. *Flora Indica: or Descriptions of Indian Plants,* 3 vols. Serampore.

RUMPHIUS, G. E., 1741–50. *Herbarium Amboinense, Plusimas Compectens Arbores, Frutices, Herbes, Plantas Terrestres et Adjacentibus Reperiunter Insulis,* 6 vols. Amsterdam.

SANTAPAU, H., 1948. Notes on Solanaceae of Bombay. *Journal of the Bombay Natural History Society, 47:* 652–662.

SASTRI, B. H., 1962. *The Wealth of India: Raw materials.* VI: L–M. New Delhi: Council of Scientific and Industrial Research.

SENGUPTA, G., 1963. *Solanum arundo* Mattei, a new record for India. *Indian Forester, 89*(7): 480–482.

SENGUPTA, G., 1969. Nomenclatural Notes on *Solanum. J. Sen mem.* Vol., 425–426.

SENGUPTA, G., 1973. Solanaceae. Materials for the Flora of Bhutan. *Record of the Botanical Survey of India, 22*(2): 154–156.

SHINNERS, L. H., 1956. Technical names for the cultivated *Capsicum* peppers. *Baileya, 4:* 81–83.

SMITH, P. G. & HEISER, C. B., 1951. Taxonomic and genetic studies on the cultivated peppers. *American Journal of Botany, 38:* 362–368.

SMITH, P. G. & HEISER, C. B., 1957. Taxonomy of *Capsicum sinense* Jacq. and the geographic distribution of the cultivated *Capsicum* species. *Bulletin of the Torrey Botanical Club, 84:* 413–420.

TERPO, A., 1966. Kritische Revision der wildwachsenden Arten und der Kultivierten Sorten der Gattung *Capsicum* L. *Feddes Repertorium Specierum Novarum Regni Vegetabilis, 72:* 155–191.

VOIGT, J. O., 1845. *Hortus Suburbunus Calcuttensis:* 508–519. Calcutta: Bishop's College Press.

WATT, G., 1908. *The Commercial Products of India.* London: John Murray.

YAMAZAKI, 1966. In H. Hara (Ed.), *The Flora of Eastern Himalaya.* Tokyo: Univ. Tokyo.

ADDITIONAL REFERENCES

BHADURI, P. N., 1933. Chromosome numbers of some solanaceous plants of Bengal. *Journal of the Indian Botanical Society, 12:* 56–64.

BITTER, G., 1920. *Die Gattung Lycianthes: Abhandlungen herausgegeben vom Naturwissenchaftlichen Verein zu Bremen, 24:* 292–520.

BLUME, C. L., 1825–1826. *Bijdragen tot de Flora van Netherlandsch Indië,* Batavia.

DEB, D. B., 1978. Some new combinations for Indian taxa of *Lycianthes* (Solanaceae). *Botanical Journal of the Linnean Society, 76:* 292–294.

HEISER, C. B. & SMITH, P. G., 1953. The cultivated *Capsicum* peppers. *Economic Botany, 7:* 214–226.

HEISER, C. B. & PICKERSGILL, B., 1975. Names for the Bird peppers (*Capsicum*—Solanaceae). *Baileya, 19:* 151–156.

HENDERSON, R. J. F., 1974. *Solanum nigrum* L. (Solanaceae) and related species in Australia. *Contributions from the Queensland Herbarium,* no. 16: 1–78.

GERASIMENKO, I. I. & REZNIKOVA, S. A., 1968. In Index to Plant chromosome numbers for 1968 (1970). *Regnum Vegetabile, 68:* 74.

KUNTZE, C. E. O., 1891. *Revisio Generum Plantarum,* Vol. 2. Leipzig.

LOUREREIO, J. de, 1790. *Flora Cochinchinensis,* Vol. 2. Lisbon.

RAO, G. R., KHAN, R. & KHAN, A. H., 1971. Structural hybridity and genetic system of *Solanum nigrum* complex. *Botanical Magazine of the University of Tokyo, 84:* 335–338.

RAO, N. N., PUNNAYA, B. N. K. & MENON, P. M., 1970. Studies on interspecific hybrids of nontuberous species of *Solanum* L. *Andhra Agricultural Journal, 17:* 1–6.

SENGUPTA, G., 1964. A rare and interesting *Solanum* (*S. hovei* Dun.) from Western India. *Bulletin of the Botanical Survey of India, 17:* 3–5.

STEBBINS, G. L., 1950. *Variation and Evolution in Plants.* New York: Columbia University Press.

TANDON, S. L. & RAO, G. R., 1966. Biosystematics and origins of the Indian tetraploid *Solanum nigrum* L. *Current Science, 20:* 324–325.

VARTAK, V. O., 1957. *Solanum esuriale* Lindl.—a new record for Bombay State. *Journal of the Bombay Natural History Society, 54:* 965.

VENKATAREDDI, B., 1969. The identity of *Solanum mccannii* Santapau. *Bulletin of the Botanical Survey of India, 11:* 209.

VISWANATHAN, T. V., 1975. On the occurrence of natural hybridization between *Solanum incanum* L. and *S. melongena* L. *Current Science, 44:* 134.

WILLDENOW, C. L., 1806. *Caroli a Linne Species Plantarum,* Editio quarta, Vol. 4, no. 2. Berlin.

4. *Solanum* in Nigeria

Z. O. GBILE

Forestry Research Institute of Nigeria, Ibadan, Nigeria

Of the 20 species, subspecies and varieties in the genus *Solanum* L. that occur in Nigeria, 15 are indigenous. The genus is of importance to Nigerians because most of its members produce edible fruits and leaves used as vegetables. The indigenous species can be divided into two ecological groups, the lowland and the highland species. The term highland species denotes those confined to or concentrated in areas attaining an elevation over 1220 m (4000 ft).

CONTENTS

INTRODUCTION

The genus *Solanum* L. is one of the largest in the plant kingdom, being represented by about 2000 species.

The classification of the West African species of the genus is very complex, difficult and often confused, particularly because it is often difficult to characterize or delimit a species due to the variability in gross morphology between different specimens of what was initially considered as constituting one species. Such varied forms which may be horticultural varieties, interspecific hybrids or indeed new species are either lumped together as a species or else placed in infraspecific taxa. As indicated by Heine (1963), a modern revision of this genus is long overdue.

The indigenous species can be divided into two ecological groups, the lowland and the highland species. The highland species are those that are restricted to the four major Nigerian highlands (Mambilla, Obudu, Vogel Peak and Jos Plateau) at an elevation of more than 1200 m (Fig. 4.1). So far, no ecological study dealing specifically with *Solanum* in Nigeria is known to have been made.

Of the 20 Nigerian taxa, 15 are indigenous to Nigeria while the remaining 5 were introduced and are now cultivated, viz. *Solanum hispidum* of Mexican and Guatemalan origin, *S. seaforthianum* var. *disjunctum* from the West Indies and Central America; *S. gilo* and *S. aethiopicum* are also cultivated, the leaves or fruits being used as vegetables. *S. wrightii* is grown as an ornamental. A few of the indigenous species are also cultivated either for ornamental or dietary purposes.

MATERIALS AND METHODS

For the gross morphological studies, about 225 herbarium specimens at the

Forestry Research Institute were examined. *S. clerodendroides* Hutch. & Dalz. and *S. cerasiferum* Dunal, stated in *Flora of West Tropical Africa* to have occurred in Nigeria, are not represented at Forest Herbarium Ibadan (FHI) and they therefore could not be examined by the author. Information on all the specimens was recorded and morphological characters of the species were scored.

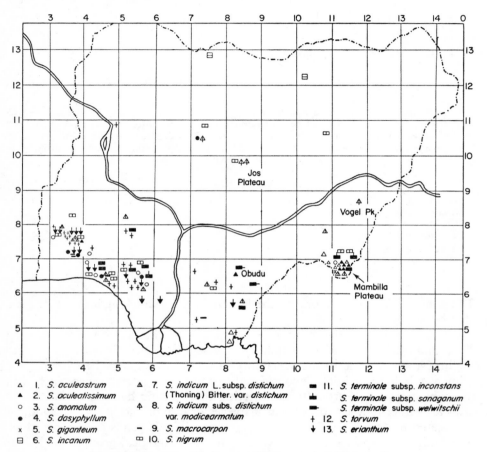

△	1. *S. aculeastrum*	△	7. *S. indicum* L. subsp. *distichum* (Thoning) Bitter. var. *distichum*	▬ 11. *S. terminale* subsp. *inconstans*
▲	2. *S. aculeatissimum*			┷ *S. terminale* subsp. *sanaganum*
○	3. *S. anomalum*	⚏	8. *S. indicum* subs. *distichum* var. *modicearmatum*	➤ *S. terminale* subsp. *welwitschii*
●	4. *S. dasyphyllum*			+ 12. *S. torvum*
×	5. *S. giganteum*	▬	9. *S. macrocarpon*	↓ 13. *S. erianthum*
⊟	6. *S. incanum*	⊞	10. *S. nigrum*	

Figure 4.1. Localities of indigenous Nigerian species of *Solanum* L.

DESCRIPTION OF SPECIES

1. Solanum aculeastrum Dunal in *DC. Prod.*, **13**, 1:366 (1852); *Fl. West Trop. Afr.*, ed. 2, **2**:332 (1963), var. *aculeastrum*. Shrub up to 3 m high. Stem and branches covered with stellate hairs and armed with spines up to 1.5 cm long. Leaves irregularly lobed to pinnatifid, covered with soft stellate hairs and often grey beneath, glabrous above. Corolla white to purple. Fruit pendant, spherical or slightly ovoid, up to 5 cm in diameter, yellow when ripe.

This is one of the Mambilla Plateau highland species. It is more or less confined to the open grassland of about 1500 m and it is often found forming a localized group in the places where cattle graze. It is probably dispersed by cattle eating the fruit.

2. *S. aculeatissimum* Jacq., *Icon. Pl. Rar.*, *1*:5, t. 41 (1781); *Fl. West Trop. Afr.*, ed. 2, *2*:334 (1963). Undershrub up to 60 cm high. Stem, branches and leaves covered with prickles of two kinds, the longer up to 1.4 cm long. Leaves with simple uniseriate, multicellular hairs. Corolla white to purple. Fruit light green with medium green stripes, dirty yellow or brownish when ripe.

S. aculeatissimum is a highland species. It is not confined only to the Mambilla Plateau but it has also extended to one of the four major highlands in Nigeria—Obudu Plateau at an altitude of about 1600 m. The two plateaux are adjoining the highlands of the Cameroun Republic. It is an annual that grows mostly in rough ground and amongst rocks.

3. *S. aethiopicum* L., *Amoen. Acad.*, *4*:307 (1759); *Fl. West Trop. Afr.*, ed. 2, *2*:332 (1963). A glabrous herb up to 60 cm high. Leaves up to 14 cm long and 7 cm broad, sinuate. Flowers white or pale violet. Fruit orbicular, in clusters of five or more, under 2 cm diameter, shining red when ripe.

S. aethiopicum is not an indigenous species. It is cultivated, the leaves being used as vegetables, and has been recorded for Ibadan and Ibuya.

4. *S. anomalum* Thonning in Besk., *Guin. Pl.*: 126 (1827); *Fl. West Trop. Afr.*, ed. 2, *2*:334 (1963).
   Syn. *S. mannii* Wright var. *compactum* C. H. Wright in *Kew Bull.*, 1894: 129.
   S. warneckeanum Dammer in Engl., *Bot. Jahrb.*, *38*: 168 (1906).
   S. pauperum of *Fl. Trop. Afr.*, 4, *2*: 217 (1906).

Shrub up to 60 cm high. Stem armed with triangular spines up to 6 mm long; branches mealy-pubescent. Leaves mostly ovate; margin sinuate, up to 15 cm long, 6 cm broad; young leaves densely stellate. Corolla white. Young fruit under 1 cm diameter, in axillary clusters, medium shining green, red when ripe.

S. anomalum has a very limited distribution, occurring mostly in Omo, Ago-Owu and the Idanre Forest Reserves in moist lowland forest of south-western Nigeria. It also occurs in the coastal area of Badagry. The annual rainfall in its main areas of distribution ranges from 1000–1500 mm. It is a high forest species.

It is sometimes cultivated for the red berries that are used as condiment for soup.

5. *S. dasyphyllum* Schum. & Thonn. in Besk., *Guin. Pl.*: 126 (1827); *Fl. West Trop. Afr.* ed. 2, *2*: 334 (1963).
   Syn. *S. duplosinuatum* Klotzsch in Peters, *Reise, nach Mossamb, Bot.*, *1*:233 (1852).
   S. afzelii Dunal in *DC. Prod.*, *13*, 1: 363 (1852).
   S. duplosinuatum Klotzsch var. *semiglabrum* C. H. Wright in *Fl. Trop. Afr.*, 4, *2*: 244 (1906).
   S. macinae A. Chev., *Bot.* 463 (1920).

Spiny undershrub up to 90 cm high. Stem pilose. Upper surface of leaf velvety, lower surface stellate. Corolla bluish to purple. Calyx armed with long sharp bristles, very accrescent in fruit, the fruit about 3 cm diameter.

S. dasyphyllum occurs mostly in waste places, often in the *Sida* community. It also occurs in wet areas. Most of the herbarium specimens were collected from villages; thus little is known about its ecological preference.

6. **S. giganteum** Jacq., *Collect.*, *4*: 125 (1790); *Fl. West Trop. Afr.*, ed. 2, 2: 332 (1963).

Armed shrub or tree up to 7 m high. Stem, undersurface of leaf and base of spine covered with white indumentum; upper surface of leaf glabrous. Leaves ovate to elliptic, up to 28 cm long and 12 cm broad; spines flat and triangular. Corolla violet to purple.

S. giganteum has a very limited distribution. It is known only from a flowering specimen collected from Ngelyaki, which is a remnant of high forest that extended farther from the stream valleys. *S. giganteum* has been recorded as one of the highland species of *Solanum*. There are also records from the Cameroun highlands which are contiguous with the Mambilla Plateau.

7. **S. gilo** Raddi in *Atti Soc. Ital. Sci. Modena*, *18*: 31 (1820); *Fl. West Trop. Afr.*, ed. 2, *2*: 332 (1963).

Syn. *S. geminifolium* Thonning in Schum & Thonn. Besk. *Guin. Pl.*: 121 (1827).
S. naumannii Engl. (1886)—*Fl. Trop. Afr.*, 4, 2: 216 (1906).

Shrub up to 120 cm high. Stem and petiole mealy-pubescent. Leaves sinuate to lobed, more densely pubescent below than above. Corolla white. Fruits globose or pear-like, up to 2.5 cm diameter, bright red.

S. gilo var. *gilo* is probably an introduced species widely cultivated in many parts of Nigeria for its edible fruits that may be cooked and eaten as vegetables.

8. **S. hispidum** Pers., *Syn.*, *1*: 228 (1805); *Fl. West Trop. Afr.*, ed. 2, *2*: 335 (1963).

Syn. *S. stellatum* Ruiz & Pavon, *Fl. Peruv.*, *2*: 40, t. 176 (1799).
S. warscewiczi Weick ex Lambertye in *Rev. Hort.*, 1865: 429.
S. antiguense Coult. In Donnell Smith, *Enum. Pl. Guatem.*, *4*: 187 (1895).
S. pynaertii De Wild, in *Miss. E. Laurent*, *2*: 437, t. 119 (1907).

Shrub up to 9 m high; old stem grooved and sparsely stellate; spines erect, up to 9 mm long. Leaf sinuate to lobed; upper surface scabrid and covered with appressed hairs; lower surface densely covered with stellate hairs and sparsely spiny on the nerves. Corolla white to blue.

S. hispidum is an introduced species. It is usually cultivated as an ornamental plant.

9. **S. incanum** L., *Sp. Pl.*, 188 (1753); *Fl. West Trop. Afr.*, ed. 2, *2*: 332 (1963).
Shrub, much branched, up to 1.8 m high. Stem spiny softly stellate; spines up to 5 mm long, straight or recurved. Leaf velvety below, densely stellate above, margin sinuate. Flowers blue to purple. Fruit globose, gold yellow and up to 2.4 cm diameter.

S. incanum has a very limited distribution, occurring only in Mixed Combretaceous Woodlands. It is often found in cultivated fields. Its fruit is inedible.

10. **S. indicum** L., *Sp. Pl.*, *187* (1753); *Fl. West Trop. Afr.*, ed. 2, *2*: 333 (1963).
subsp. **distichum** (Thonning) Bitter var. **distichum**.
Syn. *S. indicum* (Thonning) subsp. *distichum* (Thonning) Bitter var. *immunitum* Bitter in *Fedde Rep.*, Beih. 16: 14 (1923).
S. senegambicum Dunal in *DC. Prod.*, *13*, 1: 194 (1852).
S. anomalum of *Fl. Trop. Afr.* 4, 2: 232.

Small spineless shrub up to 1.5 m high. Stem densely stellate. Leaves sinuate to entire, grey, densely stellate beneath. Corolla white. Young fruit green, turning yellow and becoming orange, under 1 cm diameter.

S. indicum, subsp. *distichum*, var. *distichum* occurs mostly along the undulating montane grassland of Mambilla Plateau. Its occurrence in some other parts of southern Nigeria might result from cultivation. The fruit is edible and the seeds may be dispersed by cattle.

10a. Subsp. **distichum** var. **modicearmatum** Bitter *op. cit.* 16 (1923).
　　Syn. *S. buettneri* Dammer in Engl., *Bot. Jahrb., 38*: 59 (1905).
　　S. schroederi Dammer in Engl., *Bot. Jahrb., 48*: 250 (1912).
Shrub with characters similar to var. *distichum* but differing in being armed with spines up to 5 mm long.

　　var. *modicearmatum* occurs in the Vogel Peak area. Some species which appear to be intermediate between this variety and var. *grandemunitum* are found on the Jos Plateau.

11. **S. macrocarpon** L., *Mant. Alt.*: 205 (1771); *Fl. West Trop. Afr.*, ed. 2, *2*: 334 (1963).
Shrub up to 1.5 m high, fairly woody and unarmed. Stem glabrous. Leaves sinuate to lobed, lower and upper surfaces glabrous; petiole subsessile to distinctly petiolate. Corolla white or bluish-purple. Fruit globose, up to 3 cm diameter, green-white, yellow when ripe.

　　S. macrocarpon, which is mostly cultivated as a vegetable, was collected from the Gambari Forest Reserve and from the high forest of Ore. It is not unlikely that the specimen from the forest reserve grew in the exploited part of the forest.

12. **S. nigrum** L., *Sp. Pl., 186* (1753), *sensu lato*; *Fl. West Trop. Afr.*, ed. 2, *2*: 335 (1963).
Spreading to prostrate stout herb up to 60 cm long; branches square to terete with narrow slightly toothed wings. Leaves mostly ovate, up to 15 cm long; nerves looped near margin; petiole often winged. Inflorescence umbellate. Corolla white. Fruit small, under 1 cm diameter, green, purplish black when ripe.

　　The natural habitat of *S. nigrum* is the moist lowland forest zone of southern Nigeria. It occurs in most of the forest reserves in this zone, and also in some parts of the mixed leguminous woodland savanna in the northern part of the country. It is cultivated in the Mambilla Plateau and some parts of the northern states as a vegetable.

13. **S. seaforthianum** Andr. var. **disjunctum** O. E. Schulz in Urban, *Symb. Antill., 6*: 169 (1909); *Fl. West Trop. Afr.*, ed. 2, *2*: 332 (1963).
Climber. Leaves alternate; leaflets sessile to shortly stalked. Inflorescence paniculate. Corolla violet-purple. Fruits globose, red when ripe.

　　S. seaforthianum is believed to be a native of the West Indies and Central America, introduced for ornamental purposes.

14. *S. terminale* Forsk. subsp. *inconstans* (*C. H. Wright*) *Heine* in *Kew Bull.*, *14*: 247 (1960); *Fl. West Trop. Afr.* ed. 2, *2*: 331 (1963).
    Syn. *S. inconstans* C. H. Wright in *Kew Bull.*, 1894: 127.
    S. suberosum Dammer in Engl., *Bot. Jahrb.* 38: 182 (1906).
    S. togoense Dammer in Engl., *Bot. Jahrb.*, 38: 59 (1905).
Woody climber; main stem brownish and fissured; branches covered with brown hairs. Leaves glabrous to sparsely pubescent. Corolla blue, anthers yellow, style white, stigma pale green. Fruit ellipsoid, green, turning red when ripe.

14a. Subsp. *sanaganum* (*Bitter*) *Heine* in *Kew Bull.*, *14*: 248 (1960).
    Syn. *S. bansoense* Dammer in Engl., *Bot. Jahrb.*, 48: 237 (1912).
    S. plousianthemum Dammer (1906) Bitter in Engl., *Bot. Jahrb.*, *54*: 456 (1917).
Climber. Stem terete and glabrous. Leaves ovate, sparsely pubescent on both surfaces. Flowers large, terminal; sepals and corolla densely pubescent; corolla blue with yellow centre. Fruit ellipsoid.

14b. Subsp. *welwitschii* (*C. H. Wright*) *Heine* in *Kew Bull.*, *14*: 248 (1960).
    Syn. *S. welwitschii* C. H. Wright in *Kew Bull.*, 1894.
Woody climber; stem brownish and glabrous. Leaves obovate to elliptic, glabrous. Flowers in pendular racemes. Corolla pale blue. Fruits green.
    S. terminale subsp. *inconstans* is a moist lowland forest species, while subspp. *sanaganum* and *welwitschii* are restricted to the Mambilla and Obudu highlands respectively.

15. *S. torvum* Sw., Prod., *47*: (1788); *Fl. West Trop. Afr.*, ed. 2, *2*: 333 (1963).
    Syn. *S. mannii* C. H. Wright in *Kew Bull.*, *129*: (1894).
Shrubby weed up to 3 m high. Stem spiny and mealy-stellate. Leaf sinuate to lobed; under surface and petiole softly stellate, upper surface stellate and scabrid. Inflorescence a corymbose cyme; pedicel about 6 mm long with glandular hairs. Calyx 4 mm long, lobes triangular. Corolla up to 2.5 cm diameter, white. Fruit up to 1 cm diameter, dirty brown.
    S. torvum is a weed that grows up quickly in abandoned clearings. This is why it is common in exploited forest reserves. It generally thrives well on good soils. *S. torvum* is rare in the northern part of the country and common in the exploited parts of the high forests of the south.

16. *S. erianthum* D. Don, *Prod. Fl. Nepal*, *96*: (1825).
    Syn. *S. verbascifolium* of authors, not of Linn.; *Fl. West Trop. Afr.*, ed. 2, *2*: 335 (1963).
Unarmed small tree or shrub up to 3.5 m high, mealy-stellate pubescent. Stem ash-grey. Leaves elliptic to ovate, up to 21 cm long and 10 cm broad, pubescent above, tomentose and velvety beneath. Corolla white, turning blue-purple. Fruit green, yellow when ripe.
    S. erianthum is similar to *S. torvum* in its distribution and ecology. It is found mostly in exploited parts of forest reserves, old farmland and cocoa plantations. It is restricted to the moist lowland rain forest zone of southern Nigeria.

17. *S. wrightii* Benth. in *Fl. Hong Kong*: 243 (1861).

 Syn. *S. macranthum* hort. in *Rev. Hort.*, 1867: 132 not of Dunal.

 S. grandiflorum of authors, not of Ruiz & Pavon, *Fl. Peruv.*, 2: 35.

Small tree or shrub up to 5 m high. Stem and leaves armed with spines. Leaves deeply lobed; upper surface fairly scabrid, and covered with appressed simple tuberculate hairs, lower surface softly stellate. Corolla pinkish-white, surface covered with simple hairs. Fruit yellow, glabrous.

 S. wrightii is cultivated for ornamental purposes.

DISCUSSION AND CONCLUSION

 Of all the *Solanum* species in Nigeria (Fig. 4.1) *S. nigrum* is the most widely distributed. It is not restricted to a definite zonal or latitudinal distribution as it occurs in both the lowlands and highlands. The *S. nigrum* complex, as it is sometimes called, is taxonomically difficult (D'Arcy, 1974). It is very variable and has polyploid forms in nature having $n = 12$, 24 and 36 chromosomes. The distribution of the hexaploid *S. nigrum* in Japan has been correlated by Nakamura (1937) to a particular latitudinal zone. The hexaploid form is known from the Himalayas at altitudes reaching up to 2400 m. Bhaduri (1951) associated black berries with the diploid, hexaploid and colchicine induced tetraploids, and bright orange to the natural tetraploid forms. The Nigerian species have black fruits. The occurrence of the hexaploid form in Nigeria has been confirmed by Omidiji (1975).

 Some of the lowland species—*S. erianthum, S. torvum* and *S. macrocarpon*—are restricted to the moist lowland forest zone in the southern part of the country. *S. incanum* has not been known from any other part of Nigeria apart from the mixed combretaceous woodlands, which is a very distinct geographical region typified as much by its climate and vegetation. It has less than 2.5 cm of rainfall from October–April and an average of about 75 cm annual rainfall (Keay, 1960). *S. incanum* appears unable to tolerate the wetter zones of the south.

 The natural habitats of both the lowland and the highland species are indicated by their occurrences in the forest reserves of the country (Table 4.1). Cytological evidence is lacking for most of the indigenous species.

 In *S. terminale* subsp. *inconstans* the inflorescence is axillary while those in subspp. *sanaganum* and *welwitschii* are terminal. In subsp. *welwitschii*, the raceme is up to 20 cm while the two other subspp. have comparatively shorter ones. The fruits of subspp. *inconstans* and *sanaganum* are ellipsoid while that of subsp. *welwitschii* is globose. Subsp. *inconstans* is a lowland species while the other two subspecies belong to the highlands (Fig. 4.1).

 In view of the major differences in gross morphology and pollen morphology (see this Volume, Chapter 27), the author recommends that each of the three subspecies be given specific status. The palynological study also clearly shows that the specimens belong to distinct species, and that pollen morphological features can be used to distinguish between the species or varieties investigated.

ACKNOWLEDGEMENT

 I am very grateful to Dr S. T. Olatoye, Director of the Forestry Research Institute of Nigeria for the facilities for this study, to Mr V. Eimunjeze for his assistance and to all the herbarium staff for their co-operation.

Table 4.1. The occurrence of *Solanum* spp. in the forest reserves in Nigeria

Forest reserve	Vegetation	*Solanum* spp. occurrence
1. Ago-Are	Mixed leguminosae savanna	*S. nigrum*
2. Ago-Owu	Moist lowland forest	*S. anomalum*
3. Akure	Moist lowland forest	*S. terminale* ssp. *inconstans*
		S. erianthum
4. Aponmu	Moist lowland forest	*S. torvum*
5. Boshi Extension	Moist sub-montane forest	*S. aculeatissimum*
6. Gambari	Moist lowland forest	*S. macrocarpon*
7. Idanre	Moist lowland forest	*S. anomalum*
8. Ngel yaki (Mambilla)	Moist sub-montane forest	*S. giganteum*
		S. torvum
9. Nikrowa	Moist lowland forest	*S. nigrum*
10. Oban	Moist lowland forest	*S. terminale* ssp. *inconstans*
11. Okomu	Moist lowland forest	*S. nigrum*
		S. terminale ssp. *inconstans*
12. Olokemeji	Moist lowland forest	*S. nigrum*
		S. torvum
		S. erianthum
13. Omo	Moist lowland forest	*S. anomalum*
		S. nigrum
		S. terminale ssp. *inconstans*
		S. torvum
		S. erianthum
14. Sapoba	Moist lowland forest	*S. torvum*
15. Shasha	Moist lowland forest	*S. dasyphyllum*
		S. indicum
		S. nigrum
		S. terminale ssp. *inconstans*

REFERENCES

BHADURI, P. N., 1951. Inter-relationship of non-tuberiferous species of *Solanum* with some consideration on the origin of Brinjal (*S. melongena* L.). *Indian Journal of Genetics and Plant Breading, 11*: 75–82.

D'ARCY, W. G., 1974. *Solanum* and its close relatives in Florida. *Annals of Missouri Botanical Garden, 61*: 819–867.

HEINE, H., 1963. Solanaceae. In J. Hutchinson & J. D. Dalziel (Eds), *Flora of West Tropical Africa*, 2nd ed. by F. N. Hepper, Governments of Nigeria, Ghana, S. Leone and the Gambia.

KEAY, R. W. J., 1959. *An Outline of Nigerian Vegetation*, 3rd ed. Lagos: Federal Ministry of Information.

NAKAMARA, M., 1937. Cytological studies in the genus *Solanum* I. Autopolyploidy of *Solanum nigrum* Linn· *Cytologia (Fuji, Jub.), 1*: 57–68.

OMIDIJI, M. O., 1975. Interspecific hybridization in the cultivated, non-tuberous *Solanum* species. *Euphytica, 24*: 341–353.

5. Australian genera of the Solanaceae

L. HAEGI

Waite Agricultural Research Institute, P.M.B.1, Glen Osmond, South Australia, 5064

Only three groups of Solanaceae are well developed in Australia. *Solanum,* with about 70 to 90 endemic species and *Nicotiana,* with about 20 endemic species, are represented elsewhere in the world. *Anthocercis, Anthotroche, Duboisia* and *Isandra* are the only endemic genera and form a coherent group of about 30 species. Properties characteristic of this group clearly place it in the Solanaceae. The restricted, apparently relict distributions of many of the species support the notion that the group is an old one which diverged early from a solanaceous stock ancestral to the tribe Cestreae. Contrary to previous suggestions these genera show little affinity with the Scrophulariaceae.

CONTENTS

INTRODUCTION

In Australia the Solanaceae is represented by several genera with native species. *Lycium, Physalis* and *Datura* all have major centres of speciation elsewhere. They are each represented by only a single native or endemic species. Only three groups of Solanaceae are well developed. *Solanum* is the largest genus with about 70 to 90 endemic species and *Nicotiana* has about 20 endemic species. Both genera are well developed in other parts of the world, chiefly the Americas.

The third group comprises the only endemic genera, *Anthocercis, Anthotroche* and a genus called *Isandra**, whose taxonomic status is uncertain. Also included is the genus *Duboisia* which is predominantly Australian with two endemic species and one native which is also found in New Caledonia. There are about 30 species in this distinctive Australasian group of genera.

PREVIOUS TREATMENTS

In the past there has been uncertainty and lack of agreement among authors about the affinities of these predominantly Australian genera. *Anthocercis* was the first genus of the group discovered but Labillardière (1806) did not indicate its family position at that time. When Brown (1810) described the genus *Duboisia* he included it with some new species of *Anthocercis* in the Solanaceae. However several authors (e.g. Endlicher, 1838; Bentham, 1846; Mueller, 1859; Hooker, 1860) subsequently assigned new species of *Anthocercis* to the Scrophulariaceae.

* This name is illegitimate, being a later homonym of *Isandra* R. A. Salisb. (Liliaceae).

Shortly after the genus *Anthotroche* was discovered, Miers (1849) recognized the similarities of the three genera. Miers, a monographer of the Solanaceae, treated them as a distinct tribe in his family Atropaceae, intermediate between the Solanaceae and Scrophulariaceae. The recognition of the genera as a group suffered a reversal in Bentham's (1868) treatment in which he placed *Anthocercis* and *Duboisia* in the tribe Salpiglossideae of the Scrophulariaceae, and *Anthotroche* in the Solanaceae. In Wettstein's (1891) classical treatment of the Solanaceae the three genera were included in the tribe Salpiglossideae which was transferred to the Solanaceae. Although *Isandra* had been discovered by then, Wettstein knew too little about the genus to place it. About 50 years later Baehni (1946) separated the genera yet again in a treatment based solely on corolla aestivation. Possibly because of misinterpretation of this character *Duboisia* and *Anthocercis* were placed in separate tribes while *Anthotroche* and *Isandra* could not be placed.

Currently the group is treated as part of the tribe Salpiglossideae in the Solanaceae. D'Arcy (1975) and others have recognized that this treatment is unsatisfactory, since all species have regular corollas and *Anthotroche* has a strictly actinomorphic flower.

AFFINITIES OF THE AUSTRALIAN GENERA

Inclusion of *Anthocercis* and *Duboisia* in the family Scrophulariaceae was based on the single criterion of these two genera having didynamous stamens. Such a classification and the resultant separation of the Australian genera into two different families (Bentham, 1868) is unsatisfactory because it ignores the existence of several characters which are common to all the genera. Recent classifications have recognized that *Anthocercis* and *Duboisia* belong in the Solanaceae and are related to *Anthotroche* and *Isandra*, but the current placement of the whole group in the tribe Salpiglossideae is also based solely on *Anthocercis* and *Duboisia* having didynamous stamens. *Anthotroche* and *Isandra* which have five equal stamens are as a result ill-placed.

A classification which is more satisfactory can be constructed from a study of as many properties as possible characteristic of these genera and other genera which seem related. On this basis it becomes clear that several of the characters and the combination of characters common to all the Australian genera are also peculiar to the group. (Table 5.1). Furthermore, a classification of the group based on characters common to all its members results in their unequivocal inclusion in the Solanaceae, and suggests their placement in a separate tribe related neither to the Scrophulariaceae nor the Salpiglossideae (Table 5.1). The presence of didynamous stamens in *Anthocercis* and *Duboisia*, and also in the Salpiglossideae (and Scrophulariaceae) may therefore be explained better by convergent evolution than by close common ancestry.

Treatment of the Australian genera as a separate tribe is supported by indications that the group is an old one. While the genera are not distinguished by numerous characters, the differences are clear cut. *Duboisia* has baccate fruits but the other genera have capsular fruits. *Anthocercis* (like *Duboisia*) has didynamous stamens while *Anthotroche* and *Isandra* have five equal stamens. In *Isandra* the corolla lobes are much shorter than the tube and in the other genera the lobes are always longer than the tube. Added to this, considerable morphological diversity occurs amongst the species, including variation in habit, foliage, indumentum, corolla lobing and colour, and anther morphology.

A further indication of the age of the group is provided by the high proportion of distinct species with restricted or relict distributions. More than half of the thirty species are confined in occurrence to very small areas, or if more widespread, occur in widely disjunct small populations. This suggestion of age and the morphological evidence presented support the notion that these Australian genera diverged early from a solanaceous stock ancestral to the tribe Cestreae.

Table 5.1. Affinities of the Australian genera *Anthocercis, Anthotroche, Duboisia* and *Isandra*

Characters common to all genera	Family Position. Support for inclusion in:	Tribal Position (10 and 11 suggest closer affinity with tribes Cestreae* and Salpiglossideae than any other tribes in Solanaceae)
1. Woody shrubs	Solanaceae (Scrophulariaceae mostly herbaceous)	Supports exclusion from Salpiglossideae and Nicotianinae
2. Stem and petiole with intraxylary phloem	Solanaceae (not known in Scrophulariaceae)	Applies to whole family
3. Inflorescences non-axillary, derived from sympodial branching	Solanaceae (axillary in Scrophulariaceae)	Applies to whole family
4. Aestivation of corolla volutive (see Miers, 1853)	Solanaceae (imbricate in Scrophulariaceae)	Uniform in the Australian group, rare elsewhere
5. Corolla regularly 5-lobed	Solanaceae (zygomorphic in Scrophulariaceae)	Supports exclusion from Salpiglossideae
6. Corolla streaked along main veins and 2 veins on each side of each of these	Not applicable at family level	Peculiar to Australian group
7. Stamens epipetalous low in corolla tube	Not applicable at family level	Supports exclusion from Nicotianinae
8. Anthers dehiscing extrorsely by longitudinal slits	Not applicable at family level	Known otherwise only in *Petunia* (Nicotianinae)
9. Ovary 2-locular; placentation axile	Characteristic of both families	Applies to whole family
10. Seed reniform with reticulate surface	Not applicable at family level	Applies to all Cestreae, Nicotianinae and Salpiglossideae, but not rest of Solanaceae
11. Embryo only slightly curved	Not applicable at family level	
12. Embryo with radicle not pointing towards hilum	Solanaceae (towards hilum in Scrophulariaceae)	Applies to whole family

*There is strong evidence (D'Arcy, 1975) that the subtribe Nicotianinae, at present included in the tribe Cestreae, should be raised to tribal status. Hence reference to the affinities of the Australian group with the Nicotianinae have been included in the table.

REFERENCES

BAEHNI, C., 1946. L'ouverture du bouton chez les fleurs de Solanées. *Candollea, 10*: 399–492.

BENTHAM, G., 1846. Scrophulariaceae. In A. P. De Candolle, (Ed.), *Prodromus Systematis Naturalis Regni Vegetabilis, 10.* Paris: Victor Masson.

BENTHAM, G., 1869 (publ. 1868). *Flora Australiensis, 4.* London: L. Reeve.

BROWN, R., 1810. *Prodromus Florae Novae Hollandiae.* (Facsimile ed.: Weinheim, 1960.)

D'ARCY, W. G., 1975. The Solanaceae: an overview. *Solanaceae Newsletter, 2*: 8–15.

ENDLICHER, S. L., 1838. *Anthocercis anisantha*, in Stirpium Australasicarum Decades III. *Annalen des Weiner Museums der Naturgeschichte, 2*: 201–202.

HOOKER, J. D., 1860 (publ. 1859). *Anthocercis tasmanica*. In, *Flora Tasmaniae*: 289. London.

LABILLARDIÈRE, J. J. H. DE, 1806. *Anthocercis* in *Novae Hollandiae Plantarum Specimen, 2*: 19. Paris.

MIERS, J., 1849. Observations upon several genera hitherto placed in Solanaceae, and upon others intermediate between that family and the Scrophulariaceae. *Annals and Magazine of Natural History* (Ser. 2), *3*: 161–183.

MIERS, J., 1853. On the genera of the tribe Duboisieae. *Anthocercis*. *Annals and Magazine of Natural History* (*Ser. 2*), *11*: 368–381.

MUELLER, F., 1859. *Anthocercis* in *Fragmenta Phytographiae Australiae*, Fasc. *1*: 179. Melbourne: Government Printer.

WETTSTEIN, R. VON, 1891. Solanaceae. In H. G. A. Engler & K. A. E. Prantl, (Eds), *Die Naturliche Pflanzenfamilien, 4* (3b): 4–38. Berlin.

6. The genus *Solanum* in Australia

D. E. SYMON

Waite Agricultural Research Institute, Private Bag 1, Glen Osmond, South Australia 5064

There are about 90 species of *Solanum* considered native to Australia of which all but eight are endemic. Species occur in all except saline, aquatic and alpine habitats. More species occur in the wet or dry tropics than in the southern temperate areas. Most species are short lived, subwoody herbaceous perennials whose above-ground parts live for several years and regrow from clonal root systems. There are a few annuals and a few small trees. Tuber-bearing species do not occur in Australia.

Species are often found in disturbed sites, rocky outcrops, breakaways and dune systems (both coastal and inland); they may be prominent after fire and physical disturbance and are rarely components of climax vegetation.

CONTENTS

PRINCIPAL SUBGENERIC GROUPS REPRESENTED IN AUSTRALIA

(1) The subgenus *Solanum* (Black Nightshade) is poorly represented in Australia. Probably only two species are native, though another six to eight have become naturalized very widely. The native species *S. nodiflorum* Jacq. (whose relationships with *S. americanum* Mill. are not yet clear) is now a pantropic weed. The second species is the endemic *S. opacum* A. Br. and Bouché which has pale green, succulent aromatic fruit, is prostrate in habit and is a hexaploid. Putative ancestors have not been suggested and it is difficult to conceive of it being derived from the diploid *S. nodiflorum* from which it differs in many characters. Both species are found in mesic tropic to subtropic sites in eastern Australia. The group has recently been revised in Australia by Henderson (1974) who provides descriptions, maps, illustrations and extensive commentaries.

(2) The small subgenus *Archaesolanum* is of considerable interest. It contains a dozen or fewer species. It has a peculiar distribution from New Guinea to Tasmania along the eastern coast of Australia and from Perth in Western Australia eastwards along the south coast and thence to New Zealand. It is not an element of the arid flora though two species *S. simile* F. Muell. and *S. capsiciforme* (Domin) Baylis occur in Mallee formation (dwarf woodland) in southern Australia.

The subgenus is unique in *Solanum* in having the base chromosome number $x = 23$. Despite Bitter's name suggesting an archetypal *Solanum* the chromosome number indicates a derived condition and this has itself developed polyploidy.

Putative ancestors have certainly not been recognized in Australia nor are extra-Australian relatives apparent. They are not clonal in habit, are almost glabrous, have no prickles or stellate hairs, have large and variable leaves, anthers which split down the cells almost completely, and succulent fruits with abundant stone cell masses. Few of these characters occur in the other subgenera. The fruits are eaten by birds and are probably so distributed between Australia, New Zealand, Tasmania and New Guinea. The species often occur in subcoastal dunes.

Species of this subgenus contain high levels of solasodine and because of this one or two species are now being cultivated in USSR, eastern Europe and New Zealand as a source of the precursors for corticosteroid drugs. They are amongst the most recent additions to domesticated plants.

As a subgenus it highlights the problems of generic boundaries in the genus *Solanum*. This group of species is at least as distinct as *Lycopersicon* from many other *Solanums*, and apart from the chromosome number has no single character to distinguish it from all other groups. To raise it to generic level would justify splitting *Solanum* into a dozen ill-defined genera.

The species include: *S. aviculare* Forst. f.

S. laciniatum Ait.

S. linearifolium Gerasimenko

S. vescum F. Muell.

S. simile F. Muell.

S. symonii Eichler

S. capsiciforme (Domin) Baylis

The group was first treated comprehensively by Baylis (1963) but Gerasimenko (1970, 1971) and Korneva (1972) have recently named several new varieties or raised older varieties to specific status. It is possible that some of these names at least would be better dealt with under the Code for cultivated plants.

(3) The subgenus *Brevantherum* is represented by a single species *S. erianthum* D. Don considered native, and by the alien *S. mauritianum* Scop. *S. erianthum* was collected in Australia very early and evidence suggests that it was established before white settlement in 1788. The subgenus is concentrated in Central and South America where both species occur and *S. erianthum* is quite isolated in relation to the other stellate haired species in Australia. Both species have weedy tendencies and both are now widely distributed through the tropics. I would suggest that their distribution began with Portuguese and Spanish trade in the Pacific in the early 1500s; their subsequent distribution being augmented by birds and fruit bats both of which eat the fruits. If this is so, establishment prior to white settlement in Australia would have been possible. A number of other anomalous *Solanaceae* occur in Australia whose origin is an enigma—a single *Lycium*, a single *Datura* and a single *Physalis*. The subgenus *Brevantherum* has been revised recently by Roe (1967).

(4) By far the largest number of *Solanum* species in Australia belong to the subgenus *Leptostemonum* (about 60 species). While they occur widely they are less common in the cooler mesic areas of south eastern Australia and are absent from Tasmania and New Zealand. They fall into a number of groups some of which are difficult to equate with those published to date.

The principal groups are discussed below.

(A) A group of about 12 species with pale blue, stellate flowers and with smallish, red, succulent fruits. The group appears to be close to the subgenus *Graciliflorum*. Their present distribution is concentrated in the higher rainfall areas of the east coast including two species with disjunct distribution across arid Australia. The isolation of these populations suggests a once more extended distribution now disrupted and contracted. This is supported by similar disjunctions in the distribution of other *Solanum* species.

The species belonging to it include:

S. amblymerum Dun.	*S. nemophilum* F. Muell.
S. chenopodinum F. Muell.	*S. parvifolium* R. Br.
S. defensum F. Muell.	*S. semiarmatum* F. Muell.
S. densevestitum F. Muell. ex Benth.	*S. stelligerum* Sm.
S. discolor R. Br.	
S. ferocissimum Lindl.	

(B) Also primarily on the east coast is a varied group with large succulent green/orange/red fruits. Most are very prickly especially when young; they are relatively large leaved, often sparsely pubescent and a few reach small tree size. Their distribution is similar to that of *Archaesolanum* along the east coast with two of the smallest species in the southern arid regions, and none in Tasmania or New Zealand. The species are often localized and both the morphological and phytogeographical discordance within this group is considerable.

The species belonging to the group include:

S. dimorphospinum White	*S. inaequilaterum* Domin
S. dallachii Benth.	*S. macoorai* Bailey
S. furfuraceum R. Br.	*S. multiglochidiatum* Domin
S. hamulosum White	*S. prinophyllum* Dun.
S. hoplopetalum Bitt. & Sum.	*S. pungetium* R. Br.
S. hystrix R. Br.	

(C) Numerically the largest is a group of greenish-yellow fruited species. These are generally small in stature and have medium sized ($\pm$ 1–1.5 cm diam.) fruit. They vary from herbaceous perennials to small shrubs. The flowers may be stellate or rotate, the seeds are pale, the flowers are hermaphrodite, and the species grow in the arid to semi-arid areas with no great concentration in any area. However some show signs of disruption of a once wider spread. Others are "recent" species and have evolved on the more modern landscapes; these tend to be the herbaceous perennials. Some species approach American species like *S. elaeagnifolium* Cav. in morphology and may belong to section *Oliganthes*.

The species concerned include:

S. adenophorum F. Muell.	*S. lachnophyllum* Symon
S. cleistogamum Symon	*S. lacunarium* F. Muell.
S. coactiliferum Black	*S. lasiophyllum* Dun.
S. dianthophorum Dun.	*S. lucani* F. Muell.
S. echinatum R. Br.	*S. oldfieldi* F. Muell.

S. elachophyllum F. Muell.	*S. oligacanthum* F. Muell.
S. ellipticum F. Muell.	*S. papaverifolium* Symon
S. eremophilum F. Muell.	*S. petrophilum* F. Muell.
S. esuriale Lindl.	*S. quadriloculatum* F. Muell.
S. horridum Dun.	*S. sturtianum* F. Muell.
S. karsensis Symon	*S. tetrathecum* F. Muell.
	S. tumulicola Symon

Next occurs a collection of andromonoecious and androdioecious species. All have rotate flowers and relatively large (2–4 cm diam.) yellowish fruits with dark or black seeds. All are small or medium sized shrubs and are mainly northern and especially north-western in distribution.

(D) The andromonoecious species have a single hermaphrodite flower below a raceme of male flowers. Their relationships are African and Asian and they have affinities with species such as *S. melongena* L., *S. incanum* L., and *S. sodomeum* L. though there are no species in common with Afro-Asia. Species include:

S. diversiflorum F. Muell.	*S. melanospermum* F. Muell.
S. eburneum Symon	*S. phlomoides* Benth.

(E) The androdioecious species are unique to Australia. They are all sub-shrubs and have separate male and hermaphrodite plants. They are an oddly diverse lot in other characters and it does not look as if they have a common set of ancestors. A few could have evolved from the andromonoecious group (D) above. They are all confined to north-western Australia. A brief comment on them was published by Symon (1970) and the possible rationale of their evolution is discussed elsewhere by me in this volume. The species include:

S. asymmetriphyllum Specht	*S. cunninghamii* Benth.
S. carduiforme F. Muell.	*S. dioicum* W. V. Fitz.
S. cataphractum Benth.	*S. leopoldensis* Symon

(F) Lastly, two anomalous species.

(a) *S. pugiunculiferum* White, probably deserves Sectional status on its own. It is an annual, is very prickly but essentially glabrous, has small campanulate flowers, dryish green fruits, flat papery seeds and small oblong anthers. It is confined to heavy soil flats south of the Gulf of Carpentaria. In all, it possesses a most unusual combination of characters.

(b) *S. campanulatum* R. Br., is confined to the east coast and is very glandular pubescent, has deeply campanulate flowers, large yellow fruit, black seeds and sometimes the upper flowers are male only.

It is clear from this brief account that the genus is varied and well developed in Australia and that distinctive groups have evolved. The evolution of these groups has long separated the varied gene pools. There is no sign of hybridity between any of the major groups. It is also noticeable how many species appear to be isolated and the term "hybrid" is rarely invoked to account for variation within the smaller groups. Probably the most widespread and difficult species complex is that about *S. ellipticum* R. Br., in which geographical subspeciation has resulted in numerous more or less localized variants difficult to define in conventional taxonomic categories. It is the most widespread species in Australia. *S. petrophilum* F. Muell. also has a number of geographically isolated variants.

Species not endemic to the Australian mainland and shared with neighbouring regions are shown in Table 6.1.

Table 6.1. Australian mainland *Solanum* species
shared with neighbouring regions

Species	Tasmania	New Zealand	New Guinea	Western Pacific Islands
S. aviculare Forst. f.	.	+	+	.
S. laciniatum Ait.	+	+	.	.
S. vescum F. Muell.	+	.	.	.
S. nodiflorum Jacq.	.	+	+	+
S. dunalianum Gaud.	.	.	+	+
S. ferox L.	.	.	+	+
S. tetrandrum R. Br.	.	.	+	+
S. viride R. Br.	.	.	+	+

S. ferox and *S. dunalianum* have only recently been recorded for Australia and may be recent introductions. Like *S. erianthum, S. ferox* has its close relatives in the Americas and it too may be an early introduction to the western Pacific from tropical America. All of these species have fleshy fruits and could be spread by birds or fruit bats.

CHROMOSOME NUMBERS

An extensive survey of chromosome numbers has recently been published, Randell & Symon (1976). The results may be summarized in the following table:

	$n = 12$	$n = 23$
diploid only	50	5
diploid and tetraploid	9	–
tetraploid only	3	2
hexaploid only	1	–
	63	7

The chromosomes are small and no differentiation between the two sex forms has been noted. As morphological distinction of those species with both diploids and tetraploids has been minimal it is assumed that most of the tetraploids are autopolyploids.

FRUIT TYPES AND DISPERSAL

The fruits of most *Solanum* species are berries but some interesting variants have evolved.

1. *Bony hard berries.* A small group of species have fruits which finally dry hard, pale, bony and indehiscent. Each berry may contain several hundred seeds and their method of distribution and release is obscure. An example is *S. petrophilum* F. Muell.

2. *Trample burrs*. These species have the berry enclosed in a very prickly calyx and the mature fruits are either shed on the ground or are readily broken off. The species frequently occur on rocky outcrops and in rocky gorges and are possibly transported by wallabies. An example is *S. echinatum* R. Br.

3. *Censer mechanism*. An as yet unnamed species from north-western Australia has the berry enclosed in a firm calyx with a small orifice. The berry when mature is circumcissile near the base and dries and shrinks to form a loose plug within the calyx. The seeds are then released by shaking or knocking the tall slender stem.

4. *Tumble weed*. *S. pugiunculiferum* White is a possible tumble weed though the plants have not been seen in motion.

The scattered evidence available indicates that birds are the principal agents of dispersal, in particular the emu which once ranged very widely in Australia. Dispersal by mammals is less well documented but kangaroos, possums, dingo and man all eat fruits and disseminate seeds.

The aboriginal uses of *Solanum* species in Australia are dealt with by Peterson (this Volume, Chapter 10) and a survey of the alkaloids of the Australian species is given by Bradley *et al.* (this Volume, Chapter 12).

Finally, a taxonomic revision of the genus is well advanced and should go to press in 1978. It will include at least ten new species, illustrations of all species and maps of distribution.

REFERENCES

BAYLIS, G. T. S., 1963. A cytogenetical study of the *Solanum aviculare* complex. *Australian Journal of Botany*, *11*: 168–177.

GERASIMENKO, I. I., 1970. Conspectus subgeneris *Archaesolanum* Bitt. ex Marz. generis *Solanum* L. *Novosti Sist. Vyssh. Rast.*, 7 (publ. 1971): 270–275.

GERASIMENKO, I. I., 1971. Intraspecific variation of *Solanum laciniatum* Ait. *Rast. Resursy*, 7: 363–371.

HENDERSON, R. J. F., 1974. *Solanum nigrum* L. Solanaceae and related species in Australia. *Contributions from the Queensland Herbarium*, No. 16.

KORNEVA, E. I. *et al.*, 1972. Intraspecific variability of Bird Nightshade and Lobed Nightshade *S. aviculare* and *S. laciniatum*. *Rast. Resursy*, *8*: 507–515.

RANDELL, B. R. & SYMON, D. E., 1976. Chromosome numbers in Australian *Solanum* species. *Australian Journal of Botany*, 24: 369.

ROE, K. E., 1967. A revision of *Solanum* sect. *Brevantherum* Solanaceae. *Brittonia*, 24: 239–278.

SYMON, D. E., 1970. Dioecious *Solanum*. *Taxon*, *19*: 909–910.

7. On typifying Linnaean names of Solanaceae

F. N. HEPPER

The Herbarium, Royal Botanic Gardens, Kew

The reasons for some recent name changes in *Solanum* are discussed in order to bring out certain principles of botanical nomenclature. Lines of approach necessary for any attempt to typify a Linnaean name are provided and some of the pitfalls are mentioned.

Users of plant names often find them troublesome, especially when well known names give way to obscure ones. Despite this, internationally recognized plant names are vital and fundamental to research. We must know and be able to communicate the names of our plants: they must be unambiguous, otherwise information retrieval systems will break down. Taxonomic research papers may be indigestible fare for Solanaceae gluttons but they provide the scientific basis for local or regional Floras with which to identify our plants, and the Royal Botanic Gardens, Kew, has a long tradition of writing the world's Floras.

It is important, I feel, to be aware of the problems, principles and practices of plant taxonomists, so for the benefit of those unfamiliar with these aspects, I would like to look back two centuries to Linnaeus's work, in order to see how relevant it is to modern botanical nomenclature.

First, it might be instructive to look at three cases of name changes in order to appreciate the reasons behind them, which fall into three main categories:

1. *Priority—Solanum mauritianum* Scopoli (1788) was described one year before *S. auriculatum* Aiton (1789) and so must take precedence over it. This is in accordance with Article 11 of the *International Code of Botanical Nomenclature* (1966).

2. *Taxonomy—S. ciliatum* Lam. is morphologically distinct from *S. aculeatissimum* Jacq. although they have been confused, and the latter name has been used for the former species, which according to Nee should be called *S. capsicoides* Allioni for reasons of priority.

3. *Misinterpretation—Solanum erianthum* D. Don, with its type specimen from Nepal, replaces *S. verbascifolium* L. which was described from the West Indies (it is, however, still obscure to which species the name *S. verbascifolium* should apply).

Today when a species is given a name there must be a herbarium or preserved specimen which is the standard or nomenclatural type. This specimen is the name-carrying voucher to which the name applies although it may not be typical of the variable species as a whole. The author of a new name has before him the plant specimen he is describing and if there is only one, that becomes the TYPE (technically the HOLOTYPE). Nowadays he may have several specimens from several gatherings, which cannot be exactly identical, and he must select one of them as the type. All this and much more is set out in the *International Code of Botanical Nomenclature* which the taxonomist has to observe for his plant name to be validly published and acceptable.

This is normally quite straightforward until we try to apply the *Code* retrospectively. Linnaeus, for example, often worked from earlier published descriptions or on drawings without the backing of a specimen, although a visual element either a specimen or an illustration almost always forms the basis of his concept. We must not forget that Linnaeus gave us our familiar binomial nomenclature for plants and it is worth recalling how this came about.

It began in 1745 when he published an account of his botanical travels on the Baltic Islands of Oeland and Götland and to save space in the index he used only two words for each species. Until then the species of a genus had usually been designated by a long string of adjectives which were both unmemorizable and clumsy to write though intended to convey the major diagnostic features, while the new catch-word was easy to remember and quicker to write. How much more preferable is simple *Solanum nigrum* to the verbose *Solanum caule inermi herbaceo, foliis ovatis angulatis.* He later applied this method to the plant kingdom as a whole in his *Species Plantarum* of 1753.

However, Linnaeus continued to provide a descriptive definition or "phrase-name" in order to distinguish each species from the others in the genus by its "essential" characters. Thus F. A. Stafleu has had to point out in *Linnaeus and the Linnaeans* (1971:88) that "when taxonomic botanists agreed in 1930 to introduce the type method for linking names and plants, thereby abandoning the method of circumscription, what they actually abandoned was Linnaeus's *essentialism*". Stearn has stressed in his scholarly Introduction to the 1957 facsimile edition of *Species Plantarum*, that these phrase-names are of first importance in matters of typification. The date of publication of *Species Plantarum* (1753), was adopted as the internationally recognized starting point of botanical nomenclature, although some botanists consider it would have been preferable to have taken the second edition (1762–63) to account for changes between the two editions.

The format of *Species Plantarum* is very concise and he based many of his species on his own earlier works. For example in 1738 Linnaeus had published a splendidly illustrated book on the plants in the garden and herbarium of an Anglo-Dutch financier George Clifford. This work, *Hortus Cliffortianus*, includes long descriptions and many figures of the plants, so he did not repeat the descriptions in his 1753 *Species Plantarum* but simply gave a reference to *Hort. Cliff.*, with an asterisk (*) to show there was much fuller relevant information, and he provided an epithet such as *nigrum* and *verbascifolium*. One must therefore consult the Clifford Herbarium, which is at the British Museum (Natural History), in order to typify many Linnaean specific names. However, it is not quite so simple since Linnaeus may have consulted other material between 1738 and 1753 and this needs to be taken into consideration in deciding on the type specimen.

Clifford's were mainly foreign cultivated plants, but Linnaeus also drew on other sources for *Species Plantarum*, such as those he had seen wild in Scandinavia and had written up in his *Flora Suecica*. He also had specimens sent from all over the world for his own herbarium which was purchased by J. E. Smith in 1784 and is now housed in the specially designed basement strong-room of the Linnean Society in Burlington House, Piccadilly, London. Consultation of the herbarium sheets (or microfiche) in order to typify a name is fraught with pitfalls since one should know a good deal about the history of each specimen in order to determine whether Linnaeus had it in his possession at the time he wrote *Species Plantarum* and, if so, whether he used it.

Besides his own herbarium, Linnaeus used those of other people, such as that of Hermann made in Ceylon during the 17th century. He even based his *Flora Zeylanica* (1747) on this herbarium now in the British Museum (Natural History) and on Burman's *Thesaurus zeylanicus* which was dated 1737 although published in 1738.

Perhaps the Linnaean names most difficult to typify are a few published by earlier botanists and taken up by Linnaeus. When herbarium specimens were wanting he used illustrations, but as these were often coarse woodcuts that he tried to match with species known to himself the conclusions were not always clear or accurate.

This paper on the principles of typification has been included here since the taxonomic discussion and conclusions have been published in the *Botanical Journal of the Linnean Society*, 76: 287–292 (1978). They may be summarized as follows, and will be formally proposed in *Taxon*.

The species hitherto known as "*Solanum indicum* L." would now be named *S. anguivi* Lam., *Illustr.* (*Tabl. encyc. méth.*) 2: 23 (1794).

The species hitherto known as "*Solanum sodomeum* L." would now be named *S. hermannii* Dunal, *Hist. Solanum*: 212, t. 2 fig. B (1813).

Two other conclusions are noted: that *S. indicum* L. is a synonym of *S. ferox L.*; and, as already mentioned by J. H. Willis (1972), *S. violaceum* R. Br. (non Ortega) is a synonym of *S. brownii* Dunal, *Hist. Solanum*: 201 (1813).

Fuller information should be sought in W. T. Stearn's Introduction to the Ray Society's facsimile edition of the *Species Plantarum* (London 1957, 1959), and in F. A. Stafleu, *Linnaeus and the Linnaeans* (A. Oosthoek's Uitgeversmaatschappij, Utrecht, 1971).

II. Ethnobotany

8. Solanaceous hallucinogens and their role in the development of New World cultures

RICHARD EVANS SCHULTES

Botanical Museum, Harvard University, Cambridge, Mass., U.S.A.

The role of solanaceous plants in the magico-religious and medical aspects of New World cultures has long been recognized as great. Its full extent and significance is, however, only now being realized. Contemporary field work by anthropologists and botanists, supported by phytochemical and pharmacological investigations, is constantly emphasizing the importance of narcotic use of members of the Solanaceae in the Americas. It is wholly probable that the end of new discoveries along these avenues of research has not yet arrived.

CONTENTS

INTRODUCTION

"Besides other disagreeable symptoms, these Solanaceae and their active elements, give rise to hallucinations and illusions of sight, hearing and taste, which differ, however, from those produced by the other Phantastica. They are not of an agreeable but, on the contrary, of a terrifying and distressful kind."

Louis Lewin "Phantastica" (1964)

Solanaceous plants have played major roles in the development of many cultures and civilizations. Following the discovery of the New World, however, American representatives of this family have drastically changed life in the Old World (Heiser, 1969).

Mexico and South America provided the chilli peppers; the Andes contributed the tomato. The potato earned its place as the staff of life of much of Europe's population and eventually became one of the dozen plants that feed mankind. Tobacco is today used around the world by more people than any other species of plant. Indeed, the New World solanaceous plants have won a merited place in the panorama of human affairs.

But this place of exalted invasion into the life of the world can hardly compare with their importance in pre-conquest American civilizations. Because the use of American medicinal, toxic, and narcotic species of this family has not spread to the Old World, their great significance in the New World cultures, both ancient and contemporary, is often not widely recognized.

It has long been recognized that solanaceous drugs played significant roles in the Americas. It was not, however, really until the writings of William E. Safford, some 50 or 60 years ago, that ethnobotanical research into the biodynamic species of this family in the New World was stimulated (Safford, 1917, 1920, 1921a).

In primitive societies of the Americas, man has mastered the utilization of many of the valuable properties of solanaceous species. No aspect of their use, however, has more deeply fascinated the American Indian than their mind-altering activities. He has bent these potent and often dangerous narcotic effects to his own peculiar uses in medicine, magic, and religion. He has valued especially the many hallucinogenic species provided by this family (Cooper, 1949; Emboden, 1972; Lewin, 1964; Safford, 1917, 1920; Uscátegui, 1959).

The New World is extraordinarily rich in plants employed as hallucinogens— more than 100 species are now known to be so used in North, Middle, and South America (Schultes, 1969–70, 1970; Schultes & Hofmann, 1973). New hallucinogens are continually being identified and some of those most recently discovered have belonged to the Solanaceae.

By far the greater number of New World hallucinogens owe their psychoactivity to alkaloids (Raffauf, 1970; Schultes, 1970; Schultes & Hofmann, 1973). The commonest of these alkaloids are indolic (Fodor, 1970; Saxton, 1960; Schultes, 1976; Taylor, 1966). The tropanes, characteristic of most of the solanaceous hallucinogens, are probably second to the indoles in importance. Members of the Solanaceae have played and still play distinctive roles in primitive societies in many parts of the Western Hemisphere, in spite of their recognized dangerous side-effects. More than 30 of the 85 genera in the family have been found to contain alkaloids (Lockwood, 1973a, b). Three major groups of alkaloids: the tropanes, pyridines and steroidal alkaloids; as well as minor ones, are present: the pyrrolidines, piperidines, and monoterpene derivatives (Gibbs, 1974; Hegnauer, 1973; Raffauf, 1970).

The genera of the Solanaceae known or suspected to be employed in the New World for their hallucinogenic effects are the following: *Brunfelsia, Cestrum, Datura* (including *Brugmansia*), *Iochroma, Juanulloa, Latua, Markea, Methysticodendron, Nicotiana, Petunia* and *Solandra*. These genera are grouped, according to the family classification offered by von Wettstein (1895), as follows: Datureae— *Datura, Methysticodendron, Solandra*; Cestreae—*Cestrum, Juanulloa, Markea, Nicotiana, Petunia*; Salpiglossideae—*Brunfelsia*; Solaneae—*Iochroma, Latua*.

In the following sections a discussion of our present knowledge of the hallucinogenic effect of the New World species in these genera is given.

DATUREAE

Datura Linnaeus

Stramonium

Datura stramonium Linnaeus, *Sp. Pl.*: 179 (1753).

Dutra

Datura discolor Bernhardi in *Tromms. N. J. Pharm.*, *26*: 149 (1833).
*Datura inoxia** Miller, *Gard. Dict.*, 8 ed. (1768) *Datura*, No. 5.
Datura kymatocarpa A. S. Barclay in *Bot. Mus. Leafl.*, *18*: 256 (1959).
Datura pruinosa Greenman in *Proc. Am. Acad.*, *33*: 486 (1898).
Datura quercifolia Humboldt, Bonpland et Kunth, *Nov. Gen. et Sp.*, *3*: 7 (1818).
Datura reburra A. S. Barclay in *Bot. Mus. Leafl.*, *18*: 258 (1959).
Datura wrightii Regel *Gartenfl.*, *8*: 193 (1859).

Ceratocaulis

Datura ceratocaula Ortega *Dec. Prim.*: 11 (1797).

Brugmansia

Datura arborea L. *Sp. Pl.*, *1*: 179 (1753).
Datura aurea (Lagerh.) Lagerheim, *Gartenfl.*, *42*: 33 (1893).
Datura candida (Pers.) Pasquale, *Cat. Ort. Bot. Nap.*: 36 (1867).
Datura dolichocarpa (Lagerh.) Safford in *J. Wash. Acad. Sci.*, *11*: 186 (1921).
Datura sanguinea Ruíz et Pavón, *Fl. Peruv.*, *2*: 15 (1799).
Datura suaveolens Humboldt et Bonpland ex Willdenow, *Enum. Hort. Berol.*: 227 (1809).
Datura vulcanicola A. S. Barclay in *Bot. Mus. Leafl.*, *18*: 260 (1959).

Undoubtedly *Datura* represents the most important group of solanaceous hallucinogens—at least, insofar as its widespread use is concerned. Many species have been employed from earliest times in both the Old and the New World (Barclay, 1959a; Lewin, 1964; Safford, 1920).

The botany of *Datura* is far from well understood (Avery, Satina, Rietsema, 1959; Safford, 1921b). Although the arborescent forms native to South America are frequently, and probably not without justification (Lockwood, 1973a, b), treated as a separate genus—*Brugmansia*—they will be considered, for purposes of this paper, as members of *Datura, sensu lato*. Under this treatment, the genus may be said to comprise from 15 to 20 species, divided usually into four sections: (1) *Stramonium* (with three species in the two hemispheres); (2) *Dutra* (containing eight species); (3) *Ceratocaulis* (with one Mexican species); (4) *Brugmansia* (South American trees, representing possibly six or seven species) (Willis, 1966).

Even though the hallucinogenic use of *Datura* goes back beyond written records in the Old World, the major centre lies in the New World, where many more species play significant roles in magic, medicine and religion in sundry cultures (Barclay, 1959a, b; Safford, 1917).

Datura seems to have been used in many parts of North America, but it found

* Although more commonly written *Datura innoxia*, the author is following Barclay in Botanical Museum Leaflets, Harvard University 18 (1959) 245 ff. in using *Datura inoxia*. (Eds.)

its centre in the South-west, where the hallucinogenically most important species is *D. inoxia*, formerly known as *D. meteloides*. *D. discolor* and *D. wrightii* are likewise employed. Many tribes utilized *Datura* ceremonially, especially in California, Arizona and New Mexico. Among the tribes known to use *Datura* are the Yokuts, Tubatulabals, several groups of Yumans, Papagos, Navajos, Tewas, Luiseños and Zunis (Schultes, 1969, 1969–70, 1972a, b, c, d).

The Zunis, who call *D. inoxia a-neg-la-kya*, employ it extensively as a narcotic, anaesthetic and, in the form of a poultice, for treating wounds and bruises. When their rain priests commune at night with the feathered kingdom, they put the powdered root in their eyes and chew the root to commune with the spirits of the dead who intercede for rain. The Zunis maintain that *D. inoxia* has a divine origin and that it still belongs only to the rain priests who alone are permitted to gather it. The Luiseños employ the plant in an initiation rite, administering a decoction to boys who dance in their intoxication, screaming like animals until finally they succumb to the deadening effects of the narcotic and are carried off in a stupor; the day following, the boys are instructed in certain mysteries connected with passage from childhood to manhood.

The Yumans take *D. inoxia* to induce dreams and gain occult powers which enable them to prophesy the future. They chew the roots or drink a tea of the leaves, exercising great care not to take a lethal dose. Many other tribes use *Datura* in the belief that they may enlist supernatural help through the drug, especially in the acquisition of secret knowledge. The Yokuts, who call this species *ta-nai*, take the seeds only once in a lifetime, in an early spring ceremony designed to ensure future good health and long life to adolescent children of both sexes. Boys who are seeking supernatural powers and who are studying to be shamans must under-go the intoxication once a year.

The Indians in other parts of North America are believed to have used *Datura*. Even the Algonquins and other tribes of the eastern woodlands employed *Datura stramonium**—the jimson weed or thorn apple—as the principal ingredient in an inebriating medicine called *wysoccan*. It was given to youths, who were confined for long periods before their adolescent initiation: "... they became stark, staring mad, in which raving condition they are kept eighteen or twenty days. ... These poor creatures ... perfectly lose the remembrance of all former things, even of their parents, their treasure and their language. When the doctors find that they have drunk sufficiently of the *wysoccan* ... they gradually restore them to their senses again. ..." They are said to "unlive their former lives" and commence manhood by losing the memory of ever having been boys (Schultes, 1969–70).

The importance of *Datura* in Mexico—both pre-Conquest and modern—was even greater and goes far back into the earliest history of the region where it was and is valued both as a medicine and a narcotic. The earliest New World herbal, the *Badianus Manuscript*, written in Aztec Mexico in 1552, illustrated *D. inoxia* or *tolohuaxihuitl* and detailed its medicinal properties "contra laternum dolorem" (Barclay, 1959b; De la Cruz, 1964; Emmart, 1940). One of the earliest accurate accounts was given by Hernández (1651), who described *D. inoxia*—the *toloatzin* of the Aztecs, *toloache* of modern Mexico—and listed its many therapeutic uses,

---

* There has been much disagreement as to whether *Datura stramonium* is native to the Old World or New. Modern students tend to accept it as indigenous to the Western Hemisphere (Barclay, 1959a, b).

especially in poultices as an anodyne, warning, however, that excessive applications could drive the patient to madness and "various and vain imaginations".

The use of toloache has persisted to the present time in Mexico. The modern Tarahumares, for example, add *Datura inoxia* or *tikuwari* to *tesquino* (a fermented drink prepared from sprouted maize) to make it strong, while the roots, seeds and leaves are the basis of a beverage employed ceremonially to promote visions and, taken by Tarahumare medicine-men, to help diagnose disease. Some modern Mexican Indians, however, consider toloache to be an hallucinogen which, unlike the peyote cactus, is inhabited by a malevolent spirit (Bye, 1976).

Several other species of *Datura—D. discolor*, *D. wrightii* and probably also *D. pruinosa*, *D. quercifolia*, *D. kymatocarpa* and *D. reburra*—are similarly used in Mexico (Barclay, 1959a, b).

An extremely interesting Mexican species is *Datura ceratocaula*, a fleshy plant with thickish, forking stems and growing in marshy and shallow waters. It has very marked narcotic properties which one of its local vernacular names—*torna-loco* ("maddening plant")—recognizes. The ancient Mexicans, who referred to it as "Sister of Ololuiqui*," held it to be very sacred. Priests about to administer it as a holy medicine were said to address it reverently before using it (Safford, 1917, 1920).

In South America the indigenous species of *Datura* are arborescent (*Brugmansia*, Lockwood, 1973a). All of the species, however, seem to be cultigens unknown in the wild and, although they bear fertile seeds—albeit often irregularly—they are usually propagated vegetatively. Handsome trees with large, showy flowers and now highly valued in horticulture, at least some of them seem to be chromosomally aberrant. They show a high degree of variability. It is apparent that, mainly because of their medicinal and narcotic properties, they have long been associated with man. "The variability expressed in the tree-daturas as a group has been enhanced through their cultivation by many native peoples: in fact, their absence from any natural vegetation implies that their recent evolution has taken place entirely under man's influence." (Bristol, 1965, 1966, 1969).

Not only is there the suspicion that all tree-daturas may be cultigens, but furthermore these plants offer complex biological problems resulting directly or indirectly from close association with man over the millenia (Bristol, 1969; Schultes, 1972a-d; Schultes & Hofmann, 1973).

Although there is still disagreement concerning their classification and interrelationships, we may, for the purpose of this paper, continue to accept seven species: *Datura arborea*, *D. aurea*, *D. candida*, *D. dolichocarpa*, *D. sanguinea* and *D. vulcanicola* in the Andean highlands from Colombia to Chile; and *D. suaveolens* in the warmer lowlands (Schultes, 1970). A recent study (see Fig. 8.1) has suggested that there are only three species on the basis of floral types—*D. candida*, *D. sanguinea* and *D. suaveolens*—and a number of cultivars of these species (Bristol, 1969). A more recent and exhaustive taxonomic revision of the tree-daturas (which were treated as generically distinct as *Brugmansia*) has suggested the acceptance of five species: *B. arborea*, *B. versicolor*, *B. suaveolens*, *B. aurea* and *B. sanguinea* and a number of hybrids and cultivars (Lockwood, 1973b).

The tree-daturas are variously known in South America as *borrachero*, *haucacachu*, *huanto*, *chamico*, *campanilla*, *floripondio*, *maicoa*, *tonga*, and *toa*. There are

* Ololuiqui was the very sacred hallucinogenic morning glory, *Rivea corymbosa* (Schultes, 1941).

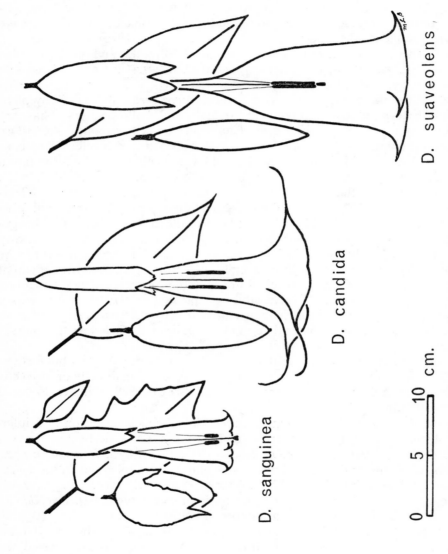

Figure 8.1. Comparison of leaf, flower and fruit in *Datura sanguinea, D. candida* and *D. suaveolens.*

many other names in the indigenous languages. The use of tree-daturas as hallucinogens tends to have a western South American distribution.

Throughout the Andes, with the exception of southern-most Chile, the mode of
preparation and use of *Datura* differs from tribe to tribe and from region to region.
Most often the drug is taken in the form of powdered seeds dropped into fermented chicha or as an infusion; or leaves and twigs may be utilized. In some
areas—amongst the Sionas of the Colombian Putumayo, for example—leaves of
Datura may be added to the yajé-drink prepared from *Banisteriopsis* to fortify its
hallucinogenic effects.

While *Datura suaveolens* is recognized as poisonous and is employed on a
limited basis as a medicine in the Amazon Valley, where it is called *toa*, the use of
this tree-datura as an hallucinogen appears to be limited to the westernmost part
of the continent, on the eastern slopes of the Andes and in the northern forested
parts of the Pacific coastal regions.

Although the tree-daturas are widely employed, the literature is surprisingly
deficient in reliable reports. There are actually only a few tribes definitely known
to use *Datura*: the Chibchas, Chocós, Inganos, Kamsás, Sionas, Kofáns in
Colombia; the Quechuas of Bolivia, Ecuador and Peru; the Mapuche-Huilliches
of Chile; and the Canelos, Piojes, Omaguas, Jívaros and Záparos of eastern
Ecuador and Peru. Amongst some of the western Amazonian Indians of Ecuador—
as with the Mapuche-Huilliches of Chile—*Datura* (probably *D. candida* and
D. sanguinea) is valued as a correctional measure for unruly children (Karsten,
1926, 1935, 1936). The Jívaros expect the spirits of their ancestors to speak to
and admonish the children during the intoxication and hallucinations (Harner,
1972, 1973). The Chibchas of pre-conquest Bogotá administered *Datura* to
wives and slaves of dead warriors or chieftans to induce in them a state
of stupor before being buried alive with their husbands (Safford, 1917). *D.
sanguinea* was a sacred hallucinogen employed in the Temple of the Sun at
Sogamoso in northern Colombia in the Chibcha empire (Castellanos, 1886;
Safford, 1917).

The Jívaros, who call the white-flowered *Datura* (probably *D. suaveolens*)
maikoa, ingest the raw sap from the green bark of the stems and are affected
within minutes. They distinguish six "kinds" of *maikoa*. An adult must always be
present to "encourage", to provide psychological support and physically to control
the drug taker should he become delirious or highly agitated. The Jívaros take
maikoa to seek an *arutam*—a soul. *Arutam* is a term for a vision that lasts for less
than a minute, while the *arutam*-soul is eternal once it is created, and a Jívaro who
possesses one cannot be killed by violence, poison, or witchcraft, although he may
die from diseases. During the course of an intoxication, the *arutam*-seeker may
awake during the night to find that the stars have gone from the sky, the earth
shakes in movement, great winds with thunder and lightning sweep the forest.
The *arutam*, usually frightening, may appear as a vision of jaguars or anacondas,
or it may take the form simply of a disembodied human head or a ball of fire
(Harner, 1972, 1973).

Intoxication from the tree-daturas is marked usually by initial effects so furious
that the partaker must be restrained, pending the onset of a deep, disturbed sleep,
during which hallucinations are induced. These are interpreted as spirit visitations

enabling the shaman to diagnose disease, discover theories and prophesy the future. It is definitely not a pleasant experience (Goodman & Gilman, 1955; Lewin, 1964).

As in North America, information on the exact species used by South American tribes for special purposes is rarely available. With the lack of voucher specimens, the species involved in each instance must usually be guessed from phyto-geographic or ecological reasoning or, perhaps, on occasion, from a vernacular name. Since all species contain similar or the same tropane alkaloids—varying only in relative concentrations—the usual lack of voucher specimens does not pose the serious problems that it might with some other hallucinogenic plants. Furthermore, in regions where there are several species, the natives recognize their close relationship and tend to take advantage of all of the species as medicines and hallucinogens.

The classification of the tree-daturas is complicated in some areas of the Andes by the presence of curious atrophied "races" of some of the species. These "races" or "strains" are valued by the natives because of their bizarre appearance which makes them attractive in magical ritual or because of their differing physiological effects, due, presumably, to varying chemical composition. These "races" are particularly numerous in the highland Valley of Sibundoy in southern Colombia (altitude about 8500 ft) where the natives propagate them vegetatively by planting pieces of stem in damp soil. Consequently, they represent virtually separate clones and have very definite native names: *borrachero buyés; borrachero dientes; borrachero ocre; borrachero biangán; borrachero amarón; borrachero salamán; borrachero quinde; borrachero munchira; borrachero andrés.* They have been botanically designated as cultivars of *Datura candida* (see Figs 8.1., 8.2., 8.3. and 8.4.). In Sibundoy, Kamsá medicine men employ all of these clones of *Datura candida*, as well as *D. dolichocarpa* and *D. sanguinea* (Fig. 8.5), as medicines and narcotics. Some of them are so highly atrophied in their vegetative parts that it would be difficult to assign them to species were not their flowers more or less identifiable (Barclay, 1959a, b; Bristol, 1966, 1969; Lockwood, 1973a, b; Schultes, 1970; Schultes & Hofmann, 1973).

There has been as yet no satisfactory explanation of why there should be such a concentration in this one high locality of so much monstrous deformity. It is clear that the interest of the natives has had a part in the preservation of these aber-rancies. But how did they arise? One idea holds that the deformities may have been induced by viral infection. Another idea is based on Blakeslee's work that indicates in the herbaceous daturas ". . . a great range of variability and the spontaneous appearance of many unusual characteristics. Of the 541 gene muta-tions encountered, 72 appeared following heating, wounding or ageing or spontan-eously in nature. Recessive genes controlling leaf shape, flower size, shape and colour and fruit form are amongst those uncovered. It is entirely possible that these simple recessive genes affecting taxonomically significant characters are present also in the tree-daturas" (Barclay, 1959a, b).

One result of intensive field studies of these cultivars has led to the belief that ". . . each cultivar is genetically distinct, quite apart from the possible influence of viruses on the leaves. Differences [occur] . . . in the morphology of the flowers and fruits, in the incidence of chromosomal inversions and in the amount of aborted pollen. Some of the unique leaves are also reflections of genetic uniqueness and not of virus infection'' (Bristol, 1966, 1969).

The principal active constituents in all species of *Datura* are the tropane alkaloids, although cuscohygrine and nicotine, which are not tropanes, have been isolated from the Old World species, *D. metel*. There is little qualitative difference among the species of *Datura*

D. inoxia contains scopolamine as its chief component, but minor alkaloids are also present: meteloidine, hyoscyamine, norhyoscyamine, norscopolamine (Gibbs, 1974).

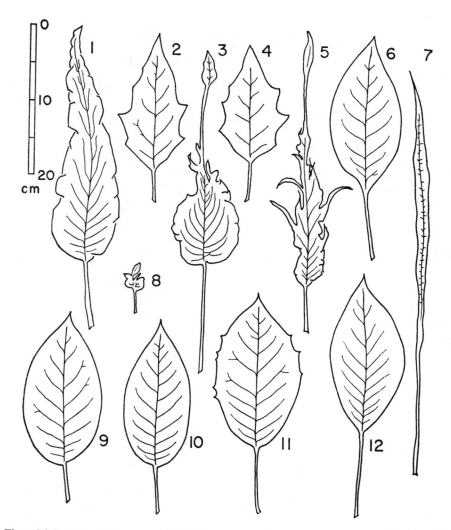

Figure 8.2. Representative leaves of all Sibundoy *Datura* cultivars, with collection numbers indicated: 1, 'Amarón, *564:* 2, 'Guamuco', *1420:* 3, 'Salamán', *1432:* 4, 'Sangre', *1309:* 5, 'Quinde', *1333:* 6, 'Andrés', *1314:* 7, 'Culebra', *1112:* 8, 'Munchira', *1268:* 9, 'Buyes', *1388:* 10, 'Biangán', *890:* 11, 'Dientes', *1447:* 12, 'Ocre', *1267.*

Recent systematic studies of the tree-daturas with authenticated botanical material indicate that scopolamine is the principal alkaloid in the aerial parts of *D. candida*, together with atropine, norscopolamine, oscine, meteloidine and noratropine. The roots contained 3α, 6β-ditigloyloxytropane-7β-ol, scopolamine, norscopolamine, 3α-tigloyloxytropane, meteloidine, oscine, atropine, noratropine and tropine. The same alkaloids are present in several cultivars of *D. candida*—the highly atrophied "races"—from Sibundoy (Evans *et al.*, 1959, 1965; Bristol *et al.*, 1969).

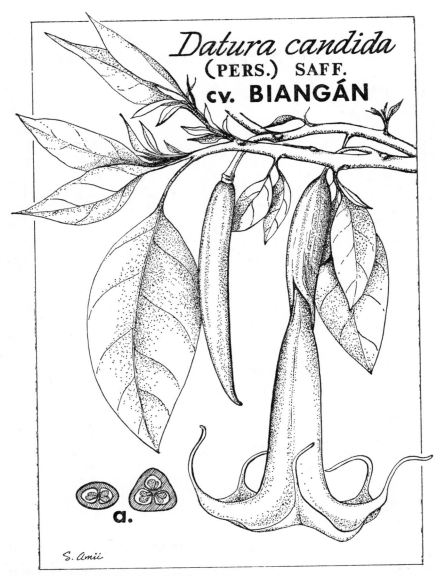

Figure 8.3. *Datura candida*. 'Normal' leaf type.

The alkaloid content varies usually between 0.3 and 0.55 % of dry plant material, with scopolamine between 31 % and 60 % of the total alkaloid content.

Methysticodendron R. E. Schultes

Methysticodendron amesianum R. E. Schultes in *Bot. Mus. Leafl., 17: 2* (1955). Known only as cultivated trees from one locality in the Colombian Andes, this

Figure 8.4. *Datura candida*. One of the cultivars from southern Colombia (Sibundoy) with deformed leaves. (See also Fig. 8.2.)

plant is an anomalous, poorly understood entity which has been described as a monotypic genus but which may represent an extremely atrophied form of a tree-datura (Fig. 8.6.).

The Kamsá and Ingano Indians of the high mountain-girt Valley of Sibundoy in southern Colombia cultivate a tree known locally as *culebra-borrachero* or, in Kamsá, *mitskway-borrachero*, both of which terms mean "snake intoxicant". It is propagated vegetatively and is reserved for very special medicinal and hallucinogenic uses. The tree has been named *Methysticodendron amesianum*.

Figure 8.5. *Datura sanguinea* R & P. cv. Guamuco. Flowering and fruiting branches, × ½

Obviously closely related to the tree-daturas, it has also been designated as a clone of *Datura candida*: *D. candida* cv. Culebra (Bristol, 1969). Vegetatively, it is extraordinarily atrophied, reduced to very narrowly ligulate, marginally irregular leaves. The flowers are even more strongly atrophied, being bilocular and tri-locular with a deeply lobate corolla and with what may be fundamental departures from a typical *Datura* morphology in their ovary and styles. If it be an atrophied tree-datura, its flowers are so altered as to make it impossible to assign it to any known species. In addition to suggestions that *Methysticodendron* is a virally

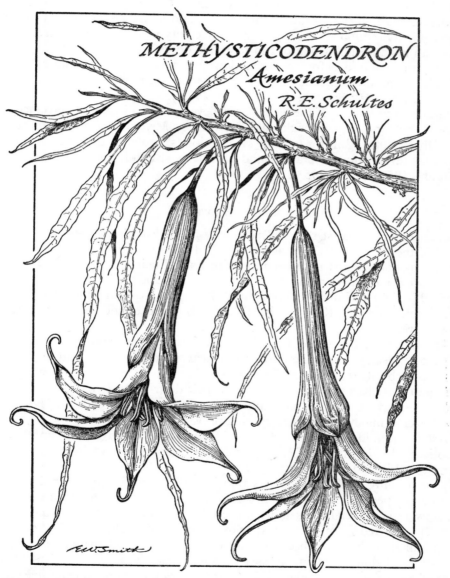

Figure 8.6. *Methysticodendron amesianum*, showing linear leaves and corolla lobes separated almost to the base. Sometimes classed as *Datura candida* cv. Culebra (see Fig. 8.2., leaf 7).

induced monstrosity, it has been theorized that it may be "the result of the action of a single pleiotropic gene mutation . . . a monstrosity of some *Datura* species of subg. *Brugmansia* . . ." (Barclay, 1959a, b). Still another possibility has been offered: that it is ". . . a remarkable case of adesmy, a tendency which is recorded from various Solanaceae" (van Steenis, 1957). Obviously, there remains much to investigate in the evolution of the tree-daturas and the related *Methysticodendron*. Consequently, it may be preferable at the present time to continue to use the generic term *Methysticodendron*.

The leaves and stems of *M. amesianum* contain 0.3 % of total alkaloids; 80 % is scopolamine, a higher percentage than that found in *Datura candida*, which may account for its reputation for higher psychoactivity amongst the Indians. Scopolamine can produce hallucinations in tolerable doses, whereas near toxic doses of atropine are required before hallucinations are experienced. The minor alkaloids of *Methysticodendron* include atropine and two as yet unidentified bases (Pachter & Hopkinson, 1960; Schultes & Hofmann, 1973).

In addition to the use of the leaves as a beverage for divination, prophecy and diagnosis of especially difficult cases of illness, *Methysticodendron* serves as a potent therapeutic agent for treating swollen joints (probably arthritis), combating chills and fevers, muscle cramps and erysipelas (Bristol, 1969; Schultes, 1955; Theilkuhl, 1957; Yépes, 1953).

Solandra Swartz

Recent reports suggest that *Solandra* may be, at least locally, one of the most sacred of the Mexican hallucinogens (Knab, 1977).

The Huichol Indians anthropomorphize an hallucinogenic plant as the hero Kieri Tewiyari. *Kieri* is a plant "from ancient times", and the story of Kieri Tewiyari is recited in the Huichol peyote ceremony. The plant grows in remote rocky places, where it retreated after Kieri Tewiyari's defeat by the deer god and the culture hero Kauyumarie.

The plant *Kieri* has white, funnel-shaped flowers and a spine-covered fruit. It lures the careless, enticing them to taste its leaves, flowers, roots and seeds. Those who succumb either die or suffer insanity through its bewitching powers, and, believing themselves to be birds, jump to their destruction from cliffs. If they do not kill themselves, they eat more and enter an eternal sleep. Only the shaman can handle such a malevolent spirit, offering prayers and gifts.

Kieri has been identified as a species of *Solandra*. There have been reports of the narcotic use by the Aztecs of *hueipatli—Solandra guerrerensis—*in central Mexico, where it is still medicinally employed.

The taxonomy of *Solandra* is very poorly understood, but there are believed to be some ten species ranging from Mexico and the West Indies to tropical America. The flowers superficially resemble those of the related genus *Datura*. Atropine has been reported from the genus; hyoscyamine and norhyoscyamine and other tropane alkaloids have also been isolated; cyanogenesis has likewise been registered. The total alkaloid content of several species is about 0.15 %. The alkaloid range of *Solandra* is not unlike the range in *Datura* but is not so extensive (Gibbs, 1974).

CESTREAE

Cestrum Linnaeus

Cestrum laevigatum Schlechtendal in *Linnaea, 7:* 59 (1832).

A species of *Cestrum*, probably *C. laevigatum*, a coastal Brazilian species commonly known as *dama da noite*, is reported to be used by seafaring personnel as a substitute for *Cannabis* (Schultes & Hofmann, 1973).

The alkaloids, parquine and solasonine, have been isolated from the toxic Chilean *C. parqui*. The characteristic constituents of the genus are saponines. Gitogenine and digitogenine have been reported from both *C. laevigatum* and *C. parqui*, and the alkaloid solasonine has been found in *C. parqui*. Solasonine is a glycosidic steroidal alkaloid. Nothing is known about the potential hallucinogenic properties of these several compounds (Gibbs, 1974; Hegnauer, 1973).

Juanulloa Ruíz et Pavón

Juanulloa ochracea Cuatrecasas in *Brittonia, 10:* 148 (1958).

Juanulloa ochracea is reputedly called *ayahuasca* in the Colombian Putumayo. This Quechua term *ayahuasca* normally refers in Peru and Ecuador to the hallucinogenic drink prepared basically from *Banisteriopsis caapi* or *B. inebrians* (Schultes, 1972a–d).

The drink made from the bark of these malpighiaceous species often has one or several additives in this region. The use of the term *ayahuasca* suggests that *Juanulloa ochracea* may on occasion be added to the hallucinogenic drink to alter or intensify its narcotic effects, or even that it may be used directly as the source of a narcotic drink. The trunk and leaves of this epiphytic shrub are employed in decoction to treat open wounds.

The alkaloid parquine has been reported from *Juanulloa*, a genus of some 12 species native from Mexico to tropical America (Raffauf, 1970).

Markea L. C. Richard

Markea formicarium Dammer in *Bot. Jb., 37:* 170 (1905).

There is evidence (*Schultes & Cabrera 12426*) that leaves of *M. formicarium* may on occasion be added to the hallucinogenic drink prepared from *Banisteriopsis caapi* and *B. inebrians* by the Makuna Indians of the middle Río Apaporis (in the region of Raudal Jerijerimo) in the Colombian Amazonia.

There are 18 species of *Markea* native to the American tropics. No phytochemical studies are apparently available for the genus.

Nicotiana Linnaeus

Nicotiana rustica Linnaeus, *Sp. Pl.:* 180 (1753).

Nicotiana tabacum Linnaeus, *loc. cit.:* 180.

Nicotiana is a genus of some 46 species of North and South America, Australia and Polynesia.

There seems to be growing evidence that *Nicotiana* may occasionally be employed for truly hallucinogenic effects (Furst, 1976; Janiger & Dobkin, 1976; Schultes, 1972a–d; Wilbert, 1972).

Everywhere in the New World, tobacco was considered a special gift to man from the gods, and it was employed in sacred magical and medicinal contexts. It was and apparently still is on occasion used to induce mystical states or the characteristically shamanistic ecstatic trance, that we commonly associate only with the vegetal hallucinogens. The Aztecs took *picietl* (*N. rustica*) to enlist the help of the supernatural world or to place themselves in "mythic time. . . when everything was possible" (Wilbert, 1972).

Tobacco was undoubtedly more widely used and for a greater variety of medicinal and magical purposes than any other plant.

Although perhaps as many as half a dozen species of *Nicotiana* were used by the American Indians, two species were very important (Goodspeed, 1954): *N. rustica*, of Mexico and North America; and *N. tabacum*, primarily of South America and the West Indies. *N. rustica*, with a much higher content of nicotine than *N. tabacum*, was often used in a more significant way in religious, magical and medicinal contexts. The Huichols, for example, consider *N. rustica* as "the proper tobacco of the shaman", and the Seneca of New York called it "real tobacco".

Tobacco was employed by American aborigines in a great variety of ways: it was smoked, drunk, licked, sucked, chewed, eaten, snuffed and even employed rectally in the form of enemas. This last method, still employed among certain South American Indians, was widely used by the ancient Peruvians and Mayas (Cooper, 1949; Mariano Ramírez, 1965).

The Warrao of the lower Orinoco in Venezuela use no other hallucinogen but tobacco. Tobacco smoke provides the shaman's means of communication with the spiritual world. Shamans may consume up to 30 large cigars, inhaling the smoke, to "travel to their respective master spirits on celestial bridges constructed by tobacco smoke" (Wilbert, 1972).

The literature of the New World cultures is replete with references to the importance of tobacco in the induction of the trance state and hallucinations, in divination, oracular witchcraft and medicinal rituals, but this literature has yet to be objectively evaluated and interpreted.

While the major active constituents in *N. rustica* and *N. tabacum* are nicotine and other pyridine and pyrrolidine alkaloids, recent reports indicate the presence in tobacco of β-carboline compounds such as harman and norharman, related to the known hallucinogenic harmine, harmaline and tetrahydroharmine. How prevalent these β-carbolines may be in commonly used types of tobacco is still not clear. Pyrosynthesis during burning is said to increase the concentration of harman and norharman from 40 to 100 times the concentration in the fresh leaves and indicates that these compounds may be formed from tryptophane. Might it not be that these β-carbolines alone or else in combination with the pyridines or other constituents of tobacco could act psychoactively as hallucinogens, especially when experienced in the proper context and against cultural patterning? (Janiger & Dobkin, 1976; Poindexter & Carpenter, 1962; Testa & Testa, 1965).

Tobacco is one of the oldest of the New World cultigens. It was spread nearly throughout the Americas by the time of the Conquest. Perhaps its ethnobotanical study is, however, just beginning and its role in primitive American societies has not yet been fully appreciated.

Petunia Jussieu

Petunia sp. Lindley in *Bot. Reg.*, *19:* t. 1626 (1833).

One of the most recently indicated South American hallucinogens is a species of *Petunia* or *shanín* which is employed in highland Ecuador to induce the sensation of flying or levitation (Haro Alvear, 1971).

Some 40 species of *Petunia*, native to South America and the warmer parts of North America, are recognized. Psychoactive principles apparently have not yet been found in *Petunia*, although the genus, a member of the Cestreae, might be expected to contain alkaloids.

SALPIGLOSSIDEAE

Brunfelsia Linnaeus

Brunfelsia chiricaspi Plowman in *Bot. Mus. Leafl.*, *23:* 255 (1973).

B. grandiflora D. Don in *Edinb. New Phil. J.;* 86 (July 1829).

B. grandiflora D. Don subsp. *schultesii* Plowman, *loc. cit.:* 259.

The genus *Brunfelsia* comprises about 40 species native to the West Indies and South America (Plowman, 1973b).

Many of the species have long played major roles in American ethnomedicine (Plowman, 1973a). The fruits of *Brunfelsia americana* are considered throughout the West Indies to be efficacious against diarrhoea. *B. uniflora* of eastern Brazil, where it is known as *manaca*, is widely esteemed in the treatment of syphilis and rheumatism, as a diuretic, emetic, abortifacient and purgative, among other uses. Another species—*B. mire*—is valued as a vermifuge. In the eastern part of the Amazon, *B. guianensis*, likewise called *manaca*, is considered to be antisyphilitic, antirheumatic, depurative and, in high doses, toxic. Natives of the western Amazon employ *B. chiricaspi* (Fig. 8.7.) and *B. grandiflora* (Fig. 8.8.) for a large number of medicinal purposes, chief of which seems to be in the treatment of rheumatism, arthritis, and fevers. The common name *chiricaspi* means "cold tree" and refers to the sensation of cold and consequent shivering brought on by drinking decoctions of these plants. *B. chiricaspi* is known only in the wild and is preferred as a medicine over the commonly cultivated *B. grandiflora* subsp. *schultesii*. Among the many vernacular names for *B. grandiflora* are *chiricaspi* (Colombia), *sanango* and *chiric-sanango* (Peru) (Pinkley, 1973).

Because of the confused state of nomenclature in *Brunfelsia*, *B. grandiflora* has often been reported in the ethnopharmacologic literature as *B. bonodora* or *B. maritima*.

Evidence for the narcotic use of *Brunfelsia* is real but still needs detailed field study in order to ascertain its extent. Annotations on herbarium collections from the western Amazon of Colombia and Peru attest to the value attached to *B. chiricaspi* and *B. grandiflora* as hallucinogens. The occasional use in Colombia and Ecuador of the vernacular name *borrachero* ("intoxicant") lends support to their rather wide recognition as an inebriant. Kofán and Jívaro medicine men of Ecuador take an infusion of *B. grandiflora* when diagnosing disease and as an additive to the yajé drink prepared from several species of the malpighiaceous *Banisteriopsis* (Schultes, 1972a–d; Schultes & Hofmann, 1973).

An obscure report that the Kachinaua Indians of the Brazilian Amazon make a psychotomimetic drink from what has been reported as *B. tastevinii* needs corroboration. In view of the similar use of *Brunfelsia* in other parts of the Amazon, this report is credible, especially since *B. tastevinii* is considered to be a synonym of *B. grandiflora*.

The chemistry of *Brunfelsia* is sorely in need of investigation (Schultes & Hofmann, 1973). Notwithstanding the importance of the several species in ethnomedicine and their observable effects on the human body, critical chemical studies have not been carried out. The older literature mentions isolation of alkaloids with such names as "franciscaine", "manacine" and "brunfelsine", but none seems to have been structurally elucidated. Preliminary tests for alkaloids have customarily been done on dried material, and volatile constituents might consequently be missed. It may be significant that a spot test (Dragendorff reagent) on fresh material of the type tree of *B. grandiflora* subsp. *schultesii* (*Schultes, Raffauf et*

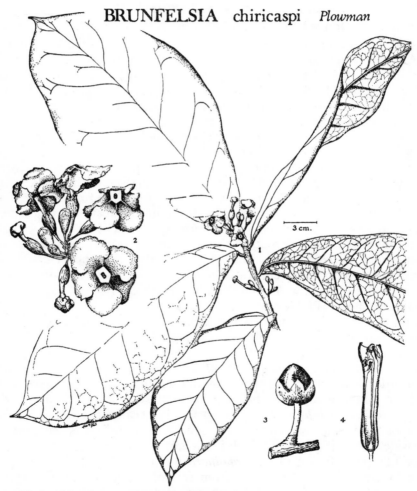

BRUNFELSIA chiricaspi   *Plowman*

Figure 8.7. *Brunfelsia chiricaspi*, showing leaf, flower and fruit.

Soejarto 24108) indicated strongly the presence of alkaloids. Cuscohygrine has been isolated from roots of a *Brunfelsia* (W. C. Evans, pers. comm.).

A recent report, on apparently vouchered material of *Brunfelsia uniflora*, *B. pauciflora* and *B. brasiliensis*, indicates no alkaloids but isolation of the non-nitrogenous scopoletin (6-methoxy-7-hydroxycoumarine). This constituent is not, however, known to be psychoactive (Mors & Ribeiro, 1957).

BRUNFELSIA grandiflora *D. Don*

Figure 8.8. *Brunfelsia grandiflora*, subsp. *grandiflora* and subsp. *schultesii:* 1, seed; 2, embryo.

SOLANEAE

Iochroma Bentham

Iochroma fuchsioides (H. & B.) Miers in *Hooker's Lond. J. Bot.*, 7: 345 (1848).

There are probably two dozen species of *Iochroma* in South America. Some of the species are ill-defined, and a taxonomic revision based in great part on field observations is overdue.

There have been vague reports that *Iochroma*, especially *I. fuchsioides*, is occasionally employed by medicine men in the Andes of Colombia to induce hallucinations (Schultes, 1970; Schultes & Hofmann, 1973). *I. fuchsioides*, the common name of which is *borrachero* ("intoxicant"), is a very frequently cultivated element in the medicinal gardens of shamans in Sibundoy, Colombia. Reports of its use as an hallucinogen have been made on herbarium collections since 1942, and its special care in cultivation argues for a unique position in the medicine man's pharmacopoeia. Recently, direct reports from a widely known medicine man of the Kamsá tribe have indicated that, in addition to its many medicinal uses (i.e., to combat dysentery and "heart weakness"), it is taken in the form of a tea in cases of "extremely difficult" divination or diagnosis (Schultes, 1977).

Chemical studies of *Iochroma* apparently have never been published.

Latua Philippi

Latua pubiflora (Griseb.) Baillon, *Hist. Plant. 9*: 334 (1888).

Latua is a monotypic genus endemic to the coastal mountains of southern Chile, apparently nowhere abundant (Fig. 8.9.).

This rare and interesting intoxicant was first reported more than a century ago. A contemporary report on its properties, made by Philippi in 1861, stated in part: "... the Indians of the Province of Valdivia possess a secret way of producing insanity with a poisonous plant for a long or short time, depending on the dose. It is considered with great secretiveness ... a missionary in Daglipulli succeeded in learning that the plant is a tall shrub called *latué* which grows in the forests of the coastal mountains. He was finally able to obtain a branch of it. ... Later, I learned the details of *latué* from Señor Juan Renous ... he did tell me of several cases of intentional and unintentional poisonings. The latter occur quite readily since ... the shrub is so very similar to *tayu* whose bark is used externally and internally ... for bruises, blows ..., etc. He related to me ... the following case. ... One of his woodcutters had suffered a strong blow ... and went into the forest to get some bark of *tayu* for it. He took instead *latué* and drank a concoction of this poison. He became insane almost immediately and wandered into the mountains. He was found three days later in an unconscious state. Several days were required for his recovery, although he suffered severe headaches for several months" (Plowman *et al.*, 1971).

There are other reports of accidental poisonings and of the malevolent use of *L. pubiflora* in the past, and contemporary inhabitants tell of deliberate poisonings resulting in insanity or death.

The effects of a decoction of *L. pubiflora* resemble those of belladonna intoxication: dilation of the pupils, dryness and later frothing of the mouth, mental disturbances, convulsions, delirium and hallucinations. Symptoms may occur at once following ingestion or even 24 hours later (Plowman *et al.*, 1971).

Local ethnopharmacology has developed several presumed antidotes to *Latua* poisoning. These antidotes include *hierba mora* (*Solanum nigrum*) of the Solanaceae, *culle* (*Oxalis* sp.) of the Oxalidaceae and the fruit of *espino negro* (*Raphithamnus spinosus*) of the Verbenaceae (Plowman *et al.*, 1971).

The employment of *Latua pubiflora* in magic and witchcraft, although well known, has only recently been thoroughly described. The *machis* or medicine-men of the Mapuche-Huilliche Indians of Chile who used *latué* (usually a woman or male transvestite) cured illness by discovering and exorcising the malevolent spirits causing the hexing of the patient. Many of the *machis* are nervous persons or even epileptic, readily disposed to trance states. They communicate with the spirit

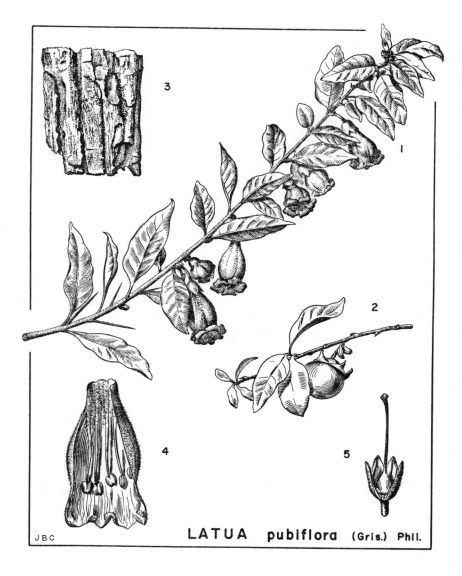

LATUA pubiflora (Gris.) Phil.

Figure 8.9. *Latua pubiflora:* 1, flowering branch; 2, fruiting branch; 3, bark; 4, dissected corolla; 5, calyx with one lobe removed showing gynoecium.

world through a number of methods. They are excellently trained in native ethno-pharmacology, and the use of narcotic plants is one of the major methods employed in their divination. The machis, even during training, take hallucinogenic plants to attain states of unreal "reality". Tobacco is perhaps the most sacred of the machis' plants, and they possess several strongly intoxicating varieties. A species of *Datura* is likewise employed. It has only recently been revealed that *Latua pubiflora* also represents one of the machis' agents for attaining the proper state for their revelations. It is still not certain how widespread is the use of *latué* amongst co-temporary medicine-men, but it may be extensive (Plowman *et al.*, 1971).

As early as 1864, chemical work was carried out on *L. pubiflora*, but this investigation failed to disclose the presence of alkaloids (Vásquez, 1864). In 1914, atropine was isolated from the plant (Pouquet, 1914). Further analyses were reported in 1918 (Miranda, 1918). Atropine was isolated in 1959 (Silva & Mancinelli, 1959), but it was not until 1962, that another alkaloid, scopolamine, in addition to atropine, was reported: 0.08 % of scopolamine, 0.18 % of atropine, with the highest concentration of the alkaloids in the leaves (Bodendorf & Kummer, 1962). The most recent analysis, reporting the same two alkaloids with vouchered botanical specimens, finds the highest concentration of alkaloids in the stem; smaller amounts of the two bases were found in the seeds and leaves (Plowman *et al.*, 1971).

REFERENCES

AVERY, A. G., SATINA, S. & RIETSEMA, J., 1959. *Blakeslee: The genus* Datura. New York: Ronald Press.

BARCLAY, A. S., 1959a. New considerations in an old genus: *Datura. Botanical Museum Leaflets, 18:* 245–272.

BARCLAY, A. S., 1959b. *Studies in the genus* Datura *(Solanaceae) I. Taxonomy of Subgenus* Datura. Ph. D. Thesis, Harvard University Cambridge, Mass.

BODENDORF, K. & KUMMER, H., 1962. Ueber die Alkaloide in *Latua venenosa. Pharmazeutische Zentralhalle für Deutschland, 101:* 620–622.

BRISTOL, M. L., 1965. *Sibundoy Ethnobotany*. Ph. D. Thesis, Harvard University, Cambridge, Mass.

BRISTOL, M. L., 1966. Notes on the species of tree daturas. *Botanical Museum Leaflets, 21:* 229–248.

BRISTOL, M. L., 1969. Tree *Datura* drugs of the Colombian Sibundoy. *Botanical Museum Leaflets, 22:* 165–227.

BRISTOL, M. L., EVANS, W. C. & LAMPARD, J. F., 1969. The alkaloids of the genus *Datura*, section *Brugmansia*. Part VI. Tree-daturas (*Datura candida* cvs.) of the Colombian Sibundoy. *Lloydia, 32:* 123–130.

BYE, R. A., 1976. *Ethnoecology of the Tarahumara of Chihuahua, Mexico*. Ph. D. Thesis, Harvard University, Cambridge, Mass.

CASTELLANOS, Juan de, 1886. *Historia del Nuevo Reino de Granada, 1:* 65–66.

COOPER, J. M., 1949. Stimulants and narcotics. In J. S. Steward [Ed] *Handbook of South American Indians. Bulletin of the Bureau of American Ethnology;* No. 143, *5:* 528–558, U.S. Gov't. Printing Office, Washington, D.C.

DE LA CRUZ, M., 1964 *Libellus de Medicinalibus Indorum Herbis*. Instituto Mexicano del Seguro Social, Mexico.

EMBODEN, W. A., 1972. *Narcotic Plants*. New York: Macmillan.

EMMART, E. W., [transl.] 1940. *Badianus Manuscript (Codex Barberini). An Aztec herbal of 1552*. Baltimore: Johns Hopkins Press.

EVANS, W. C., MAJOR, V. A. & PETHAN, M., 1965. The alkaloids of the genus *Datura*, section *Brugmansia* III. *Datura sanguinea* R. and P. *Planta Medica, 13:* 353–358.

EVANS, W. C. & WELLENDORF, W., 1959. The alkaloids of the roots of *Datura. Journal of the Chemical Society.*, 1406.

FODOR, G., 1970. Tropane alkaloids. In S. W. Pelletier (Ed.), *Chemistry of the Alkaloids*. New York: Van Nostrand Reinhold.

FURST, P. T., 1976. *Hallucinogens and Culture*. San Francisco: Chandler & Sharp.

GIBBS, R. D., 1974. *Chemotaxonomy of Flowering Plants:* 1–4, Montreal: McGill-Queen's University Press.

GOODMAN, L. S. & GILMAN, A., 1955. *The Pharmacological Basis of Therapeutics*, 2nd Ed. New York: Macmillan.

GOODSPEED, T. H., 1954. *The Genus* Nicotiana. *Origins, Relationships and Evolution of Its Species in the Light of Their Distribtuion, Morphology and Cytogenetics*. Waltham, Mass.: Chronica Botanica.

HARNER, M. J., 1972. *The Jívaro*. New York: Doubleday/Natural History Press.

HARNER, M. J. (Ed.), 1973. *Hallucinogens and Shamanism*. New York: Oxford University Press.

HARO ALVEAR, S. L., 1971. Shamanismo y farmacopea en el Reino de Quito. *Contribuciones Instituto Ecuatoriano de Ciencias Naturales*, No. 75.

HEGNAUER, R., 1973. *Chemotaxonomie der Pflanzen, 6*, Basel: Birkhauser Verlag.

HEISER, C. B., 1969. *Nightshades—the Paradoxical Plants*. San Francisco: W. H. Freeman.

HERNÁNDEZ, F., 1651. *Nova Plantarum, Animalium et Mineralum Mexicanorum Historia* . . . Rome: B. Deuersini et Z. Masotti.

JANIGER, O. & DOBKIN de RIOS, M., 1976. Nicotiana an hallucinogen? *Economic Botany, 30:* 149–151.

KARSTEN, R., 1926. *The Civilization of the South American Indians*. New York: Alfred A. Knopf.

KARSTEN, R., 1935. The head-hunters of western Amazonas. *Soc. Scient. Fennica, Commentationes Humanarum Litterarum, 7*, No. 1, Helsinki.

KARSTEN, R., 1936. Arrow-poisons and narcotics in western Amazonas. *Ethnological Studies, 2:* 68–77.

KNAB, T., 1977. Notes concerning use of *Solandra* among the Huichol. *Economic Botany, 31:* 80–86.

LEWIN, L., 1964. *Phantastica: Narcotic and Stimulating Drugs—their Use and Abuse*. London: Routledge & Kegan Paul.

LOCKWOOD, T. E., 1973a. Generic recognition of *Brugmansia. Botanical Museum Leaflets, 23:* 273–284.

LOCKWOOD, T. E., 1973b. *A Taxonomic Revision of Brugmansia (Solanaceae)*. Ph. D. Thesis, Harvard University, Cambridge, Mass.

MARIANO RAMÍREZ, C., 1965. *Témas de Hipnosis*. Editorial Andrés Bello, Biblioteca de Estudios Médicos, Santiago, Chile.

MIRANDA, J. B., 1918. Estudio químico, fisiológico y terapeútico de la *Latua venenosa* (Palo de Brujo). *Actes de la Société Scientifique du Chili, 27*, pt. 3: 10–26.

MORS, W. B. & RIBEIRO, O., 1957. Occurrence of scopoletin in the genus *Brunfelsia, Journal of Organic Chemistry, 22:* 978–979.

PACHTER, I. J. & HOPKINSON, A. F., 1960. Note on the alkaloids of *Methysticodendron amesianum. Journal of the American Pharmaceutical Association (Scientific Edition), 49:* 621–622.

PHILIPPI, R. A., 1861. Descripción de un nuevo género de plantas de la familia de las Solanáceae. *Anales de la Universidad de Chile, 18* (3): 309–311.

PINKLEY, H. V., 1973. *The Ethnoecology of the Kofan Indians*. Ph.D. Thesis, Harvard University, Cambridge, Mass.

PLOWMAN, T. C., 1973a. Four new Brunfelsias from northwestern South America. *Botanical Museum Leaflets, 23:* 245–272.

PLOWMAN, T. C., 1973b. *The South American Species of Brunfelsia (Solanaceae)*. Ph. D. Thesis, Harvard University, Cambridge, Mass.

PLOWMAN, T., GYLLENHAAL, L. O., & LINDGREN J. E., 1971. *Latua pubiflora*, magic plant from southern Chile. *Botanical Museum Leaflets, 23:* 61–92.

POINDEXTER, E. H. & CARPENTER, R. D., 1962. Isolation of harman and norharman from cigarette smoke. *Chemistry and Industry*, 27 Jan, 1962: 176.

POUQUET, D., 1914. *Contribución al Estudio Químico del Latué*. Thesis, Curso de Quimica y Farmacia, Universidad Nacional, Santiago, Chile.

RAFFAUF, R. F., 1970. *A Handbook of Alkaloids and Alkaloid-containing Plants*. New York: Wiley-Interscience.

SAFFORD, W. E., 1917. Narcotic plants and stimulants of the ancient Americans. *Annual Report of the Smithsonian Institution 1916:* 387–424.

SAFFORD, W. E., 1920. Daturas of the Old World and New: an account of their narcotic properties and their use in oracular and initiatory ceremonies. *Annual Report of the Smithsonian Institution 1920:* 537–567.

SAFFORD, W. E., 1921a. *Datura*—an inviting genus for the study of heredity. *Journal of Heredity, 12:* 178–190.

SAFFORD, W. E., 1921b. Synopsis of the genus *Datura. Journal of the Washington Academy of Sciences, 11:* 173–189.

SAXTON, J. E., 1960. The indole alkaloids. In R. H. F. Manske (Ed.), *The Alkaloids*, 7, New York: Academic Press.

SCHULTES, R. E., 1941. *A Contribution to our Knowledge of Rivea corymbosa, the Narcotic ololiuqui of the Aztecs*. Harvard Botanical Museum, Cambridge, Mass.

SCHULTES, R. E., 1955. A new narcotic genus from the Amazon slope of the Colombian Andes. *Botanical Museum Leaflets, 17:* 1–11.

SCHULTES, R. E., 1969. Hallucinogens of plant origin. *Science, 163:* 245–254.

SCHULTES, R. E., 1969–1970. The plant kingdom and hallucinogens. *Bulletin of Narcotics, 21*, no. 3 (1969): 3–16; no. 4(1969): 15–27; *22*, no. 1(1970): 25–53.

SCHULTES, R. E., 1970. The botanical and chemical distribution of hallucinogens. *Annual Review of Plant Physiology, 21:* 571–594.

SCHULTES, R. E., 1972a. An overview of hallucinogens in the Western Hemisphere. In P. T. Furst (Ed.), *Flesh of the Gods:* 3–54. New York: Praeger Publisher.

SCHULTES, R. E., 1972b. De plantis toxicariis e Mundo Novo tropicale commentationes X. New data on the malpighiaceous narcotics of South America. *Botanical Museum Leaflets, 23:* 137–147.

SCHULTES, R. E., 1972c. De plantis toxicariis e Mundo Novo tropicale commentationes XI. The ethno-toxicological significance of additives to New World hallucinogens. *Plant Science Bulletin, 18:* 34–41.

SCHULTES, R. E., 1972d. The utilization of hallucinogens in primitive societies—use, misuse or abuse? In W. Keup (Ed.), *Drug Abuse:* 17–26. Springfield, Ill.: Charles C. Thomas.

SCHULTES, R. E., 1976. Indole alkaloids in plant hallucinogens. *Journal of Psychedelic Drugs, 8:* 7–25.

SCHULTES, R. E., 1977. A new hallucinogen from Andean Colombia: *Iochroma fuchsioides*. *Journal of Psychedelic Drugs, 9:* 45–49.

SCHULTES, R. E. & HOFMANN, A., 1973. The Botany and Chemistry of Hallucinogens. Springfield, Ill.: Charles C. Thomas.

SILVA, M. & MANCINELLI, P., 1959. Atropina en *Latua pubiflora* (Griseb.) Phil. *Boletin de la Sociedad Chilena de Química, 9:* 49–50.

VAN STEENIS, C. G. G. J., 1957. Specific and infraspecific delimitation. *Flora Malesiana (Ser. 1), 5:* dxvii—ccxxxiii, dxxxvi.

TAYLOR, W. I., 1966. *Indole Alkaloids.* Oxford: Pergamon Press.

TESTA, A. & TESTA, P., 1965. Nitrogenous heterocyclic compounds in smoke condensates. *Ann. Direct. Etudes Equipm. SEITA, 2* (sect. 1): 163–191.

THEILKUHL, J. F., 1957. *Introducción al estúdio del* Methysticodendron amesianum. Thesis, Universidad Nacional de Colombia, Bogotá.

USCÁTEGUI M., N., 1959. The present distribution of narcotics and stimulants amongst the Indian tribes of Colombia. *Botanical Museum Leaflets, 18:* 273–304.

VÁSQUEZ, A., 1864. Sustancias del *Latua venenosa* de Chile, latué o arbol de los brujos. *An. Soc. Farm. Santiago, 2*(3): 71–75.

VON WETTSTEIN, R., 1895. In A. Engler and K. Prantl (Eds), *Die natürlichen Pflanzenfamilien, IV* (3), Leipzig: Verlag Wilhelm Engelmann.

WILBERT, J., 1972. Tobacco and shamanistic ecstasy among the Warrao Indians of Venezuela. In P. T. Furst (Ed.), *Flesh of the Gods:* 55–83. New York: Praeger Publisher.

WILLIS, J. C., 1966. *A Dictionary of the Flowering Plants and Ferns* [revised edit.], D. K. Airy Shaw (Ed.), 7. Cambridge: Cambridge University Press.

YÉPES, A. S., 1953. Introducción a la etnobotánica colombiana. *Publ. Soc. Colomb. Etnol.*, No. 1: 6–48.

9. Ethnobotany of Old World Solanaceae

K. L. MEHRA

National Bureau of Plant Genetic Resources, I.A.R.I. Campus, New Delhi, India

Many members of Old World Solanaceae play an important role in the social and religious life of a number of ethnic groups. An ethnobotanical account is presented on the use of certain species in magic, charms, rituals and ceremonies, worship, divination, fertility cults, etc., for the purpose of initiating and promoting further studies on these and other species.

CONTENTS

INTRODUCTION

The Old World Solanaceae comprise many highly poisonous plants, some most valuable medicinal plants and several food plant species. Although from the pre-historic period a few of its species were used by different ethnic groups in their folk-lore and therapeutic practices, no systematic ethnobotanical investigation has so far been undertaken on them. However, useful information on certain ethnobotanical aspects of many species has been rather indirectly recorded by specialists, such as archaeologists, anthropologists, folklorists, historians, explorers, travellers, missionaries, sociologists, and of course by field botanists, amongst others. Scattered information gathered from the available literature is presented here to focus attention, and thus to initiate and promote systematic ethnobotanical investigations on Old World Solanaceae similar to those conducted on New World species (Heiser, 1969; Swain, 1972; Furst, 1972). In the Old World, solanaceous plants are used in magic, charms, rituals and ceremonies, worship, divination, fertility cults, therapeutical practices, etc. Some details on these aspects are presented below.

USE IN MAGIC, CHARMS AND SUPERSTITIOUS PRACTICES

Pre-historic men and women depended on plants in manifold ways. Folk-lore, superstitions, traditions, various magic rituals, and other tribal practices in vogue, then and even to-day in the Old World, bear ample proof of the great influence magic doctors, herbal medicines, cure-deities, amulets and herbal charms exerted, and still exert, over both illiterates and civilized men and women. Many people still believe that a disease or any other human suffering can be cured not only by medical treatment or social reforms but that magical practices also play a great role in promoting or curing illness and disease.

Thus, *Withania somnifera* is highly valued among the Sotho of Basutoland, who often prescribe its use as a remedy against the intestinal parasites introduced by witchcraft (Phillips, 1917). Malaysian tribal folks place a casting net and a bunch of *Solanum ferox* and *Dracaena* on the head of an expectant mother to protect her and the unborn child from the bad effects of evil spirits (Skeat, cf. Burkill, 1935). In Basutoland, at a heathen wedding, the crushed roots of *Solanum supinum* are placed on the undigested grass taken from the stomach of an ox (or oxen) killed for the feast. This rite is supposed to protect the newly married couple from being harmed by the enemies of the family (Phillips, 1917). Similarly, the Manyika of Southern Rhodesia tie the roots of *Solanum sodomeum* to the foot of a patient to prevent him from becoming worse if a menstruating woman comes into the hut or room. The root is also tied around the neck or carried in the pocket as a protection against poisoning by an "*Umtakata*" (Watt & Breyer-Brandwijk, 1962). In India, the seeds of *Hyoscyamus niger*, made into a paste with mare's milk and tied up in a piece of wild bull's skin, are believed to prevent conception if worn by women (Dymock, Warden & Hooper, 1891). Similarly, the Southern Sotho believe that the tying of a portion of the inner part of the roots of *Solanum melongena* on the wrists of a pregnant woman ensures her good health (Watt & Breyer-Brandwijk, 1962).

Certain plants are believed to have magical potency. For instance, Malaysian magical faith prohibits the eating of the fruits of *Solanum melongena* by a mother during the 40 days rest period following the delivery of a child. This practice is perhaps connected with an Arabian superstition that the fruit is exceedingly heating, leading to melancholia and madness (Maxwell, cf. Burkill, 1935). The leaf juice of *Solanum aculeatissimum* enters into a magical preparation to be taken at childbirth in Malaysia (Hadji Bidah, cf. Burkill, 1935). Similarly, the roots of *Solanum ferox* in combination with other drugs are made into a preparation and applied to the bodies of elephants in a magical practice, to protect them from chills and to fatten them (Maxwell, cf. Burkill, 1935). In Sikkim Himalayas, the roots of *Mandragora caulescens* are used in certain magical rites (Watt, 1893).

Solanaceous plants are also prescribed for temporarily changing the climate of an area. For instance, black ripe fruits of *Solanum nigrum* symbolic of the dark clouds, are included in the rain-making medicines of the Southern Sotho and Tswana rain-doctors of Africa (Dornan, 1927–30). Similarly, the Southern Sotho in Basutoland believe that the smoke coming from the burning plants of *Lycium acutifolium* is capable of driving away an approaching hail-storm (Phillips, 1917).

The application of various recipes of herbs and herbal charms, some of which are highly effective, seems to have been practiced from the remote past. Josephus and Aelian stated that during and before their time the mandrake, *Mandragora officinarum*, was held in great superstitious awe by the Jews and the Greeks (Randolph, 1905). In the middle ages, mandrake became the mystical magic root, which existed only in fancy, and was represented by a factitious magic, in the form of a man or woman, and used by priests and charlatans as a charm (Dymock *et al.*, 1891). In Syria, Palestine and the Near-Eastern markets, mandrake roots are shaped into forms, resembling a human being, by certain trained artists, almost as a profession (Thompson, 1934). These images, called mannikins, are bathed, dressed and carefully tended in reverence (Randolph, 1905). Mannikins were kept as fetishes or talismans and were believed to bring good fortune to the possessor in many ways, viz., in divining secrets, increasing one's material prosperity, gaining the lover and the object of affection, etc. Joan of Arc was believed to have been in possession of a mandrake mannikin, which she is said to have carried in her bosom, hoping by means of it to acquire riches. This allegation was included in the articles of accusation brought against her in her trial (Thompson, 1934). In France, mandrake was known as Mandagloire or Malore (Main de gloire), and was regarded as a kind of fairy, which if well treated, would bring good luck to its owner (Dymock *et al.*, 1891). In Armenian folk-medicine, the flower buds of mandrake wrapped in linen and tied with seven strings of different colours are worn around the neck to cure jaundice. A small piece of mandrake root is burnt in a room to drive away devils and evil spirits, while the smoke inhaled is prescribed in Armenia as a cure for insanity (Thompson, 1934).

USE IN WORSHIP, RITUALS AND CEREMONIES

Primitive man lived in an intimacy with nature and began nature worship with no temples, priests or gods. This polytheistic nature worship was gradually transformed into various complex ideas and beliefs. Since from early times plants were put to many economic and other uses, sentiments of adoration for them crossed man's mind. The savage believed that the world in general was animate, and trees and plants were no exception to this rule. He thought that they had souls like his own and he treated them accordingly. In *The Golden Bough* Frazer (1954), who devoted several pages to tree worship, presented several examples of plant symbol worship in Europe, India, Australia, America, Africa, China and other parts of the world, showing different shades of beliefs and diverse types of rituals or other customs and practices associated with different plant species.

The conception of plant worship broadly falls into two categories, i.e., (i) the tree gods, whose worship became organized into a definite religion and (ii) tree demons or tree spirits, whose propitiation degraded to the level of sorcery and incantation. Furthermore, as man's conception of the deity became more definitely anthropomorphic on the one hand and less local on the other, this primitive representation of the god in the tree underwent a change, resulting in expressing more clearly the man-like form of the god, or the god became associated with certain species of plants, which thenceforth became sacred to and symbolic of Him (Gupta, 1965). Plants are thus repeatedly mentioned in connection with customs, beliefs and traditions in Old World societies. Among solanaceous plants *Datura*, mandrake, and tobacco were held sacred and have, therefore, been used in various rituals, initiation and other ceremonial practices.

Datura

In India, according to its ancient literary traditions, as mentioned in *Vamana Purana*, *Datura metel* is believed to have arisen from the chest of the God Shiva (Mukhopadya *et al.*, 1968). In ancient India, elaborate rituals were laid out for each sacred ceremony, and plants associated with specific deities formed an important and essential item of the rite. Thus, according to *Garuda Purana*, the vow of *Ananga-Trayodashi* falls on the 13th day of the moon's increase in the month of January. The God Yogeshwara (Shiva) should be worshipped on this day with offerings of *Datura* along with flowers, leaves and twigs of other sacred plants (Dutt, 1908). Similarly, on the fourth day of Bharapad (August-September) the *Vinayaka Urtam Puja* is performed in several parts of South India. The leaves of *Datura metel* are used along with the leaves and flowers of several other sacred plants as offerings to the God Shiva. The white variety is favoured for this purpose (Lancaster, 1965). *Datura* species are associated as sacred plants in Shiva worship and are not usually offered to other deities in Gujarat, Bengal and other parts of India (Desai, 1965; Mahapatra, 1965).

In Africa, the seeds of *Datura* species, called *"langboontije"* among the school boys, especially in the country schools, are used in a kind of an initiation ceremony, the aim being to administer sufficient quantity to produce inebriant effects (Watt & Breyer-Brandwijk, 1962).

Although a lot of information is presently available on the ritual use of hallucinogens in the New World (Furst, 1972; Schultes, this Volume, Chapter 8), little is known about the use of hallucinogens in the Old World. Recently, however, Johnston (1972) presented interesting details about the motivations, attitudes, and expectancies of novices approaching the final rite, the journey of fantasy, involving *Datura fastuosa* injection rites, proceeding in six stages, amongst the girls of the Tsonga puberty school *"khomba"* in Mozambique and the northern Transvaal. Specific reinforcers include the complementarity of spirit communication and dreaming among the Tsonga, indigenous beliefs concerning bluish green snakes, the unabated performance of particularly significant drumming patterns, and the authoritative suggestions of the garbed priest, who administers the drug (Johnston, 1972). An interesting aspect of these rites is that the girls are neither sickened nor poisoned by *D. fastuosa*, but see and hear that which is culturally expected of them at that time and place.

Mandrake

In Greek mythology, Aphrodite, the goddess of love and beauty, is said to have been born from the sea. Behind her stands the mandrake, *Mandragora officinarum*, its fleshy roots, sometimes having a fancied resemblance to the human form in a grotesque parody of a man or a woman, the famous "forked radish of Falstaff" (Gupta, 1965). It was so alarming and dangerous a personality and it shrieked so loudly when pulled (see Romeo and Juliet, Act IV, Sc. III, "shrieks like mandrake born out of the earth") that men dare not pluck it up themselves and they employed a dog to do so, which then expired in agony. Mandrake is also mentioned in Anthony and Cleopatra (Act VIII, Sc. V), Othello (Act III, Sc. VIII) and King Henry VI (Act III, Sc. II).

The ancient Greeks and Romans considered mandrake so potent and valuable in medicine, as a narcotic and restorative plant that the collection of its roots was made a special ceremony. Thus, Theophrastus (9.8.8) wrote "around the mandragora one must make three circles with a sword, and dig looking towards the west. Another person must dance about in a circle and pronounce a great many aphrodisiac formulae". Pliny (25.148) wrote "that those who are about to dig mandragora avoid a wind blowing on their faces, first they make three circles with a sword, and then dig looking towards west" (Randolph, 1905). More than 20 books have been written on the medicinal, spiritual and diabolic properties of mandrake (see Moldenke & Moldenke, 1952, for details). The superstitions about mandrake increased gradually and the Pythagoreans called it anthropomorphon. Columella alluded to it as being semi-homo and Hildegard believed "that it was fashioned out of the same earth whereof God created Adam, and its likeness to man is a wile of the devil, which distinguishes it above all other plants" (Thompson, 1934).

Tobacco

Bishop (1949) wrote that "the burning of incense and spices has had a place in worship from time immemorial and there is little doubt that the custom of smoking developed from the inhalation of the fumes from tobacco burnt as an offering to the great spirit. Tobacco was regarded by the aboriginal people of America as a sacred plant and as a special gift of the god to man and, when the plant reached the Old World, this attitude was adopted by Europeans along with the Indian's ideas concerning the efficacy of tobacco as a remedy against almost all bodily ills". In India, tobacco is not held sacred by the Hindus, but aboriginal tribes ascribe to it a divine origin. However, one south Indian verse certainly gives it a high place in the Indian cultural context. Once an Indian semi-god (*Devata*) Indra asked Brahma (one god out of the Trinity concept of Hindus, viz., Brahma, Vishnu, Mahesh, the God the creator, the administrator and the destroyer) "What is the best thing in the World?". Brahma replied with his four mouths, "*Tobaku, pogaku, hogesoppu* and *pogele*" or tobacco in Hindustani, Telugu, Kanarese and Tamil, respectively (Natesa Sastri, 1949). The Portuguese introduced tobacco into India in the 16th century A.D. and its early history and spread was discussed in another paper (Mehra, 1966). It spread rapidly in India and interestingly, several plant myths and traditions are in vogue amongst the tribal people. Elwin (1949, 1954) described in detail 27 plant myths built around tobacco, as regards its origin or discovery of usage. The main myths and traditions are that it was found by accident; someone used the leaf in a platter and inhaled the fumes with his food; a dancer idly picked and chewed a leaf while he was relieving himself; medicine revealed by Bara Pen (a tribal God) as a cure for sickness or a solace in bereavement or a narcotic to relax after doing hard work; and lastly there is a motif of "the girl nobody loved" who turned into a tobacco plant and was then desired by all the world (Elwin, 1949). The last tradition is widely distributed amongst the tribal people, viz., the Kond, Gadaba and Juang of Orissa, the Muria of Bastar and the Gond of Mandla (Elwin, 1949). Since Elwin's accounts are not easily available, a few plant myths are described below, based on Elwin (1949, 1954).

Formerly tobacco grew wild in a jungle and the people had no knowledge about it. Sonsai, a Gond tribal person, was a famous medicine-man. A man once came to him with a bad cough and Sonsai looked into his winnowing pan and found nothing to give him. But at night Bara Pen (a tribal God) came into his dream and told him to get tobacco leaves from the jungle as medicine. Sonsai got the leaves and the sick man recovered after smoking it. The new medicine spread thereafter.

In olden days, it was customary to entertain visitors with rice. But rice got weary of having to satisfy so many people and fled. He met tobacco in the way and explained his difficulties. Tobacco informed him not to worry and agreed to be a first item to be offered to a visitor, to be followed by rice. The master, finding the rice missing in his house, went in search of it and found tobacco in the way. He put a tobacco flower in his ear and felt drunk with its scent. Thereafter, he brought its seed, leaf and flower and the use of tobacco spread among people.

In olden days, there was no tobacco. Men had no way of getting intoxicated or having an excuse for resting, but had to go on working till the job was finished. Seeing this, Nirantali thought men had much trouble, working all the day, and felt like making something for their recreation. So, from her little girl's hair she took a white bug and threw that into a garden. When the rain fell, a tobacco plant grew from the bug. When leaves emerged on the plant, Nirantali picked them, made a cheroot and offered that to her husband, who was pleased to smoke it. Seeds and the smoking habit spread thereafter from one hand to another.

There was a king who had an ugly daughter and nobody would marry her in spite of a large dowry. When the girl grew older and nobody agreed to marry her, she committed suicide in utter desperation. At the time of her death, she was blessed by God, that in whatever form she would come back to the world, she would be loved and desired by men. After her cremation the tobacco plant grew from her ashes. Its leaves were smoked and men became addicted to its use.

Four Saora brothers and their wives lived in a forest. In those days men did not cough or sneeze due to which women could not forecast their arrival in the house, to get time to dress up properly. So they went to Piskisum and begged him to make some arrangement. The God picked a tobacco leaf and said "Take it, grind it up and throw it in front of the house, before they next come home". They did so, and men began to cough and sneeze. Thus, the habit of sneezing developed and spread, through the taking of tobacco snuff.

The Bhil tribes of India offer tobacco along with other items to the dead bodies during their funeral rites (Naik, 1956). Similarly, in the burial rites of Sema Nagas of India, a pipe and tobacco are placed in the grave of a dead warrior (Hutton, 1968). Muria Gonds of Bastar, India, ceremonially put up a small plot of tobacco in the centre of their dormitory before they set up a village (Grigson, 1938). Tobacco is freely distributed in several ceremonies and rituals especially the dancing expeditions, singing, combing and massage ceremonies, wherein both young boys and girls mostly participate, amongst the Muria tribal peoples of India (Elwin, 1947a). These peoples also refer to tobacco in the folk songs they collectively sing at the time of marriage.

To ask for a puff of tobacco is the conventional way of inviting a girl for sexual congress after marriage amongst the Muria tribals of India (Elwin, 1947a).

There are also taboos associated with tobacco smoking. Muria tribal peoples of India do not accept tobacco from a father at his child's birth ceremony (Elwin, 1947a). Amongst the Baiga people of India, no social sanction is more powerful than the fear that a friend may refuse one's tobacco at a feast or funeral (Elwin, 1947b). In fact, in several ethnic groups of India there is a common way of social sanctions against people by saying that nobody should share water and tobacco with them.

USE AS POISONS

Many Old World solanaceous plants, being highly poisonous, are used for criminal purposes in several places.

Datura metel has been used in Indo-China for the production of fatal criminal poisoning, while in the East Indies, women have been known to breed beetles and to feed them on the leaves of *Datura metel*; the poisonous excreta is used for poisoning an unfaithful lover (Watt & Breyer-Brandwijk, 1962). In India, the seeds of *Datura metel* are rarely used as a homicidal poison but are more commonly applied to produce stupefaction, aimed at committing robbery, sometimes with fatal results (Byam & Archibald, 1921–23). In Tanzania, *Datura* species are used for inebriant effects, especially to facilitate robbery. *Datura*, called by a secret name "*Lukuma*" is added to a kind of local beer, *pombe*, and offered to the victims for facilitating the robbery (Raymond, 1944).

The Bushman and possibly Hottentot in Africa use the fruit juice of *Solanum incanum* along with other ingredients for preparing their arrow poisons (Watt & Breyer-Brandwijk, 1962). Similarly, in Nigeria, the roots of *Solanum incanum* are considered poisonous and are used along with *Amorphophallus dracontioides* roots, in preparing an arrow poison (Dalziel, 1937).

Tobacco juice, as a poison for weapons, is said to be used in various parts of the world, including Borneo and the Mentawei Islands off Sumatra (Burkill, 1935).

In the interior of Angola, *Datura* plants are used as a fish poison by the indigenous inhabitants. The leaves are rubbed between the hands and thrown into the water. The effect is claimed to be so strong that the fish rise to the surface within a few minutes (Roark, 1931).

USE IN DIVINATION

The practice of divination, viz., the foretelling of future events or the discovery of what is hidden or obscure by supernatural or magical means, is widespread amongst several tribal populations in the Old World. In parts of Africa and India, certainly, it is closely linked with the magical aspects of the indigenous practice of medicine, so that many medicine men are diviners as well.

The stimulating effects of *Datura* species are said to have been used to inspire the pythoness of the Delphian oracle (Lancet, 1953).

In Uganda, the tribal priest, in order to be inspired by his god, smokes a pipe of tobacco fiercely till he works himself into a frenzy. The loud excited tones in which he then talks are recognized as the voices of the god speaking through him and they behave as incarnate temporary human gods (Frazer, 1954). Certain persons are supposed to be possessed from time to time by a spirit or deity. While the possession lasts, their own personality lies in abeyance. The presence of a spirit is revealed by convulsive shiverings and shaking of the body, and the inhalation of tobacco helps in the process (Frazer, 1954).

An appraisal of the history of tobacco and its ceremonial, ritualistic uses reveals that tobacco snuff has had significant magical applications especially in the practice of divination and countering of witchcraft. The Mano of Liberia blow tobacco snuff up the nose for convulsions or coma. If the sufferer responds by sneezing, there is little hope of recovery (Watt, 1972).

USE IN FERTILITY CULTS

The concept of plants as the symbol of fertility seems to be clearly emphasized in the Assyrian cylinders and bas-reliefs, where it is conventionally represented as a date-palm between two personages, who approach it from either side bearing in their hands a cone similar to the inflorescence of the male date palm. The ancient people had connected the idea of the fertility cult with sex in their minds and beliefs and had viewed plants likewise. The importance of fertility is evidenced by the large number of plants used as aphrodisiacs, as magical protectors and stimulants to reproduction, and as charms to aid crop production and to prevent harmful bewitchment of crops.

In Indian folk medicine, the root of *Mandragora officinarum* is called *Lakshmana*, possessed of lucky signs or marks. It is described as a promotor of conception, aphrodisiac, and a corrective of the condition called *Tri-dosha* or disorders of the three humours of the body, i.e. bile, blood and phlegm (Dymock *et al.*, 1891). Theophrastus and Dioscorides mentioned that fruits of this plant were used in love philtres, which appears to be explained by the sexual excitement and hallucinations that are observed under the influence of *Datura* poisoning or intoxication (Dymock *et al.*, 1891). Physiologus reported in the animal kingdom concerning the elephant that "there is an animal on the mountain named elephant. There is no desire for intercourse in the male animal; if he wants to beget children he goes to the east near the paradise. Here, there is a tree called mandrake; both the male and the female go there, and the female first tastes of the tree and gives the male a share of it, and plays with him until he tastes it, and by eating it the male has sexual intercourse with her". In Greece, a part of mandrake plant is suspended as a necklace on the neck of the barren woman and its fruits or a piece of root are placed on her body, in due time, that is during sexual intercourse with her husband. A similar piece of root with the bulb of *Scilla* is placed under the pillow or the mattress of the newly married couple's bed, obviously for the same purpose (Papamichael, 1975). Mandrake is carried by women of Eastern Europe as a charm against sterility (Biswas, 1955). Mandrakes are twice mentioned in the old testament (for procreation of children, Genesis, XXX, 14–18; Song of Soloman VII, 13) and, thus, received allegorical treatment from the church fathers. The use of mandrake in matters related to the procreation of children is found amongst French, German, Russian, Italian, Greek, Jewish, English and other people (Papamichael, 1975). It is supposed that mandrake acquired its reputation as a potent love charm because of its association with Circe, as it was sometimes called "the plant of Circe" after the sorceress, who in the mythological legend turned men into swine with her magical potion (Thompson, 1934). Mandrake is the *alraun* or *alruna* of German mythology, which was believed to be a gallow's mannikin sprung from the seeds of men, who were hanged, that when pulled out of the earth by a black dog, it shrieked like a child (Dymock *et al.*, 1891). The symbolic and magical use of mandrake, preserved by the Greeks of Asia Minor, in the Dodecanese, Crete, Mani and Cyprus (especially the fruits of mandrake are placed among the eggs which are for incubation) led Papamichael (1975) to asume that this custom is of eastern origin.

In southern Nigeria, the egg plant, *Solanum melongena*, is considered as a symbol of fertility and used by barren women (Dalziel, 1937). The Zulu, in

Africa, administer the root barks of *Solanum sodomeum* (probably a misnomer for *Solanum sisymbriifolium*) as a remedy for barrenness and impotence (Bryant, 1909). Recently, Johnston (1972) presented evidence that *Datura fastuosa* ingestion within the context of the Tsonga (Shangana-Tsonga of Mozambique and northern Transvaal) girl's puberty school is linked to a fertility goal (an added ingredient provides protection against barrenness by witchcraft). The women are instructed to watch out for witchcraft in their husbands' villages and it is believed that the ingested *Datura fastuosa* washes one's eyes so as to facilitate perception of the fertility-granting ancestor-god.

The Ramah Navaho use the berries of *Solanum triflorum* to increase the productivity of watermelon seed. The dried berries are stored until spring, soaked in water and planted with the watermelon seed. They believe that since *Solanum triflorum* bears several berries, this practice would increase the productivity of watermelon (Vestal, 1952). Similarly, the leaves and berries of *Solanum villosum* are soaked in water and put on watermelon seed by Ramah Navaho to ensure a good crop (Vestal, 1952).

The people of the Bara river region put the prickly stem of *Solanum melongena* on the threshold of a house, in which the first fruits of the rice harvest, i.e., the soul of rice, have been lodged; evidently the prickles are considered as protectors (Evans, cf., Burkill, 1935).

The above account shows that several members of the Old World Solanaceae play an important role in the social and religious life of many ethnic groups of the Old World. Even some species of the New World, having been once introduced into the Old World, have also become an integral part of the cultural heritage of the Old World communities. There is a need, thus, to initiate systematic ethnobotanical studies on different ethnic groups located in different parts of the Old World. It would be of interest to make cross cultural comparison to analyse the origin and diffusion of specific traits from one group to the other. It is also to be hoped that such studies would help to unravel the inter-cultural contacts that occurred at different times of man's history in different parts of the world.

REFERENCES

BISHOP, W. J., 1949. Some early literature on addiction, with special reference to tobacco. *British Journal of Addiction, 46:* 49–65.

BISWAS, K., 1955. *Common Medicinal Plants of Darjeeling and the Sikkim Himalayas.* Calcutta: W. Bengal Govt. Press.

BRYANT, A. T., 1909. Zulu medicine and medicine men. *Annals of the Natal Museum, 2:* 1–103.

BURKILL, I. H., 1935. *A Dictionary of Economic Products of the Malay Peninsula.* London: Crown Agents.

BYAM, W. & ARCHIBALD, R. J., 1921–23. *The Practice of Medicines in the Tropics.* London: Hodder & Stoughton.

DALZIEL, J. M., 1937. *The Useful Plants of West Tropical Africa.* London: Crown Agents.

DESAI, B. L., 1965. Tree worship in Gujrat, 54–57. In S. S. Gupta (Ed.), *Tree Symbol Worship in India.* Calcutta: Indian Publications.

DORNAN, S. S., 1927–30. *Bantu Studies, 3:* 185.

DUTT, M. N., 1908. *Garuda Purana.* Calcutta: Society for the Resuscitation of Indian Literature.

DYMOCK, W., WARDEN, C. J. H. & HOOPER, D., 1891. *Pharmacographia Indica, II.* Calcutta: Thacker, Spink.

ELWIN, V., 1947a. *The Muria and their Ghotul.* Oxford: Oxford University Press.

ELWIN, V., 1947b. *The Baiga.* London: John Murray.

ELWIN, V., 1949. *Myths of Middle India.* Oxford: Oxford University Press.

ELWIN, V., 1954. *Tribal Myths of Orissa.* Oxford: Oxford University Press.

FRAZER, J. G., 1954. *The Golden Bough*. London: Macmillan.

FRIEND, H., 1886. Flowers and Flower-lore. *In* H. N. Moldenke & A. L. Moldenke (Eds), 1952. *Plants of the Bible*. New York: Ronald Press.

FURST, P. T., 1972. *The Ritual Use of Hallucinogens*. New York: Praeger Publishers.

GRIGSON, W. V., 1938. *The Maria Gonds of Bastar*. Oxford: Oxford University Press.

GUPTA, S. S., 1965. *Tree Symbol Worship in India*. Calcutta: Indian Publications.

HEISER, C. B. Jr., 1969. *Nightshades. The Paradoxical Plants*. San Francisco: W. H. Freeman.

HUTTON, J. H., 1968. *The Sema Nagas*. Oxford: Oxford University Press.

JOHNSTON, T. F., 1972. *Datura fastuosa:* Its use in Tsonga girl's initiation. *Economic Botany, 26:* 340–351.

LANCASTER, S. P., 1965. *The Sacred Plants of the Hindus*. Bulletin of the National Botanic Garden, Lucknow, India.

LANCET, 1953. *Annotations, 1,* 1036.

MAHAPATRA, P. K., 1965. Tree symbol worship in Bengal. *In* S. S. Gupta (Ed.), *Tree Symbol Worship in India:* 125–139. Calcutta: Indian Publications.

MEHRA, K. L., 1966. Portuguese introduction of fruit plants in India. *Indian Horticulturist, 10:* 9–12.

MOLDENKE, H. N. & MOLDENKE, A. L., 1952. *Plants of the Bible*. New York: Ronald Press.

MUKHOPADYA, et al., 1968. *Vamana Purana*. Varanasi, India: All India Kasiraj Trust.

NAIK, T. B., 1956. *The Bhils, a Study*. Kingsway, Delhi: Bharativa Adimjati Sevak Sangh.

NATESA SASTRI, S. M., 1949. A verse on tobacco. *The Indian Antiquary, 20:* 297.

PAPAMICHAEL, M., 1975. *Birth and Plant Symbolism. Symbolic and Magical Uses of Plants in Connection with Birth in Modern Greece*. Athens: Hellenic Folk-lore Research Centre.

PHILLIPS, E. P., 1917. A contribution to the Flora of the Leribe Plateau and environs. *Annals of the South African Museum, 16:* 1–382.

RANDOLPH, C. B., 1905. The mandragora of the ancient folk-lore and medicine. *Proceedings of the American Academy, 40:* 487–537.

RAYMOND, W. D., 1944. *East African Medical Journal, 21:* 362.

ROARK, R. C., 1931. Excerpts from consular correspondence relating to insecticides and fish poison plants. *U.S.D.A. Bureau of Chemistry and Soils*.

SWAIN, T., 1972. *Plants in the Development of Modern Medicine*. Harvard Univ. Press, Camb., Mass., U.S.A.

THOMPSON, C. J. S., 1934. *The Mystic Mandrake*. London: Rider.

VESTAL, A. P., 1952. *Ethnobotany of the Ramah Navaho*. Peabody Mus. Amer. Archeol. Ethnol. Harvard Univ. 40, 4.

WATT, G., 1893. *Dictionary of the economic products of India*. Calcutta: Central Printing Office.

WATT, J. M., 1972. Magic and witchcraft in relation to plants and folk medicine. In T. Swain (Ed.) (1972), *Plants in the Development of Modern Medicine:* 67–102. Massachusetts: Harvard University Press.

WATT, J. M. & BREYER-BRANDWIJK, M. G., 1962. *The Medicinal and Poisonous Plants of Southern and Eastern Africa*. Edinburgh & London: Livingstone.

10. Aboriginal uses of Australian Solanaceae

NICOLAS PETERSON

Department of Prehistory and Anthropology, The Australian National University, Canberra, Australia

Plants of the Solanaceae family were an important source of food and drugs to the Australian Aborigines of the arid regions. As a food source they were important because of their widespread availability and storage potential. Preparation and storage methods are described and a chemical analysis of the composition of the staple plants provided. The use of Solanaceae as a source of narcotics at the time of European arrival is well documented but until recently confusion has surrounded which species were chewed and which were used in hunting to poison waterholes. This matter is clarified. An annotated list of Solanaceae of ethnobotanical significance is appended to the paper.

CONTENTS

INTRODUCTION

Forty thousand years before Captain Cook sailed into Botany Bay, the Aborigines discovered and settled Australia. From the time of their arrival until the coming of Europeans, they lived by hunting and gathering, adapting to all the major floristic zones of the continent. Broadly speaking, two types of Aboriginal economy can be distinguished: an inland economy based on the gathering of seeds and fruits supplemented by lizards and small game, and a coastal economy of roots, fruits, nuts, fish and other aquatic resources. Large marsupials, although highly valued, were nowhere a daily part of the diet.

The balance of plant to animal foods appears to have been proportional to the regional distribution of biomass between these two. Thus in the arid interior plant foods contributed up to 80% by weight of the diet, while in the tropical and temperate woodlands, where there was access to aquatic resources, vegetable foods were often only 30%–50% of the diet.

There are between 110–120 indigenous Australian species in the Solanaceae family (see Burbidge, 1963). Although found throughout the continent, the majority of species grow in the arid regions with fewer representatives in the temperate woodlands and least in the tropical forests of the north. Aboriginal use reflects this distribution. Solanaceae were rarely used in the tropical north, though in the temperate woodlands of Victoria, New South Wales and Tasmania, a few species were eaten frequently. It is only in the desert regions that the plants are an important source of food and drugs.

The information recorded here is based on a survey of the literature supplemented by observations from fieldwork in Central Australia among the Walpiri. For the temperate areas, the only source of information is the early literature, mostly dating from before the turn of the century. This is deficient in several respects: the areas covered are limited, the information sketchy—usually no more than a name in a list of utilized plants—and the identifications often doubtful. Identifications, however, are a problem throughout the continent. Most authors who have written about Aboriginal foods are not botanists, although a number are natural scientists. Consequently, while the genus is usually correct, the species name is frequently wrong: there is simple misidentification in the field; there is reclassification and change in nomenclature since the author published; and there are the confusions introduced by Europeans using Aboriginal names, the best example of which is the history of the identification of Aboriginal chewing tobaccos discussed below. Nevertheless, the more recent work in the arid regions provides reliable information from the area of greatest use.

SOLANACEAE AS FOOD

Direct evidence on the antiquity of the use of Solanaceae is currently lacking. However, it is a reasonable assumption that members of this family have been part of the diet for as long as the desert areas have been occupied. The staple *Solanum* species are visually obvious, abundant and many can be eaten without preparation.

Archaeological evidence in western New South Wales indicates that there has been complete continuity in the diet of that region for 15,000 years (Allen, 1974: 315). All the animal species recovered from the ancient living site at Lake Mungo were still present in the area during the 19th century and being used by the Aborigines, so that assuming vegetable foods other than seed were eaten in the past, as they must have been, *S. ellipticum* and *S. esuriale* were in all probability among them. Then as now the Aborigines have competed for the fruit of these and other solanums with insects, kangaroos, emus and even dingoes to a limited extent (Finlayson, 1936: 142).

Solanum species are particularly important in the desert for two reasons: they are widely available and some can be stored. Like all the desert flora, they are dependent on the highly variable rainfall. Rain falls either as a result of the monsoon or from localized storms. Unless it is particularly wet, most solanums bear in the cooler parts of the year, especially spring, autumn and early winter and, given the right combination of temperature and rainfall, they can bear fruit within 3 to $3\frac{1}{2}$ months after falls of rain (e.g. see Richards, 1882: 136–7), making June and July frequent times of abundance. Gould, however, reports the collecting of *ngaru* (which he identifies as *S. eremophilum* but which is a misidentification for *S. chippendalei*—see Appendix) in substantial quantities between December and February 1966–67 in Western Australia during the height of the antipodean summer (1969: 262), and that in a drought year (1969–70), *ngaru* ripened earlier and lasted longer (1970: 64, 65). At least one species, *S. ellipticum*, is reported to be a permanent producer, if conditions are not too dry (Cleland & Tindale, 1959: 127).

Detailed understanding of the combination of factors that control the production of fruit on solanums has yet to be achieved, but the role of fire in increasing the abundance of a number of species, in particular *S. centrale*, is known to both Aborigines and botanists (e.g. Richards, 1882: 137). The Walpiri, Ngatatjara (Gould, 1971: 22, 23) and doubtless other Aborigines, are quite explicit about the effects of their deliberate and systematic burning of the countryside. It is their most important way of manipulating the environment to increase the amount of plant production directly available to them.

S. centrale, *S. chippendalei* and other *Solanum* species grow in areas where there is also spinifex (*Triodia* sp.) which expands from isolated clumps to form a climax community of almost continuous ground cover over four to five years without burning, tying up scarce nutrients only released by fire. The solanums regenerate rapidly after fire, because they form large fire resistant clones by expanding from an initial plant via underground stems which respond quickly to the release of nutrients with the first rain (Latz, Alice Springs Herbarium).

Another way in which Aborigines are influencing the distribution and abundance of *Solanum* species has recently been observed in Central Australia (O'Connell, n.d.). The Alyawara people inadvertently introduced *S. chippendalei* south of its normal range, when substantial quantities of fruit were brought into camp by car. The discarded seeds germinated in the disturbed and enriched soil about the camp, and bore fruit.

During good years, many *Solanum* species are incredibly abundant. In August 1975, a count of fruit on *S. chippendalei* bushes in a 3×2 m area at well 49 on the Canning Stock route showed 221 fruit on 18 small bushes, about a fifth of the fruit ripe enough to eat (Kimber, n.d.: 2). Three hundred kilometres to the south-east of this stock route, in the Clutterbuck Hills, Gould observed a group of 13 Ngatatjara with three women and an older girl supplying plant food. During an average $4\frac{1}{2}$ hour working day, they were collecting just under 13.6 kg of dried *S. centrale* (*kampurapa*) and 4.5 kg of *S. chippendalei* (*ngaru*) after cleaning and drying (i.e. the fresh weight of the *S. chippendalei* would be about 22.7 kg based on my own experiments; a loss of 25 % through cleaning, in turn reduced by 80 % by drying). This group of women were collecting just under 4.5 kg of processed plant food per day during the period of his observations (1969: 262).

Apart from the abundance of solanums in good times, they are important in lean times, too. The most important is the fruit of *S. centrale* which dries out naturally on the bush in hot dry conditions and eventually falls to the ground, rock hard and very like a sultana in size, colour and shape. These dried fruit can be collected months after they have ripened, if they have not been eaten by insects, ground with water and made into an edible paste. The paste is either eaten or compacted into round cakes as much as 25.4 cm in diameter, and weighing up to 1.59 kg (Gould, 1969: 361) (see Plate 10.1). The outer surface, sometimes red ochred, dries to form a crust which keeps the inside from spoiling indefinitely.

S. chippendalei can also be stored. In 1896, Carnegie, travelling in the Western Desert, gave chase to some Aborigines to help him find water, and came upon their camp where he noticed:

> "several wooden sticks on which were skewered dried fruits, not unlike gooseberries; these were hidden in a bush, and are remarkable, for they not only show that the natives have some forethought, but that they trade in

edible goods as well as in weapons and ornaments. These fruits are from the *Solanum Sodomeum* [*S. chippendalei*], and were only seen by us near the Sturt Creek (300 miles away)" (1898: 230–231).

The reference to trading of dried fruits is dubious, since nowhere else has trading of food in this way been reported, but the drying of *S. chippendalei* is widespread in the desert. Thomson, working less than 100 km to the east of Carnegie's route, reported a very similar find in 1956:

"Deeper among the dunes, closer to Lake Mackay, another camp, larger and better organized, was located ... This camp was ... evidently more permanent and seemed to be used as a base ... from which food-gathering sorties were being made into the sand dunes, perhaps for a few days at a time. On top of some of the brushwood shelters ... reserves of prepared vegetable foods had been left. Some of this had been desiccated and was carefully stored. One of these foods was brown in colour, with the appearance and texture of a mass of pulverized preserved figs and contained numerous conspicuous pale yellow seeds, again suggesting a kind of thick fig paste. It had a half-sweet, half-tart or acid flavour, but it proved palatable and satisfying, as it was evidently obtained in large quantities even in this drought year, it was probably an important staple food. In the absence of the natives, I was not able to identify the fruit from which this food was derived, but concluded that it was one of the several species of Solanaceae that were seen in flower at this season—some of which, however, were highly poisonous. A quantity of dehydrated or desiccated material was also seen in this same camp, neatly impaled on slender twigs which had been stripped of their bark. This proved to be the dried pericarp, or ovary wall, of another species of *Solanum* which, in the absence of the inflorescence, I could not identify specifically. The discovery of a reserve of prepared and desiccated vegetable food stored in this way was of much interest, particularly in a drought year and in face of the belief that is widely held that the Australian Aborigines live from hand to mouth and make no attempt to conserve food." (Thomson, 1962: 11–12).

These two species can confidently be identified with *S. centrale* and *S. chippendalei* as the Pintupi, about whom Thomson is writing, are western neighbours of the Walpiri and share a similar subsistence pattern in which these two species are staples when available.

The black seeds of *S. chippendalei* and the placenta and mucilage immediately adjacent to them are extremely bitter. These are removed with a small wooden blade called *katjalara* by the Walpiri, and *pangara* (Gould, 1969: 261) by the Ngatatjara. The base of the fruit is stabbed two or three times, so that it opens to look like a small tulip, and the seeds and thin skin around them scraped and flicked out with a few deft twists of the wrist. Sun or fire dried on thin sticks (see Gould, 1971: 16), they can be stored indefinitely, those in Fig. 10.1 being already over three years old. To eat they are either soaked in water or pounded up with water to make a thick sweet paste.

Wallace & Wallace (n.d.:1) report that *S. coactiliferum* is used by the Pitjant-jatjara after preparation, although regarded as inedible by the Walpiri. The bitter juices are squeezed out, and the skins washed and heated before eating. No mention is made of storage, but it would seem a definite possibility, given the style of preparation. Without doubt, most Pitjantjatjara live in a more hostile environment than the Walpiri, which probably accounts for the differential interest in this plant. Other *Solanum* species used in the desert without preparation are: *S. cleistogamum*, *S. ellipticum* and *S. esuriale*.

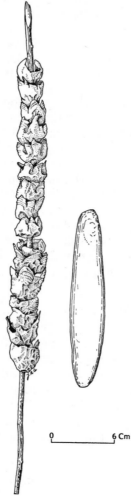

0 _____ 6 Cm

Figure 10.1. Cleaned and sun dried fruits of *Solanum chippendalei* are strung on sticks for storage throughout much of Central Australia. The small wooden blade, 8 mm thick, is used to remove the seeds and bitter mucilage.

Of the species eaten elsewhere, the most commonly referred to are the so-called kangaroo-apples of the temperate woodlands, *S. aviculare*, *S. simile*, *S. laciniatum* and *S. vescum*. The fruit of these species can only be eaten when completely ripe, and even then are reported as leaving a burning sensation in the mouth.

Physalis minima (Specht, 1958: 499), *S. dioicum* (Crawford, n.d.) and *S. lucani* (Heath, n.d.) are the only species of Solanaceae for which there is firm evidence of use in the tropical woodlands.

The limited amount of information available on the food value of these plants is summarized in Table 10.1. The four specimens of *S. chippendalei* were all close to, but in slightly different stages of, ripeness. Their importance as a source of vitamin C is clear but underemphasized, as the specimens were not preserved in oxalic acid. A further set of specimens, some ripe, some raw, and some ripe in oxalic acid, were analysed and gave the following ascorbic acid values;

ripe	110.4 mg/100 g
ripe in acid	113.43 mg/100 g
green	19.01 mg/100 g

While these specimens of *S. chippendalei* were poor in protein, a sample analysed for the Department of Health, Alice Springs, gave 10.6 g/100 g (the sample was insufficiently large for any further analysis, and it is unknown whether the inedible seeds were removed before analysis). From other analyses of desert plants, it is clear that the seeds of *Acacia* species were the main source of vegetable protein when available, but the fact that *S. centrale*, whose seeds are edible, can store as much as 8.44 g/100 g further underlines its importance in the diet.

Aboriginal nomenclature for the *Solanum* species varies from area to area, but the data are inadequate for any detailed or sophisticated analysis and comparison of Aboriginal taxonomy. However, a brief consideration of Walpiri usage draws attention to differences from the current Linnean classification. In at least one case the Walpiri classify two Linnean species together:

warakalukalu	*S. coactiliferum*
	S. quadriloculatum

The Walpiri typify *warakalukalu* as kangaroo food, unfit for humans, although Cleland & Tindale (1959: 137) report that their western neighbours, the Pintupi, eat them. This is an error, for *S. quadriloculatum* is quite poisonous (Everist, 1974: 475), and from their own linguistic evidence interpreted in the light of more recent work (Hansen & Hansen, 1974b), *S. quadriloculatum* is classified with *S. coactiliferum* by the Pintupi, too. On the other hand, the Walpiri make at least one distinction not made in the Linnean taxonomy:

S. chippendalei	*wanakitji* (level ground)
	ngayiyaki (hillsides)

The basis for this distinction other than location was never entirely clear, although the fruit of the hillside variety were said to be yellower and slightly smaller than those on the level ground. No attempt was made to verify this experimentally.

Many plants have multiple names as a result of borrowing terms from adjacent linguistic groups, or because they are referred to in some oblique manner, partially dependent on context. Thus, *S. chippendalei*, usually called *wanakitji/ngayiyaki* by the Walpiri, can be called *kulapanta*, having waste matter. *S. centrale* is known as *yakatjiri* in Walpiri, but they also use the Pintupi term *kamparrarpa* mainly, although not exclusively, for the dried fruit.

Table 10.1. *Solanum* Food composition of three Australian species

	Protein (g/100g)	Moisture (g/100g)	Fat (g/100g)	Ash (g/100g)	Carbo-hydrate	Kcals	Minerals (mg/100 g)			Ascorbic acid (mg/100g)	Collector and locality	Analyst
							sodium	potassium	calcium			
S. centrale	8.44	28.82*	5.53	3.29	53.92	299.21	119	195	46	—	Peterson & Peterson, Yuendumu, N.T., 1972	Maggiore
S. chippendalei	1.17	72.93	1.39	1.72	22.79	108.35	17.74	638.92	97.61	29.42	Kimber, from near well 49 on Canning stock route W.A. Sandy soil. 31.8.75	Maggiore
S. chippendalei	1.92	68.43	0.56	2.58	26.51	118.76	20.78	794.57	129.89	34.29	Kimber, from 30–35 miles east of Billiluna Station W.A. Slightly stony soil. 31.8.75	Maggiore
S. chippendalei	2.47	77.81	0.33	1.38	18.01	84.89	19.31	564.67	80.94	58.74	Kimber, from Granites, N.T. 1.9.75	Maggiore
S. chippendalei	1.38	81.7	0.54	0.84	15.54	72.54	36.16	418.72	66.61	12.16	Kimber, from Granites, N.T. Clay soil on edge of stone country. 23.8.75	Maggiore
S. ellipticum	2.63	86	1.77	1.25	5.70	49.25	—	355	35	—		Dadswell 1934:17–18
Tomato (for comparison)	1.0	93.7	0.3	—	4.1	21	4	252	14	22		Thomas & Corden 1970:17

All the specimens of *S. chippendalei* had their seeds removed. They were in slightly different stages of ripeness.
*The specimens analysed were naturally dry, when collected, and edible seeds left in.

SOLANACEAE AS DRUGS

Interest in the use of Solanaceae as a source of narcotics began in 1861. Then Wills of the Burke and Wills expedition recorded in his diary on 7 May that:

"Various members of the tribe . . . gave us some stuff they called bedgery or pedgery; it has a highly intoxicating effect when chewed, even in small quantities. It appears to be the dried stems and leaves of some shrub" (quoted in Johnston & Cleland, 1933–1934:202).

This was the beginning of an intense interest in identifying the plant and its active principles that was not completely resolved until the 1950s. The botanical and pharmacological confusions that plagued the discussion from the outset have been documented in great detail by Johnston & Cleland down to the 1930s in their interesting paper on *pituri* (1933–1934), so I shall only draw attention to the main points.

Pituri (i.e. Wills' *bedgery* or *pedgery*) was the indigenous word used only by the Aborigines of the Upper Mulligan river, southeastern Queensland (Roth, 1897: 51) for the plant *Duboisia hopwoodii*, which they chewed, mixed with ash. Although tobacco chewing is common throughout the desert regions, Aboriginal names for tobacco differ from group to group, and almost everywhere else only *Nicotiana* spp. were used. But like kangaroo—originally a Guugu-Yimidhir word for a particular species of macropod, but introduced into English by Captain Cook for them all—*pituri* entered English as the term for all indigenous chewing tobaccos. The inevitable result was that across the continent, whatever people chewed was likely to be identified as *D. hopwoodii*, even though this was most unlikely to be correct. The plausibility of the identification was enhanced by the widespread distribution of *D. hopwoodii* through most of the arid zone, and the rapid assimilation of the term *pituri* into Aboriginal English. From early on, however, plants also known in English as *pituri* were recorded as animal poisons. The Aborigines in many areas of Central and Western Australia would crush a few of these leaves in small pools of water used by emus and kangaroos, so that their movements would become uncoordinated, making them easy to kill. Camels, too, were reported as dying after eating *pituri* leaves, no more than 30 being required to kill an adult animal (Terry, 1974:63). The question of how an animal poison could also be used as a chewing tobacco called out for clarification.

The active alkaloid in *D. hopwoodii* was variously identified as pituria, piturine duboisine and nicotine (see Johnston & Cleland, 1933–1934:207–208) by different workers. One of their problems arose from the fact, subsequently confirmed, that *D. hopwoodii* appeared to have a different alkaloid from the only other common member of the genus, *D. myoporoides*, which is now known to contain hyoscine and hyoscyamine (see Barnard, 1952:14). Eventually in 1934, Hicks and Le Messurier established that it was not nicotine, but d-nor-nicotine, a chemical four times as strong as nicotine (Hicks, 1963:54) that was the active principal in *D. hopwoodii*. Subsequent research has shown, however, that the situation is not as simple as this. The chemical properties of the plant vary not only with seasons and state of maturity, but apparently by region. Bottomley, Nottle & White (1945:18–19) analysed 58 samples of *D. hopwoodii* from Western Australia and found only one specimen with a chemical composition like the material examined by Hicks and Le Messurier, while all the others contained nicotine. According to

Barnard (1952:14), the evidence points to the plants of Western Australia and western Queensland containing mainly nicotine, while those of South Australia and Central Australia contain nor-nicotine.

The significant point is that in most areas chewing tobacco is rarely ever *D. hopwoodii*, but nearly always one or other species of indigenous *Nicotiana*. Apart from western Queensland, *D. hopwoodii* appears to be used only when *Nicotiana* species are in short supply or unavailable, and even then it is usually mixed with the ends of a *Nicotiana* quid (Hicks, 1963:54). *D. hopwoodii* was only widely used as an animal poison.

The preparation of all *Nicotiana* is straightforward. Freshly collected leaves are either torn up and lightly masticated into a quid and then rolled with wood ash, or more usually the leaves are sun or fire dried first (see Thomson, 1961:6–7). A wide range of species are used for ash including *Acacia salicina* (Aiston, 1930:49), *A. pruinocarpa*, *A. coriacea*, *A. ligulata*, *A. aneura*, *A. kempeana*, *Grevillea striata* (Meggitt, 1966:126), *Cassias* and even *Eucalyptus* sp. The quid is then ready for chewing or storing behind the ear. It never seems to have been smoked, although smoking was introduced to the north coast by Indonesian sailors several hundred years ago (see Thomson, 1939). The literature is obscure as to what was smoked in the elaborate pipes common in Arnhem Land, but for Cape York, Thomson (1939:83) mentions *Derris trifoliata* var. *macrocarpa* and *Grewia polygama*. Sometimes, a binder is added to the quid. Johnston and Cleland (1933–1934:275 and 277) report the Aranda and Pitjantjatjara using wallaby, euro or even rabbit hair, and Roth (1897:100) refers to native flax (probably *Linum marginale*) for the Boulia area of Queensland. Other additives recorded are yellow ochre and, in more recent times, sugar (Johnson & Cleland, 1933–1934:270, 275).

The species used depends, of course, on availability, but wherever there is more than one species, there is a preference for the stronger varieties: these are generally species found high on the hillsides, such as *N. gossei*. The following remarks on "*pituri*" by a Pintupi tribesman reveal this preference and some social aspects of tobacco use:

"Pluck pitcheri [the tribesman used the word *mingkulpa*, a general word for all tobaccos in Pintupi; pitcheri is the translator's term] plants. Let us cook and eat it. Pluck pitcheri plants. Don't bring back the weak leaves—bring back the strong ones. Let us try it first. Don't bring back the weak leaves without trying it. Let us bring back ash tree to mix with the pitcheri. Let us eat it together with the ash, we who are starving for pitcheri. Let us eat it, so it can burn our throats. When walking without water, chewing pitcheri is good to keep one alert. Let us cook and eat pitcheri. One should break his lump in half and give it to another. After preparing it, let us hide it in the shelter, so the women won't grab it from us. Let us carry it in our pockets. If you keep it where people can see it, they ask you for it, and finish it all up. Not only pitcheri, but tin of tobacco and cigarettes as well. We are all eating that pitcheri which is from the white man, cigarettes and tobacco, it is very good pitcheri. That which grows on the hills is called pinatilypa [*N. gossei*?]." (Hansen & Hansen, 1974a:13–14).

Most desert groups in Central Australia have access to three or four different species, the most commonly used being *N. ingulba*, as it is abundant and widespread. Burbidge notes (1960:343) that many species are usually associated with disturbed areas, such as recent regrowth on burnt patches in the shade of trees and shrubs.

Indigenous tobaccos were traded widely, although there is little accurate information on the routes and quantities involved, and supplies of dried leaves were stockpiled in caves (see Finlayson, 1936:45) and by the Walpiri (observation by the present author, 1972). The most detailed accounts are from Roth and Howitt, referring to the areas of southwestern Queensland and northeastern South Australia:

> "If all is well, pituri [*D. hopwoodii* in this area] arrives in Boulia in the rough about the beginning of March. By "in the rough" is meant the condition, very much like half-green half-yellow tea with plenty of chips, in which it is conveyed in the dilly-bags for barter, etc. The pituri shrub itself flowers about January. The supply for the Boulia District is obtained in the neighbourhood of Carlo (*vel* Mungerebar), on the Upper Mulligan. ... the plant grows further eastwards than this, though in scattered patches only, e.g. about sixteen miles westwards of Glenormiston head station; a patch of it is also said by the Mitakoodi aboriginals to be growing in one of the gullies at Cloncurry, on the Rifle Mountain (where the old target-range used to be). From Boulia and Marion Downs, from Herbert Downs and Roxburgh, messengers are sent direct to the Ulaolinya tribes at Carlo with spears and boomerangs, "Government" and other blankets, nets, and especially red-coloured cloths, ribbons, and handkerchiefs to exchange and barter for large supplies of the drug. On its advent at Roxburgh the pituri may travel partly up the Georgina and partly along the ranges to the Kalkadoon, who may supply the Mitakoodi with it, but very little gets further eastward. From Boulia it is sent up the Burke, and so through the Yellunga and Kalkadoon, again carried to the Mitakoodi, or may be forwarded on to Warenda and Tooleybuck. Marion Downs sends it *via* Springvale, etc., to the Middle Diamantina, whence it may go up as far as Elderslie and Winton, very little, if any, ever reaching the Thomson River." (Roth, 1897: 100).

In Central Australia, the trade ran out from the hills to the people to the south and north of the ranges, trips of 150 miles being recorded from the Serpentine Lakes area to the Blackstone ranges by bush dwelling Aborigines in the 1950s (see Macaulay, 1960:5). Today, landrovers and cars are used to visit productive areas, when the chance allows, but shop bought chewing tobacco and cigarettes are replacing indigenous chewing tobaccos, at least with the younger people, in all but the remotest areas.

Many Central Australian groups have a generic term for tobacco. Thus the Southern Walpiri use the term *tjanyungu* to cover all kinds, ranked below in decreasing order of preference:

	N. gossei (*tjunpunpa*)
tjanyungu	*N. ingulba* (*yarunpa*)
	N. simulans (*tjungarayi tjungarayi*)
	Commercial tobaccos (ranking *vis-a-vis* the traditional tobaccos varies with age of peron spoken to).

The Alyawara have the following terminology and preference ranking (O.Connell, n.d.):

ingulba

[*N. gossei*] (*ingulba inbinba*)
This is known of but does not occur locally.
N. ingulba (*ingulba ngunjiga*)
Goodenia lunata (*ingulba ndarinya*)
N. benthamii (*ingulba pudura*)
N. velutina (*ingulba ndurilba*)
Very rarely used.
N. megalosiphon (*ingulba ndurilba*)
Very rarely used.

The reported use of *Duboisia myoporoides* to make an intoxicating drink in 1867 (Barnard, 1952) by making a hole in the trunk of this tree and filling it with water overnight is not well substantiated. Indeed, it is the only such report for this species without any firm location. However, a similar practice is recorded in Tasmania with *Eucalyptus resinifera*, but in this case, it was simply the sap which, if allowed to ferment, became intoxicating (Roth, 1899:94). There is also one reported use of *Datura leichhardtii* to make an intoxicating drink (see appendix), but apart from this doubtful case, there is no other reference to the use of this genus in Australia.

The only other kind of recorded use as a medicine of a member of the Solanaceae family is of *S. lasiophyllum*. In the Carnarvon, Onslow and Roeburn area of Western Australia, a decoction of roots of this plant was applied as a poultice to leg swellings (Webb, 1969:142, 145). No other medicinal uses are recorded.

ACKNOWLEDGEMENTS

I am most grateful to the people of Yuendumu whose friendly interest in explaining their way of life made it possible for myself and my wife to collect much of the information recorded here. I am indebted to my wife, Rosalind Peterson, who, with the help of Walpiri women, enthusiastically collected and documented plants used in the area. We made no special study of Solanaceae, consequently there are gaps in the information, particularly with respect to ethnotaxonomy.

I would like to thank Peter Latz, John Maconochie and Clyde Dunlop of the Alice Springs Herbarium for identifying plant specimens, and Peter Latz for generously giving me the benefit of his interest in Aboriginal plant use and knowledge of the Solanaceae. I am much indebted to David Symon for his careful reading of a first draft and for drawing my attention to references that I would otherwise have missed.

I am also grateful to Mrs P. Maggiore of the Western Australian Institute of Technology for the chemical analysis of the specimens of *S. centrale* and *S. chippendalei*; to Mrs J. Goodrum for the drawing; Dick Kimber for providing fresh specimens of *S. chippendalei* for analysis together with some interesting observations; Richard Gould for the photograph of *S. centrale* preparation; Jim O'Connell for his interest and information; and Noel and Phil Wallace for their helpful response to my enquiry.

APPENDIX

An annotated list of Solanaceae of ethnobotanical importance

The number of authors who have referred to the use of Solanaceae by Aborigines is limited. The most frequent references are to be found in papers resulting from the expeditions sponsored by the Board for Anthropological Research of the University of Adelaide to Central Australia, on which Cleland, Johnston, Tindale and Hicks all served on several occasions. Cleland was the most interested in Aboriginal plant use, but he had great difficulty with the Solanaceae. Thus, although he refers to *S. diversiflorum*, *S. eremophilum*, *S. nemophilum* and *S. phlomoides*, these species either do not occur in the areas he visited, or are misidentifications. None of them is known to be used. Similarly, the references to *N. occidentalis* and *N. suaveolens* are also incorrect.

Aboriginal names for various Solanaceae are reported in the literature, but their status is equivocal in almost all cases. Authors sometimes write as if there were a single name used by all Aborigines, whereas there were in fact two hundred languages and two to three times as many dialects. While some names were shared by neighbouring languages and dialects, many were not. Further, establishing the correct Aboriginal name for a particular species is much more difficult than many collectors appreciate. Unless plants are named spontaneously *in situ*, the likelihood of error is high, particularly where the questioner has negligible control of the language. I have, however, included some Aboriginal names in this Appendix, where they have been checked by recent fieldwork associated with professional identification of the plants, or are names common in the literature from the early period. Any names about which there is uncertainty I have placed in inverted commas.

Anthocercis myosotidea F. Muell. Distribution: mallee scrubs of western Victoria, the Murray area and Eyre peninsula of South Australia. George French Angus mentions that an intoxicating root was used by some South Australian Aborigines without identifying it. Cleland (1966:120) comments that it is difficult to know what this plant might be, and suggests that in the absence of Ranunculaceae or Umbelliferae with likely tap roots in the area, *A. myosotidea* is the only likely candidate.

Datura leichhardtii F. Muell. ex Benth. Distribution: arid zone. Not used, but Cleland & Tindale (1959:138) comment: "Once two boys ate these, and they became drunk. Their mother wanted to know what was wrong. Very 'strong' and dangerous, only makes one drunk." According to them, the Aranda call it a "*ranga*" *rakata*.

Duboisia hopwoodii F. Muell. Distribution: arid zone from western Queensland to Kimberley coast (see Barnard, 1952:5). Widely used to incapacitate emus and larger mammals by crushing leaves in drinking water. There is a somewhat doubtful report from Western Australia (Webb, 1969:141–145) of balls of ground leaves mixed with ash being strewn near water holes to catch game, particularly emus. This species was the main indigenous tobacco in the Mulligan River area, southeastern Queensland, where it was known as *pituri* (Roth, 1897:51), but was rarely used for chewing elsewhere (see Johnston and

Cleland, 1933–1934, and Barnard, 1952, for an exhaustive discussion of the extensive literature on this species).

Duboisia myoporoides F. Muell. Distribution: eastern coastline of Australia only (see Barnard, 1952:5). Almost certainly not used, although there is one secondhand report of use as an intoxicant (see Barnard, 1952).

Lycium australe F. Muell. This shrub occurs in the drier areas of southern Australia and records of its use are very limited, although it was probably regularly eaten. Koch (1898) records that it was eaten by the Aborigines at Mt. Lyndhurst (northern Flinders Ranges, South Australia), and they called it "Beeree" and "Wandneree". Black (1957) states that it was eaten by Aborigines and this statement is repeated by Ewart (1930), both may be repeating Koch. An early specimen in Melbourne from the Murray River gives the Aboriginal name as "Tylany" but makes no mention of its use.

Nicotiana benthamii Domin. Distribution: scattered areas of northern parts of the arid zone (see Burbidge, 1960:346), but mostly absent from Macdonnell and Musgrave Ranges (Latz, 1974:282). Used by the Walpiri as a chewing tobacco in the Granites area (Cleland & Johnston, 1939:26). This species was prized, where it was available (Latz, 1974:282).

Nicotiana cavicola Burbidge Distribution: central western part of Western Australia (see Burbidge, 1960:347). Used as a chewing tobacco mixed with ash from white river gum bark by people on the Fortescue River (von Brandenstein, 1970:14–15, 239–240).

Nicotiana excelsior J. M. Black Distribution: a rare plant (Latz, 1974:281), found mainly in the Petermann and Musgrave Ranges area of the arid zone (see Burbidge, 1960:346). Used as a chewing tobacco. Preferred over *N. gossei* in the Ernabella area, where flowers and flowering stalks as well as leaves are chewed, mixed with ash of *Acacia aneura* and red gum (Johnston & Cleland, 1933–1934:277). According to Cleland (1932:38), it is used by Alyawara who say that among other things it will prevent vomiting, but this may be a misidentification, as it does not fit with the known distribution.

Nicotiana glauca Graham Distribution: introduced. Apparently growing wild in parts of Western Australia. Highly prized by Pitjantjatjara according to Wallace & Wallace (n.d.), and known as *minkulpa mulapa* (true *minkulpa*).

Nicotiana gossei Domin. Distribution: a rare plant (Latz, 1974:282) found in the southern ranges of Northern Territory (see Burbidge, 1960:347). Used as a chewing tobacco. Most highly preferred by the Walpiri and Aranda (e.g. see Latz, 1974:282). Found in sheltered areas of rocky gorges, and harvested by preference when the first blossoms occur, according to one source (see Johnston & Cleland, 1933–1934:275). Contains 1.10 % nicotine (see Webb, 1948:156). The Walpiri call this species *tjunpunpa*, the Pitjantjatjara *pitulpa*.

Nicotiana ingulba J. M. Black Distribution: spinifex plains areas of Central Australia (see Burbidge, 1960:346; and Thomson, 1961). The most commonly used chewing tobacco in this area because of widespread availability. Thomson (1961) provides details of preparation and use by the Pintupi. The Walpiri call it yarunpa; the Alyawara *ingulba ngunjiga*.

Nicotiana megalosiphon Heurck and F. Muell. Distribution: widespread in Central Australia, but restricted to water courses. This species is recognized by the Alyawara but rarely, if ever, used (O'Connell, n.d.). They call it *ingulba ndurilba*. According to Latz, this species is very similar to, or possibly identical with, *N. simulans*.

Nicotiana rotundifolia Lindley Distribution: Cleland & Tindale (1959:138) give Aranda and Pintupi names for this plant, but comment that it is not used by the Aranda, without making it clear whether Pintupi use it or not. However, Burbidge (1960:371) has examined the specimen and finds it is a misidentification for *N. simulans*. *N. rotundifolia* is not found in the Northern Territory, but is confined to Western Australia (see Burbidge, 1960:347).

Nicotiana simulans Burbidge Distribution: widespread in the arid regions of Western Australia, South Australia and the southern portion of the Northern Territory (see Burbidge, 1960:346). Rarely used as a chewing tobacco.

Nicotiana suaveolens Lehmann Distribution: this species is found only in Victoria and eastern New South Wales (see Burbidge, 1960:346). Cleland and Johnston, however, report it from the Mount Leibig area, but comment that although named, it is unused (1933:123; see also Johnston & Cleland, 1933–1934:275, 277, for reports from Ernabella area of South Australia). Latz has checked the original specimens and finds they were misidentifications for *N. velutina* and *N. simulans*.

Nicotiana velutina Wheeler Distribution: northern South Australia and the adjacent parts of the Northern Territory, Queensland and New South Wales (see Burbidge, 1960:346). It is the second most common species of the genus in Central Australia (Latz, 1974:282). Cleland and Tindale record this species as being used in the Aranda and Pintupi areas, but note that it is not ranked as highly as *N. ingulba* (1959:139). According to Latz, however, the Aranda shun this plant (1974:282).

Physalis minima L. Distribution: common in the tropical north and parts of New South Wales (Maiden, 1889:52). There is some difference of opinion as to whether it is endemic or not. Maiden reports Aborigines in the Cloncurry River area of North Queensland eating the berries (1889:52), and Specht, the Aborigines of the East Alligator River where the fruit ripens during the latter half of the dry season (1958:499).

Solanum aviculare Forst. Distribution: found in the temperate woodlands of southern and eastern Australia. Smyth (1878; 2:174) records that the Aborigines of Lake Condah region eat this species. Maiden (1889:58) notes that it can only be eaten when fully ripe; otherwise it has an acrid taste and burns the throat. Ripeness is indicated by the outer skin bursting. With *S. laciniatum*, *S. vescum* and *S. simile*, this species forms a very distinct and Australasian group of *Solanum* species (Symon: pers. comm.).

Solanum centrale J. M. Black Distribution: common throughout much of the arid region. Fruit eaten when ripe or after drying naturally on plant. The dried fruit can be ground into paste and either eaten or moulded into round balls that keep indefinitely. This species is a desert staple (see Gould, 1969; Thomson, 1962:11–12; Sweeney, 1947:290). The Walpiri call it *yakatjiri* or *kamparrarpa*, the Pitjantjatjara term.

Solanum chippendalei Symon *ms*. Distribution: common throughout much of Western Australia and the Northern Territory north of Alice Springs. Fruit eaten after removal of black seeds and surrounding skin. The fruits can be stored by sun or fire drying, strung on sticks. Dried fruits are soaked before use. A staple food (see also Carnegie, 1898:230–231; Sweeney, 1947:290; Thomson, 1962:11–12; Gould, 1969). The Walpiri distinguish two varieties: one growing on the plains *wanakitji*, the other on the hillsides *ngayiyaki*. The Ngatatjara call this species *ngaru* and the Pintupi *pura*.

Solanum cleistogamum Symon Distribution: Central Australia. It is commonly eaten by the Walpiri without preparation. This is the unidentified *Solanum* referred to by Cleland & Tindale (1954:85) as *japindiri*. The Walpiri call it *yipuntri*, the Pitjantjatjara *wiriny-wirinypa*.

Solanum coactiliferum J. M. Black Distribution: Central Australia. The fruit of this species is not eaten by the Walpiri and Aranda, but is, after preparation, by the Pitjantjatjara (Wallace & Wallace, n.d.). Fruits are squeezed between stones, washed and heated before eating. The Walpiri call this species *warakalukalu*, and the Pitjantjatjara *itunpa*.

Solanum dallachii Benth. Distribution: tropical Queensland. Bailey (1901:1087) gives an Aboriginal name for this plant, "*koori*", but does not make it clear that it was eaten.

Solanum dioicum W. V. Fitzg. Distribution: north-western Australia. At Kalumburu Mission, this species grows on the flood plains amid introduced pasture plants. Fruit eaten in dry season. Known as *gilu* in that area (Crawford, n.d.).

Solanum diversiflorum F. Muell. Distribution: northwest Northern Territory and Kimberley region. Cleland and Johnston, however, give Aranda and Luritja names for this species and indicate that the outside of the fruit is eaten, after the pulp and seeds have been removed. They indicate some doubt about their identification (1933:123), and Latz confirms it is a misidentification for *S. chippendalei*.

Solanum ellipticum R. Br. Distribution: widespread in the arid areas of Australia. The fruit is eaten without preparation and is bitter to taste, unless completely ripe (cf. Cleland & Johnston, 1933:115, with Cleland & Tindale, 1954:85). Cleland & Tindale comment on this species that it "is of real importance, as it is said to furnish small edible green tomatoes all the year around and is a widely distributed species ... numerous enough to furnish food on most occasions and situations" (1959:127). The Walpiri call this species *wangki*, and the Pitjantjatjara *tawal-tawalpa*. Specimens at Kew collected in 1899 by M. Koch from Mt. Lyndhurst, South Australia, bear the Aboriginal name "*yoomeroo*" and "*yoomerov*". This would appear to be either a Kujani or Dieri term.

Solanum eremophilum F. Muell. Distribution: Central Australia. According to Cleland & Tindale (1959:138), the fruit of this species is eaten by the Aranda and Luritja, and the Ngatatjara, according to Gould (1969:262). There is no record of use by the Walpiri. Gould's description of the preparation of this species is identical with the Walpiri preparation of *S. chippendalei*, but Cleland & Tindale (1959:138) simply comment that it yields a fruit "very nice to eat" without reference to cleaning, although in the same paper, they refer to the need to clean *S. chippendalei* (which they identified as *S. melanospermum*). Both references are misidentified for *S. chippendalei*.

Solanum esuriale Lindl. Distribution: Central Australia, western New South Wales and Queensland (Maiden, 1889:58). This is among the first of the solanums to be recorded as food: in April 1836, the guides of Sir Thomas Mitchell gathered and ate them on the Lachlan river (1839,2:43). Chippendale & Murray (1963:95–96) note that it is poisonous to stock, but probably only when unripe. Ewart in Flora Victoria (1930:1005) gives an Aboriginal name, "*quena*". A specimen in the Adelaide Herbarium collected by M. Koch in September 1898 at Mt. Lyndhurst, South Australia, gives the name "*puddaddee*"; others from western New South Wales in the Sydney Herbarium bear the name "*goomi*". A specimen collected by Basedow in 1919, now at Kew, bears the name *intunbu*, which would appear to be the same word the Pitjantjatjara also use for *S. coactiliferum*.

Solanum hystrix R. Br. Distribution: Eyre peninsula of South Australia (Maiden, 1889:58). Richards (1882:136–137) records that the fruit is used, but only with the pounded and baked bark of the mallee root (*Eucalyptus* sp.). Before eating, the Aborigines remove the seeds and make the fruit and bark into a cake. Unprepared, the fruit makes the throat sore and inflames cuts. Richards notes that it prevails in burnt scrub lands, and that if there is winter rain, two crops a year are possible.

Solanum laciniatum Ait. Distribution: New South Wales, Victoria and Tasmania. Roth (1899:95) reports the fruit of this species to be one of the principle vegetable foods of Tasmania, eaten only when "dead ripe".

Solanum lasiophyllum Dun. Distribution: Western Australia and southwest of Northern Territory. Von Brandenstein (in Webb, 1969:142, 145) records that it is used in Western Australia as a medicament: a decoction of roots is applied as a poultice to leg swellings.

Solanum lucani F. Muell. Distribution: northern Western Australia and Northern Territory. The fruits of this species are eaten by the Nunggubuyu of southeastern Arnhem Land. They call them *nga: lig* (Heath, n.d.).

Solanum melanospermum F. Muell. Distribution: northern part of the Northern Territory, Queensland and Gulf Country. This species is frequently confused with *S. chippendalei* in Central Australia (see Cleland & Tindale, 1959:138). There is no record of use of this plant.

Solanum nemophilum F. Muell. Distribution: found in southeastern Queensland, but not in the arid areas to the west. All references to this species in Central Australia refer to *S. centrale*. There is no record of use of this plant in Queensland.

Solanum orbiculatum Dun. Distribution: widespread in Western Australia and southwestern Central Australia. Fruit eaten by the Walpiri, Aranda and Pintupi (see Cleland & Tindale, 1959:137) without preparation.

Solanum petrophilum F. Muell. Distribution: widespread in southern inland from western New South Wales to Western Australia (Everist, 1974:474). Cleland & Tindale (1959:138) have a slightly ambiguous reference to this species which they say is not eaten. There is no record of use. It has been suspected of poisoning stock (Everist, 1974:474). The Walpiri call this species *ngaliyirki*.

Solanum phlomoides A. Cunn. ex Benth. Distribution: Hammersley ranges of Western Australia. Cleland & Tindale (1959:138) and Cleland & Johnston

(1939:25–26) tentatively identify this species as a food plant used by the Luritja and Pintupi peoples, but it is certainly a reference to *S. chippendalei*, since the species is only in Western Australia. They suggest that the species has affinity with *S. melanospermum* (i.e. *S. chippendalei*), and mention the need to remove the seeds before eating. This interpretation is further confirmed by Cleland & Tindale (1954:85) giving the Walpiri name as *wanakitji* which in 1972–1973 was used exclusively for *S. chippendalei*.

Solanum quadriloculatum F. Muell. Distribution: Central Australia. Named by several desert groups, but not eaten by any of them: strongly poisonous (Everist, 1974:475). The Walpiri call this species *warakalukalu*.

Solanum simile F. Muell. Distribution: throughout the mallee vegetation across southern Australia. Richards (1882:137) reports South Australian Aborigines eating the fruit after it has fallen to the ground. Consumption in great numbers causes sickness and a burning taste in the mouth. Richards notes that it springs up abundantly on burnt scrublands, and records an abundant crop in mid-February after very heavy storms the previous November. Ewart (1930:1004) gives the Aboriginal name "*oondoroo*" for this plant.

Solanum stelligerum Sm. Distribution: Southeastern Australia. Found in the open forest. Fruit eaten in the New South Wales coastal areas (see Lampert & Saunders, 1973:106).

Solanum vescum F. Muell. Distribution: New South Wales, Tasmania and Victoria (Cribb & Cribb, 1975:75). Smyth reports that this species was eaten in Victoria (1878, 1:213). Called *gunyang* by some Aboriginal groups in Victoria.

REFERENCES

AISTON, G., 1930. Magic stones of the tribes east and north-east of Lake Eyre. *Papers and Proceedings of the Royal Society of Tasmania for the year 1929: 47–50.*

ALLEN, H., 1974. The Bagundji of the Darling Basin: cereal gatherers in an uncertain environment. *World Archaeology, 5: 309–322.*

BARNARD, C., 1952. The *Duboisias* of Australia. *Economic Botany, 6: 3–17.*

BAILEY, F. M., 1901. *The Queensland Flora, Part 4*, Brisbane: Queensland Govt.

BLACK, J. M., 1929. *Flora of S. Australia* (1st ed.). Adelaide: Govt Printer.

BLACK, J. M., 1957. *Flora of South Australia, Part 4*, 2nd ed. Adelaide: Govt. Printer.

BOTTOMLEY, W., NOTTLE, R. A. & WHITE, D. E., 1945. The alkaloids of *Duboisia hopwoodii*. *Australian Journal of Science, 8:* 18–19.

BRANDENSTEIN C. G., VON, 1970. *Narratives from the North-west of Western Australia*, 3 vols. Canberra: Australian Institute of Aboriginal Studies.

BURBIDGE, N. T., 1960. The Australian species of *Nicotiana* L. (Solanaceae). *Australian Journal of Botany, 8:* 342–380.

BURBIDGE, N. T., 1963. *Dictionary of Australian Plant Genera*, Sydney: Angus & Robertson.

CARNEGIE, D. W., 1898. *Spinifex and Sand: a Narrative of Five Years' Pioneering and Exploration in Western Australia*, London: C. Arthur Pearson.

CHIPPENDALE, G. M. & MURRAY, L. R., 1963. *Poisonous Plants of the Northern Territory: Extension Article No. 2*, Alice Springs: Northern Territory Administration Animal Industry Branch.

CLELAND, J. B., 1932. Botanical notes of anthropological interest from Macdonald Downs, Central Australia. *Transactions of the Royal Society of South Australia, 56:* 36–38.

CLELAND, J. B., 1966. The ecology of the Aboriginal in South and Central Australia. In B. C. Cotton (Ed.), *Aboriginal Man in South and Central Australia:* 111–158. Adelaide: Government Printer.

CLELAND, J. B. & JOHNSTON, T. H., 1933. The ecology of the Aborigines of Central Australia; botanical notes. *Transactions of the Royal Society of South Australia, 57:* 113–124.

CLELAND, J. B. & JOHNSTON, T. H., 1939. Aboriginal names and uses of plants at the Granites, Central Australia. *Transactions of the Royal Society of South Australia, 63:* 22–26.

CLELAND, J. B. & TINDALE, N. B., 1954. The ecological surroundings of the Ngalia natives in Central Australia and native names and uses of plants. *Transactions of the Royal Society of South Australia, 77:* 81–86.

CLELAND, J. B. & TINDALE, N. B., 1959. The native names and uses of plants at Haast Bluff, Central Australia. *Transactions of the Royal Society of South Australia, 82:* 123–140.

CRAWFORD, I. M., n.d. *Aboriginal Plant Foods of Northern Kimberley.* Unpublished manuscript, Canberra: Australian Institute of Aboriginal Studies.

CRIBB, A. B. & CRIBB, J. W., 1975. *Wild Food in Australia.* Sydney: Collins.

DADSWELL, I. W., 1934. The chemical composition of some plants used by Australian Aborigines as food. *Australian Journal of Experimental Biology and Medical Science, 12:* 13–18.

EVERIST, S. L., 1974. *Poisonous Plants of Australia,* Sydney: Angus & Robertson.

EWART, A. J., 1930. *Flora of Victoria,* Melbourne: Melbourne University Press.

FINLAYSON, H. H., 1936. *The Red Centre: Man and Beast in the Heart of Australia,* Sydney: Angus & Robertson.

GOULD, R. A., 1969. Subsistence behaviour among the Western Desert Aborigines of Australia. *Oceania, 39:* 253–274.

GOULD, R. A., 1970. Journey to Pulykara. *Natural History, 79:* 56–67.

GOULD, R. A., 1971. Uses and effects of fire among the Western Desert Aborigines of Australia. *Mankind, 8:* 14–24.

HANSEN, K. C. & HANSEN, L. E., 1974a. *Mingkulpa by Charlie Tjakamarra: a Pintupi Post-Primer Reader,* Darwin: Department of Education.

HANSEN, K. C. & HANSEN, L. E., 1974b. *Pintupi Dictionary,* Darwin: Summer Institute of Linguistics.

HEATH, J., n.d. *Nunggubuyu Ethnobotany: Preliminary Outline.* Unpublished manuscript, Canberra: Australian Institute of Aboriginal Studies.

HICKS, C. S., 1963. Climatic adaptation and drug habituation of the Central Australian Aborigine. *Perspectives in Biology and Medicine, 7:* 39–57.

JOHNSTON, T. H. & CLELAND, J. B., 1933–1934. The history of the Aboriginal narcotic, pituri. *Oceania, 4:* 201–233; *4:* 268–289.

KIMBER, R., n.d. Letter dated 3 September 1975.

KOCH, M., 1899. A list of plants collected on Mt. Lyndhurst run, S. Australia. *Transactions of the Royal Society of South Australia, 22:* 114. (1898)

LAMPERT, R. J. & SAUNDERS, F., 1973. Plants and men on the Beecroft Peninsula, New South Wales. *Mankind, 9:* 96–108.

LATZ, P., 1974. Central Australian species of *Nicotiana. Australian Plants, 7:* 280–283.

MACAULAY, R., 1960. Letter report to the Commissioner of Native Welfare, Perth, Western Australia, on the Shell Lakes Boundary Dam Patrol, May–June 1959. Dated 12 January 1960, pp. 1–8.

MAIDEN, J. H., 1889. *The Useful Native Plants of Australia,* Sydney: Turner & Henderson.

MEGGITT, M. J., 1966. Gadjari among the Walpiri Aborigines of Central Australia. *Oceania, 37:* 124–147.

MITCHELL, T. L., 1839. *Three Expeditions into the Interior of Eastern Australia.* 2nd ed., 2 vols. London: T. W. Boone.

O'CONNELL, J., n.d. Unpublished field notes 1974–1975, Canberra: Australian National University.

RICHARDS, A. F., 1882. *Solanum hystrix* and *Solanum simile. Transactions of the Royal Society of South Australia, 4:* 136–137.

ROTH, H. L., 1899. *The Aborigines of Tasmania.* Halifax: King & Sons.

ROTH, W. E., 1897. *Ethnological Studies among the North-west-central Queensland Aborigines.* Brisbane: Government Printer.

SMYTH, R. B., 1878. *The Aborigines of Victoria,* 2 vols. Melbourne: Government Printer.

SPECHT, R. L., 1958. An introduction to the ethnobotany of Arnhem Land. In R. L. Specht & C. P. Mountford, (Eds) *Records of the American-Australian Scientific Expedition to Arnhem Land, 3:* 479–503. Melbourne: Melbourne University Press.

SWEENEY, G., 1947. Food supplies of a desert tribe. *Oceania, 17:* 289–299.

TERRY, M., 1974. *War of the Warramullas.* Adelaide: Rigby.

THOMAS, S. & CORDEN, M., 1970. *Tables of Composition of Australian Foods.* Canberra: Australian Government Publishing Services.

THOMSON, D. F., 1939. Notes on the smoking-pipes of north Queensland and the Northern Territory of Australia. *Man, 39:* 81–91.

THOMSON, D. F., 1961. A narcotic from *Nicotiana ingulba,* used by the desert Bindibu. *Man, 61:* 5–8.

THOMSON, D. F., 1962. The Bindibu expedition: exploration among the desert Aborigines of Western Australia. *Geographical Journal, 128:* 1–14.

WALLACE, N. & WALLACE, P., n.d. Letter dated 26 January 1976.

WEBB, L. J., 1948. *Guide to the Medicinal and Poisonous Plants of Queensland,* Melbourne: C.S.I.R.O. Bulletin No. 232.

WEBB, L. J., 1969. The use of plant medicines and poisons by Australian Aborigines. *Mankind, 7:* 137–146.

Plate 10.1

(*Facing p. 189*)

EXPLANATION OF PLATE

PLATE 10.1

A Ngatatjara woman prepares *Solanum centrale* for storage. The naturally dried fruit heaped in the background are being ground into paste which is then consolidated into balls like that in the foreground. (Photograph R. Gould.)

III. Alkaloids

11. The steroid alkaloids of *Solanum*

K. SCHREIBER

Institute of Plant Biochemistry of the Academy of Sciences of the GDR, Halle/Saale, German Democratic Republic

The recent knowledge in the occurrence and chemical structure of the steroidal alkaloid glycosides and alkamines from *Solanum* and related genera are reviewed. Special emphasis is given to a discussion of the following five groups of alkaloids representing different types of structure: the solanidanes, the spirosolanes, the 16-unsubstituted 22,26-epiminocholestanes, the alkaloids with an α-epiminocyclohemiketal moiety, and the 3-aminospirostanes.

CONTENTS

INTRODUCTION

As a result of a lot of work done in a number of laboratories all over the world, *Solanum* steroid alkaloids have been isolated from more than 300 species of the plant families Solanaceae and Liliaceae, especially from plants of the genus *Solanum* and the related one *Lycopersicon* in which they generally occur as glycosides (see, e.g. Prelog & Jeger, 1953, 1960; Schreiber, 1968, 1970, 1974). All the known steroidal alkamines (about 60), the structures of which have been fully established, possess the C_{27} skeleton of cholestane. They belong to one of the following five groups representing different types of structure:

(1) hexacyclic tertiary bases with a fused indolizidine moiety, the so-called solanidanes (cf. solanidine, I),

(2) spirosolanes (cf. solasodine, II),

(3) 22,26-epiminocholestanes (cf. solacongestidine, III),

(4) the alkaloids with an α-epiminocyclohemiketal moiety (cf. solanocapsine, IV), and

(5) the 3-aminospirostanes (cf. jurubidine, V), a novel type of nitrogenous steroidal sapogenins (spirostanes) with a spiroketal moiety.

SOLANIDANE ALKALOIDS

The earliest known *Solanum* alkaloid, solanidine (I), occurs as glycosides in the potato plant, *Solanum tuberosum* L., as well as in a number of other, especially

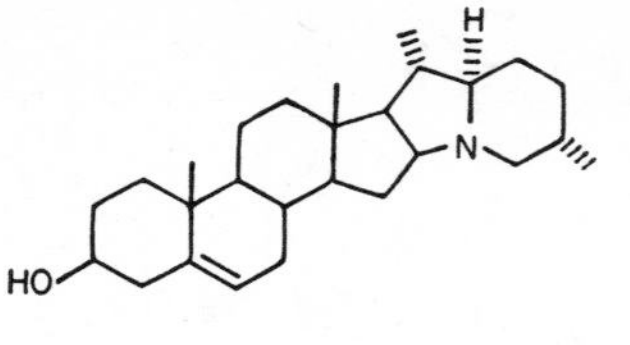

I

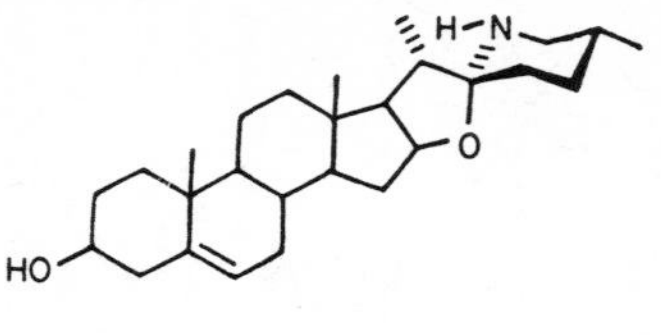

II

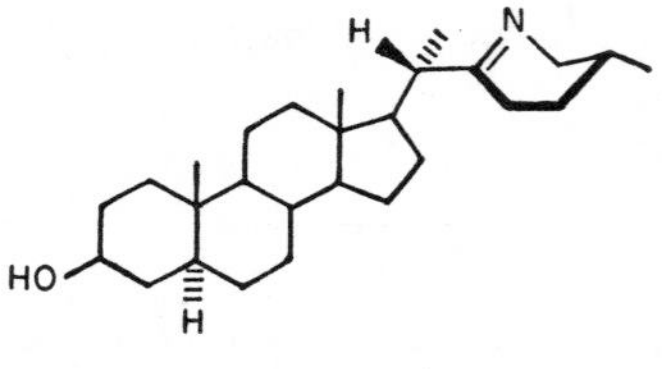

III

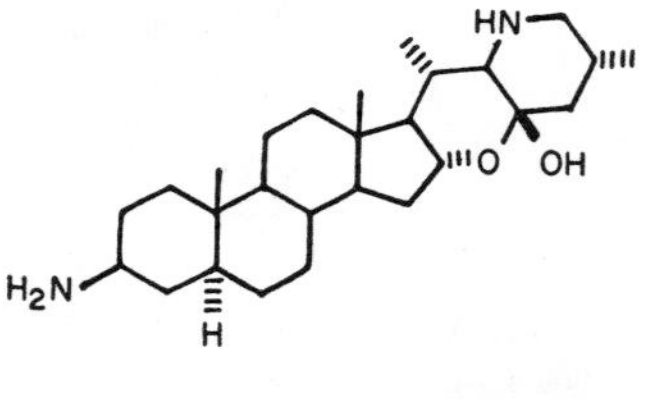

IV

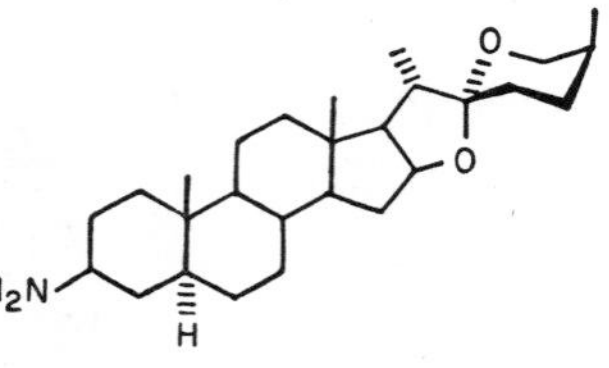

V

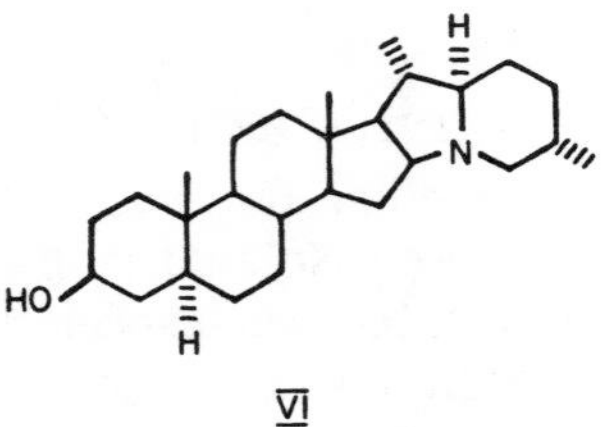

VI

tuber-bearing, *Solanum* species and in *Veratrum* species. Its dihydro-derivative, demissidine (VI), was isolated for the first time from the Mexican wild potato *Solanum demissum* Lindl. (Kuhn & Löw, 1947), but has been found, too, in some other tuber-bearing *Solanum* species, e.g. in *Solanum horovitzii* Buk. (Prokoshev, Petrochenko & Baranova, 1952; Petrochenko, 1953, 1956), and in selected plant introduction lines of *Solanum chacoense* Bitt. and *Solanum commersonii* Dun. (Osman, Herb, Fitzpatrick & Sinden, 1976), as well as, in addition to tomatidine in a number of mutants of *Lycopersicon esculentum* Mill. and *L. pimpinellifolium* (Jusl.) Mill. (Schreiber *et al.*, 1961). It should be mentioned, that *Solanum demissum* Lindl. in addition to demissidine, also contains considerable amounts of tomatidine (Schreiber & Aurich, 1963). The complete structures and stereochemistry of solanidine (I) and demissidine (VI), as illustrated in the formulae, have been established by X-ray analysis of demissidine hydroiodide (Höhne, Schreiber, Ripperger & Worch, 1966; Höhne 1972).

According to these results and in contrast with former statements (Rosen &

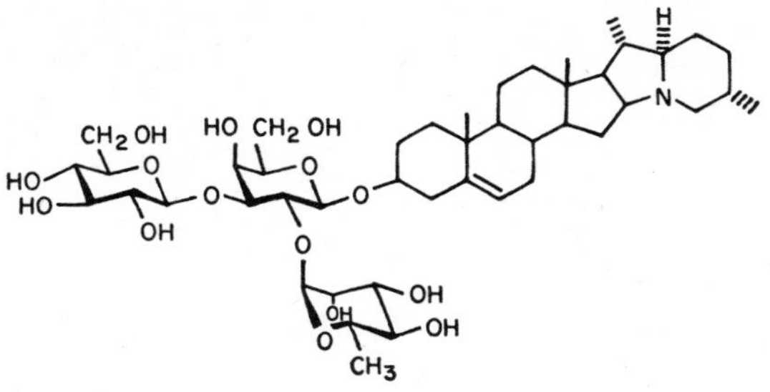

VII

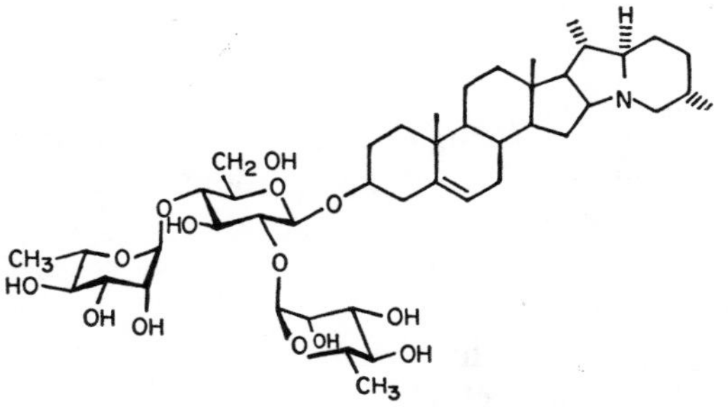

VIII

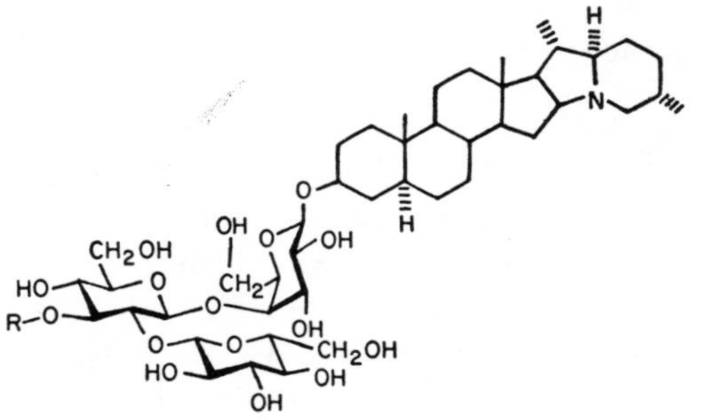

IX : R = β—D — Xylopyranose
X : R = β—D — Glucopyranose

Rosen, 1954; Sato & Ikekawa, 1961; see Fieser & Fieser, 1959, 1961), the natural solanidanes (cf. formulae I and VI) possess 20S : 22R : 25S : NS configuration. This means that the hydrogen atom at C-22 is oriented to the rear and that the six-membered ring F of the heterocyclic *trans*-indolizidine system has a chair conformation, putting the (25S)-methyl in an equatorial position. The two main glycosides of solanidine (I) isolated from *Solanum tuberosum* L. as well as a number of related wild potatoes are α-solanine (VII) and α-chaconine (VIII). According to investigations done by Kuhn & Löw (1954 and 1955a) the sugar moiety of α-solanine, which is named β-solatriose, possesses the structure of a branched α-L-rhamnopyranosyl-β-D-glucopyranosyl-β-D-galactopyranose as pictured in the formula VII. On the other hand, the trisaccharide portion of α-chaconine (VIII), the so-called β-chacotriose, is the branched bis-α-L-rhamno-

XI

pyranosyl-β-D-glucopyranose (Kuhn *et al.*, 1955b). The other naturally found minor glycosides of solanidine (I), the β- and γ-solanines and β- and γ-chaconines, are products of a partial hydrolysis of the respective α-glycosides.

The principal glycoside of demissidine (VI), the so-called demissine (IX) isolated from *Solanum demissum* Lindl., also possesses a branched tetrasaccharide moiety, the β-lycotetraose, which has been shown to contain one molecule each of D-xylose and D-galactose as well as two molecules of D-glucose (Kuhn *et al.*, 1953, 1956, 1957). Special selected plant introduction lines of *Solanum chacoense* Bitt. and *Solanum commersonii* Dun. are found to contain, instead of α-solanine and α-chaconine, demissine (IX) as well as a new glycoalkaloid, commersonine (X) (Osman *et al.*, 1976). The leptines (cf. leptine II, formula XI) represent another interesting type of related glycoalkaloid isolated by Kuhn & Löw (1957, 1961a, b) from a further special line of *Solanum chacoense* Bitt. They are mono-acetylated compounds which, by action of esterases or by mild alkaline treatment, afford the corresponding acetyl-free leptinines. The acetyl group is not located in the sugar moiety but on a second hydroxy group at C-23 of the respective aglycone, leptinidine. Leptinidine has the structure of 23-hydroxysolanidine. In connection with the aforementioned known stereochemistry of the basic solanidane skeleton it has been established by us that the 23-hydroxyl in leptinidine is in the axial β-position (Schreiber & Ripperger, 1967a).

SPIROSOLANE ALKALOIDS

Another group of widely distributed *Solanum* alkaloids is the so-called spiro-solanes (cf. formula II) which are nitrogenous analogues of the spirostanes, that is of the steroidal sapogenins. Out of the four most important ones, solasodine (II) and soladulcidine (XII) belong to the 25*R* series, tomatidenol (XIII) and the tomato alkaloid tomatidine (XIV) to the 25*S* series with a reversed *S*-configuration at the spiro C-atom 22.

Solasodine (II) was shown to occur as glycosides in more than 100 *Solanum* species (see e.g. Schreiber, 1968). Soladulcidine (XII) and tomatidenol (XIII) have been isolated as main alkaloids, after hydrolysis of the naturally occurring glycosides, from different chemovarieties of the bitter-sweet nightshade, *Solanum dulcamara* L. (cf. Schreiber, 1968). Tomatidenol (XIII) was found, too, in addition to solanidine (I), in *Solanum tuberosum* L. (Schreiber, 1957, 1963; Shih & Kuć, 1974). Tomatidine (XIV), isolated for the first time from different *Lycopersicon* species, was shown to occur also in some tuber-bearing *Solanum* species, especially of the section Acaulia (cf. Schreiber, 1968). The solasodine glycosides solasonine and solamargine as well as the tomatidenol glycosides α- and β-solamarine possess the same trisaccharide moieties as α-solanine (VII) and α-chaconine (VIII), that is the β-solatriose and the β-chacotriose, respectively. On the other hand, the tomatidine glycoside α-tomatine has the same tetrasaccharide portion as demissine (IX).

Solasodine (II) has been also isolated from *Cestrum parqui* L'Hérit. (Silva, Mancinelli & Cheul, 1962) as well as a mixture of both solasodine (II) and tomatidenol (XIII) from *Cyphomandra betacea* Sendtn. (Schröter & Neumann, 1964; Schröter *et al.*, 1964).

II : Δ⁵ → II : Δ^5
XII : 5αH

XIII : Δ^5
XIV : 5αH

Like the spirostanes, the analogous spirosolane alkaloids can be easily degraded by acetylation leading to the O(3), N-diacetate (XV), its proton-catalysed pseudo-isomerization to the nitrogenous furostane derivative (XVI), CrO_3 oxidation to compound XVII and hydrolysis leading to XVIII. Thus, for instance, solasodine (II), like diosgenin, also gives pregnadienolone acetate (XVIII) in about 65% yield. Because of these results, first obtained by Sato, Katz & Mosettig, (1951a, b, 1952) as well as by Kuhn *et al.* (1952), the *Solanum* spirosolane alkaloids have been receiving increased attention as a convenient and most promising starting material for the commercial synthesis of hormonal steroids. These possibilities have encouraged extensive studies in this field, both with a view to improving the conditions of degradation and to finding high-yielding plant species and varieties (see Schreiber *et al.*, 1967). Until quite recently solasodine (II) appeared to be the most suitable alkamine and *Solanum laciniatum* Ait., the best available plant source. In addition to this species, solasodine has been isolated in the form of its glycosides from more than 100 species. One of these, the Indian *Solanum khasianum* C. B. Clarke, has been especially recommended (Chaudhuri & Rao, 1964; Saini, Mukherjee & Biswas, 1965; Chaudhuri & Harariha, 1966; Maiti & Mookherjea, 1968; Iyengar, 1968). Finally, as already mentioned, a chemovariety of the bitter-sweet nightshade (*Solanum dulcamara* L.) was shown to contain the more advantageous spirosolane alkaloid, tomatid-5-en-3β-ol (XIII), in practical amounts. This stereoisomer of solasodine can also be degraded, in high yield up to 75%, to pregnadienolone acetate (XVIII) (Schreiber & Rönsch, 1963, 1965a, b; Bognár & Makleit, 1965).

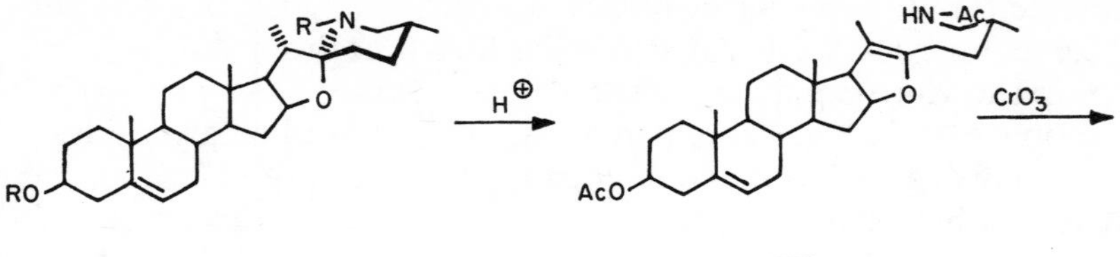

II : R=H
XV : R=CH$_3$CO

XVI

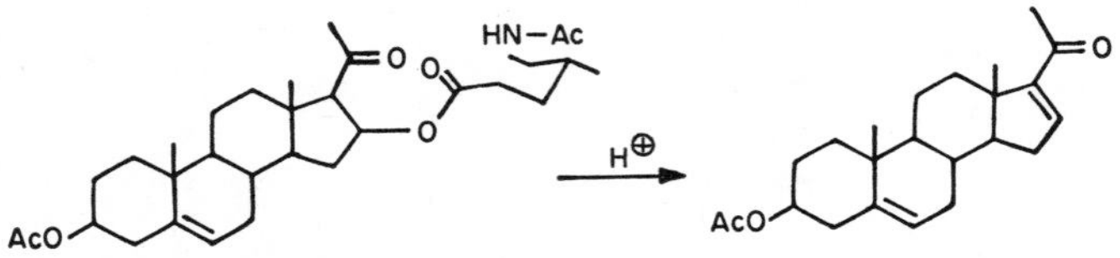

XVII

XVIII

22,26-EPIMINOCHOLESTANE ALKALOIDS

Some new C$_{27}$ steroidal alkaloids representing novel structural types have been isolated both from *Solanum* and *Veratrum*. For instance, verazine (XIX) from *Veratrum album* ssp. *lobelianum* (Bernh.) Suessenguth (Tomko & Vassová, 1964; Adam, Schreiber & Tomko 1967a, b; Tomko, Adam & Schreiber, 1967) and solacongestidine (III) from *Solanum congestiflorum* Dun. (Sato *et al.*, 1969; Adam, Voigt & Schreiber, 1971) are both 16-unsubstituted ring-E-opened alkaloids. Very likely these are biosynthetic intermediates, for example, of the spirosolanes tomatidenol (XIII) and soladulcidine (XII), which are actually identical with 16β-hydroxyverazine and 16β-hydroxysolacongestidine, respectively, which are not stable in this form but undergo spontaneously stereospecific cyclization to the spiroaminoketals (cf. Schreiber & Adam, 1960, 1961, 1963; Adam & Schreiber, 1966).

The *Solanum* alkaloids mentioned have been not only degraded to pregnane derivatives but also synthesized starting from 3β-acetoxypregna-5,16-dien-20-one (XVIII). A number of reaction sequences led both to the four spirosolane alkaloids, soladulcidine (XII), tomatidine (XIV), solasodine (II), and tomatidenol (XIII), and to the natural solanidane derivatives, demissidine (VI) and solanidine (I), simultaneously confirming the established structures. The synthesis of the 22,26-epimino-cholestanes verazine (XIX) and solacongestidine (III) have been associated with the above sequences (cf. Schreiber, 1968).

ALKALOIDS WITH α-EPIMINOCYCLOHEMIKETAL MOIETY

Some time ago we were able to complete successfully the synthesis of another steroidal *Solanum* alkaloid, solanocapsine (IV) (Ripperger, Sych & Schreiber, 1970, 1972a, b). This unusual alkaloid was isolated nearly 50 years ago from *Solanum pseudocapsicum* L., a beautiful ornamental plant, also called the Christmas cherry

(Breyer-Brandwijk, 1929). According to early investigations, solanocapsine has the empirical formula $C_{27}H_{46}N_2O_2$, H_2O; no glycoside was detected. In contrast to former structural proposals, solanocapsine was regarded by us as having the structure of a 3β-amino-5α-steroid with an α-epiminocyclohemiketal moiety (Schreiber & Ripperger, 1960a, b, 1962a, b; Ripperger & Schreiber, 1969). The structure IV shown has now been fully established by X-ray analysis of the corresponding N-bromobenzylidene derivative (Höhne, Ripperger & Schreiber, 1970). Solanocapsine has been isolated not only from *Solanum pseudocapsicum* L. but also from *S. capsicastrum* Link, *S. hendersonii* hort. and *S. seaforthianum* Andr. (Schreiber & Ripperger, 1962b; Chakravarti, 1975).

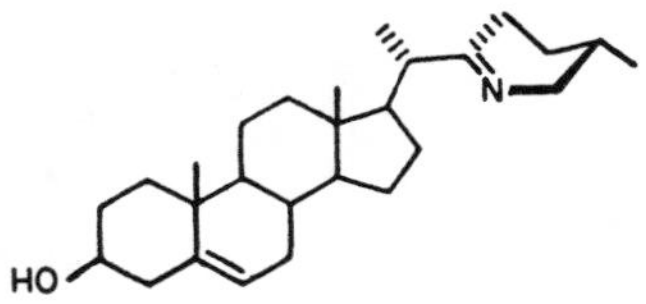

XIX

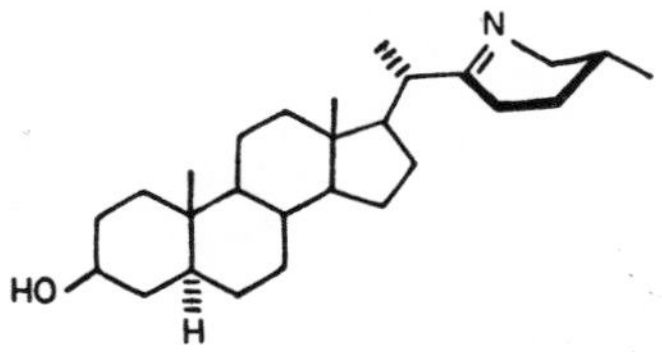

III

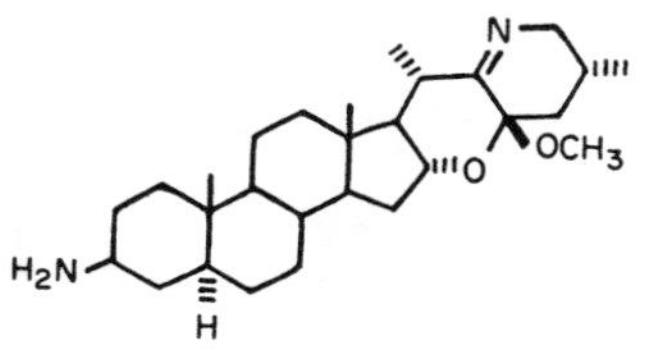

XX

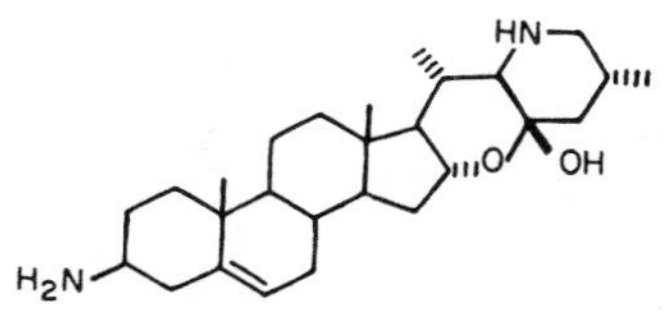

XXI

Quite recently, two other natural alkaloids with the solanocapsine skeleton have been isolated, that is solacasine, a minor compound from *Solanum pseudocapsicum* L., the methylketal of 22(N)-dehydrosolanocapsine (XX) (Mitscher *et al.*, 1976) as well as seaforthine, the 5,6-dehydrosolanocapsine (XXI) from *Solanum seaforthianum* Andr. (Chakravarti, 1975).

3-AMINOSPIROSTANE ALKALOIDS

About ten years ago we were able to isolate from roots of *Solanum paniculatum* L. a new nitrogenous steroidal sapogenin (Schreiber, Ripperger & Budzikiewicz, 1965; Schreiber & Ripperger, 1966; Ripperger, Schreiber & Budzikiewicz, 1967a; see Meyer & Bernoulli, 1961). It has been named "jurubine" in reference to the vernacular name "Jurubeba" for this species of *Solanum* which is indigenous to tropical Brazil. Acid or enzymatic hydrolysis of the saponin afforded an amino-steroid, jurubidine, the structure of which has been established as (25S)-3β-amino-5α-spirostane (V), mainly by application of physical methods such as mass and NMR spectroscopy as well as circulardichroism measurements. The 3β-amino-3-deoxyneotigogenin structure (V) of jurubidine has been confirmed by deamination of jurubidine by the aid of nitrous acid yielding neotigogenin which has been reconverted, via neotigogenone and subsequent sodium/ethanol reduction of its 3-oxime, into the starting alkaloid jurubidine.

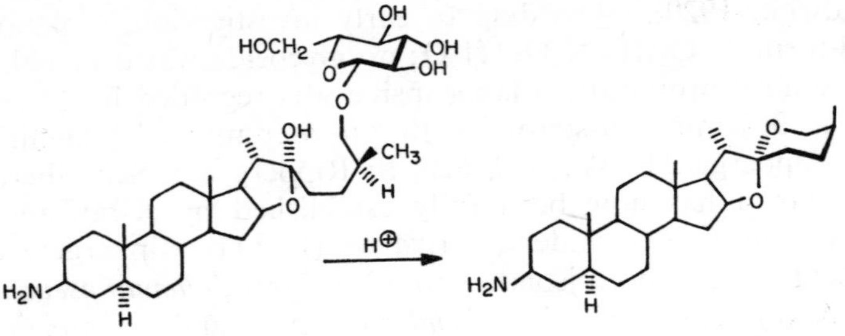

XXII   V

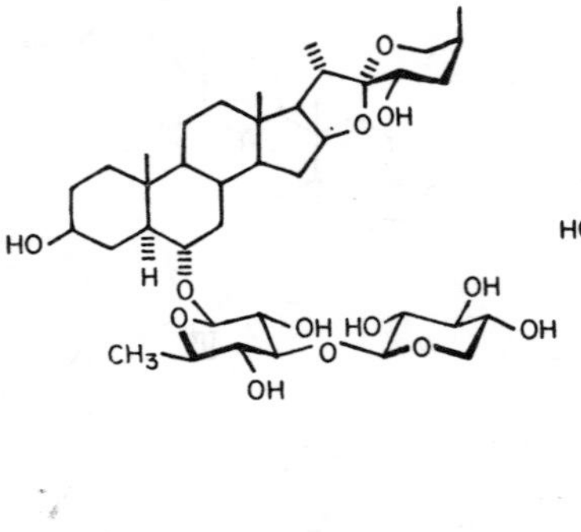

XXIII 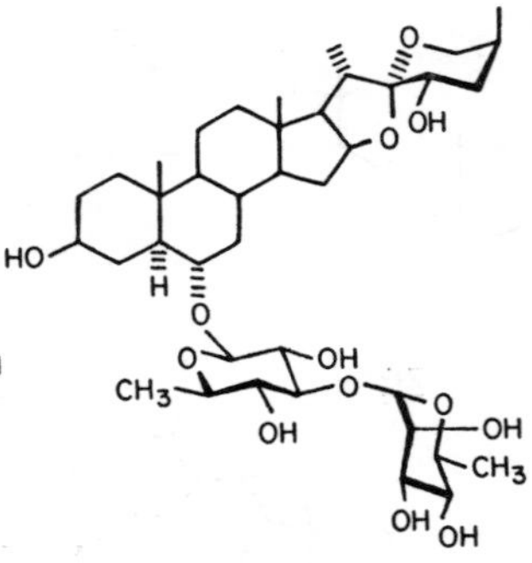   XXIV

The native glycoside jurubine contains, in addition to the aglycone jurubidine (V), one molecule of D-glucose. Extensive chemical investigations led to the result that this saponin does not have a normal spirostane skeleton but an open side-chain moiety. The structure of jurubine so elucidated, as shown in formula (XXII), is a 22α-hydroxyfurostanol glucoside in which the sugar is attached to the 26-hydroxy group. Acid hydrolysis or enzymatic cleavage with β-glucosidase yields the corresponding aglycone which spontaneously cyclizes to the spirostane derivative jurubidine (V).

After we discovered the first representative of this novel type of 22α-hydroxy-furostanol saponin, Tschesche, Lüdke & Wulff (1967, 1968, 1969) as well as Kiyosawa et al. (1968) were able to demonstrate the general occurrence also of nitrogen-free furostanol glycosides in plants containing steroidal saponins.

The isolation of 22α-hydroxyfurostanol O(26)-glucosides confirms the earlier suggestions of Marker & Lopez (1947) nearly 30 years ago that the true naturally occurring steroidal saponins biosynthesized in plants might have an open side-chain moiety, in which ring closure to the spiroketal structure encountered in the corresponding sapogenins is prevented by conjugation of the 26-hydroxy group with sugars.

Surprisingly, the leaves and fruits of *Solanum paniculatum*, the roots of which contain jurubine (XXII), possess neither this steroidal glycoalkaloid nor other (also nitrogen-free) furostanol glycosides, but contain two other new spirostane saponins, paniculonin A and B, the structures of which have been elucidated (XXIII and XXIV) (Schreiber et al., 1965; Ripperger et al., 1967b; Ripperger &

Schreiber, 1968). Both of them represent saponins of a hitherto unknown structural type. Their aglycone, paniculogenin, has three hydroxy groups at C-3, C-6 and, a novel feature, at C-23. The sugar moieties are not bonded to the 3-hydroxy group as normally observed (or mostly only assumed), but to the hydroxyl at C-6. Finally, these saponins contain, in addition to D-xylose (paniculonin A, XXIII) and L-rhamnose (paniculonin B, XXIV), the rare sugar D-quinovose, the 6-deoxy-D-glucose, not found before as a sugar component of steroidal saponins of this type.

After acid hydrolysis of the whole saponin fraction isolated from leaves and fruits from *Solanum paniculatum*, in addition to the aglycone paniculogenin, the known steroidal sapogenin neochlorogenin has also been isolated (Ripperger *et al.*, 1967a). Jurubine (XXII), as well as paniculogenin and neochlorogenin (after acid hydrolysis of the naturally occurring glycosides) have also been found both in roots and leaves of *Solanum torvum* Sw. (Schreiber & Ripperger, 1967b).

ACKNOWLEDGEMENT

I acknowledge with many thanks the collaboration of all my co-workers engaged in the investigations reported. I am especially indebted to my colleagues Dr G. Adam, Dr H. Ripperger, Dr H. Rönsch, and Dr F.-J. Sych as well as to Prof. E. Höhne for the X-ray analytical investigations.

REFERENCES

ADAM, G. & SCHREIBER, K., 1966. *Tetrahedron, 22:* 3591–3595.
ADAM, G., SCHREIBER, K. & TOMKO, J., 1967a. *Justus Liebigs Annalen der Chemie, 707:* 203–208.
ADAM, G., SCHREIBER, K., TOMKO, J. & VASSOVA, A., 1967b. *Tetrahedron, 23:* 167–171.
ADAM, G , VOIGT, D. & SCHREIBER, K., 1971. *Journal für Praktische Chemie, 313:* 45–50.
BOGNÁR, R. & MAKLEIT, S., 1965. *Acta Chimica Academiae Scientiarum Hungaricae, 46:* 205–219.
BREYER-BRANDWIJK, M. G., 1929. *Bulletin des Sciences Pharmacologiques, 36:* 541–550.
CHAKRAVARTI, R. N., 1975. *Annual Report 1974, Indian Institute of Experimental Medicine:* 18–19. Calcutta.
CHAUDHURI, S. B. & HAZARIKA, J. N., 1966. *Current Science, 35:* 187.
CHAUDHURI, S. B. & RAO, P. R., 1964. *Indian Journal of Chemistry, 2:* 424.
FIESER, L. F. & FIESER, M., 1959. *Steroids:* 847–857. New York: Reinhold.
FIESER, L. F. & FIESER, M., 1961. *Steroide:* 935–988. Weinheim: Verlag Chemie.
HÖHNE, E., 1972. *Journal für Praktische Chemie, 314:* 371–376.
HÖHNE, E., RIPPERGER, H. & SCHREIBER, K., 1970. *Tetrahedron, 26:* 3569–3577.
HÖHNE, E., SCHREIBER, K., RIPPERGER, H. & WORCH, H.-H., 1966. *Tetrahedron, 22:* 673–678.
IYENGAR, M. S., 1968. *Indian Drugs Pharmaceutical Industry,* 13–14.
KIYOSAWA, S., HUTSCH, M., KOMORI, T., NOHARA, T., HOSOKAWA, I. & KAWASAKI, T., 1968. *Chemical and Pharmaceutical Bulletin, 16:* 1162–1164.
KUHN, R. & LÖW, I., 1947. *Chemische Berichte, 80:* 406.
KUHN, R. & LÖW, I., 1954. *Angewandte Chemie, 66:* 639.
KUHN, R. & LÖW, I., 1957 *Angewandte Chemie, 69:* 236.
KUHN, R. & LÖW, I., 1961a. *Tagungsberichte Deutschen Akademie der Landwirtschaftswissenschaften zu Berlin, 27:* 7.
KUHN, R. & LÖW, I., 1961b. *Chemische Berichte, 94:* 1088.
KUHN, R., LÖW, I. & TRISCHMANN, H., 1952. *Angewandte Chemie, 64:* 397.
KUHN, R., LÖW, I. & TRISCHMANN, H., 1953. *Chemische Berichte, 86:* 1027.
KUHN, R., LÖW, I. & TRISCHMANN, H., 1955a. *Chemische Berichte, 88:* 1492.
KUHN, R., LÖW, I. & TRISCHMANN, H., 1955b. *Chemische Berichte, 88:* 1690.
KUHN, R., LÖW, I. & TRISCHMANN, H., 1956. *Angewandte Chemie, 68:* 212.
KUHN, R., LÖW, I. & TRISCHMANN, H., 1957. *Chemische Berichte, 90:* 203.
MAITI, P. C. & MOOKHERJEA, S., 1968. *Indian Journal of Chemistry, 6:* 547–548.
MARKER, R. E. & LOPEZ, J., 1947. *Journal of the American Chemical Society, 69:* 2389–2392.

MEYER, K. & BERNOULLI, F., 1961. *Pharmaceutica Acta Helvetiae, 36:* 80–96.

MITSCHER, L. A., JUVARKAR, J. V. & BEAL, J. L., 1976. *Experientia, 32:* 415–416.

OSMAN, S. F., HERB, S. F., FITZPATRICK, T. J. & SINDEN, S. L., 1976. *Phytochemistry, 15:* 1065–1067.

PETROCHENKO, E. I., 1953. *Dokladÿ Akademii Nauk SSSR, 90:* 841.

PETROCHENKO, E. I., 1956. *Uspekhi Sovremennoĭ Biologii, 42:* 19.

PRELOG, V. & JEGER, O., 19˜3. *Alkaloids, 3:* 247–312.

PRELOG, V. & JEGER, O., 1960. *Alkaloids, 7:* 343–361.

PROKOSHEV, S. M., PETROCHENKO, E. I. & BARANOVA, 1952. *Dokladÿ Akademii Nauk SSSR, 82:* 995.

RIPPERGER, H. & SCHREIBER, K., 1968. *Chemische Berichte, 101:* 2450–2458.

RIPPERGER, H. & SCHREIBER, K., 1969. *Justus Liebigs Annalen der Chemie, 723:* 159–180.

RIPPERGER, H., BUDZIKIEWICZ, H. & SCHREIBER, K., 1967a. *Chemische Berichte, 100:* 1725–1740.

RIPPERGER, H., SCHREIBER, K. & BUDZIKIEWICZ, H., 1967b. *Chemische Berichte, 100:* 1741–1752.

RIPPERGER, H., SYCH, F.-J. & SCHREIBER, K., 1970. *Tetrahedron Letters:* 5251–5252.

RIPPERGER, H., SYCH. F.-J. & SCHREIBER, K., 1972a. *Tetrahedron, 28:* 1619–1627.

RIPPERGER, H., SYCH, F.-J. & SCHREIBER, K., 1972b. *Tetrahedron, 28:* 1629–1644.

ROSEN, W. E. & ROSEN, D. B., 1954. *Chemistry and Industry,* 1581–1582.

SAINI, A. D., MUKHERJEE, M. & BISWAS, R. C., 1965. *Indian Journal of Plant Physiology, 8:* 103.

SATO, Y. & IKEKAWA, N., 1961. *Journal of Organic Chemistry 26:* 1945–1947.

SATO, Y., KATZ, A. & MOSETTIG, E., 1951a. *Journal of the American Chemical Society, 73:* 880.

SATO, Y., MILLER, H. K. & MOSETTIG, E , 1951b. *Journal of the American Chemical Society, 73:* 5009.

SATO, Y., KATZ, A. & MOSETTIG, E., 1952. *Journal of the American Chemical Society, 74:* 538–539.

SATO, Y., KANEKO, H., BIANCHI, E. & KATAOKA, H., 1969. *Journal of Organic Chemistry, 34:* 1577–1582.

SCHREIBER, K., 1957. *Angewandte Chemie, 69:* 483.

SCHREIBER, K., 1963. *Kulturpflanze, 11:* 422.

SCHREIBER, K., 1968. *Alkaloids, 10:* 1–192.

SCHREIBER, K., 1970. *Pure and Applied Chemistry, 21:* 131–152.

SCHREIBER, K., 1974 *Biochemical Society Transactions, 2:* 1–25.

SCHREIBER, K. & ADAM, G., 1960. *Tetrahedron Letters:* 5–8.

SCHREIBER, K. & ADAM, G., 1961. *Experientia, 17:* 13–14.

SCHREIBER, K. & ADAM, G., 1963. *Justus Liebigs Annalen der Chemie, 666:* 155–176.

SCHREIBER, K., ADAM, G., AURICH, O., HORSTMANN, C., RIPPERGER, H. & RÖNSCH, H., 1967. *Proceedings of the International Congress Hormonal Steroids 2nd: Excerpta Medica Foundation International Congress Series, 132:* 344–353.

SCHREIBER, K. & AURICH, O., 1963. *Zeitschrift für Naturforschung, B 18:* 471.

SCHREIBER, K., HAMMER, U., HOF, U., ITHAL, E. & RUDOLPH, W., 1961. *Tagungsberichte Deutschen Akademie der Landwirtschaftswissenschaften zu Berlin, 27:* 75.

SCHREIBER, K. & RIPPERGER, H., 1960a. *Experientia, 16:* 536.

SCHREIBER, K. & RIPPERGER, H., 1960b. *Tetrahedron Letters:* 9–11.

SCHREIBER, K. & RIPPERGER, H., 1962a. *Justus Liebigs Annalen der Chemie, 655:* 114–135.

SCHREIBER, K. & RIPPERGER, H., 1962b. *Zeitschrift für Naturforschung, B 17:* 217–221.

SCHREIBER, K. & RIPPERGER, H., 1966. *Tetrahedron Letters:* 5997–6002.

SCHREIBER, K. & RIPPERGER, H., 1967a. *Chemische Berichte, 100:* 1381.

SCHREIBER, K. & RIPPERGER, H., 1967b. *Kulturpflanze, 15:* 199–204.

SCHREIBER, K., RIPPERGER, H. & BUDZIKIEWICZ, H., 1965. *Tetrahedron Letters:* 3999–4002.

SCHREIBER, K. & RÖNSCH, H., 1963. *Abhandlungen der Deutschen Akademie der Wissenschaften zu Berlin, (Kl. Chem. Geol. Biol.), 4:* 395–408.

SCHREIBER, K. & RÖNSCH, H., 1965a. *Archiv der Pharmazie, 298:* 285–293.

SCHREIBER, K. & RÖNSCH, H., 1965b. *Justus Liebigs Annalen der Chemie, 681:* 187–195.

SCHRÖTER, H.-B. & NEUMANN, D., 1964. *Mitteilungsblatt der Chemischen Gesellschaft der DDR, 11:* 199.

SCHRÖTER, H.-B., NEUMANN, D., AURICH, O., ROMEIKE, A. & SCHREIBER, K., 1964. Unpublished data.

SHIH, M.-J. & KUĆ, J., 1974. *Phytochemistry, 13:* 997.

SILVA, M., MANCINELLI, P. & CHEUL, M., 1962. *Journal of Pharmaceutical Science, 51:* 289.

TOMKO, J., ADAM, G. & SCHREIBER, K., 1967. *Journal of Pharmaceutical Science, 56:* 1039–1040.

TOMKO, J. & VASSOVÁ, A., 1964. *Chemické Zvesti, 18:* 266–272.

TSCHESCHE, R., LÜDKE, G. & WULFF, G., 1967. *Tetrahedron Letters:* 2785–2790.

TSCHESCHE, R., LÜDKE, G. & WULFF, G., 1969. *Chemische Berichte, 102:* 1253–1269.

TSCHESCHE, R., TJOA, B. T., WULFF, G. & NORONHA, R. V., 1968. *Tetrahedron Letters:* 5141–5144.

12. Distribution of steroidal alkaloids in Australian species of *Solanum*

V. BRADLEY, D. J. COLLINS, F. W. EASTWOOD, M. C. IRVINE, J. M. SWAN

Department of Chemistry, Monash University, Victoria, Australia, 3168,

AND

D. E. SYMON

Department of Agronomy, Waite Agricultural Research Institute, South Australia, 5064

We have examined 84 species of Australian *Solanum* plants for steroidal alkaloids. We have found for 20 species, less than 0.1% alkaloid in any part of the plant; for 34 species, alkaloid(s) in the fruits only, and for 30 species, alkaloid(s) in leaf, stem and fruit (for 13 of the last group, fruits were unavailable and only leaf and stem have been assayed at this stage). We have isolated several new alkaloids, in particular from *S. callium* and from *S. dunalianum*, and biological and chemical studies on these alkaloids are continuing. The most important Australian species from the viewpoint of commercial production of solasodine would still appear to be *S. laciniatum* and *S. aviculare*.

CONTENTS

INTRODUCTION

The present world population is around 4000 million, and this includes about 1000 million women of child bearing age. During the last decade the overall rate of population increase has been 2% which has serious implications for ecological systems and for the maintenance of social and political stability.

Steroidal contraceptives are already playing a major role in family planning and population control. In 1974, some 50 million women were using oral contraceptives, including 12 million women in the U.S.A., 15 million in China and 1 million in Australia. In the developed countries, about 20–25% of fertile women choose to use oral contraceptives and available evidence suggests that the same figure may eventually apply in developing countries. Moreover, it is predicted that the figure of 50 million will rise to 100 million by 1980, and bodies such as the International Planned Parenthood Federation and the Agency for International

Development are aiming at a target of 250 million women using steroidal contraception. On the other hand there could be a relative decline in the use of oral contraceptives if sterilization programmes and abortion become more widely accepted.

In addition to large scale programmes for human fertility control, there will be an expanding use of steroids in many areas of medicine, including cardiovascular therapy, libido-promoting agents, human abortifacients, menopause regulants, anti-inflammatory agents, and in animal population control. A rapid increase in the demand for steroidal raw materials can therefore be expected and for this reason we have undertaken a survey of Australian *Solanum* species for their content of steroidal alkaloids.

Medicinal steroids are manufactured by American and European companies, and by government agencies in eastern Europe, Russia and China. Similar industries are being developed in India and Japan. The American and European companies use as their principal starting compound the steroid diosgenin, isolated from a Mexican yam, a *Dioscorea* species. China also uses an indigenous *Dioscorea* as a source of diosgenin. Russia and some of the eastern European countries use diosgenin but also produce a proportion of their steroid raw material from *Solanum laciniatum*, an Australian and New Zealand plant which Russian botanists have obtained and cultivated for this purpose. *Solanum laciniatum* contains 1–2 % of solasodine, a steroidal alkaloid which is a close chemical relative of diosgenin. Chemical processes for the conversion of diosgenin and solasodine into steroidal drugs are practically identical, and the present world shortage of diosgenin, and the likely growth in demand for steroids could be met by increased production of solasodine, provided overall costs are comparable. Already the high price of diosgenin has meant that virtually all the 19- nor steroids required for oral contraceptives are now being manufactured by total synthesis; other potential sources of steroid, at least for certain purposes, include the microbiological fermentation of stigmasterol from soybean, and cholesterol from animal spinal cord.

Table 12.1 summarizes world steroid production in 1973, Table 12.2 lists the principal steroid raw materials and Table 12.3 shows the 1973 distribution of steroid sources as between plant and animal steroids and total synthesis. These figures relate to the time when our survey commenced, and already the situation has changed a good deal. In particular the proportion of steroid made by total synthesis has certainly increased. Whether there will be an expanding role for *Solanum* steroid production over the next ten years is a question which is very difficult to answer. Certainly, with regard to 16- substituted steroids such as the corticoid β-methasone, diosgenin and solasodine are likely to remain important. At present the most favoured plants are *S. laciniatum*, *S. khasianum* and possibly *S. marginatum* for production of solasodine, while a chemovariety of *S. dulcamara* has been shown to provide good yields of the still more advantageous tomatid-5-en-3β-ol, a stereoisomer of solasodine which can also be degraded in high yield to 16-dehydropregnenolone acetate (Schreiber & Roensch, 1965a, b).

As already mentioned, investigations on *S. laciniatum* and its close relative *S. aviculare* have been carried out in Hungary, Russia, Bulgaria, Czechoslovakia, Yugoslavia, Poland, Roumania, China and India. These studies have led to processes for the commercial production of solasodine. More recently, growing trials of *S. laciniatum* and *S. aviculare* have been carried out in Egypt, New Zealand and Cuba, and have commenced in Australia and elsewhere.

Table 12.1. World steroid production—1973

Raw materials (diosgenin equivalent)	1000 tonnes
Finished products	100 tonnes
Sale value (finished products)	U.S. $1000 million

Table 12.2. Steroid raw material

Compound	Source
Cholesterol	{ Cattle spinal cord { Wool grease
Stigmasterol	Soybean oil
Bile Acids	Cattle bile
Diosgenin	*Dioscorea* (yams)
Hecogenin	Sisal (Africa)
Solasodine	*Solanum* species

Table 12.3. Distribution of steroid sources—1973*

Plant		Animal		Total synthesis	
Diosgenin		*Cholesterol and Bile Acids*			
Mexico	50 ⎤				
China	5 ⎬56%	France	9%		
India	1 ⎦	Holland ⎫			
		Germany ⎭ 1%		U.S.A.	7.5%
				France	5.0%
				Germany	0.5%
Hecogenin					
Africa	6%				
Stigmasterol					
U.S.A.	15%				
Totals	77%		10%		13%

*Hardman (pers. comm.)

METHODS

Table 12.4 shows the approximate number of species in each of the 8 genera of indigenous Australian *Solanaceae*. *Solanum* dominates with 92 species, followed by *Nicotiana*, 25, and *Anthocercis*, 20. When our survey commenced in 1973, only 18 of the 92 *Solanum* species had been subjected to general alkaloidal screening and of these 18 species only 10 had been subjected to more detailed examination. We felt that before initiating any agronomic studies on *S. laciniatum* and *S. aviculare* in Australia, we had an obligation to examine the whole range of native *Solanum* species for steroidal alkaloids and to determine the structures of any new steroids that might be discovered. This has involved collection of plant material from all over Australia, proper identification of each species and chemical investigations using standard procedures. These have included drying of plant material under standard conditions, colorimetric estimation of total steroidal alkaloids (as methyl orange complexes of the aglycones), thin layer chromatography of crude extracts for the detection of glycoalkaloids, thin layer chromatography of hydrolyzed crude extracts for detection and partial identification of alkaloid aglycones and the isolation, purification and characterization of alkaloids.

Table 12.4. Australian Solanaceae (indigenous)

Genus	No. of species (approx.)
Anthocercis	20
Anthotroche	6
Datura	1
Duboisia	3
Isandra	1
Lycium	2
Nicotiana	26
Solanum	92

RESULTS

As expected, the Archaesolanum group has proved to be the richest source of steroidal alkaloid, and indeed *S. laciniatum* and *S. aviculare* have emerged as still the most valuable sources of solasodine. The assay results for this group are shown in Table 12.5. Another interesting plant is *S. callium* C. T. White ex R. J. Henderson (formerly known as *S. aff. superficiens*). This large rain-forest shrub, found in

Table 12.5. Steroid content of Australian species of *Solanum*

Subgenus, section, and species	Alkaloid assay % Leaf	Stem	Fruit
Subgenus *Archaesolanum*			
Section Archaesolanum			
S. aviculare Forst. f.	0.3–3.1	0.2–1.3	0.8–3.5
S. capsiciforme Domin (Baylis)	0.4–1.1	0.1–0.2	0.6
S. laciniatum Ait.	1.0	0.1	3.5
S. linearifolium Her.	0.5	0.1	0.6
S. simile F. Muell.	0.5–2.5	0.1–0.6	0.1–1.6
S. symonii Eichler	0.6, 0.8	0.1, 0.2	0.6
S. vescum F. Muell.	0.3	trace	0.6
Subgenus *Solanum*			
Section Solanum			
S. americanum Mill. ($\equiv$ *nodiflorum*)	0	0	0.1
S. nigrum	trace	0	0.4
S. opacum A. Br. and Bouche	0	0	0.4
Section Leiodendra			
S. callium C. T. White ex R. J. Henderson (*S. aff. superficiens*)	1.2–3.1	0.4–1.3	0.6–3.8

Subgenus *Leptostemonum*
Section Graciliflorum

S. amblymerum Dun.	0.2	0.1	0.1
S. chenopodinum F. Muell.	trace	0	0.1
S. densevestitum F. Muell.	trace	0	—
S. discolor R. Br.	0.3	0.2	0.1 (g)
S. ferocissimum Lindl.	0	0	0.1 (r)
			0.1 (g)
S. nemophilum F. Muell.	0	0	—
S. parvifolium R. Br.	trace	0	—
S. semiarmatum F. Muell.	0	0	0
S. stelligerum Sm.	0	0	trace

Not examined *S. defensum*, *S. yirrkalensis* Symon (ms)

Section Brevantherum

S. erianthum D. Don	0.1	0.1	0.8

Section Campanulatum

S. campanulatum R. Br.	0	0	0.7

Section Irenosolanum

S. viride R. Br.	0.2	trace	1.0 (g)
			0.3 (r)

Section Pugiunculiferum

S. pugiunculiferum C. T. White	0.2	trace	1.5 (seed)
			1.2

Section Micracanthum

S. dimorphospinum C. T. White*	0.5–0.7	0.1–0.2	0.1–0.2
S. hamulosum C. T. White	0	0	0

Section Oliganthes, related groups

S. karsensis Symon	0.1	0.1	trace
S. oligacanthum F. Muell.	0.3	0.1	—
S. sturtianum F. Muell.	0	0	trace
S. dunalianum Gaud.	0.3–1.0	—	0.2
S. tetrandrum R. Br.	0	0	—

Section Melongena, Andromonoecious group

S. chippendalei Symon (ms)	0	0	0.2
S. clarkiae Symon (ms)	0	0	trace
S. diversiflorum F. Muell.	0	0	1.5 (g)
S. eburneum Symon	0	0	0.1–0.4 (r)
			0 (g), 0.1 (r)
S. melanospermum F. Muell.	0	0	0.3
S. oedipus Symon	0	0	trace
S. phlomoides Benth.	0	0	0.9 (r)
S. aff. phlomoides Symon (ms)	trace	—	—

Not examined *S. heteropodium* Symon (ms)

Section Melongena, Androdioecious group

S. asymmetriphyllum Specht	0	0	0.6 (g)
S. cunninghamii Benth.	0	0	0.5 (g)
S. dioicum W. V. Fitz.	0	0	0.3 (r)
			1.5 (g)
S. leopoldensis Symon	0	0	0.8 (r)
S. petraeum Symon (ms)	0	0	—
S. tudununggae	0	0	0.6 (g)
S. sp. (new) (Kalumburu)	0	0	—

Not examined *S. carduiforme*, *S. cataphractum*, *S. vansittartensis*

*The alkaloidal aglycone from this plant was shown to be tomatidine.
g, green; r, ripe

southern Queensland, has yielded substantial quantities of two new alkaloids. The first of these, $C_{27}H_{45}NO_2$, m.p. 163-166.5°, $[\alpha]_D^{35}+44.8°$ (CHCl$_3$,) has been shown by X-ray crystallography (B. M. K. C. Gatehouse & A. J. Jozsa, unpubl.) to be the C-25 epimer of solafloridine (Fig. 12.1), and we propose the name 25-iso-solafloridine. The C(5)–C(6) linkage is saturated and the alkaloid could also be named '5,6-dihydroetioline'. Etioline is the corresponding 5,6-unsaturated steroidal alkaloid isolated from *Veratrum grandiflorum*. The 25-*S* stereochemistry for 25-isosolafloridine has also been confirmed by circular dichroism studies ($\Delta\epsilon=$ -2.1, $\lambda=242$ nm), and the structure is fully supported by ^{1}H and ^{13}C nmr data.

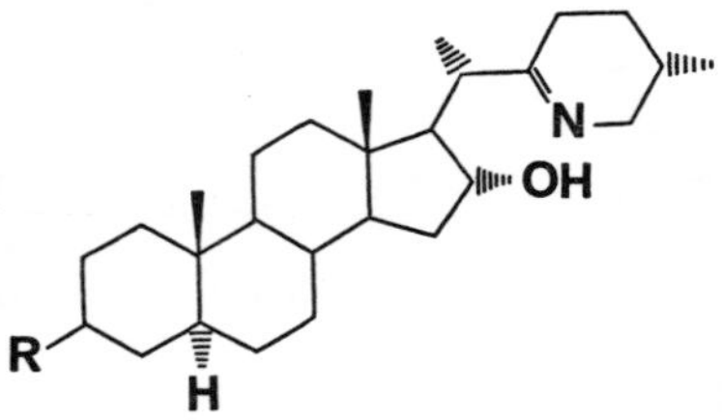

Figure 12.1. New alkaloids from *S. callium*. R=OH, 25-Isosolafloridine ($\equiv$5,6-dihydroetioline); R=NH$_2$, Solacallinidine. (Plus traces of other alkaloids not yet characterized.)

The second new alkaloid $C_{27}H_{46}N_2O$, solacallinidine, is identical with 25-iso-solafloridine except that the 3-hydroxyl group is replaced by a 3-amino group. It is of interest that 25-isosolafloridine has already been synthesized, having been prepared by Professor Schreiber and co-workers (Ripperger, Sych & Schreiber, 1972) in the course of their synthesis of solafloridine.

The remaining analytical results are summarized in Table 12.5. Some points of interest emerge. The alkaloid analysis has established a clear chemotaxonomic differentiation between *S. dimorphospinum* and *S. hamulosum*, two closely related species which have often been confused. *S. hamulosum* contains no alkaloid in any part of the plant whereas *S. dimorphospinum* is relatively rich in steroidal glyco-alkaloid, the aglycone of which proved to be tomatidine. The plant *S. dunalianum* was found on one of our field trips growing at Weipa on Cape York Peninsula; it had not previously been found in Australia but was described for New Guinea and the Solomon Islands. This plant contains new alkaloids, the main constituent, $C_{27}H_{46}N_2O$, having two nitrogen atoms, one of which is a 3-amino group. Further work on these alkaloids is in progress.

Among the dioecious and monoecious species, only one of the latter type, namely *S. diversiflorum* F. Muell. had been tested prior to our work; it was found to give a moderately strong reaction with Mayer's reagent. Our survey has revealed almost no steroidal alkaloids in the leaf or stem of any of the seven dioecious and eight monoecious species which were examined, but the fruits of all except *S. petraeum* Symon (ms), *S. clarkiae* Symon (ms) and *S. oedipus* Symon showed solasodine in the fruits, ranging up to 1.5 % in the green fruits of *S. dioicum* and *S. diversiflorum*. The evidence for the presence of solasodine rests mainly on thin layer chromatography, but in the case of *S. phlomoides* pure solasodine was isolated from the hydrolyzed extract. Apart from the three exceptions already mentioned it appears to be characteristic of both the andromonoecious and andro-dioecious groups that the leaf and stem contain virtually no steroidal alkaloid but the fruits contain appreciable amounts.

In the two dioecious species *S. dioicum* and *S. leopoldensis* for which both green and ripe fruits from the same collections were examined, the latter showed significantly less alkaloid, a feature shown by the Archaesolanum group, and also, for example, by *S. dulcamara*.

ACKNOWLEDGEMENTS

We thank the Rural Credits Development Fund of the Reserve Bank of Australia for financial support for the survey, and the Australian Research Grants Committee for maintenance for structural studies.

REFERENCES

RIPPERGER, H., SYCH, F.-J. & SCHREIBER, K., 1972. Solanum alkaloids-XCVI. Synthesis of solafloridine and 25-isosolafloridine. *Tetrahedron, 28:* 1619–1627.

SCHREIBER, K. & ROENSCH, H., 1965a. Solanum alkaloids—XLIV. Separation of Tomatid-5-en-3β-ol from *Solanum dulcamara* L. and its degradation to 3β-acetoxypregna-5,16-dien-20-one. *Justus Liebigs Annalen der Chemie, 681:* 187–195.

SCHREIBER, K. & ROENSCH, H., 1965b. Solanum alkaloids—LIII. Steroid alkaloids and sapogenins of chemically distinct varieties of *Solanum dulcamara. Archiv der Pharmazie, 298:* 285–293.

13. Variation in alkaloids in *Solanum dulcamara* L.

IMRE MÁTHÉ Jr. AND IMRE MÁTHÉ Sr.

Research Institute for Botany, Hungarian Academy of Sciences, Vácrátót, Hungary

In *Solanum dulcamara* L. there are three main steroid alkaloids, tomatidenol, solasodine and soladulcidine, which occur in varying proportions in many glycosides. This paper surveys the literature together with the authors' own findings, and tries to determine what factors are correlated with alkaloid composition.

Much information is now available on the alkaloid, sapogenin and triterpene content of various organs at different stages of development. There are many chemical races in *S. dulcamara*: the predominantly tomatidenol producing taxa tend to be pubescent and come from the humid climatic regions of western Europe, the predominantly soladulcidine producing taxa tend to be glabrous and occur in the drier climatic regions of eastern Europe, but this distinction is not sufficiently clear to support formal subspecific morphological division of this species.

The importance of *Solanum* alkaloids as precursors for steroid hormones is discussed.

CONTENTS

INTRODUCTION

Since 1951, when Sato and Kuhn (Sato, Miller & Mosettig, 1951; Kuhn, Löw & Trischmann, 1952) elaborated the degradation of spirosolane alkaloids into pregnane derivatives, much has been done to find *Solanum* species in which these alkaloids, especially solasodine, are present in industrially exploitable quantities. The practical background of these researches is the possibility that from these compounds several important medicines, among them oral contraceptives, may be produced. Although, nowadays, great efforts are made to avoid plant sources in steroid production, the importance of the study of *Solanum* alkaloids does not seem to decrease for both practical and theoretical reasons. On the practical side, there is a deficiency of available plant sources of diosgenin, the main starting material currently used, so that the replacement of diosgenin sources by other plants such as solasodine-bearing ones, may be temporarily necessary in certain countries (Manning, 1969). On the other hand, from the theoretical point of view, studies of *Solanum* alkaloids can contribute to the better understanding of the genus *Solanum*.

Several investigators have studied the chemistry and the occurrence of *Solanum* alkaloids. Instead of trying to introduce their works (there are good compilations dealing with these researches e.g. Boll, 1966; Schreiber, 1968; etc.) we wish to refer to the research of a Hungarian scientist Pál Tuzson. He not only drew attention to *S. aviculare* (it proved to be *S. laciniatum*) as a solasodine source industrially utilizable in Central Europe (Tuzson, 1954), but also he was the first, together with his co-worker Kiss (Tuzson & Kiss, 1957) and McKee (1957), to distinguish alkaloids of *S. dulcamara* L. from those of *S. tuberosum* L. Tuzson and Kiss gave the name of soladulcidine to the first compound isolated from *S. dulcamara* L. They gave in their paper a good summary of the earliest chemical work on *S. dulcamara* L. as well.

STEROID COMPOSITION OF *SOLANUM DULCAMARA* L.

The correct structure of soladulcidine, namely the recognition that it is a dihydrogenated solasodine (25R)-5α,22α-N-spirosolan-3β-ol, is attributed to Schreiber (1958), (Fig. 13.1). The presence of this compound in *S. dulcamara* was confirmed by many other authors (Sander, Alkemeyer & Hänsel, 1962; Alkemeyer & Sander, 1959a; Sander & Willuhn, 1961; Sander, 1963b, c; Bognár & Makleit, 1965; Makleit, 1966; Makleit & Bognár, 1963, 1967, etc.). Beside soladulcidine, solasodine (25R)-22α-N-spirosol-5-en-3β-ol was identified (Sander, 1963b; Boll, 1962; Boll & Andersen, 1962; Schreiber & Rönsch, 1963a, 1965a; Tomowa, 1962, 1964; Willuhn, 1968; Makleit & Bognár, 1963, 1967; Makleit, 1966, etc.). Some authors (Alkemeyer & Sander, 1959a, b; Rasmussen & Boll, 1958; Makleit & Bognár, 1963) mentioned another unknown, Δ[5] unsaturated compound present in the plant in certain cases besides solasodine. Schreiber (Schreiber & Rönsch, 1963a, b, 1965b) determined the structure of this alkaloid, called tomatidenol, the configuration of which is 25S at the C-25 of the spirosolane skeleton (Fig. 13.1) in contrast to that of solasodine and soladulcidine. The structure of tomatidenol proved to be (25S)-22β-N-spirosol-5-en-3β-ol. The co-occurrence of these two alkaloids, with opposite configurations at C-25, in the same plant is a rare thing in the genus *Solanum* (Sander, 1963a).

From the root of *S. dulcamara*, besides tomatidenol, solasodine and soladulcidine, their 15α-hydroxy derivatives were isolated independently of what kind of

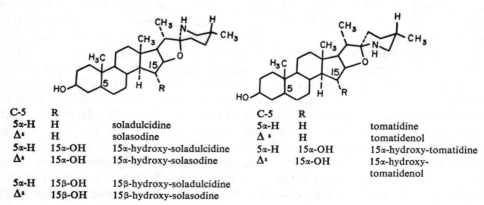

C-5	R	
5α-H	H	soladulcidine
Δ[5]	H	solasodine
5α-H	15α-OH	15α-hydroxy-soladulcidine
Δ[5]	15α-OH	15α-hydroxy-solasodine
5α-H	15β-OH	15β-hydroxy-soladulcidine
Δ[5]	15β-OH	15β-hydroxy-solasodine

C-5	R	
5α-H	H	tomatidine
Δ[5]	H	tomatidenol
5α-H	15α-OH	15α-hydroxy-tomatidine
Δ[5]	15α-OH	15α-hydroxy-tomatidenol

Figure 13.1. Steroid alkaloids isolated from *Solanum dulcamara* L.

alkaloids were present in the above-ground organs (Rönsch & Schreiber, 1965a, 1966a; Willuhn & Kun-Anake, 1970). The presence of 15β-hydroxysolasodine, and 15-oxosoladulcidine is also probable in the root (Schreiber, 1968).

Solanum alkaloids occur predominantly in the form of glycosides. Some glycosides isolated from the above-ground organs of *S. dulcamara* are well identified (Table 13.1). This Table summarizes these glycosides, regardless of how thoroughly their structures have been determined.

Table 13.1. Glycoalkaloids identified in *Solanum dulcamara* L.

Glycosides	Sugars
TOMATIDENOL	
α-solamarine	L-Rham., D-Glu., D-Gal.
β-solamarine	2L-Rham., D-Glu.
γ$_1$-solamarine	L-Rham., D-Glu.
γ$_2$-solamarine	L-Rham., D-Glu.
δ-solamarine	D-Glu., D-Gal.
Soladulcamarine	2L-Rham., L-Arab., D-Glu.
SOLASODINE	
Solasonine	L-Rham., D-Glu., D-Gal.
Solamargine	2L-Rham., D-Glu.
β-solamargine	L-Rham., D-Glu.
SOLADULCIDINE	
Soladulcidine-tetraoside	D-Xyl., 2D-Glu., D-Gal.
(α-, β-, γ-) soladulcine	L-Rham., D-Xyl., D-Glu., D-Gal.

L-Rham., L-Rhamnose; D-Glu., D-Glucose; D-Xyl., D-Xylose; D-Gal., D-Galactose; L-Arab., L-Arabinose

The sugars of the α-, β-, and γ-soladulcines are described (Schreiber, 1958) but not their structures. The presence of these glycosides was confirmed by Soviet workers (Tukalo & Ivanchenko, 1971; Aslanov, 1975). In addition the soladulcidine aglycone can be found in the form of soladulcidinetetraoside, the sugar moiety of which is most likely the same as in tomatine, β-lycotetraose (O-β-D-glucopyranosyl-(1 → 2$_{glu}$)-O-β-D-xylopyranosyl-(1 → 3$_{glu}$)-O-β-D-glucopyranosyl-(1 →4$_{gal}$)-β-D-galactopyranose) (Alkemeyer & Sander, 1959a; Sander & Willuhn, 1961; Sander *et al.*, 1962). The occurrence of another tetraoside with tomatidenol aglycone, called soladulcamarine, was discovered by Danish research workers (Rasmussen & Boll, 1958; Boll, 1962). Soladulcamarine contains among other sugars L-arabinose which is a rather unusual component of the *Solanum* glycosides. Schreiber & Rönsch (1963a), finding no L-arabinose, assume that it is a decarboxylated product of D-galacturonic acid which may be the sugar component of a glycoside highly soluble in water.

Tomatidenol is the aglycone of some well identified glycosides: in α-solamarine the sugar moiety is β-solatriose i.e. O-α-L-rhamnopyranosyl(1 →2$_{gal}$)-O-β-D-glucopyranosyl-(1 →3$_{gal}$)-β-D-galactopyranose; in β-solamarine it is β-chacotriose i.e. O-α-L-rhamnopyranosyl-(1 →2$_{glu}$)-O-α-L-rhamnopyranosyl-(1 →4$_{glu}$)-β-D-glucopyranose (Boll, 1962, 1963; Rönsch & Schreiber, 1966b). From α- and β-solamarines biosides can be derived. In *S. dulcamara* the O-α-L-rhamnopyranosyl-

(1 →4)-β-D-glucopyranosyl-, i.e. β₂-chaconyltomatidenol, derived from β-sola-marine was first found and identified by Boll (1962, 1963). He named it γ-sola-marine but later instead of this designation the use of γ_2-solamarine was suggested by Rönsch and Schreiber (1966b). These authors described two more biosides; γ_1-solamarine i.e. O-α-L-rhamnopyranosyl-(1 →2)-β-D-glycopyranosyl-(β_1-cha-conyl-), and δ-solamarine i.e. β-solabiosyltomatidenol, the structure of which is O-β-D-glucopyranosyl-(1 →3)-β-D-galactopyranosyl-tomatidenol.

The solasodine glycosides identified by Tomowa (1962, 1964), Willuhn (1966, 1968), and Rozumek and Sander (1967), are as follows; solasonine (β-solatriosyl-solasodine), solamargine (β-chacotriosylsolasodine) and β-solamargine (structure not yet certain).

The closely related biosynthetic pathways of *Solanum* alkaloids and steroid sapogenins explains their co-occurrence in *Solanum* species (including *S. dulcamara*). Beside tomatidenol, solasodine and soladulcidine alkaloids, yamogenin, diosgenin and tigogenin sapogenins have been found. (Sander, 1963b, c; Boll & Andersen, 1962; Dersch & Sander, 1962; Máthé Jr., 1974a; Schreiber & Rönsch, 1963a, 1965a; Sander & Willuhn, 1961; Willuhn, 1967; etc.).

From the biochemical point of view the identification of the triterpenes, obtusi-foliol, cycloeucalenol, cycloartenol, 24-methylencycloartenol,lophenol, 24-methy-lenlophenol and 24-ethylidenlophenol, and of the sterols, cholesterol, brassicasterol, campesterol, 24-methylencholesterol, stigmasterol, β-sitosterol and isofucosterol was of great significance (Köstens & Willuhn, 1973; Willuhn & Liebau, 1976; Willuhn & Köstens, 1974, 1976). The triterpenes were found in both free and acylated forms, the sterols in free, acylated, glycoside and acylglycoside forms.

While the sapogenins appear with the structurally corresponding alkaloids, namely tomatidenol with yamogenin, solasodine with diosgenin and soladulcidine with tigogenin, no such interdependence in the case of sterols and alkaloids could be revealed. Independently of whatever kinds of alkaloids are in the plant, the same above-mentioned sterols occur (Willuhn & Köstens, 1976). Regarding the presence of cholesterol, β-sitosterol and stigmasterol no organ-dependent variability could be found either (Máthé Jr., 1974a). These findings support the idea that the biosynthetic route of sapogenins and *Solanum* alkaloids divides only after the formation of sterols (Schütte, 1969). The saturated or unsaturated character (at C-5) of the alkaloids forms at the "final" stage (Willuhn & Köstens, 1976).

As shown by the brief survey above, the steroids of *S. dulcamara*, especially the alkaloids, show divergence from other *Solanum* species in their greater complexity because of the presence of alkaloids (sapogenins) of 25S and 25R configuration, with or without the Δ^5 double bond, and because of the differences in composition of the above ground and below ground organs. These complexities allow the possibility of multiple variability. An attempt will be made in the rest of this paper to relate this variability in steroids, especially the alkaloids, to other factors.

VARIABILITY DUE TO GEOGRAPHIC ORIGIN

At the very beginning of the investigation of *S. dulcamara* alkaloids, it was found that plants taken from different countries, from Denmark (Rasmussen & Boll, 1958; Boll & Andersen, 1962), Germany (Sander, 1963c; Schreiber & Rönsch,

1963a; etc.) and from Hungary (Tuzson & Kiss, 1957), showed differences in their alkaloid constitution. In Denmark Δ^5 unsaturated, in Hungary saturated, and in Germany both kinds of compounds (aglycones) were found. Having extended these investigations to plants coming from yet other countries (initially from various Botanic Gardens), Sander (1963c) was the first to show the existence of geographical differences in their alkaloid constitution. This was confirmed by other authors (Rozumek & Sander, 1967; Bognár & Makleit, 1965; Makleit, 1966; Makleit, Bognár & Máthé, 1967; Máthé & Máthé Jr., 1970, 1973; Máthé Jr., 1970, 1974a). According to Sander, plants in West Europe contain tomatidenol, those in East Europe contain soladulcidine-solasodine. The border between these two areas is in Germany.

Sander (1963c), observing the same *S. dulcamara* clones for four years in experimental fields and studying plants originating from humid and arid habitats but introduced to contrasting conditions, found that the alkaloid constitution remained in all cases the same as it had been. The conclusion he drew was that the chemical constitution of the vegetative organs of *S. dulcamara* is a genetically fixed character of the plant; no environment-dependent variability could be established. He distinguished tomatidenol and soladulcidine-solasodine taxa (chemische Sippen). The designation of the chemical taxa changed a little in time. While in most papers (Bognár & Makleit, 1965; Makleit *et al.*, 1967; Máthé & Máthé Jr., 1970; Máthé Jr., 1970) Sander's designation is used, Willuhn (1966) and Rönsch, Schreiber & Stubbe (1965) distinguished pure tomatidenol, solasodine and soladulcidine taxa. In the course of detailed population studies, they found plants with only one alkaloid (aglycone). The occurrence of plants with more than one alkaloid originated by the crossing of the pure taxa. Their genetically fixed character was confirmed by studying the offspring of the same mother plants (after self-fertilization). No changes could be observed (Willuhn, 1966).

Here should be mentioned a crossing experiment (Rönsch *et al.*, 1965). Pure tomatidenol and soladulcidine taxa were crossed with each other; the offspring, besides tomatidenol, solasodine and soladulcidine, also contained tomatidine. The interpretation of this experiment poses problems in the light of the afore mentioned, namely that the most frequently occurring plants have more than one alkaloid. Apart from Rönsch and Schreiber's work, tomatidine has not been identified in the above-ground organs of *S. dulcamara*.

Referring to Sander's and Willuhn's works on the genetically fixed character of chemical races of *S. dulcamara*, one thing stands out. The area of tomatidenol taxa more or less coincides with the region of the atlantic climate, that of the soladulcidine ones with the continental climate. From the climatic point of view Germany has an intermediate position; consequently, the environmental conditions of Germany favour the co-occurrence of both chemical taxa. So, the open field experiments there can not unequivocally exclude the possibility of the influence of environment on the alkaloid composition.

This uncertainty prompted us to investigate, by further studies, the origin-dependent variability. Unlike our earlier work (Máthé & Máthé Jr., 1970; Máthé Jr., 1970), individual plants grown in experimental fields in Hungary, but originating from foreign Botanic Gardens, were analysed. Plants of local origin were also investigated, (Máthé & Máthé Jr., 1973; Máthé Jr., 1974a). Parallel with this,

Hungarian *S. dulcamara* populations from more than 150 natural habitats were also studied (Máthé & Máthé Jr., 1972a).

These investigations confirmed not only the distribution differences existing between the tomatidenol and soladulcidine-solasodine chemical taxa but also revealed the transitional position of the solasodine taxa (the grouping was based on the main alkaloid content; other components in small amounts were disregarded) (Fig. 13.2). The important point established is that the east-west differences refer only to the predominant but not to the exclusive occurrence of the particular taxa. In accordance with this statement, referring in general to the whole of Europe, it was not surprising that in Hungary tomatidenol taxa were gathered mainly from humid habitats. Detailed study of these populations is going on. The data already available show that these populations are not homogeneous. In general, Hungary belongs to the area of soladulcidine-solasodine taxa, as could be suspected on the basis of the earlier sporadic data (Bognár & Makleit, 1965).

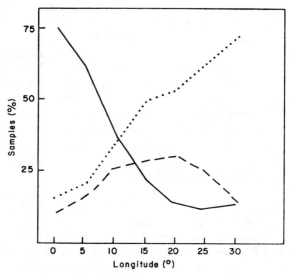

Figure 13.2. Percentage distribution of *Solanum dulcamara* L. Chemical taxa according to geographical longitude: ———, tomatidenol; – – – –, solasodine; . . . . ., soladulcidine.

QUANTITATIVE VARIATION BETWEEN ORGANS AND WITH TIME

As has already been discussed, there are qualitative differences in alkaloid composition between above and below ground organs, and also quantitative differences between vegetative and generative organs.

There are changes in alkaloid composition and content during berry development (Sander, 1963b; Willuhn, 1967, 1968; Sander & Willuhn, 1961). Besides the alkaloids normally present in above ground vegetative organs, the berries contain solasodine. The proportion of solasodine increases as the berries develop, but during ripening it is metabolized more rapidly than the other alkaloids, and by the end of ripening it is reduced to the original level (Willuhn, 1967). The total alkaloid content (dry wt. %) decreases gradually during development, more rapidly during ripening, and the ripe fruits contain only traces (Sander, 1963a; Willuhn, 1967; Máthé Jr., 1974b).

In order to determine how the alkaloid content changed during ripening, Sander (1963b) followed the formation of the soladulcidinetetraoside fraction and found little variation but a little increase. This finding does not conflict with the varying solasodine concept of Willuhn (1967). Sander excluded the possibility of translocation of alkaloids from the ripening berry to the vegetative organs, and assumed that they were metabolized by N-acylation. Initially the glycoside bond is intact. This agrees with other results such as the occurrence of 3β-hydroxy-5 α-preg-16-en-20-one in *Lycopersicon pimpinellifolium* (Schreiber, 1968).

The sapogenins, the other main steroid fraction, have a higher concentration in the inflorescence than in the other generative organs (Sander, 1963c). During the development of the fruits the sapogenin content (dry wt. %) decreases, increases as ripening starts, then decreases again. The sterol fraction decreases during development of the berry, but remains static during ripening.

Máthé Jr. (1974a) found the alkaloid content of the inflorescence to be between that of leaves and of stem, much higher than reported by other workers.

Some correlation is found between the alkaloid content (dry wt. %) of the berries and that of the vegetative organs of the same plant. Both berries and vegetative organs of the tomatidenol taxa have a higher alkaloid content than those of the solasodine-soladulcidine taxa. These relationships were revealed by geographical studies in which plants coming from West Europe but grown in Hungary, had higher alkaloid contents than those from East Europe (Máthé & Máthé Jr., 1973). Many samples of each chemical taxon were analysed for total glycoalkaloid content of the vegetative organs (Fig. 13.3). The tomatidenol with soladulcidine, solasodine, and soladulcidine taxa were similar in mean glycoalkaloid content and range; the tomatidenol taxa gave significantly higher values for both parameters.

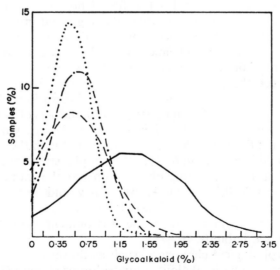

Figure 13.3. Frequency distribution of samples of the four chemical taxa of *Solanum dulcamara* L. according to the quantity of glycoalkaloids in the vegetative organs. ——, tomatidenol; – . –, tomatidenol/soladulcidine; – – – –, solasodine; . . . . . ., soladulcidine.

Variation throughout the growing season and in successive years was also studied (Máthé & Máthé Jr., 1970; Máthé Jr., 1970, 1974a, b). All plants were grown under standard conditions in the Botanic Garden in Hungary. Differences in habitat may affect the alkaloid content (Máthé & Máthé Jr., 1972a).

In general the organs of *S. dulcamara* show the following decreasing order of alkaloid content: green berry, leaf, stem, ripe berry (Alkemeyer & Sander, 1959a; Máthé Jr., 1974a; etc.).

CONNECTION BETWEEN ALKALOID COMPOSITION AND PHYTOMASS

Plants of the soladulcidine taxa had about 20 % higher dry weight above ground phytomass production than the tomatidenol plants, when grown for the same period of time under the same arid environmental conditions in Hungary. However, although smaller, the tomatidenol plants developed more rapidly during the growing season, produced more berries and had a higher proportion of ripe berries. Most of them were from humid, more shaded habitats in other countries, whereas the soladulcidine plants were native to Hungary (Máthé Jr., 1974a).

Our results may be related to those of Gauhl (1968) who studied the photosynthetic performance of *S. dulcamara* ecotypes from shaded and exposed habitats. He found that sudden strong exposure to light stimulated increase in the protein and chlorophyll content and the CO_2 uptake of leaves of ecotypes from exposed habitats, but had no stimulating effect on ecotypes from shaded habitats. Possibly these two sets of observations are complementary.

The influence of environment on the phytomass has also been studied in detail (Máthé Jr. & Máthé, 1973; Máthé Jr., 1974a; Máthé Jr. *et al.*, 1975, 1976).

CONNECTION BETWEEN ALKALOID COMPOSITION AND MORPHOLOGICAL CHARACTERS

Several early workers looked for but found no close correlation between the alkaloidal composition of the vegetative above ground organs and any morphological characters. Sander mentioned, however, that all the hairy plants studied by him were of the tomatidenol chemical type. In contrast Vo Hong Nga (1975), investigating plants from natural habitats in Hungary, found a significant positive correlation between the pubescent character and the proportion of soladulcidine in soladulcidine-solasodine bearing plants. Our observations (Máthé & Máthé Jr., 1972b, c; Máthé Jr., 1974a), based on studies of hundreds of herbarium specimens, and the chemical assay of plants originating from Hungary and abroad, support Sander's statement; the hairy character is correlated with the content of unsaturated alkaloid, tomatidenol; about 60 % of the samples studied belonged to the tomatidenol taxon. Another interesting observation was that none of the tomatidenol taxa had a strong purplish colour. However it would be premature to draw far reaching conclusions from this, because of the small sample size, and in any case the purplish colour depends primarily on the environmental conditions and is shown by plants in exposed habitats.

No other morphological characters, such as flower colour or leaf shape, were found to be correlated with alkaloid composition or content (Máthé & Máthé Jr., 1972c).

Various morphological characters of *S. dulcamara* are assigned different significance by different taxonomists; similar characters have been used to describe species (Pojarkowa, 1955), and subspecific taxa (Hegi, 1927; Soó, 1968). Detailed chemical study may help solve these uncertainties.

Studies of the alkaloid, sapogenin and sterol chemical constituents of various morphological subspecific taxa of *S. dulcamara* have not revealed any unequivocal differences among the taxa, or any significant divergence from *Solanum dulcamara* L. var. *dulcamara* (Máthé Jr., 1974a). These investigations support the view that morphological characters are of little importance in subdividing this species. The effects of genetics, time and environment on the morphological characters of populations of *S. dulcamara* has been studied in Hungary (Vo Hong Nga, Bernáth & Tétényi, 1976), as has been the geographical distribution of morphological taxa (Máthé & Máthé Jr., 1972b).

CONCLUSION

This survey has tried to discuss the variability of *Solanum dulcamara* alkaloids from various aspects. Since the beginning of the chemical study of this plant the most interesting question has been how constant is the alkaloid composition. In the course of our work we try to trace back the variation of the alkaloids to the influence of two main agents: to ecological factors and to the stage of development of the plants (although the latter depends on the former, the stage of development expresses also the variability due to genetic factors).

A diagram (Fig. 13.4) illustrates the dependence of the properties discussed in this compilation on the two above-mentioned agents. (Detailed discussion of problems other than the alkaloid composition is beyond the scope of this paper).

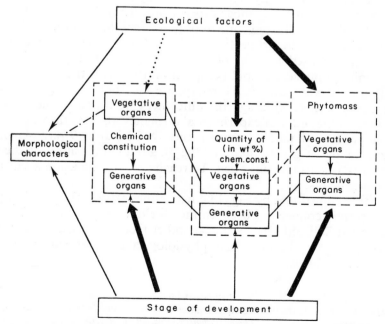

Figure 13.4. Contacts among the studied questions of *Solanum dulcamara* L. Contacts: . . . . . ., questionable; ———, proved; – . – . –, possible slight; ▬▬, strong.

Among the connections demonstrated, the direct one between ecological factors and the alkaloid composition is the most questionable. Although the problem of genetic constancy of the alkaloid composition was thoroughly examined in Germany (see above), we can not regard it as solved beyond doubt; among others, because of the following reasons.

The fact of the co-occurrence of plants with various alkaloid composition in Germany, where the genetic fixity of the chemical taxa was first studied, can also be explained by supposing that the ecological conditions here may favour the formation of both tomatidenol, solasodine and soladulcidine in the vegetative above-ground organs. Consequently, in our opinion at least, the experiments reported here could justify the supposition that no strong influence of ecological factors on the alkaloid composition exists. Even if there were any such influence, it would be very slight. Proof of its existence needs, however, studies under more extreme ecological conditions.

It should be mentioned repeatedly that the possibility of producing alkaloids other than those present in the above-ground organs exists in any plant. Independently of what kind of alkaloids occur in the above-ground organs, all of the *Solanum dulcamara* alkaloids are present in the root (Rönsch & Schreiber, 1965, 1966a; Willuhn & Kun-Anake, 1970).

The berry during development, has a varying solasodine content; the quantity of solasodine changes independently of the composition of the vegetative organs. Referring to the latter phenomenon Boll (Boll & Andersen, 1962) expresses doubt whether "chemical strains or races" of *Solanum dulcamara* exist at all. He regarded the differences in the alkaloid composition only as quantitative, not qualitative.

Recently Bernáth and Tétényi (Bernáth, Tétényi, Horváth & Zámbo, 1976) published a paper on growth chamber experiments. They found that the alkaloid composition and also the alkaloid content of tomatidenol-solasodine-soladulcidine bearing plants changed; the proportion of Δ^5 unsaturated compounds increased when plants were exposed to light of short wavelength (blue). Having studied plants with similarly mixed alkaloid composition (but with the main alkaloid as tomatidenol) for several years in Hungary, we can not exclude the tendency towards saturation either. Although the differences we found in successive years were very slight, of the order of experimental error, nevertheless the same tendency was observed in several parallel experiments (Máthé Jr., 1974a). Anyhow, at present, the question of genetic fixity of the alkaloid composition of the above-ground vegetative organs remains open. To reach a final conclusion needs further, more exact experiments.

The connections illustrated by Fig. 13.4 may either express the possibility of the existence of a certain connection between the alkaloid composition and ecological factors, manifested indirectly by way of the other factors studied, or may refer to various aspects of so far unknown basic physiological or genetic differences within the species *Solanum dulcamara*.

REFERENCES

ALKEMEYER, M. & SANDER, H., 1959a. Untersuchungen über die Inhaltsstoffe von *Solanum dulcamara* L. *Tagungsberichte Deutsche Akademie der Landwirtschaftswissenschaften zu Berlin, 27:* 23–28.
ALKEMEYER, M. & SANDER, H., 1959b. Ein Glykoalkaloid aus *Solanum dulcamara* L. *Naturwissenschaften, 6:* 207–208.

ASLANOV, S. M., 1975. Glycoalkaloids and steroid sapogenins of *Solanum persicum* (Russian). *Khimija Prirodnykh Soedinenij, 2:* 264–265.

BERNÁTH, J., TÉTÉNYI, P., HORVÁTH, I. & ZÁMBÓ, I., 1976. Effects of light intensity and spectrum composition on the formation of steroidal alkaloids. (Hungarian). *Herba Hungarica, 15:* 43–53.

BOGNÁR, R. & MAKLEIT, S., 1965. Steroidalkaloid-glykoside, X. Über den Steroidalkaloid-Glykosidgehalt von *Solanum dulcamara* L. *Acta Chimica Hungarica, 46:* 205–219.

BOLL, P. M., 1962. Alkaloidal Glycosides from *Solanum dulcamara* II. Three New Alkaloidal Glycosides and a Reassessment of Soladulcamaridine. *Acta Chemica Scandinavica, 16:* 1819–1830.

BOLL, P. M., 1963. Alkaloidal Glycosides from *Solanum dulcamara* IV. The Constitution of β- and γ-Solamarine. *Acta Chemica Scandinavica, 17:* 1852–1858.

BOLL, P. M., 1966. *Solanum* steroid alkaloids. Chemistry and botanical distribution (Danish). *Akademisk Forlag København:* 1–210.

BOLL, P. M. & ANDERSEN, B., 1962. Alkaloidal glycosides from *Solanum dulcamara* III. Differentiation of geographical strains by means of thin-layer chromatography. *Planta Medica, 4:* 421–436.

DERSCH, G. & SANDER, H., 1962. Über *Solanum dulcamara* L. 6. Mitteilung: Cytologische Untersuchung chemisch verschiedener Sippen. *Pharmazeutische Zeitung, 107:* 1540–1541.

GAUHL, E., 1968. Differential photosynthetic performance of *Solanum dulcamara* ecotypes from shaded and exposed habitats. *Carnegie Institute Year Book, 67:* 482–487.

HEGI, G., 1927. *Illustrierte Flora von Mittel-Europa, München, 5:* 2589–2592.

KÖSTENS, J. & WILLUHN, G., 1973. Steringlykoside und Acylsteringlykoside in Blättern von *Solanum dulcamara. Planta Medica, 24:* 278–285.

KUHN, R., LÖW, I. & TRISCHMANN, H., 1952. Abbau von Tomatidin zum Tigogeninlacton. *Chemische Berichte, 85:* 416–424.

MAKLEIT, S., 1966. Investigations on steroid alkaloid glycosides and steroid alkaloid aglycones. (Hungarian) Candidate Dissertation, 1–81.

MAKLEIT, S. & BOGNÁR, R., 1963. Steroid-alkaloid-glykoside, VII. Über den Δ^5 tomatidenol-/3β/-gehalt des *Solanum dulcamara* L. *Acta Chimica Hungarica, 38:* 53–54.

MAKLEIT, S., BOGNÁR, R. & MÁTHÉ, I. Jr., 1967. Steroidalkaloidglykoside XII. Untersuchung weiterer Stämme von *Solanum dulcamara* L. (Hungarian). *Herba Hungarica, 1:* 13–16.

MANNING, C. E. F., 1969. The market for steroid drug precursors with particular reference to diosgenin. (Report.) Tropical Production Institute, April.

MÁTHÉ, I. & MÁTHÉ, I. Jr., 1970. Study of the ecology and chemical constituents of *Solanum dulcamara* L. (Hungarian) *Herba Hungarica, 9:* 7–16.

MÁTHÉ, I. & MÁTHÉ, I. Jr., 1972a. The alkaloid contents of *Solanum dulcamara* L. populations in Hungary. *Herba Hungarica, 11:* 5–12.

MÁTHÉ, I. & MÁTHÉ, I. Jr., 1972b. Contributions to variability of *Solanum dulcamara* L. (Hungarian). *Botanikai Közliménynck, 59:* 129–134.

MÁTHÉ, I. & MÁTHÉ, I. Jr., 1972c. Contribution to the relationship between alkaloidal and morphological character of *Solanum dulcamara* L. (Hungarian). *Botanikai Közliménynck, 59:* 237–240.

MÁTHÉ, I. & MÁTHÉ, I. Jr., 1973. Data to the European area of the chemical taxa of *Solanum dulcamara* L. *Acta Botanica Academiae Scientiarum Hungaricae, 19:* 441–451.

MÁTHÉ, I. Jr., 1970. Investigations on the variability of the alkaloid content in *Solanum dulcamara* L. *Herba Polonica, 3:* 278–285.

MÁTHÉ, I. Jr., 1974a. Contribution to the study of alkaloid content in different organs (leaf, stem, fruit, root) of *Solanum dulcamara* L. (Hungarian). *Botanikai Közliménynck, 61:* 133–139.

MÁTHÉ, I. Jr., 1974b. Investigation with special respect to the ecological conditions of the alkaloid production of *Solanum* species native or introducible to Hungary. (Hungarian). Candidate Dissertation, 1–196.

MÁTHÉ, I. Jr. & MÁTHÉ, I., 1973. Influence of some Soil-nutrients on *Solanum dulcamara* under green-house conditions. (Hungarian). *Herba Hungarica, 12:* 29–39.

MÁTHÉ, I. Jr., SZÖCS, Z., PRÉCSÉNYI, I. & MÁTHÉ, I., 1976. Study of the *Solanum* alkaloid production of *Solanum dulcamara* L. populations in Hungary. *Herba Hungarica, 16:* 37–44.

MÁTHÉ, I. Jr., TÓTH, Gy., VAJDA, S. & MÁTHÉ, I., 1975. Study on the effects of ecological factors on *Solanum dulcamara. Acta Agronomica Academiae Scientiarum Hungaricae, 24:* 325–334.

McKEE, R. K., 1957. Solanine and Related Alkaloids. *Nature, 179:* 313–314.

POJARKOVA, A., 1955. *Solanum.* In *Flora USSR* (Russian), *22:* 12–23.

RASMUSSEN, H. B. & BOLL, P. M., 1958. Soladulcamarine, the Alkaloidal Glycoside of *Solanum dulcamara. Acta Chemica Scandinavica, 12:* 802–806.

ROZUMEK, K. E. & SANDER, H., 1967. Ein Beitrag zur chemischen Differenzierung innerhalb der Art von *Solanum dulcamara* L. *Archiv der Pharmazie, 4:* 316–321.

RÖNSCH, H., SCHREIBER, K. & STUBBE, H., 1965. Isolierung von Tomatidin aus Bittersüssen Nachtschatten. *Naturwissenschaften, 55:* 182.

RÖNSCH, H. & SCHREIBER, K., 1965. 15α-hydroxy-soladulcidin und 15α-hydroxy-tomatidin, zwei neue Steroidalkaloide aus *Solanum dulcamara* L. *Tetrahedron Letters, 24:* 1947–1952.

RÖNSCH, H. & SCHREIBER, K., 1966a. Über vier neue 15α-Hydroxy-spirosolan-Alkaloide aus Wurzeln von *Solanum dulcamara* L. *Justus Liebigs Annalen der Chemie, 694:* 169–182.

RÖNSCH, H. & SCHREIBER, K., 1966b. Solanum-Alkaloide-LXXII. Über γ₁- und δ-solamarin, zwei neue Tomatidenol-glykoside aus *Solanum dulcamara* L. *Phytochemistry, 5:* 1227–1233.

SANDER, H., 1963a. Zur Chemotaxonomie der Gattung *Solanum* (Sensu ampl.) *Botanische Jahrbücher für Systematik, Pflanzengeschichte und Pflanzengeographie, 82:* 404–428.

SANDER, H., 1963b. Über *Solanum dulcamara* L. 7. Mitteilung Abbau von Spirosolanolglykosiden in reifende Früchten. *Planta Medica, 11:* 23–35.

SANDER, H., 1963c. Chemische Differenzierung innerhalb der Art *Solanum dulcamara* L. *Planta Medica, 3:* 303–316.

SANDER, H., ALKEMEYER, M. & HÄNSEL, R., 1962. Über *Solanum dulcamara* L. 4. Mitteilung: Chemische Differenzierung innerhalb der Art und Isolierung von Soladulcidin-tetraosid. *Archiv der Pharmazie, 295:* 6–13.

SANDER, H. & WILLUHN, G., 1961. Über *Solanum dulcamara* L. 5. Mitteilung. Isolierung von Tigogenin neben Soladulcidin aus reproduktiven Organen. *Flora.-Morphologie, Geobotanik, Ökophysiologie, 151:* 150–154.

SATO, Y., MILLER, H. K. & MOSETTIG, E., 1951. Degradation of solasodine. *Journal of the American Chemical Society, 73:* 5009.

SCHREIBER, K., 1958. Die Alkaloide von *Solanum dulcamara* L. *Planta Medica, 6:* 94–97.

SCHREIBER, K. & RÖNSCH, H., 1963a. Neuere Untersuchungen zur Chemie und Biochemie der Alkaloide von *Solanum dulcamara* L. *Abhandlungen der Deutschen Akademie der Wissenschaften zu Berlin (Kl. Chem. Geol. Biol.), 4:* 395–407.

SCHREIBER, K. & RÖNSCH, H., 1963b. Über Tomatid-5-en-3β-ol aus *Solanum dulcamara* L. und dessen Synthese. *Tetrahedron Letters, 5:* 329–334.

SCHREIBER, K. & RÖNSCH, H., 1965a. Die Steroidalkaloide und sapogenine chemisch unterschiedlicher Sippen von *Solanum dulcamara* L. *Archiv der Pharmazie, 298:* 285–292.

SCHREIBER, K. & RÖNSCH, H., 1965b. Über Tomatid-5-en-3β-ol aus *Solanum dulcamara* L. und dessen Abbau zu 3β-Acetoxy-pregna-5,16-dien-20-on. *Justus Liebigs Annalen der Chemie, 681:* 187–195.

SCHREIBER, K., 1968. Steroid Alkaloids: The *Solanum*. Group. In R. H. F. Manske (Ed.), *The Alkaloids, 10:* 1–192. New York & London: Academic Press.

SCHÜTTE, M. R. In K. Mothes & M. R. Schütte, 1969. Biosynthese der Alkaloide, VEB. *Deutschen Verlag der Wissenschaften, Berlin:* 616–644.

SOÓ, R., 1968. Systematical and geobotanical handbook of Hungarian Flora and Vegetation. 3. (Hungarian). *Akadémiai Kiadó* 153–154.

TOMOWA, M., 1962. Untersuchung von Solanum-Arten auf Glykoalkaloidgehalt. 3. Mitteilung. Untersuchung über *Solanum dulcamara* L. var. *persicum* Willd. *Planta Medica, 10:* 450–454.

TOMOWA, M., 1964. Untersuchung des Glykoalkaloidgehalts einiger Solanum-Arten. 4. Mitteilung β-Solamargin in *Solanum dulcamara* L. var. *persicum* Wild. *Planta Medica, 12:* 541–542.

TUKALO, E. A. & IVANCHENKO, B. T., 1971. Glycoalkaloids of *Solanum dulcamara*. *Khimija Prirodnykh Soedinenij, 2:* 207–208.

TUZSON, P., 1954. *Solanum* Alkaloids. Akad. Doctor Dissertation 1–75. (Hungarian).

TUZSON, P. & KISS, J., 1957. *Solanum*-Alkaloide II. Soladulcidin. *Acta Chimica Academiae Scientiarum Hungaricae, 12:* 31–34.

VO HONG NGA, 1975. Chemotaxonomical investigation on the form-circle of Hungarian *Solanum dulcamara* L. Candidate Dissertation, 1–132. (Hungarian)

VO HONG NGA, BERNÁTH, J. & TÉTÉNYI, P., 1976. Response of *Solanum dulcamara* population to ecological factors I. Fixedness and instability of Morpho-phenological features. (Hungarian) *Herba Hungarica 15:* 31–44.

WILLUHN, G., 1966. Untersuchungen zur chemischen Differenzierung bei *Solanum dulcamara* L. *Planta Medica, 14:* 408–420.

WILLUHN, G., 1967. Untersuchungen zur chemischen Differenzierung bei *Solanum dulcamara* L. II. Der Steroidgehalt in Früchten Verschiedener Entwicklungsstadien der Tomatidenol- und Soladulcidin-Sippe. *Planta Medica, 15:* 58–73.

WILLUHN, G., 1968. Untersuchungen zur chemischen Differenzierung bei *Solanum dulcamara* L. III. Der Steroidalkaloidgehalt in Früchten der Solasodin-Sippe. *Planta Medica, 16:* 462–466.

WILLUHN, G. & KUN-ANAKE, A., 1970. Untersuchung zur chemischen Differenzierung bei *Solanum dulcamara*. B. Isolierung von Tomatidin aus Wurzeln der Solasodin-Sippe. *Planta Medica, 18:* 354–360.

WILLUHN, G. & KÖSTENS, J., 1974. *Solanum dulcamara* L. Triterpenoide und Sterine aus dem Petrolätherextrakt der Blätter. *Planta Medica, 25:* 115–137.

WILLUHN, G. & KÖSTENS, J., 1976. Vergleichende Untersuchung der Sterine aus den Blättern der drei Steroidalkaloid-Sippen von *Solanum dulcamara* L. *Herba Hungarica 16:* 19–22.

WILLUHN, G. & LIEBAU, A., 1976. Wasserlösliche Sterinkomplexe in Blättern von *Solanum dulcamara*. *Planta Medica, 29:* 63–65.

14. Distribution of steroidal glycoalkaloids in cells of *Solanum* and *Lycopersicon*

J. G. RODDICK

Department of Biological Sciences, University of Exeter, U.K.

Using differential centrifugation, cell fractions have been obtained from various organs of *Solanum tuberosum* and *Lycopersicon esculentum*. Analyses of fractions revealed that in cells of both species glycoalkaloids (α-solanine and α-chaconine in potato and α-tomatine in tomato) were most abundant in the soluble phase, present in smaller concentrations in the microsomal fraction and usually not detectable in lower fractions. The significance of these findings is discussed in relation to the site of synthesis and possible functions of these compounds, and also to cell membrane structure.

CONTENTS

INTRODUCTION

Many plants belonging to the genus *Solanum* and the genus *Lycopersicon* elaborate steroidal glycoalkaloids (Schreiber, 1968). Some of these compounds (e.g. α-solanine and α-tomatine) are known to exhibit *in vitro* toxicity to a wide variety of micro-organisms and animals (see McKee, 1956; Renwick, 1972; Roddick, 1974), a feature which led early workers to postulate that such compounds may play a role *in vivo* in inhibiting or reducing invasion by fungal pathogens and/or predation by phytophagous animals. Over the years, numerous arguments have been put forward both in favour of and against such rôles, but few authors (if any) have taken into consideration in their arguments and discussions the location of glycoalkaloids within the cell, even though this is a key factor in determining whether or not these compounds would actually be available to function as mentioned above.

The distribution of steroidal glycoalkaloids in the whole plant is well documented but, despite their possible involvement in plant protection and their important lytic effects on cell membranes, virtually no information exists on their distribution at the subcellular level. It has been claimed (Akahori *et al.*, 1970) that the closely related steroidal saponins of *Dioscorea tokoro* are mainly bound to cell and organelle membranes, but recent studies with young green tomato fruits (Roddick, 1976a, and see also Tables 14.1 and 2) did not reveal a similar location

223

for the tomato glycoalkaloid, α-tomatine (Fig. 14.1A). As a further check, investigations have been extended to other organs of the tomato plant and also to the glycoalkaloids of the potato plant (viz. α-solanine and α-chaconine—Fig. 14.1B,C) and this report details and discusses results obtained.

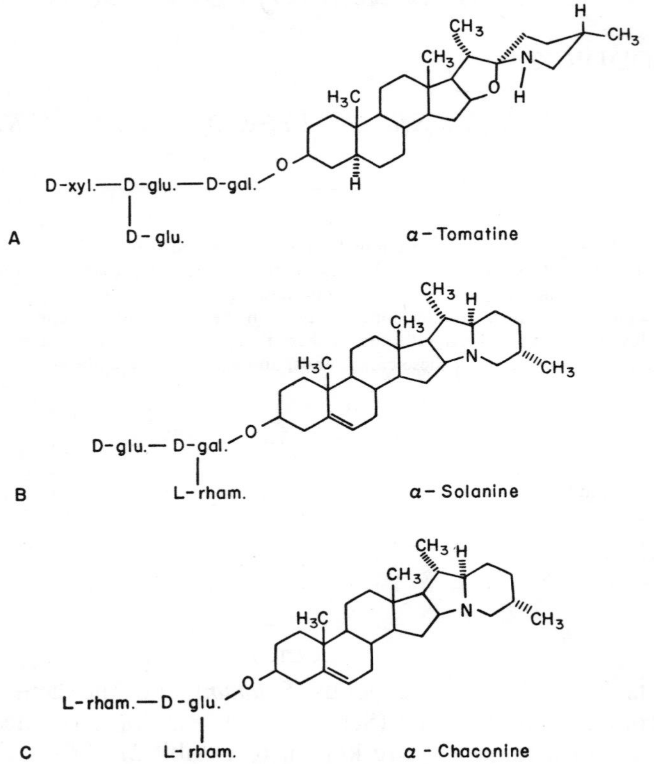

Figure 14.1. The major steroidal glycoalkaloids of *Lycopersicon esculentum* (A), and *Solanum tuberosum* (B and C). Note that α-solanine and α-chaconine differ in structure only in two sugar molecules.

MATERIALS AND METHODS

Plant material. 2–3-month-old glasshouse-grown plants of *Solanum tuberosum* L. cv. Majestic and *Lycopersicon esculentum* Mill. cv. Best of All were used.

Fractionation of tissues. Freshly-harvested plant material was homogenized in cold O.1 M phosphate buffer pH 7.4 containing D-mannitol (0.5 M), disodium EDTA (0.001 M) and dithiothreitol (0.1 %). Homogenates were strained through muslin followed by centrifugation at $500 \, g \times 10$ min, $2500 \, g \times 20$ min, $16,000 \, g \times 30$ min and $105,000 \, g \times 90$ min. Pellets were re-suspended in fresh medium, an aliquot of which was removed for protein analysis. Suspensions were then centrifuged as before and pellets extracted for alkaloids. All procedures were carried out at 4°C.

Expressed sap. Tissues were first deep-frozen for at least 24 hours then thawed and pressed in a hand-operated screw press. The expressed sap was cleared of particulate matter by centrifuging at 105,000 **g** × 90 min.

Extraction of glycoalkaloids. Glycoalkaloids from both species were extracted by essentially the same method. This is detailed elsewhere (Roddick, 1976a) but briefly involves extraction with acidified aqueous methanol, reduction of the combined extracts to aqueous under vacuum and adjustment of the pH to 10 with conc. ammonia. At this stage, potato extracts were heated at 80°C for 30 min. After cooling overnight, all extracts were centrifuged (27,000 **g** × 20 min), and the pellets washed in 1 % ammonia then dried. Pellets were then extracted with methanol and the combined methanol extracts reduced to dryness. Flask contents were taken up and made to volume, in methanol in the case of potato, and in 96 % ethanol in the case of tomato.

Supernatants were reduced in volume by rotary evaporation, ammonified and extracted as above. Cleared expressed sap was extracted as for supernatants. Tissue residues from fractionation experiments and whole plant material were extracted as for pellets (after homogenization in the latter case).

Assay of protein and glycoalkaloids. Protein was determined by the colorimetric Folin-Ciocalteau method of Lowry, Rosebrough, Farr & Randall (1951). Total potato alkaloids were estimated as α-solanine by a colorimetric method based on treatment with acidified ethanol and 1 % formaldehyde (Shih, Kuć & Williams, 1973), while α-tomatine was assayed by a radioligand method based on the precipitation of tomatine with [4-^{14}C]-cholesterol (Heftmann & Schwimmer, 1973). The presence or absence of glycoalkaloids was in all cases confirmed by TLC.

RESULTS

The organelle composition of the various fractions was determined by electron microscopy and found to be as previously described (Roddick, 1976a), except that chloroplasts were not present in roots and potato tubers and starch grains from the latter were present mainly in the 500 g pellet.

A similar pattern of α-tomatine distribution was observed in cells from small green fruits, leaves and roots of the tomato plant, the alkaloid being most abundant (on a unit protein basis) in the 105,000 g supernatants, present in small amounts in the 105,000 g pellets and usually not detectable in lower organelle fractions (Table 14.1). It should be stressed that no meaningful comparisons can be made

Table 14.1. Distribution of α-tomatine in cell fractions from
Lycopersicon esculentum

	Tomatine conc. (mg g^{-1} protein)		
	leaf	root	small green fruit
Tissue residue	1.9	1.7	3.3
500 g pellet	N.D.*	N.D.	N.D.
2500 g pellet	N.D.	N.D.	N.D.
16,000 g pellet	Trace	N.D.	N.D.
105,000 g pellet	9.9	Trace	69.2
105,000 g supernatant	105.0	2.4	141.1

*N.D. = not detectable by TLC.

between alkaloid concentrations per unit of protein in different organs. Sonication of lower organelle fractions did not release alkaloid or increase levels (Roddick, 1976a).

Tomatine concentrations calculated from fractionation data were of a similar order to values obtained by extraction of the corresponding whole organ (Table 14.2). Expressed sap from all the tested organs contained tomatine, although in differing concentrations, and the correspondence between the sap alkaloid concentration based on fresh weight of original tissue and the concentrations revealed by fractionation experiments and whole-organ extracts showed some variation, corresponding very closely in fruit tissue but less so in root tissue (Table 14.2). In some cases this may simply reflect the relative ease with which expressed sap is obtained.

Subcellular localization of potato alkaloids in all the organs tested was essentially the same as that observed for α-tomatine, with highest concentrations again occurring in the 105,000 g supernatant and smaller amounts in some of the higher cell fractions (Table 14.3). Within each organ of the potato, correspondence between whole-organ data, expressed sap values (based on fresh weight of tissue) and calculated concentrations from fractionation experiments was very close indeed (Table 14.4).

Table 14.2. α-Tomatine content of whole organs and expressed sap from
Lycopersicon esculentum

| | Tomatine conc.* | | | | | |
| | Leaf | | Root | | Small green fruit | |
	mg g^{-1}	mg cm^{-3}	mg g^{-1}	mg cm^{-3}	mg g^{-1}	mg cm^{-3}
Whole organ	1.8	—	0.5	—	0.3†	—
Expressed sap	0.8	1.2	0.1	0.1	0.3†	0.4†
Total from fractionated tissue	3.0	—	0.2	—	0.5†	—

*weight basis is fresh weight.
†based on fruit pericarp tissue only.

Table 14.3. Distribution of glycoalkaloids in cell fractions from
Solanum tuberosum

| | Glycoalkaloid conc. (mg g^{-1} protein)† | | |
	leaf	root	tuber
Tissue residue	1.0	1.8	0.1
500 g pellet	N.D.*	N.D.	N.D.
2500 g pellet	N.D.	N.D.	N.D.
16,000 g pellet	2.0	N.D.	0.4
105,000 g pellet	4.1	N.D.	0.9
105,000 g supernatant	13.6	2.4	2.9

*N.D. = not detectable by TLC.
†calculated as α-solanine.

Table 14.4. Glycoalkaloid content of whole organs and expressed sap from *Solanum tuberosum*

| | Glycoalkaloid conc.* | | | | | |
| | leaf | | root | | tuber | |
	mg g^{-1}	mg cm^{-3}	mg g^{-1}	mg cm^{-3}	mg g^{-1}	mg cm^{-3}
Whole organ	0.6	—	0.2	—	0.03	—
Expressed sap	0.6	0.9	0.2	0.3	0.04	0.06
Total from fractionated tissue	0.5	—	0.3	—	0.03	—

*calculated as α-solanine. Weight basis is fresh weight.

DISCUSSION

The cell fractionation data suggest that the glycoalkaloids of both potato and tomato are located principally, but not necessarily exclusively, in the soluble phase of the cell, and this tends to be supported by the levels of glycoalkaloids detected in expressed sap. It is not yet known however whether these compounds accumulate mainly in the vacuole, mainly in the soluble phase of the cytoplasm or evenly distributed between the two. Since the aglycones of the compounds involved (viz. solanidine and tomatidine) are virtually insoluble in aqueous solution, glycosylation would appear to be, at least in part, a solubilization process.

With the exception of potato root, glycoalkaloids were always detected in the 105,000 **g** pellet, although not always in lower fractions, and this raises the question of whether these compounds may be synthesized, either partly or wholly, in the microsomes and then transported to the soluble phase of the cell where they are stored. Evidence has already been presented that plant sterols, and possibly some steroidal sapogenins, are synthesized in these organelles (Knapp, Aexel & Nicholas, 1969; Akahori *et al.*, 1970) and glycoalkaloids are known to be derived from the former (Heftmann, Lieber & Bennett, 1967) and closely related chemically and biogenetically to the latter (Schreiber, 1968). The failure to detect glycoalkaloids in the microsomal fraction from potato root is probably a result of the small total amounts present. A similar situation has been encountered in large green tomato fruits where total tomatine is also low (Roddick, 1976a). The sporadic appearance of glycoalkaloids in the 16,000 **g** fractions could again be due to very low concentrations, in some cases below the limits of detection, or, and probably more likely, to contamination of the fractions by microsomal material (fractions obtained by differential centrifugation are by no means pure). It yet remains to be established whether the intracellular distribution of other steroidal glycoalkaloids resembles that proposed here for the potato and tomato alkaloids.

Glycosylation confers on tomatidine and solanidine not only their solubility but also, to a large extent, their toxicity (McKee, 1959; Arneson & Durbin, 1968). The toxic nature of these compounds frequently manifests itself in membrane disruption and this appears to stem partly from their surfactant properties (McKee, 1959) and partly from their ability to bind membrane sterols (Schreiber, 1957; Arneson & Durbin, 1968). Since the glycoalkaloids of potato and tomato appear to be located in the soluble phase of the cell and are present in relatively high concentrations in some organs (e.g. around 1mM in leaves), the limiting membranes of the "parent" cells must somehow be able to withstand the considerable lytic potential of such levels. That they in fact do is suggested by a recent report (Roddick, 1976b) that cells of tomato are less susceptible to lysis and general disruption by exogenous tomatine than cells of non-tomatine-containing plants. How this apparent resistance to endogenous alkaloid occurs however is not yet clear. Although the toxicity of glycoalkaloids is reduced with decreasing pH (McKee, 1959), the pH of exogenous solutions used in the above study (Roddick, 1976b) was sufficiently similar to that of cell sap to rule out this as the principal detoxification mechanism. For some glycoalkaloids at least (e.g. α-tomatine), susceptibility to lysis is thought to be correlated with the free sterol composition of the membrane (Arneson, 1967) and thus it may be that damage to cell membranes by endogenous alkaloid is prevented either by physical separation of alkaloid and membrane sterol or by modifications to the sterol component. Since glycoalkaloids only bind sterols with a free 3β-hydroxyl group (Schulz & Sander, 1957), some degree of "immunity" could be achieved by an overall reduction in the frequency of sterol molecules. Alternatively, the availability of suitable target molecules would be reduced if a proportion of the sterols existed, not in the free form, but as esters or glycosides, both of which involve bonding at C-3 and loss of the free hydroxyl group.

The rôle of steroidal glycoalkaloids in plant protection is still a subject of much debate. Most work has been done on the tomato and potato glycoalkaloids and concerned with protection against fungal pathogens (see Roddick, 1974; Frank, Wilson & Webb, 1975). There exist numerous conflicting reports relating to whether these compounds are more concentrated in resistant (cf. susceptible) cultivars of healthy plants, whether their levels actually increase following infection, and whether resistant cultivars contain more alkaloid than susceptible cultivars following infection. The concentration of glycoalkaloids in certain tissues makes it likely that they at least contribute to the creation of an unfavourable environment for the pathogen, although the extent to which they do so is not yet clear, possibly because it varies with the plant organ, pathogen, etc.

On the other hand, the fact that glycoalkaloids are continually present, and often in considerable amounts in aerial organs, has led to suggestions (Schreiber, 1957; Fraenkel, 1959) that they may be more important as repellants (rather than toxins) of would-be predators (such as certain insects and herbivores), the effects of which are more or less instantaneous.

In either role, it would probably be advantageous if glycoalkaloids were present in a readily-available soluble form, and indeed it would be difficult to imagine how they could operate effectively as toxins or deterrents if bound to, or a component of, cell or organelle membranes.

Lack of knowledge of the intracellular distribution of glycoalkaloids has meant that up till now it has not always been certain that these compounds would actually be available *in vivo* to act according to the methods discussed above. However, the findings of this work suggest that, at least from the point of view of their location within the cell, no major objections can be raised to glycoalkaloids functioning as protective agents (of whatever type) in potato and tomato plants.

ACKNOWLEDGEMENT

The author wishes to acknowledge the invaluable technical assistance of Mrs Linda Firminger.

REFERENCES

AKAHORI, A., YASUDA, F., KAGAWA, K., ANDO, M. & TOGAMI, M., 1970. Intracellular distribution of the steroidal sapogenins in *Dioscorea tokoro*. *Phytochemistry, 9:* 1921–1928.

ARNESON, P. A., 1967. *The Rôle of Tomatine in the Resistance of Tomato Plants to Fungal Infection.* PhD thesis, University of Wisconsin.

ARNESON, P. A. & DURBIN, R. D., 1968. Studies on the mode of action of tomatine as a fungitoxic agent. *Plant Physiology, 43:* 683–686.

FRAENKEL, G. S., 1959. The raison d'être of secondary plant substances. *Science, 129:* 1466–1470.

FRANK, J. A., WILSON, J. M., & WEBB, R. E., 1975. The relationship between glycoalkaloids and disease resistance in potatoes. *Phytopathology, 65:* 1045–1049.

HEFTMANN, E., LIEBER, E. R. & BENNETT, R. D., 1967. Biosynthesis of tomatidine from cholesterol in *Lycopersicon pimpinellifolium*. *Phytochemistry, 6:* 225–229.

HEFTMANN, E. and SCHWIMMER, S., 1973. A radioligand assay of tomatine. *Phytochemistry, 12:* 2661–2663.

KNAPP, F. F., AEXEL, R. T. & NICHOLAS, H. J., 1969. Sterol biosynthesis in sub-cellular particles of higher plants. *Plant Physiology, 44:* 442–446.

LOWRY, O. H., ROSEBROUGH, N. J., FARR, A. L. & RANDALL, R. J., 1951. Protein measurement with the Folin phenol reagent. *Journal of Biological Chemistry, 193:* 265–275.

McKEE, R. K., 1956. *Studies on the Toxicity of Solanine and Related Compounds to Micro-organisms and on the Influence of Solanine Content on the Resistance of Potato Tuber Tissues to Fungal Infection.* PhD thesis, University of Nottingham.

McKEE, R. K., 1959. Factors affecting the toxicity of solanine and related alkaloids to *Fusarium caeruleum*. *Journal of General Microbiology, 20:* 686–696.

RENWICK, J. H., 1972. Hypothesis. Anencephaly and spina bifida are usually preventable by avoidance of a specific but unidentified substance present in certain potato tubers. *British Journal of Preventative and Social Medicine, 26:* 67–88.

RODDICK, J. G., 1974. The steroidal glycoalkaloid α-tomatine. *Phytochemistry, 13:* 9–25.

RODDICK, J. G., 1976a. Intracellular distribution of the steroidal glycoalkaloid α-tomatine in *Lycopersicon esculentum* fruit. *Phytochemistry, 15:* 475–477.

RODDICK, J. G., 1976b. Response of tissues and organs of tomato to exogenous α-tomatine. *Journal of Experimental Botany, 27:* 341–346.

SHIH, M., KUĆ, J. & WILLIAMS, E. B., 1973. Suppression of steroid glycoalkaloid accumulation as related to rishitin accumulation in potato tubers. *Phytopathology, 63:* 821–826.

SCHREIBER, K., 1957. Natürliche pflanzliche Resistenzstoffe gegen den Kartoffelkäfer und ihr möglicher Wirkungsmechanismus. *Züchter, 27:* 289–299.

SCHREIBER, K., 1968. In R. H. F. Manske (Ed.), *The Alkaloids. Chemistry and Physiology, X.* New York & London: Academic Press.

SCHULZ, G. & SANDER, H., 1957. Über Cholesterin-Tomatid, eine neue Molekulverbindung zur Analyse und präparativen Gewinnung von Steroiden. *Hoppe-Seyler's Zeitschrift für Physiologische Chemie, 308:* 122–126.

15. The characteristics of solasodine accumulation in *Solanum khasianum* C. B. Clarke var. *chatterjeeanum* Sengupta* and *S. laciniatum* Ait. grown under field conditions in Birmingham

LINDA M. F. MILLER AND M. E. DAVIES

Department of Plant Biology, University of Birmingham, Birmingham, U.K.

Seasonal variation in solasodine content was determined for *Solanum viarum* (= *S. khasianum* var. *chatterjeeanum*)* and *S. laciniatum* grown under identical field conditions in Birmingham. The results confirm the major differences in the pattern of accumulation reported for these two species, the alkaloid being located solely in the berries of *S. khasianum* and in both berries and foliage of *S. laciniatum*. Solasodine concentrations in individual organs are similar to those reported for plants grown at other locations.

INTRODUCTION

The social and economic importance of steroidal drugs has been emphasized by Bradley *et al.* in a previous contribution to this Volume (p. 203). The increasing demand for these products has prompted several investigations into the potential of other solanaceous alkaloids, particularly solasodine, as alternatives to the currently used steroidal sapogenins. The present report is concerned with the characteristics of solasodine accumulation in two species, one Indian (*S. khasianum*) and one Australasian (*S. laciniatum*), grown under temperate European conditions.

METHODS

S. khasianum and *S. laciniatum* seeds were sown in early April 1975. After seven weeks growth under greenhouse conditions the plants were transferred to the field, planting in alternating blocks, five blocks of 20 plants for each species. At each harvest five plants of each species were selected on a random basis and divided into separate organs before being dried at 60°C for 24 hours. Each sample was ground to a fine powder and analysed for solasodine content using a modification of the colorimetric method of Lancaster & Mann (1975).

* Examination of this material and the relevant literature has convinced us that the correct name for *Solanum khasianum* C. B. Clarke var. *chatterjeeanum* Sengupta is *Solanum viarum* Dunal, as established by Babu (Babu, C. R., 1971. The identity of *Solanum khasianum* C. B. Clarke var *chatterjeeanum* Sengupta (Solanaceae). *Journal of the Bombay Natural History Society*, 67: 609–611.). It is not conspecific with *S. khasianum* C. B. Clarke var *khasianum*, *S. myriacanthum* Dunal or *S. reflexum* Schrank.—Eds.

In order to define the environmental conditions during the growth period, local climatic data is presented in Fig. 15.1.

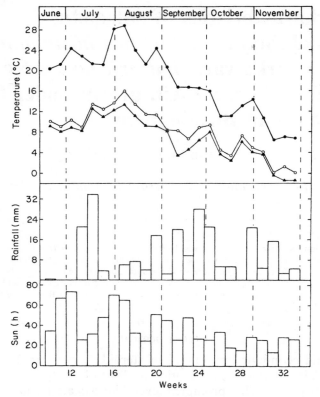

Figure 15.1. The climatic conditions experienced in Birmingham; June to November, 1975. ●, Maximum temperature; ○, minimum temperature; ▲, minimum ground temperature.

RESULTS

Growth

For both species maximum growth occurred between 14 and 21 weeks after germination and thereafter the size of the plants remained fairly constant (Fig. 15.2). In the case of *S. khasianum*, however, there was a decrease in leaf fresh weight due to leaf abscission associated with a lower ambient temperature, whilst berry set and maturation beginning at week 18 increased fruit fresh weight. During the latter part of the growing period any increases in fresh weight of *S. laciniatum* were directed largely to the thickening of the stem.

Solasodine accumulation

The berry is the sole organ which accumulates solasodine in *S. khasianum*, a maximum concentration (1.3 %) being established after 25 weeks which is maintained for the remainder of the growth period (Fig. 15.3). On the other hand, the

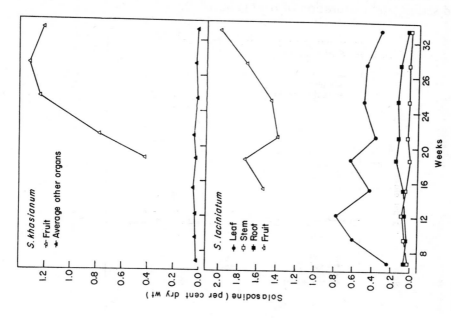

Figure 15.3. Changes in solasodine concentration. Standard errors of the means were as follows: *S. khasianum* = 0.05, *S. laciniatum* = 0.06.

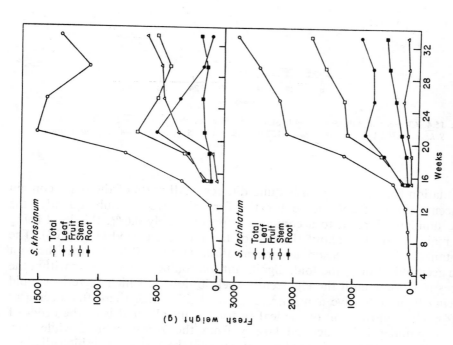

Figure 15.2. Changes in fresh weight. Standard errors of the means were as follows: total plant—*S. khasianum* = 157.8, *S. laciniatum* = 254.9; plant parts—*S. khasianum* = 19.5, *S. laciniatum* = 38.9.

yield of alkaloid per plant increases steadily throughout the growing period due to continued setting and maturation of fruit (Fig. 15.4).

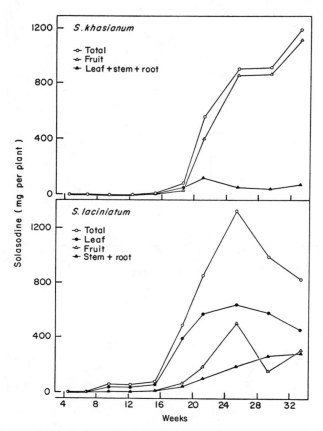

Figure 15.4. Changes in solasodine yield. Standard errors of the means were as follows: total plant—*S. khasianum* = 134.0, *S. laciniatum* = 152.6; plant parts—*S. khasianum* = 45.1, *S. laciniatum* = 31.2.

The situation for *S. laciniatum* is quite different. All parts of the plant contain low concentrations of solasodine for the first nine weeks. Subsequently, the leaves accumulate alkaloid to a content of approximately 0.6%, this level being roughly maintained throughout the rest of the growing period (Fig. 15.3). The fruit contain the highest solasodine concentrations (1.4 to 2.0%) whereas the root and stem accumulate little alkaloid. Significant solasodine accumulation within the plant does not occur until the eighteenth week, following which there is a rapid increase in alkaloid to give a maximum at week 25 which thereafter decreases (Fig. 15.4). The contribution of the leaf to total plant alkaloid is of the order of 60% the remainder being derived largely from the berries which, while they represent only a small proportion of total fresh weight, because of their high alkaloid content contribute significantly to the yield.

DISCUSSION

The present observations on the pattern of accumulation of alkaloid within these two species confirm those reported by other research workers in this field (Khanna & Murty, 1972; Saini, 1966; Lancaster & Mann, 1975). Workers from several countries including the Soviet Union and Hungary have reported on the alkaloid contents of field grown *S. laciniatum*. Some strains have been described as having a solasodine content of approximately 2.0 % of dried overground material (Bernáth & Foldesi, 1974; Murav'eva, Konoratenko & Brink, 1969). Field trials in New Zealand have reported solasodine concentrations of 0.6 % in mature leaves, 0.9 % in immature leaves and 2.0 % in berries (Lancaster & Mann, 1975). Alkaloid concentrations observed in plants cultivated in Birmingham were found to be similar to the results from New Zealand although they are consistently lower than those reported by investigators in the Soviet Union and Hungary.

No clear conclusions can be drawn from a comparison of the solasodine content of *S. khasianum* berries produced by plants grown under temperate conditions with results reported in India as these range from 0.5 to 5.4 % (Puri & Bhatnagar, 1974).

The practical problems of cultivating the two species under temperate conditions are somewhat different.

Although there is a considerable difference in stature, yields of glycoalkaloid from the two species are remarkably similar. Use of *S. khasianum* offers the advantages that solasodine is located solely in the berry and this may facilitate processing. On the other hand the requirement for heavy fruit set and maturation means that under these conditions the season would have to be artificially extended by raising plantlets under greenhouse conditions. In this respect, the potential of *S. laciniatum* would appear to be somewhat greater, in that the requirement is largely for substantial amounts of foliage—a condition which is achieved quite early in the growing season. Extending the season artificially only serves to increase the ratio of stem material to leaf which may be undesirable (Lancaster & Mann, 1975; Foldesi, Lang & Kovács, 1969). In our experience berry set and maturation appears to be a phenomenon associated with the early development of plants in the field, and hence no virtue would accrue from artificially extending the growth period.

ACKNOWLEDGEMENTS

We wish to express our gratitude to the Science Research Council for a grant in support of one of us (L.M.F.M.)

REFERENCES

BERNÁTH, J. & FOLDESI, D., 1974. The relationship between the solasodine content of *Solanum laciniatum* leaves and changes in the air temperature. *Herba Hungaria, 13:* 37–43.

FOLDESI, D., LANG, T. & KOVÁCS, T., 1969. Localisation of the active substance of *Solanum laciniatum* Ait. in individual plant organs and its accumulation in the course of the vegetative period. *Herba Hungarica, 8:* 49–61.

KHANNA, K. R. & MURTY, A. S., 1972. Effect of fruit stage and maturity on the glycoalkaloid content in *Solanum khasianum*. *Planta Medica, 21:* 182–187.

LANCASTER, J. E. & MANN, J. D., 1975. Changes in solasodine content during the development of *Solanum laciniatum*. *New Zealand Journal of Agricultural Research, 18:* 139–144.

MURAV'EVA, V. I., KONORATENKO, P. T. & BRINK, N. P., 1969. Changes in the solasodine content of *Solanum laciniatum*. *Rast. Resur., 5:* 187–190.

PURI, R. K. & BHATNAGAR, J. K., 1974. Studies on the genus *Solanum*—a review. *Pharmacos, 19:* 7–24.

SAINI, A. D., 1966. Alkaloid content of *Solanum khasianum* Clarke. *Current Science, 35:* 600.

16. The hydrolysis of tomatine by an inducible extracellular enzyme from *Fusarium oxysporum* f. sp. *lycopersici*

J. E. FORD[1], D. J. McCANCE[2] AND R. B. DRYSDALE

Microbiology Department, Birmingham University, Birmingham, U.K.

Fusarium oxysporum f. sp. *lycopersici* detoxifies α-tomatine by producing an inducible extracellular enzyme which cleaves the glycoalkaloid forming the tetrasaccharide lycotetraose and tomatidine.

The antifungal glycoalkaloid tomatine, found in a number of species of *Solanum* and *Lycopersicon* including the tomato *L. esculentum*, is composed of the spiro-solane tomatidine and lycotetraose, a tetrasaccharide containing two glucose residues, one xylose residue and one galactose residue (Fig. 16.1).

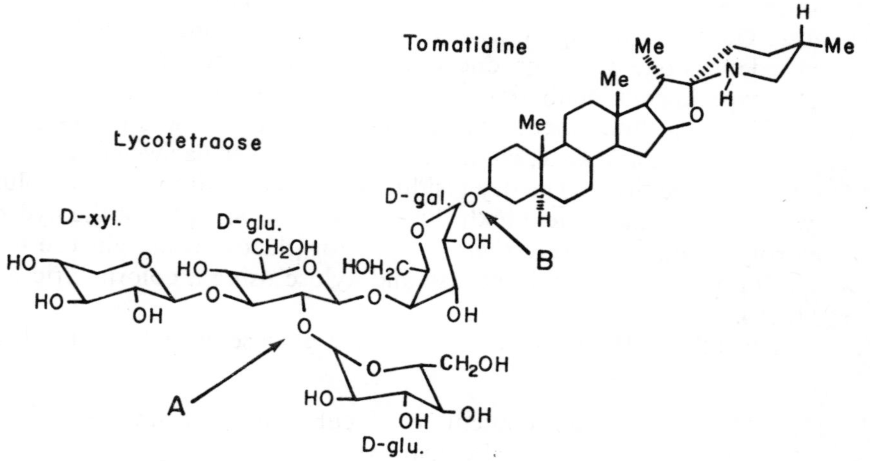

Figure 16.1. Structure of tomatine. Enzyme from *Septoria lycopersici* hydrolyses the molecule at A. B is the proposed site of attack by enzyme from *Fusarium oxysporum* f. sp. *lycopersici*.

In healthy tomato plants the concentration of tomatine is highest in the leaves and this has led to the suggestion (Arneson & Durbin, 1967) that tomatine may be important in the resistance of tomato to leaf spot caused by *Septoria lycopersici*. This fungus can detoxify tomatine by enzymatically removing one glucose residue

---

[1] Present address: Department of Pathology, Scarborough General Hospital.
[2] Present address: Microbiology Department, Guy's Hospital Medical School, London.

from the molecule forming β2—tomatine (Fig. 16.1) which is much less toxic to the fungus than the parent compound.

Tomatine may also be involved in the resistance of tomato to wilt caused by *Fusarium oxysporum* f. sp. *lycopersici* but this is not yet clearly established (McCance & Drysdale, 1975).

This paper describes the mechanism of detoxification of tomatine by an extracellular enzyme produced by *F. oxysporum* f. sp. *lycopersici*.

The enzyme was obtained by growing *F. oxysporum* f. sp. *lycopersici*, Race 1 in tomato leaf extract for five days at 25°C on a shaker. Ammonium sulphate was added to the spent medium to give 90 % saturation, the precipitate collected, redissolved in water, dialysed overnight at 4°C against citrate-phosphate buffer (pH 5.0) and then stored in 1 ml volumes at −20°C until required.

Following incubation (24 h, 25°C) of tomatine (1 mg/ml final concentration) with the enzyme preparation the incubation mixture was examined by thin layer chromatography (TLC) on silica gel using n-propanol/water (85/15) as solvent. Tomatidine was detected on the TLC plates but there was no tomatine nor any spots corresponding to the monosaccharides glucose, galactose or xylose. The incubation mixture also failed to give a positive test for glucose when assayed by a glucose oxidase method although a positive test was obtained if the incubation mixture was acid hydrolysed (1 M HCl, 100°C, 60 min) prior to the glucose oxidase assay. In addition following acid hydrolysis the incubation mixture could be shown, by TLC, to contain the monosaccharides glucose, galactose and xylose.

These results suggest that *F. oxysporum* f. sp. *lycopersici* produces an enzyme which converts tomatine to tomatidine and, presumably, the tetrasaccharide lycotetraose. Quantitative experiments were then carried out to confirm that the tetrasaccharide was one of the products of the enzyme reaction.

For these experiments incubation mixtures were prepared as described above, but following the 24 hour incubation period, the mixture was dialysed against distilled water and the dialysate, containing the low molecular weight components of the mixture, freeze-dried. The dry residue was redissolved in a small volume of distilled water and a sample acid hydrolysed as before. Samples of the hydrolysed and unhydrolysed material were then assayed for glucose using glucose oxidase, galactose using galactose dehydrogenase and xylose using a colorimetric method (Tracey, 1950).

These assays (Table 16.1) showed that no free glucose, galactose or xylose was

Table 16.1. Monosaccharide content of incubation mixture following acid hydrolysis[1]

Sugar	mg/ml	Molarity (μM)	Molar ratio[2]
Glucose	22	122.1	2.00
Galactose	12	66.6	1.09
Xylose	11	73.3	1.20

[1] Monosaccharides were not detectable in incubation mixture prior to acid hydrolysis.
[2] Glucose taken as 2.00.

present in unhydrolysed material whereas following acid hydrolysis the three monosaccharides were present in the 2 : 1 : 1 molar ratios expected for the tetrasaccharide.

TLC of the reconstituted freeze-dried dialysate also failed to reveal the presence of the three individual monosaccharides but some material was present on the plate in the region with an Rf of 0-0.22. Silica gel from the corresponding region of a duplicate plate was recovered, eluted with water and the eluate divided into two portions one of which was hydrolysed with acid as before. The hydrolysed and unhydrolysed samples were then freeze-dried, redissolved in a small volume of distilled water and assayed for the three sugars as above.

These assays (Table 16.2) showed that the unhydrolysed sample did not contain the monosaccharides but that following acid hydrolysis glucose, galactose and xylose were present in the 2 : 1 : 1 molar ratio anticipated for the tetrasaccharide.

Results in agreement with those presented in Tables 16.1 and 16.2 have also been obtained following isolation of the presumptive tetrasaccharide from the enzyme incubation mixture by chromatography on Bio-Gel P2 polyacrylamide.

Table 16.2. Monosaccharide content of presumptive tetrasaccharide recovered from TLC plate and subjected to acid hydrolysis[1]

Sugar	mg/ml	Molarity (µM)	Molar ratio[2]
Glucose	33	183.1	2.00
Galactose	19	105.4	1.15
Xylose	15	99.9	1.09

[1] Prior to acid hydrolysis monosaccharides were not detectable.
[2] Glucose taken as 2.00.

These experiments indicate that *Fusarium oxysporum* f. sp. *lycopersici* produces an enzyme which detoxifies tomatine by removing the sugars as the tetrasaccharide, leaving the much less toxic aglycone tomatidine. The enzyme is inducible. This detoxification mechanism is different from that of *Septoria lycopersici* (Arneson & Durbin, 1967) but may be similar to that reported briefly by Ver Hoeff and Liem (1975) for *Botrytis cinerea*.

REFERENCES

ARNESON, P. A. & DURBIN, R. D., 1967. Hydrolysis of tomatine by *Septoria lycopersici*: a detoxification mechanism. *Phytopathology, 57*: 1358–1360.
McCANCE, D. J. & DRYSDALE, R. B., 1975. Production of tomatine and rishitin in tomato plants inoculated with *Fusarium oxysporum* f. sp. *lycopersici*. *Physiological Plant Pathology, 7*: 221–230.
TRACEY, M. V., 1950. Pentose determination in the presence of hexoses and uronic acids. *Biochemical Journal, 47*: 433.
VER HOEFF, K. & LIEM, J. I., 1975. Toxicity of tomatine to *Botrytis cinerea* in relation to latency. *Phytopathologische Zeitschrift, 82*: 333–338.

17. Tropane alkaloids of the Solanaceae

W. C. EVANS

Department of Pharmacy, University of Nottingham, Nottingham, U.K.

The review covers the distribution of tropane alkaloids within the plant kingdom and more specifically their occurrence, chemical structure and biosynthesis in the family Solanaceae. Some clear-cut chemotaxonomic features are evident. The genus *Datura* contains a wide spectrum of tropane alkaloids as do also *Solandra*, *Anthocercis* and *Duboisia*. All species of the subtribe Hyoscyaminae contain principally mixtures of the hyoscine-hyoscyamine type alkaloids; other taxa of the family contain variously no tropane alkaloids, alkaloids restricted to the hyoscyamine type, and alkaloids of the tigloyloxytropane type. Alkaloid inheritance in various hybrids of *Datura* species is discussed.

CONTENTS

INTRODUCTION

Tropane alkaloids constitute one of the distinctive groups of secondary metabolites of the Solanaceae and many plants containing them have long been utilized for their medicinal, hallucinogenic, and poisonous properties. Such alkaloids are not, however, confined to this family and they are slowly being discovered in other families (Table 17.1). With the possible exception of the Erythroxylaceae, Euphorbiaceae, Solanaceae and Convolvulaceae, for which some phylogenetic relationship can be claimed (e.g. Bessey, 1915), it would seem that tropane alkaloid formation represents a parallel development throughout the plant kingdom. Even within the group of families mentioned above, it is only for the Erythroxylaceae and Solanaceae that these compounds are sufficiently common within the family to warrant their consideration as chemotaxonomic characters. Within the Solanaceae, certain tribes and genera are well characterized by the presence of these compounds whereas others are notable for their complete absence (Table 17.2).

Table 17.1. Distribution of tropane alkaloids in the plant kingdom

	Approx. number of:				
Family	genera	genera containing tropane derivatives	alkaloids described	Alkaloid type	References
Orchidaceae	600–700	(1)	(2)	⎡ *cis*- and *trans*- ⎤ Cuscohygrine ⎣ derivative ⎦	Ekevag and Gawell (1973)
Dioscoreaceae	11	1	2	Tropine ester, a 2-hydroxy-tropane derivative	Pinder (1952), (1957)
Proteaceae	62	4	10+	4-Benzylic derivatives of hydroxytropanes, a tropane-γ-pyrone & others	Bick *et al.* (1975) Gillard (1975) Kan-Fan & Lounasmaa (1973) Lounasmaa (1975) Motherwell *et al.* (1971)
Rhizophoraceae	16	1	6	Tropine esters, a dithiolane	Loder & Russell (1966), (1969)
Cruciferae	350	1	2	Tropine esters	Platonova & Kuzovkov (1963)
Erythroxylaceae	4	1	over 20	Hydroxytropane esters, ecgonine derivatives	Henry (1949), (for early references) Johns *et al.* (1970) Agar *et al.* (1974), (1975)
Euphorbiaceae	290	2	3	Hydroxytropane esters	Parello *et al.* (1963) Johns *et al.* (1971)
Solanaceae	68	21	over 30	See this review	
Convolvulaceae	51	2	5	Tropine & nortropine esters, cuscohygrine	Willaman & Schubert (1961) Evans & Somanabandhu (1974)

Table 17.2. Distribution of tropane-type alkaloids in the Solanaceae (Wettstein's Classification)

Tribe	Subtribe	Genera	Genera containing tropane-type alkaloids (including cuscohygrine)	No. of alkaloids isolated (approx.)
I Nicandreae		1	1	2
II Solaneae	Lyciinae	14	3	9
	Hyoscyaminae	4	4	14
	Solaninae	12	4	6 (also non-tropane bases)
	Mandragorinae	6	3	11
III Datureae		2	2	30
IV Cestreae	Cestrinae	3	0	0
	Goetzeinae	4	0	0
	Nicotianinae	12	0	0 (non-tropane bases present)
V Salpiglossideae		9	4	22 (also non-tropane bases)

STRUCTURES AND BIOGENESIS OF TROPANE ALKALOIDS

The range of tropane alkaloids within the Solanaceae arises from the esterification of various acids, as illustrated below, with hydroxytropanes of the structures I, II or III.

Acetic acid $CH_3.COOH$

Propionic acid $CH_3.CH_2.COOH$

Isobutyric acid $\dfrac{CH_3}{CH_3}{>}CH.COOH$

Isovaleric acid $\dfrac{CH_3}{CH_3}{>}CH.CH_2.COOH$

2-Methylbutyric acid $CH_3.CH_2.\underset{\underset{CH_3}{|}}{CH}.COOH$

Tiglic acid $CH_3.HC : C.CH_3.COOH$

(+)−α−Hydroxy−β−phenylpropionic acid

Tropic acid

Atropic acid

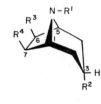

I

R^1 = H or CH_3
R^2 = OH
R^3 = H or OH
R^4 = H or OH

II

R^1 = H or CH_3
R^2 = OH

III

With esters involving a tropane-diol or triol two identical, or two different, acids may be involved in the formation of a bisacyloxytropane; in the case of ψ-tropine (III), tiglic acid is the only known esterifying acid. No cases are known of natural alkaloids in which all three hydroxyls of the triol are substituted, and esterification with tropic and atropic acids appears to be confined to C-3. Two configurations of the tropane-3,6-diol are known in the Solanaceae. The (-)-form constitutes the alkamine of valeroidine (*q.v. Duboisia*) and it has the 3S : 6S absolute configuration, and the (+)-antimer is the alkamine of the mono- and di-tigloyltropane diols (*q.v. Datura*) having the 3R : 6R configuration (Fodor & Sóti, 1965).

The biosynthetic origin of the tropane ring system in such well-known genera as

Datura and *Atropa* has been fully investigated (Fodor, 1967, 1971) and will not be discussed here; the general pathway is indicated below:

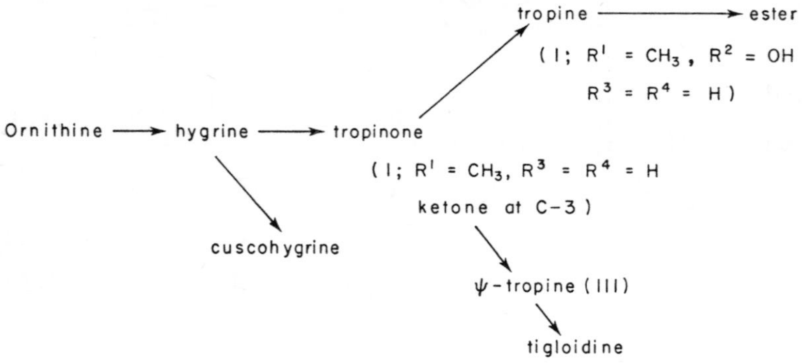

tropine   → ester

(I; R^1 = CH_3 , R^2 = OH

R^3 = R^4 = H)

Ornithine ⟶ hygrine ⟶ tropinone

(I; R^1 = CH_3, R^3 = R^4 = H

ketone at C–3)

cuscohygrine

ψ–tropine (III)

tigloidine

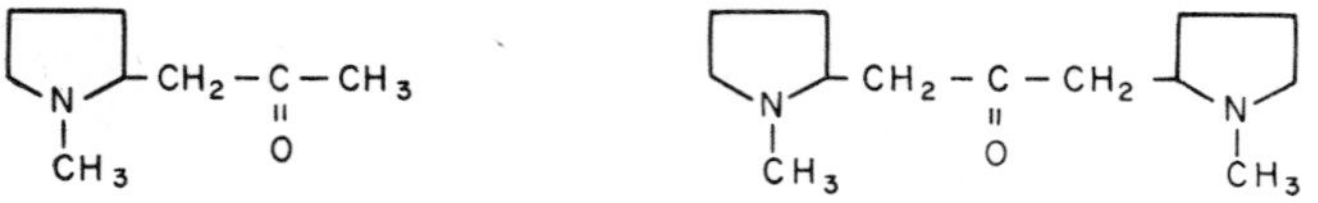

Hygrine

Cuscohygrine

The biogenesis of cuscohygrine, which is a side-product of the principal tropane pathway, has been studied in *Scopolia lurida* (O'Donovan & Keogh, 1969). The intermediates in the tropane pathway often form significant components of the alkaloid mixture of solanaceous plants and in some genera, as discussed below, they may represent the limit of alkaloid synthesis.

Current biogenetic interest is centred on the formation of the tropane-diols and triol and their tigloyl esters. A number of pathways are plausible, e.g. direct hydroxylation of tropine, esterification of tropine followed by hydroxylation, hydroxylation of a di-ester to yield a triol derivative, formation of diols and triols by different routes. Experiments which give unequivocal results for the elucidation of such pathways are extremely difficult to design (Leete, 1972; Basey & Woolley, 1973a, 1975; Beresford & Woolley, 1975; Leete & Lucast, 1975).

Hyoscine (II; R^1 = CH_3, R^2 = tropoyloxy) is formed from hyoscyamine (I; R^1 = CH_3, R^2 = tropoyloxy, R^3 = R^4 = H) via 6-hydroxyhyoscyamine (Romeike & Fodor, 1960) and the extent to which this conversion occurs often depends on the age of the plant, particular variety or species, and geographical source, and is an important factor in determining the ultimate use and pharmacological properties of the plant. The tropic acid moiety of hyoscine and hyoscyamine with its C_3 side-chain is derived entirely from phenylalanine with side-chain rearrangement (Leete & Louden, 1962). But, in spite of much research, the intermediates in this re-arrangement still remain obscure. The acid moiety of littorine (α-hydroxy-β-phenylpropionic acid) is also derived from phenylalanine but without side-chain rearrangement (Evans & Woolley, 1969); however, neither the rearrangement of this ester nor of cinnamoyl tropine to produce hyoscyamine could be demonstrated in *Datura stramonium* (Leete & Kirven, 1974). The evidence for the role of cinnamic acid in tropic acid biosynthesis is conflicting (Evans & Woolley, 1976; Prabhu, Gibson & Schraum, 1976). N-Demethylation and dehydration of hyoscine and hyoscyamine can occur in the plant to give the nor- and apo-derivatives respectively; whether other routes can be utilized for the formation of these alkaloids is not clear.

DISTRIBUTION OF TROPANE ALKALOIDS IN THE FAMILY

Wettstein's (1895) classification of the Solanaceae into five tribes with seven subtribes (Table 17.2) is used below for the consideration of the distribution of tropane alkaloids in the family. The tribes are treated in an order which lends itself to the discussion of alkaloid occurrence.

Tribe Datureae

The Datureae, comprised of the genera *Datura* and *Solandra*, constitutes a uniform chemotaxonomic group. All species of the tribe examined contain tropane alkaloids and exhibit the greatest range of alkaloid examples in the family (Table 17.2).

Safford (1921) grouped the genus *Datura* into four sections:—I, Stramonium; II, Dutra; III, Ceratocaulis; IV, Brugmansia. Some fifteen species have now been studied in considerable chemical detail and all exhibit much the same type of alkaloid spectrum; in this respect, the tree daturas (Brugmansia), sometimes afforded generic rank, exhibit no marked differences to the other sections. The aerial parts of all plants contain hyoscine and/or hyoscyamine as principal alkaloids with smaller amounts of derivatives of these bases. The roots contain, in addition, a large number of esters formed from tropane-diol (I; $R^1 = CH_3$, $R^2 = R^3 = OH$, $R^4 = H$) and tropane-triol (I; $R^1 = CH_3$, $R^2 = R^3 = R^4 = OH$). The alkaloids characterized from the genus are given in Table 17.3. Although not all the alkaloids listed have been isolated from each species, the variation between species appears largely to be one of degree with minute amounts of some bases often being difficult to separate from the more prevalent components.

Solandra is a relatively small genus of about 12 vine-like species. Subsequent to the isolation of the principal alkaloids from the leaves of *S. longiflora* by Petrie (1907, 1916) the genus received little chemical attention until five species (*S. grandiflora, S. guttata, S. hartwegii, S. hirsuta* and *S. macrantha*) were systematically examined by modern methods (Evans, Ghani & Woolley, 1972b). The alkaloids from the five species suggest a uniform chemotaxonomic group with atropine and/or hyoscyamine and their nor-derivatives constituting the principal alkaloids; a number of minor bases are present in both aerial parts and roots. In addition to the above alkaloids, the following were identified: littorine (three species), hyoscine (present in small proportion only), norhyoscine (*S. guttata* stems), tigloidine, 3α-tigloyloxytropane, 3α-acetoxytropane, valtropine, tropine, ψ-tropine, cuscohygrine. The range is, therefore, similar to that found in *Datura* spp. but is distinguished from it by the apparent absence from the roots of the mono- and ditigloyl esters of tropanediol and tropanetriol, and by the presence of valtropine, an alkaloid not detected in *Datura* but a constituent of *Duboisia* (*q.v.*).

Table 17.3. Alkaloids of *Datura*

Alkaloid	Basic moiety	Esterifying acid(s)	Occurrence in genus	Reference to isolation from *Datura*
Tropine			All spp.	Evans & Wellendorf (1959)
Hyoscyamine (and atropine)	Tropine	Tropic acid	All spp.	Feldhaus (1905)
Apoatropine	Tropine	Atropic acid	Limited	Romeike (1953)
Littorine	Tropine	2-Hydroxy-3-phenyl-propionic acid	General	Evans & Major (1968)
3α-Tigloyloxytropane	Tropine	Tiglic acid	General	Evans & Wellendorf (1959)
3α-Acetoxytropane	Tropine	Acetic acid	*D. sanguinea*	Evans & Major (1966)
Hyoscyamine N-oxide (two isomers)	Tropine N-oxide	Tropic acid	*D. stramonium*	Phillipson & Handa (1975)
Norhyoscyamine (and noratropine)	Nortropine	Tropic acid	General	Carr & Reynolds (1912)
Pseudotropine			Most spp.	Evans & Wellendorf (1959)
Tigloidine	Pseudotropine	Tiglic acid	Several spp.	Evans & Wellendorf (1959)
Tropane-3α, 6β-diol			General	Evans & Than (1962)
(−)-3α,6β-Ditigloyloxy-tropane	(+)-Tropane-3α, 6β-diol	Tiglic acid (2 mol)	All	Evans & Wellendorf (1958)
(−)-6β-Tigloyloxy-tropan-3α-ol	(+)-Tropane-3α, 6β-diol	Tiglic acid (2 mol)	*D. innoxia*	Evans & Griffin (1963)
(−)-3α-Tigloyloxy-tropan-6β-ol	(+)-Tropane-3α, 6β-diol	Tiglic acid (2 mol)	*D. innoxia*	Evans & Griffin (1963)
(±)-3α-Tigloyloxy-tropan-6β-ol	(±)-Tropane-3α, 6β-diol	Tiglic acid (2 mol)	*D. suaveolens*	Evans & Lampard (1972)
6β-Acetoxy-3-tigloyl-oxytropane	(+)-Tropane-3α, 6β-diol	Tiglic and acetic acids	*D. sanguinea*	Evans & Major (1966)
6β-(2-Methylbutyryloxy)-3α-tigloyloxytropane	(+)-Tropane-3α, 6β-diol	Tiglic+2-methylbutyric acids	*D. suaveolens*	Evans & Lampard (1972)
6β-Propanoyloxy-3α-tigloyloxytropane	(+)-Tropane-3α, 6β-diol	Tiglic+propionic acids	*D. innoxia*	Beresford & Woolley (1974a)
6β-(2-Methylbutyryloxy)-tropan-3α-ol	(+)-Tropane-3α, 6β-diol	2-Methylbutyric acid	*D. ceratocaula*	Beresford & Woolley (1974c)
6-Hydroxyhyoscyamine	(−)-Tropane-3α, 6β-diol	Tropic acid	Interspecific hybrids	Romeike (1962a)
6β-Tigloyloxynortropan-3α-ol	Nortropane-3α, 6β-diol	Tiglic acid	*D. sanguinea*	Unpublished result
Meteloidine	Tropane-3α-, 6β,7β-triol	Tiglic acid	Most species	Pyman & Reynolds (1908)
6β-Tigloyloxytropan-3α, 7β-diol	Tropane-3α, 6β,7β-triol	Tiglic acid	*D. suaveolens*	Evans & Lampard (1972)
3α,6β-Ditigloyloxy-tropan-7β-ol	Tropane-3α, 6β,7β-triol	Tiglic acid (2 mol)	All species	Evans & Partridge (1957)
6β-Isovaleryloxy-3α-tigloyloxytropan-7β-ol	Tropane-3α, 6β,7β-triol	Tiglic+isovaleric acids	*D. suaveolens*	Evans & Major (1968)
Hyoscine	Scopine	Tropic acid	All spp.	Kirchner (1905)
Apohyoscine	Scopine	Atropic acid	Most spp.	Evans & Woolley (1965)
Hyoscine N-oxide	Scopine N-oxide	Tropic acid	*D. stramonium*	Phillipson & Handa (1975)
Oscine			Tree daturas	Evans & Major (1968)
Norhyoscine	Norscopine	Tropic acid	Most spp.	Evans *et al.* (1965)
Cuscohygrine			All spp.	van Haga (1954)

Tribe Salpiglossideae

With the notable exception of the three Australian woody genera, *Duboisia*, *Anthocercis* and *Anthotroche*, which are included in the tribe as a matter of convenience, most members of the Salpiglossideae appear to be devoid of tropane alkaloids.

Anthocercis represents some twenty species and the poisonous properties of some have been noted. Cannon, Joshi, Meehan & Williams (1969), identified hyoscyamine as the principal alkaloid of the aerial parts of *A. viscosa* and *A. littorea*, and isolated a new base, littorine, from the latter together with meteloidine and atropine. A study of the minor alkaloids and root-alkaloids of both species was made by Evans & Treagust (1973); the two species contain a similar range of alkaloids, 16 being identified, and constituting a pattern indistinguishable from that of a typical *Datura* species. Bick *et al.* (1974) have found the major leaf alkaloids of *A. tasmanica* to be hyoscine and nicotine. The occurrence of these two alkaloids together is of interest (see *Duboisia* below). An investigation of other species to determine the overall alkaloid pattern of the genus would be valuable.

Three species constitute the genus *Duboisia*. *D. myoporoides* and *D. leichhardtii* are important commercial sources of tropane alkaloids and their chemistry has been extensively studied. Both the total alkaloid yield (up to 5 %, dry wt basis) and the type of alkaloid produced are very variable and depend on geographical source, time of collection, and the existence of chemical races and hybrids (Loftus Hills, Bottomley & Mortimer, 1954a). Such variations caused considerable confusion in early chemical investigations of these plants; for an account of such early work see Barnard (1952). In addition to the principal alkaloids hyoscyamine, hyoscine, and norhyoscyamine isolated by early workers, other alkaloids have also been characterized (Table 17.4) and in some cases represent the original isolation of these compounds as natural products. Although lacking ditigloyl esters (Coulson &

Table 17.4. Tropane alkaloids of *Duboisia*

Alkaloid	Hydrolysis products	Reference to isolation from *Duboisia*
Hyoscine	see "Datura"	Ladenburg (1887)
Hyoscyamine	see "Datura"	Bender (1885)
Norhyoscyamine	see "Datura"	Carr & Reynolds (1912)
Tigloidine	see "Datura"	Barger *et al.* (1937)
Valeroidine	(−)-Tropane-3α,6β-diol + isovaleric acid	
Poroidine	Nortropine + isovaleric acid	Barger *et al.* (1938)
Isoporoidine	Nortropine + 2-methylbutyric acid	
Butropine	Tropine + isobutyric acid	Rosenblum & Taylor (1954);
Valtropine	Tropine + 2-methylbutyric acid	Deckers & Maier (1953)
3α-Tigloyloxytropane	see "Datura"	
3α-Acetoxytropane	see "Datura"	
Norhyoscine	see "Datura"	Coulson & Griffin (1967, 1968)
Apohyoscine	see "Datura"	
Tropine		
[Tetramethylputrescine]		Griffin (1967)
6-Hydroxyhyoscyamine (from hybrid)	see "Datura"	Griffin (1975)

Griffin, 1968), the range of alkaloids found in these two species of *Duboisia* resembles that of *Anthocercis* and *Datura*. However, unlike *Datura*, *Duboisia* spp. may also produce other alkaloids; a race of *D. myoporoides* from New Caledonia contained nicotine and nornicotine, and one from Acacia Plateau, Queensland contained nicotine, anabasine and isopelletierine as well as hyoscine (Loftus Hills *et al.*, 1953; Mortimer, 1957). Pyridine alkaloids have also been reported in the fine roots of *D. leichhardtii* (Kennedy, 1971). This non-uniformity of alkaloid synthesis in the genus is further emphasized by *D. hopwoodii* which produces, in the leaves, principally nicotine or nornicotine in concentrations of up to over 5% (see Barnard, 1952). Repeated investigations of this species have failed to detect tropane alkaloids in the aerial parts but Kennedy (1971) reported the presence of hyoscine, hyoscyamine and cuscohygrine in the roots.

The phytochemistry of *Anthotroche* has received little attention; Bremner & Cannon (1968) isolated hyoscyamine in low yield from the leaves and stems of *A. pannosa*, six other bases being indicated by t.l.c. Further investigation is required to establish the nature of these alkaloids and to ascertain whether other members of the genus contribute towards a chemotaxonomic unit.

Another woody genus of the Salpiglossideae, but of New World origin, is *Brunfelsia*. Ethnobotanical reports on *B. hopeana* (Schultes, 1966), and the reported presence of atropine-like alkaloids in the plant (Webb, 1948) suggested this genus as worthy of further investigation for tropane alkaloids. In an investigation of *B. bonodora*, *B. calycina*, *B. hopeana* and *B. undulata* raised under glass in Nottingham, England, Somanabandhu (1974) was able to isolate cuscohygrine only and to detect a small proportion of other uncharacterized bases in various morphological parts of the above. Further investigation of this genus is therefore necessary.

Tribe Solaneae

The Solaneae is the largest tribe of the family and its division by Wettstein into four subtribes is now considered artificial (D'Arcy, 1975). It contains a number of Old World genera which are particularly noted for their hyoscyamine–hyoscine content and have found much use in medicine. Other genera of the tribe are restricted to the accumulation of cuscohygrine, tropine, ψ-tropine and tigloyl esters of the last two, and others appear to contain no tropane derivatives.

Subtribe Lyciinae

Most members of this tribe produce alkaloids of the hyoscyamine type. *Atropa belladonna* was extensively studied by early workers and found to contain hyoscyamine as the principal alkaloid throughout the plant and small amounts of hyoscine in the young plants. Cuscohygrine (originally isolated from belladonna as bellaradine), apoatropine, belladonnine (an apoatropine dimer) together with simpler volatile bases are all minor constituents. Recently, Phillipson & Handa (1975) have demonstrated hyoscyamine and hyoscine N-oxides as normal constituents of the plant. No acids, other than tropic acid and atropic acid have been detected as esterifying units in the belladonna alkaloids. Hyoscyamine and atropine are the principal alkaloid constituents of *Latua pubiflora* leaves (Silva & Mancinelli, 1959) and of *L. venenosa* (Bodendorf & Kummer, 1962) with a small proportion of hyoscine in the latter. Ghani (unpublished data) examined the roots of two species of *Acnistus*; hyoscyamine, cuscohygrine, tigloidine were identified and other minor bases detected. In our studies on the co-occurrence of littorine with hyoscyamine in the Solanaceae we could detect the former in neither belladonna nor *Acnistus* roots. Cuscohygrine has been characterized from the roots of *Margaranthus solanaceus* (Somanabandhu, 1974).

Subtribe Hyoscyaminae

The four genera (*Scopolia, Physochlaina, Przewalskia, Hyoscyamus*) produce hyoscyamine as the usual major alkaloid with some hyoscine. N-Oxides are normal constituents of *Scopolia* and *Hyoscyamus* (Phillipson & Handa, 1975). Some *Hyoscyamus* species may contain hyoscine as the principal alkaloid; the following alkaloids have been recorded in the genus: apohyoscine, norhyoscine, littorine (two species), tropine, ψ-tropine, cuscohygrine and possibly tigloidine and tigloyloxytropane (Pelt, Younos & Hayon, 1967; Paris & Saint Firmin, 1967; Ghani, Evans & Woolley, 1972). Two *Przewalskia* species examined by t.l.c. (Hsiao, Hsia & Ho, 1973) contained hyoscyamine, hyoscine and 6-hydroxyhyoscyamine. An alkaloid anisodine was detected in one species; this base is recorded as the ester of scopine and 2,3-dihydroxy-2-phenylpropionic acid. A number of *Scopolia* species constitute a rich source of hyoscyamine and atropine. As with belladonna, their perennial habit means that the roots are economically useful and they have been extensively studied in Eastern Europe; some species contain hyoscine. Cuscohygrine and tropine are present in most roots and 3α-tigloyloxytropane and ψ-tropine have been reported in the roots of *S. carniolica* (Zito & Leary, 1966). The alkaloid anisodine was first reported in *S. tangutica* (see Hsiao *et al.*, 1973). Ghani (1971) failed to detect littorine in three species examined. In an examination of two *Physochlaina* spp. (*P. alaica, P. dubia*), Russian workers have recorded the presence of hyoscyamine and hyoscine with 6-hydroxyhyoscyamine metabolism evident during the growing stages (Mirzamatov, Malikov, Lutfillin & Yunusov, 1972).

Subtribe Solaninae

Unlike the previous subtribes of the Solaneae, none of the genera of the Solaninae is noted for the production of hyoscyamine and hyoscine. Tigloyl esters of tropine and ψ-tropine represent the most advanced level of alkaloid production. Cuscohygrine occurs in sporadic species of *Solanum*; Somanabandhu (1974) examined the roots of 55 species and recorded cuscohygrine present in 12 of these and the possibility of its presence in a further seven species. In the roots of *Physalis* tropane alkaloid production appears to reach the level of monotigloyl ester synthesis and N-oxide formation. Tigloidine has been reported in all the species examined and the following alkaloids in some: 3α-tigloyloxytropane and its N-oxide, tropine, ψ-tropine, cuscohygrine. Ditigloyl esters, hyoscine, hyoscyamine and littorine could not be detected (Basey & Woolley, 1973b; Yamaguchi, Numata & Hokimoto, 1974). *P. peruviana* has been utilized to study the biogenesis of tigloidine (Beresford & Woolley, 1974b). A more complex pattern of alkaloid biosynthesis is presented by *Withania somnifera* in that pyrazole (Schroeter, Newmann, Katritzky & Swinbourne, 1966) and anaferine-type alkaloids are also involved; tropane alkaloids identified include 3α-tigloyloxytropane, tropine and ψ-tropine together with cuscohygrine (Leary, Khanna, Schwarting & Bobbit, 1963; Schwarting *et al.*, 1963). Little chemical work has been reported on the South American genus *Saracha*; the roots of *S. viscosa* yielded cuscohygrine and other unidentified bases (Somanabandhu, 1974).

Subtribe Mandragorinae

This subtribe constitutes a non-uniform group as regards alkaloid content· Mandragora roots (from *Mandragora officinalis*) have long been known for their pharmacological action; they contain hyoscyamine, hyoscyamine N-oxide, atropine, hyoscine, apoatropine, belladonnine, cuscohygrine, and probably scopine (Staub, 1962; Phillipson & Handa, 1975). An investigation of the roots of the distinct species *M. autumnalis* and *M. vernalis*, usually referred to collectively as *M. officinalis*, has indicated by t.l.c. and g.l.c. that in addition to the above alkaloids, 3α-tigloyloxytropane and 3α, 6β-ditigloyloxytropane are also present. This is the first report (Jackson & Berry, 1973) of the occurrence of a ditigloyl ester outside the Datureae and Salpiglossideae. *Salpichroa origanifolia* roots contain small quantities of cuscohygrine, ψ-tropine, tropine and possibly hyoscyamine (t.l.c. and Vitali-Morin test), (Evans, Ghani & Woolley, 1972a). No other species of this genus appear to have been investigated for alkaloids. The roots of *Cyphomandra betacea* contain tropinone and cuscohygrine in addition to non-tropane bases; the presence of hyoscyamine, tropine, ψ-tropine and tigloidine has been tentatively suggested (Evans *et al.*, 1972c). No alkaloids appear to have been reported from the other genera (*Nectouxia, Jaborosa, Trechonaetes*) of the subtribe.

Tribes Nicandreae and Cestreae

The one species (*Nicandra physalodes*) of the tribe Nicandreae contains tropinone and hygrine in the roots (Romeike, 1966a). No tropane alkaloids have been reported from the three subtribes of the Cestreae but nicotine-type alkaloids (cf. *Duboisia*) are characteristic of *Nicotiana*.

ALKALOIDS OF HYBRIDS

Hybridization followed by selection offers the possibility of producing interesting new chemical races in solanaceous plants. With tropane alkaloids, most work has centred on the inheritance of hyoscine and hyoscyamine in the genera *Datura*, *Duboisia* and to a less extent *Scopolia* with a view to producing races with a superior alkaloid content.

The classical genetical work of Blakeslee and his colleagues (see Avery, Satina & Rietsema, 1959) gives a clear picture of the possibilities of hybrid formation in the herbaceous daturas.

An interspecific cross which has received attention is that between *Datura ferox* and *D. stramonium*. *D. ferox* contains in its aerial parts predominantly hyoscine whereas, at the time of flowering *D. stramonium* contains a hyoscine : hyoscyamine ratio of about 1 : 2. The F_1 generation is characterized by hyoscine dominance and in the F_2 generation an apparent 3 : 1 segregation into hyoscine and predominantly hyoscyamine types occurs. From these plants, races having the morphological form of one parent but the alkaloid characteristics of the other can be developed. (Romeike, 1961, 1962b, 1966b). Romeike (1962b) stabilized the desirable F_1 characteristics by tetraploid formation; this race retained its vigour in subsequent generations. By hybridization of the two species followed by two back-crosses to *D. stramonium* and selection for high hyoscine-containing plants, we have produced at Nottingham races, which by the F_7 generation resembled *D. stramonium* in their general vigour and morphological form and contained the alkaloid characteristics of *D. ferox*; these hybrids have now been grown through fourteen generations.

An interspecific cross involving *D. leichhardtii* (♀) and *D. innoxia* (♂), both of section Dutra, has also been studied (Evans *et al.*, 1969a). The female parent resembles *D. stramonium* in its alkaloid content, and the male parent, when grown in a temperate climate, usually produces hyoscine as principal alkaloid with some hyoscyamine. F_1 plants possessed in the aerial parts, equal quantities of hyoscine and hyoscyamine, and in the F_2, segregation occurred with some plants resembling either parent and others intermediate between the parents. The *D. stramonium* × *D. discolor* hybrid is interesting because it represents parents from two different sections of the genus (Al-Yahya & Evans, 1975). F_1 plants showed in their leaves a hyoscine : hyoscyamine ratio between that found for either parent but more clearly resembling that of *D. stramonium*. From the above, it is apparent that in *Datura* the tendency towards hyoscine production is dominant but, for the production of hybrids containing a high hyoscine content, at least one parent (e.g. *D. ferox*) is required which itself produces little or no hyoscyamine.

With a view to obtaining improved strains of *Duboisia* for alkaloid production, studies have been made on hybrids of *D. leichhardtii* and *D. myoporoides*. (Loftus Hills *et al.*, 1954b). Even from one cross, the hyoscine : hyoscyamine ratio in the leaves of F_1 clones varied considerably and further variations were evident by the use of different strains of *D. myoporoides* as a parent. The complex interaction of factors influencing alkaloid production in this genus renders the interpretation of results difficult.

It is perhaps noteworthy that 6-hydroxyhyoscyamine, shown to serve as an intermediate in the conversion of hyoscyamine to hyoscine in *Datura*, has been reported as an accumulated metabolite in hybrids; it was isolated from the *D. ferox* × *D. stramonium* cross by Romeike, detected in the *D. leichhardtii* × *D. innoxia* F_1 hybrid, and isolated by Griffin (1975) in relatively high yield (0·5 %) from a hybrid *Duboisia*.

CONCLUSION

Based on the extent of present knowledge, the production of tropane alkaloids in the Solanaceae is as below:

Tribe I.   Nicandreae   Synthesis limited to hygrine and tropinone

Tribe II.	Solaneae	
	Subtribe Lyciinae	Alkaloids of the hyoscyamine type in three genera
	Subtribe Hyoscyaminae	All species contain hyoscine-hyoscyamine type alkaloids
	Subtribe Solaninae	Highest level of synthesis represented by two genera producing tigloyl esters
	Subtribe Mandragorinae	A non-uniform group regarding level of alkaloid synthesis
Tribe III.	Datureae	All species produce a wide range of alkaloids which is particularly evident in *Datura*
Tribe IV.	Cestreae	No tropane alkaloids reported
Tribe V.	Salpiglossideae	Tropane alkaloid production restricted to the three Australian woody genera which produce a wide range of alkaloids comparable to *Datura*; nicotine alkaloids frequently co-occur

REFERENCES

AGAR, J. T. H., EVANS, W. C. & TREAGUST, P. G., 1974. *Journal of Pharmacy and Pharmacology, 26: Suppl.*, 111P.

AGAR, J. T. H. & EVANS, W. C., 1975. *Journal of Pharmacy and Pharmacology, 27: Suppl.*, 85P.

AL-YAHYA, M. & EVANS, W. C., 1975. *Journal of Pharmacy and Pharmacology, 27: Suppl.*, 87P.

AVERY, A. G., SATINA, S. & RIETSEMA, J., 1959. *Blakeslee: The Genus* Datura. New York: Ronald Press.

BARGER, G., MARTIN, W. F. & MITCHELL, W., 1937. *Journal of the Chemical Society*, 1820.

BARGER, G., MARTIN, W. F. & MITCHELL, W., 1938. *Journal of the Chemical Society*, 1685.

BARNARD, C., 1952. *Economic Botany, 6:* 3.

BASEY, K. & WOOLLEY, J. G., 1973a. *Phytochemistry, 12:* 2197.

BASEY, K. & WOOLLEY, J. G., 1973b. *Phytochemistry, 12:* 2557.

BASEY, K. & WOOLLEY, J. G., 1975. *Phytochemistry, 14:* 2201.

BENDER, K., 1885. *Pharmazeutische Zentralhalle für Deutschland, 26:* 38.

BERESFORD, P. J. & WOOLLEY, J. G., 1974a. *Phytochemistry, 13:* 1249.

BERESFORD, P. J. & WOOLLEY, J. G., 1974b. *Phytochemistry, 13:* 2143.

BERESFORD, P. J. & WOOLLEY, J. G., 1974c. *Phytochemistry, 13:* 2511.

BERESFORD, P. J. & WOOLLEY, J. G., 1975. *Phytochemistry, 14:* 2205, 2209.

BESSEY, C. E., 1915. *Annals of Missouri Botanical Garden, 2:* 109.

BICK, I. R. C., BREMNER, J. B., GILLARD, J. W. & WINZENBERG, K. N., 1974. *Australian Journal of Chemistry, 27:* 2515.

BICK, I. R. C., GILLARD, J. W. & WOODRUFF, M., 1975. *Chemistry and Industry:* 794.

BODENDORF, K. & KUMMER, H., 1962. *Pharmazeutische Zentralhalle für Deutschland, 101:* 620.

BREMNER, J. B. & CANNON, J. R., 1968. *Australian Journal of Chemistry, 21:* 1369.

CANNON, J. R., JOSHI, K. R., MEEHAN, G. V. & WILLIAMS, J. R. 1969. *Australian Journal of Chemistry, 22:* 221.

CARR, F. H. & REYNOLDS, W. C., 1912. *Journal of the Chemical Society:* 946.

COULSON, J. F. & GRIFFIN, W. J., 1967. *Planta Medica, 15:* 459.

COULSON, J. F. & GRIFFIN, W. J., 1968. *Planta Medica, 16:* 174.

D'ARCY, W. G., 1975. In J. G. Hawkes (Ed.), *Solanaceae Newsletter,* No. 2: 8. University of Birmingham.

DECKERS, W. & MAIER, J., 1953. *Chemische Berichte, 86:* 1423.

EKEVAG, U. E. & GAWELL, L. M., 1973. *Acta Chemica Scandinavica, 27:* 1982.

EVANS, W. C., GHANI, A. & WOOLLEY, V. A., 1972a. *Phytochemistry, 11:* 469.

EVANS, W. C., GHANI, A. & WOOLLEY, V. A., 1972b. *Phytochemistry, 11:* 470.

EVANS, W. C., GHANI, A. & WOOLLEY, V. A., 1972c. *Journal of the Chemical Society:* 2017.

EVANS, W. C. & GRIFFIN, W. J., 1963. *Journal of the Chemical Society:* 4348.

EVANS, W. C. & LAMPARD, J. F., 1972. *Phytochemistry, 11:* 3293.

EVANS, W. C. & MAJOR, V. A., 1966. *Journal of the Chemical Society (C):* 1621.

EVANS, W. C. & MAJOR, V. A., 1968. *Journal of the Chemical Society (C):* 2775.
EVANS, W. C., MAJOR, V. A. & THAN, M. P., 1965. *Planta Medica, 13:* 353.
EVANS, W. C. & PARTRIDGE, M. W., 1957. *Journal of the Chemical Society:* 1102.
EVANS, W. C. & SOMANABANDHU, A., 1974. *Phytochemistry, 13:* 519.
EVANS, W. C., STEVENSON, N. A. & TIMONEY, R. F., 1969a. *Planta Medica, 17:* 120.
EVANS, W. C. & THAN, M. P., 1962. *Journal of Pharmacy and Pharmacology, 14:* 147.
EVANS, W. C. & TREAGUST, P. G., 1973. *Phytochemistry, 12:* 2505.
EVANS, W. C. & WELLENDORF, M., 1958. *Journal of the Chemical Society:* 1991.
EVANS, W. C. & WELLENDORF, M., 1959. *Journal of the Chemical Society:* 1406.
EVANS, W. C. & WOOLLEY, J. G., 1965. *Journal of the Chemical Society:* 4936.
EVANS, W. C. & WOOLLEY, V. A., 1969. *Phytochemistry, 8:* 2183.
EVANS, W. C. & WOOLLEY, J. G., 1976. *Phytochemistry, 15:* 287.
FELDHAUS, J., 1905. *Archiv der Pharmazie (und Berichte der Deutschen Pharmazeutischen Gesallschaft), 243:* 328.
FODOR, G., 1967. In R. H. F. Manske (Ed.), *The Alkaloids, IX:* 295. London: Academic Press.
FODOR, G., 1971. In R. H. F. Manske (Ed.), *The Alkaloids, XIII:* 383. London: Academic Press.
FODOR, G. & SÓTI, F., 1965. *Journal of the Chemical Society:* 6830.
GHANI, A., 1971. PhD Thesis, Univ. Nottingham, Eng.
GHANI, A., EVANS, W. C., & WOOLLEY, V. A., 1972. *Bangladesh Pharmaceutical Journal, 1:* 12.
GILLARD, J., 1975. PhD Thesis, Univ. Tasmania, Austr.
GRIFFIN, W. J., 1967. *Australasian Journal of Pharmacy, 48:* 520.
GRIFFIN, W. J., 1975. *Naturwissenschaften, 62:* 97.
van HAGA, P. R., 1954. *Nature, 174:* 833.
HENRY, T. A., 1949. *The Plant Alkaloids.* London: Churchill.
HSIAO, P., HSIA, K. & HO, L., 1973. *Acta Botanica Sinica, 15:* 187.
JACKSON, B. P. & BERRY, M. I., 1973. *Phytochemistry, 12:* 1165.
JOHNS, S. R., LAMBERTON, J. A., SIOUMIS, A. A., 1970. *Australian Journal of Chemistry, 23:* 421.
JOHNS, S. R., LAMBERTON, J. A., SIOUMIS, A. A., 1971. *Australian Journal of Chemistry, 24:* 2399.
KAN-FAN, C. & LOUNASMAA, M., 1973. *Acta Chemica Scandinavica, 27:* 1039.
KENNEDY, G. S., 1971. *Phytochemistry 10:* 1335.
KIRCHNER, A., 1905. *Archiv der Pharmazie (und Berichte der Deutschen Pharmazeutischen Gesellschaft), 243:* 309.
LADENBURG, A., 1887. *Pharmaceutical Journal, 17:* 1049.
LEARY, J. D., KHANNA, K. L., SCHWARTING, A. E. & BOBBIT, J. M., 1963. *Lloydia, 26:* 44.
LEETE, E., 1972. *Phytochemistry, 11:* 1713.
LEETE, E. & KIRVEN, E. P., 1974. *Phytochemistry, 13:* 1501.
LEETE, E. & LOUDEN, M. L., 1962. *Journal of the American Chemical Society, 84:* 1510, 4507.
LEETE, E. & LUCAST, D. H., 1975. *Phytochemistry, 14:* 2199.
LODER, J. W. & RUSSELL, G. B., 1966. *Tetrahedron Letters:* 6327.
LODER, J. W. & RUSSELL, G. B., 1969. *Australian Journal of Chemistry, 22:* 1271.
LOFTUS HILLS, K., BOTTOMLEY, W. & MORTIMER, P. I., 1953. *Nature, 171:* 435.
LOFTUS HILLS, K., BOTTOMLEY, W. & MORTIMER, P. I., 1954a. *Australian Journal of Applied Science, 5:* 258, 276.
LOFTUS HILLS, K., BOTTOMLEY, W. & MORTIMER, P. I., 1954b. *Australian Journal of Applied Science, 5:* 283.
LOUNASMAA, M., 1975. *Planta Medica, 27:* 83.
MIRZAMATOV, R. T., MALIKOV, V. M., LUTFILLIN, K. L. & YUNUSOV, S. U., 1972. *Khimija Prirodnykh Soedinenij, 8:* 493; *9:* 566.
MORTIMER, P. I., 1957. *Australian Journal of Science 20:* 87.
MOTHERWELL, W. D. S., ISAACS, N. W., KENNARD, O., BICK, I. R. C., BREMNER, J. B. & GILLARD, J., 1971. *Journal of the Chemical Society (D):* 133.
O'DONOVAN, D. G. & KEOGH, M. F., 1969. *Journal of the Chemical Society (C):* 223.
PARELLO, J., LONGEVIALLE, P., VETTER, W. & McCLOSKEY, J. A., 1963. *Bulletin Société Chimique de France:* 2787.
PARIS, R. R. & SAINT FIRMIN, A., 1967. *Compte Rendu Hebdomadaire des Séances de l'Academie des Sciences (Ser. D), 264:* 825.
PELT, J. M., YOUNOS, Ch. & HAYON, J. C., 1967. *Annales Pharmaceutiques Françaises, 25:* 59.
PETRI, J. M., 1907. *Proceedings of the Linnean Society of New South Wales, 32:* 789.
PETRI, J. M., 1916. *Proceedings of the Linnean Society of New South Wales, 41:* 815.
PHILLIPSON, J. D. & HANDA, S. S., 1975. *Phytochemistry 14:* 999.
PINDER, A. R., 1952. *Journal of the Chemical Society:* 2236.
PINDER, A. R., 1957. *Chemistry and Industry,* 1240.

PLATONOVA, T. F. & KUZOVKOV, A. D., 1963. *Meditsinskaya Promÿshlennost'*, *SSSR*, *17:* 19 (through *Chemical Abstracts*, *60:* 8350).

PRABHU, B. V., GIBSON, C. A. & SCHRAMM, L. C., 1976. *Lloydia*, *39:* 79.

PYMAN, F. L. & REYNOLDS, W. C., 1908. *Journal of the Chemical Society*, *93:* 2077.

ROMEIKE, A., 1953. *Pharmazie*, *8:* 729.

ROMEIKE, A., 1961. *Kulturpflanze*, *9:* 171.

ROMEIKE, A., 1962a. *Naturwissenschaften*, *49:* 281.

ROMEIKE, A., 1962b. *Kulturpflanze*, *10:* 140

ROMEIKE, A., 1966a. *Naturwissenschaften*, *53:* 82.

ROMEIKE, A., 1966b. *Kulturpflanze*, *14:* 129.

ROMEIKE, A. & FODOR, G., 1960. *Tetrahedron Letters:* 1.

ROSENBLUM, E. I. & TAYLOR, W. S., 1954. *Journal of Pharmacy and Pharmacology*, *6:* 410.

SAFFORD, W. E., 1921. *Journal of the Washington Academy of Sciences*, *11:* 173.

SCHROETER, H. B., NEWMANN, D., KATRITZKY, A. R. & SWINBOURNE, F. J., 1966. *Tetrahedron*, *22:* 2895.

SCHULTES, R. E., 1966. *Lloydia*, *29:* 293.

SCHWARTING, A. E., BOBBIT, J. M., ROTHER, A., ATAL, C. K., KHANNA, K. L., LEARY, J. D. & WALTER, W. G., 1963. *Lloydia*, *26:* 258.

SILVA, M. & MANCINELLI, P., 1959. *Boletín de la Sociedad Chilena de Quimica*, *9:* 49.

SOMANABANDHU, A., 1974. PhD Thesis, Univ. Nottingham, Eng.

STAUB, H., 1962. *Helvetica Chimica Acta*, *45:* 2297.

WEBB, L. J., 1948. *CSIRO Australian Bulletin*, No. 232: 152.

WETTSTEIN, R., 1895. Solanaceae. In A. Engler & K. Prantl (Eds), *Die natürlichen Pflanzenfamilien*, *IV* (36): 4. Leipzig: Englemann.

WILLAMAN, J. J. & SCHUBERT, B. G., 1961. *Alkaloid-bearing Plants*. U.S. Department of Agriculture, Technical Bulletin, No. 1234.

YAMAGUCHI, H., NUMATA, A. & HOKIMOTO, K., 1974. *Yakugaku Zasshi*, *94:* 1115.

ZITO. S. W. & LEARY, J. D., 1966. *Journal of Pharmaceutical Sciences*, *55:* 1150.

IV. Flavonoids, terpenes and proteins

18. Flavonoids of the Solanaceae

J. B. HARBORNE

Department of Botany, University of Reading, U.K.

AND

T. SWAIN

Biology Centre, Boston University, Mass., U.S.A.

Since an earlier summary of flavonoid patterns in the family, published in 1967, relatively little new information has emerged. Flavonoids have been fully characterized in several crop plants, notably the potato, tobacco and petunia, but little is known of the patterns in wild species. The anthocyanin pigments are characteristically acylated, usually have rutinose as the 3-sugar and are frequently based on petunidin. The two flavonols, kaempferol and quercetin, are universal, occurring often as the 3-glucosides and 3-rutinosides; more complex triglycosides are also common, especially in *Solanum*. Flavones appear to be rare, being regularly present only in *Capsicum*. Other phenolic constituents are abundant, especially caffeyl esters and coumarins.

From the systematic point of view, flavonoids are of interest in indicating that the Solanaceae is somewhat isolated from most of the neighbouring families, although there are links with the Convolvulaceae. Undoubtedly, flavonoid patterns have a contribution to make in systematic studies at the lower levels of classification in the family, but progress so far has done no more than indicate the potentialities of such an approach.

CONTENTS

INTRODUCTION

The most conspicuous class of flavonoid present in the Solanaceae is undoubtedly the anthocyanins, which are responsible for the characteristic purple colours of both flowers and fruits. For example, in the deadly nightshade, *Atropa belladonna*, anthocyanins are responsible for the colour of both the brown-violet bell-shaped corollas and the attractive, but deadly, black berries. Anthocyanins, in fact, provide a range of colour from purple to orange in many genera and have been especially studied in *Nicotiana*, *Petunia* and *Solanum*. Another class of flavonoids universally distributed in the family is the flavonol glycosides, which occur in both flowers and leaves. Other compounds, such as flavones and dihydroflavonols, have been recorded on occasion but they appear to be much less common. Biosynthetically related C_9 compounds, hydroxycinnamic acids and hydroxy-coumarins, are by contrast very often present and their occurrence will be briefly mentioned in view of their close biogenetic relationship to the flavonoids.

Because of genetical interest, flavonoid pigments have been intensively studied in at least two genera, *Petunia* (Meyer, 1964) and *Solanum* (Harborne, 1960a) and economic importance has ensured that the flavonoids of *Nicotiana* have been fully identified (Tso, 1972). Thus, a reasonable background of information on flavonoid patterns is available in the family. In spite of this, flavonoid markers, which have been so widely exploited in systematic studies in other taxa (see e.g. Harborne 1975), have rarely been applied to like problems in this family. Some progress has been made in this respect in only a few cases, e.g. *Capsicum* (Ballard, McClure, Eshbaugh & Wilson, 1970) and tuberous *Solanum* (Harborne, 1964 and unpubl. results).

In order to extend the information on leaf flavonoids in the family, a representative survey has been carried out recently on plant material supplied from the University of Birmingham collection and the results are incorporated in this review. In the present account, the major classes of flavonoid will be considered in turn and an assessment will be made of their potential value in future chemosystematic studies in the family.

ANTHOCYANINS

Anthocyanins have been investigated in some nine genera of the Solanaceae (Table 18.1). They have been detected variously in flowers, where they contribute to intense purple colours, in berries, where the concentration may be so high as to make the skin appear black, in leaves, stems and in tubers. Undoubtedly the most

Table 18.1. Anthocyanins identified in the Solanaceae

Genus and species	Organ examined	Pigment(s) present*
Atropa belladonna	flower, fruit	Petanin
Browallia speciosa		Dp 3 (di-*p*-coumarylglucoside)-5-glucoside
Brunfelsia calycina		Negretein
Cestrum purpureum	flower	Pg and Cy 3-rutinosides
Iochroma coccinea		Pg 3-glucoside
I. tubulosa		Mv 3-glucoside
Lycopersicon esculentum	stem, leaf	Petanin
Nicotiana sanderae		Dp 3-rutinoside
N. setchellii	flower	Cy 3-glucoside, Cy 3-sophoroside
N. tabacum		Pg and Cy 3-rutinosides
Petunia hybrida	flower	Peonanin, delphanin, petanin, negretein, Cy 3-gentiobioside, Cy 3,7-diglucoside, Cy and Pt 3-sophorosides, Cy, Pn, Dp, Pt and Mv 3-glucosides
Solanum spp. and cultivars	Petal, stem, fruit, tuber	Pelanin, cyananin, peonanin, delphanin, petanin, negretein, Pg, Cy, Dp and Pt 3-rutinosides

* Pg, Pelargonidin; Cy, cyanidin; Pn, peonidin; Dp, delphinidin; Pt, petunidin; Mv, malvidin; pelanin, cyananin, peonanin, delphanin, petanin and negretein are the 3-(*p*-coumarylrutinoside)-5-glucosides of Pg, Cy, Pn, Dp, Pt and Mv respectively.

For literature references, see Harborne (1967) and Harborne *et al.* (1975).

characteristic anthocyanidin of the family is petunidin, particularly when it occurs as the acylated glycoside petanin, the 3-(*p*-coumaryl-rutinoside)-5-glucoside (Fig. 18.1). This pigment occurs widely throughout the genus *Solanum*, and, with other petunidin derivatives, is also found in *Atropa* and *Lycopersicon*. Related pigments based on delphinidin and malvidin are also common, the former occurring for example in the skin of the aubergine, *Solanum melongena* and the latter, in quantity, in the tuber of purple black forms of *S. tuberosum*.

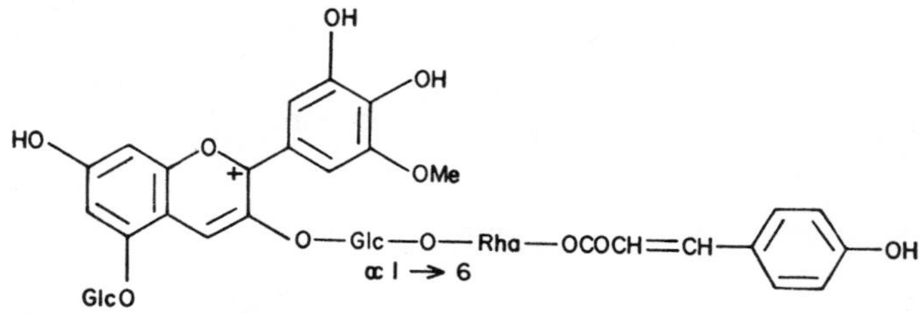

Petanin
Atropa, Lycopersicon, Solanum

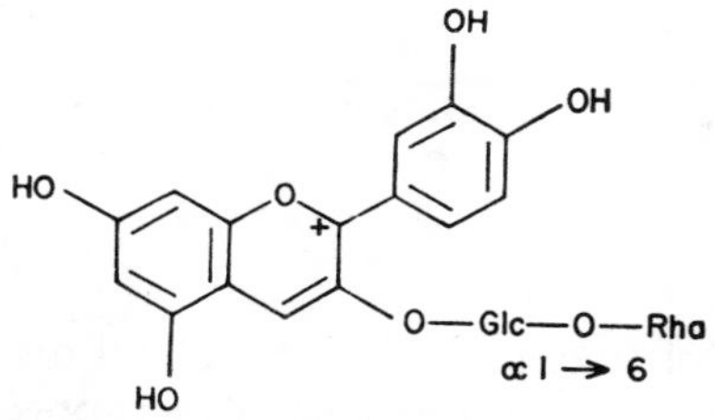

Cyanidin 3 – rutinoside: *Cestrum*
pelargonidin analogue: *Nicotiana*
delphinidin analogue: *S. melongena*

Figure 18.1. The two main anthocyanin types of the Solanaceae.

Anthocyanidin 3-rutinosides, a simpler glycosidic pattern than that of petanin (see Fig. 18.1), is also of importance in the family, occurring regularly in *Cestrum*, *Nicotiana* and *Solanum*. Three much rarer types have been found almost exclusively in *Petunia*: the 3-sophorosides, 3-gentiobiosides, and 3,7-diglucosides.

Much genetical and biosynthetic work has been done on these pigments in *Petunia* and *Solanum* and the pathway of synthesis and genetical control in now relatively well understood (see e.g. Hess, 1968). In the case of *Solanum* (Harborne, 1960), one of the most interesting discoveries was the presence of a pleiotropic gene, (**Ac**), controlling three aspects of pigment synthesis. The anthocyanin in the dominant genotype is petanin, while that of the recessive form (**ac**) is delphinidin 3-rutinoside (Fig. 18.2). The gene **Ac** is thus responsible for three distinct biosynthetic steps: methylation of delphinidin to petunidin, addition of glucose to the 5-hydroxyl; and acylation of rutinose with *p*-coumaric acid. Because of the close genetic linkage, it is likely that the three events occur sequentially, probably at a late stage in biosynthesis, and function to protect the final product from enzymic destruction. This view is supported by experiments with other plants; for example there are reports of a similar genetic situation regarding the linkage of acylation with glycosylation in *S. melongena* (Abe & Gotoh, 1959) and in *Petunia* (Meyer, 1964).

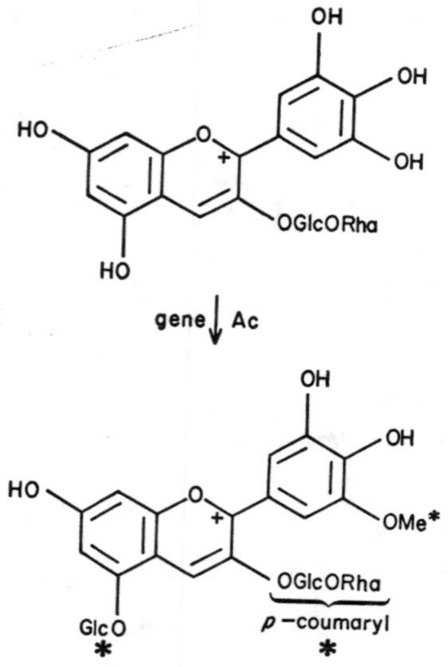

Figure 18.2. Pleotropic effect of **Ac** gene in *Solanum*.

The anthocyanin pattern of the Solanaceae is distinctive and distinguishes it in varying degrees from related families. Some features, however, appear in the anthocyanin patterns of these other groups. Thus, families with acylated pigments include the Convolvulaceae (*p*-coumaryl or caffeyl 3-sophoroside-5-glucosides), Labiatae (*p*-coumaryl or caffeyl 3,5-diglucosides) and Polemoniaceae (*p*-coumaryl 3,5-diglucosides). Again, the 3-rutinoside-5-glucoside pattern is found in Gesneriaceae (e.g. *Saintpaulia*) and the 3-rutinoside type in Bignoniaceae (e.g. *Campsis*). Finally, petunidin occurs as a major pigment in certain Boraginaceae (e.g. petunidin 3,5-diglucoside in *Anchusa*).

Within the Solanaceae, anthocyanins have not been used to any extent so far in systematic studies. That they are potentially useful is apparent from the work on *Nicotiana* flowers (see Table 18.1), where some species contain different anthocyanidins (cf. *N. tabacum* with pelargonidin and cyanidin, *N. sanderae* with delphinidin) while others have different glycosidic patterns (cf. *N. tabacum* with 3-rutinoside, *N. setchellii* with 3-sophoroside). More detailed studies of pigment patterns in this and other genera might yield results of systematic interest.

Pigment variation is readily apparent in the flowers, fruits and tubers of *Solanum* species and appears to be useful in taxonomic studies of this genus. However, the only serious attempt to employ pigment patterns has been at the level of cultivars of *S. tuberosum*. The pigments of potato tubers, as revealed by simple 1-dimensional TLC of sprout extracts, can be used as a partial means of cultivar identification

(Table 18.2). The method, of course, does not separate the cultivars completely since the same pattern may be present in more than one. However, when used in combination with other screening techniques, it does provide a positive and relatively rapid additional means of variety testing (Brown & Moss, 1976).

Table 18.2. Anthocyanins identified in the sprouts of some potato cultivars

Pigment pattern	Cultivar*
Negretein	Pentland Crown
Petanin	Ulster Chief, Ulster Prince
Negretein + petanin	Arran Peak, Congo
Negretein + petanin + peonanin	Pentland Ivory
Petanin + peonanin	Record, Ulster Lancer
Peonanin	Pentland Dell, Maris Piper, Thynia
Peonanin + pelanin	King Edward, Stormont Enterprise
Pelanin	Desirée

* There are quantitative variations where mixtures occur which further separate cultivars with the same qualitative pattern.
Data from unpublished results of J.B.H.

FLAVONOL AND FLAVONE GLYCOSIDES

Many surveys have shown that the majority of angiosperm families have a particular pattern of leaf flavonoids, depending on the degree of advancement of several morphological features (Harborne, 1977). These patterns include presence/absence of: proanthocyanidins (previously referred to as leucoanthocyanidins or condensed tannins); flavonoids with trihydroxylation in the B-ring; common flavonols kaempferol or quercetin; common flavones (apigenin or luteolin, with or without C-glycosidic attachment); and various rarer compounds (e.g. flavones or flavonols with an extra hydroxyl at 6- or 8-position or with unusual methylation) (Bate-Smith, 1962; Harborne, 1975, 1977; Swain, 1975). In the case of the Solanaceae, it is clear that only one simple pattern occurs in the leaves, based on only two compounds, the common flavonols kaempferol and quercetin. Other flavonoids are, by contrast, rare or absent. Thus, proanthocyanidins are completely absent from the family. Flavonoids with B-ring trihydroxylation (such as the flavonol myricetin) are also absent from leaves, although myricetin does appear occasionally in flowers as a by-product of delphinidin synthesis. Flavones, although detected in certain genera (Table 18.3), are rare. Other flavonoid features are almost completely absent; two exceptions are the 3,3'-dimethyl ether of quercetin, present in tobacco, and dihydroquercetin, found in *Petunia*. It may be noted that flavones with extra hydroxylation or methoxylation at the 6-position, which occur widely and characteristically in many related tubiflorous families (Harborne & Williams, 1971) are apparently completely lacking from the Solanaceae.

Information on the leaf flavonoids in the family is based on a survey of 26 species and 14 genera by Bate-Smith (1962), a more extensive survey of tuberous *Solanum* (Harborne, 1964) and a recent representative survey of 32 species from 24 genera (Table 18.3; Harborne, unpublished results). The number of taxa sampled is still relatively small and wider surveys are highly desirable to confirm the conclusions reached. The relative infrequency of flavones is particularly surprising in view of the fact that most families related to the Solanaceae predominantly contain flavones. In the case of *Solanum*, the flavone luteolin 7-glucoside (Fig. 18.3) was only found in two of over 60 wild taxa surveyed: in the tetraploid *S. stoloniferum* and in the triploid hybrid *S.* × *vallis-mexici* known to be derived therefrom. Even in *S. stoloniferum*, the flavone was lacking from 4 of 38 clones examined. It was always accompanied by flavonol glycosides (Harborne, 1964) and never occurred

Table 18.3. % Frequencies of leaf flavonoids in Solanaceae

Aglycones	32 spp. of 24 genera*‡	Glycosides	
Kaempferol	50	Monoglycoside	21
Quercetin	70	Diglycoside	14
Luteolin/Apigenin	6	Triglycoside	32
	100 spp. *Solanum*†		
Kaempferol	20	Monoglycoside	5
Quercetin	95	Diglycoside	94
Luteolin	3	Triglycoside	40

* Genera examined were: *Anthocercis, Athenaea, Bassovia, Brunfelsia, Capsicum, Cestrum, Datura, Dunalia, Fabiana, Iochroma, Jaborosa, Juanulloa, Lycianthes, Lycopersicon, Markea, Methysticodendron, Nicotiana, Nothocestrum, Poecilochroma, Salpichroa, Solandra, Solanum, Vestia* and *Withania* (J. B. Harborne, unpubl. results).

† from Harborne (1964).

‡ Bate-Smith (1962) in his survey of 26 spp. of 14 genera only looked at aglycones, finding kaempferol and/or quercetin in all but 4 spp. The possible presence of luteolin was recorded in *Physalis*.

alone as found in other related families. The rareness of flavones in the Solanaceae was also apparent during the most recent survey of 24 genera, when it was found once, in *Capsicum*. Here, however, flavones predominate and there are no co-occurring flavonols (Harborne, unpubl. results).

While the pattern of flavonols in the Solanaceae is simple at the aglycone level, it is clearly more complex than in related families in the glycosides that are present. Thus, while the expected simple 3-glucosides and 3-rutinosides of kaemp-

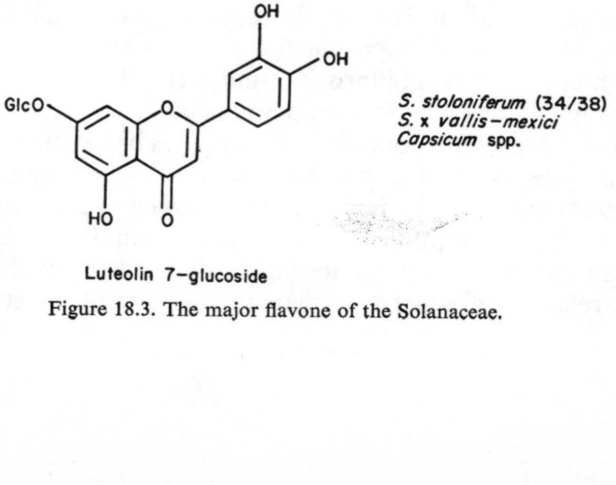

S. stoloniferum (34/38)
S. x *vallis-mexici*
Capsicum spp.

Luteolin 7-glucoside

Figure 18.3. The major flavone of the Solanaceae.

ferol and quercetin are widespread, more complex triglycosides are also of very regular occurrence in the family (Table 18.3). One of the most interesting of these triglycosides is quercetin 3-(2^G-glucosylrutinoside) (Fig. 18.4) which occurs in 37% of wild *Solanum* species and in 41% of cultivated taxa. This compound is of variable occurrence in leaves of *S. chacoense* and *S. stoloniferum,* and genetic experiments (Harborne, 1964) have shown that its occurrence is controlled by a single dominant gene **G1**. Its synthesis is accompanied by that of the 3-sophoroside and it is apparent that the gene **G1** actually controls the transfer of glucose in the $\beta1 \rightarrow 2$ position to two substrates, quercetin 3-glucoside (giving the 3-sophoroside) and quercetin 3-rutinoside (giving the 3-triglycoside (Fig. 18.4)) (see Table 18.4). There

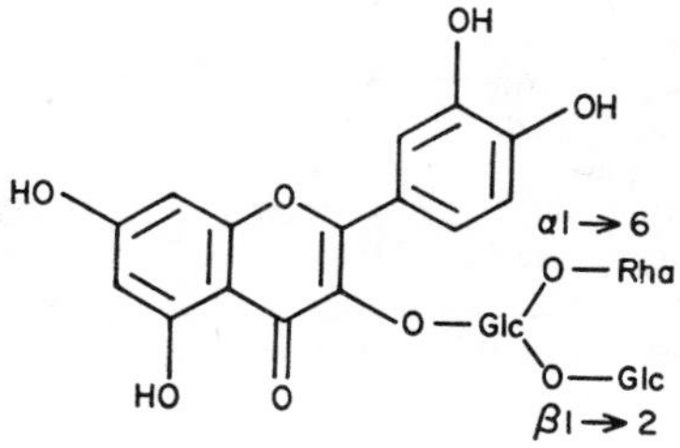

Quercetin 3-(2^G-glucosylrutinoside)
Solanum 37% wild spp., 41% cult. spp.
variable in *S. chacoense, S. stoloniferum*

Figure 18.4. A flavonol with branched trisaccharide in *Solanum.*

Table 18.4. Concentration of flavonol glycosides in flowers of *Solanum chacoense*

	μmoles of flavonol/g. fr. wt of flower	
Quercetin glycoside	Dominant	recessive*
3-(2^G-glucosylrutinoside)	7.64	0.27
3-sophoroside	2.25	0.0
3-rutinoside	1.86	7.48
3-glycoside	0.0	0.31

* Dominant form was an F_2 plant which could have been either **Gl Gl** or **Gl gl**; recessive form was **gl gl** (Harborne, 1964).

is another possibility, i.e. that an extra sugar moiety might be attached in the 7-position. In fact, at least five such flavonol 3,7-glycosides are known (Table 18.5); in this case the 7-sugar may be either glucose or rhamnose. One of these 3,7-glycosides, namely quercetin 3-sophoroside-7-glucoside, is accompanied in *Petunia* flowers by a flavonol glycoside which has yet another structural peculiarity—the presence of an acyl group on the 3-sugar. This compound, called petunoside (Fig. 18.5) is the $\beta1 \rightarrow 2$ ferulyl derivative of kaempferol 3-sophoroside and its structure was elucidated by Birkofer, Kaiser & Kosmol in 1965. Acylation which is very common among the anthocyanins of the Solanaceae, is thus represented in this case among the flavonols of the family.

One of the flavonol 3,7-glycosides listed in Table 18.4 is unique in having not three but four monosaccharide units in its structure. This is kaempferol 3-sophorotrioside-7-rhamnoside, which has three glucoses attached to the 3-position in a linear $\beta 1 \rightarrow 2$ sequence and rhamnose attached at position 7. This compound was first isolated and identified from potato seed in 1964 (Harborne, 1964), its structure being rigorously established more recently by means of GC-MS of its permethylated derivative (Schmid & Harborne, 1973).

Table 18.5. Flavonol 3,7-diglycosides in the Solanaceae

Aglycone	3-sugar	7-sugar	occurrence
Kaempferol	Sophorotriose	Rhamnose	*Solanum* seed
Kaempferol	Sophorose	Rhamnose	
Quercetin	Rutinose	Glucose	*Nicotiana* leaf/flower
Quercetin	Robinobiose	Glucose	*Atropa* leaf
Quercetin	Sophorose	Glucose	*Petunia* flower

For references, see Harborne *et al.* (1975).

Acylation of Solanaceae flavonols

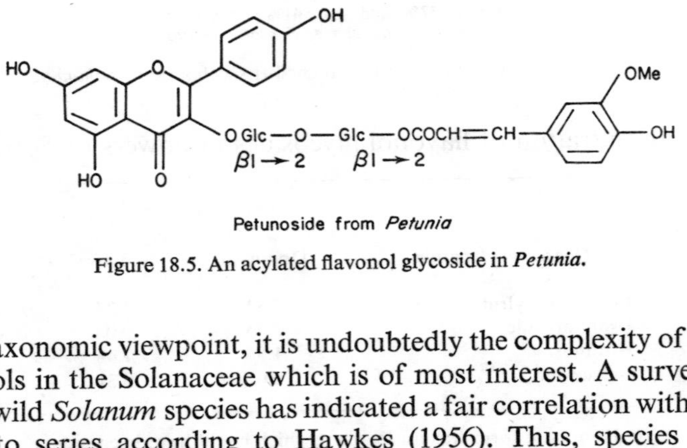

Petunoside from *Petunia*

Figure 18.5. An acylated flavonol glycoside in *Petunia*.

From the taxonomic viewpoint, it is undoubtedly the complexity of glycosylation of the flavonols in the Solanaceae which is of most interest. A survey of flavonol glycosides in wild *Solanum* species has indicated a fair correlation with classification of species into series according to Hawkes (1956). Thus, species of the series *Conicibaccata* are distinguished from other taxa by containing 3-sophorosides at the expense of the more usual 3-rutinosides. Again, in some series (e.g. *Acaulia*) only the 3-rutinoside occurs while in others (e.g. *Demissa*) the 3-(2^G–glucosylrutinoside) predominates. While over ten flavonoids occur in wild *Solanum* leaves, only six are present in cultivated forms. Here again, they are of taxonomic interest, since the pattern in both flower and leaf varies with the cultivar examined. For example, the flowers of Arran Victory contain the quercetin 3-triglycoside while those of Majestic and Arran Viking lack it.

One attempt to apply leaf flavonoid data to the solution of taxonomic problems outside *Solanum* is in the genus *Capsicum* where the leaf constituents are, as mentioned above, flavones rather than flavonols. Here, the flavones occur both as O-glycosides (e.g. luteolin 7-glucoside) and as C-glycosides (e.g. vitexin, orientin, etc.). A combined chemical and biological analysis of selected taxa allowed Ballard *et al.* (1970) to recognize three groups. These were: I, *C. baccatum* group, white flowered with a simple flavone pattern (*O*- and *C*-glycosides of luteolin and apigenin); II, *C. eximium* group, purple to white flowered, having a more complex flavonoid pattern, especially the presence of chrysoeriol (luteolin 3′-methyl ether); and III, *C. cardenasii/pubescens* group, purple flowered containing (with one exception) only flavone *O*-glycosides. The chemical results support the view that group II *Capsicum* arose by hybridization of taxa from groups I and III, and, while other explanations are possible, this provides a basis for more extensive studies using a range of other techniques. The flavonoids of *Capsicum* have not all yet been fully identified and their chemistry deserves further attention. A rare glycoside of apigenin, the 7-apiosylglucoside, has been reported here (Rangoonwala & Friedrich, 1967) and this needs confirmation and taxonomic exploration.

Other experiments reporting the use of leaf flavonoids in systematic investigations are those of Averett (this Volume, p. 493) at the populational level in *Chamaesaracha* and related genera and those of Whalen (this Volume, p. 581) on *Solanum* section *Androceras*. Both these plant groups again contain the usual kaempferol and quercetin glycosides. Natarella & Sink (1974) have studied leaf phenolics, without chemical identification, in *Petunia* in order to throw light on the origin of the garden hybrids. Their results support the view that the cross between *P. axillaris* and *P. violacea* was a key to the evolution of *P. hybrida*, rather than *P. inflata*, which had been proposed as a parental species on cytological grounds.

RELATED C₉ COMPOUNDS

Hydroxycinnamic acids are closely related to flavonoids both in their biosynthesis and also in their distribution in higher plants. In the case of the Solanaceae, there is a special link between the two classes of polyphenols, since many of the anthocyanins contain *p*-coumaric acid acylating the sugar moieties. Flavonol glycosides are similarly acylated with ferulic acid, especially in the case of *Petunia* (Fig. 18.5). It is worth considering briefly therefore the distribution of hydroxycinnamic acid derivatives in the family and also the hydroxycoumarins, which are derived biosynthetically from *o*-hydroxycinnamic acids by dehydration and ring closure.

Hydroxycinnamic acids are widespread in leaves and flowers of Solanaceae, usually in a range of combined forms. Caffeic acid (3,4-dihydroxycinnamic acid) is particularly common and generally occurs either as the quinic ester, chlorogenic acid, or as the glucose ester, 1-caffeylglucose (Fig. 18.6). In the berries, especially of tuberous *Solanum*, other conjugates occur, including the glucose ester of *p*-coumaric acid in cultivated species and the 3-glucoside of caffeic acid in wild species.

The fact that the Solanaceae in general have caffeic acid combined both as chlorogenic acid and caffeylglucose is taxonomically important, since this distinguishes it from other related families (see Table 18.6). Surveys have shown (Harborne, 1966) that other combined forms of caffeic acid, i.e. isochlorogenic and rosmarinic acid, progressively replace chlorogenic and caffeylglucose in these other families. In the Boraginaceae, for example, both these latter compounds are absent, rosmarinic acid being the only caffeic acid ester present.

Three hydroxycoumarins which are characteristic of the Solanaceae are aesculin, cichoriin and scopolin (Fig. 18.7). There are many other related structures known, including the 7-xylosylglucoside of scopoletin (fabiatrin) first isolated from leaves of *Fabiana imbricata* and the furanocoumarin, marmesin, recently reported in leaves of tomato, *Lycopersicon* (Mendez, 1971). Of all these coumarins, scopoletin (usually present as the 7-glucoside scopolin) is undoubtedly the most widespread. Its distribution is relatively sporadic and it is of little taxonomic interest. By

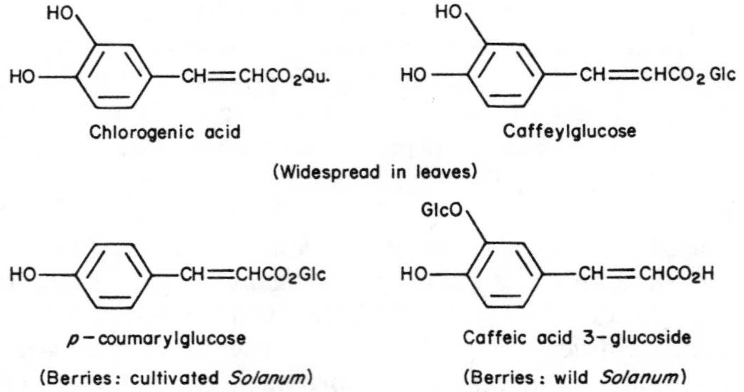

Figure 18.6. Cinnamic acid conjugates in the Solanaceae.

Table 18.6. Hydroxycinnamic acid conjugates: distribution patterns

Family	Chlor.	Cag.	Isochl.	Rosm.
Solanaceae	+ +	+	−	−
Convolvulaceae	+	−	+ +	−
Polemoniaceae	+	−	+	−
Hydrophyllaceae	+	−	+	+
Boraginaceae	−	−	−	+

Chlor., Chlorogenic acid; Cag, 1-caffeylglucose; Isochlor, iso-chlorogenic acid; Rosm., rosmarinic acid.

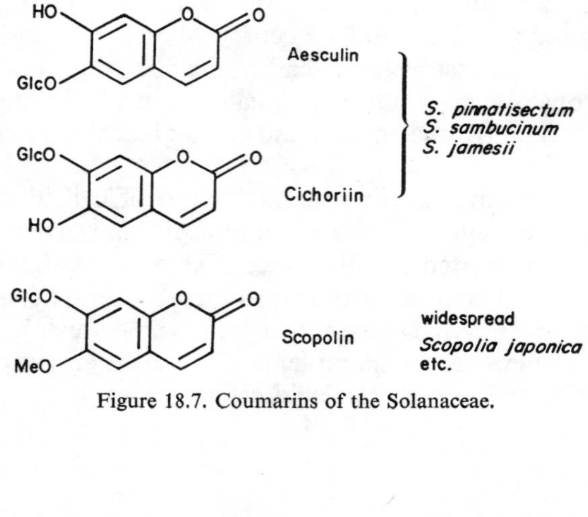

Figure 18.7. Coumarins of the Solanaceae.

contrast, aesculin and cichoriin are of restricted occurrence and hence of more taxonomic value. They occur, for example, in the tuberous *Solanum* in just three closely allied species *S. pinnatisectum*, *S. jamesii* and *S. sambucinum*, being present in both leaves and flowers (Harborne, 1960b). Originally these three species were placed together in the same series *Pinnatisecta* (Correll, 1952), although later Hawkes (1956) transferred *S. sambucinum* from here to the *Cardiophylla* series. Since *S. cardiophyllum* lacks the two coumarins, the phytochemical evidence argues against such a transfer. Indeed, Hawkes (1956) in making the rearrangement, regarded *S. sambucinum* as being intermediate in character between the two series rather than being a true member of *Cardiophylla*. In a yet later treatment, this author placed all three coumarin-containing taxa in the same series, together with *S. cardiophyllum*, at the same time describing *S.* × *sambucinum* as a hybrid between *S. pinnatisectum* and *S. cardiophyllum* (Hawkes, 1963).

CONCLUSION

As has been discussed here, the Solanaceae has a particular flavonoid pattern which generally distinguishes it from other plant families with which it is most usually associated in systematic treatments. The pattern is based on: presence of acylated anthocyanins, especially derivatives of petunidin; the flavonols kaempferol and quercetin, often present as triglycosides; infrequency of flavones; presence of caffeic esters, chlorogenic and 1-caffeylglucose; and the sporadic occurrence of the coumarin, scopoletin. Other rarer flavonoids are encountered on occasion. There are several anthocyanidin glycosides specific to *Petunia* and some distinctive methyl ethers of quercetin in *Nicotiana*. How far such unusual types are present elsewhere in the family remains for future exploration.

In terms of its flavonoid chemistry, therefore, the Solanaceae shows no close relationship with any of the families of the Tubiflorae *sensu lato*. This suggests, together with other phytochemical evidence (e.g. the richness of alkaloid types, etc.) that it has a somewhat isolated position in the angiosperm system, with possible links only with the Convolvulaceae and Nolanaceae. This would support more recent treatments of the Tubiflorae, where the Solanaceae is placed in a separate order Solanales, containing Convolvulaceae, Nolanaceae and (sometimes) Polemoniaceae.

With regard to chemosystematics within the family, there is not the richness of structural diversity in the flavonoid series which is met with in such families as the Ericaceae (Harborne & Williams, 1973), Primulaceae (Harborne, 1968) or Leguminosae (Harborne, 1971). Nevertheless, there is sufficient variation for flavonoids to provide a reasonable number of chemical characters for taxonomic purposes. It is also clear from the few preliminary studies that have been undertaken that there is considerable potential in the flavonoids providing useful data in future taxonomic revisions of genera and tribes. The scene is at least now set for a much wider exploitation of flavonoid chemistry in the study of the systematics and phylogeny of this important family.

ACKNOWLEDGEMENTS

We are grateful to Prof. J. G. Hawkes and Dr R. N. Lester for the supply of fresh leaf material of the Solanaceae used in the flavonoid survey.

REFERENCES

ABE, Y. & GOTOH, K., 1959. Biochemical and genetical studies in eggplant. *Botanical Magazine, Tokyo, 72:* 432–437.

BALLARD, R. E., McCLURE, J. W., ESHBAUGH, W. H. & WILSON, K. G., 1970. A chemosystematic study of selected taxa of *Capsicum. American Journal of Botany, 57:* 225–233.

BATE-SMITH, E. C., 1962. The phenolic constituents of plants and their taxonomic significance. I. dicotyledons *Journal of the Linnean Society (Botany), 58:* 95–173.

BIRKOFER, L., KAISER, C. & KOSMOL, H., 1965. Neue Flavonglykoside aus Petunia hybrida. *Zeitschrift für Naturforschung, 20b:* 605–607, 923–924.

BROWN, E. & MOSS, J. P., 1976. The identification of potato varieties from tuber characters. *Journal National Institute of Agricultural Botany, 14:* 49–69.

CORRELL, D. S., 1952. Section Tuberarium of the genus Solanum of North America and Central America. *U.S. Department of Agriculture, Monograph,* No. 1.

HARBORNE, J. B., 1960a. Anthocyanin production in the cultivated potato. *Biochemical Journal, 74:* 262–269.

HARBORNE, J. B. 1960b. The coumarins of *Solanum pinnatisectum. Biochemical Journal, 74:* 270–273.

HARBORNE, J. B., 1964. The flavonol glycosides of wild and cultivated potatoes. *Biochemical Journal, 84:* 100–106.

HARBORNE, J. B., 1966. Caffeic acid ester distribution in higher plants. *Zeitschrift für Naturforschung, 21b:* 604–605.

HARBORNE, J. B., 1967. *Comparative Biochemistry of the Flavonoids:* 383pp. London: Academic Press.

HARBORNE, J. B., 1968. Flavonoid patterns in the Primulaceae. *Phytochemistry, 7:* 1215–1221.

HARBORNE, J. B., 1971. Distribution of flavonoids in the Leguminosae. In J. B. Harborne *et al.* (Eds), *Chemotaxonomy of the Leguminosae:* 31–72. London: Academic Press.

HARBORNE, J. B., 1975. Biochemical systematics of flavonoids. In J. B. Harborne, T. J. Mabry & H. Mabry (Eds), *The Flavonoids:* 1056–1095. London: Chapman & Hall.

HARBORNE, J. B., 1977. Flavonoids and the evolution of the angiosperms. *Biochemical Systematics and Ecology, 5:* 7–22.

HARBORNE, J. B. & WILLIAMS, C. A., 1971. 6-Hydroxyluteolin and scutellarein as phyletic markers in higher plants. *Phytochemistry, 10:* 367–378.

HARBORNE, J. B. & WILLIAMS, C. A., 1973. A chemotaxonomic survey of flavonoids and simple phenols in leaves of the Ericaceae. *Botanical Journal of the Linnean Society, 66:* 37–54.

HARBORNE, J. B., MABRY, T. J. & MABRY, H., 1975. *The Flavonoids:* 1204 pp. London: Chapman & Hall.

HAWKES, J. G., 1956. A revision of the tuber-bearing Solanums. *Annual Report Scottish Society for Research and Plant Breeding,* 38–110.

HAWKES, J. G., 1963. A revision of the tuber-bearing Solanums, 2nd ed. *Scottish Plant Breeding Station Record.*

HESS, D., 1968. *Biochemische Genetik.* Berlin: Springer Verlag.

MENDEZ, T. C., 1971. Phenolic compounds of the tomato plant. *Experientia, 27:* 758.

MEYER, C., 1964. Die Genetik des B-ringes bei Petunia-anthocyanen. *Zeitschrift für Vererbungslehre, 95:* 171–183.

NATARELLA, N. J. & SINK, K. C., 1974. Chromatography of phenolics of species ancestral to *Petunia hybrida. Journal of Heredity, 65:* 85–92.

RANGOONWALA, R. & FRIEDRICH, H., 1967. Flavone glycosides of Capsicum. *Naturwissenschaften, 54:* 368.

SCHMID, R. D. & HARBORNE, J. B., 1973. Mass spectrometric identification of a kaempferol tetraglycoside from *Solanum* seed. *Phytochemistry, 12:* 2269–2273.

SWAIN, T., 1975. Evolution of flavonoid compounds. In J. B. Harborne *et al.* (Eds), *The Flavonoids:* 1096–1129. London: Chapman & Hall.

TSO, T. C., 1972. *Physiology and Biochemistry of Tobacco Plants:* 410 pp. New York: Dowden, Hutchinson & Ross.

19. The biosynthesis of chlorogenic acid in Solanaceae

F. PARMENTIER

Lab. Alg. Scheik., RUCA, Antwerp, Belgium

Evidence for a biosynthetic pathway from L-phenylalanine through cinnamic acid, p-coumaric acid and 5-0-p-coumarylquinic acid to chlorogenic acid has been obtained in *Cestrum* and *Solanum*. Some properties of the enzymes are known.

Hydroxy-cinnamic acids occur widely in the plant kingdom particularly as esters of (−) quinic acid, (−) shikimic acid and D-glucose. The oldest known is chlorogenic acid, 5-0-caffeyl-quinic acid. The phenylpropanoid skeleton plays a central role in the plant phenol biosynthesis: *p*-coumaric acid and caffeic acid are believed to be the ultimate precursors of most plant phenolics, e.g. lignin, flavonoids etc.

Although these compounds could act as germination inhibitors (Van Sumere, 1960) and plant growth regulators (Gross, 1975) no specific role can as yet be attributed to them.

L-phenylalanine, formed in plants via the shikimate pathway (Haslam, 1974) is the link to the biosynthesis of hydroxy-cinnamic acids in Solanaceae. The conversion to t-cinnamic acid is catalysed by the enzyme L-phenylalanine ammonia lyase (PAL, E.C.4.3.1.5.). It has been isolated and purified from potato tuber (Havir & Hanson, 1968).

It is possible to draw the following scheme for chlorogenic acid biosynthesis and to discern in fact three possible routes (Fig. 19.1):

(1) C→CQ→pCQ→CaQ

(2) C→pC→pCQ→CaQ

(3) C→pC→Ca →CaQ

Levy & Zucker (1960) proposed the first route for the biosynthesis of chlorogenic acid in potato slices. Hanson (1966) however indicated that the cinnamyl ester of quinic acid is not an intermediate. W. Steck (1968) proposed route 2 to be effective in tobacco among others. Nagels & Parmentier (1974, 1976 and unpubl. results) demonstrated conclusively that in *Cestrum poeppigii* the second hypothesis is a major route for the biosynthesis of chlorogenic acid. The evidence follows from kinetic trapping experiments: when radioactive t-cinnamic acid is fed to *Cestrum* leaves the labelling of *p*-coumaric acid, *p*-coumarylquinic acid, chlorogenic acid reaches a maximum at 2, 5 and 6 h respectively. Also the specific activity of the first two acids is always higher than that of chlorogenic acid. The labelling of cinnamylquinic and caffeic acid is slower than that of chlorogenic acid and excludes them as lying on a direct pathway to chlorogenic acid.

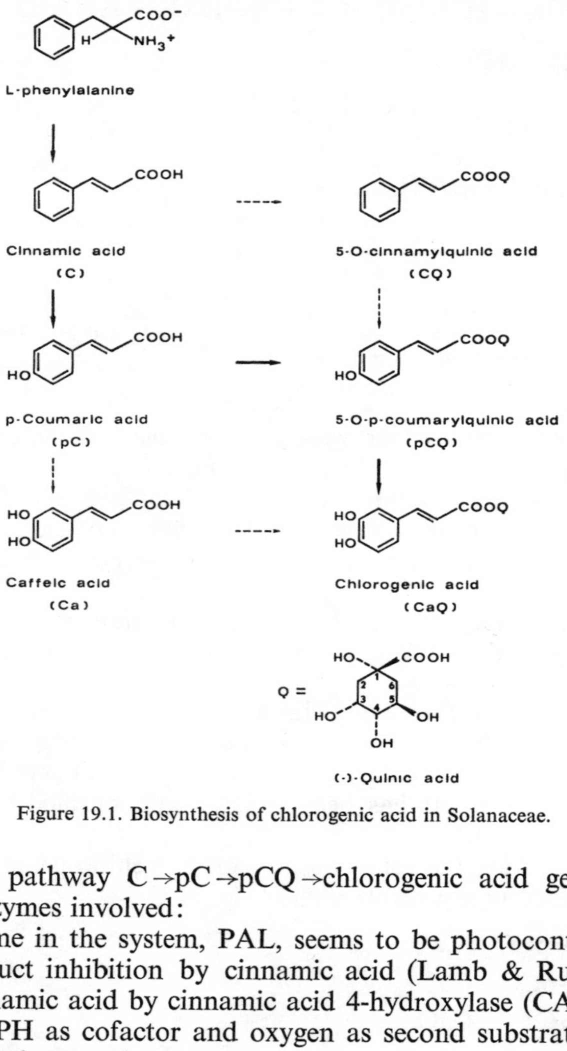

Figure 19.1. Biosynthesis of chlorogenic acid in Solanaceae.

The proposed pathway C→pC→pCQ→chlorogenic acid gets support from studies of the enzymes involved:

The first enzyme in the system, PAL, seems to be photocontrolled. It is very sensitive to product inhibition by cinnamic acid (Lamb & Rubery, 1976). The oxidation of cinnamic acid by cinnamic acid 4-hydroxylase (CA4H, E.C.1.14.13. 11.) needs NADPH as cofactor and oxygen as second substrate. Both enzymes were studied in *Solanum tuberosum*.

The esterfication of the phenylpropanoic acid could involve an acid: Co A ligase and an acid quinate: Co A transferase. The former has been studied in *Solanum tuberosum* (Rhodes & Wooltorton, 1975). The enzyme is specific for Co A and ATP and is activated by Mg^{++}. The fact that the enzyme has the highest affinity for *p*-coumaric acid is consistent with the above proposed pathway of chlorogenic acid biosynthesis.

For the second step in esterification we will have to look for a *p*-coumaryl quinate: Co A transferase.

The last enzyme to fit in the proposed pathway will be a *p*-coumarylquinate hydroxylase.

ADDENDUM

p-coumaryl quinate: Co A transferase, isolated from fruits of *Lycopersicon esculentum*, has now been studied. The enzyme has the highest affinity for *p*-coumaryl Co A, another fact that sustains the proposed pathway (Rhodes & Wooltorton, 1976).

REFERENCES

GROSS, D., 1975. *Phytochemistry, 14:* 2105.

HANSON, K. R., 1966. *Phytochemistry, 5:* 491.

HASLAM, E., 1974. In *The Shikimate Pathway.* London: Butterworths.

HAVIR, E. A. & HANSON, K. R., 1968. *Biochemistry, 7:* 1904.

LAMB, C. J. & RUBERY, P. H., 1976. *Phytochemistry, 15:* 665.

LEVY, C. C. & ZUCKER, M., 1960. *Journal of Biological Chemistry, 235:* 2418.

NAGELS, L. & PARMENTIER, F., 1974. *Phytochemistry, 13:* 2759.

NAGELS, L. & PARMENTIER, F., 1976. *Phytochemistry, 15:* 703.

RHODES, M. J. C. & WOOLTORTON, L. S. C., 1975. *Phytochemistry, 14:* 2161.

RHODES, M. J. C. & WOOLTORTON, L. S. C., 1976. *Phytochemistry, 15:* 947.

STECK, W., 1968. *Phytochemistry, 7:* 1711.

VAN SUMERE, C. F., 1960. In J. B. Pridham (Ed.), *Phenolics in Plants in Health and Disease.* Oxford: Pergamon Press.

20. The diterpenes of *Nicotiana* species and *N. tabacum* cultivars

W. W. REID

Department of Chemistry, Queen Elizabeth College, University of London, U.K.

A study of the diterpenes of *Nicotiana* has shown these are synthesised within the trichomes and excreted on to the leaf surface. The diterpenes are of either the labdane or duvane classes. *N. tabacum* cultivars, now considered to be hybrids, generally contain one or other group arising from the duvane producing parent *N. sylvestris* or the labdane producing parent *N. tomentosiformis*. These diterpenes have plant growth inhibiting properties and are precursors of tobacco flavour.

CONTENTS

INTRODUCTION

The leaves of cultivated tobacco are covered with a gummy exudate which is synthesized in the trichomes and contains large amounts of diterpenes (Michie & Reid, 1968). It has been generally considered in the tobacco industry that this gum is the source of the aroma, particularly of aromatic tobaccos. Some 30 years ago Wolf, in his pioneer work on aromatic tobaccos (Bentley & Wolf, 1945; Wolf, 1946) clearly recognized the importance of the role of the trichomes and exudate relative to aroma. The amounts of these diterpenes are considerable in fresh leaf $(0.5 \rightarrow 10.0$ g/100 g dry tissue). However these compounds are very susceptible to light and oxygen, and only small amounts are found after curing (Reid, 1974–1977). It was not until 1962 (Roberts & Rowland, 1962) that the first macrocyclic diterpene, a derivative of the duvane structure, was characterized (4,8,13-duvatriene-1,3-diol) in very small amounts from cured tobaccos. The first labdane derivatives were isolated in small amounts from Turkish tobacco in 1961 (Giles & Schumacher, 1961).

Note: The original terminology for the diterpenes has been retained in this article. The preferred nomenclature recently internationally adopted is labdanoids in place of labdanes and thunberganoids in place of duvanes.

THE CHEMISTRY OF *NICOTIANA* DITERPENES

Since the start of this study this subject has become increasingly complex. A preliminary account has been published (Colledge, Reid, & Russell, 1975) and a more detailed paper is in preparation (Colledge *et al.*, in prep.). In summary the current position is as follows. An examination of the diterpenes of some 20 species of *Nicotiana* and 30 cultivars of *N. tabacum* shows these fall into either the labdane or duvane classes. The majority of labdane producing species yielded cis-abienol or (12 Z)-labda-12,14-diene-8α-ol (1), the component N1 mentioned *loc. cit.*, together with the related diol (2) (13 E)-labda-12-ene-8α,15-diol. *N. glutinosa* was exceptional in that it produced four labdanes but no (1) or (2). The major metabolite of this species was identified as sclareol or (13 R)-labda-14-ene-8α,13-diol (3) and its (13 S) epimer (4). A similar mixture from the same species was reported about the same time (Bailey, Vincent & Burden, 1974). The minor metabolites of *N. glutinosa* were assigned the structures (5) (6) (7). All compounds were isolated as crystalline samples with satisfactory mp, ultraviolet, infrared, n.m.r. and mass spectra. An exception was components (3) and (4) isolated as a eutectic mixture. The duvane producing varieties produced a mixture of the α and β isomers of 4,8,13-duvatriene-1,3-diol (8).

THE DITERPENES OF HYBRIDS

An early example of the effect of hybridization was with *N. sylvestris* and *N. glutinosa*. The former is a *duvane* producer whereas the latter, as we have seen above, produces four *labdane* diterpenes in which sclareol and its 13-epimer were the major components. The hybrid contained both the *duvanes* and the *labdanes* in about equal proportions. Of more interest was the relationship of diterpene formation to the origin and evolution of *N. tabacum*, the commercial tobacco plant.

N. tabacum ($n = 24$) is believed to have arisen by chromosome doubling after hybridization of *N. sylvestris* Spegazzini & Comes ($n = 12$) with a species in the Tomentosae section of *Nicotiana*. Goodspeed & Clausen suggested that *N. tomentosa* Ruiz & Pavon ($n = 12$) was the species, but Clausen amended it to *N. tomentosiformis*, Goodspeed ($n = 12$). Goodspeed favoured *N. otophora* Grisebach ($n = 12$). For a summary see Goodspeed (1954). Recently (Gray, Kung, Wildman & Sheen, 1974) from an analysis by isoelectric focusing of the polypeptide composition of Fraction 1 protein isolated from *N. tabacum* and the putative progenitor species, showed that *N. tabacum* arose from the hybridization of *N. sylvestris* ♀ and *N. tomentosiformis* ♂. Reid (1974b) showed that *N. tomentosiformis* was a *labdane* producer and *N. sylvestris* a *duvane* producer. In later work (Reid, 1975c) an examination was made of the diterpenes of *N. sylvestris*, *N. tomentosiformis*, *N. otophora*, *N. tabacum* PB and two hybrids, *N. otophora* × *N. sylvestris* (produced at Bergerac) and an amphidiploid *N. tomentosiformis* × *N. sylvestris* developed by Dr L. G. Burk, U.S. Department of Agriculture, Oxford Tobacco Research Station. The results may be summarised as follows:

(1) *cis*-abienol was only found in *N. tomentosiformis*, the amphidiploid *N. sylvestris* × *N. tomentosiformis* and some cultivars of *N. tabacum*.

(2) *N. otophora* and *N. otophora* × *N. sylvestris* showed identical chromatographic patterns and *cis*-abienol was not present.

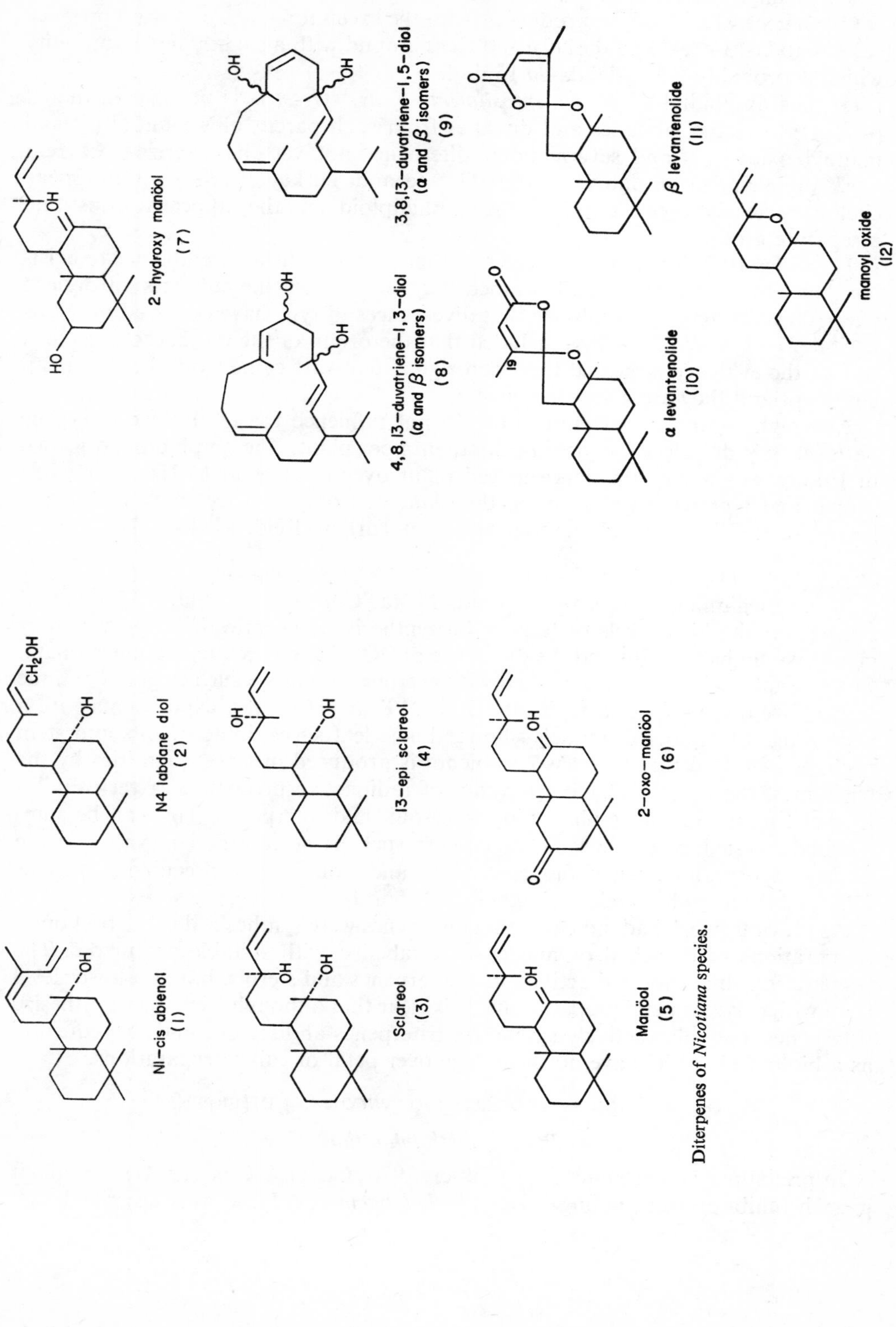

2-hydroxy manöol
(7)

3,8,13-duvatriene-1,5-diol
(α and β isomers)
(9)

β levantenolide
(11)

4,8,13-duvatriene-1,3-diol
(α and β isomers)
(8)

α levantenolide
(10)

manoyl oxide
(12)

CH₂OH

N4 labdane diol
(2)

13-epi sclareol
(4)

2-oxo-manöol
(6)

N1-cis abienol
(1)

Sclareol
(3)

Manöol
(5)

Diterpenes of *Nicotiana* species.

(3) Improvements in technique suggested that species and cultivars producing labdanes might in addition produce small amounts of duvanes.

(4) In general all labdane producers (with the exception of *N. glutinosa*) produce (13 E)-labda-13-ene-8α,15-diol (2), and a compound with a slightly higher mobility which is probably (13 E)-labda-8α,15-diol.

(5) The amphidiploid *N. tomentosiformis* × *N. sylvestris* is unusual in that it produces *both* the labdanes and duvanes from each parent plus manöol (5) and manoyl oxide (12) and several polar diterpenes not yet characterized. Current work on *N. tabacum* cultivars from Greece and Turkey shows their diterpene chemistry closely resembles that of the amphidiploid, and they appear to constitute a separate group.

The older established classical oriental tobaccos have little resemblance to existing American varieties. If it is assumed they derive from the cultivars introduced over 300 years ago, it is probable that divergences of type have occurred over this period. Wolf & Wolf (1948) concluded that the origin of these tobaccos was lost, but all the evidence indicated that the original cultivars came from the American continent and they are a separate group.

Although conditions of cultivation do not influence the qualitative diterpene pattern, they do affect the yield of diterpenes per plant. The amphidiploid grown in France was a vigorous large leaved plant over ten feet high. However when grown under extreme water stress, the plant was only two feet high with small leaves, resembling oriental varieties grown in Turkey (Reid, 1974–1977).

BIOSYNTHESIS OF DITERPENES IN *NICOTIANA TABACUM* PB

At the beginning of this work (Reid, 1974a, Colledge & Reid, 1974) it was hoped that the high levels of terpene biosynthesis which prevail in the secretory glands would have facilitated the direct use of ^{14}C labelled acetate and mevalonate. For example, tobacco stems washed with acetone to remove surface gum continue to produce diterpenes at a high rate (Reid, 1974a). However, experiments with a wide range of biological models—flower buds, leaf slices, intact plants and stem peelings—showed that whereas *cis*-abienol is produced in large quantities by the trichomes, the pattern of incorporation of radioactive precursors is very similar to that found with other plant mono-, sesqui- and diterpenes. This can be summarised by saying that specific precursors such as acetate and mevalonate are poorly incorporated when fed externally, and non-specific precursors such as CO_2 and sucrose show relatively good incorporation.

As previous work had indicated that diterpenes were synthesized in the trichomes preparations of detached organelles were labelled with suitable precursors. The preparations incorporated activity into diterpenes and sterols, but at a lower level than with calyx or petal preparations. It is clear that although trichomes synthesize diterpenes, they also actively synthesize triterpenes and sterols, and consequently as a biological model have no advantage over petal or calyx preparations.

THE BIOLOGICAL PROPERTIES OF *NICOTIANA* DITERPENES

Plant growth inhibition

In preliminary communications (Cutler, 1970; Cutler & Cole, 1974) a new plant growth inhibitor from young leaves of *N. tabacum* cv. Hicks was described, and shown to inhibit the growth of wheat coleoptiles to 10^{-5}M. The inhibitors were subsequently shown to be β4,8,13-duvatriene-1,3-diol (DVT) and the α isomer which was ten times less active in bioassays (Springer, Clardy, Cox, Cutler & Cole, 1975). Studies on *N. tabacum* cv. PB (Deletang, 1973) showed that an inhibitor exists in the flower buds which keeps root growth in check. Removal of those buds and young leaves during cultivation not only induces the expansion of the remaining leaves but also promotes root growth. There is a subsequent increase in nicotine synthesis in the roots, which eventually gives leaves with a high nicotine content. As the extracts prepared by Deletang were found to contain small amounts of diterpenes (Reid, 1974–1977), biological assays were carried out on 10 tobacco diterpenes (Cutler, Reid & Deletang, 1977).

The diterpenes *cis*-abienol (1) and the related diol (2) obtained from *N. tabacum* cv. PB significantly inhibited the growth of wheat coleoptiles ($P = 0.01$) at 10^{-3} and 10^{-4}M but the diol (2) was inexplicably more active at 10^{-4} than at 10^{-3}M. The synthetic derivative iso-dihydroabienol was inactive. Diterpenes from *N. glutinosa*, 2-hydroxymanöol, the eutectic mixture of sclareol and episclareol and pure sclareol were all active at 10^{-3}M and 10^{-4}M, while manöol and manoyl oxide were equally active at 10^{-3}M. α-Levantenolide was active at 10^{-3}M but the β isomer was inactive. β-4,8,13-duvatriene-1,3-diol was the only diterpene assayed that was active at 10^{-5}M (Cutler *et al.*, 1977).

Antifungal activity

In the course of a study of the reaction of *N. glutinosa* to infection by tobacco mosaic virus (Bailey *et al.*, 1974) the diterpenes sclareol and 13-epi sclareol were isolated as a eutectic mixture. These diterpenes did not prevent germination of fungal spores but markedly inhibited the radial extension of colonies growing on agar. The inhibition was shown to be due to an effect on the morphology of the fungi in which the degree of hyphal branching was increased.

Extracts of *N. glutinosa* produced several areas of fungal growth inhibition on silica gel plates when assayed by means of *Cladosporium cucumerinum*. The material responsible for the main inhibitory area was identified as a mixture of sclareol and 13-epi sclareol. In continuation of this work Bailey & Burden (1976) showed that two major inhibitory components from *N. sylvestris* were the α and β duvatriene diols. Of the major diterpenes of *N. tabacum* cv. PB *cis*-abienol was active whereas the diol was inactive. Assay of silica gel chromatograms of diterpene fractions of *N. tabacum* cv. PB showed several strong zones of inhibition due to unidentified terpenoid compounds (Bailey & Reid, 1976).

REFERENCES

BAILEY, J. A. & BURDEN, R. S., 1976. Private communication.
BAILEY, J. A. & REID, W. W., 1976. Unpublished results.
BAILEY, J. A., VINCENT, G. G. & BURDEN, R. S., 1974. Diterpenes from *N. glutinosa* and their effect on fungal growth. *Journal of General Microbiology, 85:* 57–64.
BENTLEY, N. J. & WOLF, F. A., 1945. Glandular leaf hairs of oriental tobacco. *Bulletin of the Torrey Botanical Club, 72:* 345–360.
COLLEDGE, A. & REID, W. W., 1974. The biosynthesis of *cis*-abienol and phytosterols in *Nicotiana tabacum* PB. *Annales du Tabac, Sect. 2, 11:* 177–183.
COLLEDGE, A., REID, W. W. & RUSSELL, R. A., 1974. A survey of surface diterpenoids of green leaves. *Annales du Tabac, Sect. 2, 11:* 159–164.

COLLEDGE, A. REID, W. W. & RUSSELL, R. A., 1975. The diterpenoids of *Nicotiana* species and their potential technological significance. *Chemistry and Industry:* 570–571.

COLLEDGE, A., REID, W. W. & RUSSELL, R. A. Paper in preparation.

CUTLER, H. G., 1970. A growth inhibitor from young expanding tobacco leaves. *Science, 170:* 356–357.

CUTLER, H. G. & COLE, R. J., 1974. Properties of a plant growth inhibitor extracted from immature tobacco leaves. *Plant and Cell Physiology, 15:* 19–28.

CUTLER, H. G., REID, W. W. & DELETANG, J., 1977. Plant growth inhibiting properties of diterpenes from tobaccos. *Plant and Cell Physiology, 18:* 711–714.

DELETANG, J., 1973. Influence exercée par les bourgeons floraux sur la rhizogènes de *Nicotiana tabacum*. *Compte Rendu Hebdomadaire des Sciences de l'Academie des Sciences:* 5822–5825.

GILES, J. A. & SCHUMACHER, J. N., 1961. Turkish tobacco. I. Isolation and characterization of α- and β-levantenolides. *Tetrahedron, 14:* 246–251.

GOODSPEED, T. H., 1954. *The genus* Nicotiana. Waltham, Massachusetts: Chronica Botanica.

GRAY, J. C., KUNG, S. D., WILDMAN, S. G. & SHEEN, S. J., 1964. Origin of *Nicotiana tabacum* L. detected by polypeptide composition of Fraction 1 protein. *Nature, 252:* 226–227.

MICHIE, M. J. & REID, W. W., 1968. Biosynthesis of complex terpenes in the leaf cuticle and trichomes of *N. tabacum*. *Nature, 218:* 578.

REID, W. W., 1974a. The biosynthesis of terpenoids in two varieties of *N. tabacum*. *Annales du Tabac, Sect. 2, 11:* 164–175.

REID, W. W., 1974b. Note on the diterpenes of cultivars of *N. tabacum*. *Annales du Tabac, Sect. 2, 11:* 176–177.

REID, W. W., 1975a. The diterpenes of *N. tomentosiformis* hybrids and *N. tabacum* cultivars. *Annales du Tabac, Sect. 2, 12:* 25–28.

REID, W. W., 1975b. A survey of some species of *Nicotiana* 1975. *Annales du Tabac, Sect. 2, 12:* 29–33.

REID, W. W., 1975c. The diterpenes of *Nicotiana* as precursors of aroma constituents of commercial tobaccos. *Annales du Tabac, Sect. 2, 12:* 33–37.

REID, W. W., 1974–1977. Unpublished work.

ROBERTS, D. L. & ROWLAND, R. L., 1962. Macrocyclic diterpenes α- and β-4,8,13-duvatriene-1,3-diols from tobacco. *Journal of Organic Chemistry, 27:* 3989–3995.

SPRINGER, J. P., CLARDY, J., COX, R. H., CUTLER, H. G. & COLE, R. J., 1975. The structure of a new type of plant growth inhibitor extracted from immature tobacco leaves. *Tetrahedron Letters, 32:* 2734–2740.

WOLF, F. A., 1946. Further consideration of the glandular leaf hairs of tobacco and of their significance. *Bulletin of the Torrey Botanical Club, 73, 3:* 224–234.

WOLF, F. A. & JONES, F. E., 1944. Comparative structure of green leaves of oriental tobacco at different levels on the stalk in relation to their quality upon curing. *Bulletin of the Torrey Botanical Club, 71:* 512–528.

WOLF, F. A. & WOLF, F. T., 1948. The origin of tobaccos of the oriental type. *Bulletin of the Torrey Botanical Club, 75:* 51–55.

21. Characterization of proteins from potatoes, and the "Index of European Varieties"

H. STEGEMANN

Institut für Biochemie, Biologische Bundesanstalt, Braunschweig, West Germany

Attention is drawn to basic features for applying electrophoretic methods to chemotaxonomy. Constancy or change of protein patterns from potato tubers and leaves are illustrated, as well as properties of, and immunological relationships among, protein fractions of *Solanum tuberosum*.

CONTENTS

INTRODUCTION

The use of organic molecules as a tool in plant taxonomy is now well established. This is in spite of some scepticism from a few traditionally orientated botanists. Their caution is partly a reaction to others who have made exaggerated claims for this method, especially when small molecules are used for differentiation.

Until 1960, only a few scientists used proteins for differentiation among species, taking advantage of the surface characteristics of the proteins, which govern the formation of specific antibodies. Among them outstanding work was done by Mez (1925), and Mez & Ziegenspeck (1926) in the early 20th century in Königsberg, and more recently for *Solanum* species by Hawkes and co-workers (Gell, Hawkes & Wright, 1960; Hawkes & Lester, 1966, 1968).

However, one has to keep in mind, that the surface of a folded macromolecule gives only part of the information about the chain, or agglomerate of chains. Charge and size and the sequential composition of the whole chain are other important attributes. This also means, that small molecules cannot be as characteristic as larger ones. In addition, a molecule directly coded by the nucleic acid of the gene will be less influenced by environmental factors than a molecule which is a secondary gene product like all the low molecular compounds. Therefore proteins are to be preferred, or even individual nucleic acids, if there were a faster means of characterizing them by hybridization.

PRE-REQUISITES FOR USING PROTEINS FOR PLANT TAXONOMY

The following points are important if proteins (and to some extent other biomolecules) are to be used for chemotaxonomy:

(1) The method for separation and identification should be inexpensive, easy to perform and not interfered with by concomitant substances of variable concentration.

(2) The separation should be a one-step-procedure and start with the crude sap in order to minimize unwanted changes during processing.

(3) At first one variety of cultivar must be checked under different conditions: (e.g. differences in maturity, climate, fertilizer application, pesticide treatment and so on) to ensure that protein or enzyme patterns are unaffected, or to provide precise knowledge of such shifts. Subsequently, other varieties may be investigated.

(4) The protein pattern (spectrum) of a given specimen should have a number of individual characters easily distinguishable from each other. The larger the number, the better the chance for differentiation. Reliable comparisons of samples to internal standards are needed.

TECHNIQUES FOR GEL-ELECTROPHORETIC SEPARATIONS

The best method for the separation of proteins and the unequivocal comparison of bands, is gel electrophoresis in slabs, preferably as polyacrylamide gel electrophoresis (PAGE). Here, separation according to size (sodium dodecylsulfate (SDS), and porosity gradient) and charge (different buffers, and pH-gradients in ampholytes) can be chosen at will. A further advantage of using slab gel electrophoresis is to have only two variables in differentiating samples, namely migration rate and intensity of the bands. This means that a computer-readable scheme could be developed if a better densitometric tracing or fourier-calculation could be found.

The apparatus tested by us for the different PAGE-procedures including immuno-chemical methods, and used for about ten years, is the PANTA-PHOR (Stegemann, 1972) from Labor-Müller, D-3510 Hann. Münden, West Germany. Detailed experimental instructions for technicians and students can be obtained from the present author.

Patterns from one figure cannot always be compared to those of another since the processing of the tubers was sometimes quite different. A comprehensive publication is in preparation.

CONSTANCY OF PROTEIN PATTERNS IN MATURE TUBERS, AND THE APPEARANCE OF A PROTEIN COMPLEX DURING IMMATURITY

Between 1960 and 1965 we found that the relative amount of storage proteins in the tubers is not influenced by environmental conditions, although the absolute amount may be (Stegemann, 1970, 1975; Stegemann & Loeschcke, 1977). The same is true for some patterns of isozymes like esterases and peroxidases, but not for others such as phosphorylases (Gerbrandy, Shankar, Shivaram & Stegemann, 1975). A minor change in the patterns is seen when sprouting is strong (Stegemann, Francksen & Macko, 1973). The "mapping-technique" (two-dimensional iso-electric focusing and electrophoresis) between pH 5 and 7 was also used to analyse proteins from plants before and after passage through meristem culture (Stegemann, Loeschcke, Bode & Huth, 1970). We found no change in the protein pattern in standard PAGE. Examples of the lack of influence of even extremely different salt contents in the soil, or different growth regulators, is given in Fig. 21.1. (The lower absolute amount of proteins in columns 9–11 is due to growth of the potato plants in pots instead of in the field, and the possible shortage of nutrients.)

Storage proteins in tubers (and to some extent in seeds) of different potato cultivars differ mainly in charge (Macko & Stegemann, 1969; Olteanu, Tanasescu & Moldoveanu, 1973), rather than in the molecular weights of their monomers (Fig. 21.2), which is in contrast to our findings in maize seeds (unpubl. results). However, if the tubers are very immature (about two months before harvest) the individual patterns are less pronounced, and one protein-band is characteristic of immaturity (Stegemann et al., 1973). It is compared with a leaf protein (fraction-1-protein or ribulose-1,5-diphosphate-carboxylase-oxygenase) which behaves similarly in standard PAGE at pH 8.9 and pH 7.9 (Fig. 21.3). Both show irregular migration rates since both rates are not influenced by the different pH values. The protein band splits after focusing in ampholytes (pH 3–10) and gives the same picture in the two-dimensional gel-electrophoresis for all cultivars tested. That means, it is dependent on the physiological state of the tuber, but not on the cultivars tested. This protein complex does not give a cross-reaction if its antibody system is tested against the fraction-1-protein from leaves. It disappears if the tuber becomes mature (Stegemann, 1977) (see Fig. 21.4). The process of disappearance is shown in Fig. 21.5, where the amount decreases with later harvest and with increasing tuber size. It has a molecular weight of about 2.7 and 4.9×10^5 daltons as checked by electrophoresis in a porosity gradient (Poro-PAGE). The corresponding value for the leaf protein is 3.15×10^5 daltons (Fig. 21.6, left). The monomers, as calculated after SDS-PAGE, have MW of 19, 10, 9 and 3.8×10^4 daltons for the proteins from immature tubers, compared with the protein from young (51, 40, 14×10^3) and older (51, 49, 40, 14×10^3 daltons) leaves, (Fig. 21.6, right). Virus particles in tubers are not detectable in standard electrophoresis (small-pore gel, low concentration of virus); in infected leaves, however, they can be separated by Poro-PAGE (3–7%).

The relationships among the detected protein bands are important in order to understand the pattern. SDS-electrophoresis showed that these bands are composed of only a few monomers. The more pronounced bands, in the cathodic part of the gel after focusing or electrophoresis, are weak antigens, and are similar but quantitatively different in different cultivars (Fig. 21.7). Bands in the anodic part are stronger antigens and are very distinctive for each cultivar (Stegemann, 1977). However, they are more or less related immunologically giving rise to coalescent precipitin lines in the two-dimensional tendem-electrophoresis after isolation or enrichment by preparative electrophoresis.

USE OF ELECTROPHORESIS TO CHARACTERIZE CULTIVARS

A grouping of the cultivars in the "Index Europäischer Kartoffelsorten" (Stegemann & Loeschcke, 1976) is based on these protein and esterase patterns. When we started our investigations in the sixties, we did not anticipate much practical use of our findings in the near future. We were aware that isozymes could be used as genetic markers, leading to better planning for breeding programmes in spite

Table 21.1. Excerpt from the tables of the "Index of European Potato Varieties" showing three of the 530 cultivars listed. The plates showing the protein spectra are not given here

(a)

Name of cultivars	A	B	CH	CS	D	DDR	DK	E	F	GB	I	N	NL	PL	S	Z	Typ.
Achat																	8
Ackersegen	4	6			11		4		9							P	8
Admirandus											X	5					4

Registration

National registers (code as for automobiles)
Reference to plate numbers of the spectra
Protein grouping of the varieties

(b)

Property evaluation (1—9)

Name	1	2	3	4	5	6	7	11	12	13	14	15	16	21	22	23	24	31	32
Achat	4	3	2	3	5	4	2	3	3	5				3	0	3	9	+	5
Ackersegen	7	2	5	4	4	5	6	6	1					2	0	3	3	A1	6
Admirandus	5	3	5	4	4	6	5									5	4		

Key for column numbers

Properties	Susceptibility to
(1) Maturity	(11) Leaf roll-virus
(2) Readiness to sprout	(12) Y-Virus
(3) Colour of skin	(13) A-Virus
(4) Colour of flesh	(14) M-Virus
(5) Shape of tuber	(15) X-Virus
(6) Deepness of eyes	(16) X (Erstling)
(7) Starch content	
	(21) Scab
	(22) Wart disease
	(23) Late blight/tuber
	(24) Late blight/foliage
	(31) Root-eelworm
	(32) Spraing

(c)

Genetic data

Name	Z	A	♀	♂	W	R
Achat	D-18	67	Fina	Rheinhort	A, D	R
Ackersegen	D-5	29	Hindenburg	Allerfrüheste Gelbe		r
Admirandus	NL-6	66	Jakobi	USDA X 96–56		

Breeders: Country and Number in the list of breeders
Parents of the varieties
Wild species involved
Genotype of resistance to Phytophthora infestans/leaves

of the complexity of potato genetics (Howard, 1970). One has to keep in mind that resistance to diseases is the normal feature of the plant and that only a very few pathogens succeed in breaking down these barriers. Therefore it should be possible to correlate even polygenic properties with bands in isozyme- or other protein-patterns. There are already hints of this in the animal field where a relationship is found between certain isozymes and the milk production of cattle (Geldermann, 1976).

As early as 1971 the identification of potato cultivars by electrophoresis was undertaken by German testing stations as a routine, to provide decisions within a day in disagreements about cultivar designations. Now it is even accepted in court cases. For these reasons we have been encouraged to compile our data for the 530 registered cultivars grown in most European countries, and also to list the national registration, the evaluation of disease susceptibility and the genetic data. An example is seen in Table 21.1, where the data of three cultivars are listed. The index also contains lists of addresses of breeders, the addresses of institutions dealing with cultivar-registration, and plates with all the electrophoretic spectra of proteins and esterases using the cultivar "Maritta" as an internal standard. For more details see Stegemann & Loeschcke (1976, 1977). Primitive cultivars from the gene bank of C.I.P. (Lima, Peru) are being tested for identity, to reduce the number of clones which has to be kept for breeding. The revised edition is scheduled for 1978.

ACKNOWLEDGEMENTS

I wish to thank Frau Hella Francksen and Ellen Krögerrecklenfort for their skilled cooperation.

REFERENCES

GELDERMANN, H., 1976. *Untersuchung der Milchleistungsvererbung beim Deutschen Schwarzbunten Rind mit Hilfe von Marker-Genen.* Habilitations-Schrift, Universität Göttingen.

GELL, P. G. H., HAWKES, J. G. & WRIGHT, S. T. C., 1960. The application of immunological methods to the taxonomy of species within the genus *Solanum. Proceedings of the Royal Society (B), 151:* 364–383.

GERBRANDY, S. J., SHANKAR, V., SHIVARAM, K. N. & STEGEMANN, H., 1975. Conversion of Potato Phosphorylase Isozymes. *Phytochemistry, 14:* 2331–2333.

HAWKES, J. G. & LESTER, R. N., 1966. Immunological Studies on the Tuber-bearing Solanums. II. Relationships of the North American species. *Annals of Botany, 30:* 269–290.

HAWKES, J. G. & LESTER, R. N., 1968. Immunological Studies on the Tuber-bearing Solanums. III. Variability within *S. bulbocastanum* and its Hybrids with Species in Series Pinnatisecta. *Annals of Botany, 32:* 165–187.

HOWARD, H. W., 1970. *Genetics of the potato, Solanum tuberosum:* 1–126. London: Logos Press.

MACKO, V. & STEGEMANN, H., 1969. Mapping of Potato Proteins by Combined Electrofocusing and Electrophoresis. Identification of Varieties. *Hoppe-Seyler's Zeitschrift für Physiologische Chemie, 350:* 917–919.

MEZ, C., 1925. In Boas, Neuberg & Rippel (Eds), *Der serodiagnostische Stammbaum des Pflanzenreiches. Naturwissenschaft und Landwirtschaft,* Heft 4: 25–47. Freising: Verlag Datterer.

MEZ, C., 1926. Die Bedeutung der Sero-Diagnostik für die stammesgeschichtliche Forschung, *Botanisches Archiv, 16:* 1–23.

MEZ, C. & ZIEGENSPECK, H., 1926. *Botanisches Archiv, 16:* 483–485.

OLTEANU, G., TANASESCU, D. & MOLDOVEANU, N., 1973. Electrophoretic characterization of seeds of some potato lines and varieties. *Revue Roumaine de Biochimie, 10:* 43–47.

STEGEMANN, H., 1970. Neue Sortenbestimmung von Kartoffeln. *Der Kartoffelbau, 21:* 338–339.

STEGEMANN, H., 1972. Apparatur zur thermokonstanten Elektrophorese oder Fokussierung und ihre Zusatzteile. *Zeitschrift für Analytische Chemie, 261:* 388–391.

STEGEMANN, H., 1975. Properties of and physiological changes in storage proteins. In J. B. Harborne & C. F. Van Sumere (Eds), *The Chemistry and Biochemistry of Plant Proteins:* 71–88. London & New York: Academic Press.

STEGEMANN, H., 1977. Plant proteins evaluated by two-dimensional methods. In D. Graesslin & B. J. Radola (Eds), *International Symposium on Electrofocusing and Isotachophoresis 1976:* 385–394. Berlin: de Gruyter.

STEGEMANN, H., FRANCKSEN, H. & MACKO, V., 1973. Potato Proteins: Genetic and Physiological Changes, Evaluated by One- and Two-dimensional PAA-gel-techniques. *Zeitschrift für Naturforschung, 28b:* 722–732.

STEGEMANN, H. & LOESCHCKE, V., 1976. Index Europäischer Kartoffelsorten/Index of European Potato Varieties (bilingual), based on electrophoretic spectra. *Mitteilungen der Biologischen Bundesanstalt für Land- u. Forstwirtschaft, 168:* 1–215.

STEGEMANN, H. & LOESCHCKE, V., 1977. Das Europäische Kartoffelsortiment und seine Indexierung. *Potato Research, 20:* 101–110.

STEGEMANN, H., LOESCHCKE, V., BODE, O. & HUTH, H., 1970. Gel-elektrophoretische Untersuchungen von Kartoffelknollen nach Aufzucht aus Meristem-Kultur. *Jahresbericht der Biologischen Bundesanstalt für Land- und Forstwirtschaft in Braunschweig,* A 65 (1970), P 62 (1971).

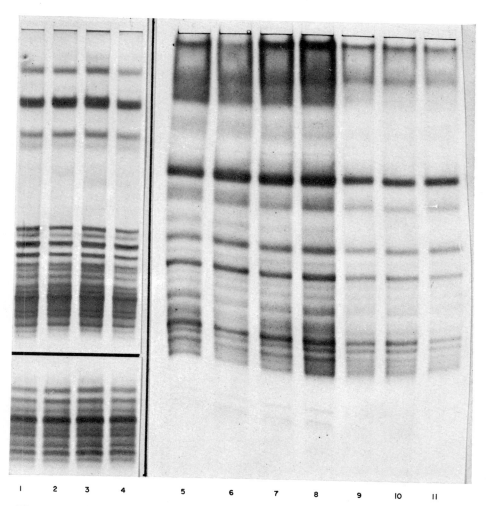

Figure 21.1. Pattern identity of potato tuber proteins after different growth conditions.

Left: (Columns 1–4) cv. Grata grown in pots with 0 to 133 % KCl, and 133 to 0 % K_2SO_4. Proteins (top), esterases (bottom). PAGE at pH 7.9 in Tris/borate and 5 % Cyanogum. (In cooperation with Dr Karl Müller, Universität, Göttingen).

Right: cv. Feldeslohn, plants treated with growth regulators; Hedonal (column 6) MH 30 (7), NCl (8, 10), doubled NCl (11), control (5,9). Field (5–8) and pot (9–11) trials. PAGE at pH 8.9. (In cooperation with Dr C. Pätzold, Forsch. Anst. Landw., Braunschweig).

(*Facing p. 284*)

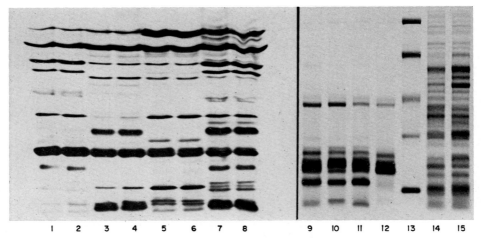

Figure 21.2. Charge (left) and size (right) distribution of tuber proteins from different cultivars. *Left:* cv. Maritta (columns 1, 2), Voran (3, 4), Saturna (5, 6) and Bintje (7, 8) after 3 h focusing in a gradient pH 6 (top) to pH 10 (bottom) from 1 % Servalyt, 400–1000 volt.

Right: cv. Maritta (9), Sieglinde (10), Voran (11), Voran not reduced by mercaptoethanol (12), marker proteins (13) with 100; 67; 37; 25.7 and 14.3×10^3 dalton. Maize proteins from dry, defatted seeds of 2 cultivars (14, 15) for comparison. SDS-PAGE at pH 7.1, Tris/borate buffer.

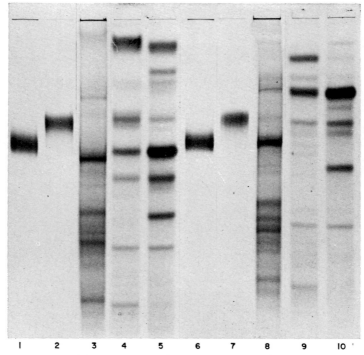

Figure 21.3. Similarity of fraction-1-protein from young (columns 1, 6) and older (2, 7) leaves, and proteins from immature tubers cv. Grata (3, 8), with respect to the migration rate at pH 8.9 (1–5) and at pH 7.9 (6–10). Comparison with proteins from mature tubers cv. Grata (4, 9) and cv. Voran (5, 10). PAGE in buffer Tris/borate.

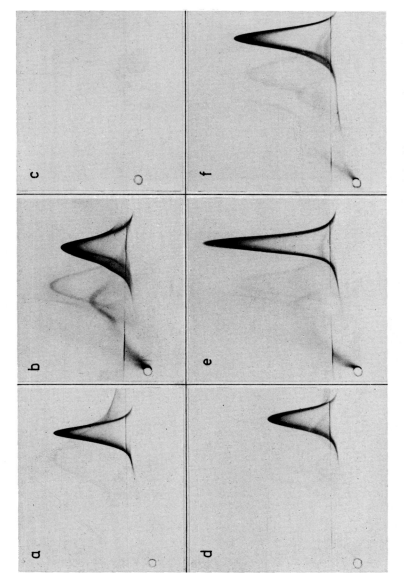

Figure 21.4. Immuno-identity and disappearance of the proteins characteristic of immature tubers with maturity.

Left to right: electrophoresis of an isolated protein fraction (a), and total sap (b) from immature tubers of cv. Grata, total sap of mature Grata (c), sap from immature tubers of Sieglinde (d), Hansa (e) and Maritta (f) in 1 % agarose, in Tris/borate buffer pH 8.9; anode at the right.

Bottom to top: electrodiffusion in the same system, containing antiserum (rabbit) against sample (a). Anode at the top.

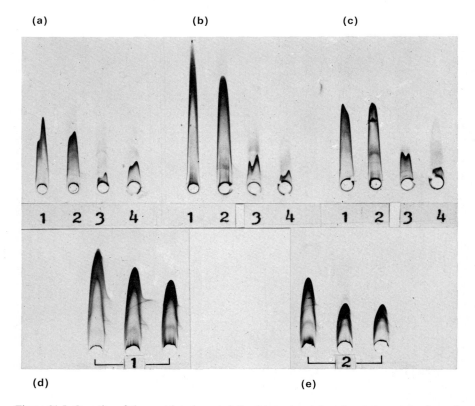

Figure 21.5. Quantity of the proteins characteristic of immature tubers in relation to the date of harvest and tuber size. Rocket technique with antiserum as in Fig. 21.4.

Dates of harvest: (1) 24.6, (2) 6.7, (3) 19.7, (4) 2.8.1975. Cultivars: (b) Maritta, (c) Sieglinde, (a), (d), (e) Grata. Tuber size in (d) and (e), 2, 11, 16 and 6, 18, 33 g respectively.

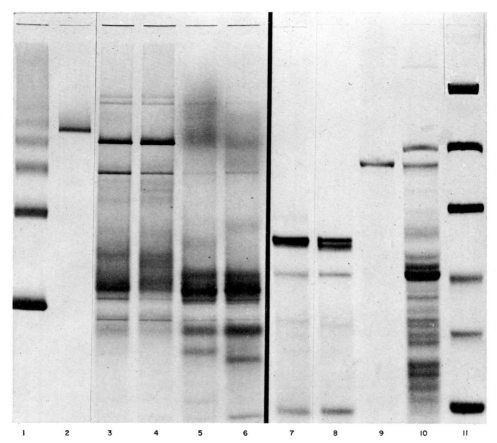

Figure 21.6. Size distribution of intact (left) and monomeric (right) proteins as revealed in porosity gradient PAGE and SDS-PAGE, respectively.

Bovine albumin marker in multiples of 67,000 daltons (column 1), fraction-1-protein from older leaves of cv. Grata (2), proteins from immature tubers of cv. Sieglinde (3) and of cv. Maritta (4), and proteins from mature tubers of the same cultivars (5, 6). Poro-PAGE in 6–26% Cyanogum and Tris/borate pH 8.9.

Monomers from fraction-1-protein of young (7) and older (8) leaves of cv. Grata, and from immature tubers of cv. Grata, isolated fractions (9) and SDS-treated sap (10). MW-markers: 160; 100; 67; 37; 25.7 and 14.3×10^3 dalton (11). SDS-PAGE in 5% Cyanogum and 0.025 M Na-phosphate with 0.1% SDS at pH 7.1.

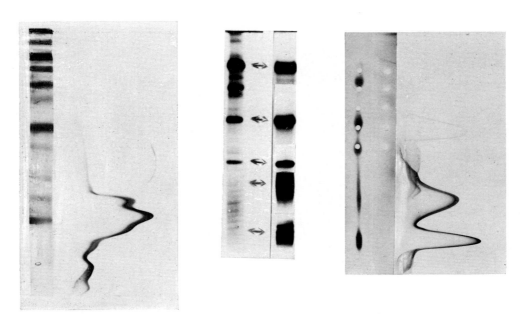

Figure 21.7. Immuno- precipitation of protein fractions from mature tubers cv. Voran.
Left: first dimension (top to bottom): charge separation by focusing in 1 % Servalyt and 4 % Cyanogum, gradient pH 4 (bottom) to pH 6. Second dimension: electrodiffusion into 1 % agarose (left to right) containing rabbit antiserum against all of the tuber proteins. Buffer 0.125 M Tris/borate pH 8.9, anode right.

Centre: separation in standard analytical PAGE (left) and fractions No. 36, 23, 18, 17 and 14 (top to bottom, right side) in the same system after preparative PAGE in 5 % Cyanogum and Tris/borate pH 8.9, anode at the bottom.

Right: first dimension (top to bottom): corresponding fractions after electrophoresis in 1 % agarose at pH 8.9 applied with the tandem-technique (left column), anode at the bottom. Second dimension: electrodiffusion as above.

22. The use of protein characters in the taxonomy of *Solanum* and other *Solanaceae*

RICHARD N. LESTER

Department of Plant Biology, University of Birmingham, U.K.

The reactions of rabbit antisera to seed extracts of eight diverse species of *Solanum*, absorbed with and then reacted with extracts from each of the eight species, are presented. The absorption experiments are described using set theory, and similarity coefficients based on negative and on positive results are defined. The eight *Solanum* species were shown to be serologically distinct, but their affinities agreed fairly well with the division of the genus into two halves, *Solanum* and *Stellatipilum*. Zone- and immuno-electrophoresis of species in section *Androceras* provided useful information on their relationships. Some other papers on serology of the Solanaceae are reviewed.

CONTENTS

INTRODUCTION

For almost a century, serology, the study of the sera or juices of animals and plants, was an uneasy combination of art and science, and even the science was largely empirical (Smith, 1976). However, since the middle of the twentieth century, knowledge of the structure and synthesis of proteins has advanced immensely, and procedures for the isolation, identification and total structural analysis of proteins from diverse organisms are now in common use. Even complex and obscure immunological phenomena such as the synthesis of antibody molecules and their reactions with antigenic determinants are being elucidated. This new knowledge is providing a sound theoretical basis for comparative serology.

The use of amino acid sequences of homologous proteins, such as cytochrome c, to assess the relationships of diverse organisms is well known. Boulter *et al.* (1972) have shown that the cytochrome c of tomato (Solanaceae) is most similar to that of sesame (Pedaliaceae) and castor (Euphorbiaceae), but has different amino acids at several positions. Provisional data on the amino acid sequence of plastocyanin from *Solanum tuberosum* show that it is more like that of *Lycopersicon esculentum* than *Solanum crispum* (Boulter, pers. comm.).

285

The separation and partial characterization of proteins from plants, animals, or other organisms by zone electrophoresis is often valuable in biosystematic studies. The resolving power of this method has been elegantly demonstrated in the characterization of cultivars of potato by Stegemann (this Volume, p. 279) For taxonomic purposes proteins with identical electrophoretic mobilities are deemed to represent the same unit character. This method is useful for comparing closely related species, but comparisons become meaningless at generic or higher levels.

PROTEIN CHARACTERS OF THE SOLANACEAE

Over fifty years ago Green (1926) showed serological similarity between *Solanum tuberosum* and *Lycopersicon esculentum*, and also within other groups of species (*Citrus* spp., *Prunus* spp., and *Pyrus* spp.) which are easily grafted together, using seed proteins in ring precipitation tests. The first major investigation of the Solanaceae was by Chester (1937) using leaf proteins of ten commonly cultivated genera. By the Schulz-Dale anaphyllaxis method he showed cross reactions between *Capsicum*, *Datura*, *Hyoscyamus*, *Lycopersicon*, *Nicandra*, *Nicotiana*, *Petunia*, *Physalis*, *Schizanthus* and *Solanum*. Sequential treatment of the uterine horns with first a cross-reacting antigen system, then the reference system, enabled some differentiation between genera to be made, and indicated that globulins were more species-specific than albumins. The next major study was by Hammond (1947, 1955) using the semi-quantitative methods of Boyden on the globulin fractions of seed extracts. Although many rabbits failed to produce good antisera, and some species such as *Solanum dulcamara* produced gross antisystematic reactions (Hammond, 1947), his results with three good antisera indicated the similarity to *Lycopersicon esculentum* of *L. pimpinellifolium*, to *Nicotiana* of *Petunia*, and to *Hyoscyamus* of *Physalis* and *Petunia* (Hammond, 1955).

Extensive studies of seed extracts were made by Tucker (Hawkes & Tucker, 1968; Tucker, 1969) using immunodiffusion techniques with antisera to *Atropa*, *Capsicum*, *Datura*, *Hyoscyamus*, *Nicotiana*, *Salpiglossis* and *Solanum*. He found fairly strong cross reactions between nearly all members of the Solanaceae, but weak reactions with *Schizanthus* and the *Scrophulariaceae*. Strong spurs were produced by each reference reaction against all cross reactions, but antiserum to *Capsicum* was anomalous in that arcs of some cross-reacting antigen systems (e.g. *Salpichroa*) crossed the arcs of the reference reaction. Attempts were made to assess the overall relationships of 15 genera, but with little success: the assumption that the genera producing strongest precipitates form the core of the Solanaceae is open to question. Absorption of the antisera allowed further discrimination. Thus antiserum to *Nicotiana* absorbed by extract of *Solanum* still reacted with *Datura*, *Hyoscyamus* and *Salpiglossis* indicating that they were more like *Nicotiana* than *Solanum*. Antiserum to *Hyoscyamus* absorbed by *Atropa* did not react with *Capsicum* and *Withania*; this was said to show the affinity of these two genera to *Atropa*, but really it shows that *Capsicum* and *Withania* are no more like *Hyoscyamus* than *Atropa* is. Analyses with antisera to species from several sections of *Solanum* showed that the diversity within this genus is as great as that between many other genera in the Solanaceae. Furthermore, species of *Lycopersicon* and *Cyphomandra* were shown to be relatively similar to *Solanum tuberosum*. In this and some other cases Tucker's work supported the classification of Bentham and Hooker rather than that of Wettstein.

Many studies have been made of potato proteins using immunodiffusion, immunoelectrophoresis and absorption. The classic work of Gell, Hawkes & Wright (1960), continued by Lester (Lester, 1965; Hawkes & Lester, 1966, 1968) provided much useful information, especially on the relationships of various Mexican wild diploid potatoes. The hybrid origin of *Solanum sambucinum* from *S. pinnatisectum* × *S. cardiophyllum*, which was postulated by Hawkes & Lester (1968) has been proved recently by Krasheninnik & Gavrilyuk (1973). Astley (1975) showed great similarities between the antigenic proteins of *S. oplocense*, *S. sucrense* and *S. tuberosum* subsp. *andigena*; the slight differences between individual accessions did not shed light on the evolutionary relationships of these species.

Many attempts to distinguish *Solanum sparsipilum*, *S. stenotomum* and several other diploid potatoes in Series Tuberosa by electrophoresis have been frustrated by the great variability of patterns of the tuber proteins (Desborough & Peloquin, 1967; Cribb, 1972; Astley, 1975; Huamán, 1975; Schmiediche, 1977). This genetic diversity, and probably heterozygosity, is not unexpected in a group of species most of which are outbreeding and the interspecific hybrids of which are generally, highly fertile. Only the clonally propagated potatoes such as cultivars or other single genotypes show uniformity (Stegemann, this Volume, p. 279). This contrasts strongly with the situation in most other *Solanum* species which are highly self-fertile but with strong interspecific barriers such as *S. melongena* (*sensu lato*) and the *S. aethiopicum* complex, in which even morphologically diverse plants produce similar patterns of proteins (Pearce & Lester, this Volume, p. 615). This difference may be one cause of the different species concepts applicable in the different sections of *Solanum*.

Studies of *Lycopersicon* species are treated fully elsewhere in this book by Rick, who has previously described the genetic control of peroxidase isozymes in *L. esculentum*, *L. e.* var. *cerasiforme* and *L. pimpinellifolium* (Rick, Zoebel & Fobes, 1974), and of four enzyme systems in *L. cheesmanii* which showed a wide range of variation between 54 populations from diverse parts of the Galapagos archipelago, but great uniformity within each population (Rick & Fobes, 1975).

Electrophoresis of seed proteins from several species of *Lycopersicon* indicate relationships similar to those discerned by cytogenetic and other methods (West, 1973; Kostova, Khristova & Georgieva, 1976). Leaf proteins were found by West (1973) to be very similar in *L. esculentum* and *L. pimpinellifolium*, and even in *Solanum pennellii* but distinct from *L. peruvianum* and *L. p.* var. *glandulosum* which were similar to each other. Immunoelectrophoretic analysis of seed proteins showed similarities between *L. esculentum* and *L. peruvianum* var. *glandulosum* on the one hand, and *Solanum tuberosum* and *S. brevidens* on the other. *S. ochranthum* appeared to be more like the *Lycopersicon* group than the *Solanum* group.

Martin (1970) found remarkable uniformity in the seed proteins of ten species of *Capsicum* by immunoelectrophoresis with antisera to *C. frutescens*, *C. baccatum* var. *pendulum*, *C. chacoense* and *C. pubescens*, but could find no reliable differences between any of them.

Electrophoresis of seed proteins from 20 species of the *Solanum nigrum* complex produced species specific patterns from morphologically and genetically distinct species, but greater variation was found in species with infraspecific morphological variation and/or incomplete genetic isolation (Edmonds & Glidewell, 1977 and Edmonds, this Volume, p. 529.).

Nicotiana is one of the most intensively studied genera of the Solanaceae, and many attempts have been made to prove the origins of the two amphiploids *N. tabacum* and *N. rustica*. Electrophoretic analysis of tetrazolium oxidase showed uniformity in five morphologically distinct cultivars of *N. tabacum*, but considerable differences between these and *N. glutinosa*, *N. rustica* and *N. sylvestris* (Harn, Cheong & Young, 1975). Different peroxidase isozymes are present in different tissues of *N. tabacum* and other species (Sheen & Rebagay, 1970). A comprehensive study of several isozymes by Sheen (1972) supported the ancestry of *N. tabacum* from the cross *N. sylvestris* × *N. tomentosiformis*, which was 86% similar, rather than from *N. sylvestris* × *N. otophora*, which was only 64% similar. Similar conclusions were obtained by measuring the amounts of nuclear DNA (Narayan & Rees, 1974).

Valuable information has been obtained by Kawashima, Tanabe & Iwai (1976) from tryptic digests of the large and the small subunits of Fraction I protein, which are coded by chloroplast DNA and nuclear DNA respectively. The 38 fractions from the large subunit of *Nicotiana sylvestris* were identical to those from *N. tabacum*, which agrees with it being the maternal progenitor. Of the 25 fractions from the small subunit of *N. tabacum*, 23 could be accounted for by *N. sylvestris*; the other two were shown by *N. tomentosiformis* but not by *N. otophora* or *N.* sp. K-12. Of the artificial amphidiploid hybrids only *N. sylvestris* × *N. tomentosiformis* had the same small subunits as *N. tabacum*, again proving the ancestry. Similar analyses of 30 species of *Nicotiana* showed variation in 13 fractions which enabled a chemical taxonomic classification of the genus to be made which was very similar to orthodox views (Kawashima *et al.*, 1974). Extensive serological studies of Fraction I proteins from 61 species of *Nicotiana* with antisera to three species (*N. glutinosa*, *N. tabacum* and *N. gossei*) have been reported by Gray (1977), using immunodiffusion with spur analysis. The results are considered to show clearly that *N. glauca* is different from other species in Section *Paniculatae* and that *N. sylvestris* is distinct from other species in Section *Alatae* but indistinguishable from *N. tabacum*. Unfortunately the basic data do not seem to be self-consistent, since no spur was formed by the reference reaction of *N. glutinosa* against *N. tabacum* antigen system indicating that *N. glutinosa* is not antigenically distinct from *N. tabacum*; likewise the reference reaction of *N. tabacum* did not produce a spur against *N. gossei*, which is surprising.

Natarella & Sink (1975) used electrophoresis to characterize leaf proteins and peroxidase enzymes from four species of *Petunia*, and 11 cultivars of *P. hybrida*. No two taxa were identical. *P. axillaris* appeared to be one of the parents of *P. hybrida*, and *P. inflata* rather than *P. violacea* the other. Several studies have been made of peroxidase isozymes from leaves of *Datura*. Ferri & Guzman (1970) found identical bands in *D. metel*, *D. innoxia* and *D. stramonium*, but one band less in *D. ferox*. Conklin & Smith (1971) found however that these and six other species all had different patterns, which they used together with cytogenetic data to divide the genus into five groups. Different tissues of the same plant have different peroxidases (Guzman, Ferri & Trippi, 1971; Scogin, 1975).

PRINCIPLES OF IMMUNOLOGY

The use of immunological techniques to assess the relationships between a wide range of organisms has been shown to be of value by many workers (Fairbrothers, 1969; Smith, 1976). This approach, variously called immunotaxonomy, sero-taxonomy, comparative serology, etc., relies on the immune response of higher vertebrate animals to provide antisera or antibody systems capable of reacting with the antigen systems of the sera or protein extracts of the organisms being studied to produce a precipitate. Both the immune response and the precipitin reaction are very complicated and have been poorly understood until recently. A good knowledge of immunology is necessary for correct experimental design and interpretation of results and there are many pitfalls for the ignorant.

To a certain extent immunological techniques can be used to recognize antigenic proteins and thus define them. The combined technique of immuno-electro-phoresis has been used extensively on species of *Phaseolus* and *Vigna* by Kloz and his co-workers (Kloz, 1971), and was found to be particularly valuable in bio-systematic studies of wild potato species (Hawkes & Lester, 1968), since it permitted the immunological recognition of proteins which were probably homologous, despite significant differences in their electrophoretic mobility. However although this method resolves the major antigenic proteins in the extract, it does not detect which antigenic determinant sites are specific and which are common to two or more taxa.

The basis of comparative serology is the production of an antibody system, (Abs) in a host animal by immunization with the extract of the organism under investigation (i.e. antigen system, Ags), and the subsequent analysis of the reactivity of the antibody system with each cross reacting (or heterologous) antigen system from further organisms in comparison with its reactivity with the reference (or homologous) antigen system originally used for immunization.

Although there are many technical problems yet to be solved in observing and comparing cross reactions and reference reactions, even more essential is a correct understanding of, and a precise description of, what is involved. Major contributions have been made by Moritz & Jensen (1961), Kirsch (1967), and Moore & Goodman (1968). The present account is based on these, and the reader is referred back to them for fuller and more precise descriptions.

The reference antigen system of an organism (A) consists of several proteins (a_1, a_2, a_3, etc.). The configurations of amino acids on the surfaces of these proteins present antigenic determinant sites (α_1, α_2, α_3, etc.) and the sum of these constitute the set of antigenic determinants of antigen system A. Immunization results in an antibody system (symbolized as A′ by Moritz & Jensen, and modified to A^{ab} by Kirsch) consisting of a set of antibodies (α_1^{ab}, α_2^{ab}, α_3^{ab}, etc.). The reference reaction measures the extent to which the set of antigenic determinants of A maps onto the set of antibodies of A^{ab}. The cross reactions of antigen systems B, C, D, etc. measures the extent to which their respective sets of antigenic determinants map onto the set of antibodies A^{ab}. There are several techniques which compare the mapping of cross-reacting and reference antigenic determinants on to the complete reference set of antibodies (Butler & Leone, 1968).

In some ways the ideal situation in immunotaxonomy would be to have a series of antibody systems, each containing a single species of antibody (e.g. α_1^{ab}) which could be used to test for the presence or absence of the corresponding antigenic determinants in the taxa being compared. A practical approximation to this is the absorption (or saturation, Moritz & Jensen, 1961) of antibody systems by different cross-reacting antigen systems. When the reference antibody system, A^{ab}, is absorbed fully by a cross-reacting antigen system, B, only those antibodies which are capable of reacting with antigenic determinant sites which are members of the set of A but are the complement of the set of B (i.e. $A \cap B'$) are left. The absorbed antibody system may be referred to as $A^{ab}_{A \cap B'}$ or $A^{ab}_{B'}$ (Kirsch, 1967; modified from Moritz & Jensen, 1961). The use of absorbed antibody systems, and the analysis of the results will be described in the present paper, together with results from immuno-electrophoresis and zone electrophoresis.

EXPERIMENTAL

Seeds of eight species representing eight sections of the genus *Solanum* were produced in the Birmingham University Solanaceae Collection (Table 22.1). Proteins were extracted from defatted seed flour using phosphate buffered saline (Lester, 1969). Antisera were produced by immunizing rabbits with proteins which

Table 22.1. List of species studied, the section to which they belong and the abbreviation, accession and antiserum used

Section	Species	Abbreviation	Accession	Antiserum
Tuberarium	*S. tuberosum* L.	TB	S.952	R4b/4
Solanum	*S. scabrum* Mill.	SC	S.243	R.28/2
	(=*S. nigrum* L. var. *guineense* L.)			
Brevantherum	*S. mauritianum* Scop.	MR	S.049	R.35/3
Pseudocapsicum	*S.* × *hendersonii* Hort.	HN	S.167	R.43
Jasminosolanum	*S. seaforthianum* Andr.	SF	S.051	R.45
Oliganthes	*S. olivare* Paill. + Boiss.	OL	S.279	R.47
Archaesolanum	*S. simile* F. Muell.	SM	S.211	R.51
Androceras	*S. citrullifolium* A.Br.	CT	S.168	R.54

had been further processed by ammonium sulphate precipitation and dialysis (Hawkes & Tucker, 1968). Immunodiffusion and immunoelectrophoresis experiments were made in Ionagar in a borate-phosphate buffer (Sangar, Lichtwardt, Kirsch & Lester, 1972). Absorption of antisera was achieved using the technique of Dray & Young (1959), loading the antigen system three times and subsequently the antibody system twice (Fig. 22.2). This technique facilitated the production of all 512 tests under uniform conditions, and provided a complete cubic $8 \times 8 \times 8$ matrix. The reactions were observed after 24 hrs, and again after washing and staining. The results were recorded as weak (+), moderate (+ +), and strong (+ + +) positives, and definite negatives (−). Uncertain positive-negatives were recorded as (±), and were given values of half a negative or half a positive reaction in subsequent calculations. Polyacrylamide gel electrophoresis was used following the standard procedure of Davis (1964).

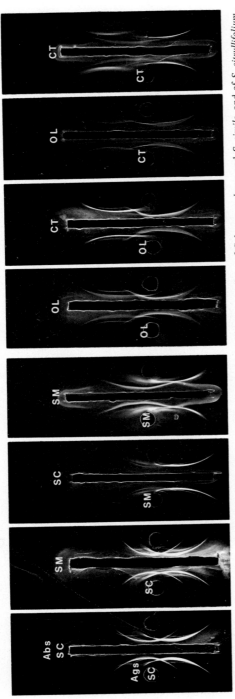

Figure 22.1. Immuno-electrophoretograms of the reference reaction and reciprocal cross reactions of *Solanum scabrum* and *S. simile*, and of *S. citrullifolium* and *S. olivare*.

Figure 22.2. Double diffusion with absorption. The centre wells were loaded several times with the absorbing antigen system before adding the antibody system there, and the test antigen systems to the peripheral wells. a, MR_{MR}^{ab} shows complete absorption and no reaction with any of the eight test antigen systems. b, and c, MR_{HD}^{ab} and MR_{SF}^{ab} show no reaction with their respective absorbing antigens, but positive reactions with MR and some of the other antigen systems.

(*Facing p. 291*)

RESULTS

Immuno-electrophoresis

The spectra produced by the reaction of all eight antigen systems against all eight antibody systems were assessed for the number of arcs in each spectrum and the disposition of the arcs. Attempts were made to analyse these data in several ways, but with little success.

In Table 22.2 the number of arcs produced by the antigen systems with each antibody system are set out. Most antisera produced about eight arcs in the reference reaction, and some as many as eleven. Unfortunately in some cases such as with TB[ab], SC[ab] and MR[ab] more arcs were produced by some cross reacting antigen systems than by the reference antigen system; this was apparently illogical and invalidated further analyses of these data. A further anomaly was that SC consistently produced more arcs than did MR with all antibody systems. Such apparently antisystematic reactions are found occasionally with plant extracts (Moritz, 1964) but only rarely has an explanation and solution been found (Lester, 1969). Obviously the present anomalies need further investigation.

Table 22.2. Number of arcs produced by immuno-electrophoresis of the eight antigen systems with the eight antibody systems

Ags	Abs							
	TB[ab]	SC[ab]	MR[ab]	HN[ab]	SF[ab]	OL[ab]	SM[ab]	CT[ab]
TB	10	7	4	5	5	6	9	8
SC	11	7	4	6	5	6	7	6
MR	6	5	4	4	4	5	6	5
HN	8	6	5	8	5	5	8	6
SF	10	6	5	5	8	7	4	6
OL	11	7	5	5	5	10	9	8
SM	9	8	4	3	5	6	9	7
CT	8	7	3	5	5	5	8	8

The disposition of the arcs in the reference reactions was found to be characteristic for each of the eight species, but similar patterns were sometimes produced in cross reactions with other antibody systems (Fig. 22.1).

Absorption

For any one antibody system, only the reference antigen system produced complete absorption and prevented any further reactions. In all other cases the absorbed antibody system produced no reaction with the cross reacting antigen system used for absorption (indicating that absorption was complete) but it produced a positive reaction with the reference antigen system. It produced various reactions with the several other cross reacting antigen systems (Fig. 22.2, Table 22.3).

Analysis of the results obtained by absorption experiments presents many problems, as has been discussed by Moritz & Jensen (1961), Moore & Goodman

Table 22.3. Reactions with absorbed antibody systems. (For explanation see text). A, anti-*S. tuberosum*; B, anti-*S. scabrum*; C, anti-*S. mauritianum*; D, anti-*S. hendersonii*; E, anti-*S. seaforthianum*; F, anti-*S. olivare*; G, anti-*S. simile*; H, anti-*S. citrullifolium*

(A) TB^{ab}

Test antigen system	Absorbing antigen system								$S_{no.pos.}$	$S_{ppt.pos.}$	$S_{abs.pos.}$
	TB	SC	MR	HN	SF	OL	SM	CT			
TB	−	++	+++	++±	++	+++	++	++	1.00	1.00	1.00
SC	−	−	++	+±	+	++	+	+	0.86	0.52	0.69
MR	−	−	−	±	+	+	+	+	0.64	0.33	0.49
HN	−	±	++	−	±	+	+	+	0.71	0.36	0.54
SF	−	−	+	±	−	++	++	+	0.64	0.39	0.52
OL	−	±	+±	+	+	−	++	+	0.79	0.42	0.61
SM	−	−	++	++	+	++	−	+	0.71	0.48	0.60
CT	−	+	+±	+	++	±	++	−	0.79	0.48	0.64
$S_{abs.neg.}$	1.00	0.63	0.13	0.25	0.19	0.19	0.13	0.13			
Coefficient of correlation with $S_{abs.neg.}$									0.65	0.49	0.59

(B) SC^{ab}

Test Ags	Absorbing Ags								$S_{no.pos.}$	$S_{ppt.pos.}$	$S_{abs.pos.}$
	TB	SC	MR	HN	SF	OL	SM	CT			
TB	−	−	++	+	±	+	−	−	0.50	0.25	0.38
SC	+++	−	+++	+++	++	+++	++	++	1.00	1.00	1.00
MR	±	−	−	±	±	+	−	−	0.36	0.14	0.25
HN	±	−	++	−	−	±	−	−	0.29	0.17	0.23
SF	±	−	++	±	−	+	−	±	0.50	0.25	0.38
OL	−	−	++	−	−	−	−	−	0.14	0.11	0.13
SM	±	−	++	+	±	++	−	±	0.64	0.36	0.50
CT	±	−	++	±	+	+	±	−	0.64	0.31	0.48
$S_{abs.neg.}$	0.56	1.00	0.13	0.44	0.56	0.19	0.81	0.75			
Coefficient of correlation with $S_{abs.neg.}$									0.87	0.96	0.91

(C) MR^{ab}

Test Ags	Absorbing Ags								$S_{no.pos.}$	$S_{ppt.pos.}$	$S_{abs.pos.}$
	TB	SC	MR	MR	SF	OL	SM	CT			
TB	−	+	−	±	+	−	±	−	0.43	0.21	0.32
SC	±	−	−	−	+	−	−	−	0.21	0.11	0.16
MR	++	++	−	++	++	++	++	++	1.00	1.00	1.00
HN	+	+	−	−	+	±	±	±	0.64	0.32	0.48
SF	−	±	−	−	−	−	−	±	0.14	0.07	0.11
OL	+	+	−	±	++	−	++	+	0.79	0.54	0.67
SM	±	±	−	−	±	±	−	±	0.36	0.18	0.27
CT	±	+	−	±	+	−	+	−	0.57	0.29	0.43
$S_{abs.neg.}$	0.44	0.25	1.00	0.69	0.19	0.75	0.50	0.56			
Coefficient of correlation with $S_{abs.neg.}$									0.97	0.92	0.96

(D) HN[ab]

Test Ags	Absorbing Ags								$S_{no.pos.}$	$S_{ppt.pos.}$	$S_{abs.pos.}$
	TB	SC	MR	HN	SF	OL	SM	CT			
TB	−	±	−	−	+	±	+	±	0.50	0.25	0.38
SC	−	−	−	−	+	±	±	−	0.29	0.14	0.22
MR	+	+	−	−	+	+	+	+ +	0.86	0.50	0.68
HN	+ +	+ +	+ +	−	+ +	+ +	+ +	+ +	1.00	1.00	1.00
SF	−	−	−	−	−	±	+	±	0.29	0.14	0.22
OL	+	+	±	−	+	−	+ +	+	0.79	0.46	0.63
SM	−	−	−	−	+	±	−	−	0.21	0.11	0.16
CT	±	+	−	−	+	±	+	−	0.57	0.29	0.43
$S_{abs.neg.}$	0.56	0.44	0.81	1.00	0.13	0.44	0.19	0.50			
Coefficient of correlation with $S_{abs.neg.}$									0.79	0.76	0.78

(E) SF[ab]

Test Ags	Absorbing Ags								$S_{no.pos.}$	$S_{ppt.pos.}$	$S_{abs.pos.}$
	TB	SC	MR	HN	SF	OL	SM	CT			
TB	−	−	+	±	−	±	−	−	0.33	0.29	0.31
SC	−	−	+	±	−	±	−	−	0.33	0.29	0.31
MR	−	−	−	−	−	±	−	−	0.08	0.07	0.08
HN	−	−	+	−	−	±	−	−	0.25	0.21	0.23
SF	+	+	+	+	−	+ +	±	+	1.00	1.00	1.00
OL	−	−	+	−	−	−	−	−	0.17	0.14	0.16
SM	−	−	±	±	−	+	−	±	0.42	0.36	0.39
CT	−	−	+	−	−	−	−	−	0.17	0.14	0.16
$S_{abs.neg.}$	0.88	0.88	0.19	0.69	1.00	0.50	0.94	0.81			
Coefficient of correlation with $S_{abs.neg.}$									0.86	0.85	0.86

(F) OL[ab]

Test Ags	Absorbing Ags								$S_{no.pos.}$	$S_{ppt.ops.}$	$S_{abs.pos.}$
	TB	SC	MR	HN	SF	OL	SM	CT			
TB	−	−	−	−	±	−	−	−	0.07	0.04	0.06
SC	−	−	−	−	−	−	−	−	0.00	0.00	0.00
MR	−	±	−	−	±	−	−	−	0.14	0.08	0.11
HN	−	±	±	−	−	−	−	−	0.14	0.08	0.11
SF	−	−	−	−	−	−	−	−	0.00	0.00	0.00
OL	+ +	+ + +	+	+ +	+ +	−	+ +	+	1.00	1.00	1.00
SM	−	±	−	±	±	−	−	−	0.21	0.12	0.17
CT	+	+ +	±	+	+	−	±	−	0.71	0.46	0.59
$S_{abs.neg.}$	0.75	0.56	0.75	0.69	0.56	1.00	0.81	0.88			
Coefficient of correlation with $S_{abs.neg.}$									0.80	0.78	0.79

(G) SM[ab]

Test Ags	Absorbing Ags								$S_{no.pos.}$	$S_{ppt.pos.}$	$S_{abs.pos.}$
	TB	SC	MR	HN	SF	OL	SM	CT			
TB	−	+	+ +	+ +	+	+ +	−	+	0.86	0.64	0.75
SC	+	−	+ +	+ +	+	+ +	−	+	0.86	0.64	0.75
MR	±	±	−	+	±	+	−	−	0.50	0.25	0.38
HN	±	±	+	−	−	±	−	±	0.43	.21	0.32
SF	+	−	+ +	+	−	+ +	−	+	0.71	0.50	0.61
OL	±	−	+	+	±	−	−	±	0.50	0.25	0.38
SM	+ +	+ +	+ +	+ + +	+ +	+ +	−	+	1.00	1.00	1.00
CT	+	+	+	+ +	+	+	−	−	0.86	0.50	0.68
$S_{abs.neg.}$	0.31	0.50	0.13	0.13	0.38	0.19	1.00	0.38			
Coefficient of correlation with $S_{abs.neg.}$									0.89	0.89	0.90

(H) CT^{ab}

Test Ags	Absorbing Ags								$S_{no.pos.}$	$S_{ppt.pos.}$	$S_{abs.pos.}$
	TB	SC	MR	HN	SF	OL	SM	CT			
TB	−	+	+	+	+	+	+	−	0.86	0.38	0.62
SC	−	−	±	+	+	±	−	−	0.43	0.19	0.31
MR	+	+	−	+	+	±	+	−	0.79	0.34	0.57
HN	−	±	+	−	±	±	±	−	0.43	0.19	0.31
SF	−	−	±	+	−	+	−	−	0.36	0.16	0.26
OL	±	+	+	+ +	+ +	−	+	−	0.79	0.47	0.63
SM	±	±	+	+ +	+	±	−	−	0.64	0.34	0.49
CT	+ +	+ +	+ +	+ + +	+ + +	+ +	+ +	−	1.00	1.00	1.00
$S_{abs.neg.}$	0.63	0.38	0.25	0.13	0.19	0.38	0.44	1.00			
Coefficient of correlation with $S_{abs.neg.}$									0.64	0.59	0.63

(1968), and Kirsch (1967). Theoretically a logical analysis might be possible, using set theory, and this possibility has been thoroughly explored by Kirsch. However in practice, such an analysis is fraught with problems, the main one being that the intersect of the set of determinants of antibody system A^{ab} with antigen system B is not identical to the intersect of B^{ab} with A, that is to say that the subset of antibody determinants induced by antigenic determinants of antigen system A and capable of reacting with the antigenic determinants of antigen system B may not be identical to the subset of antibody determinants induced by antigenic determinants of antigen system B which are capable of reacting with the antigenic determinants of antigen system A. This imperfect commutativity between A∩B and B∩A is partly due to variation between individual host animals in their immune responses, and tends to frustrate attempts to use pure logic to determine the degrees to which members are held in common between the sets of antigenic determinants of a series of organisms being compared.

Nevertheless, after many futile attempts to make sense of the present data, a method of analysis has been devised somewhat empirically, which is self consistent, which is in concordance with immunological principles, and which produces results which appear to be taxonomically acceptable.

If we consider the results obtained with any one antibody system, such as anti-*S. mauritianum* (MR^{ab}) (Table 22.3C), it can be seen that, although each of the cross reacting antigen systems used for absorption does not react with the antibody system absorbed by it, the reference antigen system still does react, showing that *S. mauritianum* is distinct from all othe other seven species. The use of all eight antibody systems shows in this way that all eight antigen systems being compared are distinct, but it does not show degrees of difference. The amount of precipitate produced by the reference antigen system with any absorbed antibody system might be taken as a measurement of the dissimilarity between the reference antigen system and the one used for absorption, but these reactions were generally large and uniform and therefore provided no discrimination.

The amount of precipitate produced by a cross reacting antigen system C with antibody system A^{ab} after absorption by antigen system B, compared with that produced by the reaction of B and A$^{ab}_C$ may indicate the relative affinities of the cross reacting antigen systems B and C to the reference antigen system A.

Thus MR$^{ab}_{HN}$, (anti *S. mauritianum* antibody system absorbed by *S. hendersonii* antigen system) produced no reaction with SC, SF or SM, indicating that HN is more like MR than are SC, SF and SM. Furthermore HN produced positive reactions with the absorbed antibody systems $MR_{SC'}^{ab}$, $MR_{SF'}^{ab}$ and $MR_{SM'}^{ab}$, confirming that HN is more like MR than are SC, SF or SM. Likewise, absorbed antiserum to *S. simile*, $SM_{SF'}^{ab}$, produced no reaction with HN, and SF produced a reaction with the absorbed antibody system $SM_{HN'}^{ab}$. If all the results were as consistent as these it would be easy to produce a logical scheme of relationships. Unfortunately this is not so, and all attempts to employ simple logic end in frustration.

However, at least in a complete cubic matrix of absorption experiments in which antibody systems to all taxa are absorbed by antigen systems of all taxa, and then tested with antigen systems of all taxa, further statistical analysis is possible using the relative numbers of negative reactions or positive reactions in the appropriate columns or rows of the matrix.

Each negative reaction indicates that the absorbing antigen system is more like the reference system than is the test antigen system. Thus the affinity of B to A is indicated by the degree to which antigen system B absorbs antibody system A^{ab} and removes from it antibodies capable of reacting with the antigen systems of other taxa, e.g. C, D, etc. Hence the number of negative reactions produced by the absorbed antibody system A_B^{ab} to a range of test antigen systems may be taken as an indication of the similarity of B to A, and the Absorption Similarity Coefficient based on Negative Reactions ($S_{abs.\ neg.}$) may be defined as the number of negative reactions produced by the absorbed antibody system with a range of test antigen systems, divided by the number of negative reactions produced by the reference antibody system absorbed by the reference antigen system to the same test antigen systems.

Using (somewhat loosely) the formulary of Moritz & Jensen (1961) with modifications by Kirsch (1967) it may be stated that $S_{abs.\ neg.}$ for $A^{ab} \cap B =$

$$\frac{(A + A_{B'}^{ab}\text{---}:\text{non}) + (B + A_{B'}^{ab}\text{---}:\text{non}) + (C + A_{B'}^{ab}\text{---}:\text{non}) + \text{etc.}}{(A + A_{A'}^{ab}\text{---}:\text{non}) + (B + A_{A'}^{ab}\text{---}:\text{non}) + (C + A_{A'}^{ab}\text{---}:\text{non}) + \text{etc.}}$$

Thus the absorption experiments using MR^{ab} (Table 22.3C), show that absorption by MR results in negative reactions with all eight test antigen systems, and so $S_{abs.\ neg.} = 1.00$. Absorption with OL caused 6 negative reactions and two positive ones ($S_{abs.\ neg.} = 0.75$), HN gave $5\frac{1}{2}$ ($S_{abs.\ neg.} = 0.69$), for CT $S_{abs.\ neg.} = 0.56$, etc. These analyses provide a ranking order of the degree of similarity to the reference antigen system of each antigen system used for absorption. Thus for the eight antibody systems the ranking orders of the antigen systems used for absorption are as follows:

Abs	*Ranking order of Ags*
TB^{ab}	TB, SC, HN, SF + OL, MR + SM + CT
SC^{ab}	SC, SM, CT, TB + SF, HN, OL, MR
MR^{ab}	MR, OL, HN, CT, SM, TB, SC, SF
HN^{ab}	HN, MR, TB, CT, SC + OL, SM, SF
SF^{ab}	SF, SM, TB + SC, CT, HN, OL, MR
OL^{ab}	OL, CT, SM, MR + TB, HN, SC + SF
SM^{ab}	SM, SC, CT + SF, TB, OL, HN, MR
CT^{ab}	CT, TB, SM, OL + SC, MR, SF, HN

Some of the indicated similarities are both reciprocal and taxonomically feasible such as SC and SM, but others such as CT and TB are definitely not. Inspection of the values of $S_{abs. neg.}$ in Table 22.4 shows that this matrix is very asymmetrical, and further analysis to produce a phenogram may not be valid. It is noticeable that some antibody systems such as SF^{ab} and OL^{ab} have consistently high values for $S_{abs. neg.}$, whereas others such as TB^{ab} and SM^{ab} have low values. If these values were standardized mathematically, they might perhaps produce a symmetrical matrix suitable for further analysis.

Table 22.4. Values for $S_{abs. neg.}$, the Absorption Similarity Coefficient based on Negative Reactions

Antibody system	Absorbing antigen system							
	TB	SC	MR	HN	SF	OL	SM	CT
TA^{ab}	1.00	0.63	0.13	0.25	0.19	0.19	0.13	0.13
SC^{ab}	0.56	1.00	0.13	0.44	0.56	0.19	0.81	0.75
MR^{ab}	0.44	0.25	1.00	0.69	0.19	0.75	0.50	0.56
HN^{ab}	0.56	0.44	0.81	1.00	0.13	0.44	0.19	0.50
SF^{ab}	0.88	0.88	0.19	0.69	1.00	0.50	0.94	0.81
OL^{ab}	0.75	0.56	0.75	0.69	0.56	1.00	0.81	0.88
SM^{ab}	0.31	0.50	0.13	0.13	0.38	0.19	1.00	0.38
CT^{ab}	0.63	0.38	0.25	0.13	0.19	0.38	0.44	1.00

An alternative analysis is based on the production of positive reactions. If a positive reaction is produced, it indicates that some antibodies of A^{ab} which are capable of reacting with antigenic determinants of test antigen system C have been left despite the previous availability for reaction of the antigenic determinants of B which was used for absorption. A positive reaction indicates the existence, and to some extent the size, of $C \cap (A \cap B')$. The similarity of C to A may be assessed both by the number of absorbed antibody systems $A_{B'}^{ab}$, $A_{D'}^{ab}$, $A_{E'}^{ab}$, etc. which still react positively with C, and also by the sum of the amounts of precipitate produced in all these reactions.

The Absorption Similarity Coefficient based on the Numbers of Positive Reactions ($S_{no. pos.}$) may be defined as the number of positive reactions produced by the test antigen system with a range of absorbed antibody systems, divided by the number of positive reactions produced by the reference antigen system with the same absorbed antibody systems. This may be expressed as

$$S_{no. pos.} \text{ for } A^{ab} \cap B = \frac{(B + A_{A'}^{ab}\text{---}: sic) + (B + A_{B'}^{ab}\text{---}: sic) + (B + A_{C'}^{ab}\text{---}: sic) + etc.}{(A + A_{A'}^{ab}\text{---}: sic) + (A + A_{B'}^{ab}\text{---}: sic) + (A + A_{C'}^{ab}\text{---}: sic) + etc.}$$

Likewise the Absorption Similarity Coefficient based on the Amount of Precipitate ($S_{ppt. pos.}$) may be defined as the total precipitate score of the reactions produced by the test antigen system with a range of absorbed antibody systems, divided by the total precipitate score of the reactions produced by the reference antigen system with the same absorbed antibody systems. This may be expressed as

$$S_{ppt. pos.} \text{ for } A^{ab} \cap B = \frac{^TBA_{A'}^{ab} + {}^TBA_{B'}^{ab} + {}^TBA_{C'}^{ab} + etc.}{^TAA_{A'}^{ab} + {}^TAA_{B'}^{ab} + {}^TAA_{C'}^{ab} + etc.}$$

The two Absorption Similarity Coefficients based on positive reactions, $S_{no.\ pos.}$ and $S_{ppt.\ pos.}$, indicated similar ranking orders, although they produced very different values (Table 22.3A–G). Since both these coefficients are merely different ways of quantifying the same attributes, their mean has been taken as the Absorption Similarity Coefficient based on Positive Reactions ($S_{abs.\ pos.}$), (Table 22.5).

Table 22.5. Values for $S_{abs.pos.}$, the Absorption Similarity Coefficient based on Positive Reactions

Test antigen system	Antibody system							
	TB^{ab}	SC^{ab}	MR^{ab}	HN^{ab}	SF^{ab}	OL^{ab}	SM^{ab}	CT^{ab}
TB	1.00	0.37	0.32	0.38	0.31	0.05	0.75	0.62
SC	0.69	1.00	0.16	0.22	0.31	0.00	0.75	0.31
MR	0.49	0.25	1.00	0.68	0.08	0.11	0.38	0.57
HN	0.54	0.23	0.48	1.00	0.23	0.11	0.32	0.31
SF	0.52	0.37	0.11	0.22	1.00	0.00	0.61	0.26
OL	0.61	0.13	0.67	0.63	0.16	1.00	0.38	0.63
SM	0.60	0.50	0.27	0.16	0.39	0.17	1.00	0.49
CT	0.64	0.48	0.43	0.43	0.16	0.59	0.68	1.00

For example in absorption experiments using MR^{ab} (Table 22.3C), MR produced a positive reaction with all seven preparations of MR^{ab} absorbed with cross reacting antigen systems, and had a total precipitate score of 14. Antigen system OL scored $5\frac{1}{2}$ positive reactions and a total precipitate score of $7\frac{1}{2}$. Hence for MR$^{ab} \cap$OL, $S_{no.\ pos.} = 0.79$, $S_{ppt.\ pos.} = 0.54$, and hence $S_{abs.\ pos.} = 0.67$.

The ranking orders for the reactivity of the various test antigen systems with all the absorbed preparations of each of the antibody systems are as follows:

Abs	Ranking order of Ags
TB^{ab}	TB, SC, CT, OL, SM, HN, SF, MR
SC^{ab}	SC, SM + CT, TB + SF, MR, HN, OL
MR^{ab}	MR, OL, HN, CT, TB, SM, SC, SF
HN^{ab}	HN, MR, OL, CT, TB, SF + SC, SM
SF^{ab}	SF, SM, TB + SC, HN, OL + CT, MR
OL^{ab}	OL, CT, SM, MR + HN, TB, SC + SF
SM^{ab}	SM, TB + SC, CT, SF, OL + MR, HN
CT^{ab}	CT, OL + TB, MR, SM, SC + HN, SF

The similarities of SM and SC, and of CT and OL, are both reciprocal and taxonomically possible, but others such as TB and CT are taxonomically unlikely.

The matrix of values of $S_{abs.\ pos.}$ (Table 22.5) is very asymmetrical, and the values for TB^{ab} and SM^{ab} are much larger than those for SF^{ab} and OL^{ab}. This is the converse of the situation shown by $S_{abs.\ neg.}$, as might be expected. These discrepancies between antisera are largely due to differences in the immune responses of different rabbits, such that the numbers and kinds of antibodies in antibody system A^{ab} which will react with antigen system B are not identical to those in antibody system B^{ab} which will react with antigen system A, thus A$\cap$B does not map with a one-to-one correspondence on to B$\cap$A.

Although the analysis of negative reactions and the analysis of positive reactions did not give uniform or symmetrical matrices (Tables 22.4 and 5), they did give similar ranking orders for each antibody system. This is shown by the correlation coefficients of $S_{abs. neg.}$ with $S_{no. pos.}$, $S_{ppt. pos.}$ and $S_{abs. pos.}$ given in Tables 22.3A–H. The reference reactions were excluded from the calculations since they were defined as unity, and were the basis for calculating the cross reactions. Most of the correlation coefficients were greater than 0.75 ($P < 5\%$).

For each antibody system the corresponding values of $S_{abs. neg.}$ and $S_{abs. pos.}$ were averaged to give the Mean Absorption Similarity Coefficient ($S_{abs. mean.}$). As can be seen in Table 22.6 this provided a much more uniform matrix which was relatively symmetrical in that the ranking order of the various antigen systems with any one antibody system corresponded closely to the ranking order of the various antibody systems with the appropriate antigen system. Thus the matrix was much more self consistent.

Table 22.6. Values for $S_{abs. mean}$, the average of the values for the Absorption Similarity Coefficients based on Negative and on Positive Reactions, for the antigen systems with each antibody system

Antigen system	Antibody system							
	TB^{ab}	SC^{ab}	MR^{ab}	HN^{ab}	SF^{ab}	OL^{ab}	SM^{ab}	CT^{ab}
TB	1.00	0.47	0.38	0.47	0.60	0.40	0.53	0.63
SC	0.66	1.00	0.21	0.33	0.60	0.28	0.63	0.35
MR	0.31	0.19	1.00	0.75	0.14	0.43	0.26	0.41
HN	0.40	0.33	0.59	1.00	0.46	0.40	0.23	0.22
SF	0.36	0.47	0.15	0.18	1.00	0.28	0.50	0.23
OL	0.40	0.16	0.71	0.54	0.33	1.00	0.29	0.51
SM	0.37	0.66	0.39	0.18	0.67	0.49	1.00	0.47
CT	0.39	0.62	0.50	0.47	0.49	0.74	0.53	1.00

An alternative way of producing this matrix is to add the numbers of negative or positive reactions in the raw data without calculating $S_{no. neg.}$ ($= S_{abs. neg.}$) and $S_{no. pos.}$, i.e. out of all the reactions produced with any one antibody system, the number of negative reactions caused by an absorbing antigen system is added to the number of positive reactions produced by it when it is used as a test antigen system with the whole range of absorbed antigen systems, and then divided by the corresponding total of negative and positive reactions for the reference antigen system. This procedure omits any assessments of the amounts of precipitate, and it is much simpler than the full procedure used in this paper, yet it produces a similar matrix of values for $S_{abs. mean}$.

Further analysis of the matrix was made by averaging mutual values to produce a half matrix, and then clustering using the unweighted pair-group method using arithmetic averages (UPGMA or group average method, Sneath & Sokal, 1973) to produce a phenogram (Fig. 22.3).

This complete analysis resulting in a phenogram of all taxa against which antibody systems were raised is only possible with a cubical matrix of results produced by absorbing each antibody system in turn with each antigen system and then testing the products with each antigen system, in all permutations. However the use of the phenogram can be extended to include other taxa against which antibody systems are not available. For each new taxon an additional row of values of $S_{abs.\ mean}$ can be added to Table 22.6, and the highest of these values can be used to place the taxon on the appropriate arm of the phenogram (e.g. Roberts, 1977).

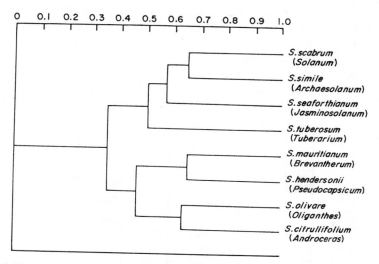

Figure 22.3. Phenogram of similarity shown by absorbed antisera for eight species from different sections of *Solanum*.

Analysis of spur and absorption data should be possible by computer, but an appropriate similarity coefficient must be used. The Simple Matching Coefficient, or something very similar, was used by Cristofolini & Chiapella (1977 and pers. comm.), but this treats all negatives and positives as equally important. For each pair of otu's (antigen systems) the similarities in rows (of antigen systems being used for absorption) and the similarities in columns (of antigen systems being used as test reagents) were compared, and all negative matches and positive matches were taken to indicate similarity. However, as discussed earlier, only the negative matches in the columns and the positive matches in the rows should be used to estimate similarity in absorption experiments, by using Jaccard's or a comparable similarity coefficient.

Zone- and immuno-electrophoresis of section Androceras (*Roberts, 1977*)

Polyacrylamide gel electrophoresis (PAGE) produced characteristically distinct patterns of protein bands for *Solanum rostratum*, *S. citrullifolium*, *S. fructo-tecto* and *S. heterodoxum* (Fig. 22.4a, c, e, g). Several bands were common to two or more species. Their identity was proved by electrophoresis of mixed extracts (Fig. 22.4b, d, f), and they were used to assess the relationships of the species. *S. rostratum* had four bands in common with *S. heterodoxum*, and *S. citrullifolium* and *S. fructo-tecto* also had four bands in common. Species from other sections of the genus *Solanum*, such as *S. hirtum*, had totally different electrophoretic patterns.

Immuno-electrophoresis of the antigen system of *S. rostratum* S.97 with the antibody system raised against it produced 14 arcs (Fig. 22.5). These were valued according to intensity (strong = 4, medium = 3, weak = 2), giving a total score of 38 or 100%. The spectra of the other species in section *Androceras* and also *S. sisymbrifolium* (section *Cryptocarpum*) produced similar spectra and scores (Table 22.7), but that of *S. hirtum* was very different. *S. citrullifolium* and *S. heterodoxum* produced characteristic spectra which were no more similar to those of *S. rostratum* and *S. fructo-tecto*, than were those of *S. sisymbrifolium*. Antibody system to *S. sisymbrifolium* indicated that all species in section *Androceras* are equally closely related to this species (Table 22.7).

The distinctness of the four species as shown by PAGE is supported by the failure of Coles to produce hybrids between them (Susan M. Coles, pers. comm.). However the affinities indicated by the common bands do not agree with the

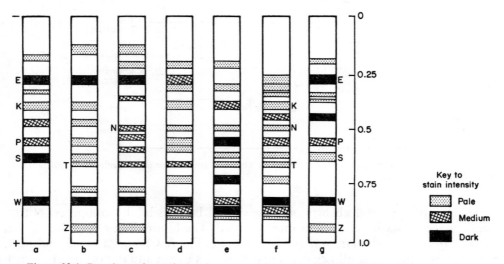

Figure 22.4. Drawings of protein bands separated by polyacrylamide gel electrophoresis of: a, *S. rostratum* S.097; c, *S. citrullifolium* S.195; e, *S. fructo-tecto* S.025; and g, *S. heterodoxum* S.593. b, d, f, Mixtures of the two adjacent samples to prove identity of bands.

Table 22.7. Immuno-electrophoresis scores produced by antigen systems of various species with antibody systems against *S. rostratum* S.097 (R.114), and *S. sisymbrifolium* S.1099 (R.119). (For explanation see text)

Code	Species	No.	RS[ab]	SS[ab]
RS	*S. rostratum*	S.097	100	73
RS	*S. rostratum*	S.399	92	70
FR	*S. fructo-tecto*	S.025	87	77
CT	*S. citrullifolium*	S.127	66	70
CT	*S. citrullifolium*	S.195	61	67
HT	*S. heterodoxum*	S.593	58	63
SS	*S. sisymbrifolium*	S.136	68	83
SS	*S. sisymbrifolium*	S.1099	63	100
HR	*S. hirtum*	S.1142	50	47

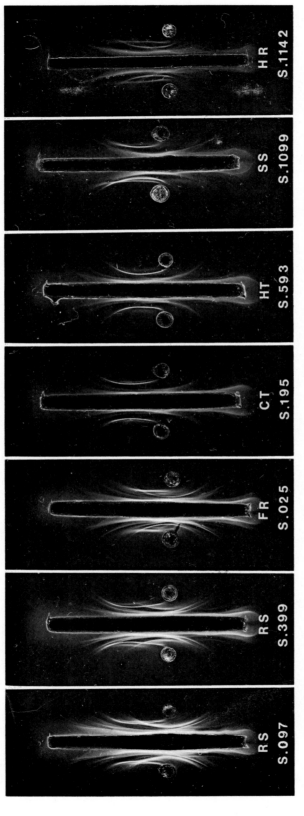

Figure 22.5. Immuno-electrophoretograms of antigen systems of several species with antibody system to *S. rostratum* S.097 (R.114). (For explanation see text and Table 22.7.)

(*Facing p. 300*)

relationships assessed by Whalen (this Volume, p. 581), who placed *S. fructo-tecto* and *S. rostratum* in series *Androceras*, but *S. citrullifolium* and *S. heterodoxum* in series *Violaceiflorum*, of section *Androceras*. The relationships shown by immuno-electrophoresis agree well with Whalen's arrangement, but the apparent affinity of series *Violaceiflorum* to *S. sisymbrifolium* rather than to series *Androceras* deserves further investigation.

DISCUSSION

The self-consistency of the matrix of $S_{abs.\ mean}$ allows some confidence in the validity of this approach of producing a cubic matrix of absorption experiments, despite the problems seen at earlier stages of analysis. The production of such a matrix would be extremely laborious, if it were not for the practical simplicity of the absorption technique of Dray & Young (1959); indeed complete cubic matrices have seldom been presented before (e.g. Cristofolini & Chiapella, 1977). The procedure of analysis of the matrix as set out above is very easy, even though a complete understanding of all the theoretical aspects of absorption experiments and their description in terms of set theory is very difficult (Kirsch, 1967). Basically the absorption experiments measure in two ways, the degree to which the set of antibodies in a given antibody system are precipitated by the antigenic determinants in each of the cross reacting antigen systems.

The taxonomic implications of the results of these experiments, as set out in the phenogram (Fig. 22.3), are very interesting. They indicate that all the eight species are relatively distinct, no phenon having a value higher than 0.65 for $S_{abs.\ mean}$, but they also indicate that certain pairs of species can be considered to be more closely related to each other than to any of the other species. Thus *S. mauritianum* and *S. × hendersonii* are clustered, and so are *S. olivare* and *S. citrullifolium*, and all four taxa are clustered together at a lower level. *S. scabrum* and *S. simile* form another group, to which is linked *S. seaforthianum* and subsequently *S. tuberosum*. All species showed distinct patterns in immuno-electrophoresis (Fig. 22.1).

Although only eight species were chosen from more than a thousand species in the genus *Solanum*, they represent eight distinct sections or subgenera (Table 22.1), distributed widely across the genus (Seithe, this Volume, p. 307). The basic dichotomy shown by the present results agrees fairly well with the division of the genus into Chorus subgenerum *Solanum* and Chorus subgenerum *Stellatipilum*. Members of sections *Solanum*, *Jasminosolanum* and *Petota* (*Tuberarium*) of the subgenus *Solanum*, and members of subgenus *Archaesolanum* all lack any stellate hairs, but have finger hairs. Their inclusion together in Chorus subgenerum *Solanum* is in agreement with the present immunological results. However the particularly close immunological relationship of subgenus *Archaesolanum* with section *Solanum* of subgenus *Solanum* is surprising since these two groups have few morphological characters in common, and have never been closely associated in any of the classifications of this genus. Whereas the rest of the genus has a base number of $n=12$, subgenus *Archaesolanum*, which is closeknit and probably of recent origin, is distinct in having $n=23$.

Solanum × hendersonii in the small section *Pseudocapsicum* presents an interesting problem. On the basis of its lack of both spines and truly stellate hairs, this section is placed in subgenus *Solanum*, although it does have ramosely branched hairs. It is peculiar in producing steroid alkaloids such as solanocapsine which are slightly different from those of most other species of *Solanum* (Schreiber, this Volume, p. 193). The present indication of an affinity to section *Brevantherum* of subgenus *Stellatipilum*, which also lacks spines but has truly stellate hairs is interesting and should be investigated further.

Both sections *Oliganthes* and *Androceras* have stellate hairs, spines, and long thin anthers, and the grouping of these two sections is generally accepted, even though members of section *Androceras* have such distinct flower and fruit characters that they have in the past been placed as a separate genus.

The similarities and differences shown by polyacrylamide gel electrophoresis and immuno-electrophoresis between the species within section *Androceras* are in broad agreement with the views of Whalen (this Volume, p. 581).

ACKNOWLEDGEMENTS

I am very grateful to Miss Sarah Marsh for her careful cultivation of the plants, to Dr Bill Tucker for providing some of the antisera, and to Phil Roberts for the data on section *Androceras*.

REFERENCES

ASTLEY, D., 1975. *Studies on the Taxonomic Relationships of* Solanum sucrense *Hawkes—a Bolivian Weed Potato Species*. Ph.D. Thesis, University of Birmingham.
BOULTER, D., RAMSHAW, J. A. M., THOMPSON, E. W., RICHARDSON, M. & BROWN, R. H., 1972. A phylogeny of higher plants based on the amino acid sequences of cytochrome c and its biological implications. *Proceedings of the Royal Society* (B) *181:* 441–455.
BUTLER, J. E. & LEONE, C. A., 1968. Determination of immunologic correspondence for taxonomic studies by densitometric scanning of antigen-antibody precipitate in agar-gel. *Comparative Biochemistry and Physiology, 25:* 417–426.
CHESTER, K. S., 1937. Anaphyllaxis tests for differentiating the proteins of normal Solanaceae. *American Journal of Botany, 24:* 451–454.
CONKLIN, M. E. & SMITH, H. H., 1971. Peroxidase isozymes: A measure of molecular variation in ten herbaceous species of *Datura*. *American Journal of Botany, 58:* 688–696.
CRIBB, P. J., 1972. *Studies on the Origin of* Solanum tuberosum L. *Sub-species* andigena *(Juz. et Buk.) Hawkes —the Cultivated Tetraploid Potato of South America*. Ph.D. Thesis, University of Birmingham.
CRISTOFOLINI, G. & CHIAPELLA, L. F., 1977. Serological systematics of the tribe Genisteae (Fabaceae). *Taxon, 26:* 43–56.
DAVIS, B. J., 1964. Disc Electrophoresis II. Methods and application to human serum. *Annals of the N.Y. Academy of Sciences, 121:* 404–407.
DESBOROUGH, S. & PELOQUIN, S. J., 1967. Esterase isozymes from *Solanum* tubers. *Phytochemistry, 6:* 989–994.
DRAY, S. & YOUNG, G. O., 1959. Two antigenically different γ-globulins in domestic rabbits revealed by isoprecipitins. *Science, 129:* 1023–1025.
EDMONDS, J. M. & GLIDEWELL, S. M., 1977. Acrylamide gel electrophoresis of seed proteins from some *Solanum* (Section *Solanum*) species. *Plant Systematics and Evolution, 127:* 277–291.
FAIRBROTHERS, D. E., 1969. Plant Serotaxonomic (Serosystematic) Literature, 1951–1968. *Bulletin Serological Museum*, No. 41: 1–11.
FERRI, M. V. & GUZMAN, C. A., 1970. The isoperoxidases from leaves of some species of the genus *Datura*. *Phyton, 27:* 137–140.
GELL, P. G. H., HAWKES, J. G. & WRIGHT, S. T. C., 1960. The application of immunological methods to the taxonomy of species within the genus *Solanum*. *Proceedings of the Royal Society* (B): 151: 364–383.
GRAY, J. C., 1977. Serological relationships of Fraction I proteins from species in the genus *Nicotiana*. *Plant Systematics and Evolution, 128:* 53–69.
GREEN, F., 1926. The precipitin reaction in relation to grafting. *Genetics, 11:* 73–82.
GUZMAN, C. A., FERRI, M. V. & TRIPPI, V. S., 1971. Isoperoxidases in organs of two species of the genus *Datura* (Solanaceae). *Phytochemistry, 10:* 2389–2391.
HAMMOND, H. D., 1947. *A Study of Taxonomic Relationship Within the Solanaceae as Revealed by the Photon'er Serological Method*. M.Sc. Thesis, Rutgers University, New Brunswick, N.J.

HAMMOND, H. D., 1955. A study of taxonomic relationship within the Solanaceae as revealed by the photron'er serological method. *Bulletin Serological Museum*, No. 14: 3–5.

HARN, C., CHEONG, Y. S. & YUNG, J. K., 1975. Studies on the electrophoretic variation in tetrazolium oxidase isozyme of *Nicotiana* species. *Korean Journal of Botany*, *18*: 150–154.

HAWKES, J. G. & LESTER, R. N., 1966. Immunological studies on the tuber-bearing Solanums. II. Relationships of the North American species. *Annals of Botany*, *30*: 269–290.

HAWKES, J. G. & LESTER, R. N., 1968. Immunological studies on the tuber-bearing Solanums.III. Variability within *S. bulbocastanum* and its hybrids with species in series Pinnatisecta. *Annals of Botany*, *32*: 165–186.

HAWKES, J. G. & TUCKER, W. G., 1968. Serological assessment of relationships in a flowering plant family (Solanaceae) In J. G. Hawkes (Ed.), *Chemotaxonomy and Serotaxonomy*: 77–88. Systematics Association. Special Volume, No. 2. London & New York: Academic Press.

HUAMÁN, Z., 1975. *The Origin and Nature of* Solanum ajanhuiri *Juz. et Buk.—A South American Cultivated Diploid Potato*. Ph.D. Thesis, University of Birmingham.

KAWASHIMA, N., TANABE, Y. & IWAI, S., 1974. Similarities and differences in the primary structure of Fraction I proteins in the genus *Nicotiana*. *Biochimica et Biophysica Acta*, *371*: 417–431.

KAWASHIMA, N., TANABE, Y. & IWAI, S., 1976. Origin of *Nicotiana tabacum* detected by primary structure of Fraction I protein. *Biochimica et Biophysica Acta*, *427*: 70–77.

KIRSCH, J. A. W., 1967. *Comparative Serology of Marsupials*. Ph.D. Thesis, University of Western Australia, Nedlands.

KLOZ, J., 1971. Serology of the Leguminosae. In J. B. Harborne, D. Boulter & B. L. Turner (Eds), *Chemotaxonomy of the Leguminosae*: 309–365. London & New York: Academic Press.

KOSTOVA, R., KHRISTOVA, J. & GEORGIEVA, R., 1976. (Application of protein electrophoresis in studies on the genetic relationships among species of the genus *Lycopersicon*.) *Genetika i Selektsiya 9*: 3–14.

KRASHENINNIK, N. V. & GAVRILYUK, I. P., 1973. (The hybrid origin of *Solanum sambucinum*.) *Dokladȳ Vsesoyujnoĭ Akademii Sel'sko-khojyaĭstvennȳkh Nauk im V. I. Lenina*, *1*: 25–26.

LESTER, R. N., 1965. Immunological studies on the tuber-bearing Solanums. I. Techniques and South American species. *Annals of Botany*, *29*: 609–624.

LESTER, R. N., 1969. An apparently asystematic reaction of vicilin from *Pisum* with an antiserum to *Lotus* seed proteins. *Archives of Biochemistry and Biophysics*, *133*: 305–312.

MARTIN, S. D., 1970. *A Biochemical Study of the Genus* Capsicum *L*. M.Sc. Thesis, University of Birmingham.

MOORE, G. W. & GOODMAN, M., 1968. A set theoretical approach to immunotaxonomy: analysis of species comparisons in modified ouchterlony plates. *Bulletin of Mathematics and Biophysics*, *30*: 279–289.

MORITZ, O., 1964. Some special features of serobotanical work. In C. A. Leone (Ed.), *Taxonomic Biochemistry and Serology*: 275–290. New York: Ronald Press.

MORITZ, O. & JENSEN, U., 1961. An attempt at a formular description of procedures and results in comparative serology. *Bulletin Serological Museum*, No. 25: 1–5.

NARAYAN, R. K. J. & REES, H., 1974. Nuclear DNA, heterochromatin and phylogeny of *Nicotiana* amphidiploids. *Chromosoma*, *47*: 75–83.

NATARELLA, N. J. & SINK, K. C., Jr., 1975. Electrophoretic analysis of proteins and peroxidases of selected *Petunia* species and cultivars. *Botanical Gazette*, *136*: 20–26.

RICK, C. M. & FOBES, J. F., 1975. Allozymes of Galapagos tomatoes: Polymorphism, geographic distribution and affinities. *Evolution*, *29*: 443–457.

RICK, C. M., ZOEBEL, R. W. & FOBES, J. F., 1974. Four peroxidase loci in red-fruited tomato species: Genetics and geographic distribution. *Proceedings of the National Academy of Sciences of the U.S.A.*, *71*: 835–839.

ROBERTS, P. A., 1977. *Interrelations of* Globodera *and some* Solanum *Species*. M.Sc. Thesis, University of Birmingham.

SANGAR, V. K., LICHTWARDT, R. W., KIRSCH, J. A. W. & LESTER, R. N., 1972. Immunological studies on the fungal genus *Smittium* (Trichomycetes). *Mycologia*, *64*: 342–358.

SCHMIEDICHE, P., 1977. *Biosystematic Studies of the Cultivated Frost-resistant Potato Species* Solanum × juzepezukii *Buk. and* Solanum curtilobum *Juz. et Buk.* Ph.D. Thesis, University of Birmingham.

SCOGIN, R. L., 1975. Isoenzymes of *Datura meteloides* (Solanaceae): Developmental patterns and tissue specificities. *Aliso*, *8*: 275–280.

SHEEN, S. J., 1972. Isozymic evidence bearing on the origin of *Nicotiana tabacum* L. *Evolution*, *26*: 143–154.

SHEEN, S. J. & REBAGAY, G. R., 1970. On the localization and tissue difference of peroxidases in *Nicotiana tabacum* and its progenitor species. *Botanical Gazette*, *131*: 297–304.

SMITH, P. M., 1976. *The Chemotaxonomy of Plants*. London: Edward Arnold.

SNEATH, P. H. A. & SOKAL, R. R., 1973. *Numerical Taxonomy*. San Francisco: W. H. Freeman.

TUCKER, W. G., 1969. Serotaxonomy of the Solanaceae: A Preliminary Survey, *Annals of Botany*, *33*: 1–23.

WEST, H. R., 1973. *A Chemotaxonomic Study of the Genus* Lycopersicon *(Tourn.) Mill*. M.Sc. Thesis, University of Birmingham.

V. Anatomy and fine structure

23. Hair types as taxonomic characters in *Solanum*

ALMUT SEITHE

Krefeld, West Germany

An analysis of the eight different hair types in *Solanum* is given, including prickles and bristles. Variations according to the parts of the plant on which they are found and the time in the life of the individual on which they develop, are described. Developmental relationships between hairs are investigated, thus differentiating between branchlet hairs on the one hand and stellate ones on the other, but stressing their common origin. Multicellular glands are shown to be very distinct in their origins from cover hairs and each of these is considered to be a main hair class. Prickles and bristles are analysed in terms of the basic types from which they are derived.

The taxonomic value of hair types is discussed, and tentative conclusions are drawn on the value of hairs in attempts to understand phylogenetic trends in the genus.

CONTENTS

INTRODUCTION

The significant role of hair types in elucidating phylogenetic relationships has been clearly demonstrated in the genus *Rhododendron* (Seithe, 1960). By means of this study it was possible to divide the genus into three large groups of species, each characterized by a particular combination of variants of two hair classes.

Similar taxonomic problems are found in the genus *Solanum*, where the wide range of hair types promises useful results in taxonomic and phylogenetic studies. Thus previous investigations showed (Seithe, 1962) that in *Solanum* several hair types on a plant can be considered as expressions of one *hair class* which are modified by their location and also the time when they begin their development.

DESCRIPTION OF BASIC HAIR TYPES

There are eight clearly recognizable hair, gland, prickle and bristle types to be found on mature plants. These are as follows:

1. *Finger hairs* (Fig. 23.1: 12, 13). These are simple uniseriate unbranched hairs consisting of more than one cell.

2. *Branchlet hairs* (Fig. 23.1 : 14, 15, 16). Hairs of this type are uniseriate, many-celled and dendroid, with the branches emerging at different levels.

3. *Stellate hairs* (Fig. 23.1 : 19–28). These may be uniseriate or multiseriate. In most cases the rays emerge at the same level (19, 20), but sometimes there is a second series or storey of rays emerging beneath the first ring (21, 24, 25). There is always a central hair or thorn protruding at right angles from the surface of the leaf and this can be very long, unicellular or multicellular (19, 20), more or less short (22, 23), or even very short indeed (27, 28).

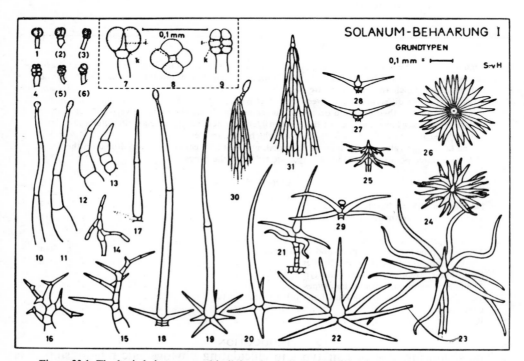

Figure 23.1. The basic hair types: multicellular glands (1–9); glandular finger hairs (10, 11); finger hairs (12, 13); branchlet hairs (14, 15, 16); stellate hairs (17, 19–28) glandular stellate hairs (18, 29); prickles (30, 31); (bristles see Fig. 2: 33).

Square gland: 1, *Solanum aviculare*, 2nd leaf: margin; 2, *S. aviculare*, transition form to storey gland, 2nd leaf; 3, *S. aviculare*, transition form to storey gland, 2nd leaf. Storey gland: 4, *S. aviculare*, 2nd leaf: lower surface; 5, *S. aviculare*, transition form to square gland, 4th leaf; 6, *S. melongena*, transition form to square gland, leaf: lower surface. Square gland: 7, *S. capense*, longitudinal section, i=intercellular space, k=cuticle; 8, *L. pimpinellifolium*, transverse section of the head. Storey gland: 9, *L. pimpinellifolium*, longitudinal section. Glandular finger hair: 10, *S. aviculare*, petiole of cotyledon; 11, *S. aviculare*, 3rd leaf: midrib. Finger hair: 12, *S. aviculare*, 4th leaf: midrib; 13, *S. aviculare*, 4th leaf- midrib. Branchlet hair: 14, *S. brachystachys*, leaf: lower surface; 15, *S. capsicastrum*, pedicel; 16, *S. congestiflorum*, leaf: lower surface: midrib. Stellate hair: 17, *S. sisymbrifolium*, calyx; also without the ray. Glandular stellate hair: 18, *S. sisymbrifolium*, leaf: midrib; longitudinal section. Stellate hair: 19, *S. sisymbrifolium*, leaf: upper surface; 20, *S. barbisetum*, leaf: upper surface; 21, *S. albicaule*, twig; longitudinal section; 22, *S. melongena*, ovary; 23, *S. cernuum*, leaf: lower surface; 24, *S. crotonoides*, leaf: upper surface; view from above; 25, *S. crotonoides*, leaf: upper surface; view from the side; 26, *S. swartzianum*; leaf: lower surface; view from above; 27, *S. swartzianum*, leaf: lower surface; longitudinal section; 28, *S. swartzianum*, corolla outside; longitudinal section. Glandular stellate hair: 29, *S. carense*, leaf: lower surface. Young prickle: 30, *S. ciliatum*, leaf: lower surface; 31, *S. marginatum*, 2nd leaf: lower surface: midrib.

4. *Gland-tipped finger hairs* (Fig. 23.1: 10, 11). These are several-celled finger hairs with a unicellular glandular tip.

5. *Gland-tipped stellate hairs* (Fig. 23.1: 18, 29). The central hair or thorn of the stellate hairs here develops a unicellular glandular tip.

6. *Multicellular glands*. Very short hairs, normally with one stalk cell, develop a multicellular glandular head (Fig. 23.1: 1–9). This head may consist of four cells in the same plane with one small cell beneath them (Fig. 23.1: 1, 7, 8); or it may be composed of two or three storeys of cells, each with two or more cells

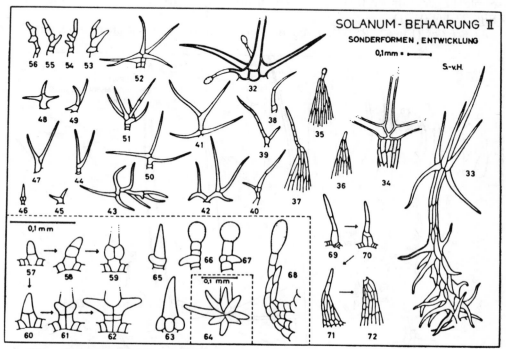

Figure 23.2. Special hair forms of the floral region, prickles, bristles and stellate hairs; further developmental steps of stellate hairs and a young prickle.
Glandular stellate hair: 32, *S. megalonyx*, leaf: upper surface. Bristle: 33, *S. cernuum*, calyx exterior; only the upper half; 34, *S. ferox*, leaf: midrib; longitudinal section of upper part. Prickle: 35, *S. ferox*, leaf: midrib; young stage; 36, *S. ferox*, leaf: midrib; young stage; 37, *S. marginatum*, young 2nd leaf: lower surface: midrib. Stunted stellate hair: 38, *S. gemellum*, calyx margin; 39, *S. gemellum*, calyx margin; 40, *S. gemellum*, calyx margin; 41, *S. melongena*, corolla lobes: tips; 42, *S. melongena*, corolla lobes: tips; 43, *S. melongena*, corolla lobes: tips; 44, *S. carense*, corolla exterior: thin parts; 45, *S. sisymbrifolium*, corolla exterior: thin parts; 46, *S. sisymbrifolium*, corolla exterior: thin parts; 47, *S. sisymbrifolium*, corolla exterior: thin parts; 48, *S. sisymbrifolium*, corolla exterior: thin parts; 49, *S. sisymbrifolium*, corolla exterior: thin parts; 50, *S. megalonyx*, corolla exterior: thin parts; 51, *S. megalonyx*, corolla exterior: thin parts; 52, *S. megalonyx*, corolla exterior: thin parts. Branchlet hairs, stunted: 53, *S. bulbocastanum*, corolla lobes: tips; 54, *S. bulbocastanum*, corolla lobes: tips; 55, *S. bulbocastanum*, corolla lobes: tips; 56, *S. bulbocastanum*, corolla lobes: tips. Young stellate hair: 57, *S. gilo*, seedling; longitudinal section; 58, *S. gilo*, seedling; longitudinal section; 59, *S. gilo*, seedling; longitudinal section; 60, *S. gilo*, seedling; longitudinal section; 61, *S. gilo*, seedling; longitudinal section; 62, *S. gilo*, seedling; longitudinal section; 63, *S. melongena*, leaf; view from the side; 64, *S. melongena*, corolla outside; view from below; 65, *S. carense*, leaf. Young glandular stellate hair: 66, *S. carense*, leaf; 67, *S. carense*, leaf. Young prickle: 68, *S. aculeastrum*, 6th leaf: lower side: midrib; 69, *S. bicorne*, leaf: lower side: midrib; 70, *S. bicorne*, leaf: lower side: midrib; 71, *S. bicorne*, leaf: lower side: midrib; 72, *S. bicorne*, leaf: lower side: midrib.

(Fig. 23.1: 4, 5, 6, 9). The former are known as *square glands* and the latter as *storied glands*.

7. *Prickles* (Fig. 23.1: 30, 31). The prickles occur as various types which grade into each other. The base can be longish so that the prickle is flattened on two sides, rather like a rose prickle, and is triangular in lateral view. The apex may be straight or curved. Other prickles have a smaller, more rounded base and look like awls or needles. All prickles are stiff and shining, and may sometimes be covered with stellate hairs and multicellular glands. Very young prickles bear a gland-tipped or a simple finger hair or their apex. These are very thin-walled and are lost later (Fig. 23.1: 30, 31).

8. *Bristles* (Fig. 23.2: 33). Even in their older rigid or stiffened state, bristles

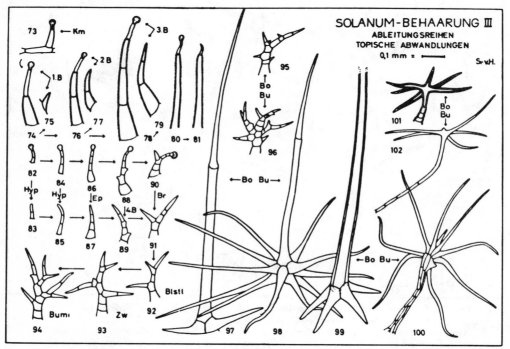

Figure 23.3. Lines of derivation with finger hairs and branchlet hairs, topographically induced variants with branchlet hairs and stellate hairs.

Glandular finger hair: 73, *S. nigrum*, cotyledon; 74, *S. nigrum*, 1st leaf. 75, (Finger hair): *S. nigrum*, 1st leaf. 76, Glandular finger hair: *S. nigrum*, 2nd leaf. 77, Finger hair: *S. nigrum*, 2nd leaf. 78, Glandular finger hair: *S. nigrum*, 3rd leaf. 79, Finger hair: *S. nigrum*, 3rd leaf. 80, Glandular finger hair: *S. muricatum*, pedicel; rare exception. 81, Finger hair: *S. muricatum*, leaf: upper surface; "bayonet hair". 82, Glandular finger hair: *S. capsicastrum*, hypocotyl. 83, Finger hair; *S. capsicastrum*, hypocotyl. 84, Glandular finger hair: *S. capsicastrum*, hypocotyl. 85, Finger hair: *S. capsicastrum*, hypocotyl. 86, Glandular finger hair: *S. capsicastrum*, epicotyl. 87, Finger hair: *S. capsicastrum*, epicotyl. 88, Glandular finger hair: *S. capsicastrum*, 4th leaf: lower surface: midrib. 89, Finger hair: *S. capsicastrum*, 4th leaf: lower surface: midrib. 90, Glandular branchlet hair: *S. capsicastrum*, leaf: very rare exception. Branchlet hair: 91, *S. capsicastrum*, leaf: margin; 92, *S. capsicastrum*, pedicel; 93, *S. capsicastrum*, twig; 94, *S. capsicastrum*, leaf: lower surface: midrib; 95, *S. congestiflorum*, leaf: upper surface; 96, *S. congestiflorum*, leaf: lower surface. Stellate hair: 97, *S. megalonyx*, leaf: upper surface; 98, *S. megalonyx*, leaf: lower surface; 99, *S. concinnum*, leaf: upper surface; 100, *S. concinnum*, leaf: lower surface; 101, *S. macropus*, leaf: upper surface*; 102, *S. macropus*, leaf: lower surface.*
*The stellate part is unicellular!

show clearly their original flexible structure. They bear a stellate hair on their apex which can break off later. The base of this hair gradually passes into the bristle itself, which, like the prickle, is also formed from epidermal and subepidermal layers. The bristles can be regarded as very long-stalked stellate hairs.

TOPOGRAPHICAL VARIANTS

Variants of the basic hair types are found in different parts of the plant. The most marked changes are seen in the stellate and branched hairs and the least marked in the unbranched ones. We shall therefore begin with a discussion of the stellate hairs which show the greatest range of topographical variants.

The differences between the upper and lower leaf surfaces are particularly well-marked in this respect. Thus the stellate hairs on the upper surface bear a longer central thorn than on the lower surface and sometimes have fewer, shorter rays and shorter stalks, though on the whole these can be larger (Fig. 23.3: 97, 98, 99, 100, 101, 102). Certain differences are not seen, as when a species has not advanced to the development of stellate hair stalks or the phylogenetic stage of the long

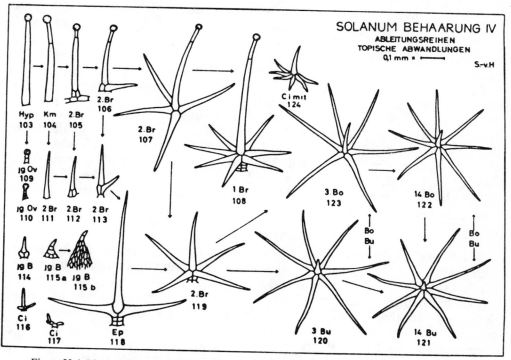

Figure 23.4. Lines of derivation with finger hairs and stellate hairs; topographically induced variants with stellate hairs.

S. capense. Glandular finger hair: 103, hypocotyl; 104, cotyledon; 105, 2nd leaf: margin. Glandular stellate hair: 106, 2nd leaf: margin; 107, 2nd leaf, margin; 108, 1st leaf margin. Glandular finger hair: 109, ovary; 110, ovary. (Finger hair): 111, 2nd leaf: margin; 112, 2nd leaf: margin. (Stellate hair): 113, 2nd leaf: margin. Young stellate hair: 114, leaf. Young prickle: 115a, leaf; 115b, leaf. Stunted stellate hair: 116, corolla inside; 117, corolla inside. Stellate hair: 118, epicotyl; 119, 2nd leaf: margin; 120, 3rd leaf: lower surface; 121, 14th leaf: lower surface; 122, 14th leaf: upper surface; 123, 3rd leaf: upper surface; 124, corolla inside: middle parts.

central thorn has already been passed. In this latter case, of course, extreme differences in thorn length cannot be seen, though they are mostly still seen on young plants (Fig. 23.4: 123, 120, 122, 121).

The variants of hairs in the flowers and pedicels are also remarkable. Thus stellate hairs occur on the midribs of the corolla lobes with a central thorn longer than on the lower leaf surface, and even sometimes multicellular (Fig. 23.4: 124). Sometimes there is a trend to more than one storey of rays. On the thin parts of the corolla, however, the stellate hairs become small and appear stunted (compare Fig. 23.2: 41 with Fig. 23.1: 22; and Fig. 23.2: 50-52 with Fig. 23.3: 98). The number of rays is reduced and while the tendency towards two storeys remains the hairs begin to resemble branchlet hairs.

On the calyx and pedicel the stellate hair stalks can reach a surprising length; the stellate part is similar to that on the lower leaf surface and only the central thorn tends to be a little longer. Stellate hairs on anthers and ovaries also show the same tendency to longer stalks and central thorns which often also bear a glandular tip.

The branchlet hairs undergo similar variation, being sometimes larger but less branched on the upper leaf surface than on the lower (Fig. 23.3: 95, 96). In the floral region they also become simpler and smaller (Fig. 23.3: compare 92 with 94).

The finger hairs, also, are often larger on the upper than on the lower surface of the leaf, except on the midrib, and are also smaller in the floral region.

Variations of the multicellular glands are fewer than those of the stellate, branchlet and finger hairs since the stalk remains as a single cell* and can only differ in the number of cell divisions in the glandular head (Fig. 23.1: 1–9). The typical form of the square glands prevails on the leaf margin but also occurs on the upper and lower surfaces of the leaf and on the stem. On the leaf midrib, however, it is often replaced by the storied gland. The two types are connected by transitional forms (Fig. 23.1: 2, 3, 5, 6). Storied glands occur, in addition, on the calyx. They can also be found on all organs in place of square glands and are sometimes smaller than these latter.

TEMPORAL VARIANTS

In addition to the positional or topographical variants, others are connected with the time in which the hairs begin to be formed during the ontogenetic development of the plant. Thus, we often find amongst the regular distribution of normal stellate hairs smaller ones with shorter stalks and fewer shorter rays which have developed later than the normal ones. These simplifying tendencies can result in completely unbranched hairs which arise through later development. Corresponding simplifications can be found also in branchlet and finger hairs.

Parallels with hair variation in *Rhododendron* indicate that most of the hair types found on any given plant are closely connected, since they are all conditioned by the same genotype. The differences seem to be entirely due to the internal environment during their development.

However, not all differences in hair type on one plant can be interpreted as simple variants. In *Rhododendron*, for instance, the hair types on one plant belong

* Very occasionally two cells, as in *S. polyadenium* (Eds).

to two distinct hair classes—the cover hair and the gland hair. In *Solanum*, too, a line of demarcation between two hair classes can be recognized, though the equivalent of a "cover hair class" includes forms ranging from simple to branched and stellate, with or without a glandular tip (Fig. 23.1: 10–29); the second hair class is restricted to the multicellular gland with a single-celled stalk (Fig. 23.1: 1–9). These two classes never intergrade.

The coherence of the types within a hair class becomes clearer when we examine the indumentum of young plants. To do this we must first examine their onto-genetic development.

THE ONTOGENETIC DEVELOPMENT OF *SOLANUM* HAIRS

All hairs in *Solanum* begin as a single cell cut off by a periclinal division of an epidermal cell (Fig. 23.2: 57). This develops into a uniseriate hair and so into a finger hair. In a branchlet hair the cells of the finger-like hair soon grow into short lateral protruberances, which then become divided transversely. These new cells can themselves branch, so producing second degree ramifications (Fig. 23.3: 89, 91, 94).

In the stellate hair development, anticlinal divisions soon begin in one cell of a finger-like hair and these daughter cells then develop into rays, though they mostly remain unicellular (Fig. 23.2: 59, 61, 62). The gland tips are formed simply by a swelling of the end cell (Fig. 23.2: 66, 67). Sometimes the stalk of a stellate hair soon becomes multiseriate through several anticlinal divisions following the first periclinal division of the epidermis cell (Fig. 23.2: 60–62).

The multicellular glands form their glandular heads by longitudinal divisions as well as transverse ones. In the square glands, only the upper of the original two cells divides. After two divisions the resulting four cells swell up into small bladders (Fig. 23.1: 7, 8). The head cells of the storied glands do not swell up so much but divide once or even twice more, transversely (Fig. 23.1: 9). The stalk remains unicellular in both cases. Since the multicellular heads begin their development from one cell there is a natural connection with the unicellular gland tip, but only in the earliest stages of ontogeny.

Returning now to the hypothesis of the existence of two basic classes of hairs let us now consider the indumentum of the young seedlings. This differs in an interesting manner from that of the adult plants (contrast Fig. 23.4: 103, 104 with 122, 121). Nearly all the species investigated bear gland-tipped finger-hairs on their cotyledons as well as some multicellular glands. They thus bear repre-sentatives of the two hair classes but in the former case these are not always identical with those of the mature plants. However, the indumentum of the mature plants is reached gradually, so that on successive leaves one observes the gland-tipped finger hairs being replaced by simple finger hairs which are often two cells shorter (Fig. 23.3: compare 74, 75; 76, 77; 78, 79; and 80, 81. Also Fig. 23.4: 104, 111). These again, at a later stage, are replaced by branchlet hairs or by stellate hairs (Fig. 23.3: 82–94; Fig. 23.4: 103–108, 119–121).

Sometimes we find only gland-tipped finger hairs on the hypocotyl, cotyledons and epicotyl. These are followed on the first true leaf by a few gland-tipped finger hairs and many simple hairs. On the second leaf stellate hairs may appear in those species genetically conditioned to develop stellate hairs (see Fig. 23.4).

These first stellate hairs still have a long, sometimes multicellular central thorn and a few short rays, or sometimes only a divided cell at the base of the finger hair (Fig. 23.4: 112, 113). Later on, the plant may produce its typical stellate hairs, such as those with a short unicellular thorn and eight long rays (Fig. 23.4: 121).

This transition from one hair type to another is so regular and so clearly apparent that it would seem to illustrate Haeckel's Law of Ontogeny repeating Phylogeny. If this were so, then one would expect to find that the species with simpler gland-tipped hairs were more primitive than those with multicellular branched or stellate forms, at least in a similar group of species that were considered on other criteria to be closely related.

THE DEVELOPMENT OF GLAND-TIPPED AND STELLATE HAIRS

To understand the relationship between the gland-tipped and stellate hairs it is important to follow their early stages of development during the ontogeny of the plant. For instance, the rays of the stellate hairs often begin to develop before the capacity for gland tip development is lost, so that one can find what appear to be gland-tipped finger hairs with one or a few rays (Fig. 23.4: 105, 106). These are followed developmentally by fully developed gland-tipped stellate hairs and then by normal, stellate hairs (Fig. 23.4: 107, 108, 119–123).

The loss of the gland tip and the development of the rays are not invariably associated with each other since the two processes can proceed at different rates. Thus, if the gland tip is lost much earlier than the time when the rays develop, many simple finger hairs can be seen as intermediate stages. If the development of the rays begins earlier, the simple finger hairs are lacking. If both developments proceed synchronously we can find all four hair types side by side: gland-tipped finger hairs, finger hairs, gland-tipped stellate hairs and stellate hairs.

The time when the gland is lost and that when ray development begins, as well as the speed of each developmental process seems to be fixed genetically for each species. At the one extreme there are species such as *S. sisymbrifolium* where both processes begin very late and are never wholly finished; at the other extreme there are species where both developments begin even on the hypocotyl and are already finished on the epicotyl (*S. marginatum*).

It is of interest to note here that those species whose hairs develop into branchlet hairs seem always to lose their gland tips *before* the branching begins (see Fig. 23.3: 82–92, but note that 90 is a very rare exception).

The direct development from gland-tipped finger hairs to stellate hairs shows that this process cannot have passed through a branchlet hair stage as might perhaps have been thought by observing the hairs *only* on adult plants. Thus there are two completely distinct parallel developmental stages:

(a) gland-tipped finger hairs directly to stellate hairs, and
(b) gland-tipped finger hairs directly to branchlet hairs.

It can thus be inferred that the species to which these hairs belong represent two distinct parallel lines of evolution. If we wish to clarify species relationships we must understand the evolution of their characters and try to distinguish between genuine phyletic relationships and convergent evolution, however.

THE DEVELOPMENT OF PRICKLES AND BRISTLES

Bristles are evidently long-stalked stellate hairs, and in agreement with this postulate are found only on species in the subgenus Stellatipilum. However, prickles, with their apical finger hair or gland-tipped finger hair (Fig. 23.1: 30, 31) could be related to the branchlet hairs, too. Nevertheless, we find them occurring almost exlusively on Stellatipilum species. Within the chorus subgenerum *Solanum* prickles are found only in section Aculeigerum; even so, these are very small "hooklets", whilst those of Stellatipilum species are often very large and not infrequently covered with stellate hairs towards their bases.

It is interesting to see that the small hooklets of section Aculeigerum begin their development with a finger hair and not with a gland-tipped finger hair, consistent with the early loss of the gland tip in the evolution of the branchlet hairs.

The question therefore arises as to how we should interpret the bristles and prickles as a whole. The bristles can be understood as a developmental step in the stellate hair where the stalk has had time to become longer and thicker. Later in their development the stellate tip breaks off and only the long thick stalk remains.

The prickles (Fig. 23.1: 30, 31) are more difficult to understand, for they are developed on mature plants where the stages of gland-tipped finger hairs and finger hairs have often already passed. However, when the youngest stages of prickle and stellate hair development are compared on the same plant they can scarcely be distinguished (Fig. 23.4: 114, 115). At first, when the central thorn appears and the rays begin to grow out we can see whether a stellate hair is developing (Fig. 23.2: 65–68). The young prickles do not prolong their apical cell, however, nor do they develop rays. Instead, they soon initiate cell divisions in quick succession, aided by the subepidermal cell layers.

The tendency to develop either a stellate hair or a prickle seems to be determined according to the "pattern scheme" described by Bünning (1948), known previously for leaf stomata. In certain instances the stimulus for extending the cells to form stellate hairs is replaced by another stimulus which promotes cell divisions and turns the hairs into prickles.

The connection between the degree of differentiation of prickles and stellate hairs is interesting. Species which bear stellate hairs without gland tips in the seedlings, also produce simple finger hairs on the prickle apices (*S. marginatum*) (Fig. 23.1: 31). In other species we find gland-tipped finger hair bristles (Fig. 23.1: 30) on young seedlings followed later by simple finger hair prickles and a parallel succession of gland-tipped stellate hairs and stellate hairs. Finally, there are species which retain gland-tipped finger hair prickles and gland-tipped stellate hairs even on the adult plants.

EVOLUTIONARY TRENDS IN *SOLANUM* HAIR TYPES

Stellate hairs: The following trends can be postulated:

(i)	From gland-tipped to glandless central thorn	(Fig. 23.4: 107, 119)
(ii)	From multicellular to unicellular central thorn	(Fig. 23.1: 19, 20)
(iii)	From long to short central thorn	(Fig. 23.1: 20, 28)
(iv)	From gland-tipped to glandless rays	(Fig. 23.2: 32, 33)
(v)	From a few to many rays	(Fig. 23.1: 20, 22)
(vi)	From short to long rays	(Fig. 23.1: 20, 22)
(vii)	From uniseriate to multiseriate stalk	(Fig. 23.1: 21, 23)
(viii)	From short to long stalk	(Fig. 23.1: 29, 23)

Branchlet hairs: The following trends can be postulated:

(i)	From a few to many branches	(Fig. 23.3: 91, 94)
(ii)	From short to long branches	(Fig. 23.3: 91, 93)
(iii)	From unicellular to multicellular branches	(Fig. 23.3: 91, 93)
(iv)	From branches of first degree only, to additional ones of second degree	(Fig. 23.3: 93, 94)

Multicellular glands: Here there are two trends:

(i)	From a few to many gland cells	(Fig. 23.1: 2, 5)
(ii)	From homogeneous to heterogeneous gland cells	(Fig. 23.1: 4, 1)

The individual parts of a hair can show different degrees of differentiation. This seems to point to the action of modifying genes varying the expression of the major genes which determine the hair class.

THE TAXONOMIC IMPORTANCE OF HAIR TYPES

To make use of hair types as taxonomic characters we must look for the highest and lowest degree of differentiation of a particular hair class in each species. We must also observe the hair development on all juvenile and all adult organs of a plant.

To estimate species differences in terms of the indumentum it would be important to know something of the genetic control of such differences. Thus, the loss of an upper or lower layer of hairs may be due to one mutational step, whilst for an alteration in the form many modifying mutations might be necessary, even though the external appearance can change more in the first case than in the second. We know nothing about whether reverse mutations take place or are possible, and without crossing experiments we can only work on hypotheses. Since genetic analyses of hair inheritance have not been made it seems advisable to look on the variation of hair classes more as a basis for the separation of taxa than as a basis for uniting them.

THE USE OF HAIR CHARACTERS BY DUNAL AND OTHER TAXONOMISTS

It we now look at the classification proposed by Dunal in 1852 (see Fig. 23.5) we see that he often grouped together species with branchlet hairs and those with stellate hairs. However, we have seen that these two hair types diverge at an early stage in the ontogeny of the plant, and we should therefore expect species with one or the other type not to be very closely related. For instance, in Dunal's taxon Anthoresis we find species with branchlet hairs included with those bearing stellate hairs. Bitter suggested the separation of the stellate haired species of Dunal's taxon Anthoresis from those species with branchlet hairs (Bitter, 1917).

Bitter's method was to form new taxonomic categories by grouping in the first place a few species that were very definitely closely related. He then tried to find characters which could help associate the taxa of lower rank with those of higher rank. The only suitable characters for this purpose were the form of the anthers and the presence or absence of prickles. However, since he found transitional states he was not able to make a perfectly logical and satisfactory division of the genus by these methods.

We have seen that it is now possible to distinguish very clearly the branchlet from the stellate hairs in the cover hair class (as contrasted with the multicellular glands class). On this basis, therefore, we can unhesitatingly divide the genus into two large parts, *Solanum* and *Stellatipilum* respectively.

The studies of *Rhododendron* carried out previously by the present writer made it seem inadvisable to alter the ranking height of already existent taxa, especially of subgenera. Thus the newly proposed taxon, *chorus subgenerum* was introduced in conformity with the International Code of Botanical nomenclature, at one higher order than *subgenus*, to unite those subgenera which belong together.

A problem remains however, as to where those species with gland-tipped finger

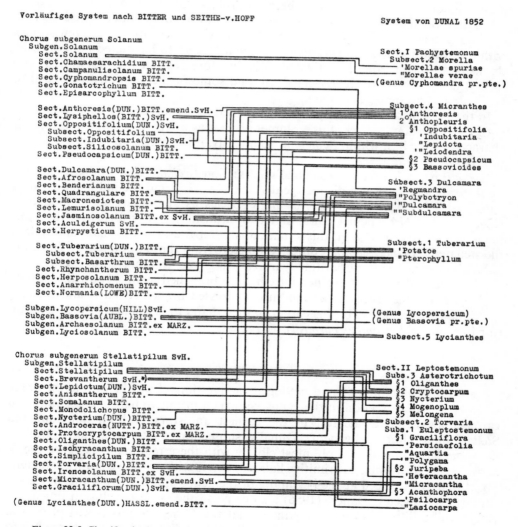

Figure 23.5. Classification proposed by Dunal (1852) compared with the taxa proposed by Bitter and Seithe-v.Hoff.

hairs should be placed, since these hair types can be the ancestors of branchlet hairs as well as of stellate hairs. It is certain that they do not form a separate group comparable, for instance, with the Azaleas in the genus *Rhododendron*. In most cases the other characters of these species show that their nearest relatives are species with branchlet hairs. Only in a very few cases do near relatives of stellate haired species bear only unbranched hairs on adult plants, but when this is so the finger hairs are distinguished from normal ones by being composed of very long cells, thus looking like the central thorns of stellate hairs. Indeed, if we search carefully on such plants we generally find a few stellate hairs too (for instance on *S. atropurpureum*, section Simplicipilum).

We can thus arrange the sections and subgenera as shown in Fig. 23.5. It is unfortunately a provisional system because Professor Bitter died in 1927 whilst he was still in the middle of his work on *Solanum*, and so could not complete his task of establishing a definitive system. Thus, he sometimes describes only a few species within a section, often enumerating others which seem to be related but which he thinks need to be further investigated. It is therefore possible that not all the 2000 species now known are rightly classified, and some further sections may need to be created to accommodate them.

In the present scheme, *Lycopersicon* is classed as a subgenus of *Solanum*. Some authors (for example, Rick, this Volume) place it as a separate genus, but on hair characters it fits best within *Solanum*. Admittedly, it is a matter of opinion whether to separate it or not.

In contrast to *Lycopersicon*, Dunal's subsection *Lycianthes* was in part elevated by Bitter to the level of a distinct genus (see Fig. 23.5) and the indumentum data agree with this decision. The *Lycianthes* species as at present understood bear stellate hairs *and* branchlet hairs, united by transitional hair forms on the same plant. As we have just seen, this situation is never found in *Solanum*, and indeed the stellate hairs of *Lycianthes* differ completely from those of *Solanum*. Thus, the hair axis does not grow out distinctly into a straight central thorn, but is often bent at an angle and the rays are not infrequently branched again. Thus, the stellate hairs of *Lycianthes* seem to be modified depressed branchlet hairs, and furthermore the hair branching begins earlier in these hairs than the loss of the gland tip (for instance in *Lycianthes purpusii*).

CONCLUSIONS

Returning to taxonomic and phylogenetic questions it will be of interest to see whether there is a section in the genus with such primitive hair characters that it can be postulated as nearest to the earliest ancestors of both Solanum and Stellatipilum chori subgenera. Dunal's subsection (Bitter's section) *Morella* (= *Maurella*) seems to comply most closely with these requirements, since the *S. nigrum* complex combines primitive hair characters with primitive floral characters. So Morella was placed at the beginning of the genus and *S. nigrum* was made the generic type species. Danert (1956) also came to the same conclusion independently, whilst Bitter also, was convinced that the *S. nigrum* group was of central importance. The name for section Morella (Maurella) had to be changed into section Solanum according to the International Code.

On the basis of our present knowledge of *Solanum* hairs we cannot say much more about the phylogeny of the genus. However, another point may be worthy of mention. At the very earliest stages of development of each plant, both the multicellular glands and the gland-tipped finger cells begin their development as a two-celled body with a single stalk cell and a unicellular gland tip. Thus the first gland-tipped finger hairs found on the hypocotyl (Fig. 23.3: 73) represent the basic structure from which each of the two main hair classes is later derived. This might indicate that the two hair classes had a common genetic origin (possibly by genome duplication) in the past and that further mutations later modified them to develop into the two distinct hair classes that we have been describing in this chapter.

In other solanaceous genera the same picture seems to be apparent. Young seedlings bear gland-tipped finger hairs, either alone or together with simple finger hairs, sometimes also with multicellular glands. Simple finger hairs or multicellular glands occur very seldom. As expected, the more primitive hair types are followed by more differentiated ones on the older plants. The gland-tipped finger hairs are replaced by simple finger hairs; finger hairs are followed by branchlet hairs; and unicellular glands are succeeded by multicellular glands. True stellate hairs, that is to say, with rays arising at one or two levels, are not found, except in the previously discussed genus *Lycianthes*. As with *Solanum*, then, the more primitive hair types form the unifying element between the genera, and hair types with a higher degree of differentiation seem to be limited to a few genera and within these to certain species only.

REFERENCES

BITTER, G., 1917. Solana africana II. *Botanische Jahrbücher für Systematik, Pflanzengeschichte und Pflanzengeographie, 54:* 424–425.

BÜNNING, E., 1948. *Entwicklungs—und Bewegungsphysiologie der Pflanze:* 169–179. Göttingen.

DANERT, S., 1956. Zur Systematik von *Solanum tuberosum* L. *Die Kulturpflanze, 6:* 83–129.

DUNAL, F., 1852. Solanaceae. In A. De Candolle, *Prodromus, 13* (1): 1–690.

SEITHE-v. HOFF, A., 1960. Die Haarformen der Gattung *Rhododendron* und die Möglichkeit ihrer taxonomischen Verwertung. *Botanische Jahrbücher für Systematik, Pflanzengeschichte und Pflanzengeographie, 79:* 297–393. Tafel 6–11.

SEITHE-v. HOFF, A., 1962. Die Haararten der Gattung *Solanum* L. und ihre taxonomische Verwertung. *Botanische Jahrbücher für Systematik, Pflanzengeschichte und Pflanzengeographie, 81:* 261–335. Tafel 25–28.

ABBREVIATIONS USED IN FIGURES

Ci	corolla inside		Bumi	leaf: lower surface: midrib
Cimit	corolla inside: middle part		Ep	epicotyl
1.B	1st leaf following the cotyledons		Hyp	hypocotyl
Blstl	pedicel (and peduncle)		jg	young (leaf or other organ)
Bo	leaf: upper surface		Km	cotyledon
Br	leaf: margin		Ov	ovary
Bu	leaf: lower surface		Zw	twig

24. An attempt to use stomatal characters in systematic and phylogenetic studies of the Solanaceae

J. BESSIS AND M. GUYOT

University of Dijon, France

The study of the epidermis of 56 species belonging to 45 genera of Solanaceae has enabled us to recognize most of the different stomatal types commonly described in the Dicotyledons (anomocytic type, meso-perigenous and mesogenous anisocytic types, mesogenous bicytic paracytic and diacytic types).

The distribution of these stomatal types can be used to establish systematic divisions and to suggest phylogenetic relationships within the Solanaceae. It seems that the curved embryo Solanaceae are relatively homogeneous as regards stomata, the anisocytic types being the main one. On the contrary, the straight embryo Solanaceae have very diversified stomatal types. Species with stomata of different types are particularly important because they enable us to trace the different stages of evolution. This reinforces the hypothesis of a polyphyletic origin of the Solanaceae.

CONTENTS

INTRODUCTION

In this study we shall not review the bibliography of attempts to use stomata for taxonomic and phylogenetic purposes since the first works of Vesque (1889) and Tognini (1897).

Observations of M. Guyot on the structure and development of Umbelliferae stomata have led us to attribute a certain taxonomic and phylogenetic value within a family to the stomatal apparatus. These observations have enabled us to define the phylogenetic relationships between the different stomatal types (Fig. 24.1).

There would seem to have been an evolution of the stomatal apparatus from the perigenous anomocytic type, to the meso-perigenous anomocytic and anisocytic types and from the latter to the mesogenous bicytic (paracytic and diacytic) and towards the mesogenous anisocytic type.

It appeared to us that an analysis of these stomatal types in various families of Dicotyledons should be undertaken with these new data based on stomatal ontogeny. Solanaceae is a particularly interesting family in this respect: on the one hand, we can observe almost all stomatal types and on the other hand,

Wettstein (Engler & Prantl, 1895), who wrote the chapter on the Solanaceae, considers them as a rather heterogeneous and probably polyphyletic family. This is why we thought it interesting to find out whether a study of the stomata in the Solanaceae would be of systematic and phylogenetic interest.

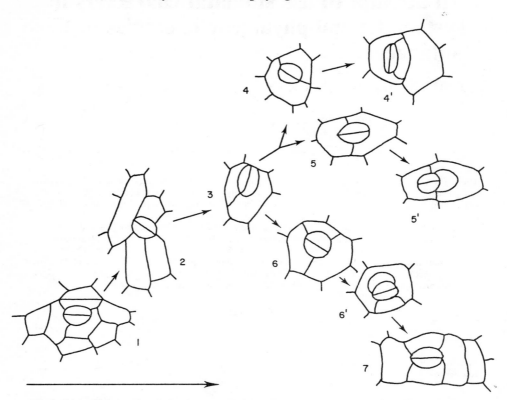

Figure 24.1. Phylogenetic relationships between the various types of stomata found in the Umbelliferae. (Guyot, 1971). 1. Perigenous anomocytic stoma. 2. Meso-perigenous anomocytic stoma. 3. Meso-perigenous anisocytic stoma. 4, 4'. Bicytic paracytic stomata. 5, 5'. Bicytic diacytic stomata. 6. Mesogenous anisocytic stoma. 6'. Tetracytic stoma linked up to anisocytic type. 7. Tetracytic stoma (this type has not been observed in the studied genera).

MATERIALS AND METHODS

Observations were made on the lower epidermis of leaves by light microscopy and scanning electron microscopy. In the case of observations by light microscopy, epidermal peels were stripped, either from fresh leaves, or from hydrated portions of leaves from herbarium specimens. For observations made by SEM, fragments of leaves were either cryodessicated and gold-coated, or directly observed in a cryo-unit of the SEM which makes possible a study of cell surfaces in a frozen state (at liquid nitrogen temperature), with a minimum of damage. Leaves may thus be observed in their nearly natural condition, still containing water.

RESULTS

The stomatal specifications of the different species are shown in Table 24.1.

Table 24.1. Stomatal specifications for 56 species in 45 genera of the Solanaceae. Codes as in Fig. 24.1 (rare occurrences in parenthesis)

NICANDREAE
Nicandra physaloides L. — 3, (4)

SOLANEAE
SOLANINEAE

Nothocestrum latifolium L.	6, (3)
Withania somnifera Dun.	6, (6')
Withania frutescens Pauq.	6, (6')
Physalis alkekengi L.	3, (6)
Physalis ixocarpa Brot.	3, (6)
Saracha edulis Tell.	2, 3, (6)
Saracha umbellata G. Don.	2, 3, (6)
Capsicum baccatum L.	6, 3
Capsicum luteum Lam.	6, (3)
Solanum dulcamara L.	6, 3
Solanum laciniatum Ait.	6
Solanum linearifolium H.	6
Solanum pseudocapsicum L.	6

LYCIINEAE

Grabowskia duplicata Arn.	3, (6), 4
Lycium chinense Mill.	3, 6, 6'
Dunalia australis Griseb.	3, 6
Acnistus parviflorus Gris.	6, 3
Iochroma coccinea Scheid.	3, 6
Discopodium eremanthum Chiov.	2, 3
Latua pubiflora Baill.	3, 6
Margaranthus solanaceus Schlect.	3, 6
Atropa belladona L.	6, 3

HYOSCYAMINEAE

Scopolia carniolica Jacq.	6
Physochlaina orientalis M.B.Don.	6, 3
Hyoscyamus albus L.	6, 3

MANDRAGORINEAE

Cyphomandra betacea Send.	6
Salpichroa origanifolia	6
Jaborosa integrifolia Lam.	3, 6, 4
Trechonaetes sativa Miers.	2, 3
Mandragora officinarum L.	6, (3)

DATUREAE

Solandra guttata Dun.	3, (6), (4)
Solandra hirsuta G. Don	3
Datura arborea L.	6, (3)
Datura inermis Jacq.	6, (3)
Datura stramonium L.	6, (3)
Dyssochroma viridiflora Miers.	4, (3)

CESTREAE
CESTRINEAE

Markea ulei	1, 2, (3)
Juanulloa aurantiaca Otto. Dietr.	1, 2, (3)
Cestrum nocturnum L.	1, 2, (3)

GOETZEINEAE

Goetzea elegans Wydl.	1, 2, (3)
Espadea amoena Rich.	1, 2, (3)
Coeloneurum ferrugineum	1, 2, (3)

Table 24.1—*continued*

NICOTIANINEAE	
Nicotiana glauca Grah.	3, (6)
Fabiana imbricata Ruiz et Pav.	3, (6)
Vestia lycioides Willd.	1, 2, (3)
Petunia violacea Lindl.	2, 3, (5), (6)
Nierembergia frutescens Durieu.	2, 3, 6
SALPIGLOSSIDEAE	
Schizanthus retusus Hook	2, 3, 5
Schwenckia americana L.	5
Browallia americana L.	2, 3, 5
Streptosolen jamesonii Miers.	2, (5)
Duboisia myoporoides R.Br.	2, 3, 6
Anthocercis littorea Labill.	2, 3, 6
Brunfelsia calycina Benth.	4

The distribution of stomata in the species studied may be commented on as follows:

Besides their characteristic stomatal type, all species studied present perigenous anomocytic stomata (type 1). The first stomata to appear are indeed always of this type and thus do not present any systematic interest. This type of primitive stoma has therefore only been indicated in Table 24.1 when it was particularly numerous.

Some species, such as *Schwenckia americana*, *Brunfelsia calycina* or *Cyphomandra betacea*, present homogeneous stomatal types, all three having mesogenous, diacytic, paracytic or anisocytic stomata. In some other species, on the contrary, different types of stomata coexist, as is the case, among others, in *Grabowskia duplicata*, *Jaborosa integrifolia*, *Petunia violacea*, etc.

SYSTEMATIC AND PHYLOGENETIC CONCLUSIONS

When we have studied several species of the same genus, we have up to now observed almost always the same type of stomata.

On the other hand, the different tribes of Solanaceae show great stomatal heterogeneity. It also appears that the two series of Wettstein's classification can be clearly recognized: one containing Nicandreae, Solaneae and Datureae, and the other, Cestreae and Salpiglossideae. The first series corresponds to curved embryo Solanaceae, the second to straight embryo Solanaceae. The first series appears relatively homogeneous, with mainly anisocytic stomata. On the contrary, species with very diversified stomata can be found in the second series: anomocytic, mesoperigenous anisocytic, mesogenous bicytic and anisocytic.

By using the principles which have helped us to establish phylogenetic relationships in the Umbelliferae, we can suggest hypothetical connections within the Solanaceae also. The principles on which these phylogenetic hypotheses have been based are mainly the following: the most primitive species would possess anomocytic stomata (types 1 and 2) and evolution would have taken place from these species to those with mesogenous stomata, either bicytic or anisocytic (3–6). The connections which then appear do not mean that there are direct links between these species, but they reveal evolutionary tendencies which have governed species differentiation. In this phylogenetic field (Fig. 24.2), the species with diversified stomata are particularly important because they show intermediate stages in this evolution. On the contrary, the species with stomata of homogeneous type can be considered as ends of a phylum.

The family Solanaceae, therefore, clearly appears polyphyletic and it is remarkable that it presents both genera with primitive stomata and genera with evolved stomata. This evolution would have occurred in a parallel way, at the level of the embryogeny and at the level of epidermis development, since, as we have shown, curved embryo Solanaceae would correspond to an evolution in the anisocytic direction and straight embryo Solanaceae to an evolution in the bicytic direction.

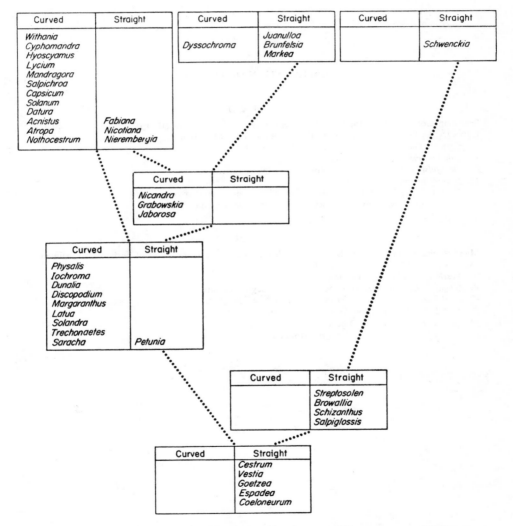

Figure 24.2. Hypothetical phylogenetic relationships between different genera of Solanaceae, grouped into parallel columns according to whether the embryo is curved or straight.

REFERENCES

GUYOT, M., 1971. Phylogenetic and systematic value of stomata of the Umbelliferae. In V. H. Heywood (Ed.), *The Biology and Chemistry of the Umbelliferae*. 197–214. London: Academic Press.
ENGLER, A. & PRANTL, K., 1895. *Die Natürlichen Pflanzenfamilien, 4*. Leipzig.
TOGNINI, F., 1897. Contribuzione allo studio della organogenia comparata degli stomi. *Atti dell'Istituto Botanico della Universita e Laboratorio Crittogamico di Pavia* (2ème série), *4:* 1, 56.
VESQUE, J., 1899. De l'emploi des caractères anatomiques dans la classification des végétaux. *Bulletin Société Botanique de France, 36:* 41, 89.

EXPLANATION OF PLATES

PLATE 24.1

Scanning electron micrographs of lower epidermis of fresh leaves (B, F); air-dried and coated leaves (A, C, D), or frozen leaves in the cryo-unit of the SEM (E).
A. *Cestrum nocturnum*. Anomocytic stomata. × 480
B. *Solanum dulcamara*. Anisocytic stomata. × 500
C. *Brunfelsia calycina*. Paracytic stomata. × 500
D. *Brunfelsia calycina*. × 1000
E. *Saracha umbellata*. Young epidermis showing stomata mother-cells. × 470
F. *Physochlaina orientalis*. Anomocytic and anisocytic stomata × 600

PLATE 24.2

Micrographs of epidermal peels from fresh leaves (C, D), or from leaves of herbarium specimens (A, B, E, F) × 400.
A. *Vestia lycioides*. Anomocytic stomata
B. *Grabowskia duplicata*. Anomocytic and anisocytic stomata
C. *Cyphomandra betacea*. Anisocytic stomata
D. *Mandragora officinarum*. Anisocytic stomata
E. *Schwenckia americana*. Diacytic stomata
F. *Dyssochroma viridiflora*. Paracytic stomata

Plate 24.1

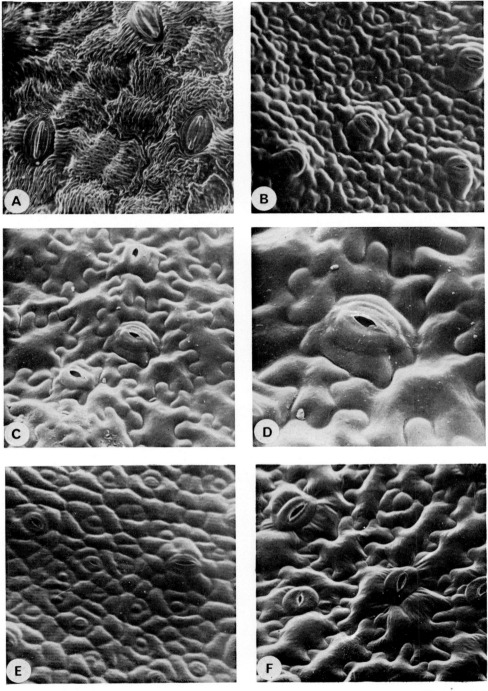

(*Facing p. 326*)

Plate 24.2

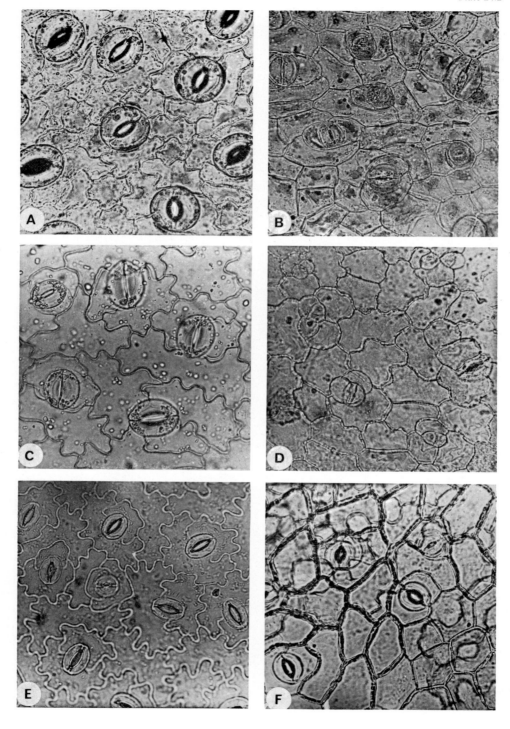

25. Pollen morphology of the Salpiglossideae (Solanaceae)

JOHNNIE L. GENTRY, JR.

Department of Botany-Microbiology, University of Oklahoma, Norman, U.S.A.

Pollen morphology of 12 genera and 23 species of the tribe Salpiglossideae (Solanaceae) was examined by scanning and transmission electron microscopy. The genera studied were *Anthocercis* Labill., *Anthotroche* Endl., *Browallia* L., *Brunfelsia* L., *Duboisia* R.Br., *Hunzikeria* D'Arcy, *Melananthus* Walpers, *Protoschwenkia* Soler., *Salpiglossis* Ruiz & Pavón, *Schizanthus* Ruiz & Pavón, *Schwenckia* L. and *Streptosolen* Miers. Tricolporate monad pollen grains characterize the majority of taxa in the tribe. The exine sculpturing is reticulate, striate, rugulate, foveolate or spinulate. Exine stratification indicates the presence of ektexine and endexine layers. The former consists of foot layer, columellae and tectal extensions which form the various surface sculpturing patterns. Of these ektexine units, the foot layer is highly pleomorphic and appears to have taxonomic significance. The endexine is well developed throughout nearly all taxa. Tetrahedral tetrads were encountered in *Salpiglossis*. Tetrads are reported for the first time in *Salpiglossis parviflora*. Scanning and transmission electron microscopy indicate that the tetrads are acalymmate with cohesion resulting from both fusion and inter-association of adjacent ektexine components. The diverse pollen morphology in *Salpiglossis* indicates that a re-evaluation of the genus is in order. Pollen morphology confirms a closer relationship of *Browallia* to *Streptosolen* than to *Brunfelsia*. *Schizanthus*, with spinulate exine sculpturing and channelled ektexine is palynologically distinct. Pollen morphology does not support the separation of the Salpiglossideae from the closely related tribe, the Cestreae.

CONTENTS

INTRODUCTION

The solanaceous tribe Salpiglossideae as currently delimited includes 12 genera and some 130 species. The genera are *Anthocercis* Labill., *Anthotroche* Endl., *Browallia* L., *Brunfelsia* L., *Duboisia* R. Br., *Hunzikeria* D'Arcy, *Melananthus* Walp., *Protoschwenkia* Soler., *Salpiglossis* Ruiz & Pavón, *Schizanthus* Ruiz & Pavón, *Schwenckia* L. and *Streptosolen* Miers. Wettstein placed nine genera in the tribe in 1891.

The tribe is considered to be the most advanced in the family and is characterized by zygomorphic corollas, reduction in the number of fertile stamens to four and didynamous, or the fertile stamens to two. There is some morphological overlap with the tribe Cestreae, subtribe Nicotianinae. The geographical distribution of the tribe is mostly South American. However, three genera, *Anthocercis*, *Anthotroche* and *Duboisia* are endemic to the Australian region.

Only a few accounts of the pollen morphology of the Salpiglossideae have been published (Erdtman, 1952; Basak, 1967; Heusser, 1971 and Rao & Leong, 1974). These studies were based on observations using light microscopy. The study by Basak (1967) recognized eight pollen types in the family. Two members of the tribe, *Browallia* and *Schizanthus* were proposed as distinctive pollen types.

The present study utilized scanning and transmission microscopical techniques to provide fundamental pollen morphological data and to investigate the systematic value of pollen characters. This report serves as a basis for future pollen studies in the family.

MATERIALS AND METHODS

Pollen of 23 species of the Salpiglossideae was examined by scanning electron microscopy (SEM) and transmission electron microscopy (TEM). Flower buds were removed from herbarium specimens and processed by the acetolysis method of Erdtman (1969). A few unacetolyzed samples were viewed with the SEM. Most of the pollen was critical point dried with a Pelco Model H critical point dryer.

Pollen for SEM was mounted on specimen stubs with rubber cement as well as without an adhesive. Vacuum evaporation was done with carbon followed by a gold-palladium mixture or only with gold and samples for SEM were examined and photographed with a Jelco JSM-2, International Scientific Instrument Super-II and Cambridge Stereoscan S-4 scanning electron microscopes. Pollen for TEM was examined with a Philips 200 microscope after processing by previously described methods (Skvarla, 1973).

The voucher specimens and collection data of the species examined are listed in Table 25.1.

RESULTS AND DISCUSSION

Pollen of two species of *Brunfelsia*, *B. undulata* (Sect. *Brunfelsia*) and *B. dwyeri* (Sect. *Franciscea*), exhibit major differences in exine sculpturing, aperture morphology, and shape. *Brunfelsia undulata* has foveolate, 3-porate pollen (Plate 25.1, figs 1–3). The circular polar view (Plate 25.1, fig. 1) is in marked contrast to the suboblate shape as noted in equatorial view (Plate 25.1, fig. 2). In *B. dwyeri* the apertures are more highly developed (i.e. colporate) (Plate 25.1, figs 4, 5), the shape is more spheroidal (Plate 25.1, fig. 5) and the sculpturing is rugulate-foveolate rather than foveolate (Plate 25.1, fig. 6). The ektexine (Plate 25.3, fig. 15) is composed of a thick, perforate tectum, short and thick columellae and an extremely thin foot layer. The slightly undulating endexine is approximately equal in thickness to the foot layer and columellae. Exine stratification of *B. undulata* is similar to *B. dwyeri*. This study indicates a possible correlation between the pollen morphology and the sections, *Brunfelsia* and *Franciscea*, recognized by Plowman (1973).

Pollen of *Streptosolen*, a monotypic genus, is rugulate and usually 7–8 colporate (Plate 25.2, figs 7, 8). In *Browallia*, a genus of two species, the pollen is rugulate or striate-rugulate and 5–7 colporoidate or colporate (Plate 25.2, figs 9, 10). The unacetolyzed pollen of *Browallia* does not reveal the true nature of the colpi (Plate 25.2, figs 11, 12). The region of the colpi is indicated by a relatively smooth and slightly raised area at the equator (Plate 25.2, fig. 12). Although SEM reveals similar exine sculpturing and aperture morphology in both genera, TEM indicates somewhat different ektexine patterns. In *Streptosolen* (Plate 25.3, fig. 13) the imperforate tectum forming the rugulae is supported by granular columellae, a thin, barely perceptible foot layer, and an endexine that is thick, uniform and lamellate in the colpus. In contrast, in *Browallia* (Plate 25.3, fig. 14) the ektexine is composed primarily of columellae which are highly variable in thickness. The columellae are linked by basal plate-like components. The latter also appear to bridge the apparent mesocolpial gap within the exine and connect with a problematic foot layer and smooth, uniform endexine. The ektexine is approximately five times thicker than the endexine. These two genera are readily distinguished from other members of the tribe by the number of apertures and exine sculpturing. The results of this investigation suggests that *Browallia* is more closely related to *Streptosolen* than to *Brunfelsia*.

Table 25.1. Material examined

Taxon	Location	Collector and number		Herbarium	Plate and fig. no.
Anthocercis anisantha Endl.	Australia	P. G. Wilson	557	US	25.14: 67, 68
A. littorea Labill.	Australia	E. Pritzel	266	F	25.13: 63–66
A. tasmanica Hook.	Tasmania	D. E. Symon	8820	MO	25.14: 69–71
Anthotroche truncata Ising	Australia	D. E. Symon	8188	MO	25.15: 72–74
Browallia americana L.	Costa Rica	W. & M. Burger	8003	F	25.3: 14
B. americana L.	Costa Rica	J. L. Gentry & W. C. Burger	2740	F	25.2: 9, 10
B. americana L.	Costa Rica	A. Smith	A532	F	25.2: 11, 12
Brunfelsia dwyeri D'Arcy	Panama	R. Wilbur & Weaver	11358	F	25.1: 4–6; 25.3: 15
B. undulata Swartz	Panama	T. Plowman	3163	F	25.1: 1–3
Duboisia hopwoodii F. Muell.	Australia	E. Pritzel	857	US	25.15: 75, 76
D. myoporoides R. Br.	Australia	M. S. Clemens	42559	US	25.15: 77
Hunzikeria texana (Torr.) D'Arcy (=*Leptoglossis texana* (Torr.) Gray	United States: Texas	E. A. Mearns	1446	US	25.8: 40, 43
H. texana (Torr.) D'Arcy	United States: Texas	E. A. Mearns	1331	US	25.8: 41, 42
H. texana (Torr.) D'Arcy	Mexico	A. Perkins & J. Hall	3313	F	25.8: 44
Melananthus fasciculatus (Benth.) Soler.	Brazil	L. B. Smith	6453	US	25.6: 27–30; 25.7: 32, 33
M. guatemalensis (Benth.) Soler.	Honduras	P. C. Standley	22146	F	25.6: 31
Protoschwenkia mandonii Soler.	Bolivia	O. Buchtien	178	F	25.5: 21–24; 25.7: 38, 39
Salpiglossis acutiloba I. M. Johnston	Peru	F. W. Pennell	13063	F	25.9: 45–47
S. lomana (Diels) Macbr.	Peru	P. C. Hutchinson	1298	F	25.9: 48, 49
S. lomana (Diels) Macbr.	Peru	Y. Mexia	7776	F	25.9: 50
S. parviflora Phil.	Chile	E. Werdermann	195	F	25.11: 56, 57
S. schwenkioides (Benth.) Wettst.	Peru	P. C. Hutchinson	1032	F	25.10: 51–55
Schizanthus lilacinus Kuntze	Chile	E. Werdermann	817	F	25.12: 58–62
Schwenckia americana L.	Guatemala	A. & A. R. Molina	25273	F	25.7: 34, 35
S. curviflora Benth.	Brazil	F. C. Hoehne	1079	US	25.4: 19, 20
S. lateriflora (Vahl) Carv.	Brazil	M. C. Gonig	s.n.	F	25.5: 25, 26; 25.7: 36, 37
S. paniculata (Raddi) Carv.	Brazil	E. Pereira	5653	US	25.4: 16–18
Streptosolen jamesonii (Benth.) Miers	Peru	E. Edwin & J. Schunke	3777	F	25.2: 7, 8; 25.3: 13

Pollen of *Melananthus* (Plate 25.6, figs 27–31) is 3-colporate with reticulate exine sculpturing. Both *M. fasciculatus* (Plate 25.6, figs 27–30) and *M. guatemalensis* (Plate 25.6, fig. 31) have identical exine sculpturing. In thin section (Plate 25.7, figs 32–33) the ektexine is composed of a thick, perforate tectum, short columellae and an undulating continuous foot layer. The smooth endexine is approximately the same thickness as the foot layer. The pollen morphology of *Melananthus* indicates an affinity with *Schwenckia* (see Discussion below).

Pollen of *Schwenckia* is reticulate and 3-colporate (Plates 25.4, figs 16–20 and 25.5, figs 25, 26). *Schwenckia lateriflora* (Plate 25.5, figs 25, 26) is distinguished from other species (Plate 25.4, figs 16–20) by the possession of free standing lacunar columellae. In TEM (Plate 25.7, figs 34–37) exine structure is similar to *Melananthus* (Plate 25.7, figs 32, 33). However, as noted in *S. americana* (Plate 25.7, fig. 35) the foot layer appears to be perforated by radially oriented channels. Within the limitations of this study foot layer channels have been noted in only one other genus, *Schizanthus* (Plate 25.12, fig. 62).

Protoschwenkia (= *Schwenckia mandonii*), a segregate genus of *Schwenckia*, has pollen which is reticulate and 3-colporate (rarely 4-colporate) (Plate 25.5, figs 21–24). Exine structure as noted by TEM (Plate 25.7, figs 38, 39) is similar to other species of *Schwenckia* (Plate 25.7, figs 34–37). *Schwenckia* and *Protoschwenkia* are closely related and both have been considered to be congeneric. The palynological evidence can be interpreted as supporting this congeneric grouping as well as the generic status of *Protoschwenkia* as proposed by Solereder (1898) and Carvalho (1966).

The genus *Hunzikeria* (D'Arcy, 1976) has a very distinctive pollen morphology (Plate 25.8, figs 40–43). Exine sculpturing is reticulate and with irregularly spaced "bulges" (Plate 25.8, figs 42, 43) which are inflations of the tectum. The pollen is 3-colporoidate or 3-colporate. Exine structure (Plate 25.8, fig. 44) including the perforate tectum supported by short, granular columellae, is similar to that noted in *Streptosolen* (Plate 25.3, fig. 13).

The pollen of *Salpiglossis* (Plates 25.9, figs 45–50; 25.10, figs 51–55 and 25.11, figs 56, 57) shows the greatest range of morphological variation in the tribe. Exine sculpturing is reticulate (*S. acutiloba*, Plate 25.9, figs 45, 46; *S. lomana*, Plate 25.9, figs 48, 49), striate (*S. schwenckioides*, Plate 25.10, figs 51–54) and rugulate (*S. parviflora*, Plate 25.11, fig. 56). The irregular pattern of the muri in *S. acutiloba* (Plate 25.9, fig. 45) is readily distinguished from the more regular pattern of *S. lomana* (Plate 25.9, fig. 48). In addition to the typical monad pollen, mature tetrahedral tetrad pollen was noted for *S. parviflora* (Plate 25.11, fig. 56). Previously the only record of *Salpiglossis* tetrads was recorded for *S. sinuata* Ruiz & Pavón (Erdtman, 1945). TEM of *Salpiglosiss* reveals a thick perforate tectum and short columellae in all species. Most unusual is the foot layer. As noted in all species examined (Plates 25.9, figs 47, 50; 25.10, fig. 55 and 25.11, fig. 57) the foot layer is highly irregular and appears to be discontinuous. This observation is in need of subsequent investigation but it seems that the foot layer in *Salpiglossis* pollen is the most pleomorphic of all genera in the tribe. In contrast to the foot layer the underlying endexine is usually of a constant thickness. However, like the foot layer, it apparently also is somewhat discontinuous (Plate 25.9, fig. 50).

Cohesion of the tetrahedral tetrads in *S. parviflora* results from both fusion and interassociation of the ektexine of adjacent pollen units (Plate 25.11, fig. 57). This type of cohesion is considered as acalymmate (Van Campo & Guinet, 1961).

Electron microscopy reveals that the species of this genus can be readily distinguished by their palynological characters. This diverse pollen morphology may provide useful taxonomic characters in a re-evaluation of the genus.

Schizanthus, a genus of eight species, is characterized by 3-colporate, spinulate pollen (Plate 25.12, figs 58–61). In a light microscopy study, Basak (1967) described the sculpturing as finely reticulate. The ektexine is composed of a thick tectum, short but thick columellae and a well developed foot layer (Plate 25.12, fig. 62). The tectum and foot layer are perforated by radially and transversely oriented channels. These channels are not known to occur in any other member of the tribe, with the exception of *Schwenckia americana* (Plate 25.7, fig. 35). The genus is unique in the Salpiglossideae because of the distinctive pollen type. It stands apart from the other genera and is readily distinguished by the exine sculpturing and by the channelled tectum and foot layer. The relationships of *Schizanthus* to other genera of the Salpiglossideae are uncertain.

The pollen of *Anthocercis* (Plates 25.13, figs 63–66 and 25.14, figs 67–71), *Anthotroche* (Plate 25.15, figs 72–74) and *Duboisia* (Plate 25.15, figs 75–77) is striate, 3-colporate or 3-colporoidate. Basak (1967) described the apertural condition as colpate, colporate or colporoidate for *Duboisia*. As seen in scanning electron micrographs, the ridges (lirae) appear to be formed by "bulges" which are evident on each side. These "bulges" are extensions of the underlying "internal" tectum and are not known to occur in the other genera herein examined. The colpus membrane in these taxa is coarsely and densely granular. Although the sculpturing is basically similar in all three genera, subtle differences may prove to be of taxonomic importance at the specific rank (Plates 25.13, fig. 66; 25.14, figs 68, 70 and 25.15, figs 73, 77). In TEM of these genera the ektexine consists of an irregular, moderately thick foot layer, slender columellae, and a thick but perforate tectum. Arising from this tectum is a second group of columellae which are terminated by a smooth ridge (i.e., the lirae of SEM). The endexine is continuous but highly irregular and approximately equal in thickness to the foot layer (in Plate 25.15, fig. 74 of *Anthotroche truncata* the endexine-foot layer separation is obscured as a result of excessive electron staining).

There is a correlation between pollen morphology and geographical distribution. These genera, all endemic to the Australian region, form a natural group of about 30 species. The pollen morphological features of *Anthocercis*, *Anthotroche* and *Duboisia* are sufficiently distinct from the other members of the Salpiglossideae to warrant recognition as a distinctive tribe within the Solanaceae.

In conclusion, preliminary study of the pollen in the Salpiglossideae revealed a rather heterogeneous morphological assemblage. Some of the genera and most species can be differentiated on the basis of the pollen characters revealed by electron microscopy. This work and additional investigations (Gentry & Skvarla, unpubl.) does not support the separation of the Salpiglossideae from the closely related tribe, the Cestreae.

ACKNOWLEDGEMENTS

I am grateful to John J. Skvarla of the University of Oklahoma for the transmission electron microscopy work and general support during the preparation of this paper. Gratitude is expressed to William F. Chissoe of the University of Oklahoma and Christine Niezgoda and Fred Huysmans of the Field Museum of Natural History for technical assistance. Also, I want to thank the curators of the herbaria that permitted me to remove polliniferous materials for this study. This work was supported by NSF Grant BMS72-02149 to the Field Museum of Natural History in support of the Scanning Electron Microscope facility, the K. P. Schmidt Fund, Field Museum of Natural History and an Arts and Sciences Faculty Research Grant, the University of Oklahoma to J. L. Gentry.

REFERENCES

BASAK, R. K., 1967. The pollen grains of Solanaceae. *Bulletin of the Botanical Society of Bengal, 21:* 49–58.

CARVALHO, L. d'A. F. DE, 1966. O gênero *Protoschwenkia* no Brasil (Solanaceae). *Sellowia, 18:* 67–72.

D'ARCY, W. G., 1976. New names and taxa: Solanaceae. *Phytologia, 34:* 283.

ERDTMAN, G., 1945. Pollen morphology and plant taxonomy. IV. *Svenska Botanisk Tidskrift 39:* 279–285.

ERDTMAN, G., 1952. *Pollen Morphology and Plant Taxonomy. I. Angiosperms.* Stockholm: Almquist & Wiksell.

ERDTMAN, G., 1969. *Handbook of Palynology.* Copenhagen: Munksgaard.

HEUSSER, C. J., 1971. *Pollen and Spores of Chile.* Tucson: The University of Arizona Press.

PLOWMAN, T., 1973. *The South American Species of* Brunfelsia *(Solanaceae).* Ph.D. Dissertation, Harvard University.

RAO, A. N. & LEONG, L. F., 1974. Pollen morphology of certain tropical plants. *Reinwardtia, 9:* 153–176.

SKVARLA, J. J., 1973. Pollen. In P. Gray (Ed.), *Encyclopedia of Microscopy and Microtechnique:* 456–459. New York: Van Nostrand.

SOLEREDER, H., 1898. Zwei beiträge zur systematik der Solanaceen. *Berichte der Deutschen Botanischen Gesellschaft, 16:* 242–249.

VAN CAMPO, M. & GUINET, P., 1961. Les pollens composes. L'exemple des mimosacees. *Pollen et Spores, 3:* 201–218.

WETTSTEIN, R. VON, 1891. Solanaceae. In A. Engler & K. Prantl (Eds), *Die Natürlichen Pflanzenfamilien, 4:* 4–38, 1895.

EXPLANATION OF PLATES

PLATE 25.1

Figs 1–6. SEM of *Brunfelsia*. 1–3. *B. undulata*: 1, polar view showing portions of 3 apertures, × 1280; 2, equatorial view showing 2 apertures with partial pore membranes, × 1730; 3, aperture view showing fragmented endexine, × 6330. 4–6. *B. dwyeri*: 4, polar view, × 1520; 5, equatorial view showing colporate aperture, × 1600; 6, exine sculpturing, × 8000. Unless indicated, the scale equals 1 μm.

PLATE 25.2

Figs 7–12. SEM of *Streptosolen* and *Browallia*. 7–8. *S. jamesonii*: 7, equatorial view showing 3 apertures, × 1800; 8, exine sculpturing, × 10,000. 9–12. *B. americana*: 9, subequatorial-subpolar view showing 3 apertures, the pollen grain in a slightly collapsed condition, × 1900; 10, polar view; exine sculpturing and portions of 2 apertures; one aperture with a partial colpial membrane, × 3000; 11, polar view of unacetolyzed pollen grain, the pollen grain angular in shape and with a polar "cap", × 1130; 12, equatorial view of unacetolyzed pollen showing apertural regions (arrows), × 1235. Unless indicated, the scale equals 1 μm.

PLATE 25.3

Figs 13–15. TEM of *Streptosolen*, *Browallia* and *Brunfelsia*. 13. *Streptosolen jamesonii*: oblique section through colpus. Endexine on left is incomplete because of sectioning angle. Layers of thin fibrillar lamellae occur in aperture region, × 8000. 14. *Browallia americana*: section through mesocolpium. Arrows indicate elements of columellae base which extend across cavus-like mesocolpium and connect with thin problematic foot layer, × 21,000. 15. *Brunfelsia dwyeri*: section through mesocolpium showing a slightly undulate endexine, thin foot layer, short columellae and thick tectum, × 6000. The scale equals 1 μm.

PLATE 25.4

Figs 16–20. SEM of *Schwenckia*. 16–18. *S. paniculata*: 16, polar view, × 2850; 17, equatorial view showing colporate aperture, × 2800; 18, exine sculpturing, × 10,000. 19, 20: SEM of *S. curviflora*: 19, polar view, × 1895; 20, exine sculpturing, × 12,000. The scale equals 1 μm.

PLATE 25.5

Figs 21–24. SEM of *Protoschwenkia mandonii*: 21, subpolar view, × 4225; 22, equatorial view showing colporate aperture and smooth colpus membrane, × 4000; 23, equatorial view of mesocolpial area, × 4380; 24, subequatorial view showing partial colpus, × 12,000. Figs 25, 26. SEM of *Schwenckia lateriflora*: 25, equatorial view of mesocolpial area; × 3275; 26, exine sculpturing showing free standing lacunar columellae, × 6400. The scale equals 1 μm.

PLATE 25.6

Figs 27–31. SEM of *Melananthus*. 27–30. *M. fasciculatus*: 27, polar view, × 3375; 28, equatorial view, × 3400; 29, exine sculpturing, × 15,000; 30, exine sculpturing, × 20,000. 31. SEM of *M. guatemalensis*: equatorial view of mesocolpium, × 2965. The scale equals 1 μm.

PLATE 25.7

Figs 32–39. TEM of *Melananthus*, *Schwenckia* and *Protoschwenkia*. 32–33. *Melananthus fasciculatus*: 32, tangential section, × 5000; 33, note prominent and continuous foot layer, × 14,000. 34, 35. *Schwenckia americana*: 34, oblique section showing part of a colpus, × 6000; 35, note channelled foot layer, × 23,000. 36, 37. *Schwenckia lateriflora*: 36, near equatorial view showing thickened endexine in colpial regions, × 5000; 37, note continuous foot layer, × 6000. 38, 39. *Protoschwenkia mandonii*: 38, section is near-medial longitudinal, × 3000; 39, exine structure is similar to *Melananthus fasciculatus* and *Schwenckia lateriflora*, × 12,000. The scale equals 1 μm.

PLATE 25.8

Figs 40–44. 40–43. SEM of *Hunzikeria texana*: 40, polar view, × 1600; 41, equatorial view showing colporate aperture and coarse and granular colpus membrane, × 1600; 42, equatorial view of meso-colpial area, × 1600; 43, exine sculpturing in mesocolpial area (note inflations of the tectum), × 5000. 44. TEM of *H. texana* showing an undulate endexine and an absence of a foot layer, × 16,240. Unless indicated, the scale equals 1 µm.

PLATE 25.9

Figs 45–50. 45–47. *Salpiglossis acutiloba*: 45, SEM polar view (note irregular pattern of muri), × 2000; 46, SEM of partial colpus, × 3100; 47, TEM showing a poorly delimited foot layer, × 8460. 48–50. *S. lomana*: 48, SEM subpolar view, × 1310; 49, SEM of equatorial view in mesocolpial area, × 2500; 50, TEM showing a prominent and discontinuous foot layer, × 11,730. Unless indicated, the scale equals 1 µm.

PLATE 25.10

Figs 51–55. *Salpiglossis schwenkioides*: 51, SEM polar view, × 2525; 52, SEM of subpolar view show-ing 2 colporate apertures, × 1980; 53, SEM of exine sculpturing in mesocolpial area, × 5330; 54, SEM of equatorial view showing colporate aperture, × 3780; 55, TEM showing a discontinuous foot layer and slightly undulate endexine. The scale equals 1 µm.

PLATE 25.11

Figs 56, 57. *Salpiglossis parviflora*: 56, SEM of tetrahedral tetrad, × 1050; 57, TEM of two adjacent monad units (note fusion of ektexine to form a "bridge-like" connection), × 8050. Unless indicated, the scale equals 1 µm.

PLATE 25.12

Figs 58–62. EM of *Schizanthus lilacinus*: 58, SEM polar view, × 2585; 59, SEM of equatorial view showing colporate aperture, × 2130; 60, SEM of equatorial view in mesocolpial area, × 2125; 61, SEM of partial colpus and exine sculpturing (note finely granular colpus membrane), × 4165; 62, TEM showing thick tectum and longitudinal channels, × 31,000. The scale equals 1 µm.

PLATE 25.13

Figs 63–66. SEM of *Anthocercis littorea*: 63, polar view showing granular colpus membrane, × 2000; 64, equatorial view showing colporate aperture, × 2915; 65, equatorial view of mesocolpial area; × 2480; 66, exine sculpturing in mesocolpium, × 13,710. Unless indicated, the scale equals 1 µm.

PLATE 25.14

Figs 67–71. EM of *Anthocercis*. 67, 68. SEM of *A. anisantha*: 67, polar view showing granular colpus membrane, × 2000; 68, exine sculpturing in mesocolpium, × 20,000. 69, 70. SEM of *A. tasmanica*: 69, subpolar view showing granular colpus membrane, × 1600; 70, exine sculpturing in mesocolpium, × 20,000. 71. TEM of *A. tasmanica* in mesocolpium. The section is similar to *Anthotroche truncata* (fig. 74) and *Duboisia hopwoodii* (fig. 75) in the extensions of the "internal" tectum, × 13,000. Unless indicated, the scale equals 1 µm.

PLATE 25.15

Figs 72–74. EM of *Anthotroche*: 72, 73, SEM of *A. truncata*: 72, polar view, × 1885; 73, exine sculpturing in mesocolpium, × 20,000. 74. TEM of *A. truncata*: because of excessive electron staining the endexine cannot be distinguished from the foot layer, × 16,000. Figs 75–77. EM of *Duboisia*: 75, TEM of *D. hopwoodii*, section through colpus with a thickened endexine, × 7700; 76, SEM of *D. hopwoodii*, polar view showing granular colpus membrane, × 2000; 77, SEM of *D. myoporoides*, exine sculpturing in mesocolpium, × 20,000. Unless indicated, the scale equals 1 µm.

Plate 25.1

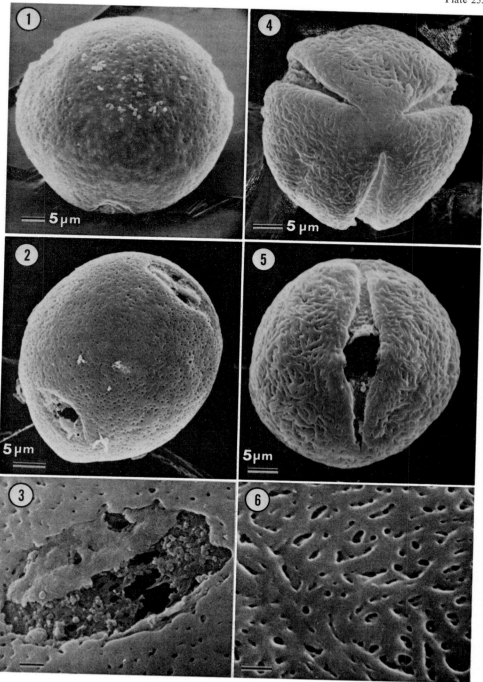

Plate 25.2

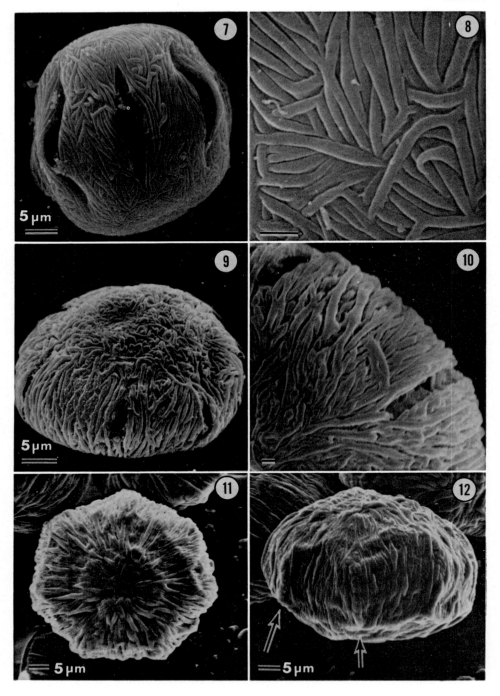

Plate 25.3

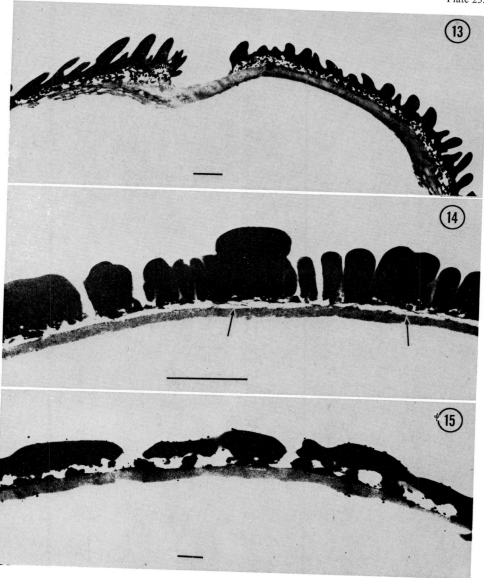

Plate 25.4

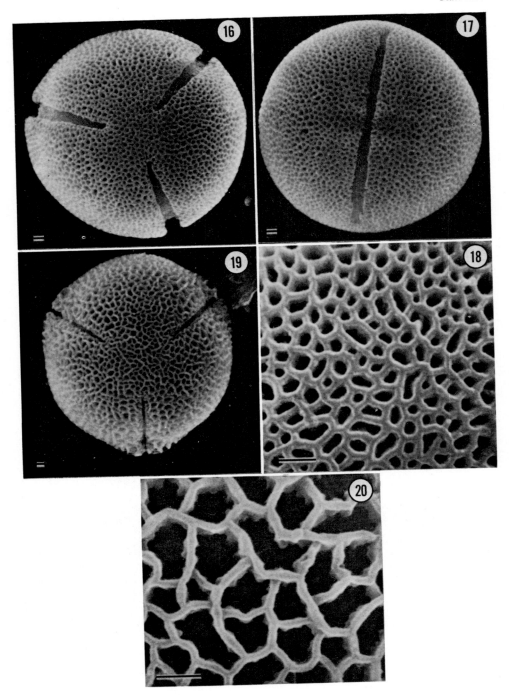

Plate 25.5

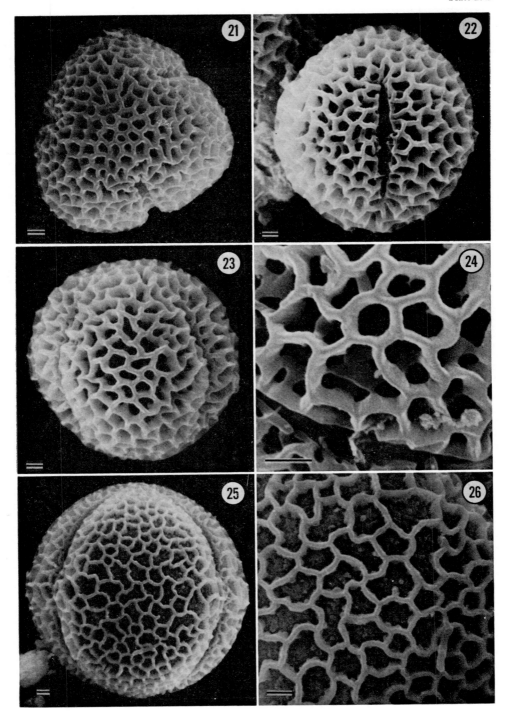

Plate 25.6

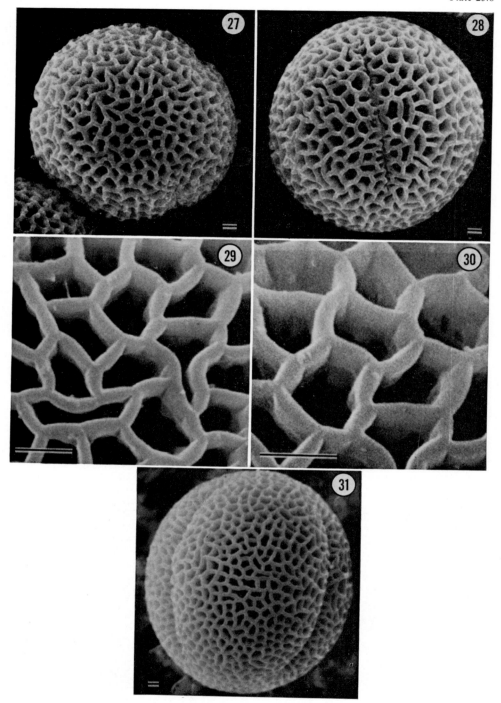

Plate 25.7

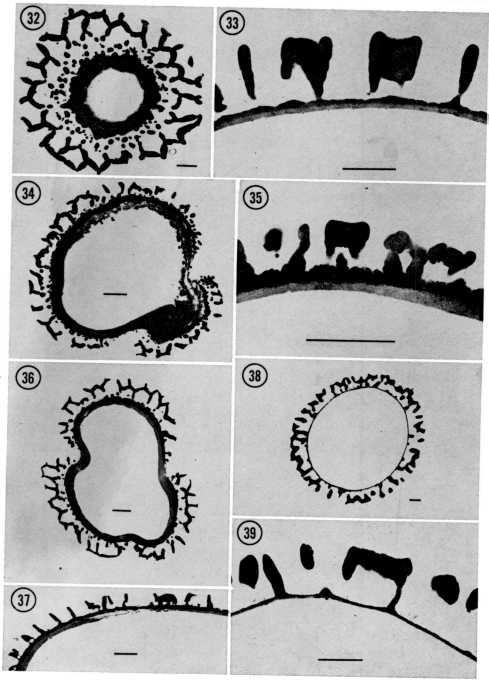

Plate 25.8

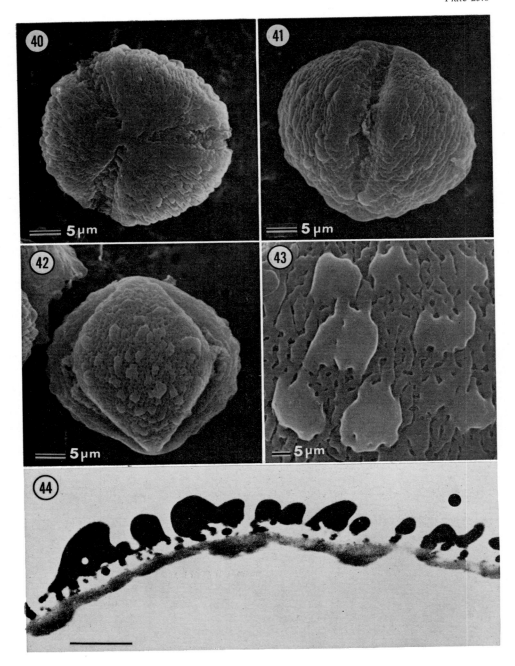

Plate 25.9

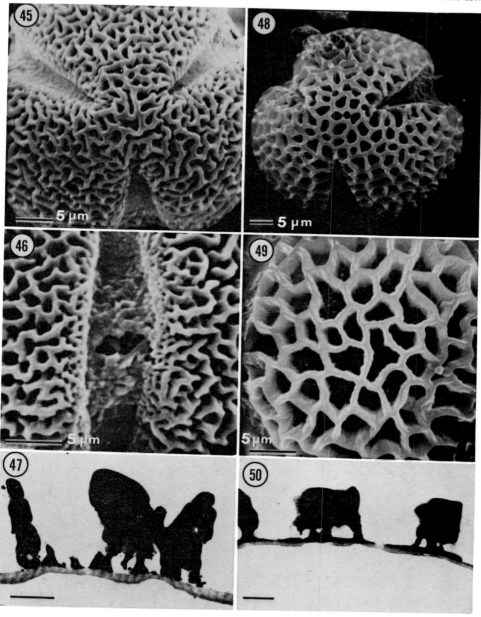

Plate 25.10

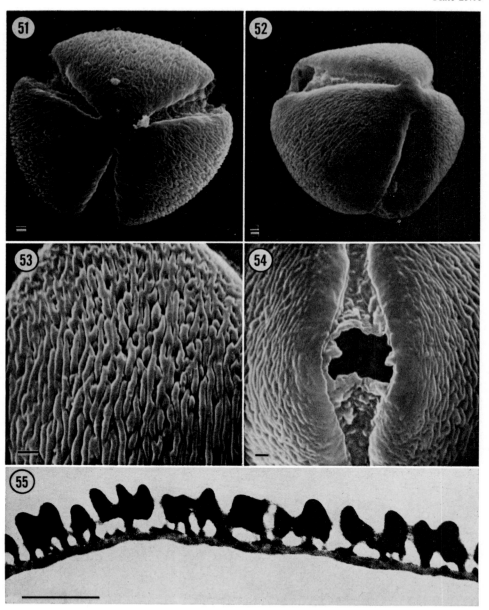

Plate 25.11

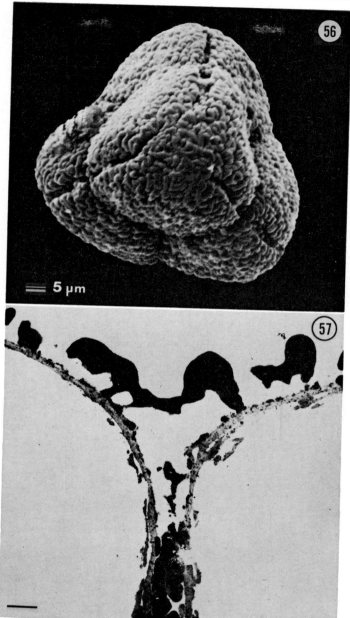

5 µm

Plate 25.12

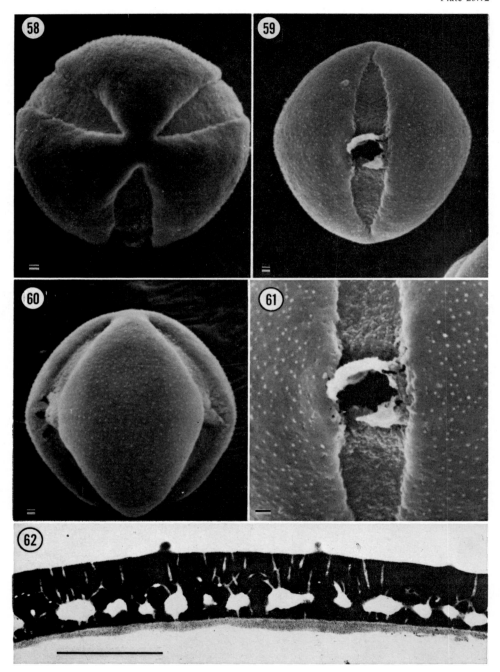

Plate 25.13

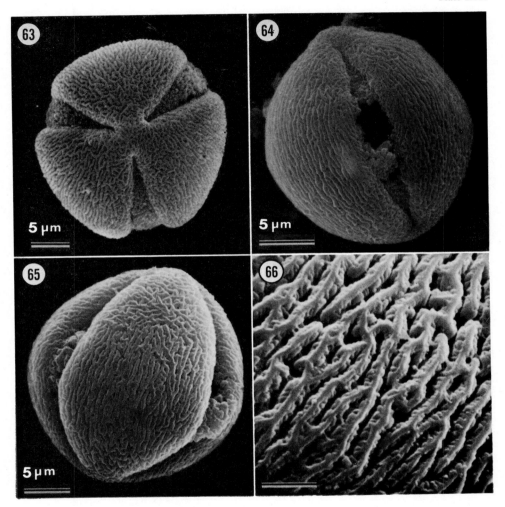

Plate 25.14

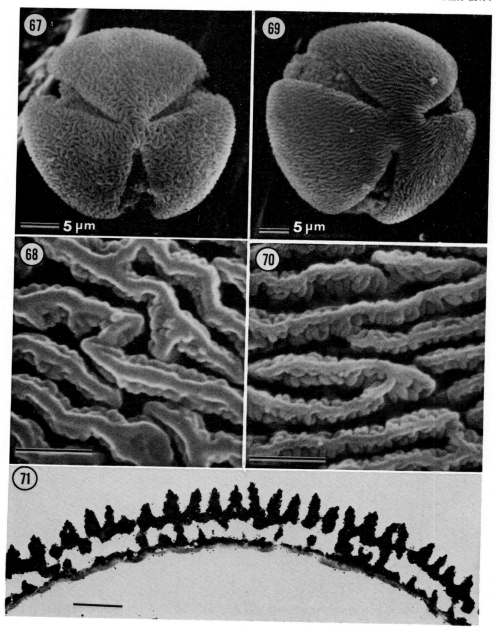

5 µm

5 µm

Plate 25.15

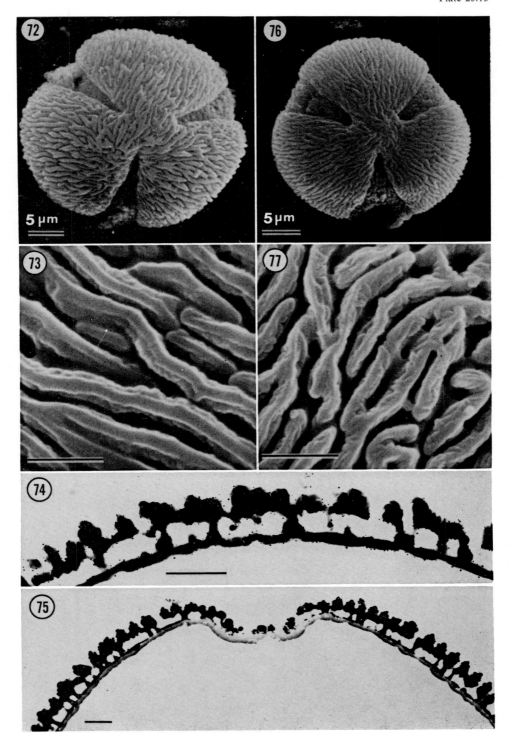

5 µm

5 µm

26. The pollen morphology of Nigerian *Solanum* species

Z. O. GBILE

Forestry Research Institute of Nigeria, Ibadan, Nigeria

AND

M. A. SOWUNMI

Department of Archaeology, University of Ibadan, Ibadan, Nigeria

The pollen grains of 19 taxa of Nigerian *Solanum* were found to have very similar apertural status, i.e. 3-colporate, the colpi and ora characteristics also being generally similar. The exine pattern is similar, though there are differences in the degree of distinctiveness. There are major differences in overall size and shape. The pollen morphological features are sufficiently distinct to permit the identification of the various species, subspecies and varieties.

CONTENTS

INTRODUCTION

The pollen grains of 19 species, subspecies and varieties of the genus *Solanum* were studied (Table 26.1); of these, 15 are indigenous to Nigeria while the remaining four were introduced and are now cultivated (see this Volume, p. 113).

In this paper a pollen key is provided (Appendix 1). The palynological investigation was carried out to ascertain whether pollen morphological features could provide additional parameters which could be used along with gross morphological and other characters in the clarification and improvement of the taxonomy of this group.

MATERIALS AND METHODS

Polliniferous materials were all obtained from herbarium specimens in the herbarium of the Forestry Research Institute of Nigeria. The materials were acetolyzed and mounted in glycerine jelly, and the slides sealed with paraffin wax (Erdtman, 1969). The pollen grains were examined with a Leitz Ortholux microscope, and the photographs taken with an Orthomat camera attached to the

Table 26.1. Pollen grain morphology of Nigerian *Solanum* species

Species/subspecies/variety	Polar axis (P) (μm)	Equatorial diameter (E) (μm)	P/E (%)	Shape class	Amb	Exine thickness (μm)	Exine thickness at aperture (μm)	Sexine pattern
S. aculeastrum Dunal var. *aculeastrum*	28.1 (27.2–29.2)	25.0 (24.0–26.0)	112.5	Subprolate	Circular to triangular	1.2	3.1	Indistinct
S. aculeatissimum Jacq.	28.2 (27.6–30.0)	22.6 (21.6–23.2)	124.6	Subprolate	Triangular	1.2	2.7	Faintly distinct
S. aethiopicum Linn.	24.2 (22.8–25.2)	23.3 (21.6–25.2)	103.8	Prolate spheroidal	Triangular	1.3	2.4	Distinct
S. anomalum Thonn.	29.5 (28.0–30.8)	23.8 (22.0–24.4)	124.1	Subprolate	Triangular	1.2	3.2	Distinct
S. dasyphyllum Schum. & Thonn.	29.2 (28.4–30.4)	27.2 (26.8–27.2)	107.5	Prolate spheroidal	Triangular	1.2	4.0	Very distinct
S. giganteum Jacq.	30.2 (28.8–31.2)	25.4 (24.4–26.8)	119.0	Subprolate	Triangular	1.3	3.6	Faintly distinct
S. gilo Raddi var. *gilo*	27.2 (24.4–29.6)	22.8 (21.2–24.4)	119.1	Subprolate	Circular	ca. 1.3	ca. 2.8	Distinct
S. hispidum Pers.	21.6 (20.8–22.8)	22.0 (21.2–22.8)	98.4	Oblate spheroidal	Triangular	1.1	2.7	Indistinct
S. indicum subsp. *distichum* (Thonn.) Bitter var. *distichum*	25.4 (24.4–27.2)	23.1 (21.6–24.0)	110.2	Prolate spheroidal	Circular	1.3	3.1	Distinct
S. indicum subsp. *distichum* var. *modicearmatum* Bitter	22.4 (21.2–23.2)	21.6 (20.0–22.4)	103.5	Prolate spheroidal	Triangular	1.2	2.4	Faintly distinct
S. incanum Linn.	26.2 (23.6–28.4)	22.4 (21.2–24.0)	117.2	Subprolate	Triangular	1.1	3.3	Indistinct
S. macrocarpon Linn.	27.7 (26.0–29.6)	27.6 (26.8–28.0)	100.4	Prolate spheroidal	Circular to triangular	1.1	3.0	Most distinct
S. nigrum Linn.	21.0 (20.4–22.0)	21.6 (20.8–23.2)	97.6	Oblate spheroidal	Triangular	1.0	2.4	Indistinct
S. seaforthianum Andr. var. *disjunctum* O. E. Schulz	16.1 (15.2–16.4)	14.7 (14.4–15.2)	109.2	Prolate spheroidal	Triangular	1.2	2.5	Indistinct
S. terminale Forsk. subsp. *inconstans* (C. H. Wright) Heine	20.7 (19.6–22.4)	21.9 (21.2–24.0)	94.3	Oblate spheroidal	Triangular	1.0	2.9	Indistinct
S. terminale subsp. *sanaganum* (Bitter) Heine	*ca.* 26.2 (*ca.* 24.8–28.4)	26.9 (25.2–28.0)	97.4	Oblate spheroidal	Circular	1.6	4.4	Faintly distinct
S. terminale subsp. *welwitschii* (C.H. Wright) Heine	22.4 (20.8–23.6)	22.0 (20.8–23.2)	101.8	Prolate spheroidal	Triangular	1.2	2.5	Indistinct
S. torvum Sw.	25.9 (24.8–28.0)	21.2 (19.2–23.2)	122.0	Subprolate	Circular	ca. 1.2	ca. 2.9	Distinct
S. erianthum Linn.	17.2 (16.4–18.4)	16.7 (16.0–17.2)	102.9	Prolate spheroidal	Triangular	0.9	1.9	Indistinct

microscope. The figures recorded for the various pollen grain dimensions were measurements from ten grains (mean and range); where, due to the paucity of grains, less than ten grains were measured, "ca." precedes the figures.

The terminology is after Erdtman (1969).[1]

POLLEN MORPHOLOGY

(P, polar axis; E, equatorial diameter.)

In order to avoid repetition, the main morphological features for all specimens studied are given here and the details for each species/variety are given separately (Table 26.1).

Pollen grains isopolar, radially symmetrical, 3-colporate, peritreme to gonio-treme, sides straight to convex, oblate spheroidal to subprolate; (P16.1–30.2 µm, E14.7–27.6 µm). Sexine granulate, subtectate, densely but faintly baculate, reticulate; aspidote.

Colpi comparatively long, extending for most of the length of the polar axis, usually distinctly constricted at and around the middle and widening towards the ends where they are widest, with distinct granulate margins. Ora conspicuous, markedly elongated transversely (length 1.2–3.6 µm, width 5.8–17.5 µm), often tapering towards the ends, may be sharply constricted in the middle, with distinct granulate margins that are often irregular in outline, being thickest around the colpi; occasionally there is a distinct annulus around the central parts; sometimes with granulate membranes (e.g. *S. incanum*, Plate 26.2, fig. 7). *S. seaforthianum* var. *disjunctum* is unique in being entirely syncolpate (Plate 26.2, fig. 29): only 3% of the grains (as seen in polar view) in *S. nigrum* are syncolpate (Plate 26.2, fig. 17).

Exine 0.9–1.6 µm thick, thicker towards end at the aperture (1.9–4.4 µm thick) where it is often less dense; aspidote (see Plates 26.1, fig. 9 and 26.2, fig. 3). Sexine minutely reticulate, reticulation often only faintly discernible, though sometimes quite distinct (cf. Table 26.1). Tectum supported by a densely baculate zone (bacules not often distinctly seen, but more clearly visible in the colpi margins), provided with dense granules (tiny verrucae?) that are often very conspicuous in the colpi and ora (e.g. *S. macrocarpon*, Plate 26.2, figs 11, 12).

DISCUSSION AND CONCLUSION

Regarding two of the most important and diagnostic pollen morphological features, i.e. apertural status and sexine pattern, the pollen grains of the *Solanum* species studied are similar to one another. The basic pollen type is one with three compound apertures (i.e. 3-colporate), comparatively long colpi and conspicuous ora that are markedly elongated transversely, and a minutely reticulate tectum on which are minute granules. Variations occur with regard to shape, size and details of exine structure.

Although it is generally recognized that the overall shape and size of pollen grains are subject to a great deal of variation, and are affected by factors such as method of preparation and age of pollen slides, pressure of the cover slip ("the Cushing effect"), and the level of maturity of the pollen grains (cf. Erdtman &

[1] For glossary of terms used, see Appendix 2.

Praglowski, 1959; Faegri & Deuse, 1960; Cushing, 1961; and Sowunmi, 1972), these characters are considered to be of relevance and importance in the present context since due consideration was given to the factors mentioned above. Thus, the pollen grains were all subjected to the same chemical treatment, measurements were carried out only a few days (three to four days) after slide preparation, and tiny balls of plasticine were placed one each in two places opposite each other beneath the cover slip thus appreciably reducing the pressure on the grains. Finally, only mature but unopened anthers from opened flowers were used. Consequently, the shapes and sizes observed and recorded are comparable and useful parameters. A consideration of the shape classes (Table 26.1) and observation of the shapes as depicted in some of the photomicrographs reveal shape variations from oblate spheroidal through to prolate spheroidal, as well as size variations from 14.7 μm in *S. seaforthianum* var. *disjunctum* to 27.6 μm in *S. macrocarpon*. A multivariate analysis of variance carried out shows conclusively that the shape and size differences within the genus are highly significant, particularly the polar axis dimension. There are slight differences in exine thickness, but these become more marked at the aspides (Table 26.1).

Not much can be deduced from these differences, however, particularly as there were no correlations between exine thickness on the one hand and pollen size or exine structure on the other. Thus, *S. terminale* subsp. *sanaganum* which has the greatest exine thickness (1.6 and 4.4 μm respectively) does not have the largest grains, neither do the smallest grains occur in *S. erianthum* which has the thinnest exine (0.9 and 1.9 μm respectively). Similarly indistinct sexine patterns are found alike in grains with either a thick or a thin exine, while grains of different exine thicknesses have a distinct pattern.

Generally, however, the species with smaller grains tend to have a faint and indistinct sexine pattern, while those with larger grains have a distinct and clearly discernible pattern.

In view of the major differences in gross and pollen morphology, the authors recommend that each of the three subspecies of *S. terminale*, subsp. *inconstans*, subsp. *sanaganum* and subsp. *welwitschii*, be given specific status. The palynological study clearly shows that the specimens belong to distinct species, and that pollen morphological features can be used to distinguish between the species or varieties investigated.

ACKNOWLEDGEMENT

Our thanks go to Dr S. O. Adamu and Mrs K. Olaofe for help with the statistical analysis, Mr R. Osoba, Mrs M. Adedeji and Mrs J. Ajayi for the photomicrographs. We are also grateful to the palynology assistants, Messrs. Omotoso and R. Afosi and to Mrs A. Aduye, Mr M. A. Odunowo, and Miss P. Okonofua for typing the manuscript.

REFERENCES

CUSHING, E. J., 1961. Size increase in pollen grains mounted in thin slides. *Pollen et Spores*, 3(2): 265–274.
ERDTMAN, G., 1969. *Handbook of Palynology* (An introduction to the study of pollen grains and spores): 486 pp. Copenhagen: Munksgaard.

ERDTMAN, G. & PRAGLOWSKI, J. R., 1959. Six notes on pollen morphology and pollen morphological techniques. *Botaniska Notiser, 112*(2): 175–184.

FAEGRI, K. & DEUSE, P., 1960. Size variations in pollen grains with different treatments. *Pollen et Spores, 2*(2): 293–298.

SOWUNMI, M. A., 1972. Pollen morphology of the Palmae and its bearing on taxonomy. *Review of Palaeobotany and Palynology, 13:* 1–80.

APPENDIX 1

Pollen key to Nigerian Solanum *species*

1.	(a) Pollen grains syncolpate	*seaforthianum* var. *disjunctum*
	(b) Pollen grains not syncolpate or only rarely syncolpate	2
2.	(a) Pollen grains rarely syncolpate	nigrum
	(b) Pollen grains not syncolpate	3
3.	(a) Pollen grains oblate spheroidal	4
	(b) Pollen grains not as above	6
4.	(a) Exine at aperture >4 μm thick	*terminale* subsp. *inconstans*
	(b) Exine at aperture <4 μm thick	5
5.	(a) P/E% not >100	*terminale* subsp. *inconstans*
	(b) P/E% sometimes >100	*hispidum*
6.	(a) Pollen grains subprolate	7
	(b) Pollen grains not as above	12
7.	(a) Sexine pattern distinct	8
	(b) Sexine pattern indistinct or only faintly distinct	10
8.	(a) Exine at aperture >3.0 μm thick	*anomalum*
	(b) Exine at aperture <3.0 μm thick	9
9.	(a) P/E% not <110	*torvum*
	(b) P/E% sometimes <110	*gilo* var. *gilo*
10.	(a) Sexine pattern indistinct	*incanum*
	(b) Sexine pattern faintly distinct	11
11.	(a) Equatorial diameter >25.0 μm	*giganteum*
	(b) Equatorial diameter <25.0 μm	*aculeatissimum*
12.	(a) Exine <1.0 μm thick	*erianthum*
	(b) Exine >1.0 μm thick	13
13.	(a) Sexine pattern distinct	14
	(b) Sexine pattern indistinct or only faintly distinct	17
14.	(a) Exine at aperture >3.5 μm thick	*dasyphyllum*
	(b) Exine at aperture <3.5 μm thick	15
15.	(a) Amb circular	*indicum* subsp. *distichum* var. *distichum*
	(b) Amb not as above	16

16. (a) P/E% not < 100   *aethiopicum*
    (b) P/E% sometimes not < 100   *macrocarpon*
17. (a) Sexine pattern indistinct   18
    (b) Sexine pattern faintly distinct   *indicum* subsp.
                                                            distichum var.
                                                            modicearmatum

18. (a) Equatorial diameter up to 25 μm   *aculeastrum* var.
                                                            aculeastrum
    (b) Equatorial diameter < 25 μm   *terminale* subsp.
                                                            welwitschii

APPENDIX 2

Glossary of palynological terms used

Aperture: any weak, preformed part of the general surface of a pollen grain or spore, which may be engaged, directly or indirectly, in forming an opening for normal exit of material in connection with the germination of a pollen grain or spore.

Amb: Outline of a (polar) pollen grain or spore viewed with one of the poles (q.v.) exactly uppermost.

Aspidote: Provided with aspides, i.e. shield-shaped exine (q.v.) area surrounding an aperture. The aspides protrude as rounded domes from the general surface of the pollen grains.

Bacules: More or less perpendicular (radial) rods.

Colpus: Longitudinal, usually more or less distinctly delimited aperture.

Colporate: Pollen grains which have compound apertures consisting of an outer colpal part and an inner oral (q.v.) part.

Equator: The line of demarcation between the two faces of isopolar pollen grains (or spores).

Exine: The outer, very resistant layer of a pollen or spore wall (sporoderm).

Goniotreme: Goniotreme pollen grains are radially symmetrical, their amb angular, their apertures situated in the angles.

Isopolar: In isopolar pollen grains or spores there are no apparent differences between the proximal and the distal face.

Oblate spheroidal: Isopolar radially symmetrical pollen grains with the ratio polar axis: equatorial diameter 1.00–0.88 are (or can be) said to be oblate spheroidal.

Ora (sing. os): term for those inner or central parts of compound apertures which are not conformable to the outer or marginal contours of an aperture.

Os: (see ora).

Peritreme: radially symmetrical pollen grains with a more or less circular amb.

Polar axis: A perpendicular imaginary line connecting the poles of a pollen grain or spore.

Pole: The centre of the surface of the inner, proximal part or face and of the outer, distal face.

Prolate spheroidal: Isopolar, radially symmetrical pollen grains or spores with the ratio polar axis: equatorial diameter 1.14–1.00.

Sexine: The outer, usually sculptured layer of the exine.

Subprolate: Isopolar radially symmetrical pollen grains or spores with the ratio polar axis: equatorial diameter 1.33–1.14.

Tectum: A more or less homogeneous layer usually distinctly separated from the nexine by a baculate zone.

Verruca (plur. verrucae): Wartlike process.

EXPLANATION OF PLATES

PLATE 26.1

Figs 1–3, *S. aculeastrum* var. *aculeastrum*; 2, note os margin; 3, note aspides. Figs 4–6, *S. aculeatissimum*; 4, note os margin. Figs 7–10, *S. aethiopicum*; 7, note constriction of colpus in the middle. Figs 11–13, *S. anomalum*. Figs 14–17, *S. dasyphyllum*; 15, note minute but distinct sexine pattern; 17, note aspides. Figs 18, 19, *S. giganteum*; 19, note aspides. Figs 20–23, *S. gilo* var. *gilo*; 23, note distinct sexine pattern. All × 1250.

PLATE 26.2

Figs 1, 2, *S. hispidum*; 1, note os margin. Figs 3–5, *S. indicum*, subsp. *distichum* var. *distichum*. Fig. 6, *S. indicum* subsp. *distichum* var. *modicearmatum*. Figs 7–10, *S. incanum*; 7, note granular os membrane. Figs 11–14, *S. macrocarpon*; 12, note exine stratification of colpus margin. Figs 15–18, *S. nigrum*; 17, syncolpate grain; 18, nonsyncolpate grain. Figs 19–21, *S. seaforthianum* var. disjunctum; 19, syncolpate grain. Figs 22, 23, *S. terminale* subsp. *inconstans*. Figs 24–26, *S. terminale* subsp. *sanaganum*. Fig. 27, *S. terminale* subsp. *welwitschii*. Fig. 28, *S. torvum*. Fig. 29, *S. erianthum*.

Plate 26.1

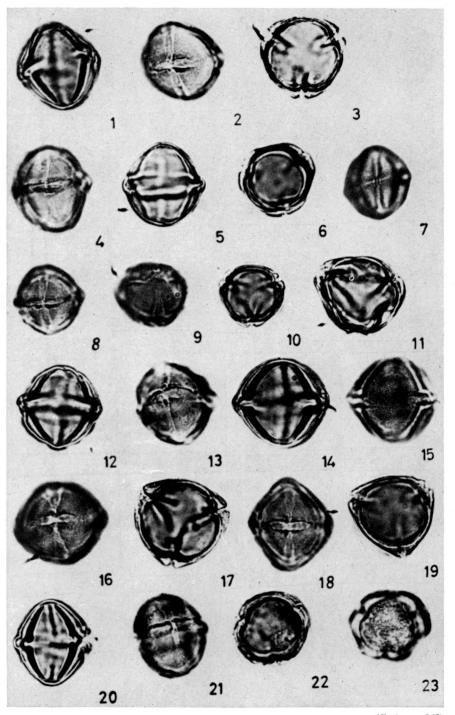

Plate 26.2

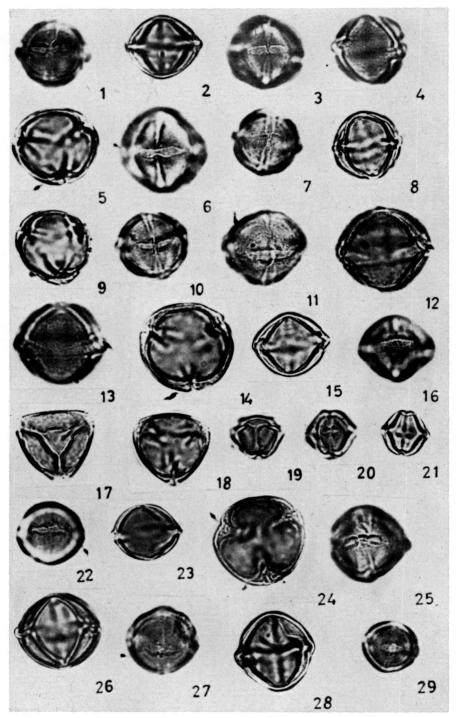

VI. Morphology and morphogenesis

27. A review of branching patterns in the Solanaceae

ALAN CHILD

Driffield, East Yorks, U.K.

The range of branching patterns in the reproductive phase of the *Solanaceae* is reviewed, following the schemes of Eichler and the papers of Danert. A possible evolutionary pathway is traced and the particular types of branching are surveyed and correlated with genera and subgeneric taxa of *Solanum* L., which are known to the author. The relevance of branching patterns to the taxonomy of the family and to the identification of individual taxa is briefly discussed.

CONTENTS

INTRODUCTION

The late Siegfried Danert (1958), by painstaking developmental studies showed the morphological basis for the observed patterns of branching (ramification) in the reproductive phase of Solanaceous plants. Subsequently (Danert, 1967), applied his findings to a study of the infrageneric variation within the complex genus *Solanum*.

Most Solanaceae exhibit a general branching pattern which suggests a regression from a monocarpic growth form where all vegetative growth is ended by reproductive shoots, which may be subtended by foliar organs or not. We still see this growth form with hapaxanthic plants like *Hyoscyamus* and many of the taxa of the tribes Cestreae and Salpiglossideae which are monocarpic ephemerals. The regression involves a return to a perennial and often woody growth form more like that of primitive angiosperms. After the shoot has been terminated by a flower or floral shoot new vegetative shoots emerge from the axils of the top leaf or leaves, and one or more will continue growth by acrotonic promotion (Danert, 1958), i.e. it is conditioned by hormone-dependent apical dominance. The top shoot, or shoots, are thus the strongest. After the continuation shoot has produced a specific number of foliar organs, it is again ended by a reproductive shoot and further vegetative continuation is initiated, and so on, throughout the growing period. This is sympodial growth and each complete section of leaves and flowers may be termed a sympodial unit, a shoot generation, or an anthoclade (Goebel, 1931). Its structure, i.e. the number of leaves, disposition of inflorescence and the degree of suppression or uneven growth of various parts of the axis and its relevant organs, is specific to the taxon involved and varies only slightly depending on the order of branching and the physiological age of the plant. Once the reproductive phase has been initiated, anthocladial growth, by repeated production of leaves and flowers, gives a more or less constant vegetative/reproductive balance throughout the growing season. It is of interest that the majority of anthocladial plants appear to be photoperiodically neutral.

Before anthocladial branching patterns are discussed, the intermediate stage between the hypothetical ancestral type of sympodial branching and the purely anthocladial may be observed with the woody members of the subfamily Cestroideae as, for example *Cestrum*, *Sessea* and the Australian *Anthocercis* (see Fig. 27.1D). Here, all long vegetative shoots end with a corymbose cyme, the lateral subsidiary branches of which are subtended by foliar organs. Vegetative continuation is then from below the reproductive section and thus the reproductive and vegetative regions are more separated than is the rule with anthocladial forms of branching. Some *Solanum* groups show a terminal reproductive structure analogous to this pattern, but usually the subtending leaves are lacking. The *Cestrum* branching pattern is closely correlated with that of hapaxanthic (i.e. monocarpic) taxa in the subfamily Cestroideae and in many Scrophulariaceae and Boraginaceae. It may therefore be considered as undifferentiated or "original" in comparison with the more evolved anthocladial forms.

Some woody members of the Solanaceae show an isolation between the vegetative and reproductive zones, but the latter behave in an anthocladial manner. This growth pattern is more typical of woody plants. Here vegetative growth is by extension shoots, and the floral region is more or less confined to lateral long or short spur shoots which have been laid down previously. The spur shoots are sympodial and resume vegetative growth in the same direction as the parent shoot, from a lateral bud below the flower or inflorescence.

The subtribe Lyciinae has several spur-flowering taxa, but there appear to be transitional forms and genera which contain both spur and purely anthocladial flowering species. Thus argument for a diphyletic origin is eroded. Spur flowering in this context could be a xero- or halophytic adaptation, as in *Grabowskia*, *Lycium*, *Dunalia* sect. *Pauciflora* and *Latua*. *Iochroma* is more mesophytic and seems intermediate in character. *Iochroma*, *Dunalia* and *Acnistus* were united by Sleumer, but his treatment seems unjustified, even only on the basis of branching patterns. *Acnistus* is associated near *Witheringia* by A. T. Hunziker.

Von Wettstein's tribe Datureae also shows arborescent, highly adapted forms with spur-flowering features. In contrast to the foregoing, however, there are climbers or shrubs ecologically specialized as epiphytes or semi-epiphytes of moist Andean or meso-American forests, some species having large chiroptophilous flowers. Genera here are: *Solandra*, *Trianaea*, *Markea* and *Juanulloa*. The two latter genera have been artificially attached to the *Cestroideae* on the basis of embryological features. Their branching pattern is different from that subfamily.

ANTHOCLADIAL BRANCHING

Eichler (1875) in his Blüthendiagrammen shows three types of anthocladial branching (Fig. 27.1A–C).

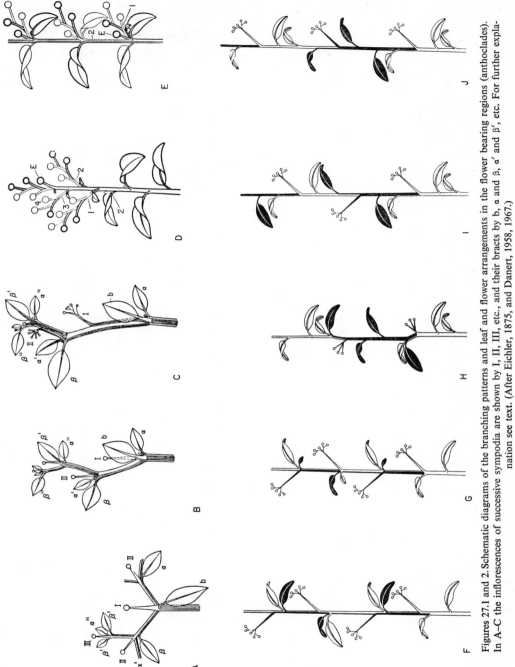

Figures 27.1 and 2. Schematic diagrams of the branching patterns and leaf and flower arrangements in the flower bearing regions (anthoclades). In A–C the inflorescences of successive sympodia are shown by I, II, III, etc., and their bracts by b, α and β, α′ and β′, etc. For further explanation see text. (After Eichler, 1875, and Danert, 1958, 1967.)

Fig. 27.2

(1) *Dichasially forked sympodial units* as exemplified by *Datura*, New World species of *Physalis*, *Margaranthus*, *Capsicum* (early order anthoclades), *Nectouxia* (Danert, 1958) and first order branchings of other genera (Fig. 27.1A). Here, the main axis ends with a terminal flower. (If "twinned" flowers, are seen here, they are the expression of a reduced lateral reproductive shoot, initiated below the terminal flower, but which grows with it concaulescently, forming a two flowered inflorescence.) As in most Solanoideae subsidiary inflorescence axes are not subtended by foliar organs. In *Datura* we see the terminal flower and two lateral shoots which have gained ascendance by apical dominance (acrotony). These grow out from the axils of the leaves preceding the flower and form the next shoot generation. The subtending leaves, however, have been carried up their product for some distance (recaulescence) and also carry a subsidiary or accessory bud in their axils. In this way, the continuation shoots apparently arise as extra-axillary growths.

If we divide each sympodial unit into hypopodium, mesopodium and epipodium (Danert, 1958), we see here that the centre section (mesopodium) has failed to elongate; therefore the subtending leaf from the previous shoot generation has been carried up to the level of the first foliage organ or α-bract of its daughter shoot. Thus we have two apparently opposite leaves of different size (the subtending leaf being the larger) and, in some instances, different shape. Such a structure is a geminate, anisophyllous cluster, consisting of foliar organs from consecutive shoot generations. In the genera quoted here, only two leafy bracts are inserted between the successive reproductive zones, but they appear separated; the β-bract, which subtends the top continuation shoot, is carried up above the flower of its own sympodial unit and is geminate with the α-bract of the daughter shoot. Thus the terminal flower sits, after anthesis, in the fork formed by the continuation shoots. Thus we have difoliate dichasial sympodial units. Later orders of branching go over to monochasial continuation where only the top axillary shoot is dominant, depending on physiological influences.

Before monochasial branching is considered the term bract as used in this context should be explained. A bract is a foliar organ, not necessarily morphologically different from a true leaf, but which is initiated at 180 degrees to its partner instead of at the two fifths divergence which is the characteristic phyllotaxy for the *Solanaceae*. According to Danert's developmental studies, the first two foliar organs of a lateral shoot are bracts, though the divergence angle may not be faithfully retained at full foliation owing to torsional influences. The pause in the initiation sequence on the differentiating part of the meristem (plastochrone) causes a divergence from the normal leaf mosaic. Sympodial units with only bracts may be therefore thought of as an adaptation, thus more evolved, in comparison to those with several leaves between the reproductive zones. Danert (1967) commented on the correlation between the number of leaves on a sympodial unit and the number of flowers or reproductive branches in an inflorescence. Where we end up with only two bracts on a sympodial unit and only 2-flowered inflorescences, or even only the terminal flower, we see the end of a reduction series and a final adaptation to anthocladial growth.

(2) *Eichler's second schema* shows difoliate sympodial units reduced to monochasial branching (Fig. 27.1B). Here, in contrast to (1), the terminal flower or inflorescence is soon pushed laterally and the dominant lateral shoot (from the axil

of the top leaf or β-bract) soon assumes the orthotropic attitude of the parent shoot. This gives rise to the widely held opinion that the inflorescence is lateral and the shoot system is monopodial. The cincinnus (monochasial sympodial unit) is betrayed by the somewhat zig-zag line of the total axis and the deflected nature of the inflorescence axis (peduncle). In *Atropa*, as shown by Eichler, the α- and β-bracts of consecutive shoot generations are geminate, as described for *Datura*, and owing to epipodial suppression, the terminal flower appears at the level of the geminate cluster, thus being falsely axillary. Here then, we have leaf recaulescence and meso- and epipodial suppression of the relevant axis. This highly differentiated branching pattern is common in the tribe *Solaneae* and elsewhere (e.g. *Nolana*). We see it, or minor variations, in *Nicandra*, *Jaltomata*, *Athenaea*, *Chamaesaracha*, *Witheringia*, *Capsicum*, *Lycianthes*, *and Salpichroa*, among others. In *Atropa*, *Nicandra* and sometimes *Capsicum*, *Lycianthes* and *Salpichroa* only the terminal flower or a pair of flowers is formed. In the latter case the distal flower represents a reduced subsidiary reproductive shoot, initiated between the terminal flower and the continuation shoot. This subsidiary floral shoot grows concaulescently with the base of the terminal flower axis and with it forms the reduced inflorescence. In *Witheringia* and *Acnistus* the inflorescence is more or less compound and more than one subsidiary inflorescence is inserted after the initiation of the terminal flower (Fig. 27.1E). Owing to the reduction of the axis of the subsidiary inflorescences (rhachides) it appears subumbellate, and owing to the suppression of the common axis (peduncle) it appears more or less sessile. In *Witheringia* there is a continuum from many flowered to few flowered inflorescences. The peduncle is often reduced to a pad-like structure at the level of the geminate leaf cluster.

(3) Eichler's representation of *Solanum nigrum* L. (Fig. 27.1C) shows similar behaviour to *Atropa*, but here the epipodium is elongated and the inflorescence axis is leaf remote, above the geminate bract cluster as described for *Datura*. The peduncle or common reproductive axis is well developed here, but the rhachides are suppressed and this gives a typically subumbellate inflorescence. The lateral deflection of the peduncle takes place early and the dominant continuation shoot is more or less orthotropic. A penultimate lateral shoot, from the axil of the α-bract, may grow out, never so strongly as the continuation shoot, but it is often significant in early order branchings. It goes over to the reproductive phase after producing two bracts and one or two leaves, but it is then not homologous with the top shoot as with true dichasial branching.

THE GENUS *SOLANUM* L.

Branching patterns of the infrageneric taxa examined

Section *Solanum*. The type section for the genus above shows an evolved form of branching as just examined in Eichler's schema (Fig. 27.1C). On higher order branchings, owing to epipodial suppression, the peduncle may be deflected at the level of the geminate cluster, but the peduncle is still elongated, thus differing from *Witheringia*. Lower order shoots on the same plant will show typical morelloid branching as per Eichler, with epipodial stretching (Fig. 27.1C).

Section *Gonatotrichum* Bitter. A morelloid group, but the inflorescence is more sessile owing to suppression of the peduncle (Fig. 27.1E).

Section *Campanulisolanum* Bitter. A shrubby morelloid taxon, hardly bearing a clear separation from section *Solanum*; the sympodial units may, however, show a few leaves above the two bracts, before the reproductive zone is inserted.

Section *Leiodendra* Dun. (*Oppositifolium* Seithe; *Geminatum* Walp.) This arborescent group shows a highly evolved anthocladial pattern similar to the morelloid Solanums but there is marked meso- and epipodial suppression, the total axis is condensed and the peduncle appears leaf-opposed or at the level of a geminate cluster (depending whether the sympodial units are di- or trifoliate) (Fig. 27.1F,G). Danert (1967) also speaks of "fasciculate" terminal inflorescences topping the long shoots (cf. *Cestrum*) but quotes no species. The author has observed no such pattern for the *Leiodendra*.

Section *Pseudocapsicum* (Dun.) Bitter. As in the two preceding taxa, the meso-podium is suppressed, giving the familiar geminate cluster (Fig. 27.1F), but another two leaves may be interposed before the reduced sessile inflorescence (only one or rarely two fertile flowers) stops the shoot generation (Fig. 27.1H). This inflorescence reduction represents an evolved character. This section is bio-chemically deviant but is near the morelloids and *Leiodendra* morphologically.

Subgenus *Leptostemonum* pro majore parte. This subdivision of the genus, with stellate trichomes, often prickly emergences and usually attenuate anthers, for the most part shows taxa with evolved anthocladial patterns, as well as reductions in the reproductive region in flower number and specialization in breeding biology. All these sections have 2-3-foliate sympodial units often with gemination of the subtending leaf and smaller first leaf of the daughter shoot (anisophyllous geminate sympodia). The peduncle is leaf-opposed or at the level of the geminate cluster (Fig. 27.1F, G, I). In floral biology a continuum may be observed in reduction forms: from the section *Torva* Nees, with many flowered corymbose cymes where most flowers are hermaphrodite and fully fertile, to section *Oliganthes* (Dun.) Bitt., with simpler, fewer-flowered inflorescences, with most flowers fertile, and section *Micracanthum* (Dun.) Bitter emend. Seithe. Then follow the sections *Lasiocarpum* D'Arcy, *Melongena* Nees and *Acanthophora* Dun. with mostly simple few-flowered inflorescences with only the terminal flower and/or the few proximal flowers of the subsidiary inflorescence hermaphrodite; the distal flowers are usually smaller, brachystylous (andromonoecious) and are often inserted with a spatial gap from the hermaphrodite flowers. Species with prickly calyces are usually unarmed with respect to the brachystylous flowers. These latter are functionally male, and in extremely specialized cases, as with some Australian *Melongena* relatives, andro-dioecy is the rule, the "female" plant has one-flowered inflorescences of herma-phrodite flowers, and the male plant has many-flowered inflorescences, with smaller brachystylous flowers (Symon, see this Volume, p. 385).

Section *Androceras* (Nutt.) Bitter ex Marzell. This leptostemonoid taxon shows unequal anther length and enantiostyly; the branching is of trifoliate geminate anisophyllous sympodial units.

Section *Protocryptocarpum* Bitter ex Marzell. This group stands a little aside from its leptostemonoid relatives. The type species, *S. sisymbrifolium*, has 2-3-foliate sympodial units and lateral, leaf-remote cincinni with most flowers fertile. The top leaf is recaulescently grown up its daughter shoot, but not so far as the first leaf of the latter; the sympodium therefore has elongated evenly (Fig. 27.1J).

Because we have followed on from Eichler's schema with *S. nigrum*, we have examined *Solana* with derived or adapted forms of branching where reduction in leaf number, suppression of parts of the axis, and reduction in the reproductive shoots are the tendencies expressed. If we correlate branching with the hapaxanthic ancestry and the woody Cestroideae, we may agree with Danert (1967) that the "original" and least specialized forms are those where plurifoliate sympodial units and richly flowered pleiochasial inflorescences occur together. He further suggested that dichasia are more "original" than monochasia and where this results in the production of complex terminal inflorescences, then we may certainly draw comparison with primitive attributes.

Dichasial branching (Fig. 27.2A) shows two distinct types in the genus *Solanum*. Sections *Brevantherum* Seithe and *Holophyllum* Walpers (*Anthoresis* Dun. pro majore parte) represent the first type and have pleiochasial, multilateral corymbose to subumbellate cymes with a long common axis or peduncle but with reduced rhachides. In contrast to most *Solana*, the inflorescence remains erect and shoot continuation is delayed until well after anthesis. Furthermore, with these two groups, the inflorescence axis remains erect, not being pushed aside by the delayed continuation shoots. When these shoots finally grow out, they remain at a shallow angle to the vertical plane. The fruiting inflorescence thus remains more or less vertical in the furcation of the continuation shoots and therefore conspicuous to animal vectors for seed dispersal. The delayed shoot continuation may be hormone-conditioned and a relic from the original branching pattern. Even on later orders of branching, where dichasia give way to monochasia, the delay and orthotropy of the inflorescence axis are still expressed. Here the sympodial units are plurifoliate, in contrast to *Datura* with difoliate dichasia in Eichler's schema, several leaves being interposed before another inflorescence supervenes. Apart from the indumentum these two taxa are morphologically similar; whether they are cladistically related or products of parallel evolution remains speculative. Section *Holophyllum* has both simple and dendritic hairs and is composed of shrubs to small trees with narrow lanceolate to ovate leaves and small flowers. *S. crispum* is an ornamental shrub from this group. The subsidiary inflorescences are sometimes subtended by small foliar organs, which is an undifferentiated character (cf. *Cestrum* and others). Section *Brevantherum* has stellate-multangular and echinoid hairs and generally broader leaves than the preceding taxon. The two bracts of dormant lateral shoots are precocious and form lunate pseudostipules at the leaf axils and are held adjacent to the main axis. With both groups the smallish ripe fruits are held erect and are coloured, being adapted for dispersal by birds or fruit bats.

Series *Giganteiformia* Bitter. This taxon, attached to section *Torvum* by Bitter (1921), branches like Brevantherum, but is prickly, with thick stellate indumentum and an thersintermediate between the leptostemonoid flowers of *Brevantherum*. The status of this group needs further assessment, but whether it is a link between the two main divisions of the genus remains speculative.

Section *Archaesolanum* Bitter ex Marzell. This represents the second type of dichasial branching (Fig. 27.2A). In this aneuploid group the sympodial units are plurifoliate, but the inflorescence, sitting in the furcation formed by the continuation shoots, which are not delayed, is simple to once-forked and is always laterally deflected. Later orders of branching go over to monochasial, the inflorescence is then soon pushed laterally and the continuation shoot is more or less orthotropic. Sometimes the common inflorescence axis has grown concaulescently with the continuation shoot up to the level of the subtending leaf, or shortly below it.

Section *Cyphomandropsis* Bitter. This section branches in a similar manner to *Archaesolanum*. It is probably better placed as a series of the genus *Cyphomandra* (D'Arcy, 1972).

Monochasial branching with plurifoliate sympodial units is often correlated with a climbing or viney habit (Fig. 27.2B). Some modifications are to be noted.

Section *Dulcamara* (Dun.) Bitt. *sensu stricto*. This conforms to the general pattern and the pleiochasial inflorescence is leaf remote or opposed at the level of the subtending leaf (concaulescent peduncle with the continuation shoot). The subtending leaf is drawn up the daughter shoot recaulescently, but gemination is rare.

Section *Jasminosolanum* Seithe. Many species climb by haptotropic petioles. Here the epipodium is stretched below the insertion point of the subsidiary inflorescences. The whole structure forms a panicle with an elongated axis quite unlike that of other Solanums. The proximal subsidiary inflorescence is subtended by a foliar organ which has a vegetative bud in its axil. This remains independent of the continuation shoot which emerges a node below and which does not assume the orthotropic attitude until the large inflorescence has been surpassed and pushed aside. Perhaps this pattern, again, is a relic from the original fasciculate reproductive termination, but here the axis has stretched and the subtending leaves have been lost. (cf. Fig. 27.1D).

Section *Aculeigerum* Seithe. The present concept of this taxon comprises climbing plants with short hooked prickly emergences on stems, petioles and leaf nervures (cf. *Rubus*), simple to dendritic, never stellate hairs, and blunt to attenuate anthers. It is apparently diphyletic on the basis of branching. The type species, *S. wendlandii*, as well as *S. molinarum*, *S. juciri* and *S. steironematophyllum* branch as for *Jasminosolanum* with the deviant paniculate inflorescence on a long central axis. *S. nemorense* and *S. megistophyllidium*, however, have difoliate geminate sympodial units and reproductive cincinni at the level of the geminate cluster (Fig. 27.1G). With the latter species, in particular, the two leaves of the cluster are markedly anisophyllous and dimorphic. Clearly, if branching patterns are regarded as being of fundamental taxonomic importance, then the similarities in all other characters indicate a convergent relationship here rather than a cladistic one. These interesting and beautiful plants need to be collected and grown for biosystematic studies before a final decision can be made on their anomalous behaviour and phyletic connections.

Subgenus *Tuberarium* sensu lato (the tomato-potato group). The loss of floral bracts seems to be almost universal with the subfamily Solanoideae, but many species of this group show a retention of floral foliar organs as well as many other specialized features. D'Arcy (1972) in a provisional conspectus for the Solanums has suggested a possible common ancestry for sections *Aculeigerum* (in the sense of the type species), *Jasminosolanum* and this group. Certainly the presence of compound leaves and bracteate (in part at least) inflorescences would support such an assumption.

Sections *Petota* Dum. (*Potatoe* Walp. and *Tuberarium* (Dun.) Bitt.) and

Basarthrum Bitter. These have plurifoliate (3–7) leaved sympodial units (Fig. 27.2C). The pseudostipular bracts as discussed for *Brevantherum*, are prominent in these two sections. The bracts may be lunate or auriculate and clasping the main axis loosely. On the main axis the clasping main foliage leaf bases arise almost a node below their apparent axils and betray this in many species by the presence of decurrent lines, ridges or wings on the stem. In the potatoes and *Basarthrum*, a reduced leaf and pseudostipules are to be found some distance up the peduncle. The great *Solanum* taxonomist George Bitter (1912) regarded these as the top leaf of the shoot generation and commented that with the potatoes the shoot continuation was from the axil of the penultimate leaf of the shoot, whereas with *S. ochranthum* (of Series *Juglandifolium*), and the tomatoes, the shoot continuation arose from the top leaf of the sympodium and this leaf was carried up its daughter shoot recaulescently. Danert (1958) has shown developmentally that the small leaf is the subtending leaf of the first subsidiary inflorescence formed after the terminal flower, but which has remained below the furcation owing to the subsequent elongation of the axis. (See Fig. 27.2D–F.)

Series *Juglandifolia* D'Arcy. Here two distinct growth forms are included; plurifoliate sympodial units in the climbing mesomorphic species, with the top leaf recaulescently carried up its daughter shoot and lateral pleiochasial inflorescences; and 2-3-foliate sympodial units in the more xeromorphic shrubby species. This branching pattern emphasises the apparent affinity of *S. juglandifolium*, *S. ochranthum*, *S. lycopersicoides* and *S. rickii* with the tomatoes rather than the potatoes.

Section *Lycopersicon* (Mill.)von Wettst. (including *Eriopersicon* C. H. Muller and *Neolycopersicon* Correll). The tomatoes have 2-3-foliate sympodial units. The inflorescence is leaf remote, pushed laterally and simple to forked. The subtending leaf is recaulescent with its daughter shoot. Pseudostipules are present, except with the cultigen *S. lycopersicon*. Danert regarded the pseudostipules at the base of the inflorescence axis as the displaced bracts of the subsidiary lateral bud in the axil of the subtending leaf above the inflorescence. Branches of the reproductive zone and individual flowers may be subtended by bracteoles in most of the wild species (see specially Fig. 27.2D). These bractlike structures are similar to the pseudostipules but are more symmetrical. In the commercial tomato various anomalies may be seen such as dichasial branching (with a narrow angle furcation and concaulescence of the axes involved) and also the reversion of a reproductive shoot to the vegetative phase, as well as various geminate formations of leaves. Space precludes a fuller treatment in this paper.

To summarize, then, the potatoes and Section *Basarthrum* have 3-7-leaved sympodial units; the two climbing species in Series *Juglandifolia* are plurifoliate whilst their more xeromorphic relatives and the tomatoes produce 2-3-foliate sympodial units. In all cases the subtending leaf of the next shoot generation is carried recaulescently up its product, but not as far as the first daughter leaf.

Section *Anarrhichomenum* Bitter. (Fig. 27.2G). This taxon appears to be related to *Tuberarium sensu lato*. Most species have compound leaves, pseudostipules and similar floral characters. The meso-American species appear to branch similarly to the potatoes, the sympodial units are plurifoliate but the peduncle is soon pushed laterally and is leaf opposed. This leaf is probably the subtending leaf of the next shoot generation and the peduncle has grown up the latter concaulescently (see *S. inscendens*). The more highly adapted South American species from moist Andean forests flower on short spur shoots, more or less independently of the long extension shoots. All species of this group are stem climbers, with adventitious roots at the nodes loosely adhering to trees and rocks. Ecological adaptation has resulted in an adpressed habit, darkish subcoriaceous simple to mostly imparipinnate leaves and smallish spur inflorescences, or perhaps more elaborate structures finally ending the long shoots.

Section *Pteroidea* Dun. (*Polybotryon* (Dun.) Bitter). These plants are herbs to shrubs, or climbers with adventitious roots, occupying a similar ecological niche to the section *Anarrhichomenum*, but probably not closely related to that taxon. Most floristic works describe the inflorescences as "axillary". But they are most probably terminal, the lower part of the common reproductive axis, the petiole, and the continuation shoot, all growing together concaulescently for the distance of one internode; thus the clasping leaf base of the unifoliate sympodial unit starts an internode below and the peduncle and stem axis run with it (Fig. 27.2H). The pleiochasial inflorescence often has a subsidiary shoot inserted some distance from the other rhachides, and thus the inflorescence appears "twinned" or two headed, as the later sectional name implies. These interesting species need collecting for cultivation, when the true ontogeny of the shoot system as well as the nature of other interesting features, can be pronounced upon more positively and the possible phyletic relationships can be ascertained.

Section *Regmandra* (Dun.) Ugent. In *S. tuberiferum* the branching is clearly as has been suggested for *Polybotryon* above (Fig. 27.2H). The common axis of the inflorescence is adnate to the petiole of the leaf of the unifoliate sympodial unit, as well as the continuation shoot, for the distance of one internode. The leaf base clasps the axis decurrently. The extreme concaulescence shown here probably represents a highly evolved form of ramification for these rather anomalous small herbs, which show intermediate characters between *Morella*, *Herpysticum* and *Polybotryon*.

Genus *Cyphomandra* Mart. ex Sendtner. The tree tomatoes have 3–4 leaved sympodial units. On early orders of branching dichasial forking is the rule, and the often spectacularly long, simple to forked scorpioid cyme sits in, or just above the furcation of the continuation shoots. Owing to mesopodial suppression, the subtending leaf is geminate to the first leaf of the continuation shoot, or even forming a group of three with the first two leaves, with extreme suppression of the mid-section. Depending on species and physiological status, dichasia give way sooner or later to monochasia. Then only the top continuation shoot is dominant and the inflorescence may be leaf remote, due to epipodial stretching; or level with the second leaf and thus falsely axillary, due to epipodial suppression; or level with the trigeminate cluster, due to extreme suppression of both the mid and upper part of the sympodial unit; or even carried up the next continuation shoot concaulescently, nearly to the level of the similarly recaulescently shifted subtending leaf. All these types of branching can be found on different shoot generations on the same plant in the cultigen *C. crassifolia*. In *C. allophylla*, as well as rarely on early order branchings of *C. crassifolia*, a leaf may appear at the first branch of the inflorescence, though it is not clear whether this is a displaced leaf or a relict subtending leaf for the first subsidiary reproductive shoot (cf. *Jasminosolanum*).

All species of this genus seem to behave alike. In *C. hartwegii* plants continue dichasial branching for several shoot generations, giving the flat-topped growth habit of this species. Here the first generation shoot continuation is subtended by a pinnatisect cauline leaf which is geminate to the smaller oblique first leaf of the daughter shoot; subsequent shoot generations are subtended by an oblong or lozenge shaped leaf, giving the characteristic leaf mosaic of this species.

As mentioned previously, the section *Cyphomandropsis* Bitt. branches similarly to *Archaesolanum* with plurifoliate dichasial to monochasial sympodial units. The flowers are similar to *Cyphomandra*, but the fruits and seeds are different.

CONCLUSIONS

An attempt has been made to survey the branching patterns of a number of Solanaceae. We may see large terminal inflorescences, plurifoliate sympodial units, bracteate subsidiary inflorescences or dichasial continuations as closer to those of a hypothetical hapaxanthic ancestry. We then see various adaptations to the anthocladial pattern, such as reduction and simplification of leaf and floral number, irregular phyllotaxy due to torsion and failure of shoot sections to elongate, and recaulescent or concaulescent adnations of the various axes and organs. All these confuse the issue of the real ontogeny of the vegetative and reproductive structures.

Clearly, a better understanding of growth sequence and initiation, particularly of the reproductive zone, is necessary as a diagnostic aid. For phylogenetic significance, however, the branching patterns must be considered in conjunction with all the observed morphological, physiological and ecological characters of a plant.

ACKNOWLEDGEMENTS

I wish to thank Dr R. N. Lester, Miss S. Marsh and all others who have co-operated in the supply of plants and seeds etc. for a study of various taxa of the Solanaceae in the living state.

REFERENCES

BITTER, G., 1912. Solana nova vel minus cognita. I. *Rep. spec. nov., 10:* 529–565.

BITTER, G., 1921. Solana africana. III. *Botanische Jahrbücher für Systematik, Pflanzengeschichte und Pflanzengeographie,* 57: 248–286.

DANERT, S., 1958. Die Verzweigung der Solanaceen in reproduktiven Bereich. *Abhandlungen der Deutschen Akademie der Wissenschaften zu Berlin (math-naturwiss.) (Kl. 1957),* No. 6: 1–183.

DANERT, S., 1967. Die Verzweigung als infragenerisches Gruppenmerkmal in der Gattung *Solanum* L. *Die Kulturpflanze,* 15: 275–292.

D'ARCY, W. G., 1972. Solanaceae studies II: typification of subdivisions of *Solanum. Annals of Missouri Botanical Garden,* 59: 262–278.

EICHLER, A. W., 1875. *Blüthendiagramme,* I. Leipzig.

GOEBEL, K., 1931. *Blütenbildung und Sprossgestaltung.* Jena.

28. Growth regulator interactions on morphogenesis in *Solanum* species

H. DAVID HAMMOND

Department of Biological Sciences, State University College at Brockport, New York, U.S.A.

Attention is first drawn to the comparative dearth of studies in comparative plant physiology, in contrast to the animal kingdom, where many data have been applied to systematic and evolutionary problems. Studies by the author and his students are discussed, where the effects of such growth regulators as gibberellin, benzyl-adenine, N-dimethylaminosuccinamic acid (B₉), quercitin, and the effect of daylength on leaf shape and lobing, spine formation, and amount of branching, on two species of *Solanum*, *S. seaforthianum* Andrews and *S. capsicoides* All. were investigated. Morphogenetic work in the Solanaceae by others is briefly discussed.

CONTENTS

INTRODUCTION

Some years ago, in the course of a book review (Hammond, 1970) it was noted that, "for the most part, biosystematic and evolutionary studies deal with the mature organism, the endpoint of ontogenetic processes," and went on to say, "when we can characterize and differentiate related species by a knowledge of the differences in the ongoing ontogeneses, and be able to say how the ontogenetic processes in these populations change with time, then we will really be able to say we know what evolution is."

Comparative physiology in the plant world has not, it would seem, reached the same level of prominence as has comparative animal physiology. The reasons are obvious: it is intrinsically more difficult, and at the same time not as exciting to most people. Nevertheless, we will surely go beyond cataloging the chemical constituents from the various species of the genera and families of plants—as interesting and necessary as this is—to a study of the processes that lead to the morphological changes that we see as plants mature, in a comparative way.

There are exciting beginnings, such as the discovery of the C₄ photosynthetic pathway, found in many tropical and/or arid-region plants along with the more common C₃ photosynthesis, and the discovery of the CAM (Crassulacean Acid Metabolism) type of photosynthesis. Already, taxonomic and evolutionary conclusions have been reached on the basis of the generic distribution of these modes of photosynthesis and the associated peculiarities of leaf structure—Kranz-type bundle sheaths in C_4 photosynthesis, succulence and CAM metabolism—for the Gramineae, Chenopodiaceae, Crassulaceae, and other families (Brown & Smith, 1975; Mooney, Troughton & Berry, 1974; Pearcy & Troughton, 1974; Smith & Turner, 1975; Stebbins, 1974; Troughton, Card & Hendy, 1974).

Such studies, however, usually do not involve experimental interference with the ongoing photosynthetic activities in the same sense as with studies of the effects of growth regulators on the developmental morphology of both vegetative reproductive parts, the timing of flowering, and so forth. We base a not inconsiderable portion of our taxonomy on vegetative structures. Leaf shape, lobing, disposition on the axis, possession of hairs and spines of various sorts, are characters used for differentiating at the species, or higher level. Whether the stem is woody or herbaceous, shrubby or viney, decumbent or upright, etc., are matters considered in species descriptions. Likewise, the inflorescence type, how it is borne on the plant, the season of bloom, etc., are characteristics used in distinguishing species. Such morphological differences are the end result of subtle changes within populations through the years, in the patterns of cell division, onset of formative processes, their duration, and their cessation. All of these processes are, of course, under the control of genetic factors the expression of which is mediated in part by the production and utilization in the plant of a number of growth regulators, some of which are known to us and have been much studied.

The present writer's interest was sparked, some years ago, when growing seedlings of a number of different *Solanum* species. There appeared a marked difference in the way the plants of the different species developed. In some, such as *S. capsicoides**, the first few leaves were heart-shaped, with a cordate base, and entire margin. The leaves were hispid, but not spiny. It was only after five or six leaves had developed that the characteristic lobing and the spines spaced along the major veins appeared. Even *S. khasianum* Clarke, the huge mature leaves of which possess the most terrifying spines, did not possess spines on the earliest leaves. On the other hand, *S. sisymbrifolium* Lam., was spiny and had highly dissected leaves from the seedling stage on. The juvenile stage, so noticeable in the first-mentioned species, was scarcely to be seen in the last-mentioned species.

It seemed that it might be possible to study such phenomena comparatively, by treating plants of various species with combinations of growth regulators to see if such things as leaf morphology would be affected similarly in different species. It has been known for some years that the several known growth regulators have definite effects on leaf shape, retention or elimination of juvenile characteristics, branching patterns, and so on.

The morphogenetic effects of the known growth regulators are becoming well-known and any up-to-date plant physiology or morphogenesis reference book discusses these in detail (e.g., Leopold & Kriedmann, 1975; Bidwell, 1974).

---

* The work herein reported was originally reported as having been done with *Solanum aculeatissimum* Jacq. However, Michael Nee of the Botany Dept., Univ. Wisconsin, Madison, informs the author that the American *S. capsicoides* All. (=*S. ciliatum* Lam.) has been confused with the former, an African species, and that we actually worked with *S. capsicoides*. Our plants had the bright, red-orange, round fruits and yellowish, flat, winged-all-around seeds of the latter species.

We can briefly summarize these below:

Exogenously supplied auxin causes stem (and grass coleoptile) elongation. It causes cell enlargement, as in potato tubers, tissue culture callus, etc., plays a role in xylem differentiation, stimulates rooting of cuttings, and along with cytokinins is involved in bud formation. Apical dominance is, in part, a function of auxin levels and movement, along with cytokinins. Auxin plays a major role in the leaf abscission phenomenon.

Applications of cytokinins bring about increased mitosis, cytokinesis, and stimulate the doubling of nuclear DNA. Cytokinins bring about swelling growth, as in radishes, inhibit root and shoot elongation and stimulate leaf enlargement. As mentioned above, cytokinins induce bud differentiation in callus, leaves, roots, etc., which can be inhibited by high auxin levels. Cytokinins can induce or enhance flowering in some species, and subsequent fruit development is affected by their levels in tissues.

Gibberellins are involved in stem elongation. Dwarf varieties of many species are deficient in gibberellin. Vines, or clamberers, contrariwise, may possibly have very high gibberellin levels. Gibberellin can induce flowering in some photoperiod-sensitive and cold-requiring species. It is involved in sex regulation in monoecious species, stimulates fruit-set, and is involved in the rate of fruit growth in some cases. One of the more interesting gibberellin effects is its effect in causing extension of or reversion to the juvenile condition. This has been seen in *Hedera helix*, *Ipomoea*, *Caryota* (a palm), *Marsilea*, and others (Allsopp, 1962; Fisher, 1976; Leopold & Kriedmann, 1975; Rogler & Hackett, 1975a, b). The change from the juvenile to the mature state would seem to involve changes in the proportions of endogenous growth regulators.

Abscisic acid is involved in stress response by plants, especially water deficiency. Abscisic acid levels rise in stressed plants. Adaptation to osmotic stress, mineral deficiency and cold, are all brought about by exogenously supplied abscisic acid. It is also involved, as the name implies, in abscission of leaves, abscission and dehiscence of fruits, and dormancy features, but not in all species, and usually only under short-day conditions.

Phenolic growth inhibitors, mostly synthesized via the shikimic acid pathway, and ordinarily occurring as the glycosides, are common in plants. They include phenolic acids, such as ferulic, caffeic, and chlorogenic acids, lactones such as coumarin, and flavonoids such as quercitin. Their mode of action is unclear, but monophenols stimulate IAA decarboxylation, and also seem to retard growth in general and promote senescence in leaves.

The experiments that have been done with several graduate students have been factorial type experiments, employing two or more growth regulators and/or environmental factors at two or more levels or states. With this approach, many different sets can be made from a given lot of plants, depending on what analysis is required. The results can be subjected to an analysis of variance and the effects, if any, of each growth substance or factor, taken separately, can be ascertained, as well as the interactions among the factors.

We hypothesized that gibberellin applications would cause plants to revert to juvenile conditions, that is, that leaf-lobing would be lost, or reduced, spines diminished or lost, and so on. Contrariwise, it was thought that applications of a cytokinin, with its cell-division inducing properties, might enhance lobing and spine formation. Growth inhibitors, which might be anti-gibberellins, such as B_9, and the phenolic compound quercitin, were also studied. In one study, the consequences of the interaction of photoperiod and growth regulators was studied as well.

FIRST STUDY

The first experiments were done with Amako Ahaghotu. Since these have been reported in detail (Ahaghotu & Hammond, 1971), they need only be summarized here. We studied the effects of gibberellin and benzyl-adenine in factorial combination on plant height, leaf area, leaf fresh and dry weights, and leaf chlorophyll and protein (as measures of senescence), of *Solanum capsicoides* All. Significance was reported at the 95% confidence level. The gibberellin levels used were 0.3 ppm (8.7×10^{-7}M), 0.03 ppm (8.7×10^{-8}M), 0.003 ppm (8.7×10^{-9}M), and zero. The benzyl-adenine concentrations were 300 ppm (10^{-3}M), 30 ppm (10^{-4}M), 3.0 ppm (10^{-5}M). The gibberellin and benzyl-adenine combinations were sprayed on the plants twice a week for six weeks, (using sodium lauryl sulphate, 0.05%, as wetting agent), the reasoning being that more or less constant higher levels of growth regulators over the normal endogenous levels might have morphogenetic effects different from a one-time application. At the end of the six weeks the plants were measured for differences in the structures and attributes mentioned above.

The highest level of benzyl-adenine was toxic and the plants in that set died except for the ones also receiving the highest gibberellin level.

As was to be expected, the main effect of gibberellin was to increase plant height. The effect was directly proportional to concentration. Benzyl-adenine had no significant effect on height. Gibberellin also had the effect of causing leaves to senesce and drop off, while benzyl-adenine caused their retention. The number of leaves remaining on the plants receiving the high level of gibberellin was only 65% of those on the plants receiving no gibberellin. The number of leaves on plants receiving no benzyl-adenine was only 38% of those receiving the middle level of that substance. Benzyl-adenine also increased the number and size of the basal side-branches, although we failed to get quantitative data on that phenomenon. The end result was that plants receiving gibberellin were tall and relatively leafless, whilst those receiving benzyl-adenine were shorter and bushier.

In terms of leaf area, both growth substances increased leaf area separately and together. In other species, gibberellin application may decrease leaf area.

Since the other things we measured are not especially relevant to the present discussion, they will not be reviewed here.

SECOND STUDY

The second study, done with Zenora Williams (Williams, 1967), overlapped the first in part, to see if the initial results would be confirmed, and then also looked into the effect of spine production and leaf-lobing in the same species, *S. capsicoides*. In this study, the effects of quercitin, a flavonoid compound, were also observed, as well as those of gibberellin and benzyl-adenine (Table 28.1). Zero concentrations for each of these substances were those with the subscript 3, which identifies the controls for each substance and for the overall control ($GA_3BA_3Q_3$). As before, the results from factorial combinations of the growth regulators were subjected to an analysis of variance and multiple comparison (Edwards, 1963).

Six to eight-week plants, 8 to 10 cm tall, selected for uniformity, were transplanted two to a container. After a week, the spraying regimens were commenced and the plants sprayed weekly for six weeks. At the end of this time the height of the plants from the cotyledonary node to the tip of that expanding leaf that was *c*.0.5 cm long was obtained. The leaf areas were obtained from tracings on graph paper. The spines on the upper and lower sides of the leaves were counted, and also the number of lobes per leaf. The effect on senescence, as measured by the amount of leaf abscission, was noted for each treatment, and the degree of lateral branching noted.

With respect to height, gibberellin caused increased inter-node elongation, as expected. Benzyl-adenine and quercitin, singly and additively together, inhibited elongation, a well-known effect for cytokinins, at least. The effects, in each case, were proportional to concentration (Table 28.2). Only the benzyl-adenine effect was statistically significant, however.

In this experiment, only benzyl-adenine significantly increased leaf area, at a concentration of $10^{-5}M$ (Table 28.3). Although quercitin seemed to reduce leaf area, the difference was not statistically significant. Gibberellin seemed to bring about a slight increase in leaf area, but again this was not significant.

Likewise, benzyl-adenine had the greatest effect of the three substances on sequential leaf senescence. It prevented leaves from dropping off. Neither gibberellin nor quercitin significantly affected leaf retention (Table 28.4).

Benzyl-adenine and quercitin were both inhibitory of spine production. Both the number and size were greatly reduced. Gibberellin, on the other hand, was

Table 28.1. Concentrations** of gibberellic acid, benzyl-adenine and quercitin received by twenty-seven sets of *Solanum capsicoides* plants

				Sets				
1	2	3	4	5	6	7	8	9
GA_1	GA_1	GA_1	GA_1	GA_1	GA_1	GA_1	GA_1	GA_1
BA_1	BA_2	BA_3	BA_1	BA_1	BA_2	BA_2	BA_3	BA_3
Q_1	Q_1	Q_1	Q_2	Q_3	Q_3	Q_3	Q_2	Q_3
10	11	12	13	14	15	16	17	18
GA_2	GA_2	GA_2	GA_2	GA_2	GA_2	GA_2	GA_2	GA_2
BA_1	BA_2	BA_3	BA_1	BA_1	BA_2	BA_2	BA_3	BA_3
Q_1	Q_1	Q_1	Q_2	Q_3	Q_3	Q_3	Q_2	Q_3
19	20	21	22	23	24	25	26	27
GA_3	GA_3	GA_3	GA_3	GA_3	GA_3	GA_3	GA_3	GA_3
BA_1	BA_3	BA_1	BA_1	BA_2	BA_2	BA_3	BA_3	BA_3
Q_1	Q_1	Q_2	Q_3	Q_3	Q_3	Q_2	Q_2	Q_1

**Concentrations GA$_1$ $10^{-5}M$ GA$_2$ $10^{-7}M$ GA$_3$ none
BA$_1$ $10^{-3}M$ BA$_2$ $10^{-5}M$ BA$_3$ none
Q$_1$ $10^{-3}M$ Q$_2$ $10^{-5}M$ Q$_3$ none

Table 28.2. Effects of growth regulator combinations on heights of *S. capsicoides* plants in centimeters from cotyledonary node to leaf 0.5 cm in width

	GA₁			GA₂			GA₃		
	BA₁	BA₂	BA₃	BA₁	BA₂	BA₃	BA₁	BA₂	BA₃
Q₁	6.7	7.5	6.5	7.7	7.2	6.2	6.7	7.5	7.0
	7.5	8.0	7.5	8.2	7.7	7.5	7.5	8.2	7.5
	9.2	10.8	8.0	9.0	8.7	9.0	8.5	9.2	8.5
	11.0	14.0	12.0	12.5	11.0	12.2	9.0	10.5	11.7
	15.0	18.0	16.0	14.0	14.7	15.5	11.5	12.2	13.2
	18.0	20.0	18.7	15.2	18.7	19.5	12.5	15.5	15.7
	67.4	78.3	68.7	66.6	68.0	69.9	55.7	63.1	63.6
Q₂	6.2	7.2	7.7	7.2	7.7	7.5	7.5	7.0	6.2
	7.2	8.7	8.8	8.0	9.0	8.2	7.7	8.0	6.5
	7.7	9.0	10.5	8.5	10.0	9.7	9.0	8.5	8.0
	9.0	16.0	18.5	9.9	14.2	14.2	10.2	11.2	12.5
	11.5	24.5	26.5	16.5	19.5	21.2	12.2	13.5	16.5
	12.0	27.5	29.7	18.0	20.5	23.2	13.7	19.0	18.5
	53.6	92.9	101.7	68.1	80.9	84.0	60.3	67.2	68.2
Q₃	8.5	7.2	8.7	8.0	7.5	6.2	7.0	6.7	7.0
	9.0	8.7	9.5	8.7	8.0	7.0	7.5	7.2	7.7
	10.2	9.2	10.7	9.7	9.5	8.0	7.7	8.2	9.0
	10.5	17.7	17.2	11.2	16.5	19.5	10.5	12.5	15.0
	13.2	25.0	25.5	13.5	20.0	28.2	13.2	15.5	17.0
	15.0	28.0	28.2	14.0	22.2	30.0	14.5	19.5	19.0
	66.4	95.8	99.8	65.1	83.7	98.9	60.4	69.6	74.9

Table 28.3. Effects of growth regulator combinations on the leaf area in square centimeters in *S. capsicoides*

	GA₁			GA₂			GA₃		
	BA₁	BA₂	BA₃	BA₁	BA₂	BA₃	BA₁	BA₂	BA₃
Q₁	2.96	6.84	3.12	3.32	2.44	2.48	4.08	3.92	3.36
	14.08	18.96	16.00	12.00	22.56	14.36	13.52	17.20	18.00
	13.20	20.56	21.52	18.44	18.16	15.88	14.52	12.00	22.56
	14.36	20.88	18.96	14.52	10.12	20.44	9.36	12.50	21.32
	44.60	67.24	59.60	44.28	63.28	53.16	41.48	45.68	65.24
Q₂	3.04	5.24	3.52	3.20	3.84	2.36	2.88	3.12	3.92
	7.36	21.88	31.20	15.44	14.52	18.44	12.00	26.00	10.44
	9.36	19.44	29.56	12.96	20.20	15.56	13.40	31.48	13.96
	15.44	20.20	31.88	13.52	19.20	17.40	9.36	34.60	12.88
	35.20	66.76	96.16	45.12	57.76	53.76	37.64	95.80	41.20
Q₃	3.22	3.20	2.24	2.88	7.36	2.96	2.68	3.20	3.80
	10.12	23.04	28.44	13.52	21.88	29.44	15.32	32.32	12.00
	9.84	27.60	22.52	18.00	20.00	31.20	8.36	32.60	14.36
	8.36	22.20	27.68	16.08	21.56	31.68	10.56	24.84	10.88
	31.64	76.04	80.88	50.48	70.80	95.78	36.92	92.96	41.04

mildly stimulatory of spine production and could partially overcome the inhibitory effects of the other two (Table 28.5). All the factor effects and their interactions, save GA/Q, were statistically significant.

Both benzyl-adenine and quercitin (highest concentration only) brought about a decrease in the number of lobes. For the former, it may be that it stimulates more

Table 28.4. Effect of growth regulator combinations on the sequential senescence of leaves in *S. capsicoides*
(Note: The numbers are the leaves which dropped off, per treatment)

	GA_1			GA_2			GA_3		
	BA_1	BA_2	BA_3	BA_1	BA_2	BA_3	BA_1	BA_2	BA_3
Q_1	6	8	13	9	11	10	4	7	8
	11	12	19	9	15	17	7	11	9
	—	—	—	—	—	—	—	—	—
	17	20	32	18	26	27	11	18	17
Q_2	7	6	14	4	10	8	3	8	18
	7	10	18	10	12	8	6	11	20
	—	—	—	—	—	—	—	—	—
	14	16	32	14	22	16	9	19	38
Q_3	6	11	12	5	9	17	7	10	11
	7	12	19	6	10	20	8	11	12
	—	—	—	—	—	—	—	—	—
	13	23	31	11	19	37	15	21	23

Table 28.5. Effect of growth regulator combinations on spine production on *S. capsicoides* plants
(Note: The numbers are the average spine number per leaf, per treatment)

	GA_1			GA_2			GA_3		
	BA_1	BA_2	BA_3	BA_1	BA_2	BA_3	BA_1	BA_2	BA_3
Q_1	8	10	17	9	10	18	8	10	12
	6	8	16	7	8	16	6	6	11
	5	7	16	5	8	15	4	6	10
	3	5	15	3	7	14	2	6	10
	—	—	—	—	—	—	—	—	—
	22	30	64	24	33	63	20	28	43
Q_2	5	16	19	7	8	19	9	11	15
	4	12	19	4	9	19	7	10	12
	4	9	20	3	7	20	5	9	10
	2	6	20	3	7	22	4	7	9
	—	—	—	—	—	—	—	—	—
	15	43	78	17	31	80	25	37	46
Q_3	5	18	19	8	11	22	12	10	11
	3	12	18	7	10	21	8	9	13
	3	10	17	5	9	20	7	8	10
	3	9	19	4	8	23	5	7	9
	—	—	—	—	—	—	—	—	—
	14	49	73	24	38	86	32	34	43

Table 28.6. Effects of growth regulator combinations on the degree of lobing of leaves in *S capsicoides*
(Note: The numbers are the average number of lobes per leaf, per treatment)

	GA$_1$			GA$_2$			GA$_3$		
	BA$_1$	BA$_2$	BA$_3$	BA$_1$	BA$_2$	BA$_3$	BA$_1$	BA$_2$	BA$_3$
Q$_1$	7	7	9	5	7	7	5	5	7
	7	7	5	5	5	7	3	5	7
	5	5	7	5	5	5	3	5	7
	22	24	30	20	22	26	11	20	26
Q$_2$	5	7	5	5	7	7	5	7	7
	5	7	7	5	5	5	5	5	7
	0	7	7	5	5	5	5	6	7
	0	5	9	3	5	5	3	5	7
	10	26	28	18	22	22	16	22	28
Q$_3$	5	7	5	5	7	9	7	7	7
	5	5	7	3	7	9	5	5	5
	3	7	9	3	5	9	5	5	5
	0	7	11	0	5	11	5	7	7
	13	26	32	11	24	38	22	24	22

Table 28.7 Effect of growth regulator combinations on the production of lateral branching in *S. capsicoides*
(Note: The numbers are the average number of side branches per treatment)

	GA$_1$			GA$_2$			GA$_3$		
	BA$_1$	BA$_2$	BA$_3$	BA$_1$	BA$_2$	BA$_3$	BA$_1$	BA$_2$	BA$_3$
Q$_1$	20	5	8	19	6	6	19	17	6
	23	14	9	29	13	10	19	17	9
	28	21	10	37	14	8	25	17	8
	17	10	9	32	5	4	23	12	11
	22	15	7	38	8	9	25	15	10
	21	17	8	36	7	9	23	16	9
	131	82	51	191	53	46	134	94	53
Q$_2$	13	7	11	18	9	9	18	9	9
	16	10	15	24	16	12	24	16	12
	14	8	17	27	12	10	27	12	10
	14	9	17	22	10	9	22	10	19
	11	8	13	27	11	7	27	11	7
	15	5	14	25	14	7	25	14	7
	83	47	87	143	72	54	143	72	54
Q$_3$	12	12	10	18	5	4	30	8	1
	11	13	7	27	17	6	28	11	5
	11	14	8	28	15	8	30	11	6
	12	11	9	22	9	9	27	9	3
	10	14	6	37	16	7	32	12	4
	14	15	5	38	17	9	35	15	5
	70	79	45	170	79	43	182	66	24

of the marginal meristem cells to keep dividing longer, thus reducing lobing. For the latter chemical, it is possible that it is preventing the leaf-expansion effects of gibberellin (Table 28.6).

Profuse lateral branching was caused by benzyl-adenine, since cytokinins are involved in the release of buds from apical dominance. Quercitin had no significant effect and, as in the first study, gibberellin inhibited side branching (Table 28.7).

THIRD STUDY

In the final study, with Norma Williams (Williams, 1968), a different species, *Solanum seaforthianum* Andrews, was used. This is a spineless, scrambling, clambering species, tropical, with hanging clusters of beautiful lavender flowers followed by little red fruits like those of *S. dulcamara*. Its leaves are pinnately lobed with seven to nine lobes. Seedling leaves have only a few lobes and resemble mature *S. dulcamara* leaves, however. In this study the effects of photoperiod were studied along with the effects of gibberellin, benzyl-adenine and N-dimethyl aminosuccinamic acid (B_9), and their interactions were studied. Light has, of course, important morphogenetic effects. In seedlings of woody species, for example, short days can result in the formation of resting buds. *Proserpinaca palustris* responds to 8 h light plus gibberellin by producing adult leaves, but juvenile leaves in 8 or even 12 h light alone.

Cuttings from a number of plants were rooted and were the subjects used. These were randomly distributed into the eight experimental groups of five plants each. Two photoperiods, 8 and 16 h were used, at about 1200 foot candles, in growth chambers. One level each of gibberellin (100 ppm), benzyl-adenine (30 ppm), and B_9 (0.25%) were used. Every combination of the four variables was used. The plants were sprayed with the chemicals twice a week for eight weeks, with 0.5% sodium lauryl sulphate as wetting agent. The effects were studied on those parts formed subsequent to the commencement of spraying. A leaf 1.5 cm long at the time treatment began was marked to denote the starting point. A 4-way analysis of variance was carried out on the results. Height of plants, length of leaf, leaf lobing, time of flowering, and the number of flowers were all measured.

Results showed that the 16 h photoperiod gave plants about half the height of 8 h photoperiod plants—an average of about 5 cm as opposed to about 10 cm. Gibberellin, in both photoperiods, greatly increased height—about four to six fold. As expected, B_9 and benzyl-adenine tended to inhibit growth. This effect was much more pronounced in the 16 h photoperiod. In fact, the gibberellin-B_9 interaction in the 8h photoperiod actually caused increased growth over gibberellin alone. Also, benzyl-adenine was not as inhibitory of growth in the 8 h photoperiod as in the 16 h photoperiod (Table 28.8).

Neither photoperiod nor chemical treatment had much effect on lobing in this species. Nevertheless, there were indications of increased lobing with a longer photoperiod, a depressing effect of gibberellin, and an increasing effect of both B_9 and benzyl-adenine, especially in the 16 h photoperiod (Table 28.9).

Leaf length was greatly affected by photoperiod, being nearly twice as long (7.1 cm av. vs. 3.8 cm av.) in the 8 h photoperiod. At 8 h, all growth regulators had a depressing effect on leaf length; at 16 h benzyl-adenine and gibberellin stimulated an increase in leaf length (Table 28.10).

Finally, the effect on flowering was rather interesting (Table 28.11). Photoperiod had a marked effect. For the 16 h photoperiod, the controls had a mean number of 3; in the 8 h photoperiod, the mean was almost 9. B_9 alone drastically inhibited flowering at either photoperiod. Benzyl-adenine also drastically reduced flowering in the 8 h photoperiod, but enhanced it at 16 h. Gibberellin enhanced flowering in the 16 h photoperiod, and reduced flowering in the 8 h photoperiod. Interestingly, B_9, when present with both gibberellin and benzyl-adenine, greatly enhanced

Table 28.8. Mean height of *S. seaforthianum* plants, (cm) per treatment

Treatments	— — —	— B_9	— BA —	— BA B_9	GA — —	GA — B_9	GA BA —	GA BA B_9
				16 h				
	4.5	1.6	5.5	3.0	33.5	5.5	22.0	13.5
	2.8	2.6	11.0	7.0	27.5	34.0	19.5	24.5
	4.0	0.8	6.2	3.5	30.0	29.0	28.0	29.5
	7.0	3.0	9.5	4.5	31.0	32.0	29.0	30.0
	8.0	5.0	5.2	4.5	39.0	24.5	31.5	24.5
Plant heights $\bar{x} =$	5.3	2.6	7.5	4.5	30.2	25.0	26.0	24.3
				8 hr				
	11.0	6.5	20.00	12.0	42.0	59.0	27.5	60.0
	7.7	4.0	39.0	8.5	48.0	40.0	54.0	22.5
	8.5	8.5	8.5	14.5	36.5	52.5	39.0	34.0
	8.5	7.0	26.0	23.0	61.0	61.0	31.0	52.0
	27.0	7.0	8.2	35.0	17.0	45.5	37.0	27.0
$\bar{x} =$	12.5	6.6	20.3	18.6	40.9	51.6	37.3	39.1

Table 28.9. Mean number of lobes in *S. seaforthianum* per treatment

Treatments	— — —	— B_9	— BA —	— BA B_9	GA — —	GA — B_9	GA BA —	GA BA B_9
				16 h				
	7.5	8.4	8.2	8.1	7.0	6.8	7.2	7.4
	7.0	8.5	8.4	7.3	5.8	8.0	5.4	6.9
	8.0	6.3	8.8	9.0	8.0	8.0	7.1	7.3
	5.0	8.5	8.3	9.3	7.4	7.8	7.7	6.6
	8.0	6.7	8.8	6.7	7.9	6.9	8.2	7.3
Lobe numbers $\bar{x} =$	7.1	7.7	8.5	8.1	7.2	7.5	7.1	7.1
				8 h				
	7.0	8.0	7.7	7.4	6.7	8.6	7.1	7.5
	8.0	8.4	8.5	8.2	8.5	7.6	8.6	6.4
	10.0	6.7	8.2	8.4	8.2	9.1	5.7	7.0
	7.0	8.3	7.8	8.6	9.2	8.2	6.8	8.1
	7.0	5.2	7.1	8.5	8.0	6.6	7.7	6.9
$\bar{x} =$	7.8	7.3	7.9	8.2	8.1	8.0	7.2	7.2

flowering in the 16 h photoperiod. In the 8 h photoperiod, the three together greatly inhibited flowering but, B_9 plus benzyl-adenine alone gave the same mean number of flowers as the control. *S. seaforthianum* is probably a short-day plant, but one might then have expected the cytokinin to stimulate flowering in the 8 h photoperiod. Nor did benzyl-adenine plus gibberellin stimulate flowering in the 8 h photoperiod.

Table 28.10. Mean length of leaves of *S. seaforthianum*, (cm) per treatment

Treatments	— — —	— — B_9	— BA —	— BA B_9	GA — —	GA — B_9	GA BA —	GA BA B_9
					16 h		—	
	3.6	3.4	3.7	2.9	4.7	5.7	6.3	4.9
	2.3	3.1	4.3	3.2	4.5	5.7	3.5	5.3
	4.0	1.9	4.2	3.7	5.0	6.2	3.8	5.6
	4.2	4.0	4.8	4.6	5.3	4.3	5.0	5.5
	4.9	3.8	4.6	3.8	5.4	4.4	5.8	5.6
Leaf length $\bar{x} =$	3.8	3.2	4.3	3.6	5.0	5.3	4.9	5.4
					8 h			
	6.0	5.1	4.6	4.9	5.4	6.5	7.1	6.4
	7.8	5.5	5.4	6.2	5.2	6.1	7.6	3.2
	8.3	5.7	3.5	5.9	6.3	7.4	6.2	5.6
	7.2	5.2	4.9	7.2	6.8	7.7	7.2	6.6
	6.1	2.0	2.5	6.2	4.8	4.6	6.1	5.9
$\bar{x} =$	7.2	4.7	4.2	6.2	5.7	6.5	6.8	5.5

Table 28.11. Mean number of flowers, *S. seaforthianum*, per treatment (five plants per set)

Treatment	— — —	— — B_9	— BA —	— BA B_9	GA — —	GA — B_9	GA BA —	GA BA B_9
					16 h			
	2	0	2	1	2	0	2	4
	2	0	2	2	6	2	3	9
	6	0	6	1	6	2	2	7
	3	0	11	1	7	2	15	15
	1	1	1	1	15	10	16	20
Flower numbers $\bar{x} =$	2.8	0.2	4.4	1.2	7.2	3.2	7.6	11.0
					8 hr			
	3	1	2	1	1	2	1	1
	8	2	2	5	1	2	2	1
	8	2	2	9	5	6	6	0
	20	1	0	21	5	5	5	5
	5	1	0	7	0	0	5	5
$\bar{x} =$	8.8	1.4	1.2	8.6	2.4	3.0	3.8	2.4

DISCUSSION

From the above, we can see that measurable changes in a number of morphological features occur when growth-regulating chemicals are applied exogenously. Obviously, these effects persist only so long as the substances are applied, or persist in the plant. Aside from B_9, however, these (or related) substances naturally occur in plants. Mutations occurring in the genetic information coding for the formation of one or more of these substances can have profound morphogenetic effects. Gibberellin-deficient dwarfs are well-known in *Zea* and *Pisum*, for instance. If such mutants can persist in the environment, we have the potential for new varieties, races, or, ultimately, species, differing in such characters, perhaps, as those we have experimented with above. Since species are differentiated, for example, on the basis not only of the presence or absence of spines, but also their size and shape, in such genera as *Solanum* (Morton, 1976), small changes in the time or amount of production of a growth regulator could have noticeable effects on our taxonomy.

Another approach is to analyse for what a plant is producing for itself during various ontogenetic stages or during fruit development. Information obtained from different species that are in the same development stage could be quite instructive in terms of tracing evolutionary trends in such things as fruit size and shape, tuber formation and dormancy characteristics, etc. Not too much has been done in the Solanaceae in a comparative context (but see Rick & Fobes, 1975). However, here and there, people are studying the endogenous levels of growth regulators in developing fruits, tubers, of the changing amounts of various enzymes either under natural conditions or after treatment with growth regulators. For example, a Cornell Univ. group (Abdel-Rahman *et al.*, 1975), have studied changes in endogenous plant hormones in cherry tomato fruits during development and maturation. Were such a study to be extended to other tomato species, such as *L. pimpinellifolium*, or *L. hirsutum*, valuable information on whether or not discernible differences in the rise and fall of the concentrations of the various regulator substances could be found and, if so, they could be correlated with final typical fruit size, shape, and colour. Marigo & Boudet (1975), report on their work on the growth effects of polyphenols in tomato. They found that application of quinic acid, phenylalanine, cinnamic acid, or other polyphenol precursors caused an important increase of the phenolic content and a reduced growth of the plants.

Mizrahi *et al.* (1975) at Purdue Univ., have studied abscisic acid and benzyl-adenine effects on the ripening of normal, Rutgers var., tomatoes versus the *rin* mutant, where the fruits do not ripen. They found that exogenous benzyl-adenine had no effect on the ripening of either variety. Abscisic acid caused Rutgers fruits to ripen twice as fast, but had no effect on the mutant. They concluded that endogenous levels of abscisic acid and benzyl-adenine do not account for the lack of ripening in *rin* fruit.

Tuberization in potatoes is another subject of intense practical interest and much study has gone into the rise and fall of growth regulator levels and changes in enzyme activities as tubers are formed and grow to maturity. Space does not permit a review of these studies. Our knowledge of the evolution of the tuber-forming phenomenon will be much advanced by a comparative study of the changes in growth regulator levels and any correlative changes seen in enzymes in the various stolon and tuber-forming species of *Solanum*.

In sum, I have attempted, in this short review, to get over the idea that the comparative physiological and biochemical approach can tell us much about the mechanisms of evolution within a genus or family. Our perceptions of the course of this evolution will ultimately result in changes in our taxonomic summaries of these changes. Future research must take this approach.

ACKNOWLEDGEMENTS

Assistance with the statistical analysis by Miss Vivian Prothro, of the Data Processing Center, Dr Donald King, and the late Dr Lillian Blake, both of the Psychology Dept. and other assistance from Drs Marie C. Taylor and John P. Rier, Jr., Botany Dept., all at Howard University, Washington, D.C., is gratefully acknowledged.

REFERENCES

ABDEL-RAHMAN, M., THOMAS, T. H., DOSS, G. J. & HOWELL, L., 1975. Changes in endogenous plant hormones in cherry tomato fruits during development and maturation. *Physiologia Plantarum, 34* (1): 39–43.

AHAGHOTU, A. N. D. & HAMMOND, H. D., 1971. Morphogenesis in *Solanum aculeatissimum* as affected by N⁶-Benzyladenine and Gibberellic Acid. *Journal of West African Science Association, 16* (2): 119–131.

ALLSOPP, A., 1962. The effects of gibberellic acid on morphogenesis in *Marsilea drummondii. A. Br. Phytomorphology, 12:* 1–10.

BIDWELL, R. G. S., 1974. *Plant Physiology:* 643 pp. New York: Macmillan.

BROWN, W. V. & SMITH, B. N., 1975. The genus *Dichanthelium* (Gramineae). *Bulletin of the Torrey Botanical Club, 102* (1): 10–13.

EDWARDS, A. L., 1963. *Experimental Design in Psychological Research:* 455 pp. New York: Holt, Rinehart, and Winston.

FISHER, J. B., 1976. Induction of juvenile leaf form in a palm (*Caryota mitis*) by gibberellin. *Bulletin of the Torrey Botanical Club, 103* (4): 153–164.

HAMMOND, H. D., 1970. Review of: Principles and Methods of Plant Biosystematics, by Otto T. Solbrig. Macmillan, Toronto. In *Plant Sciences Bulletin, 16* (4): 12. Dec. 1970.

LEOPOLD, A. C. & KRIEDMANN, P. E., 1975. *Plant Growth and Development*, 2nd ed.: 545 pp. New York: McGraw-Hill.

MARIGO, G. & BOUDET, A. M., 1975. Rôle des pólyphénols dans la croissance. Définition d'un modèle expérimental chez *Lycopersicum esculentum. Physiologia Plantarum, 34* (1): 51–55.

MIZRAHI, Y., DOSTAL, H. C., McGLASSON, W. B. & CHERRY, J. H., 1975. Effects of abscisic acid and benzyladenine on fruits of normal and *rin* mutant tomatoes. *Plant Physiology, 56* (4): 544–546.

MOONEY, H., TROUGHTON, J. H. & BERRY, J. A., 1974. Arid climates and photosynthetic systems. *Carnegie Institute Yearbook, 73:* 793–805.

MORTON, C. V., 1976. *A revision of the Argentine species of* Solanum. Córdoba: Acad. Nac. Ciencias (Arg.).

PEARCY, R. W. & TROUGHTON, J. H. 1974. C₄ photosynthesis in tree-form Euphorbias in wet tropical sites in Hawaii. *Carnegie Institute Yearbook, 73:* 809–812.

RICK, C. M. & FOBES, J. F., 1975. Allozyme variation in the cultivated tomato and closely related species. *Bulletin of the Torrey Botanical Club, 102* (6): 376–384.

ROGLER, C. E. & HACKETT, W. P., 1975a. Phase change in *Hedera helix:* Induction of the mature to juvenile phase change by gibberellin A3. *Physiologia Plantarum, 34* (2): 141–147.

ROGLER, C. E. & HACKETT, W. P., 1975b. Phase change in *Hedera helix:* Stabilization of the mature form with abscisic acid and growth retardants. *Physiologia Plantarum, 34* (2): 148–152.

SMITH, B. N. & TURNER, B. L., 1975. Distribution of Kranz syndrome among Asteraceae. *American Journal of Botany, 62* (5): 541–545.

STEBBINS, G. L., 1974. *Flowering Plants. Evolution Above The Species Level:* 399 pp. London: Arnold.

TROUGHTON, J. H., CARD, K. A. & HENDY, C. H. 1974. Photosynthetic pathways and carbon isotope discrimination by plants. *Carnegie Institute Yearbook, 73:* 768–780.

WILLIAMS, N. P., 1968. *Morphogenesis in* Solanum seaforthianum *as affected by N⁶ benzyladenine, B₉₉₅, gibberellic acid, and light.* M.S. Thesis. Howard University, Washington, D.C.

WILLIAMS, Z. N., 1967. *The effect of gibberellic acid, N⁶-benzyladenine, and quercitin on morphogenesis in* Solanum aculeatissimum *Jacq.* M.S. Thesis. Howard University. Washington, D.C.

29. Observations on the isolation, morphology and culture of potato leaf and meristem protoplasts

R. H. A. COUTTS

Department of Microbiology, University of Birmingham, Birmingham, U.K.

Protoplasts were isolated from stolon apices of *Solanum tuberosum* L. cv. King Edward after their excision from sprouting tubers, and also from fully expanded leaflets of plants grown from these same tubers.

Protoplasts from the apices were meristematic in appearance, of varying size, and showed characteristic traversing of the vacuoles by cytoplasmic strands. They were extremely fragile and did not survive for more than 3 days when cultured *in vitro*.

Protoplasts released from leaflet pieces were morphologically similar to leaf protoplasts isolated from other species. They regenerated a *de novo* cell wall after three days in culture, and exhibited expansion, aggregation and chloroplast systrophy. The protoplasts survived for up to eight days in culture but did not divide. Cell death at this stage may be related to the presence of either ammonium nitrate or chloride in the medium, as has been described in recent investigations.

CONTENTS

INTRODUCTION

Protoplasts have been isolated and cultured from a diverse range of plants and tissues (Takebe, Otsuki & Aoki, 1968; Cocking, 1972). However, until recently such techniques had not been used extensively to study Solanaceae species. The potential of using single cells of solanaceous origin for genetical studies, either involving hybridization or rapid cloning of particularly valuable strains, has been outlined previously (Bajaj, 1974), as has the possible production of disease-free plants (Kassanis, 1967). Clearly, before such studies can be undertaken in potato, methods have to be sought for the successful isolation and culture of protoplasts from different tissues of plants of interest. This communication outlines procedures for the isolation and culture of leaf mesophyll and meristem protoplasts from potato, and some difficulties encountered in the study. Little success was obtained in attempts to culture isolated protoplasts; however, recent work with similar systems has been more successful in this area (Lorenzini, 1973; Upadhya, 1975).

MATERIALS AND METHODS

Plant material

Potato (*Solanum tuberosum* cv. King Edward) tubers were sprouted in darkness in vermiculite and a number of the stolon apical meristems were excised. The same tubers were then transferred to pots, and grown under greenhouse conditions (Coutts, 1973a). Fully expanded leaves, usually 2nd to 4th leaf, were used for the isolation of protoplasts, taken from eight to ten week old plants.

Protoplast isolation

Apical meristems

Meristems were removed from the sprouted tubers and surface sterilized in 10% sodium hypochlorite solution, plus 0.5% "Teepol" detergent for 15 mins. The meristems (10-20) were then washed three to four times in sterile distilled water. The areas of tissue adjacent to the apices (3 to 4 mm below) were then excised aseptically, and shredded finely using a scalpel. These tissue pieces were then suspended in an enzyme mixture (5 mls), consisting of 2% w/v P1500 Cellulase, 2% w/v 5000 Cellulase, 0.5% Macerozyme (All Japan Biochemicals Co., Japan), dissolved in an osmoticum of 20% w/v sucrose at pH 5.6. The enzyme mixture was Millipore sterilized (22 μm filter) before use. The mixture was then shaken in a 50 ml Erlenmeyer flask continuously for 18 h at 25°C, at 20 reciprocating cycles/minute in reduced light. Protoplasts were released by gentle disruption of the material after the incubation, using a pasteur pipette, and were further clarified by decanting the material through muslin into centrifuge tubes. Intact protoplasts were then brought to the surface of the sucrose by centrifugation (100g, 10 min), and collected in a pasteur pipette. Protoplasts were then washed three times in fresh sterile 20% w/v sucrose by repeated centrifugation as above, and observed in a light microscope.

Leaves

Leaf protoplasts were isolated by a similar overnight mixed enzyme treatment of surface sterilized, peeled leaf pieces. Leaflets were removed and allowed to become flaccid for 1 h prior to surface sterilization in 70% v/v ethanol for 30 sec, followed by 6% v/v sodium hypochlorite for 10 min, and three washes in sterile distilled water. Leaflets were then aseptically stripped of their lower epidermis and placed face down in 12 cm petri dishes, containing 15 ml of an enzyme mixture consisting of 1% w/v P1500 Cellulase, 0.5% w/v Macerozyme, and 1.0% w/v potassium dextran sulphate (Meito Sangyo Co. Ltd., Japan) dissolved in an osmoticum of 13% w/v mannitol, at pH 5.8. The mixture was filter-sterilized as previously. The petri plates were left at 25 °C, under ambient light conditions for 18 h, before protoplasts were isolated by gentle squeezing of the leaves with forceps and decantation through muslin into centrifuge tubes. Protoplasts were sedimented by centrifugation (100 **g**, 10 min), and washed two times in 13% w/v mannitol plus 1 mM calcium chloride, by resuspension and centrifugation. Finally, protoplasts were freed from unconverted cells and debris by centrifugation through 20% sucrose (as previously) on which intact protoplasts floated.

Protoplast culture and electron microscopy

Leaf protoplasts were cultured as surface layers in screw-capped centrifuge tubes at 20°–30°C under 2500 lux constant illumination provided by fluorescent warm-white tubes. Protoplasts were cultured on 10 ml aliquots of either the medium of Nagata & Takebe (1970) omitting mannitol and raising the sucrose concentration to 20% w/v, or the medium of Morrel & Muller (1964) together with sucrose at the same concentration.

Freshly prepared leaf protoplasts were examined in the electron microscope, using fixation, embedding and observation techniques outlined previously (Coutts, 1973b).

RESULTS AND DISCUSSION

Meristem protoplasts

Released protoplasts were highly meristematic in nature (Plate 29.1, fig. 1) with characteristic cytoplasmic stranding through large vacuoles, common in such protoplasts. Protoplasts varied in size from 50–200μm, and were frequently contaminated with unconverted cells and debris. Attempts to free protoplasts from such contamination often led to their complete loss owing to their extreme fragility. Protoplast yields were low compared to other protoplast types (Cocking, 1972), and attempts to isolate protoplasts from callus propagated from meristems met with similar problems (Coutts, 1973a). No attempts were then made to culture these protoplasts. The osmotic stabilizer used in this study may have been at too high a concentration for successful isolations to be made. Similarly, pre-plasmolysis of tissue prior to enzyme digestion commonly used now (Grout & Coutts, 1974) may have been advantageous, assuming that sucrose may only partially penetrate the tissue, as it did in similar tissue systems (Ambid, Delmestre & Fallot, 1974) and that vacuum infiltration may have increased this penetration. It is also possible that a modified sucrose or ficoll density gradient used previously in other work, may have assisted in separating cells from protoplasts, and protoplasts of different size from one another. Recently, successful application of such techniques have been utilized by French workers in isolating protoplasts from potato tubers and *Helianthus* species (Lorenzini, 1973; Ambid *et al.*, 1974).

Leaf protoplasts

As seen previously, plant age was not a critical factor in the production of protoplasts from potato leaflets (Upadhya, 1975); young expanded leaves gave the most reproducible yields of protoplasts, however. Difficulties were experienced in stripping the lower epidermis off potato leaflets; allowing the leaves to become flaccid improved peeling but the use of pectin-glycosidase to dissolve the cutin and epidermal cells (Upadhya, 1975) would undoubtedly assist in this context. Released protoplasts were morphologically similar to other leaf mesophyll types (Coutts, 1973a, Gigot, Kopp, Schmitt & Milne, 1975) (Plate 29.1, fig. 2) with cytoplasm arranged peripherally around a large central vacuole. Chloroplasts were however, slightly smaller than those of tobacco leaf protoplasts but had the same ordered arrangement as in intact cells. Spontaneous fusion bodies of several protoplasts were common (Plate 29.1, fig. 3) possibly due to insufficient digestion of cellulosic cell-walls. Contamination with cells (Plate 29.1, fig. 4) and debris was a major problem and the possible toxicity of macerozyme cannot be ignored (Cocking, 1972; Upadhya, 1975). Efforts to culture leaf protoplasts in media known to support development and sustained division of protoplasts, or callus were largely unsuccessful. Intact protoplasts developed new cell-walls in three to five days (Plate 29.1, figs 5, 6) in Nagata & Takebe's medium (1970). Protoplasts showed characteristic enlargement, aggregation (due to common wall formation), systrophy of chloroplasts and chlorophyll loss (Plate 29.1, figs 5, 6), but no divisions were noted in either of the media investigated (see earlier). After seven to ten days of culture (Plate 29.1, fig. 7) cells looked similar to 3-day cultured samples, and after this period many cells turned brown and died. These results have since been confirmed by other workers (Upadhya, 1975), and systematic study of growth media has led to the conclusion that ammonium ions either with nitrate or chloride (present in both media used in this study) are toxic to potato leaf protoplasts even at low concentrations (100 mg/l). This result is in contrast to that found by Morrel & Muller (1964) for callus cells in which the addition of ammonium apparently stimulated cell proliferation. However, Upadhya (1975) has been successful in obtaining a callus from cultured protoplasts and is attempting to differentiate this tissue.

Thin sections of freshly isolated leaf protoplasts examined in the electron microscope showed them to be similar to other leaf protoplast types, showing all the morphological features characteristic of mature plant cells (Plate 29.2, fig. 8a). Only in the chloroplasts could any detectable differences from whole leaf cells and other protoplast types so far examined, be seen. Regions of stroma were cut out to form sub-chloroplasts as seen previously (Coutts, 1973a); (Plate 29.2, fig. 8a, b) but no para-crystalline arrays of supposed fraction 1 protein were found, common in tobacco protoplasts (Otsuki, Takebe, Honda & Matsui, 1972; Coutts, 1973a; Gigot et al., 1975) and thought to be due to plasmolysis and dehydration (Shumway, Weier & Stocking, 1967; Wrischer, 1973). Amorphous bodies also seen in tobacco protoplast chloroplasts were smaller in potato protoplasts and less frequent, as were starch granules; oil droplets in the cytoplasm (Plate 29.2, fig. 8a, b) were also uncommon, when compared to other protoplast type.

Since the date of the Solanaceae Conference (July 1976), for which this paper was originally prepared, a report of whole plant regeneration from potato mesophyll protoplasts has been published by Shepard & Totten (1977).

ACKNOWLEDGEMENT

The author would like to thank Professor E. C. Cocking for his advice and encouragement throughout this work, and the tenure of an S.R.C. C.A.S.E. Award from 1970–1973.

REFERENCES

AMBID, C., DELMESTRE, M. H. & FALLOT, J., 1974. Isolement et survie de protoplastes issus de tubercules de Topinambour *Helianthus tuberosus* L. *Comptes Rendu Hebdomadaire des Séances de l'Academie des Sciences*, (Sèrie D): *279* 1429–1432.

BAJAJ, Y. P. S., 1974. Potentials of protoplast culture work in agriculture. *Euphytica, 23:* 633–649.

BAJAJ, Y. P. S. & DIONNE, L. A., 1967. Growth and development of potato callus in suspension cultures. *Canadian Journal of Botany, 45:* 1927–1931.

COCKING, E. C., 1972. Plant cell protoplasts—isolation and development. *Annual Review of Plant Physiology, 23:* 29–50.

COUTTS, R. H. A., 1973a. *The Use of Protoplasts in the Study of Viral Infection and Replication.* Ph.D. thesis, University of Nottingham.

COUTTS, R. H. A., 1973b. Viruses in isolated protoplasts, a potential model system for studying nucleo-protein replication. *Coll. Int. C.N.R.S.* (Paris), *212:* 353–365.

GIGOT, C., KOPP, M., SCHMITT, C. & MILNE, R. G., 1975. Subcellular changes during isolation and culture of tobacco mesophyll protoplasts. *Protoplasma, 84:* 31–41.

GROUT, B. W. W. & COUTTS, R. H. A., 1974. Additives for the enhancement of fusion and endocytosis in higher plant protoplasts: an electrophoretic study. *Plant Science Letters, 2:* 397–403.

KASSANIS, B., 1967. Plant tissue culture. In K. Maramarosch & H. Koprowski (Eds), *Methods in Virology, 1:* 537–566. London & New York: Academic Press.

LORENZINI, M., 1973. Obtention de protoplasts de tubercule de Pomme de terre. *Comptes Rendu Hebdomadaire des Séances de l'Academie des Sciences*, (Sèrie D) *276:* 1839–1842.

MORREL, G. & MULLER, J. F., 1964. La culture *in vitro* du meristeme apical de la Pomme de terre. *Comptes Rendus Hebdomadaire des Séances de l'Academie des Sciences* (Sèrie D), *258:* 5250–5259.

NAGATA, T. & TAKEBE, I., 1970. Cell wall regeneration and cell division in isolated tobacco mesophyll protoplasts. *Planta, 92:* 301–308.

OTSUKI, Y., TAKEBE, I., HONDA, Y. & MATSUI, C., 1972. Ultrastructure of infection of tobacco meso-phyll protoplasts by tobacco mosaic virus. *Virology, 49:* 188–194.

SHEPARD, J. F. & TOTTEN, R. E., 1977. Mesophyll cell protoplasts of potato. *Plant Physiology, 60:* 313–316.

SHUMWAY, L. K., WEIER, T. E. & STOCKING, C. R., 1967. Crystalline structures in *Vicia faba* chloroplasts. *Planta, 76:* 182–189.

TAKEBE, I., OTSUKI, Y. & AOKI, S., 1968. Isolation of tobacco mesophyll cells in intact and active state. *Plant Cell Physiology, 9:* 115–124.

UPADHYA, M. D., 1975. Isolation and culture of mesophyll protoplasts of potato (*Solanum tuberosum* L.). *Potato Research, 18:* 438–445.

WRISCHER, M., 1973. Protein crystalloids in the stroma of bean plastids. *Protoplasma, 77:* 141–150.

EXPLANATION OF PLATES

PLATE 29.1

Fig. 1. Freshly isolated protoplasts of *Solanum tuberosum* var. King Edward meristems showing large vacuoles (V), cytoplasmic stranding (CS) and plasmalemma (P). Fig. 2. Freshly isolated potato mesophyll protoplast showing peripheral ordered arrangement of chloroplasts (C) and cell-wall free plasmalemma (P). Fig. 3. Spontaneous fusion body of several freshly isolated protoplasts; note common plasmalemma (CP), binding the cells together. Fig. 4. Freshly isolated potato leaf mesophyll cells, showing similar features to fig. 2, together with undigested cell-wall (CW). Fig. 5. Potato leaf protoplasts, three days after culture—note deformation of protoplasts to cellular type structure with a new cell-wall (CW), systrophy of chloroplasts (C) and aggregation due to a common wall formation. Fig. 6. Potato leaf protoplasts five days after culture, essentially similar to those in fig. 5 with some characteristic stratification of chloroplasts (C) in the cytoplasm. Fig. 7. Potato leaf protoplasts ten days after culture, almost identical to those seen at three, five days (figs 5, 6) but debris from dead cells was more common than in previous days samples. Scale 25 μm in all figs.

PLATE 29.2

Fig. 8a, b. Freshly isolated potato leaf mesophyll protoplasts showing fine structural details of chloroplasts (C), plasmalemma (P), sub-chloroplasts (SC), nucleus (N), vacuoles (V), and oil-droplets (O). Scale 2 μm.

Plate 29.1

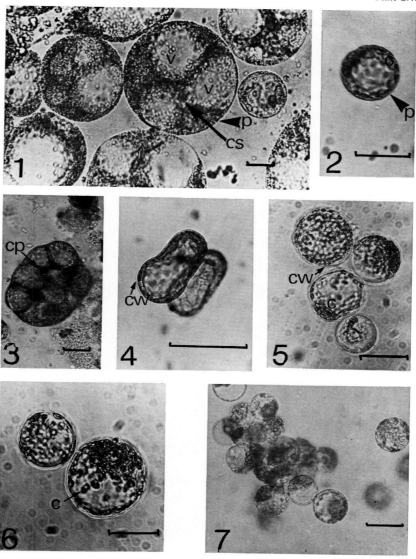

(*Facing p. 376*)

Plate 29.2

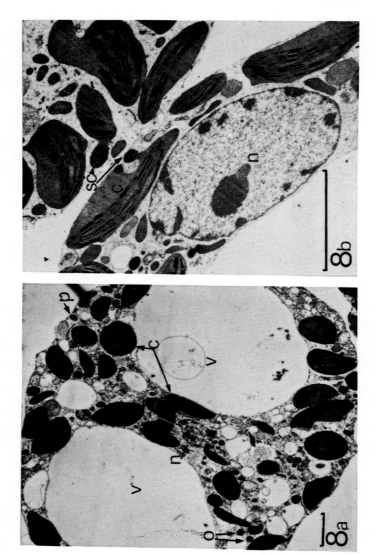

30. Rapid clonal propagation of *Solanum curtilobum* cv. Mallku by aseptic shoot meristem culture

ROGER J. WESTCOTT, BRIAN W. W. GROUT AND GRAHAM G. HENSHAW

Department of Plant Biology, University of Birmingham, Birmingham, U.K.

An explant from an axillary bud, consisting of the meristematic dome and the smallest pair of obvious leaf primordia, was cultured on Murashige and Skoog medium containing 1.0 mg dm³ 2iP and 0.01 mg dm³ NAA.

Numerous adventitious meristems arose superficially on the small amount of callus produced at the base of the explant. Each of these meristems developed into a leafy shoot when the culture was transferred to medium with 0.1 mg dm³ gibberellic acid as the only growth hormone.

Nodal explants from these shoots produced rooted plantlets, by axillary bud growth, four weeks after transfer to fresh medium. The shoots can be further propagated by "nodal cuttings", and when rooted, transplanted into soil with 95% survival. The propagation potential of such a system has been illustrated.

CONTENTS

INTRODUCTION

Potato germ plasm collections have been established in Europe, U.S.A. and Peru (at the International Potato Centre) to conserve the genetic material of both modern and primitive cultivars together with related wild species. The material in the gene banks is stored either as true seed or as tubers. There is a strong preference for propagation of clonal material by tubers in local breeding programmes, but problems of maintenance, especially in a disease-free state, have led to the use of true seed in many collections. The use of tissue cultures as an alternative method for germ plasm storage (Henshaw, 1975) may combine some of the advantages of these two systems. Tissue cultures have low space requirements, a high propagation potential in many instances, and the possibility of disease elimination. Disease-free cultures would be suitable for international distribution of germ plasm material. The possibility of some degree of genetic instability has been reported in various tissue cultures (D'Amato, 1975) and this would be an obvious disadvantage with regard to their use as storage material for genetic conservation. However, this risk appears to be reduced if organ cultures, such as those derived from isolated shoot tips, are used (Morel, 1975).

Investigations into the feasibility of using tissue culture techniques for potato germ plasm storage have been started with a single European *S. tuberosum* cultivar (Dr McIntosh) and twelve clones of various Andean potato species. The clones were chosen to cover the full range of ploidy levels from diploid to pentaploid and include representatives of *Solanum stenotomum* (2x), *S.goniocalyx* (2x), *S.chaucha* (3x), *S.juzepczukii* (3x), *S.tuberosum* subsp. *andigena* (4x), *S.tuberosum* subsp. *tuberosum* (4x) and *S.curtilobum* (5x).

Successful shoot tip cultures were obtained from the *S.tuberosum* cultivar using methods previously reported (Morel, Martin & Muller, 1968), to produce a single leafy stem. The establishment of similar cultures from the Andean potatoes was sporadic and unreliable, which would be a major disadvantage with regard to the practical aspects of establishing a germ plasm collection.

A more complex growth medium, containing a cytokinin, was found to be successful in producing multiple shoots from a single shoot tip explant of each of the Andean clones, but gave limited success with the *S.tuberosum* cultivar. A study was made of the development of these multiple shoots in one of the Andean cultivars, *S.curtilobum* cv. Mallku, to see whether they arose from differentiated tissue associated with the original shoot tip explant, or adventitiously from the surface of callus produced at the base of the explant.

Stem pieces from the regenerated shoots, containing a single node, ("nodal-cuttings"), produced a single shoot from the axillary bud when transferred onto medium containing gibberellic acid. After rooting, these shoots could be transplanted into soil with a high rate of success. The propagation potential of such a system has been illustrated using data obtained with the *S.tuberosum* cultivar.

MATERIALS AND METHODS

The potato plants used in the tissue culture experiments were grown either from tubers or cuttings in a glasshouse with a minimum air temperature of 12°C.

Stem pieces including one node were excised from the plant and the leaf and petiole removed. The pieces with the axillary buds were immersed in a sodium hypochlorite solution (0.5% available chlorine) for 5 min and then rinsed three times with sterile distilled water. Shoot tips which comprised the apical dome and no more than two pairs of obvious leaf primordia were excised under sterile conditions.

Shoot tips were cultured on filter paper bridges in test tubes with 3.5 cm³ modified* Murashige and Skoog medium containing 3% w/v sucrose, 1 mg dm³ 2iP (6-γ-γ-(dimethylallaylamino) purine), and 0.01 mg dm³ NAA (naphthalene acetic acid). This will be referred to as MSNP medium.

Plantlets formed in culture were dissected into stem segments containing one node and transferred to modified Murashige and Skoog medium containing 3% w/v sucrose and 0.1 mg dm³ gibberellic acid (GA₃). After four weeks the plantlets from these cultures were either transferred to soil or further subcultured as described above. All cultures were grown at 22° ± 1°C with a 16 h photoperiod. Illumination was at 4000 lux at bench level supplied by high pressure mercury vapour lamps.

---

* Murashige and Skoog medium without sucrose, IAA and kinetin as supplied by Flow Laboratories Ltd., Irvine, Scotland.

At transplanting, plantlets were transferred into modified John Innes Compost No. 2 in 3 cm fibre pots. These were kept in a plastic propagator for seven days in the glasshouse. The plants were then treated in an identical manner to normal cuttings and retained for tuber production in either the glasshouse or the field.

Material for scanning electron microscopy was prepared by the method of Grout, Chan & Simpkins (1976). Material for sectioning was dehydrated and embedded in paraffin wax (m.p. 55°C). Sections were cut at 15μm and stained with safranin and light green (Purvis, Collier & Walls, 1960).

RESULTS

The isolated shoot tip (Plate 30.1, fig. 1) does not grow whilst maintained on MSNP medium, but after two weeks callus tissues begin to proliferate from the base of the explant (Plate 30.1, fig. 2). This tissue comprises large, naked cells with no apparent tissue organization (Plate 30.1, fig. 3).

The rate of callus growth declines after four weeks of culture, and the callus cells become organized into spherical aggregates (Plate 30.1, fig. 4). A thin "cuticle" develops on surface regions of some of the aggregates between week 6 and week 8 of the culture period (Plate 30.2, fig. 5).

After week 8 of culture there is an obvious decrease in the mean size of the "cuticle"-covered cells, presumably as a result of cell division, and their arrangement becomes more regular (Plate 30.2, fig. 6).

Between week 8 and week 12 of culture these more organized regions begin to assume a dome and primordia arrangement (Plate 30.2, fig. 7) and, after week 12, typical shoot meristems can be found randomly distributed over the callus surface (Plate 30.2, fig. 8).

The formation of these meristems is not synchronous, however, and new meristems continue to arise throughout the culture period, taking four to six weeks from "cuticle" deposition to the attainment of normal structure.

Examination of sections throughout the culture period does not reveal any vascular or cambial connections between the original meristem of the explant and the newly forming ones, nor are they connected between themselves.

The efficiency of propagation by subculturing nodal explants is summarised in Table 30.1. There is no apparent difference in the ability of nodes from all parts of the shoot to produce new shoots from the axillary buds.

Table 30.1. Shoot production by nodal explants from regenerated plantlets of *S. curtilobum* cv. Mallku

Position of node from apex	Number excised	Shoot production after 30 days	Survival after transplanting
Apex	7	7	7
1	8	7	5
2	8	6	6
3	7	5	6
4	8	6	5
5	7	6	6
6	7	6	5
7	2	2	2
	54	45	42

DISCUSSION

In the development of the multiple shoot cultures described in this investigation the dome and primordia of the original explant are persistent. They have no apparent influence on the development of further meristems at the callus tissue surface, but the possibility cannot be discounted. The lack of any differentiated morphological connection between the original meristem and the developing ones and the apparently random spatial relationships between them, however, leads to the conclusion that the meristems on the callus surface are adventitious.

This type of multiple shoot production has been routinely obtained for the full range of potato genotypes which have been used in the germ plasm investigations. These multiple shoot cultures have two major advantages over the more commonly used meristem culture which produces only a single shoot. The first is the reliability of production from a wide range of potato genotypes, although further work must be undertaken to extend its application in this respect. The second is the high propagation potential readily achieved using nodal explants from the regenerated shoots.

A typical multiple shoot culture would produce 20 adventitious shoots within three months of isolation of the original explant. Each of these shoots would provide at least five nodes suitable for further propagation. The single shoot culture might only provide five nodes after three months and, using nodal propagation, it would take a total of five months from the isolation of the original explant to produce 100 nodes (Table 30.2).

Table 30.2. A comparison of the number of nodes available for subculture* using multiple and single shoot cultures

Time after isolation (months)	Multiple shoot	Single shoot
3	100	5
4	500	25
5	2500	125
6	12,500	625

* Assuming 5 nodes per shoot.

The possible disadvantages lie in the opportunities for genetic change which may be introduced during the callus stage, although proliferation of callus is limited in this system. The efficiency of virus elimination by such a culture process must also be investigated.

The tissue cultures described in this study have obvious potential for use in a system of germ plasm storage, but rigorous evaluation must be made of the genetic stability and disease status of plants produced in this way before they can be accepted as useful material for inclusion in a potato gene bank. Such an evaluation programme is at present being carried out in collaboration with the International Potato Centre, Peru.

REFERENCES

D'AMATO, F., 1975. The problem of genetic stability in plant tissue and cell culture. In O. H. Frankel & J. G. Hawkes (Eds), *Crop Genetic Resources for Today and Tomorrow:* 333–348. London: Cambridge University Press.

GROUT, B. W. W., CHAN, K. W. & SIMPKINS, I., 1976. Aspects of growth and metabolism in a suspension culture of *Acer pseudoplatanus* (L.) grown on a glycerol carbon source. *Journal of Experimental Botany*, *27*: 77–86.

HENSHAW, G. G., 1975. Technical aspects of tissue culture storage for genetic conservation. In O. H. Frankel & J. G. Hawkes (Eds), *Crop Genetic Resources for Today and Tomorrow:* 349–358. London: Cambridge University Press.

MOREL, G., 1975. Meristem culture techniques for the long-term storage of cultivated plants. In O. H. Frankel & J. G. Hawkes (Eds), *Crop Genetic Resources for Today and Tomorrow:* 327–332. Cambridge University Press.

MOREL, G., MARTIN, C. & MULLER, J. F., 1968. La guerison des pommes de terre attientes de maladies à virus. *Annales de Physiologie Végétale 10:* 113–139.

PURVIS, M. J., COLLIER, D. C. & WALLS, D. (Eds), 1966. *Laboratory Techniques in Botany*. London: Butterworths.

EXPLANATION OF PLATES

PLATE 30.1

Fig. 1. Freshly isolated shoot tip explant comprising the apical dome and one large leaf primordium. Fig. 2. Callus tissue at the base of the explant after two weeks in culture. Fig. 3. A portion of the explant shown in fig. 2 at increased magnification to show large, naked callus cells. Fig. 4. Callus tissue organized into spherical aggregates four weeks after culture initiation.

PLATE 30.2

Fig. 5. A cuticle-like covering developing over a callus cell aggregate eight weeks after initiation of the culture. Fig. 6. A region of smaller, more regular cells which arise below the "cuticle" covered regions. Fig. 7. Early development of a shoot meristem at the callus surface. Fig. 8. A section through an entire callus showing numerous developing shoot meristems.

Plate 30.1

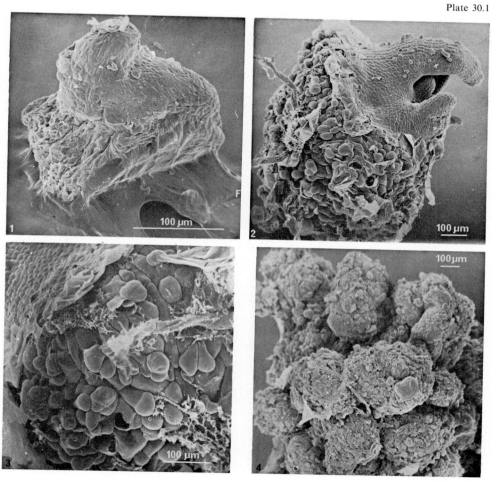

Plate 30.2

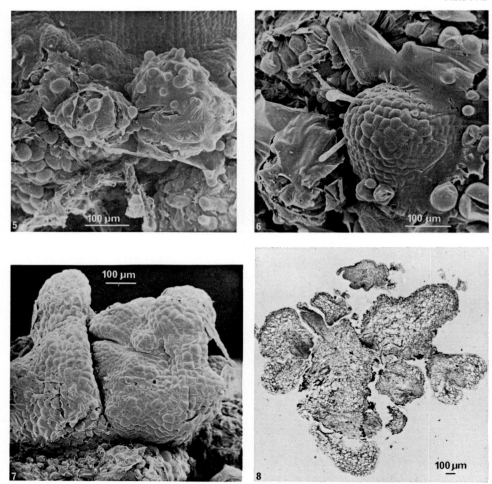

VII. Floral biology, incompatibility and haploidy

31. Sex forms in *Solanum* (Solanaceae) and the role of pollen collecting insects

D. E. SYMON

Waite Agricultural Research Institute, The University of Adelaide, Private Bag, Glen Osmond, South Australia, 5064, Australia

True monoecism and dioecism are not known in *Solanum* since purely female flowers have not been found. In all cases of sexual reduction male flowers are produced with the retention of hermaphrodite flowers on the same (andromonoecism) or in a few cases on separate plants (androdioecism).

A survey of the literature and observation on the pollen collecting insects associated with *Solanum* suggests that these sex forms have evolved from the need to provide pollen for specialized pollen vectors.

CONTENTS

SEX FORMS AND TRENDS IN THE INFLORESCENCE

The inflorescence of *Solanum* is a variously developed or reduced cyme. The more elaborate, branched, many flowered inflorescences which may appear paniculate or corymbose can probably be considered as relatively primitive. In Australia they are found only in mesic tropical environments and this appears to be the case elsewhere. In arid areas the inflorescences are reduced in complexity to simple cymes with few or many flowers and may finally be reduced to a single or few flowers. The trends to be discussed are illustrated in Fig. 31.1.

Studies on sex forms in *Solanum* have had an erratic history. As Harris (1903) writes, the flowers of *Solanum* are generally characterized as being hermaphrodite and many major reference works make no mention of different sex forms. Many recent general Floras give no indication of this condition. As we shall see later, the reason for this may be that the phenomenon is not uniformly spread throughout the genus. Individual botanists, e.g. Dunal (1813) and many others since, have encountered the different forms of flowers, some with evident surprise (Martin, 1972), as "an unusual female sterility", some considering it an abnormality.

Before proceeding with any further discussion of the sex forms some of the terms must be clarified. In one recent paper the term "heterostyly" was used. Although this is strictly correct since the plants do have styles of different lengths, the term is better restricted to plants with styles of different lengths but in which all the flowers are fertile, such as the well known cases in *Primula* or *Oxalis*—an evolutionary development involved with outbreeding mechanisms.

To avoid the use of the term heterostyly Hossain (1973) has used the term stylar-heteromorphy. This is considered preferable but is still inadequate. The short-styled flowers in *Solanum* are usually female sterile and are evolutionary stages in the development of a special dioecism. In *Solanum* hermaphrodite, andromonoecious and androdioecious forms occur. Table 31.1 summarizes the terms used for different sex forms, those marked with an asterisk (*) being found in *Solanum*.

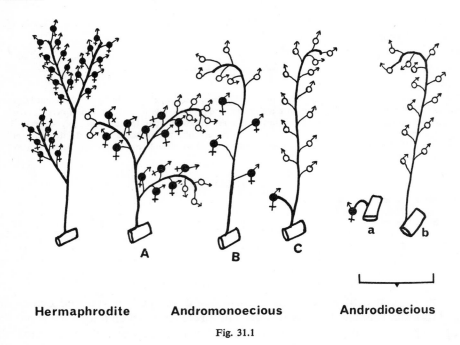

Hermaphrodite   Andromonoecious   Androdioecious

Fig. 31.1

Table 31.1. Terms for the different sexual forms of flowers and their occurrence in *Solanum*

*Hermaphrodite	♀̇	Bisexual flowers on one plant; common in *Solanum*
Monoecious	(♀+♂)	Unisexual flowers on one plant; not known in *Solanum*
Dioecious	♀, ♂	Unisexual flowers on different plants; true dioecism; not known in *Solanum*
*Andromonoecious	(♀̇+♂)	Hermaphrodite and male flowers on one plant; common in *Solanum*
*Androdioecious	♀̇, ♂	Hermaphrodite and male flowers on separate plants; rare in *Solanum*
Gynomonoecious	(♀̇+♀)	Not known in *Solanum*
Gynodioecious	♀̇, ♀	Not known in *Solanum*

Table 31.2 shows that the different sex forms are concentrated in the subgenus *Leptostemonum* according to the classification of the genus as presented by D'Arcy (1972).

Table 31.2. Distribution of sex forms in the genus *Solanum*

Subgenus *Solanum*		
10 Sections	all hermaphrodite	H
1 Section (Benderianum)	andromonoecious	A
Subgenus *Archaesolanum*		
1 Section	hermaphrodite	H
Subgenus *Bassovia*		
3 Sections	all hermaphrodite	H
Subgenus *Brevantherum*		
5 Sections	all hermaphrodite	H
Subgenus *Lyciosolanum*		
1 Section	hermaphrodite	H
Subgenus *Potatoe*		
9 Sections	all hermaphrodite	H
Subgenus *Leptostemonum*		
22 Sections		
Acanthophora		A
Aculeigerum		.
Androceras		.
Anisantherum		.
Aquartia		.
Cryptocarpum		A
Eriophyllum		H
Graciliflorum		.
Irenosolanum		.
Ischyracanthum		A
Lasiocarpum		A
Lathyrocarpum		A
Leprophora		.
Melongena	(7 series most andromonoecious) (a few androdioecious)	A + AD
Micracantha		H
Monodolichopus		H
Nycterium		A
Oliganthes	8 series mixed	A & H
Persicariae		.
Subinermia		H
Somalanum		H
Torva	3 series mixed	A & H

H, Hermaphrodite; A, andromonoecious; AD, androdioecious.

The information in Table 31.2 is derived largely from the descriptions of the type species of the various subgenera and sections. The very uneven quality and detail in descriptions raises problems. Several recent authors give quite inadequate detail of species which I know from personal experience to be andromonoecious. In some sections of the genus where there are relatively recent comprehensive treatments namely subgenus *Solanum* (Edmonds, 1972), subgenus *Tuberosum* (Correll, 1962; Hawkes & Hjerting, 1969), subgenus *Brevantherum* (Roe, 1972), and the closely related *Lycopersicon* (Muller, 1940; Luckwill, 1943), different sex forms appear to be completely absent and I have found neither comment on the state nor description of it.

The andromonoecious condition carefully described by Hossain (1973) in *S. torvum* (a species from Central America and now a pantropic weed) may be considered as an early phase in the reduction of the gynoecium. Numerous lower flowers are hermaphrodite and only distal flowers of the inforescence are male; up to twenty fruits may be set in one inflorescence (Fig. 31.1A). Whilst there is little difference in the sepals, petals, stamens, or pollen grains of the two forms of flowers, the ovaries of the males are slightly reduced, the styles are greatly reduced and their lengths fall clearly into a different size class with no overlap. Hossain points out that the males form a small proportion of the inflorescence, that they are among the later formed flowers, and that they occupy distal ends of the branches.

The second phase of differentiation is shown by such species as *S. sodomeum* and *S. campanulatum*, the former of North African/Mediterranean origin and now widespread as a weed, the latter an eastern Australian species. In both of these the inflorescence is usually an unbranched raceme-like cyme with several lower flowers fertile, and several upper flowers male (Fig. 31.1B). Poorly developed inflorescences may be reduced to very few flowers. There is some tendency for the lower flowers to be larger with rather more prickly calyces and for the upper male flowers to be smaller and less prickly. Several fruits may set and the male flowers are not numerous. This phase is probably more widespread than is currently recognized.

The third phase is better known and is exemplified by *S. melongena*: Egg Fruit. In this group the lower flowers are distinctly larger and have larger prickly calyces; the fruit in these cases is large, usually solitary and the pedicels may be massive (Fig. 31.1C). In the case of *S. melongena*, domestication may well have reduced the male flowers still further so that solitary hermaphrodite flowers may occur. We can note here that this condition is not necessarily due to domestication. Half a dozen Australian species have a single lower hermaphrodite flower with few to many male flowers above.

The distribution of the second and third phases is extensive and includes species in all continents. The trend is the same throughout with a reduction in the number of hermaphrodite flowers to few or one only and generally an increase in their size. The male flowers may be few to many but they are generally smaller, have less prickly calyces and their ovary, style and stigma become vestigial.

Only in Australia has there been further development of this trend with the evolution of true androdioecism (Symon, 1970). In about six species the fruitful inflorescence is reduced to a large solitary hermaphrodite flower, and all have relatively larger fruits enclosed in prickly calyces. The separate male plants have simple or forked long cymes of up to 50 male flowers with vestigial ovaries (Fig. 31.1.a, b). The two forms occur in the ratio of about 1 : 1 but because of their clonal habit this is not easy to count.

I think it must be concluded that the androdioecious species are "advanced" and are derived from andromonoecious species.

BREEDING SYSTEMS

Fortunately, enough is known of the reproductive biology to make some speculations on the evolutionary advantages of production of male flowers. The hermaphrodite flowers still produce pollen though seemingly in reduced amounts or in reduced quality; hence they still have the potential for selfing. The male plants set no seed so that half the individuals of the species do not contribute directly to the seed supply. The clonal habit is developed though this is widespread in the genus, and growth habits range from small to large shrubs.

Obviously, true dioecism ensures outcrossing and the maintenance of considerable heterozygosity in the population, which is generally considered to be an evolutionary advantage. *Solanum* has kept its fruitful flowers hermaphrodite and, in some cases at least, these are self fertile so that outcrossing is not always achieved, nor is it always essential. Self sterility alleles are widespread in *Solanum*, and these too assist in achieving outcrossing. Fryxell (1957) lists 14 species of *Solanum* of which 12 were self incompatable; however, the sample is both small and biased as most of those listed belong to the tuber bearing sections of the genus. Hardin *et al.* (1972) list a few more incompatible species. The present author has grown many species of *Solanum* but has not tested them specifically for incompatibility; many of the Black Nightshade group (section *Solanum*) are self fertile and all species of the subgenus *Archaesolanum* set fruit freely. In the subgenus *Leptostemonum* where the different sex forms are well developed many of the species in which only one individual was grown either set no seed or set fruits very poorly even when growing well and subject to local pollen vectors. Some of the andromonoecious species e.g. *S. sodomeum, S. marginatum, S. melongena* (all introduced aliens in Australia) will set seed when isolated, whereas other local species will not. Whatever may be the final figure of incompatible species in the genus, there is evidence that the incompatability is widespread and effective and obviously reduces any need to develop true dioecism as a mechanism for outbreeding. Unfortunately not enough is known of compatibility in most of the wild species of *Solanum* to know whether incompatibility and andromonoecism are linked.

The final switch to androdioecism achieved in the Australian species is intriguing. What is the advantage of the complete separation of the male plants? Half the population becomes non-seed bearing and the fruitful plants still retain anthers and pollen. The trend begun with *S. torvum* can be traced through a progressive line of development. Presumably each of these changes has some selective advantage for the species.

If in this series the hermaphrodite flowers are self fertile or retain an effective incompatibility system then the advantage to the plant in producing male flowers is not in achieving heterozygosity for one would expect simple dioecism to be established. There is some tendency in this direction, as there is some evidence that pollen fertility is reduced in the hermaphrodite flowers, but this only at the end of the series.

INSECT POLLINATORS

As *Solanum* species produce little or, more often, no nectar the only attraction the flowers have for insects is the pollen they produce. The flowers are rarely scented but a faint sweet scent can sometimes be detected by the human nose. If the flowers were female only and were without anthers they would possibly not attract pollinating vectors.

Heinrich & Raven (1972) suggest that a flower must provide sufficient reward to attract foragers but it must limit this reward so that animals will go on to visit other plants of the same species. The flowering regime of both the andromonoecious and androdioecious species in which a few fresh male flowers are produced on each raceme each day over a long period of time is very similar to that noted by Jansen (1971) for some of the bee flowers in Costa Rica.

The anthers of many species of *Solanum* are poricidal, i.e. the pollen is shed through small pores at the apex of the anther and the anthers are either loosely or firmly erect forming a cone about the style. In those species whose anthers open in the more conventional introrse longitudinal slits the anthers remain in a cone and the whole anther cone then has a single opening and can be considered effectively poricidal. Very little pollen is ever displayed at the orifice and some manipulation of the anther by vibration or "milking" is necessary to release pollen. Michener (1962) seems to have been the first to describe in detail the ability of bees to "buzz" or vibrate the anthers and to release pollen, J. H. Barrett having drawn his attention to it in New Guinea. The insects curl their bodies over the cone of anthers, the flower often becoming inverted with their weight. The insects release pollen with rapid wing vibrations, buzzing and antennal probing of the anthers. The pollen is caught on the anterior ventral regions of the abdomen and is relocated in their corbiculae (pollen baskets). His observations were soon supported by those of Linsley (1962) and Linsley & Cazier (1963, 1972). Michener lists 18 species of bees he had observed extracting pollen from poricidal anthers and suggests that they are frequently the females, and though they belong to several different taxonomic groups the genus *Apis* (Honey Bee) is absent. My observations in South Australia confirm those of Michener and in addition indicate that the smaller bees do not invert the flowers and usually "buzz" a single anther at a time.

The evidence that many species of Bees and Syrphids can manipulate the poricidal anthers of the *Solanum* flowers is now extensive, though close species-dependance is less well documented. The classical book by Muller (1883) deals with a few European species of *Solanum* only, and records two common Syrphids on Potato, Bittersweet and Black Nightshades, all of which were also visited by bees and bumblebees. Linsley (1962) reports *Ptiloglossa arizonensis* (family *Collitidae*) active as an early morning pollinator of *Solanum elaeagnifolium*. In addition *Bombus sonorus* and *Bombus morrisonii* (family *Apidae*) were the only bees which exploited the pollen of this *Solanum* and they manipulated the anthers somewhat violently. *Exomalopsis solani* (family *Anthophoridae*) visits *Solanum elaeagnifolium* and other species at other sites. Linsley & Cazier (1963) give further details of the pollinating activity of species of bees active on two species of *Solanum*. They describe the collection and transport of pollen. The constancy among the pollinators varied but *Ptiloglossa jonesii* had pure loads of *Solanum* pollen, *Caupolicana yarrowi* had mixed loads, *Psaenythia mexicanorum* had pure loads, *Bombus sonorus* was very constant, 85% having pure loads, and *Protoxaea gloriosa* had mixed loads, while three species of *Centris* had 59%, 37% and 16% *Solanum* pollen respectively. Most of the bees were active at the flowers of *Solanum* early in the morning and pollen collecting tapered off or ceased from mid- to late morning. Linsley & Cazier (1972), continuing their studies on the biology of bees and their competition for various pollen, after describing further the activities of bees, give some information on 'homing' ability of the bees. Of 344 males marked and recaptured, 6% returned from 1.6 km, 6% from 5 km and 3% from 8 km, and a single bee (of 159 marked) returned in 24 hours from 16 km. Jansen (1971) also gives records of the remarkable distances from which the female Euglossine bees will return and obtained recaptures from 23 km in Costa Rica. He also gives evidence of the successive daily visits by these bees to plants producing relatively few fresh flowers each day, a flowering regime that would be matched by the male plants of the androdioecious species of *Solanum* in northwestern Australia. He points out that effective outcrossing may occur at very low plant densities, because of the powerful flying ability of the bees. Jansen also notes that, as with other highly host specific interactions of flowers and bees, floral morphology may become specialized to exclude all other types of bees. Although the Euglossinae are long tongued bees, Jansen considers their behaviour to be similar to that of the large lowland bees *Xylocopa*, *Ptiloglossa*, *Centris* and *Bombus*.

Macior (1964) studied the pollination ecology of *Dodecatheon meadia* (Primulaceae) which has a cone of anthers similar to those of *Solanum*. In this case the pollen is freed by introrse dehiscence of the anthers and is discharged at the tip of the staminal cone. Macior recorded the activity of bees in collecting pollen and, in addition to his observations, cinematographic films of the insects' activities were made. His observations were extended to nearby *Solanum dulcamara* and *Lycopersicon esculentum*, both of which have a closely similar arrangement of their anthers. Two species of solitary bees and seven species of *Bombus* attended the *Dodecatheon*, the same solitary bees and five species of *Bombus* attended the *Solanum dulcamara* and two species of *Bombus* attended the *Lycopersicon*. No other insects were found to collect pollen in a like manner from these morphologically similar flowers. The two solitary bee species were not observed on 143 other plant species in the vicinity of the research area. Analysis of the pollen loads did reveal some small amounts of foreign pollen. Thirtythree females of *Augochloropsis* attending *Solanum dulcamara* flowers were examined and one had no pollen, five had one foreign pollen grain and 27 had *Solanum* pollen exclusively. *Lasioglossum* collected abundant pollen from *Solanum* and other flowers, and mixed loads were more common. Nectar collection appeared of little importance in all three genera. He concluded that the normal manner of pollination appears to be a behavioural pattern of the bee closely related to the form, function and position of the flower.

Michener (1965) collected *Trichocolletes hackeri* and *Amphylaeus mediosticus*, (family *Colletidae*), and *Lasioglossum musicum* (family *Halictidae*) from *Solanum furfuraceum* at the Bunya Mountains, Queensland. The author has collected or observed *Lasioglossum* sp. and *Nomia flavoviridis* (family *Halictidae*) and *Amegilla salteri* (family *Anthophoridae*) on six species of *Solanum* in South Australia.

Cheel (1908) collected *Amegilla bombiformis* and *Amegilla cingulata* (family *Anthophoridae*) from a *Solanum* in Sydney, New South Wales. Houston (1975, pers. comm.) reports that *Amegilla bombiformis* is common on *Solanum* in South-eastern Queensland, and *A. salteri* (family *Anthophoridae*) is common across most of Australia and that *Nomia flavoviridis* (family *Halictidae*) is common on *Solanum* in arid areas of Australia; in addition species of *Stenotritus* and *Hylaeus* (family *Colletidae*) were active on *Solanum* in South-eastern Queensland. All of these are considered to be general collectors of pollen and none is known to be specialized.

Linsley & Cazier (1970) state that it is clear that solanaceous plants, especially species of *Solanum* and to a lesser extent *Datura*, play a central role among pollen sources for caupolicanine bees (family *Colletidae*) in southeastern Arizona. The anthers of *Solanum* are tubular and only large bees that can vibrate the anthers and tip or invert the flowers are consistently able to obtain large quantities of pollen in a short time. Small bees, including halictines and *Exomalopsis* scavenge pollen from the anther tips. The authors describe in detail the vigorous early morning activity of pollen collection.

Some details of insects attending *Solanum* flowers are given by Herndon in Hardin *et al.* (1972). Four sympatric *Solanum* species were studied. All were entomophilous and bagged flowers failed to set seed. Carpenter bees, bumble bees and mining bees were all found to carry *Solanum* pollen, though none was constant for any of the four species; it is interesting to note however that the larger bees preferred the blue flowered species and the smaller mining bees preferred the yellow flowers of *Solanum rostratum*. Hand pollinations within the four species indicate self incompatibility mechanisms, while cross pollinations were successful within each species. Bowers (1975) obtained 11 genera of bees from the flowers of *Solanum rostratum* including five species of *Bombus*. None was oligolectic and no other insects attending this *Solanum* were listed by her.

Proctor & Yeo (1973) may be quoted: "*S. dulcamara* has a rather similar mechanism (cone of anthers) but the pollen is shed from pores at the tips of the anthers. It produces little nectar and has evidently become specialised as a pollen flower. It is freely visited by bees which release pollen by rapid vibration of their wings as they hang from the anther cone". Liu *et al.* (1975) investigated the pollen sources of four species of *Bombus* in Ontario to determine which plants provided foodstuffs for these bumblebees. Analysis of the pollen in the nests showed that in three species *Solanum dulcamara* comprised 66%, 10% and 70% respectively of the pollen in the larval meconia, despite a wide range of other plant species present. They state that the bumblebees often worked the majority of flowers rapidly and commenced foraging earlier in the day and finished later in the evening than did other genera of bees. They add that the basis for food preferences by insects is not fully understood and that little is known regarding the mechanisms which attract a bee searching for pollen; they considered pollen searching to be a behavioural act distinct from nectar gathering. Hurd & Linsley (1975) give details of the activity and flowers visited for pollen or honey by bees of the family *Oxaeidae* in the Americas. Five of the 14 bee species studied were known to use *Solanum* species as a source of pollen. The authors considered this family of bees to be narrowly polylectic, frequently using *Cassia* and *Kallstroemia* as well as *Solanum* as their principal sources of pollen.

Free (1970) described tomatoes being vibrated to improve pollination as a commercial practice. He states that bees and bumblebees visit *S. tuberosum* but that "little pollen is available and no nectar".

Considerable information on the pollination ecology of *Lycopersicon* in Peru and California is provided by Rick (1950). "The tomatoes by virtue of their self incompatibility depend entirely on pollen transfer for their reproduction. The great activity of the collected bees in vibrating the flower with rapid leg movements accompanied by a high pitched hum leaves no doubt that they were collecting pollen and were responsible for cross-pollination. The rate of flower visiting in Peru is so intense that flowers of fertile plants are often completely divested of their pollen. The vectors are readily collected in large numbers in short periods of time".

Eleven species of bees in six different families were collected by Rick. One species, *Augochlora nigromarginata*, was collected almost exclusively on two closely related species of tomato indicating some degree of specificity; other bees were collected more widely. "Such specialization would seemingly follow a considerable evolution and suggest that the activity of the vectors is of fundamental importance, not only to the self incompatible species, for whose reproduction it is essential, but to the self compatible ones as well". The most active vector of tomato pollen in California where tomatoes are not native belongs to the same genus as two very active vectors in Peru.

That pollen is important in the economy of the Syrphid flies and essential for solitary bees is stated in many texts on these insects. Linsley (1958) states on solitary bees "all species of bees are dependent on flowers for their basic food ... the larval food is pollen and nectar". Houston (1969), reporting on the nests and behaviour of some Euryglossine bees in Australia, states that the larval provisions were a fluid suspension of pollen which were sometimes oily in nature and on which the egg was laid. Standifer (1967) shows clearly that the per cent protein in 25 plant pollens varied by a factor of three (ranging from 11 to 36%) and had a significant effect on the longevity of honey bees fed on them. *Solanum* pollen was not included and honey bees do not work *Solanum* flowers extensively. Regarding the Syrphid flies Schneider (1969) states "Adults feed not only on honeydew of aphids but also on nectar and pollen. The uptake of pollen is a prerequisite to normal ovarial function and fecundity. Landscapes or seasons without flowers hamper their ecological efficiency".

We see then that pollen is of great importance to both groups of insects, and the attraction is to the pollen-seeking female bees. The *Solanum* flowers produce no nectar so that both males and females get this from other species.

How restricted is the relationship between the plant genus and the bees and Syrphid flies? Linsley (1958) on solitary bees states "the collection of pollen invokes the most consistently specific response to the flower species ... Consistency in the collection of pollen is characteristic of bees in general...." The term oligolecty is used for those bees which collect pollen from a few species and this is reflected in physiological and morphological adaptations which sharply limit the number of kinds of pollen sources normally utilized by bees with this inherited character. Another authority, Malyshev, cited in Linsley, considers the majority of solitary bees to be oligolectic. There are fewer oligolectics in the tropics, and more in arid areas where competition for pollen is more intense. However restricted their pollen sources bees must visit other flowers for nectar. He also states that the solitary bees, though cosmopolitan, have their greatest abundance in warm arid areas having a semi-desert character. Several genera are world wide and have retained a remarkable degree of homogeneity.

Schneider (1969) states that the Syrphidae are amongst the best known visitors to flowers, that they are swift and indefatigable fliers and show preferences for white and blue colours but also that yellow star-shaped paper flowers 15 cm diameter were particularly effective traps at the end of autumn. White, blue or yellow are the common colours of *Solanum* flowers and the conspicuous yellow cone of anthers is one of the most consistent features of *Solanum* flowers over hundreds of species. A glimpse at the importance of the anther cone is given by

Rick (1947) when describing a mutant of tomato with a gene suppressing hair formation. The absence of hairs on the anther edges allows the anthers to spread widely. The result was that selfing was greatly reduced and out-crossing greatly increased due to disruption of the normal mechanism for pollination.

Macior (1971) lists a few of the well known co-adaptations between pollinators and floral mechanisms. The Yucca moth, wasps that pollinate figs, and the pseudocopulation of some orchids are examples of extremely close insect/plant dependence. It is now evident that there is a special relationship between a wide group of bees and genera of Solanaceae (and other families) with poricidal anthers. Though there is some evidence of plant/insect dependence the greater part suggests that bees visit many different flowers for pollen and nectar but they are almost the only group capable of manipulating the cone of anthers of *Solanum*. The plants are more dependent on the bees than the bees on the plants. However the evidence that *Solanum* pollen is of major importance in the feeding of many bees may indicate a closer dependence than at first evident.

CONSEQUENCES OF SPECIALIZED POLLINATORS

Macior (1971) draws attention to the fact that recent taxonomists of *Dodecatheon* found it difficult to separate species on the basis of floral morphology, all the flowers being closely similar with a pendant cone of anthers and reflexed corolla. Macior further suggests that speciation is effected by geographic and edaphic isolation and not at all by changes in form or operation of the pollinating mechanism. The parallel with *Solanum* is striking; its abundant speciation is notorious, with tremendous diversity of plant form, and fruiting characters, but with remarkable uniformity in its flowers. It is of interest to note that the closely related genus *Lycianthes*, which has similar poricidal anthers is also characterized by numerous, localised, "difficult" species.

Grant (1950) states that the likely consequences of a close relationship between plant and pollinator could be that: "The principal evolutionary role of the isolating mechanism based on pollinating behaviour of bees is to augment the efficiency of geographical speciation in bee plants. It follows that angiosperms which are pollinated by bees will possess a more effective means of evolutionary divergence than angiosperms which are pollinated by inconstant insects (promiscuous plants). More rapid species formation should thus occur in "bee plants" than "promiscuous plants". Grant continues . . . "A survey of the flora of California, which is an area very rich in endemic bees, did in fact reveal a significantly higher number of species per genus in bee plants than in promiscuous entomophilous angiosperms, 5.9 and 3.4 respectively".

The bulk of evidence suggests that, while bees gather pollen from a variety of sources and while several species of bee may work one species of *Solanum*, they are without doubt the principal pollinators of *Solanum* and are voracious collectors of *Solanum* pollen. Few if any other insects play so important a role and, further, *Solanum* is dependent upon them. Some advantage may accrue then from the production of extra pollen *per se*. One can see several possible relationships:

(1) The abundant male flowers are produced over a long period of time; this

has been observed in the field where male plants of *S. dioicum, S. leopoldensis* and *S. asymmetriphyllum* have been seen in flower long after the main flowering of the hermaphrodite plants has finished. Four androdioecious species have been cultivated by the author and in each case the male plants have begun flowering up to two weeks before the hermaphrodite plants. Racemes on male plants have continued to produce flowers for over a month. Each flower lasts a few days and there may be up to three out at one time. This results in a succession of flowers on the raceme. In contrast, the inflorescences of the hermaphrodite plants consist of a single flower lasting about the same time (see Fig. 31.1) and the total pollen production is very much less. Androdioecism ensures ample pollen when flowers are produced and the greater likelihood of fertilization of the more limited solitary hermaphrodite flowers.

(2) If incompatibility is not present the presence of male plants ensures at least a proportion of outcrossing while still retaining the possibility of selfing.

(3) As pollen is essential in the economy of the pollinating vector and it may be advantageous to the plant to supply extra pollen, some plants supply nectar in return for services by insects. Others supply fleshy fruits attractive to agents of seed dispersal. I would suggest that the plants produce pollen over and above that needed for fertilization alone, in order to maintain the numbers of the pollinators. Supplying this pollen in hermaphrodite flowers may be too expensive or too short lived. If all the flowers were hermaphrodite this could load the plant with more fruits than it could normally mature and these few flowers might produce inadequate pollen to maintain the pollinating vectors. Syrphid flies and bees are both dependent on pollen to complete their life cycles. The maintenance of adequate populations of pollen vectors is essential for the plants.

That there is an advantage to the plant in limiting fruiting capacity of the relatively large-fruited *Solanum* species in Australian sites is certainly borne out by observations that frequently only one fruit is set per branch and one fruit set seems to inhibit or reduce further fruiting. However, in those species with the potential for producing several fruits (phase 2, Fig. 31.1) the male flowers commonly exceed the hermaphrodite flowers by 2 to 20 times.

This may be adequate in the third phase (Fig. 31.1C) in many cases. Some of the Australian species may have up to 50 male flowers above the hermaphrodite flowers. The final split to the fourth phase is difficult to explain but may provide even more pollen necessary in the somewhat torrid sites in which the species grow.

The north-west of Australia contains some of the most difficult environments in the country. At its southern margin high temperatures combine with light or erratic summer rains to form a very dry climate which grades to a more normal monsoon climate in the north with a concentrated wet season followed by an arid period, this producing great climatic extremes both within and between seasons. The production of pollen is also likely to fluctuate greatly both within and between seasons. On any raceme of the andromonoecious species the lower hermaphrodite flower is the first to open. If the flowering season is short many of these will have opened before the population of vectors is fully developed or active. The production of male flowers and ultimately androdioecious plants to supply pollen and enhance vector activity by the time the hermaphrodite flowers are produced and to continue this over a long period of time is a logical development.

CONSEQUENCES IN THE GENUS *SOLANUM*

If the thesis is correct that male flowers are advantageous to *Solanum* and have developed from the close insect-plant dependence, there are a number of interesting consequences:

(i) It helps explain the morphological constancy of the flowers, which is a feature of the genus *Solanum*. The purple-blue stellate corolla with prominent yellow anthers is extremely widespread in the genus, even in those sections such as the Maurellas and tuber-bearing species where the production of male flowers has not been recorded. In contrast, the morphology of other parts of the species is very diverse. As the apparently simple floral structure is closely related to a narrow range of pollinating insects it must be considered a specialized mechanism.

(ii) It helps explain the very large number of species in the genus *Solanum*. One can add in parenthesis that *Cassia* (*Caesalpiniaceae*), which also has poricidal anthers, is another large pantropic genus with many species and is frequently linked with *Solanum* in many of the papers cited.

(iii) It helps explain the evolution of localized *Solanum* species since successful species (depending on restricted vectors) will be the product of the ecological limitations of the bees and of the plant population, though there are obviously many other factors involved in this.

(iv) It helps explain the relative uniformity and undistinguished character of *Solanum* pollen grains. Large or highly sculptured or ornamented grains would be more likely to clog up the narrow orifice of the anthers and to flow less freely.

(v) There is evidence that with the development of andromonoecism there has been an increase in fruit size and a reduction in fruit number. I know of no andromonoecious/androdioecious species with small fruits or few seeds.

(vi) Monoecism and dioecism have often developed as mechanisms ensuring outcrossing and the maintenance of heterozygosity. In *Solanum*, at least, it is more likely that these conditions have arisen in response to a close insect-plant relationship centred on pollen production, and are only partly a mechanism for achieving outcrossing, as this is frequently controlled by incompatibility systems.

REFERENCES

BOWERS, K. A. W., 1975. The pollination ecology of *Solanum rostratum* (Solanaceae). *American Journal of Botany, 62:* 633.

CHEEL, E., 1908. Notes and Exhibits. *Proceedings of the Linnean Society of N.S.W., 33:* 287.

CORRELL, D. S., 1962. *The Potato and Its Wild Relatives.* Texas Research Foundation.

D'ARCY, W. G., 1972. *Solanaceae* studies. II. Typification of subdivisions of *Solanum. Annals of Missouri Botanical Gardens, 59:* 262.

DUNAL, M. F., 1813. *Histoire Naturelle . . . des* Solanum. Montpellier.

EDMONDS, J. M., 1972. A synopsis of the taxonomy of *Solanum* Sect. *Solanum* (Maurella) in South America. *Kew Bulletin, 27:* 95.

FREE, J. B., 1970. *Insect Pollination of Crops.* London & New York: Academic Press.

FRYXELL, P. A., 1957. Mode of reproduction of higher plants. *Botanical Review, 23:* 135.

GRANT, V., 1950. The flower constancy of bees. *Botanical Review, 16:* 379.

HARDIN, J. W., DOERKSEN, G., HERNDON, D., HOBSON, M. & THOMAS, F., 1972. Pollination ecology and floral biology of four weedy genera in southern Oklahoma. *Southwestern Naturalist, 16:* 403.

HARRIS, J. A., 1903. Polygamy and certain floral abnormalities in *Solanum. Transactions of the Academy of Science of St Louis:* 185.

HAWKES, J. G. & HJERTING, J. P., 1969. *The Potatoes of Argentina, Brazil, Paraguay and Uruguay.* Oxford: Clarendon Press.

HEINRICH, B. & RAVEN, P. H., 1972. Energetics and pollination ecology. *Science, 176:* 597.

HOSSAIN, M., 1973. Observations on stylar heteromorphism in *Solanum torvum* Sw. (*Solanaceae*). *Botanical Journal of the Linnean Society, 66:* 291.

HOUSTON, T. E., 1969. Observations on the nest and behaviour of some Euryglossine bees (*Hymenoptera: Colletidae*). *Journal of the Australian Entomological Society, 8:* 1.

HURD, P. D. & LINSLEY, E. G., 1975. The bee family *Oxaeidae* with a revision of North American species. *Smithsonian Contributions to Zoology,* No. 220.

JANSEN, D. H., 1971. Euglossine bees as long distance pollinators of tropical plants. *Science, 171:* 203.

LINSLEY, E. G., 1958. The ecology of solitary bees. *Hilgardia, 27:* 543.

LINSLEY, E. G., 1962. The colletid *Ptiloglossa arizonensis* (Timberlake) a matinal pollinator of *Solanum. Pan Pacific Entomologist, 38:* 75.

LINSLEY, E. G. & CAZIER, M. A., 1963. Further observations on bees which take pollen from plants of the genus *Solanum. Pan Pacific Entomologist, 39:* 1.

LINSLEY, E. G. & CAZIER, M. A., 1970. Some competitive relationships among matinal and late afternoon foraging activities of caupolicanine bees in south-eastern Arizona. *Journal of the Kansas Entomological Society, 43:* 251.

LINSLEY, E. G. & CAZIER, M. A., 1972. Diurnal and seasonal behaviour patterns among adults of *Protoxaea gloriosa* (Hymenoptera, Oxaeidae). *American Museum Novitates, 2509:* 1.

LIU, J. T., MACFARLANE, R. P. & PENGELLY, D. H., 1975. Relationships between flowering plants and four species of *Bombus* (*Hymenopera: Apidae*) in southern Ontario Can. *Entomologist, 107:* 577.

LUCKWILL, L. C., 1943. The genus *Lycopersicon. Aberdeen University Studies,* 120.

MACIOR, L. M., 1964. An experimental study of the floral ecology of *Dodecatheon meadia. American Journal of Botany, 51:* 96.

MACIOR, L. M., 1971. Co-evolution of plants and animals: systematic insights from plant insect interactions. *Taxon, 20:* 17–27.

MARTIN, F. W., 1972. Sterile styles in *Solanum mammosum* L. *Phyton, 29:* 127.

MICHENER, C. D., 1962. An interesting method of pollen collecting by bees from flowers with tubular anthers. *Revista de Biologia Tropicale, 10:* 167.

MICHENER, C. D., 1965. A classification of the bees of the Australian and Pacific regions. *Bulletin of the American Museum of Natural History, 130:* 1.

MULLER, H., 1883. *The Fertilisation of Flowers.* London: Macmillan.

MULLER, C. H., 1940. A revision of the genus *Lycopersicon. U.S.D.A. Miscellaneous Publications, 282:* 1.

PROCTOR, M. & YEO, P., 1973. *The Pollination of Flowers.* London: Collins.

RICK, C. M., 1947. A hair suppressing gene that indirectly affects fruitfulness and the proportions of cross pollination in the tomato. *Genetics, 32:* 101.

RICK, C. M., 1950. Pollination relations of *Lycopersicum esculentum* in native and foreign regions. *Evolution, 4:* 110–122.

ROE, K. E., 1972. A revision of *Solanum* section *Brevantherum* (*Solanaceae*). *Brittonia, 24:* 239.

SCHNEIDER, F., 1969. Bionomics and physiology of aphidophagous Syrphidae. *Annual Review of Entomology, 14:* 103.

STANDIFER, L. N., 1967. A comparison of the protein quality of pollens for growth stimulation of the hypopharangeal glands and longevity of honey bees, *Apis mellifera* L. (Hymenopera: Apidae). *Insectes Sociaux, 14:* 415.

SYMON, D. E., 1970. Dioecious *Solanums. Taxon, 19:* 909.

32. Floral biology of *Capsicum* and *Solanum melongena**

L. QUAGLIOTTI

Institute of Plant Breeding and Seed Production, University of Turin, Italy

---

Abnormalities in *Capsicum* flower morphology are reviewed and the physiological control of flowering and fruiting is discussed. The genetical aspects of male sterility and the production of haploids are then briefly commented on.

In *Solanum melongena* a discussion of flower morphology is followed by a review of flowering physiology and heterostyly; pollination biology is also discussed as well as the physiological control of fruit setting and development. The phenomenon of male sterility is reviewed.

CONTENTS

A. *Capsicum*

INTRODUCTION

The typical flower of the *Capsicum* genus is pentamerous and heimaphrodite; pedicel 10–20 mm long; calyx campanulate, shortly 5–7 dentate, ribbed, about 2 mm long, usually enlarging and enclosing base of flower; corolla rotate-campanulate gamopetalous, 5–7 partite, 10–20 mm in diameter.

There is such a wide variability between different species and different cultivars that the floral characters can be used as valid criteria for systematic classification. Thus Lippert, Smith & Bergh (1966) have distinguished the species: *C. annuum*, *C. frutescens*, *C. chinense*, *C. galapogense*, *C. chacoense*, *C. schottianum*, *C. microcarpum* or *C. pendulum*, *C. praetermissum*, *C. eximium*, *C. pubescens*, *C. cardenasii* and *C. scolnikianum*, on the basis of the characters indicated in Table 32.1.

Moreover, the intraspecific differentiation between varieties is so accentuated where *Capsicum annuum* and *C. frutescens* are concerned that they can be recognized from the exceptions to the characteristics indicated. In the *C. annuum* var. *grossum* group, for example, there are populations such as Pomodoro giallo and

*Work supported by a grant from the National Research Council, Rome.

Pomodoro rosso in which the flowers are produced in pairs or in more numerous clusters, rather than singly.

The dimensions of the flowers and their parts may also show considerable intraspecific differences (Table 32.2).

Abnormalities in *Capsicum* flowers have been described by many authors (Terracciano, 1878; Penzig, 1922; Cochran, 1934).

The most frequent are: petaloidy and pistilloidy of calyx lobes; non-abscission of the corolla; formation of staminodes; proliferation of stamens; absence of filaments; protrusion of the style above the tightly closed petals; non-abscission

Table 32.1. Morphological characters distinguishing *Capsicum* species (Lippert et al., 1966)

Corolla colour:	white (a–d–e–f–g), white to greenish-white (c), white to lavender (h–i), purple (k), blue (l), yellow (m)
Corolla throat spots:	none (a–b–c–d–e–k), yellow (f–h–i), green to yellow (g), greenish-yellow (l)
Corolla shape:	rotate (a–b–c–d–e–f–g–h–i–k), campanulate (l–m)
Anther colour:	yellow (d–e–f–g–h–i), blue (b–c), blue to purple (a), purple (k), pale blue (l)
Calyx teeth:	present (a–c–e–g–h–i–k–l–m), none (b–d–f)
Number of flowers per node:	1 (a–c–d–h–k), 1–2 (g–l), 2–3 (b–i), 3–5 (c), 5–7 (f)

Key.   a, *C. annuum*   g, *C. microcarpum* or *C. pendulum*
        b, *C. frutescens*   h, *C. praetermissum*
        c, *C. chinense*   i, *C. eximium*
        d, *C. galapogense*   k, *C. pubescens*
        e, *C. chacoense*   l, *C. cardenasii*
        f, *C. schottianum*   m, *C. scolnikianum*

Table 32.2. Size of flower components in *Capsicum annuum* L.

HIROSE (1965)

Var.	Cv.	Corolla diam. (mm)	Pedicel length (cm)	No. of stamens	Anthers length (mm)	Filament length (mm)	Style length (mm)
longum	Husimiama	15.6	11.6	5–6	2.0	2.1	5.0
grossum	Ozisi	24.2	19.2	6	2.7	3.1	4.6
acuminatum	Kagawahontaka	17.4	18.2	5–6	2.9	1.8	6.4

QUAGLIOTTI (1970)

Var.	Cv.	Corolla diam. (mm)	Pedicel length (cm)	No. of sepals	No. of petals	No. of stamens	No. of carpels
abbreviatum	Quarantino	29.6	21.8	5.9	5.9	5.8	2.3
longum	Corno di bue giallo	29.6	16.5	5.9	5.7	5.7	2.7
grossum	Quadrato d'Asti giallo	27.5	14.9	6.4	6.2	6.2	3.4
acuminatum	Spagnolino	18.5	14.7	5.7	5.5	5.5	2.3

of the style, which remains attached to the ovary, increasing in size as the fruit grows, and finally, ripening with the fruit itself and taking on the colour typical of the fruit; fasciation of the style.

Mutants for morphological floral features have been indicated by von Der Pahlen (1967), in Argentina. The mutation indicated by the term 'fasciflora' (symbol fc) is controlled by a single recessive gene and takes the form of a marked increase in number of the floral components. Single flowers may be altered or two or three flowers may be joined together and borne on a single pedicel; the mutant plants are completely female-sterile, show pollen viability of 27–55% and produce seedless fruits.

The polypetalous mutant of Bansal & Dalmir Singh (1972) was induced by γ rays and resulted from an induced translocation in at least two chromosomes. The plants show pollen fertility of 70%, two styles and stigmas per gynoecium and bear few, thin and shrivelled fruits.

Sterile mutants (in which the reproductive parts were transformed to vegetative ones) were isolated in M_1 and M_2 generations following treatment with ethyl methanesulphonate and with N-nitroso-N-methylurea. Segregation studies show that the mutation is recessive and that it is controlled by a single gene (Bansal, 1973).

Non-flowering individuals, showing different types of aberrant vegetative growth, were found by Kormos & Kormos (1956) among the progenies of inter-varietal crosses in Hungary; the non-flowering character appeared to be mono-factorially transmitted.

Morphological non-hereditary floral anomalies artificially induced in different ways have been described several times. Jayakaran (1972), in plants sprayed with the morphactin EMD-IT 7839, obtained fused campanulate corolla, fused calyx and reduced or missing stamens. The flower buds remained closed, abscissed early and in some cases had an extra ovary.

Floral abnormalities including anomalies in number, size and position of floral parts and internal and external malformations were described by Aillaud, Gondran & Pichenot (1972) as occurring after inoculation with cucumber mosaic virus (CMV); the same types of abnormality were found after grafting on *Datura stramonium*, suggesting that the abnormalities in grafted plants are the results of viral infection and not a direct effect of grafting.

Outstanding differences in flower morphology have been identified by Pochard (1970a) in eleven primary trisomics studied in *Capsicum annuum*. The colour of the anthers is very characteristic (purple, bluish, brownish yellow, yellowish or violet-purple), as also the colour of the corolla (greenish or yellowish), the size of the flower (very large or small), its shape (deeply serrated or bell-shaped) and the width of the petals (wide or narrow).

PHYSIOLOGY

Flower differentiation is affected by numerous environmental factors, particularly the photoperiod, for which the pepper seems to show a tendency to prefer the short or normal period. The data collected, however, refer mainly to *Capsicum frutescens* and on the whole are scarce. Actually the photoperiod—as long as it does not fall below 12 hours—does not seem to constitute a problem worthy of special attention for cultivation purposes in our latitude, either in the open air (flowering from June to October) or where cultivation is carried out under glass.

From the bibliographical data collected we learn that according to Auchter & Harley (1924), *Capsicum frutescens* L. is a short-day species; Deats (1925) also obtained flowering in a shorter time under moderately short-day conditions. Later, Cochran (1942) showed that the differentiation of the flower primordia did not take place at all in plants (again of *C. frutescens* L.) kept at a photoperiod of 0 or 6 hours; it was found in the shortest time (about 23 days from emergence) in individuals cultivated in 12 hours of light and in longer times (26, 27, 28 and 31 days respectively) in those kept at a photoperiod of 8–10 hours, of 14, 18, and 24 hours. A strict chronological relation also existed between the macroscopically noticeable appearance of the flower primordia and the putting out of flower buds; in fact, in every case, the period between the two manifestations was from 12 to 13 days.

The data of Artjugina (1958) and Rusenova-Kondareva (1963) seem to confirm the findings quoted with reference to *C. frutescens*. In the first case, individuals of eight Russian cultivars gave a higher yield (since the number of fruits was greater) if kept at a photoperiod of 12 h/d instead of the normal one of 14 h/d. In the second case, of the four Hungarian varieties tested, kept at 10–15–24 h/d, two gave a higher and earlier yield at the minimum photoperiod and two others produced more with the 15 h/d treatment.

According to Studencova (1964), in many cultivars both American and European, the plants grow more vigorously, flower and ripen earlier and produce higher yields in 12-hour days than in natural 18-hour or in 24-hour days.

As regards the influence of light intensity, it appears that the limit of 3000 lux is the minimum for induction of normal growth.

According to Balazs (1964) in experiments carried out in artificial light, not further defined, growth was very stunted at 1900 lux, whereas it was normal from 3000 to 10,000 lux, with a 10-day shortening of the cycle at the higher level as compared with the lower level.

For hydroponic cultivation Vargovà (1974) in Czechoslovakia, used 200 W bulbs or 40 W fluorescent tubes to complement daylight with illumination lasting 17 hours.

The effect of light intensity is obvious even from the trials with shading carried out with black plastic nets, capable of reducing total daylight by 30, 50 and 70% (Table 32.3). It has been found that in general the varieties all react in the same way,

Table 32.3. Effect of shading on the flowers per plant (no.) and on the fruit set (%) in *Capsicum* (cv. Corno di bue giallo)

	Reduction of incident solar radiation (%)							
	0		30		50		70	
	No.	%	No.	%	No.	%	No.	%
Mattei *et al.* (1971)	61.8	32.3			45.0	26.9		
	65.4				62.4			
Quagliotti *et al.* (1974)	122	8.1	107	11.4	89	9.8	70	5.7

The number of flowers for the first trial and the percentage of flowers set for both do not differ to a statistically significant extent.

reducing the production of flowers progressively as incident solar radiation is reduced (the natural average intensity of light was about 0.59 cal cm^{-2} min^{-1} for Turin and 0.69 for Rome).

It is also true that these differences in the number of flowers do not correspond to a difference in the number of marketable fruits, since the percentage of fruit set is always extremely low and the small fruit set at the end of the season are not usable (Mattei, Quagliotti, Bigotti & di Pietro, 1971; Quagliotti, Lepori & Bigotti, 1974).

The most important factor determining flower differentiation is temperature, especially night temperature. The results of Deli & Tissen (1966), for example, are interesting. From plants exposed for 25 days (from the time when the third true leaf was 1 cm long) at a 17 hour photoperiod with an intensity of 8000 lux and a night temperature of $+12$ °C, they obtained a higher flowering than that of plants kept at 16,000 lux and $+18$ °C.

Also of interest are De Donato's results (1966) from the "Quadrato d'Asti' cultivar. Treatment at $+6$ °C for 15 hours daily in darkness for 14 days from the time the cotyledons expanded, induced not only a greater degree of vegetative growth than that of the control (kept at a night temperature above $+14$ °C) but also induced the formation of a greater number of flowers.

Earliness of flowering is an interesting character from the practical point of view; the existence of a positive correlation between the length of the total growth period and that of the period from sowing to flowering has been shown, and of a negative correlation between the length of this period and the yield, the early-maturing individuals giving higher yields (Jo, Yu and I, 1973).

Opening of the flowers outdoors is adversely affected on dull or wet days but in long spells of bad weather flowers do also appear. Anthesis is complete early in the morning (by 08.00–09.00 hrs) and most flowers shed their pollen at about 09.00 hrs.

The rhythm of flowering and the total number of flowers produced per plant may vary enormously according to the genotype. In similar environmental conditions the data given in Table 32.4 can be taken as typical even though, since they were taken from plants whose flowers were removed after opening, they will certainly be found to be higher than those for plants allowed to produce fruit in the normal way.

The regulation of flowering (for the purpose of delaying and concentrating it)

Table 32.4. Number of flowers per plant in *Capsicum* cultivars

Cultivar	Flower number	
Zairaikozisi	2,665	Hirose (1965)
Ryokkowase	714	
Piccante di Cajenna	2,205	Quagliotti (1969)
Quadrato d'Asti rosso	485	
Quadrato d'Asti giallo	424	
Corno di bue giallo	538	
Corno di bue rosso	531	

by application of growth regulators has been studied by Rylski (1972). All the chemicals tested, except CCC, delayed flowering in different ways.

Alar induced growth retardation of the main axis, without preventing subsequent flowering at lower nodes, and caused loss of apical dominance. GA_3 induced accelerated apical development and consequently abortion of flowers on lateral shoots. The effects of NAA were degeneration of flower buds at the first flowering node and the delayed production of flowers at the second and third nodes. Ethephon delayed flower initiation, favoured the abortion of flowers already differentiated, decreased the growth of the main axis and induced sprouting of the lateral buds.

The pistil comprises an ovary with a longitudinal diameter of 2–5 mm and a transversel diameter of 1.5–5 mm, containing 2–4 carpels; a style 3.5–6.5 mm long; a capitate and lobed, papillate stigma which has a mean diameter slightly greater than that of the style. Cases of fasciated stigmas are quite frequent.

Receptivity varies with temperatures during and after anthesis. It is highest on the day of anthesis, when the anthers are fully developed but are still indehiscent and the corolla is still closed but is about to open, and it lasts for a maximum of 4–7 days (Popova, 1959, 1963; Cochran & Dempsey, 1966) and 5–9 days after emasculation at white bud stage (Markus, 1969).

The relative position of the stigma and the anthers may vary considerably according to the cultivars and the time of flowering, producing various kinds of flowers, which can be classified by the length of their styles into short-styled, medium-styled and long-styled (Quagliotti, 1970).

As far as pollination and fruit setting are concerned, this morphological distinction corresponds to very different behaviour. It is often held (Baldini, 1951; Daskaloff & Popova, 1962) that long-styled flowers are more easily pollinated, even though, according to Cochran (1938) and Erwin (1931) a style longer than the stamens—characteristic of primitive forms of small long berries—does not favour self-pollination at all. According to Srivastava (1964) short-styled flowers are more fertile but it is obvious that whether or not pollen falls on the stigma of the flower it was produced by, depends on the position (dangling or upright) of the flower itself at the time of the dehiscence of the anthers.

The flower position is a typical variational characteristic. It also appears (see Kiss, 1971a) that the frequency with which the stigma protrudes above the anthers is greatest in the small-fruited varieties, intermediate in the red peppers for spice and least in the varieties with big berries; the character seems to be connected with the phylogenetically most primitive forms of *Capsicum*. No substantial differences were noted in flowers of the same cv. developing at different nodes: the higher percentage of true-to-type progeny from seeds of earlier, compared with later, ripening fruits seems associated with insect visitation.

There are from 5 to 7 stamens per flower; they have filaments from 1.8–3.5 mm long and anthers 1.2–2 mm wide and 2–4 mm long. Dehiscence occurs laterally, along a line that runs the whole length of the anther. A normally fertile flower may contain 1–1.5 mg of pollen.

The pollen grain, when dry, has a minimum diameter of 17–20 µm and a maximum diameter of 34–38 µm; swollen grains have a diameter of 27–33 µm.

According to Novak & Betlach (1970a) the tapetal layer of anthers develops on a similar behaviour pattern in many varieties of *C. annuum*; it is differentiated from the archesporial complex during the early development of the anthers; during the further production of tapetum cells, the scheme of the cellular poly-nuclear type is followed, as cytokinesis does not follow karyokinesis. The high rate of polyploidy (from 4 *n* to 8 *n*) is characteristic of the whole layer.

For electron microscopic study of microsporogenesis in male-fertile and cyto-plasmic male-sterile flowers, see Horner & Rogers (1974).

Observers are not in agreement with regard to gametogenesis. Cochran (1938) for example found that in *Capsicum frutescens* var. *grossum*, the embryo-sac has a monosporic development and that the division of the generative nucleus takes place in the pollen tube, while Lengel (1960) observed a bisporic type development of embryo-sac and trinucleate pollen grains in *C. frutescens* var. Japanese Varie-gated Ornamental.

Environmental temperature has considerable importance for the physiology of the male gametophyte. The optimal temperature for pollen germination, lower than that for egg-plant and the same or slightly higher than that for the tomato, is $+20-+25$ °C.

It is above all the maximum limit of this range that can be modified according to the cultivar. In other words, the depressive effect of high temperature is apparent at different levels for the various genotypes. According to Cochran (1938) the vitality of the pollen is in all cases appreciably reduced by temperatures above 30 °C, and even the formation of pollen is compromised by high thermal levels. It seems that the divisions of the mother cells of the pollen are upset by this climatic factor, since the percentage of sterile pollen was found to be strictly related to the temperature of the air about 15 days prior to anthesis.

Since it has been observed (Hirose, 1957) that in the Tabasco cv.—in which dehiscence of the anthers occurs relatively late in the morning (roughly between 10 and 12 o'clock)—pollen still germinates to a relatively satisfactory degree at $+35$ °C and even at $+40$ °C, it can be assumed that there is some relation be-tween the hour of opening of the anthers and the optimal temperature for the germination of pollen.

The chemical composition of the anthers containing fertile pollen is different from that of the anthers with pollen which excessively low environmental tempera-ture has rendered sterile. The varying content can be registered chromatographic-ally, by certain free aminoacids; for example the fertile anthers contain much more proline than the sterile ones, while the latter are richer in asparagine (Fujishita, 1965).

We have information from many sources regarding the storability of pollen. Generally speaking it has been found that temperature is more important than the moisture in the air: at temperatures of 20°–30°C for practical purposes the viability does not last longer than a few hours (24–48 h), whereas at 0 °C or a little higher, pollen can be stored for several days (up to 5–6); if however storage is made in both refrigerator and dryer, there is a good chance of maintaining vitality for fairly long periods (Avetikyan & Stepanyan, 1973). Stored pollen, besides being less efficacious in inducing the production of seed, also seems to give rise to weaker and less productive plants (Popova, 1959, 1963; Hirose, 1965; Dempsey, 1966).

The study of pollen viability seems rather difficult and the results are often not comparable. Similar results were obtained by Novák & Betlach (1967) using colour reaction to either 2,3,5-triphenyltetrazolium, chloride and neotetrazolic blue.

Germination and tube growth were studied on a 1% agar and 10% sucrose medium by Kiss (1971b) who obtained the best results with pollen collected at 09.00 hrs, from dehiscent anthers.

Mass germination on 1% agar and 15% sucrose substrate was noted 2 hours after sowing by Avetikyan & Stepanyan (1973) who found 72–91% fertility of fresh pollen. The behaviour of the pollen *in vitro* was often found to be very variable. None of the media tested by Lepori & Scapin (pers. comm.), some of which gave excellent pollen germination and tube growth, supplied data that corresponded sufficiently well to the germination *in vivo*. Hence they are not acceptable as reliable tests of the viability of the pollen.

The behaviour of the pollen tubes in interspecific crosses has been studied by Georgieva & Molkhova (1974). The following cases are reported: an incompatibility barrier appeared on the stigma; a weaker barrier occurred at the base of the style (reducing the number of pollen tubes entering the ovary); the pollen tubes ceased to grow in the ovary and failed to enter the micropyle; the tubes entered the micropyle but fertilization failed to occur.

With regard to pollination, it should be pointed out that there is not a perfect synchronization in ripening between androecium and gynoecium; at the budding stage the flower's own pollen is often still unripe when the stigma is ready to receive pollen. This fact allows us (Marfutina, 1974) to obtain hybrid seeds without emasculation of the flowers. Selfing is the most frequent mechanism of the reproductive process and self-pollination often occurs by gravitation. Wind and insects are however responsible for cross pollination, which is not so rare as it was originally thought to be. The degree of crossing found, though varying noticeably from case to case was nonetheless always of considerable significance (Table 32.5).

Table 32.5. Percentages of cross-pollination in *Capsicum annuum* L.

Odland & Porter (1941)	9–32	natural
Murthy & Murthy (1962)	58–68	natural
Hristov & Gencev (1965)	77–92	artificial (without emasculation)
Franceschetti (1971)	1–46	natural
Lorenzetti & Cirica (1974)	5–24	natural

Wind direction was shown to influence the degree of cross pollination (Murthy & Murthy, 1962). The most important factor however seems to be the fertilizing activity of insects. Of these, the following have been recognized in Italy: *Coccinella septem-punctata, Hippodamia tredecim-punctata, Apis mellifera, Macrosteles sex-notatus, Coleoptera (Halticinae), Diptera (Brachycerinae), Hymenoptera (Formicidae), Hemiptera (Anthocoridae, Lygaeidae, Nabidae, Aphididae, Typhlocybidae), Lepidoptera (Sphingidae)*.

Concerning the action of bees, according to Lorenzetti & Cirica (1974), they are fairly active in depositing pollen on pepper flowers, especially when there is little competition from more attractive flora. According to Breuils & Pochard (1975), the activity of bees consists essentially in the gathering of pollen, which is carried out thoroughly and effectually although nectars too seem to serve to some degree as an attraction. As pollinating agents, bees were not found to be particularly effectual in transporting pollen to the male-sterile seed producer plants, since they tend rather to visit pollinating parents.

The pollinating action of large flies (*Calliphora*, *Lucilia*, etc.), used under insect-proof shelter, is also slight, since they do not seem to visit the flowers systematically, but tend merely to settle on the upper parts of the vegetation. The best results were obtained in all cases with hand pollination, which produced a number of seeds three times that obtained with the above-mentioned insects.

Concerning the genetic effects of the reproductive mechanism, it seems certain that the high proportion of crossing found gives rise to considerable frequency of heterozygous loci and a conspicuous amount of genetic variability. The manifestations of heterosis which appeared as a result of inter-varietal crossing were generally conspicuous and concerned many characters: viz. number of fruits per plant, number and weight of seeds, earliness, yield per plant, adaptability. On the depressive effects of inbreeding, information (Suzuki, 1951) is scarce, but seems to confirm the existence of the problem.

As for the procedure to be followed to ensure the success of artificial pollination, the most interesting findings are as follows:-

— the best time for emasculation is immediately before the opening of the flower (Daskaloff & Popova, 1962);

— emasculation of the female plants is best done without removing the corolla, that is by eliminating only the androecium (Hristov & Gencev, 1965);

— it is advisable to pollinate immediately after emasculation to obtain a high yield of seeds (Hristov & Gencev, 1965; Gikalo & Studenceva, 1967);

— maximum setting is obtained with pollen gathered from flowers in anthesis and from anthers in dehiscence (Markus, 1969; Kiss, 1971b; Padda & Singh, 1971); the pollen gathered from flowers about to open (the day before anthesis) or that have only just opened (the day after anthesis) gave reduced setting, while pollen collected two or three days before anthesis was completely ineffectual (Dempsey, 1966);

— production of seeds is favoured by the use of a large quantity of pollen (Daskaloff & Popova, 1962);

— vitamins of the B group sprayed on the pistils immediately after pollination increase the fruit set and the number of seeds per fruit (Popova, Kamenova & Mikhailov, 1971).

The pollen grains, after they are placed on the stigma, remain inactive for some time, in any environmental condition. The minimum time lapse between pollination and fertilization, which Cochran (1938) observed, was 42 hours, at temperatures ranging from $+21°$ to $+27°C$. But according to Hirose (1965) 6 hours are sufficient.

The fruit set, especially in *Capsicum annuum* var. *grossum*, is very small in proportion to the number of flowers produced; the values given are: 8% (Bashir, 1952), 10–11% (Nagarathnam & Rajamani, 1963), 20–25% (Breuils & Pochard, 1975), 27–32% (Mattei *et al.*, 1971), 6–11% (Quagliotti *et al.*, 1974). According to Kato & Tanaka (1971) the percentage of total fruit setting was low at peak fruit production and high when fruiting was reduced. Fruit setting on the main stem was constantly around 80%, while on lateral branches averaged 30 and ranged from 10 to 100%.

That the time of flowering is important for fruit setting was found also by Berenyi (1971): at the beginning of flowering 50–75% of the flowers set fruit, while later on the percentage decreased progressively; practically, the useful fruit setting finished in Hungary around 5–10 August.

Many authors who have examined this problem from different points of view have been in agreement in considering that the environmental factor that plays the predominant role in reducing fruit set is temperature. According to Cochran (1936), by keeping the temperature of the greenhouse within certain limits the flower drop percentage may vary from 0 to 100%. For example, setting in plants kept at 32°–38 °C was nil, while by decreasing the temperature to +21°–27 °C and again to +16°–21 °C, the production of fruits set increased progressively; flower drop, according to this author, is the result of an excess of transpiration at the higher temperature. Dorland & Went (1947) hold a different opinion. They consider it is a question of insufficient sugar translocation which probably takes place at high night temperatures. With regard to the latter factor the optimal level for flowering and setting, they say, both for peppers and tomatoes, falls gradually in proportion to the length of the vegetative cycle: for short growing periods the optimal night temperatures are 15.5°–20.5 °C; for longer periods it is only around +8.5 °C. Again with regard to the effects of night temperature, the data provided by Wells (1967) are of great interest. From plants of *Capsicum annuum* L. cultivated at a day temperature of +26 °C and a night temperature of +10°, +16°, +21° or +27 °C the maximum flowering was obtained in individuals kept at +16° and +21 °C and the maximum setting and the highest level of fruiting in those kept at +10° and +16 °C. Low night temperature also appears to be responsible for the special elongated, pointed form of the berries at the base of the style, as opposed to the normal, deeply furrowed form, obtained at a temperature of +16 °C or higher. According to Rylski & Halevy (1974) plants grown at low night temperatures (8°–10 °C) show higher fruit set than those grown at high night temperatures (18°–20 °C). When flowers were emasculated, fruit set was attained only at low night temperatures; when plants were subjected to low night temperature after anthesis, they produced a high percentage of parthenocarpic fruit set and hence often deformed, low-quality berries. The effect of low temperatures in inducing fruit set, starts from anthesis on and not earlier as there is a lack of pollen at such temperatures. These authors suggest reducing the night-temperature to 10 °C at the anthesis of the first flower till the blooming of the flowers at the 5th–6th node.

Growth regulators (GA$_3$; NOA; NAA; CPA; 2,4–D; 2,4,5–T) produced fruit setting even at 18°–20 °C (night). High day temperatures (30°–35 °C average) cause abortion of young flower buds.

Increase in fruit set of both *Capsicum annuum* and *C. frutescens* was obtained with growth regulators; by Chattopadhyay & Sen (1974) with 2,4,5–trichlorophenoxyacetic acid (10 ppm) and β-naphthoxyacetic acid (40 ppm) and by Srivastava (1964) with 2,4–dichlorophenoxyacetic (2 ppm) and β-naphthoxyacetic acid (75 ppm); in the sprayed plants and more short-styled flowers were obtained.

Parthenocarpy is fairly frequent in peppers in consequence of ineffective pollination, both in the case of artificial crossing or of failure to self under isolation in the case of greenhouse cultures. In the first case it is obviously a disadvantage because no seeds are produced, in the second it is an unfavourable result because it involves an irregular growth of the fruit. Seedless fruits developed without fertilization can easily be obtained with growth regulators, such as naphthalene acetic acid 0.05% (Cheong-Ying Wong, 1938). Parthenocarpic fruits are also produced in pollen sterile plants (Novák, Betlach & Dubovski, 1970b; Shifriss & Frankel, 1969).

Table 32.6. The most important sources of male-sterility in *Capsicum*

	Inheritance	Origin
Martin & Grawford (1951)	simple recessive	
Peterson (1958)	S ms ms	
Hirose (1965)		spontaneous
Hirose & Fujime (1975)	a pair of recessive genes	
Rusenova-Kondareva (1963)	only through the mother line	interspecific hybridization
Daskaloff (1971		X-rays
(1973a)	nuclear genes	γ-rays
Pochard (1970b)	recessive monofactorial	γ irradiation of vegetative monoploid buds treatment with EMS of seeds with monoploid embryos
Shifriss *et al.* (1969)	recessive, single gene	spontaneous
(1971)	S ms ms	spontaneous
(1973)	recessive, single gene	spontaneous

GENETICAL ASPECTS

Male-sterility in sweet pepper has been largely studied in the last 25 years owing to its great importance for hybrid seed production.

The most important male sterile sources are set out in Table 32.6. The most interesting aspects studied, apart from the hereditary pattern, are:

(a) the phenotype characteristics, especially at the level of the floral organs, of the male-sterile plants: vegetative changes of the type "xantha" (Pochard, 1970b); reduction in size of anthers, that are of blue-violet colour and shrunken, etc. (Peterson, 1958; Hirose, 1965; Shifriss, 1973; Shifriss & Frankel, 1969; Novák *et al.*, 1971; Daskaloff, 1971);

(b) microsporogenesis, which has been found to show irregularities of a different kind: the tapetal cells vacuolate and enlarge at meiosis; the sporogenous cells are surrounded by callose and undergo meiosis, forming microspore tetrads that do not develop any further, but are compressed to form a central mass (Horner & Rogers, 1973). Approximately 30 deformed microspores are in the male sterile anthers, compared with 500 pollen grains in anthers of fertile plants (Novák & Betlach, 1970b). Development of the microsporocytes stops at the tetrad stage in some cases; in others it is interrupted at various stages and the flowers show marked changes in structure of the tapetum and stamens (Pochard, 1970c). Tetrads that appear normal, gradually degenerate without developing into pollen grains (Hirose & Fujime, 1975);

(c) biochemical components of the sterile anthers (Markova & Daskaloff, 1974);

(d) the stability of the phenomenon of sterility in various environmetal conditions: maximum sterility under warm conditions and minimum in cool environment (Peterson, 1958): total sterility under field conditions and a small amount of fertile pollen grains in the glasshouse (Daskaloff, 1973b);

(e) practical utilization in the production of hybrid seed (Ohta, 1961a, b; Nassi, 1973; Breuils & Pochard, 1975; Daskaloff, 1971; Shifriss & Rylski, 1973);

(f) the relation between virus and cytoplasmic male-sterility; interactions between specific viruses and S cytoplasm result in reduced fertility; according to Ohta (1968, 1969, 1971, 1973) a cytoplasmic entity for male-sterility could have originated from an exogenous virus that lost its infectivity during evolution and became a kind of plasmon of RNA nature.

Androgenesis in vitro. Quite a number of studies have now been made on haploidy in pepper. They mostly involve the utilization of spontaneous monoploids from twin seedlings or those induced by pollen treatment (Campos, 1959; Campos & Morgan, 1960; Dumas de Vaulx & Pochard, 1974).

Anther cultures deserve special mention here. According to Novák (1974), it is possible to induce callus proliferation both from somatic anther tissue and microspores of *C. annuum* and *C. frutescens*. Quite different results in cultivars and species confirm the importance of genetic control in tissue differentiation ability and *in vitro* callus differentiation.

From anthers cultured by Harn, Kim, Choi & Lee (1975) haploid callus developed from the microspores; in combination with other growth regulators, 2,4-D was effective in inducing haploid callus. The latter and embryoids were obtained more frequently in the anthers at the late uninucleate stage.

B. *Solanum melongena* L.

INTRODUCTION

The typical flower is pentamerous and hermaphrodite; solitary or in 2–5 flowered cymes, opposite or subopposite leaves; pedicel 1–3 cm long rounded, violet and ferruginous pubescent; calyx 2–2.5 cm long, spiny, woolly and persistent; forming a distinct cup-shaped structure at its base; corolla curved, hairy beneath and glabrous within, pale purplish, but deep purple at veins; stamens 1–1.2 cm long, alternating with corolla lobes, free, erect, yellow with very short filaments flat at the base; long, narrow anthers forming a cone which surrounds the style and opening into two terminal pores; two-locular ovary, style relatively short or long with capitate, lobed, green stigma.

The number of elements that make up the floral verticils has sometimes also been used for subspecific classification. According to Fiori (1919) the *esculentum* Dun. variety has solitary flowers and 6–9 sepals, petals and stamens; the *insanum* L. variety has flowers in clusters of 2–3 with 5–6 sepals and petals and 5 stamens; the *ovigerum* Lam. variety has solitary flowers with 3–6 sepals and petals and 5–9 stamens.

The conspicuous variability between cultivars would seem to permit the use of certain flower morphology characters for the description of cvs and groups of cvs. In general the number of sepals, petals, stamens and carpels is much higher than that of the spontaneous forms (20 or more stamens are sometimes found); further-more these flowers, which can be defined as tetracyclic, show some tendency to eucyclicity, to have, that is, the same typical number of components in all the flower verticils (Table 32.7).

Table 32.7. Morphology of the flowers of *Solanum melongena* (Quagliotti, 1966)

Cultivar	Corolla diam. (mm)	No. of sepals	No. of petals	No. of stamens
Grossissima di Firenze	45	7.7	7.1	7.1
Mostruosa di New York	41	7.4	6.8	6.6
Violetta tonda precoce	40	7.0	6.6	6.7
Violetta lunga	37	6.0	5.8	5.8

The flowering habit is peculiar: the flowers are set right on the stem and the first one appears as a rule on a definite internode that may vary from the fifth to the thirteenth according to the cultivar (Kakizaki, 1924).

The peduncle sometimes bears two or three flowers but usually only one of them is fertile.

Floral anomalies have been described by Nagarajan & Xavier (1962) and by Quagliotti (1962a). They consist of malformations of the calyx (hypertrophy of some of the sepals, the growing together of two or more sepals), the corolla (irregularity of the dimensions of the petals, cohesion of two or more petals), the androecium (growing together of one or more anthers and the corresponding petals, atrophy of one or more stamens, growing together of stamens whose anthers are joined laterally, absence of filaments, hypertrophy of an anther theca which is forked and provided with two dehiscence pores), the gynaeceum (enlargement of the style which bends towards the ovary and in part grows together with a petal, fascination of the style, syncarpy of ovary consisting of several carpels each with its own distinct style).

PHYSIOLOGY

Flower formation in egg-plant, which is usually classified as a day-neutral plant, is much affected by the nutritional factor; the fertilizer level seems to cause much more difference in time of flower initiation and development than it does in those plants which respond to vernalization and photoperiod (Eguchi, Matsmura & Ashizawa, 1958).

The first opening of the flower takes place at 06.00–0.900 hrs in summer; dehiscence of anthers at 05.00–08.00 hrs. The flowers close in the evening but open again for 8–10 days (after the first day). High humidity and high temperature in the morning hours tend to hasten the opening of the flower and the dehiscence of the anthers. The corolla is visible in the calyx cup 3–4 days before it opens, and the day before it reaches or exceeds the tip of the calyx lobes; on opening the exposed portion of the corolla bulges outwards and cleavages appear between the petals; after complete opening they become reflexed (Pal & Singh, 1943).

Heterostyly*: The difference in position of the stigma from the anther tips depends on the occurrence of styles of different lengths: short (3–4.5 mm), medium (6–7 mm) and long (7.5–9 mm). The length of the stamens (9–10 mm) does not change in the three types of flowers whose frequency varies noticeably in the different cultivars (Murtazow, Petrov & Doikova, 1971; Quagliotti, 1962b).

The diameter of the pedicel is directly correlated with the length of the style and with the small phloem area in the pedicel (Smith, 1931).

The proportion of long-styled flowers, as compared with short or medium styled, and the total number of flowers produced, are higher with increasing concentrations of CO_2 (Imazu, Yabuchi & Oda, 1967).

The influence of flower heterostyly on fruit setting has been studied in 34 cultivars by Uncini (1971): he found that the secondary flowers of the inflorescence are characterized by a hypostigmatic condition and lack of fruit setting, whereas main flowers—like the solitary ones—show a variable percentage of fruit setting, which largely depends on the peristigmatic condition. This kind of flower is more favourable to selfing than the epistigmatic kind as it is less sensitive to the effects of visitor insects.

The high percentage of abortion in short-styled (hypostigmatic) flowers is generally confirmed, whereas the efficiency in fruit setting of medium-styled (peristigmatic) ones is not always accepted; the position of the flower in the inflorescence also appears to be important in controlling normal setting (Smith, 1931; Krishnamurthi & Subramanian, 1954; Khot & Kanitkar, 1956; Oganesjan, 1966; Prasad & Prakash, 1968; Murtazov, Petrov & Doikova, 1971).

Pollination: In most flowers the style projects out of the cone of anthers, bending slightly downwards. In such long-styled flowers the stigma is first touched by insects. As the anthers dehisce at terminal pores, insects, wind or anything touching or shaking them, makes pollen fall. However, there is often so much dispersion of pollen that the stigma can not receive a sufficient amount of it; or it may occur that in flowers isolated before blooming, most of the pollen is preserved in anthers and remains alive until the stigma almost loses its receptivity (Kakizaki, 1924).

According to Pal & Singh (1943) 30 to 40% of the fruit set depends on pollination by contact, gravity and wind; the rest by insects. The fact that a great number of flowers do not hang downwards but are borne at an angle with the vertical, precludes the possibility of gravity being an important factor in pollination. It has been observed that little pollen comes out of the anthers in the normal course of dehiscence; as the lip of apical pore curves inward, away from the stigma, this catches the pollen discharged from the pore. The low setting for self-pollination is thus explained by the scantiness of the pollen and the obstructing lip of the apical pores.

Nor does the wind seem to be an effective means of pollination, as the flowers mostly fail to give fruits without the agency of insects. A large number of insects (*Xilocopa, Apis, Anthophora, Polistes*) visit the flowers of egg-plant in the morning hours; they touch the stigma with their bodies, carrying pollen from other flowers. Many trials indicate that insects play the major role in pollination; consequently fertilization largely depends on crossing (Kakizaki, 1924; Magtang, 1936; Pal & Singh, 1943).

* Technically andromonoecy with brachystyly. (*Eds*)

The amount of natural cross-pollination is estimated by Kakizaki (1924) ranging from 0.2 to 46.8%, by Sambandam (1964), of 0.7–15% between flowers of the same plant and 1.9–10.9% between plants, by Franceschetti (personal communication) of 9–32%. It has also been recorded that, at a distance of more than 50 m, there is no crossing. In some cases selfing is favoured by the lack of suitable pollinators, as has been indicated, for instance, in Czechoslovakia (Frydrych, 1964).

Selfing can be greatly increased by hand pollination; this has important practical application in breeding, as it is very easy to carry out and may induce high production of seed from a single fruit.

The pollen grain is almost round in shape, with a diameter of 20–23 μm. A satisfactory medium for germination contains agar-agar and sucrose 5–10% (Mishra, 1962). Concerning the effect of surfactants, Regupathy & Subramaniam (1973) found that Triton X–100 inhibits egg-plant pollen germination and that Teepol concentrations exceeding 25 p.p.m. significantly reduce pollen tube growth.

Comparisons of the number of stigmatic lobes (pleiomery of the stigma) and fruit size indicates that the number of lobes is positively correlated with large fruits (Sambandam & Muthiah, 1969).

Flowering is staggered and lasts over a long period of time; for example from the 100th–120th and 250th day after sowing, which means, for cultures in the field in north Italy, from mid-June to the first ten days of November, with peaks half way through July and at the beginning of September.

The rate of flowering shows great varietal differences: the "Mostruosa di New York" begins flowering earlier and produces many more flowers than the "Grossissima di Firenze" (Table 32.8).

Table 32.8. Effects of the cultivar, season and thinning on production per plant (no.) in *Solanum melongena* (Quagliotti, 1974)

	Mostruosa di New York		Grossissima di Firenze	
	1968	1969	1968	1969
Control	104	200	37	60
Flowers totally removed weekly	256	535	117	188
Flowers thinned fortnightly	131	325	42	112
Flowers thinned once every four weeks	103	244	38	74

Considerable stimulus to flower production is determined by the removal of flowers (Table 32.8); intensification of flowering is found to be accompanied by a noticeable decrease in fruit setting (Quagliotti, 1967, 1974, 1975).

Earliness of flowering, which is to some extent connected with that of the ripening of the fruit, varies considerably from one cultivar to another; in normal cultures in north Italy, sown in the hotbed and transplanted in the field, flowering may begin after 55 days ("Nana nera precocissima") or even after 110 days ("Mostruosa bianca di New York") from sowing. In general the long-fruited cultivars are earlier than those with globular and oval fruit (Toderi, 1965).

With regard to fruiting earliness, the period of ripening begins earlier and ends later in "Mostruosa di New York" than in "Grossissima di Firenze". Nonetheless, if the time between flowering and ripening of the berry is focussed for the estimate of earliness, it has been found shorter in the "Grossissima di Firenze". The difference is minimal at first and increases as the flowering period proceeds (Quagliotti, 1975).

Significant positive correlations were found between days to floral initiation and opening and to fruit maturity in 16 cultivars by Nsowah (1970).

According to Pal & Osvald (1967) the fruit set of cross-pollinated plants is much higher than that of self-pollinated ones.

Flower abortion is favoured by natural day light reduction and high (30 °C) night temperature (Saito & Ito, 1973); the percentage of fruit set increases with increasing (200–300–900–3000 ppm for 8 hours daily) levels of CO_2 (Imazu, Yabuchi & Oda, 1967). The number of seeds per fruit is closely connected with the type of pollination: it is highest with free pollination, lower in selfed plants and lowest in exclusively crossed plants (Pal & Taller, 1969).

To obtain the highest levels of fruit setting and seed production under artificial pollination the following points seem important:

—suitable types of bags for isolation: embroidery cloth gives the best results (Bhore, Bhapkar & Chavan, 1965);
—the earliest, single and long-styled flowers are most suitable (Frydrych, 1964);
—emasculation is best done on buds which would open the next day (Pal & Singh, 1943);
—pollination performed without prior emasculation, at the beginning of bud opening, can give up to 97% hybrid seed (Oganesjan, 1966);
—a large quantity of pollen (or supplementary pollination as opposed to a single application) increases the fruit set and the number of seeds per fruit and decreases the frequency of parthenocarpy (Popova, 1959, 1961);
—pollen can be stored for two days and pollination may be performed one day after emasculation (Oganesjan, 1966);
—stigma receptivity and the fertilizing capacity of pollen are highest at the time of opening of the floral buds; satisfactory fertilization, however, can be obtained over a two-day interval (Tatebe, 1938; Mikaeljan, 1964);
—pollen remained viable for some days, when stored at 20°–22 °C in 60–65% R.U. (Popova, 1958).

Male sterility: Chemicals seem to be useful for practical emasculation: 2.4-D at 10 ppm applied at ten-day intervals induces complete pollen sterility and a satisfactory number of seeds per fruit in open pollination (Jyotishi & Hussain, 1968; Jyotishi & Chandra, 1969). Good effects can also be obtained by spraying buds, two or three weeks before anthesis (early meiotic stages), with 0.29% of sodium 2,3-dichloroisobutyrate (Nasrallah & Hopp, 1963). Various, more or less damaging effects, including indehiscence of anthers and pollen sterility may result from 0.01% maleic hydrazide injected in the main axis of the inflorescence when flowers are very young (Pal & Olah, 1970).

Androgenesis *in vitro*: The first information we have on anther culture comes from Raina & Iyer (1973) about callus developed from the cut end of the filament and occasionally from the connective tissue; the pollen callus was haploid, but the plants obtained after differentiation and serial transfer were diploid.

REFERENCES FOR SECTION A

AILLAUD, G., GONDRAN, M. & PICHENOT, M., 1972. Étude morphologique comparative d'anomalies florales, soit induites sur un cultivar de *Capsicum annuum* L. par inoculation au C.V.1, soit apparues dans les lignées de *Capsicum annuum* L. modifié per greffage. *Bulletin Société Botanique de France, 119:* 5–6, 303–323.

ARTJUGINA, Z. D., 1958. The effect of day-length upon the growth development and yield of sweet pepper. *Report. Agricultural Science, 8:* 138–140. *Plant Breeding Abstracts,* 4273/1959.

AUCHTER, E. C. & HARLEY, C. P., 1924. Effect of various lengths of day on development and chemical composition of some horticultural plants. *Proceedings American Society for Horticultural Science, 21.*

AVETIKYAN, N. L. & STEPANYAN, S. S., 1973. The duration of capsicum pistil and pollen viability. *Izvestiya Sel'skokhozyaistrennykh Nauk, 9:* 58–63. *Plant Breeding Abstracts,* 8266/1976.

BALAZS, S., 1964. The influence of light on the development of peppers. *Kísérletügyi Közlemények* (Sect. C) *57 C* (3): 3–23. *Horticultural Abstracts,* 6791/1966.

BALDINI, E., 1951. Contributo ad una descrizione e classificazione sistematica delle razze di peperone coltivate in Italia. *Rivista dell' Orto-floro-frutticoltura Italiana, II:* 227–241.

BANSAL, H. C. & DALMIR SINGH, 1972. Translocation heterozygote induced in *Capsicum annuum* L. *Current Science, 41:* 23, 853.

BANSAL, H. C., 1973. Induced mutation affecting flower development in *Capsicum annuum* L. *Current Science, 42:* 4, 139–140.

BASHIR, C. M., 1952. Some pollination studies on chillies. *Agriculture Pakistan, 3:* 125–128. *Horticultural Abstracts,* 3951/1954.

BERENYI, M., 1971. Some results of studies on fruit set in capsicums. *Zöldségtermesztési Kutatò Intézet Bulletinje, 6:* 83–95. *Horticultural Abstracts,* 6897/1973.

BREUILS, G. & POCHARD, E., 1975. Essai de fabrication de l'hybride de piment "Lamuyo—INRA" avec utilisation d'une stérilité male genique (ms 509). *Annales de l'Amélioration des Plantes, 25* (4): 399–409.

CAMPOS, F. F., 1959. Haploid parthenogenesis in *Capsicum frutescens* L. following crosses with untreated and X-rayed pollen. *Dissertation Abstracts, 20. Plant Breeding Abstracts,* 3011/1960.

CAMPOS, F. F. & MORGAN, D. T. Jr., 1960. Genetic control of haploidy in *Capsicum frutescens* following crosses with untreated and X-rayed pollen. *Cytologia, 25:* 362–372.

CHATTOPADHYAY, T. K. & SEN, S. K., 1974. Studies on the effects of different growth regulators on reproductive physiology and morphology of chilli (*Capsicum annuum* L.). *Vegetable Science, 1:* 42–46. *Plant Breeding Abstracts,* 3858/1975.

CHEONG-YING WONG, 1938. Induced parthenocarpy of watermelon, cucumber and pepper by the use of growth promoting substances. *Proceedings of the American Society for Horticultural Science, 36:* 632–635.

COCHRAN, H. L., 1934. Abnormalities in the flower and fruit of *Capsicum frutescens. Journal of Agricultural Research, 48* (8): 737–748.

COCHRAN, H. L., 1936. Some factors influencing growth and fruit-setting in the pepper (*Capsicum frutescens* L.). *Cornell University Agricultural Experiment Station, Memoirs, 190:* 1–39.

COCHRAN, H. L., 1938. A morphological study of flower and seed development in pepper. *Journal of Agricultural Research, 56* (6): 395–417.

COCHRAN, H. L., 1942. Influence of photoperiod on the time of flower primordia differentiation in the Perfection pimiento (*Capsicum frutescens* L.). *Proceedings of the American Society for Horticultural Science, 40:* 493–497.

COCHRAN, H. L. & DEMPSEY, A. H., 1966. Stigma structure and period of receptivity in pimientos (*Capsicum frutescens* L.). *Proceedings of the American Society for Horticultural Science, 88:* 454–457.

DASKALOFF, S., 1971. Two new male sterile pepper (*Capsicum annuum* L.) mutants. *Compte Rendu de l'Académie Bulgare des Sciences Agriculture, 4* (3): 291–294.

DASKALOFF, S., 1973a. Three new male sterile mutants in pepper (*Capsicum annuum* L.). *Comptes Rendues Academic Agricole G. Dimitrov, 6* (1): 39–41.

DASKALOFF, S., 1973b. Investigations on induced mutation in pepper (*Capsicum annuum* L.) II. A study and use of male sterile mutants in hybrid seed production. *Scientific Session Institute of Genetics and Plant Breeding,* Sofia: 217–228.

DASKALOFF, H. & POPOVA, D., 1962. Studies on flowering biology and pollination of red pepper (*Capsicum annuum* L.). *News Central Scientific Research Institute for Plant Industries, Sofia: 13:* 5–18. *Plant Breeding Abstracts,* 1949/1963.

DEATS, M. E., 1925. The effect on plants of the increase and decrease of the period of illumination over that of the normal day period. *American Journal of Botany, 12:* 384–392.

DE DONATO, M., 1966. Osservazioni intorno all'azione della temperatura notturna e di trattamenti con 6-benzil-aminopurina sull'accrescimento e sulla fioritura del peperone (*Capsicum annuum* var. *grossum*) nei primi stadi di sviluppo. *Rivista dell'Orto-floro-frutticoltura Italiana, 91* (L, 2): 140–146.

DELI, J. & TIESSEN, H., 1966. Interaction of temperature and light intensity on flowering of Calwonder peppers. *Proceedings 17th International Horticulture Congress,* Md. 1. *Horticultural Abstracts,* 1192/1968.

DEMPSEY, A. H., 1966. Effect of storage and stage of flower development on viability of pepper pollen. *Horticultural Science*, *1* (2): 56–57.

DORLAND, R. E. & WENT, F. W., 1947. Plant growth under controlled conditions VIII Growth and fruiting of the chili pepper (*Capsicum annuum*). *American Journal of Botany*, *34:* 393–401.

DUMAS DE VAULX, R. & POCHARD, E., 1974. Essai d'induction de la parthénogenèse haploïde par action du protoxyde d'azote sur fleurs des piments (*Capsicum annuum* L.). *Annales l'Amélioration des Plantes*, *24* (3): 238–306.

ERWIN, A. T., 1931. Anthesis and pollination of the Capsicums. *Proceedings of the American Society for Horticultural Science*, *28:* 309.

FRANCESCHETTI, U., 1971. Natural cross pollination in pepper (*Capsicum annuum* L.). *Eucarpia Meeting on Genetics and Breeding of Capsicum*, Turin: 346–353.

FUJISHITA, N., 1965. Cytological, histological and physiological studies on pollen degeneration. II. On the free amino-acids with reference to pollen degeneration caused by low temperature in fruit-vegetable crops. *Journal of the Japan Society for Horticultural Science*, *34:* 113–120. *Horticultural Abstracts*, 2833/1966.

GEORGIEVA, I. D. & MOLKHOVA, E., 1974. A cytological study of the behaviour of the pollen tubes in some crosses within the genus *Capsicum*. *Genetika i Selekisiya*, *7* (4): 302–310. *Plant Breeding Abstracts*, 4821/1975.

GIKALO, G. S. & STUDENCEVA, L. I., 1967. The technique of producing hybrid pepper seed. *Breeding and Seed Growing*, *1:* 1–56. *Plant Breeding Abstracts*, 4826/1967.

HARN, C., KIM, M. Z., CHOI, K. T. & LEE, Y. I., 1975. Production of haploid callus and embryoid from the cultured anther of *Capsicum annuum*. Sabrao J. *7* (1): 71–77. *Plant Breeding Abstracts*, 4625/1976.

HIROSE, T., 1957. Studies on the pollination of red pepper. I. Flowering and the germinability of the pollen. *Scientific Reports of the Saikyo University (Agriculture)* 9: 5–12. *Plant Breeding Abstracts*, 585/1959.

HIROSE, T., 1965. Fundamental studies on the breeding of pepper. *Technical Bulletin, 2: Laboratory of Olericulture, Faculty of Agriculture, Kyoto Prefectural University, Japan.* 1–180.

HIROSE, T. & FUJIME, Y., 1975. A New Male Sterility in Pepper. *Hortscience*, *10* (*3*): 314.

HORNER, H. T. & ROGERS, M. A., 1973. Microsporogenesis in normal and cytoplasmic male sterile pepper (*Capsicum annuum* L.). *Abstracts of American Journal of Botany*, *60* (4): 7.

HORNER, H. T., & ROGERS, M. A., 1974. A comparative light and electron microscopic study of microsporogenesis in male-fertile and cytoplasmic male-sterile pepper (*Capsicum annuum*). *Canadian Journal of Botany*, *52* (3): 435–441.

HRISTOV, S. & GENCEV, S., 1965. A study of some problems of the floral biology of pepper (*Capsicum annuum* L.) in relation to heterosis and hybrid-seed production. *Sofia*, *2:* 605–615. *Plant Breeding Abstracts*, 4698/1966.

JAYAKARAN, M., 1972. Suppression of stamens in *Capsicum annuum* by a morphactin. *Current Science*, *41* (23): 849–850.

JO, Y. K., YU, I. U. & I, S. S., 1973. A study on characteristics and correlation between pepper varieties. *Research Reports of the Office of Rural Development, Horticulture*, *15:* 1–7, *Plant Breeding Abstracts*, 4003/1974.

KATO, T. & TANAKA, M., 1971. Studies on fruit set and development in capsicums I. Fruiting behaviour. *Journal of the Japanese Society of Horticultural Science*, *40* (4): 359–366.

KISS, A., 1971a. Studies on floral structure in the Tápiószele international collection of pepper varieties. *Agrobotanika*, *11:* 97–104.

KISS, A., 1971b. Studies of pollen viability in the Tápiószele varietal collection of red pepper. *Agrobotanica*, *12:* 53–60. *Plant Breeding Abstracts*, 1384/1973.

KORMOS, J. & KORMOS, J., 1956. Experimental data on the morphogenesis of flowering plants. *Plant Breeding Abstracts*, 2023/1958.

LENGEL, P. A., 1960. Development of the pollen and the embryo sac in *Capsicum frutescens* L. var. Japanese variegated ornamental. *Ohio Journal of Science*, *60* (I): 8–12.

LIPPERT, L. F., SMITH, P. G. & BERGH, B. D., 1966. Cytogenetics of the vegetable crops. Garden pepper, *Capsicum* sp. *Botanical Review*, *32* (1): 24–55.

LORENZETTI, F. & CIRICA, B., 1974. Quota d'incrocio, struttura genetica delle popolazioni e miglioramento genetico del peperone (*Capsicum annuum* L,). *Genetica agraria*, *XXVIII* (2): 191–203.

MARFUTINA, V. P., 1974. Obtaining hybrid seeds of sweet pepper without emasculation of the flowers. *Plant Breeding Abstracts*, 10288/1975.

MARKOVA, M. & DASKALOFF, S., 1974. Biochemical studies of a cytoplasmic male-sterile form of pepper (*Capsicum annuum* L.). *Compte Rendu de l'Academie d'Agriculture de France*, *7* (4): 27–31.

MARKUS, F., 1969. Correlation between the age and fertilization of red pepper pistil. *Acta Agronomica Hungaria*, *18* (1–2): 155–164. *Horticultural Abstracts*, 3988/1970.

MARTIN, J. A. & GRAWFORD, J. H., 1951. Several types of sterility in *Capsicum frutescens*. *Proceedings of the American Society of Horticultural Science*, *57:* 335–338.

MATTEI, F., QUAGLIOTTI, L., BIGOTTI, P. G. & DI PIETRO, A., 1971. Effects of different solar radiation levels on some morphological and physiological characters of *Capsicum annuum* L. *Eucarpia Meeting on "Genetics and Breeding of* Capsicum*", Turin, 1971: 302–321.*

MURTHY, N. S. R. & MURTHY, B. S., 1962. Natural cross pollination in chilli. *Andhra Agricultural Journal, 9: 161–165. Plant Breeding Abstracts, 3513/1963.*

NAGARATHNAM, A. K. & RAJAMANI, T. S., 1963. Studies on fruit setting in Chillies (*Capsicum annuum* Linn.). *Madras Agricultural Journal, 50: 138–139.*

NASSI, M. O., 1973. La maschiosterilità nel peperone e le sue possibili applicazioni per le colture in serra ed in pieno campo. *Colture Protette, 2: 19–21.*

NOVÁK, F. J., 1974. Induction of haploid callus in anther cultures of *Capsicum* sp. *Zeitschrift für Pflanzenzüchtung, 72: 46–54.*

NOVÁK, F. & BETLACH, J., 1967. Determination of pollen viability by tetrazolium salts. *Plant Breeding Abstracts, 3588/1970.*

NOVÁK, F. & BETLACH, J., 1970a. Development and karyology of the tapetal layer of anthers in sweet pepper (*Capsicum annuum* L.). *Biologia Plantarum, 12* (4): 275–280.

NOVÁK, F. & BETLACH, J., 1970b. Meiotic irregularities in pollen sterile sweet pepper (*Capsicum annuum*). *Cytologia, 35: 335–343.*

NOVÁK, F., BETLACH, J. & DUBOVSKI, J., 1971. Cytoplasmatic male sterility in sweet pepper (*Capsicum annuum* L.) I Phenotype and inheritance of male sterile character. *Zeitschrift für Pflanzenzüchtung, 65: 129–140.*

ODLAND, M. L. & PORTER, A. M., 1941. A study of natural crossing in pepper (*Capsicum frutescens*). *Proceedings of the American Society of Horticultural Science, 38: 585–588.*

OHTA, Y., 1961a. The use of cytoplasmic male sterility in *Capsicum* breeding. *Seiken Ziho, 12: 59–60.*

OHTA, Y., 1961b. Grafting and cytoplasmic male sterility in *Capsicum*. *Seiken Ziho, 12: 35–39.*

OHTA, Y., 1968. Effect of virus inoculation and thermoshock treatment on cytoplasmic male sterility in *Capsicum*. *Seiken Ziho, 20: 63–67.*

OHTA, Y., 1969. Cytoplasmic male sterility and virus infection in *Capsicum annuum* L. *Japanese Journal of Genetics, 45* (4): 277–283.

OHTA, Y., 1971. Nature of a cytoplasmic entity causing male sterility in *Capsicum annuum* L. *Eucarpia Meeting on "Genetics and Breeding of* Capsicum*", Turin: 229–238.*

OHTA, Y., 1973. Identification of cytoplasms of independent origin causing male sterility in red peppers (*Capsicum annuum* L.). *Seiken Ziho, 24: 105–106.*

PADDA, D. S. & SINGH, J., 1971. Studies on some important aspects of floral biology in chillies. *Indian Journal of Agricultural Research, 5* (3): 217–218.

PAHLEN, A., 1967 'Fasciflora', a new mutant pepper (*Capsicum annuum* L.). *Plant Breeding Abstracts, 3053/1969.*

PENZIG, O., 1922. Pflanzen-teratologie: 77. Berlin.

PETERSON, P. A., 1958. Cytoplasmically inherited male sterility in *Capsicum*. *The American Naturalist, XCII,* No. 863: 111–119.

POCHARD, E., 1970a. Description des trisomiques du piment (*Capsicum annuum* L.) obtenus dans la descendance d'une plante aploïde. *Annales de l'Amélioration des Plantes, 20* (2): 233–256.

POCHARD, E., 1970b. Obtention de trois nouvelles mutations de stérilité mâle chez le piment (*Capsicum annuum* L.) par traitements mutagenes appliqués à un matérial monoploïde. *Eucarpia Meeting, Versailles: 93–95.*

POCHARD, E., 1970c. Etude comparée de la gamétogénèse chez les plantes normales, les plantes possédant la stérilité cytoplasmique Peterson et les mutant mâle-stériles mr 9, mc 509, et mc 705 (*Capsicum annuum* L.). *Eucarpia Meeting, Versailles: 169–173.*

POPOVA, D., 1959. Lebensfähigkeit des Pollens und Empfänglichkeitsdauer der Narbe beim Paprika. *Académie des Sciences de Bulgarie. Institut de Culture des Plantes, VIII: 215–226.*

POPOVA, D., 1963. Studies on the effect of the age of the pollen and egg-cell in the pollination of blossoms on the heterosis in green peppers (*Capsicum annuum*). *Comptes Rendu de l'Academie Bulgare des Sciences, 16* (3): 317–320.

POPOVA, D., KAMENOVA, V. & MIKHAILOV, L., 1971. Studies on the effect of vitamins during pollination on the F1 generation in *Capsicum annuum*. *Genetica, 1971, 7* (9): 31–35.

QUAGLIOTTI, L., 1969. Notizie sulla biologia del genere *Capsicum*, prodromiche al miglioramento genetico del peperone. *Annali della Facoltà di Scienze Agrarie, Torino, V: 119–168.*

QUAGLIOTTI, L., 1970. Osservazioni biometriche sul fiore di peperone (*Capsicum annuum* L.). *Il Coltivatore e Giornale Vinicolo Italiano I: 7–15.*

QUAGLIOTTI, L., LEPORI, G. & BIGOTTI, P. G., 1974. Responses to solar radiation by two varieties of pepper (*Capsicum annuum* L.). *Corean Journal of Breeding, 6* (1): 29–33.

RUSENOVA-KONDAREVA, I., 1963. The effect of day length on the development and yield of some sweet pepper varieties. *News Institute for Plant Industries, Sofia, 18: 5–16. Plant Breeding Abstracts, 6323/1964.*

RYLSKI, I., 1972. Regulation of flowering in sweet pepper (*Capsicum annuum* L.) by external application of several plant growth regulators. *Israel Journal of Agricultural Research*, 22 (4): 31–40.

RYLSKI, I. & HALEVY, A. H., 1974. Temperature dependence of fruit set and fruit development in sweet pepper (*Capsicum annuum* L.). *XIX International Horticultural Congress, Warsavia*, Abstract No. 122.

SHIFRISS, C., 1973. Additional spontaneous male-sterile mutant in *Capsicum annuum*. *Euphytica*, 22 (3): 527–529.

SHIFRISS, C. & FRANKEL, R., 1969. A new male sterility gene in *Capsicum annuum* L. *Journal of the American Society for Horticultural Science*, 94 (4): 385–387.

SHIFRISS, C. & FRANKEL, R., 1971. New sources of cytoplasmic male sterility in cultivated peppers. *Journal of Heredity, 62:* 254–256.

SHIFRISS, C. & RYLSKI, I., 1973. Comparative performance of F1 hybrids and open pollinated "Bell" pepper varieties (*Capsicum annuum* L.) under suboptimal temperature regimes. *Euphytica*, 22 (3): 530–534.

SRIVASTAVA, V. K., 1964. Studies on fruit set in chillies (*Capsicum frutescens* L.) as influenced by application of plant regulators. *Science and Culture, 30* (4): 210.

STUDENCOVA, L. I., 1964. The reaction of pepper and egg-plant varieties to change in day-length. *Plant Breeding Abstracts*, 5850/1965.

SUZUKI, Y., 1951. Decline of fruit productivity by artificial selfing in *Capsicum annuum*. *Plant Breeding Abstracts*, 205/1957.

TERRACCIANO, N., 1878. Intorno alla transformazione degli stami in carpelli nel *Capsicum grossum*, e di un caso di prolificazione fruttipara nel *Capsicum annuum*. *Nuovo Giornale Botanico Italiano, 10:* 28–34.

VARGOVÀ, E., 1974. The influence of substrates and illumination of plants on the growth and development of forced pepper grown in nutriculture. *XIX International Horticultural Congress, Warsavia:* 141.

WELLS, O. S., 1967. The effect of night temperature on fruit set of the pepper (*Capsicum annuum* L.). *Dissertation Abstracts Section B, 27:* 4206. *Horticultural Abstracts*, 3465/1968.

REFERENCES FOR SECTION B

BHORE, D. P., BHAPKAR, D. G. & CHAVAN, V. M., 1965. Best method of selfing in brinjal (*Solanum melongena* L.). *Poona Agricultural College Magazine, 55:* 20–23.

EGUCHI, T., MATSMURA, T. & ASHIZAWA, M., 1958. The effect of nutrition on flower formation in vegetable crops. *Proceedings of the American Society for Horticultural Science, 72:* 343–351.

FIORI, A., 1933. *Nuova Flora Analitica Italiana:* 944. Firenze.

FRYDRYCH, J., 1964. Biology of flowering in the eggplant (*Solanum melongena* L.). *Bulletin Vyzkumny Ustov Zelinarsky, Olomouc CSSR, 8:* 27–37.

IMAZU, T., YABUCHI, K. & ODA, Y., 1967. Studies on the carbon dioxide environment for plant growth— II. The effect of the CO_2 concentration on growth, flowering and fruit set of eggplant (*Solanum melongena* L.). *Journal of the Japanese Society of Horticultural Science, 36:* 275–280.

JYOTISHI, R. P. & HUSSAIN, S. M., 1968. Use of 2,4-D as an aid in hybrid seed production in brinjal (*Solanum melongena* L.). *Horticultural Abstracts*, 4006/1970.

JYOTISHI, R. P. & CHANDRA, A., 1969. Induction of pollen sterility in brinjal (*Solanum melongena* L.) by foliar spray of 2,4-D. : 16–20. *Plant Breeding Abstracts*, 4198/1971.

KAKIZAKI, Y., 1924. The flowering habit and natural crossing in the eggplant. *Japanese Journal of Genetics, 3:* 29–36.

KHOT, B. D. & KANITKAR, U. K., 1956. Structure of flowers as related to setting in brinjals. *Poona Agricultural College Magazine, 47* (I): 17–22.

KRISHNAMURTHI, S. & SUBRAMANIAN, D., 1954. Some investigations on the types of flowers in brinjals (*Solanum melongena* L.) based on stile length, their fruit set under natural conditions and in response to 2,4-dichloro-phenoxy-acetic as a plant growth regulator. *Indian Journal of Horticulture, 11:* 63–67.

MAGTANG, M. V., 1936. Floral biology and morphology of the eggplant. *Philippine Agriculturalist, 25:* 30–64.

MIKAELJAN, S. G., 1964. Viability of the pistil and pollen of eggplant. *News of the Academy of Sciences of the Armenian SSR: Biological Science, 17* (8): 79–84. *Plant Breeding Abstracts*, 5397/1965.

MISHRA, 1962. Preliminary pollen studies in four varieties of brinjal (*Solanum melongena* L.) and their F1 hybrids. *Science and Culture 28* (9): 439–440. *Horticultural Abstracts*, 3178/1963.

MURTAZOV, T., PETROV, K. & DOIKOVA, M., 1971. Some features of flower position and flowering in eggplant in relation to breeding and seed production. *Nauchni Trudove. Vissh Selskostopanski Institut "Vasil Kolarov", 20* (2): 53–61. *Plant Breeding Abstracts*, 798/1973.

NAGARAJAN, M. & XAVIER, T., 1962. Polycarpy in brinjal *Solanum melongena* L. *Madras Agricultural Journal, 49:* 311. *Plant Breeding Abstracts*, 3883/1963.

NASRALLAH, M. E. & HOPP, R. J., 1963. Effect of a selective gametocide on eggplant (*Solanum melongena* L.). *Proceedings of the American Society for Horticultural Science, 83:* 575–578.

NSOWAH, G. F., 1970. Effects of sowing date on flowering and yield in varieties of eggplant (*Solanum melongena* L.). *Ghana Journal of Agricultural Science, 3:* 99–108.

OGANESJAN, A. G., 1966. Obtaining hybrid seeds of eggplant without prior emasculation of the flower. *Report of the Scientific Transactions, Research Institute of Agriculture of the Armenian SSR:* 151–154. *Plant Breeding Abstracts*, 8789/1970.

PAL, B. P. & SINGH, H. B., 1943. Floral characters and fruit formation in the egg-plant. *Indian Journal of Genetics and Plant Breeding, 3:* 45–58.

PAL, G. & OSVALD, Z., 1967. A study of fertilization after removing different amounts of various parts of the pistil. *Acta Agronomica Academiae Scientiarum Hungaricae, 16* (1–2): 33–40. *Plant Breeding Abstracts*, 1615/1968.

PAL, G. & TALLER, M., 1969. Effects of pollination methods on fertilization in eggplant (*Solanum melongena* L.). *Acta Agronomica Academiae Scientiarum Hungaricae, 18:* 307–315. *Plant Breeding Abstracts*, 4197/1971.

PAL, G. & OLAH, E., 1970. Effect of maleic hydrazide on flower formation in eggplants (*Solanum melongena* L.). *Acta Agronomica Academiae Scientiarum Hungaricae, 19:* 33–36.

POPOVA, D., 1958. Dauer der Lebensfähigkeit der Blütenstaubs und der Narbenrezeptivität der Eierfrucht. *Academie de la Science de Bulgarie: Bulletin de l'Institute de la Culture des Plantes, VI:* 125–134.

POPOVA, D., 1959. Untersuchungen über den Einfluss der Bluten staubmenge bei der Bestäubung und Befruchtung von *Solanum melongena. Academie de la Science de Bulgarie. Bulletin de l'Institute de la Culture des Plantes, VII:* 163–173.

POPOVA, D., 1961. Contribution à l'étude de quelques questions relatives à la biologie de la floraison de l'aubergine (*Solanum melanogena* L.). *Academie de la Science de Bulgarie. Bulletin de l/Institute de la Culture des Plantes, XII:* 187–207.

PRASAD, D. N. & PRAKASH, R., 1968. Floral biology of brinjal (*Solanum melongena* L.). *Indian Journal of Agricultural Science, 38:* 1053–1061.

QUAGLIOTTI, L., 1959. Possibilità di miglioramento genetico della melanzana (*Solanum melongena* L.) in Piemonte. *Sementi Elette, V* (5): 38–45.

QUAGLIOTTI, L., 1962a. Anomalie fiorali in *Solanum melongena* L. *Allionia, 8:* 117–120.

QUAGLIOTTI, L., 1962b. Su alcuni aspetti di eteromorfismo stilare in *Solanum melongena* L. *Sementi Elette*, 8 (4): 29–38.

QUAGLIOTTI, L., 1966. Contributo alla descrizione biometrica del fiore di *Solanum melongena* L. *Annali della Facoltà di Scienze agrarie, Torino, III:* 71–84.

QUAGLIOTTI, L., 1967. Osservazioni sul ritmo annuale di fioritura in alcune cultivar di melanzana (*Solanum melongena* L.). *Sementi Elette, XIII* (1): 8–20.

QUAGLIOTTI, L., 1974. Effetto della asportazione dei fiori sulla fioritura e sulla fruttificazione della melanzana. *Annali della Facoltà di Scienze agrarie, Torino, IX:* 319–334.

QUAGLIOTTI, L., 1975. Some aspects of earliness in eggplant. *Acta Horticulturae*, No. 52: 181–188.

RAINA, S. K. & IYER, R. D., 1973. Differentiation of diploid plants from pollen callus in anther cultures of *Solanum melongena* L. *Zeitschrift für Pflanzenzüchtung, 70* (4): 275–280.

REGUPATHY, A. & SUBRAMANIAM, T. R., 1973. Effect of surfactants on pollen germination and pollen tube growth in eggplant (*Solanum melongena* L.) *Phytoparasitica, 1* (2): 115–116. *Horticultural Abstracts*, 8682/1974.

SAITO, T. & ITO, H., 1973. Studies on flowering and fruiting in eggplants—VIII. Effects of early environmental conditions and cultural treatment on flower development and drop. *Journal of the Japanese Society of Horticultural Science, 42* (2): 155–162.

SAMBANDAN, C. N., 1964. Natural cross pollination in eggplant (*Solanum melongena* L.). *Economic Botany*, 2: 128–131.

SAMBANDAN, C. N. & MUTHIAN, S., 1969. On the relation between pleiomery on the stigma and fruit size in eggplant (*Solanum melongena* L.). *AUARA, 1:* 123–125. *Plant Breeding Abstracts*, 9215/1971.

SMITH, O., 1931. Characteristic associated with abortion and intersexual flowers in the egg-plant. *Journal of Agricultural Research, 43* (I): 83–94.

TATEBE, T., 1938. On pollination in the eggplant. *Journal of the Japanese Horticultural Association, 9:* 69.

TODERI, G., 1965. Contributo alla descrizione di alcune cultivar di melanzane (*Solanum melongena* L.) e risultati produttivi nel Veneto. *Sementi Elette, II* (1): 12–24.

UNCINI, L., 1971. Prime osservazioni sui rapporti fra lunghezza dello stilo e allegagione in 34 varietà di melanzana (*Solanum melongena* L.). *Atti dell'Istituto Sperimentale di Orticoltura, II*, No. 42: 83–92.

33. The genus *Nicotiana*: evolution of incompatibility in flowering plants

K. K. PANDEY

Genetics Unit, Grasslands Division, D.S.I.R., Palmerston North, New Zealand

Studies in the genus *Nicotiana* have been significant in the elucidation of genetic mechanisms controlling intra- and interspecific incompatibility in flowering plants. They have also thrown new light on the nature and evolution of genetic polymorphism.

CONTENTS

INTRODUCTION

The study of incompatibility in the genus *Nicotiana* has played a major role in the general elucidation of the problem of self- and cross-incompatibility in flowering plants. It was in the horticultural species *Nicotiana sanderae* that East and Mangelsdorf (1925)—independently of Filzer (1926) working on *Veronica syriaca* (Scrophulariaceae)—first proposed the one-locus gametophytic system of self-incompatibility. This system has since been found to be widespread among the homomorphic angiosperms (Lewis, 1954; Pandey, 1957; de Nettancourt, 1972).

Genetic studies in five of the six self-incompatible (S.I.) species in the genus *Nicotiana* (excluding *N. tomentosa*) have suggested that they all have a single *S* locus, multiallelic, gametophytic, system of self-incompatibility (Pandey, 1969a).

INTERSPECIFIC INCOMPATIBILITY AND *S*-GENE POLYMORPHISM

The realization that the *S* locus controls not only intraspecific incompatibility but also interspecific incompatibility in related species was brought home again by studies in *Nicotiana*. Andersen and de Winton, as early as 1931, showed that *N. alata* has two classes of S.I. alleles. One when present in the style allows the style to accept the pollen from a sister self-compatible (S.C.) species *N. langsdorffii* while the presence of the other causes the style to reject that pollen. The important property of the latter class of alleles, called the S_{FI} alleles (Pandey, 1964) is their duality. They act intraspecifically as normal self-incompatibility alleles but also have the additional property of determining the acceptance or rejection of *N. langsdorffii* pollen interspecifically. The presence of a similar allele was also reported by Mather in 1943 in the S.I. species *Petunia violacea* in relation to the pollen of the sister S.C. species *P. axillaris*. (A contradictory report by Takahashi (1974) in *Nicotiana* was due to his misidentification of the materials used (Pandey, 1977).)

Recent studies in *Nicotiana* have not only confirmed the findings of Anderson and de Winton but have shown that such dual purpose S.I. alleles occur in all five S.I. species investigated (excluding *N. tomentosa*) and that such duality is not limited in reaction to the pollen of *N. langsdorffii* alone. It may occur with the pollen of other species as well (Pandey, 1967, 1969a, 1973). Thus, self-incompatibility alleles of a particular species may be classified into groups according to whether they condition acceptance or rejection of pollen from another species (Pandey, 1973). The behaviour of each allele varies characteristically in relation to the pollen of different species. In Table 33.1 four *S* alleles of the S.I. species *N. bonariensis* have been classified as 'accepting' or 'rejecting' in relation to the pollen of three other related S.I. species.

Table 33.1. *S* Allele specificity in the style in relation to interspecific incompatibility

	Stylar alleles of *N. bonariensis* interacting with pollen of:		
Allele classes	*N. langsdorffii*	*N. noctiflora*	*N. glauca*
Accepting	S_1, S_2, S_3	S_3, S_4	S_2, S_3, S_4
Rejecting	S_4	S_1, S_2	S_1

Step-wise cross-compatibility pattern

The presence of intriguing parallel, allelic variation regarding interspecific incompatibility through the whole spectrum of S.I. species in *Nicotiana* raised the question of the extent and nature of this polymorphism in the genus. This led to an extensive interspecific cross-compatibility investigation in *Nicotiana* (Pandey, 1968, 1969b, c). At present it has been extended to 37 species of *Nicotiana* (Table 33.2), more than half the 64 species in the genus. These comprise all six S.I. species,

Table 33.2. Thirty seven species of *Nicotiana* used in crosses

19 Australian species All S.C.	*N. hesperis, N. suaveolens, N. maritima, N. excelsior, N. megalosiphon, N. simulans, N. gossei, N. exigua, N. velutina, N. benthamiana, N. rosulata, N. goodspeedii, N. ingulba, N. occidentalis, N. debneyi, N. umbratica, N. cavicola, N. rotundifolia, N. amplexicaulis.*
South Pacific islands S.C.	*N. fragrans*
17 South American species 6 S.I. species	*N. alata, N. bonariensis, N. petunioides, N. noctiflora, N. tomentosa, N. forgetiana.*
9 S.C. species	*N. repanda, N. glutinosa, N. paniculata, N. sylvestris, N. rustica, N. tabacum, N. longiflora, N. plumbaginifolia, N. corymbosa.*
2 S.C. species with rare S.I. races	*N. glauca, N. langsdorffii.*

and 31 S.C. species, the latter including all 20 species indigenous to Australasia and 11 of the species indigenous to South America. Among the South American species are the two S.C. species, *N. langsdorffii* and *N. glauca*, which have rare S.I. races.

Figure 33.1 summarizes the results of crosses between species. In the figure the order of arrangement of the species as males and females is based on cross-compatibility behaviour. For clarity all species behaving alike as males or as females have been grouped together. Since all S.I. species generally showed a similar cross-compatibility relationship with all S.C. species only one, *N. bonariensis*, has been included here. The results of crosses between the S.I. species themselves, many of which show polymorphism, have been considered in detail in a separate figure.

The results showed that all 20 S.C. Australasian species were reciprocally compatible between themselves and, with the single exception of *N. fragrans*, behaved alike as a group in crosses with other species. As female parents they were all universally compatible with all other species. As male parents they were fully

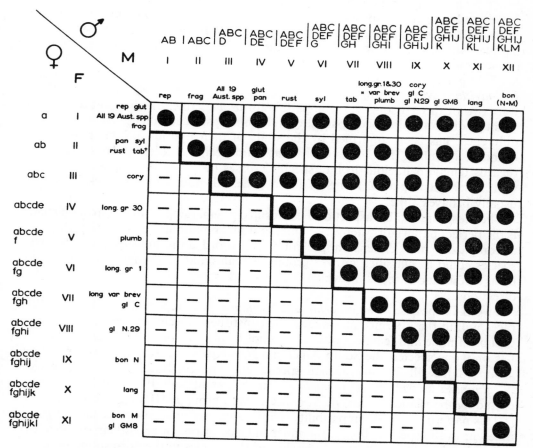

Figure 33.1. Summary of the results of crosses among 37 species of *Nicotiana*, and proposed specificity elements in the *S* gene complex controlling interspecific incompatibility. ●, compatible; —, incompatible.

compatible with five S.C. species (*N. repanda*, *N. glutinosa*, *N. paniculata*, *N. sylvestris* and *N. rustica*), polymorphically incompatible with one S.C. species (*N. tabacum*) and incompatible with four S.C. (*N. plumbaginifolia*, *N. longiflora*, *N. glauca*, and *N. langsdorffii*) and all S.I. species. The single exception, *N. fragrans*, is a lone inhabitant of the South Pacific islands, away from the rest of the species in this group which are indigenous to the Australian continent. This species differs from the other species of the group in only one respect: its pollen is rejected by *N. corymbosa* while pollen of the Australian species is accepted.

All Australian species without exception thus behaved alike as a group. They apparently have evolved from a common source probably after the Australian continent became separate from the South American continent. They all seemingly have one type of *S* gene. The island species *N. fragrans* has slightly diverged from the Australian species with regard to pollen specificity. It probably separated from that group relatively early.

Considering South American species, all crosses between S.C. species were either reciprocally or unilaterally compatible, that is, compatible one way and incompatible in the reciprocal direction. In crosses between S.C. and S.I. species those involving S.I. species as the male parents, with rare exceptions, were compatible, whereas the reciprocals involving the S.C. species as male parents were incompatible.

An important point to note here is that in two cases, *N. glauca* (C, N29) and *N. longiflora* (var. *grandiflora* strains 30 and 1, and var. *breviflora*), where there were two or three distinct self-compatible forms available, the different forms had slightly different cross-compatibility relationships. The allelic basis of these differences has been confirmed by the study of their hybrids and their progenies (Pandey, unpubl.). Thus it is clear that intraspecific *S* allele polymorphism as regards female behaviour in interspecific crosses occurs in the self-compatible as well as self-incompatible forms of the *S* gene. All pollen of a species, however, usually behaves alike.

The most remarkable aspect of these results, as emerging from this arrangement, is the perfect pattern of "stepwise" unilateral incompatibility relationship that has a complexity and dimension previously unrecorded in flowering plants. The species could be classified into 12 groups as males and 11 groups as females, and form a continuous series, in which S.C. and S.I. species are polarized in opposite directions.

Two S.C. species, *N. langsdorffii* and *N. glauca*, both reportedly having rare S.I. races, fall in the intermediate range. The constituent species in the corresponding groups of male and female, however, are not identical. For example the group MI has only one species, the South American *N. repanda*, whereas the corresponding group FI has all 20 Australasian species and another South American species *N. glutinosa* as well as *N. repanda*. As male parents, all 19 Australian species, excluding the island species, *N. fragrans*, form a group (MIII) of their own.

This pattern of cross-compatibility relationships results from two characteristics inherent in the species: firstly, the independence of pollen and stylar specificities, shown by the non-identical membership of the species in the corresponding male and female groups; and secondly, there is an inverse relationship between the pollen and stylar reactions, i.e. the species with more stylar compatibilities have less pollen compatibility, and conversely, species with more pollen compatibility have less stylar compatibility. This is best exemplified by the species at the opposite ends of the scale. For example the pollen of S.I. *N. bonariensis* which lies at the one extreme as male (MXII), is able to grow into all types of style, whereas styles of the same species, lying at another extreme as female (FXI), reject all pollen other than that of the self species. The S.C. *N. repanda* forms the opposite group I. Pollen of this group is rejected by the species of all other groups, while styles of this group accept pollen from all species (Pandey, 1968, 1969b).

The results of crosses among the S.I. species are somewhat complicated by the widespread occurrence of polymorphism (P), that is, certain *S* genotypes are compatible while certain others of the same species are incompatible with a common pollen parent.

Even among the S.I. species, despite the diversity of compatibility relationships and extensive polymorphism, when the species are arranged as males and females according to their crossing behaviour, as in the figure 2, an overall pattern of relationship emerges. All species combinations showing incompatibility (I), or polymorphic incompatibility (P), fall in one block as distinct from the block including only compatibility (C).

The most interesting feature of the pattern which emerges from the S.I. species crosses is its general similarity with that shown in the earlier figure. It is remarkably clear that despite the variability of the S.I. × S.I. results the unilateral stepwise compatibility relationship so characteristic of the crosses involving the S.C. species, i.e. S.C. × S.C., S.I. × S.C. and S.C. × S.I., also occurs in S.I. × S.I. crosses. Furthermore, while the S.I. × S.C. form of unilateral incompatibility is the most common, all other forms of incompatibility (S.I. × S.I.—; S.C. × S.I.—; S.C. × S.C.—), nevertheless also occur (Pandey, 1969b). These results show that the two principles discussed earlier—(1) the independence of pollen and stylar compatibilities, and (2) the inverse relationship between pollen and stylar compatibilities— are universal and hold true for the S.I. species as well as the S.C. species. Thus the pollen of *N. tomentosa* which lies at the one extreme as male parents is rejected by all other S.I. species, while the styles of this species which lie at another extreme as female parents, accept pollen of all other S.I. species. At the opposite end from *N. tomentosa* is *N. bonariensis*, the pollen of which is compatible with styles of all S.I. species, while its style is either totally or polymorphically incompatible with the pollen of all S.I. species (Fig. 33.2). What is the significance of this pattern?

Two separate but linked specificities

Looking at the whole picture of interspecific incompatibility in *Nicotiana* one thing becomes perfectly clear. The picture shows that fundamentally, the underlying cause of interspecific incompatibility is the same throughout the genus, irrespective of the self-incompatibility or self-compatibility of the species involved. The class of specificity which governs the incompatibility at the interspecific level is common to the whole genus, and shows one type of pattern. This specificity is evidently independent of that which is responsible for the incompatibility at the intraspecific level. The latter is usually limited to individual species, or to very closely related species having very similar polygenic backgrounds (Pandey, 1968).

Let us now look at further evidence from the study of *S* allele polymorphism discussed earlier. Consider for example the behaviour of the so-called S_{FI} alleles in *N. alata* and *N. forgetiana* in crosses with the S.C. *N. langsdorffii*. The study of the hybrids between the two S.I. species, containing the S_{FI} alleles, and the S.C. species and tests of their selfed and crossed progeny, clearly showed that the S_{FI} class of self-incompatibility alleles responsible for unilateral incompatibility behaviour, behave as perfectly normal self-incompatibility alleles in crosses within the species. The distinction between the S_{FI} alleles which reject *N. langsdorffii*

	bon.	ala.	pet.	for.	noc.	lan.	gla.	tom.
bonariensis	I	P	P	P	P	P	P	I
alata	C	I	I	P	I	P	I	I
petunioides	C	I	I	I	C	P	I	I
forgetiana	C	C	I	I	I	P	P	I
noctiflora	C	C	C	C	I	P	P	I
glauca	C	C	C	C	C	P	P	I
langsdorffii	C	C	C	C	C	C	P	I
tomentosa	C	C	C	C	C	C	C	I

Figure 33.2. Results of crosses among the self-incompatible species of *Nicotiana*. I, Incompatible; C, compatible; P, polymorphically incompatible—as females certain *S* genotypes being compatible while others are incompatible with the same pollen parent.

pollen and the more numerous S_I alleles which accept this pollen becomes apparent only when plants of the S.I. species are crossed with *N. langsdorffii* (Pandey, 1964). That is the S_{FI} class of alleles have two levels of specificity: one which is expressed intraspecifically and the other, quite independent of it but genetically very closely linked, which controls interspecific incompatibility. The evidence for the two levels of specificity is sharp and clear. The characteristic integrity of the S_{FI} and S_I classes of alleles have been confirmed, through the segregational study of the bud-selfed and crossed progeny in all five S.I. species of *Nicotiana*, where the two classes of alleles have been found to occur. The association of the intraspecific specificity characterizing an *S* allele is firmly linked with a particular pattern of interspecific incompatibility behaviour. In over 1200 plants in five S.I. species tested so far not a single case has been found where the relationship between the intraspecific and interspecific specificities, characteristic of a particular allele, has been found to be broken (Pandey, 1969a).

A third piece of evidence in favour of the linked but separate intra- and inter-specific specificities comes from the site of pollen inhibition. The interspecific incompatibility in *Nicotiana* is characteristically expressed in the stigma while the intraspecific incompatibility is usually expressed in the style (Pandey, 1964, 1969b). This is dramatically demonstrated in the expression of polymorphism of self-incompatibility alleles in *Nicotiana*. For example alleles of the S_{FI} class in *N. alata* acting intraspecifically, inhibit the pollen in the style, but the same alleles acting interspecifically inhibit the *N. langsdorffii* pollen in the stigma. The two levels of specificity although governed by the same complex, have their own distinct sites of inhibition.

Thus the identity of two separate but parallel forms of specificity, one called "Primary Specificity" controlling interspecific incompatibility and the other called "Secondary Specificity" controlling intraspecific incompatibility, is supported by evidence from at least three different sources.

Evolution

Now let us very briefly consider evolution. Some aspects have been dealt with in earlier publications (Pandey, 1968, 1969a, b), so only certain salient features pertinent for the present discussion will be touched upon here.

From the point of view of evolution it would be logical to assume that a specificity which transcends the barrier of species, and covers a whole genus, or perhaps wider taxa, and involves disjunct flora, would be more ancient than the specificity which is restricted to a single species or to very closely related forms of a species. This would suggest that *interspecific incompatibility is primitive and that the intraspecific incompatibility developed later and is superimposed upon the interspecific specificity*. This is indeed paradoxical. Why should interspecific incompatibility evolve first in the evolution of flowering plants?

Evidence from several sources suggests that the intraspecific homomorphic gametophytic incompatibility developed very early in the evolution of angiosperms. This would mean that interspecific incompatibility probably evolved even earlier, possibly during the evolution of gymnosperms. The structure of the gymnosperm "flower" may then be the key to the above mentioned paradox. With the evolution of the early anemophilous gymnosperms, plants must have faced the problem of protecting their "naked" ovules from the pollen of other distinct but related species whose pollen grains were still capable of growing into the female gametophyte and effecting fertilization, thus either virtually sterilizing the female gametes or producing zygotes inferior to those of their own species. The ability of the female gametophyte to be able to accept the pollen of its own species, while being able to reject the pollen of others which happened to be deposited on it, is thus so basic to the survival of the gymnosperm evolutionary line of development that the genetic mechanism evolved to overcome this problem may well have been the primitive foundation on which the *S* complex, as we know it in angiosperms, was eventually developed (Pandey, 1969b). The primitiveness of interspecific incompatibility compared with intraspecific incompatibility is hence evolutionarily related to the primitiveness of gymnosperms in relation to angiosperms.

The gene complex for interspecific specificity, i.e. Primary Specificity, probably first evolved in gymnosperms, through duplication and redifferentiation, initially from a single primeval specificity cistron, the chain of cistrons specifying diverse forms of primary specificity. The associated gene complex for intraspecific specificity, i.e. Secondary Specificity, probably developed early during the evolution of entomophilous angiosperms, as a means of promoting cross-fertilization, and of encouraging heterozygosity. It probably arose through a complete or partial duplication of the primary specificity cistrons. Redifferentiation of the duplicated segment then produced a large number of allelic forms.

Normally, in intraspecific pollination the Secondary Specificity is dominant over the Primary Specificity, so that only the Secondary Specificity is expressed. In interspecific pollination, however, where the two parental species are sufficiently distinct polygenically, the dominance of the Secondary Specificity over the Primary Specificity is lost, with the resultant expression of the primitive Primary Specificity.

In an interspecific pollination if the primary specificities expressed in the pollen are evenly matched by those present in the style the pollination is incompatible. Compatibility occurs when pollen has an additional primary specificity which is absent in the style.

As a result of past isolation during speciation, when the selection pressure maintaining the primary specificity, controlling interspecific incompatibility, was considerably relaxed, there was a loss of activity of some of the primary specificity elements. This resulted in a certain degree of variability of the primary specificity in S.I. species. Later when related S.I. and S.C. species came into contact in certain populations, alleles which prevented introgression usually had a selective advantage, thereby producing polymorphism of the self-incompatibility alleles on the level of Primary Specificity. However, polymorphism of the S_I and S_C alleles can occur naturally without selection through mutational erosion of the S complex (Pandey, 1968, 1969a, b).

Once a mutation involving a loss of the Secondary Specificity occurs, leading to self-compatibility, and there is a subsequent adaptation for self-pollination, selection pressure on the S locus as a whole is relaxed. Under these conditions the erosion of the primary specificity can proceed at an accelerated rate. Hence different self-compatible species may have different forms of the self-compatibility gene according to the extent of erosion of the Primary Specificity (Fig. 33.3). Thus, among the S.C. species, *N. repanda*, whose pollen is rejected by all species, has the most eroded form of the S gene in the pollen. It is also interesting to note that *N. fragrans*, the lone, isolated Pacific Islands species, has a more eroded form of the S_C gene in the pollen than all Australian species as a group.

The independent erosion of the pollen and stylar primary specificity would tend to produce self-compatible species having different forms of the S complex which when intercrossed, would show the unilateral pattern of cross-incompatibility relationships observed in *Nicotiana*.

Similarly, a limited erosion of the Primary Specificity in self-incompatibility alleles would, on the one hand, produce S gene polymorphism, and on the other, the stepwise unilateral incompatibility pattern (Fig. 33.3). Thus, among the S.I. species, *N. tomentosa*, whose pollen is rejected by styles of all other S.I. species, has the most eroded form of the S_I gene as regards primary specificity.

Probably the most interesting corollary to the idea that self-incompatibility is derived from an older interspecific incompatibility system is that it offers an explanation for another evolutionary paradox: self-incompatibility has been observed in lower plants, ferns and angiosperms but not in gymnosperms, although interspecific incompatibility has been found to occur widely in gymnosperms (Hagman, 1975). Why is there such an evolutionary gap? The key to the resolution of this dilemma, as has been discussed above, lies in the structural peculiarity of the angiosperm flower and in the close association of its evolution with the evolution of gymnosperms and angiosperms—the anemophilous gymnosperms with style-less "pistil" and naked ovules: entomophilous angiosperms with styles, stigmas, and protected ovules contained in ovaries. As stated earlier, the very survival of this evolutionary line probably depended upon the development of interspecific incompatibility.

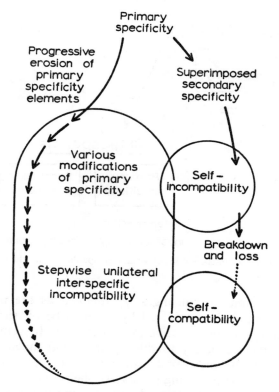

Figure 33.3. Evolution and erosion of the *S* gene complex giving rise to *S* gene polymorphism and various forms of interspecific incompatibility relationships.

The development of interspecific incompatibility in gymnosperms was presumably based on ancient specificity units, the presence of which can be traced from the earliest of plants where mechanisms exist by which two apparently similar cells are conjugationally attracted or repelled by each other (Pandey, 1969b). In gymnosperms the primeval specificity units were thus adapted to produce what has been called Primary Specificity suppressing growth of foreign pollen in anemophilous naked ovules. As mentioned before, evolution of self-incompatibility occurred later as a secondary development with the evolution of entomophilous angiosperms (Fig. 33.4).

Incongruity

Now, against this fascinatingly interwoven evolutionary picture of the S gene complex, controlling inter- as well as intraspecific incompatibility, there is a recent view expressed by Abdalla & Hermsen (1972) and Hogenboom (1973, 1975) who do not believe that there is any relationship between intraspecific incompatibility and interspecific incompatibility. Interspecific incompatibility, which these

SEED PLANTS

Gymnosperms

> Primary
> specificity

Inter specific
incompatibility

Angiosperms

> Primary   Secondary
> specificity + specificity

Inter specific   Intra specific
incompatibility   incompatibility

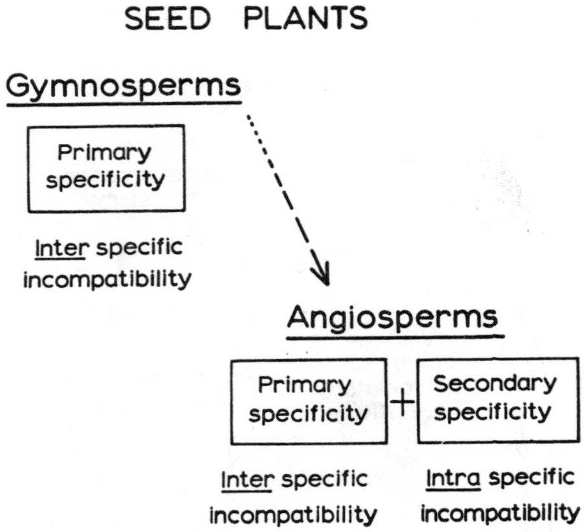

Figure 33.4. Relationship of inter- and intraspecific incompatibility with the evolution of gymnosperms and angiosperms.

authors call by the name of "incongruity" (defined as the non-functioning of pistil-pollen relationship), is considered by them to be physiologically and evolutionarily independent of intraspecific incompatibility. It seems to me that there are far too many facts contrary to this opinion, and much of what has been described here would be incomprehensible with this concept of incongruity. However, very briefly, the following points may help to clarify the situation.

Mather (1975) has recently pointed out the natural logic of evolutionary divergence producing interspecific barriers based, wherever possible, on the intraspecific foundation systems already in existence. Thus it is not unreasonable to assume that a sexual barrier between species may be primarily built on the by-products of a parent species system as it stretched and broke, unit by unit, during evolutionary divergence, and became unworkable. This is not to deny the development and existence of incongruity as an independent phenomenon, but only to suggest that incongruity is likely to be involved more seriously at a later, second phase of the formation of an interspecific breeding barrier, the first phase being based mainly on the activity of the ancient S gene complex and its physiologically associated genetic elements.

The notion of incongruity as being the sole repository of the control of inter-specific incompatibility has been mainly based on observations from plants having a polygenically disturbed background caused by inbreeding or interspecific hybridization. It has long been known that *S* genes function correctly only in their own specific genetic background. In a changed polygenic background, however, other major and minor genes affecting pollen and stylar physiology, which are normally repressed, are able to express themselves and to interfere with the *S* gene action (Mather, 1943; Pandey, 1959, 1960, 1968, 1970).

The major reason that interspecific incompatibility among angiosperms should be based on the *S* complex is that interspecific incompatibility, i.e. Primary Specificity, as stated earlier, evolved first and the intraspecific incompatibility, i.e. Secondary Specificity, was superimposed upon it later. The two have many physiological bases in common. Incongruity itself would also develop naturally with evolutionary divergence and may under certain conditions of selection for interspecific incompatibility strengthen or later replace the interspecific sexual barrier based on Primary Specificity. There is no doubt that the effectiveness of incongruity in relation to Primary Specificity would increase with increasing divergence (Fig. 33.5). It is therefore most interesting to find that the effect of Primary Specificity is as strong as it is throughout a genus. Indeed, through the classical work of Lewis & Crowe (1958) it has been traced far beyond the genus and even across the family, amply attesting to the primary and deep-rooted nature of this system.

Complex interspecific compatibility relationships have been observed not only in the genus *Nicotiana*, but also to a limited extent in certain other genera, e.g. *Petunia* (Stout, 1952), *Solanum* (Rick, 1960; Grun & Aubertin, 1966; Pandey, 1968), *Lycopersicon* (McGuire & Rick, 1954; Martin, 1961; Chmielewski, 1962, 1968) and *Phlox* (Levin, 1973). *Complex interspecific compatibility patterns and the very close and consistent association of the self-incompatibility alleles with the inter-specific incompatibility behaviour, covering the whole genus or wider taxa, are facts absolutely incomprehensible in terms of neutral or negative role embodied in the*

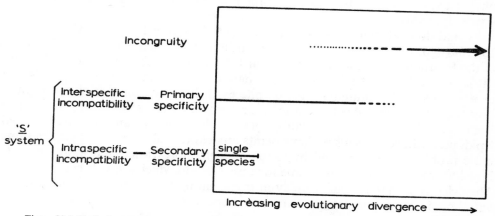

Figure 33.5. Evolutionary relationships between intra- and interspecific incompatibility—controlled by the *S* gene complex, and "incongruity".

concept of incongruity. Hogenboom (1975) has stated that while congruity is selected, incongruity is haphazard and is a natural by-product of evolutionary divergence. How then can we understand the intricate but systematic pattern of incompatibility relationships discovered between disjunct flora—for example that between Australasian and South American species of *Nicotiana*? A positive and intricately patterned inhibitory system cannot conceivably arise from a background of negative or neutral events.

The above, cogent observations are highly significant in the understanding of the evolution of interspecific incompatibility in flowering plants and cannot be ignored, or summarily dismissed by generalizations, in favour of a more simplistic hypothesis. Linkage, offered as an explanation for associated patterns, implies co-ordinated selection at an early phase of evolution, a concept included in the author's hypothesis for *S* gene evolution.

CONCLUSION

The widespread polymorphism of the *S* locus encompassing the whole genus, family or even wider taxa, and intricate relationship patterns discovered at three distinct physiological levels: that is (1) at the level of spontaneous and induced mutations of the *S* gene, (2) at the level of intraspecific incompatibility, and (3) at the level of interspecific incompatibility, are, in my opinion, inherent in the genetic architecture of the ancient *S* locus. The multi-dimensional polymorphism is an inevitable consequence of the evolution, and later erosion, of such supergenes. The highly specific and ordered nature of interspecific incompatibility affecting genera and even families quite obviously could not have arisen through *ad hoc* selection, for selection of this nature and on such a scale was not possible; nor could it have arisen through incongruity, for here no positive selection is even implied. But because Primary Specificity was present as a relic of an earlier development it was open to wide modification and disruption, the latter gaining in intensity with evolutionary divergence. From the point of view of evolution, the *S* locus, with parallel development at almost every major level of plant phylogeny, probably illustrates one of the most powerful links in the evolution of flowering plants.

I could do no better than to end by a quotation from Professor Mather, who, 13 years ago, while summarizing the symposium on "Genetic diversity and fitness" at the International Genetics Congress in 1963, made the following point. He said "It is no accident that all our speakers have been dealing with observations and experiments made on animals, for despite extensive study of their chromosomes, breeding systems and genetico-ecological relations, virtually no investigations have been carried out with wild plants comparable with the analysis of flies and lepidoptera. In some ways perhaps plants are less tractable material, yet the same basic problems are there and by virtue of their more varied and more versatile breeding systems plants offer scope for investigation which animals cannot equal. They clearly have their part to play in investigating the organization of heritable variation in wild populations. May we hope that their neglect will not long continue?"

Looking at the role of *S* gene polymorphism in the breeding behaviour of flowering plants as summarized today, one can only marvel how truly the prediction of Mather has been fulfilled! Earlier observations of Anderson and de Winton, and Mather, were only, as it were, the tip of an iceberg.

ACKNOWLEDGEMENT

I thank Mr Graham Petterson for expert technical assistance.

REFERENCES

ABDALLA, M. M. F. & HERMSEN, J. G. TH., 1972. Unilateral incompatibility: hypothesis, debate and its implications for plant breeding. *Euphytica, 21:* 32–47.

ANDERSON, E. & DE WINTON, D., 1931. The genetic analysis of an unusual relationship between self-sterility and self-fertility in *Nicotiana. Annals of Missouri Botanical Gardens, 18:* 97–116.

CHMIELEWSKI, T., 1962. Cytological and taxonomical studies on a new tomato form. Part I. *Genetica Polonica, 3:* 253–264.

CHMIELEWSKI, T., 1968. Cytogenetical and taxonomical studies on a new tomato form. Part II. *Genetica Polonica, 9:* 97–124.

EAST, E. M. & MANGELSDORF, A. J., 1925. A new interpretation of the hereditary behaviour of self-sterile plants. *Proceedings of the National Academy of Sciences of the United States of America, 11:* 166–171.

FILZER, P., 1926. Die Selbsterilitat von *Veronica syriaca. Zeitschrift für Induktive Abstammungs-u. Vererbungslehre, 71:* 156–230.

GRUN, P. & AUBERTIN, M., 1966. The inheritance and expression of unilateral incompatibility in *Solanum. Heredity, 21:* 131–138.

HAGMAN, M., 1975. Incompatibility in forest trees. *Proceedings of the Royal Society of London* (B), *188:* 313–326.

HOGENBOOM, N. G., 1973. A model for incongruity in intimate partner relationships. *Euphytica, 22:* 219–233.

HOGENBOOM, N. G., 1975. Incompatibility and incongruity: two different mechanisms for the non-functioning of intimate partner relationships. *Proceedings of the Royal Society of London* (B), *188:* 361–375.

LEVIN, D. A., 1973. Polymorphism for interspecific cross-compatibility in *Phlox. Proceedings of the National Academy of Sciences of the United States of America, 70:* 1149–1150.

LEWIS, D., 1954. Comparative incompatibility in angiosperms and fungi. *Advances in Genetics 6:* 235–285.

LEWIS, D. & CROWE, L. K., 1958. Unilateral incompatibility in flowering plants. *Heredity, 12:* 233–256.

MARTIN, F. W., 1961. Complex unilateral hybridization in *Lycopersicon hirsutum. Proceedings of the National Academy of Sciences of the United States of America, 47:* 855–857.

MATHER, K., 1943. Specific differences in *Petunia*. I. Incompatibility. *Journal of Genetics, 45:* 215–235.

MATHER, K., 1963. Genetical diversity: Synthesis. *Genetics Today, 3:* 541–552.

MATHER, K., 1975. Comment: Incompatibility and incongruity. *Proceedings of the Royal Society of London* (B), *188:* 374–375.

McGUIRE, D. C. & RICK, C. M., 1954. Self-incompatibility in species of *Lycopersicon* sect. Eriopersicon and hybrids with *L. esculentum. Hilgardia, 23:* 101–124.

NETTANCOURT, D. de, 1972. Self-incompatibility in basic and applied researches with higher plants. *Genetica Agraria, 26:* 163–216.

PANDEY, K. K., 1957. Genetics of incompatibility in *Physalis ixocarpa* Brot. A new system. *American Journal of Botany, 44:* 879–887.

PANDEY, K. K., 1959. Mutations of the self-incompatibility gene (*S*) and pseudo-compatibility in angiosperms. *Lloydia, 22:* 222–234.

PANDEY, K. K., 1960. Incompatibility in *Abutilon hybridum. American Journal of Botany, 47:* 877–883.

PANDEY, K. K., 1964. Elements of the *S*-gene complex. I. The S_{FI} alleles in *Nicotiana. Genetical Research, 5:* 397–409.

PANDEY, K. K., 1967. *S*-gene polymorphism in *Nicotiana. Genetical Research, 10:* 251–259.

PANDEY, K. K., 1968. Compatibility relationships in flowering plants: Role of the *S*-gene complex. *American Naturalist, 102:* 475–489.

PANDEY, K. K., 1969a. Elements of the *S*-gene complex. IV. *S.* allele polymorphism in *Nicotiana* species. *Heredity, 24:* 601–619.

PANDEY, K. K., 1969b. Elements of the *S*-gene complex. V. Interspecific cross-compatibility relationships and theory of the evolution of the *S* complex. *Genetica, 40:* 447–474.

PANDEY, K. K., 1969c. Interspecific cross-compatibility relationships in *Nicotiana. Proceedings of the XI International Botanical Congress, Seattle:* 165.

PANDEY, K. K., 1970. Elements of the *S*-gene complex. VI. Mutations of the self-incompatibility gene, pseudo-compatibility and origin of new self-incompatibility alleles. *Genetica, 41:* 477–516.

PANDEY, K. K., 1973. Phases in the *S*-gene expression, and *S*-allelic interaction in the control of interspecific incompatibility. *Heredity, 31:* 381–400.

PANDEY, K. K., 1977. Genetic control of interspecific incompatibility between *Nicotiana alata* and *N. langsdorffii:* Correction of Takahashi's observations. *Japan Journal of Genetics, 52:* 431–433.

RICK, C. M., 1960. Hybridization between *Lycopersicon esculentum* and *Solanum pennellii:* phylogenetic and cytogenetic significance. *Proceedings of the National Academy of Sciences of the United States of America, 46:* 78–82.

STOUT, A. B., 1952. Reproduction in *Petunia. Memoirs of the Torrey Botanical Club, 20:* 1–202.

TAKAHASHI, H., 1974. Genetical and physiological analysis of interspecific incompatibility between *Nicotiana alata* and *N. langsdorffii. Japan Journal of Genetics, 49:* 247–256.

34. Incompatibility and incongruity in *Lycopersicon*

Institute for Horticultural Plant Breeding, Wageningen, The Netherlands

A survey is given of the different breeding barriers within and between *Lycopersicon* species. The nature, evolution and genetics of these phenomena are discussed.

The discussion starts from a pistil-pollen relationship based on matching genic systems in pistil and pollen, resulting from co-evolution. Incompatibility prevents normal functioning of the relationship within populations, though the potential for functioning of both partners is complete. It results from the positive selection pressure for genes promoting fitness.

Isolating mechanisms between populations, including those acting in the progamic phase, are incompletenesses in the relationship. The partners do not have a complete potential for functioning together, from lack of information in one partner about the other. This incongruity results from evolutionary divergence.

Results are reviewed to demonstrate that incompatibility and incongruity are different mechanisms. The genetics of style characters in one species, on which information in the pollen of another species is lacking, is discussed on the basis of results with *L. peruvianum* and *L. minutum*.

Results of hybridization of *L. esculentum* and *L. peruvianum*, after pre-selection within the wild species, are presented.

Prospects of using interpopulational barriers in plant breeding, as an extra mechanism for controlled pollination, will be discussed.

CONTENTS

BREEDING BARRIERS IN *LYCOPERSICON*

Lycopersicon is a small genus in the Solanaceae, closely related to *Solanum*. The taxonomy of *Lycopersicon* is not without controversies, but a tentative classification of species may be given. The subgenus *Eulycopersicon* C. H. Mull. includes the species *L. esculentum* Mill. and *L. pimpinellifolium* Mill. The subgenus *Eriopersicon* C. H. Mull. comprises the species *L. peruvianum* (L.) Mill., *L. chilense* Dun. and *L. hirsutum* Humb. and Bonpl. (Muller, 1940; Luckwill, 1943; Rick, 1953, 1956, 1961, 1963; Rick & Lamm, 1955). In addition, there is a species *L. minutum* (Chmielewski & Rick, 1962; Chmielewski, 1968a), which cannot be placed in one of the subgenera and has recently been reclassified as two separate species: *L. chmielewskii* and *L. parviflorum* (Rick, Kesicki, Fobes & Holle, 1976).

435

All species have twelve pairs of chromosomes and are closely related. Yet the genus shows not only a multiplicity of forms, as a result of a high degree of differentiation, but also nearly all possible breeding barriers, within as well as between species (see Table 34.1) (Smith, 1944; Lamm, 1950; Bohn, 1951; De Zerpa, 1952; Rick, 1953, 1961, 1963; McGuire & Rick, 1954; Rick & Lamm, 1955; Rick & Butler, 1956; Lewis & Crowe, 1958; Martin, 1961a, b, 1966; Chmielewski, 1962, 1966, 1968a, b). Furthermore, certain barriers sometimes occur in only a part of a species. Therefore *Lycopersicon* is highly suitable for studying breeding barriers.

Table 34.1. Survey of *Lycopersicon* species and of breeding barriers within and between them

♀	♂					
	L. esc.	*L. pim.*	*L. min.*	*L. hir.*	*L. chi.*	*L. per.*
L. esculentum	+	+	+	+	EA	EA
L. pimpinellifolium	+	+	+	+	EA	EA
L. minutum	+, UI, EA	+, UI	+	EA	EA	EA
L. hirsutum	+, UI	+, UI	+, UI	+, SI, UI	?	EA
L. chilense	UI	UI	UI	?	SI	EA
L. peruvianum	UI	UI	UI	UI	EA	SI

+, No serious barrier; SI, self-incompatibility; UI, unilateral incongruity; EA, embryo abortion; ?, no research results known.

Unilateral incongruity generally occurs between species of the subgenus Eulycopersicon and those of Eriopersicon (MacArthur & Chiasson, 1947; Bohn, 1951; De Zerpa, 1952; McGuire & Rick, 1954; Lewis & Crowe, 1958; Martin, 1961b), although exceptions were found in *L. hirsutum* var. *glabratum* (Chmielewski, 1966; Martin, 1966). Within *L. hirsutum* unilateral incongruity occurs between different populations (Martin, 1961a, 1963). *L. minutum* holds an intermediate position (Chmielewski, 1962, 1968a).

Table 34.1 shows different degrees of isolation between species. The most isolated species are *L. chilense* and *L. peruvianum*. The barriers isolating them not only render it difficult to produce interspecific hybrids, but also in later generations crossability problems arise in different forms (McGuire & Rick, 1954; Martin, 1961b).

Recently, as a result of an evolutionary study on pistil-pollen relationships in *Lycopersicon*, in *L. peruvianum* the self-incompatibility was broken as well as the unilateral incongruity with *L. esculentum* (Hogenboom, 1972 a–e). As far as the author is aware this was the first time that a unilateral incongruity between species was broken by inbreeding and artificial selection. Besides existing situations of self-incompatibility occurring in only part of a species, whether or not combined with unilateral incongruity with *L. esculentum* occurring in only part of the species, this new development gave an excellent opportunity to study the genetics and evolution of the different barriers and also their possible interrelation. Because of the controversy on this point results of this study on pistil-pollen relationships in *Lycopersicon* were of general interest. It resulted in new insights into breeding barriers and their interrelation (Hogenboom, 1973, 1975).

The wild *Lycopersicon* species are interesting sources of disease resistances, temperature tolerances and other characters and, in general, sources of variation for tomato breeders. As the barriers and the lack of knowledge of their nature and genetics result in a rather meagre exploitation of the very valuable genetic material of wild species for improvement of the cultivated tomato, such studies are also of direct practical interest.

NATURE AND EVOLUTION OF THE BARRIERS

The nature and evolution of breeding barriers may be well understood when starting from the basis of the pistil-pollen relationship. For normal functioning of this relationship (i.e. to achieve fertilization) a complex of events takes place: pollen germination, formation of the pollen tube, penetration in pistil tissues, growth in a certain direction and finding the way to and into the ovule and into the embryosac for fusion with the egg. All this is based on a chain of processes in both pistil and pollen, of which those in the pollen (tube) must be accurately coordinated and interact with those in the pistil.

The pistil characters relevant to fertilization are based on a number of genes and gene complexes, each governing a process. (We may speak of barrier genes, governing barrier processes.) The pollen carries all genetic information necessary to behave adequately and react in all steps of the complex of physiological and structural situations between pollination and fertilization. (We may speak of penetration genes governing penetration processes.) Normal functioning of the fertilization process requires that the genic systems in pistil and pollen and their activities be fully matching (Hogenboom, 1973, 1975). Each of the couples of genes or gene complexes in pistil and pollen governs a part of the chain of processes and interactions leading to fertilization. The matching genic systems in pistil and pollen result from co-evolution of pistil and pollen (Hogenboom, 1975).

This description of an intimate relationship based on matching genic systems shows that there are two possible systems for non-functioning. One possibility is that *the relationship is rendered non-functional from the outside*. This is how *incompatibility* works. Functioning of the relationship is inhibited though both partners have the complete potential for functioning and coordination. When the products of *S*-gene action in pollen tube and style are identical, a repressor is formed which inhibits further functioning of the pistil-pollen relationship (Lewis, 1960, 1965; Ascher, 1966; Pandey, 1967; Linskens, 1968). This old system, preventing self-fertilization, is itself not a part of the pistil-pollen relationship. It results from the positive selection pressure for genes promoting fitness.

Another possibility is that *the relationship itself is non-functional because the partners are incompletely matching* for the genetic information regulating interaction and coordination. In that case we speak of *incongruity*: there is a lack of genetic information in one partner about some character in the other (Hogenboom, 1973, 1975). Such incomplete pistil-pollen relationships, acting as isolating mechanisms, result from evolutionary divergence. Each species develops its own pistil-pollen relationship. As a result of divergence the pollen of one species may lack essential information on one or more pistil characters of another species. This lack may be small or great, depending on the degree of divergence. It will be clear that it may be unilateral, but will generally be bilateral and that the phenomenon may occur irrespective of the presence or absence of an incompatibility system, between populations with and without any contact.

Whereas the incompatibility system, rendering the pistil-pollen relationship non-functional by similarity of partners for *S*-alleles, is the same in many species, incongruity may in each case be of a very different nature, depending on the direction of the evolutionary divergence. Whereas incompatibility may occur between certain pistils and certain pollen grains, incongruity will generally occur between all the pistils of one population and all the pollen of another.

On the basis of these insights, resulting from the study of breeding barriers in *Lycopersicon* (Hogenboom, 1972 a–e), a better and more generally valid picture of phenomena acting in interpopulational crosses was obtained and the overestimation of the role of the incompatibility mechanism was demonstrated. The fact that unilateral incongruity occurs between populations of self-incompatible plants, between populations of self-compatible plants and also between populations of self-incompatible and populations of self-compatible plants both in the cross self-incompatible × self-compatible and in the reciprocal cross, not only holds for *Lycopersicon*, but is a general phenomenon (see references in Hogenboom, 1972d). It shows that interspecific pollen tube growth inhibition occurs irrespective of the presence or absence of incompatibility. The same holds for bilateral incongruity.

Breaking the self-incompatibility in *L. peruvianum* as well as its unilateral incongruity with *L. esculentum* and studying the genetics of these break-downs (Hogenboom, 1972b–e) yielded a reinterpretation of the different results of crosses between self-compatible and self-incompatible species, in connection with the evolution of self-compatible species, earlier explained on the basis of a series of mutational steps of the incompatibility gene (Lewis & Crowe, 1958). Inbreeding of cross-fertilizing plants will immediately result in segregation of lower barrier capacities in pistils, but pollen information will only change to adapted levels in the long run (Hogenboom, 1973, 1975). This explains why self-fertilizing species of recent origin often behave differently from those of ancient origin in interspecific crosses. The work with *L. peruvianum* represents the first two steps of this three-step process: through inbreeding and artificial selection the total barrier capacity of the pistil of some *L. peruvianum* inbred lines was lowered to the level of genetic information of the *L. esculentum* pollen. In *Lycopersicon* two other evolutionary events led to a similar result: in *L. hirsutum* (Chmielewski, 1966; Martin, 1966) and in the former *L. minutum* (Rick *et al.*, 1976) the ancestral outcrossing form shows unilateral incongruity with *L. esculentum*, while in the derivative self-fertilizing type bilateral congruity with *L. esculentum* segregates.

In the work on *Lycopersicon*, reinterpretations could be given for the complex genetical situations regarding crossability after interspecific crosses, for the phenomenon of *S*-gene polymorphism and for the complex patterns of crossability between species (Hogenboom, 1973, 1975). These patterns, which have not only been found in *Lycopersicon* (McGuire & Rick, 1954; Chmielewski, 1962, 1968a; Martin 1961a, 1963) but also in *Nicotiana* (Pandey, 1968, 1969) and *Petunia* (Stout, 1952), result from stepwise unilateral relations between populations. In our opinion such patterns occur normally in taxa with a certain degree of differentiation, resulting in each population having its own pistil-pollen relationship and its own barrier and penetration capacities. They will be found irrespective of the presence or absence of an incompatibility system. In general, the self-fertilizers will be found to be best usable as female parents, the cross-fertilizers as male parents.

From the description of the pistil-pollen relationship it follows that incompatibility is not necessarily a complicated system, as the chain of processes between pollination and fertilization can easily be impeded by disturbing only one link. As a consequence incompatibility is rather easily broken. Large scale self-pollination to screen the pollen population for individuals with a spontaneously mutated *S*-allele succeeded in *L. peruvianum* and in other species (Hogenboom, 1972b). Simple treatments of the plant (e.g. high temperature) may have the same effect temporarily.

It also follows from the same description and from the evolutionary basis of incongruity that incongruity may be a simple incompleteness of information in certain cases but will mostly be complicated or very complicated. In general, its breakdown is therefore complicated too. Breaking the unilateral incongruity between *L. peruvianum* and *L. esculentum* was achieved stepwise (Hogenboom, 1972d). In inbred material of *L. peruvianum* different patterns of interaction were found between *L. peruvianum* pistil and *L. esculentum* pollen. In style and ovary separate barriers for *L. esculentum* pollen occurred. The results indicated that the incongruity is built up of a number of different processes.

The reasoning on matching genic systems and incongruity between partners may also be applied to the genomes meeting after fertilization. Between these partners, too, accurate coordination is necessary for a normal development of the zygote. Here, too, evolutionary divergence may result in inadequate interaction so that certain processes do not take place. This may result in embryo-abortion, or hybrid weakness or sterility. This relationship between genomes can, however, hardly be analysed.

GENETICS OF THE BARRIERS

The genetics of *incompatibility* has been thoroughly studied. It is mostly based on a one or two locus multiple allelic system. As in other Solanaceae, in *Lycopersicon* a homomorphic gametophytic system occurs, controlled by multiple alleles at one locus (Lamm, 1950, 1953; McGuire & Rick, 1954; Martin, 1961b, 1963). It is assumed that each *S*-allele consists of three cistrons: one controlling the specificity of the protein acting in the incompatibility reaction (this one is the same in pistil and pollen but different for each *S*-allele) and two regulatory cistrons: one controlling the activity of the specificity cistron in the pollen, the other that in the style. These two are the same for all *S*-alleles. The parts are functionally integrated (Lewis, 1958, 1960; Lundqvist, 1965).

This picture of the genetics of incompatibility, which is the same for many species, is rather simple. As the *S*-allele parts are mutationally independent, incompatibility can be rather easily broken. Mutation of one of the cistrons is enough, as has been shown in a number of genera, and other genetic situations have the same result (see references in Hogenboom, 1972c).

The study of the genetics of *incongruity* has long been hampered by the hypothesis that inhibition of growth of pollen tubes from another species is based on the incompatibility system (*S*-alleles). The validity of this hypothesis was already earlier disputed and other mechanisms for pollen tube growth inhibition suggested

(Stout, 1952; Bellartz, 1956; Swaminathan & Murty, 1957; Sampson, 1962; Martin 1963; Grun & Aubertin, 1966; Hogenboom, 1972d, e, 1973, 1975; Hermsen, Olsder, Jansen & Hoving, 1974). As research on interpopulational barriers extends, the hypothesis will more and more prove untenable.

The genetics of incongruity, depending on the degree of evolutionary divergence between populations, may be anything between very simple and very complicated, and each case may show another mechanism.

An important aspect of the study of the genetics of interpopulational barriers is that it should be carried out within a species, determining the genetic difference between plants which, for example, inhibit pollen tube growth of another species and plants which do not. This was first done in *Solanum* by Grun & Aubertin (1966), who studied the genetics of inhibition of *S. verrucosum* pollen tubes in *S. chacoense* pistils. They found that this inhibition was based on two or more independent dominant genes.

A similar approach in *Lycopersicon*, which was made possible artificially, showed similar results (Hogenboom, 1972d, e). In *L. peruvianum* the inhibition of *L. esculentum* pollen tubes was shown to be governed by one or more independent dominant genes. These genes are part of the barrier capacity of the *L. peruvianum* pistil about which the *L. esculentum* pollen is not informed. The genes may be assumed to govern processes which are inhibitive to growth of *L. esculentum* pollen tubes as a result of evolutionary divergence. The absence of the inhibition in certain *L. peruvianum* plants was a result of inbreeding in this heterozygous species and segregation of homozygous recessives for the barrier genes concerned.

In *L. minutum*, a self-compatible species part of which shows unilateral incongruity with *L. esculentum* and another part bilateral congruity, we are now obtaining results showing a similar situation: inhibition of *L. esculentum* pollen tubes is based on a number of dominant genes.

Hermsen *et al.* (1974), working in *Solanum*, came to the same conclusion: inhibition of *S. verrucosum* pollen in pistils of dihaploid *S. tuberosum* is governed by more than one dominant gene. They found no relation between (in)compatibility in dihaploid *S. tuberosum* and its genotypes for (non)inhibition of *S. verrucosum* pollen.

RELATIONSHIP BETWEEN THE BARRIERS

The conclusion from the foregoing, drawn earlier from the research on breeding barriers in *Lycopersicon* and from a reinterpretation of earlier research (Hogenboom, 1973, 1975), is that pollen tube growth inhibition in interpopulational relationships is based on incongruity and that in such relationships incompatibility plays no role or only a secondary one.

Incompatibility and incongruity have no relationship, they are separate and independent mechanisms, based on different principles and with different evolutionary backgrounds. The only character that incompatibility and incongruity may have in common is that the result of their occurrence may be the same: inhibition of pollen tube growth.

Arguments for the distinction of the two mechanisms do not only arise from the genetical analyses mentioned above, but also from anatomical and physiological research on pollen tube inhibition (cf. Hogenboom, 1972e).

The importance of the distinction of the two mechanisms is that it led to the insight that the study of incongruity is the study of missing links in pistil-pollen or other intimate relationships and is therefore a good means to unravel how such relationships function.

HYBRIDIZATION OF *L. ESCULENTUM* AND *L. PERUVIANUM*

Further knowledge of interspecific breeding barriers may lead to a better approach in breaking them. One may think of the inactivation of barrier genes on which in the pollen of the other species information is lacking or, in certain cases (as impeded pollen germination), of neutralizing the effect of such barrier genes by substituting absent penetration genes, e.g. by adding pollen of the female parent containing the products of these penetration genes (cf. Knox, Willing & Ashford, 1972).

The discussion on the evolution of self-compatible species (Hogenboom, 1973, 1975) indicates how incongruity between species can be broken. The differences in barrier and penetration capacities may be removed by inbreeding in cross-fertilizers and artificial selection. In *Lycopersicon* in this way the development of new idiotypes was enabled from the cross *L. peruvianum* ♀ × *L. esculentum* ♂ (Hogenboom, 1972d). This broadening of the possibilities for novel variation by plasmatic interaction (Rick, 1967), using the possible plasmon differentiation in *Lycopersicon* (Andersen, 1964), may have interesting consequences for tomato breeding.

The embryo abortion in crosses between *L. esculentum* and the most interesting wild species *L. peruvianum* is a very serious barrier. It can partly be avoided with the help of embryo culture (Smith, 1944), but in general the success of this technique has not been great. In *L. peruvianum* the breakdown of self-incompatibility enabled inbreeding by selfing. The genetic variability thus revealed resulted in the finding of plants without unilateral incongruity with *L. esculentum*, enabling the cross *L. peruvianum* × *L. esculentum* to take place. It further appeared that in this cross the degree of embryo abortion was in certain plants much lower than in the reciprocal cross. In some *L. peruvianum* × *L. esculentum* crosses three hybrid plants were obtained from about 15 pollinations, in spite of the incomplete fertilization (Hogenboom, 1972d). These and other results show how by inbreeding, selection and intercrossing in *L. peruvianum*—i.e. by pre-breeding within the wild species—material may be developed with which large-scale *L. peruvianum* × *L. esculentum* hybrid production is possible. It was also found that from the cross *L. peruvianum* × *L. esculentum* relatively fewer triploid F_1 plants were obtained than from the reciprocal cross.

Interesting material has been obtained from the hybridization of *L. esculentum* with *L. peruvianum* and is still being obtained in the segregating generations. Because of the very complex crossing relationships found in the interspecific F_1— probably owing to different types of interspecific interaction—and to preserve as much as possible of the genetic material present, the next generations are produced by pollinating all plants with a mixture of pollen from all plants.

Interesting disease resistances are found in the material against *Cladosporium fulvum*, tobacco mosaic virus, *Fusarium oxysporum*, *Meloidogyne* species and *Pyrenochaeta lycopersici*. Furthermore, lines have been obtained which open up prospects of adapting *L. esculentum* to considerably lower glasshouse temperatures.

Also types appear with interesting habit characters like different degrees of side shoot formation and interesting types of male sterility. These characters are under study at our institute.

PROSPECTS OF USING INCONGRUITY IN PLANT BREEDING

Plant breeders now use self-incompatibility, male sterility and femaleness in hybrid seed production. Incongruity may be an interesting extra mechanism. For its exploitation one should select a related species that has a unilateral incongruity with the crop species and also a higher barrier capacity. Now from this species one selects one of the extra barrier genes and by back-crossing this is introduced into the female parent to be developed. This parent will then have a shortage of penetration genes so that self-fertilization is impossible. The penetration gene(s) corresponding to the introduced barrier gene must be introduced into the male parent to be developed.

The ideal is, of course, to select a one-barrier-gene for one-penetration-gene relationship, although somewhat more complicated situations are also usable. At the moment we are trying to introduce such a system into *L. esculentum* by introducing (a) barrier gene(s) and the corresponding penetration gene(s) from *L. minutum*. The results are too preliminary to draw conclusions but it seems that we have not yet succeeded in finding a one-gene for one-gene situation. More difficulties may be encountered if the action of the genes introduced is too much influenced by their new genetic background.

REFERENCES

ANDERSEN, R., 1964. Evidence of plasmon differentiation in Lycopersicon. *Tomato Genetics Coop. Report, 14:* 4–6.

ASCHER, P. D., 1966. A gene action model to explain gametophytic self-incompatibility. *Euphytica, 15:* 179–183.

BELLARTZ, S., 1956. Das Pollenschlauchwachstum nach arteigener und artfremder Bestäubung einiger Solanaceen und die Inhaltsstoffe ihres Pollens und ihrer Griffel. *Planta, 47:* 588–612.

BOHN, G. W., 1951. Fertility relations in *L. hirsutum* and its hybrids with *L. esculentum. Tomato Genetics Coop. Report 1:* 3–4.

CHMIELEWSKI, T., 1962. Cytogenetical and taxonomical studies on a new tomato form, Part I. *Genetica Polonica, 3:* 253–264.

CHMIELEWSKI, T., 1966. An exception to the unidirectional crossability pattern in the genus *Lycopersicon. Genetica Polonica, 7:* 31–39.

CHMIELEWSKI, T., 1968a. Cytogenetical and taxonomical studies on a new tomato form, Part II. *Genetica Polonica, 9:* 97–124.

CHMIELEWSKI, T., 1968b. New hybrids with *L. peruvianum* obtained by means of a periclinal chimaera. *Tomato Genetics Coop. Report, 18:* 9.

CHMIELEWSKI, T. & RICK, C. M., 1962. *Lycopersicon minutum. Tomato Genetics Coop. Report, 12:* 21–22.

GRUN, P. & AUBERTIN, M., 1966. The inheritance and expression of unilateral incompatibility in *Solanum. Heredity, 21:* 131–138.

HERMSEN, J. G. TH., OLSDER, J., JANSEN, P. & HOVING, E., 1974. Acceptance of self-compatible pollen from *Solanum verrucosum* in dihaploids from *S. tuberosum.* In H. F. Linskens (Ed.), *Fertilization in Higher Plants:* 37–40.

HOGENBOOM, N. G., 1972a. Breaking breeding barriers in *Lycopersicon.* 1. The genus *Lycopersicon,* its breeding barriers and the importance of breaking these barriers. *Euphytica, 21:* 221–227.

HOGENBOOM, N. G., 1972b. Breaking breeding barriers in *Lycopersicon.* 2. Breakdown of self-incompatibility in *L. peruvianum* (L.) Mill. *Euphytica, 21:* 228–243.

HOGENBOOM, N. G., 1972c. Breaking breeding barriers in *Lycopersicon.* 3. Inheritance of self-compatibility in *L. peruvianum* (L.) Mill. *Euphytica, 21:* 244–256.

HOGENBOOM, N. G., 1972d. Breaking breeding barriers in *Lycopersicon*. 4. Breakdown of unilateral incompatibility between *L. peruvianum* (L.) Mill. and *L. esculentum* Mill. *Euphytica, 21:* 397–404.

HOGENBOOM, N. G., 1972e. Breaking breeding barriers in *Lycopersicon*. 5. The inheritance of the unilateral incompatibility between *L. peruvianum* (L.) Mill. and *L. esculentum* Mill. and the genetics of its breakdown. *Euphytica, 21:* 405–414.

HOGENBOOM, N. G., 1973. A model for incongruity in intimate partner relationships. *Euphytica, 22:* 219–233.

HOGENBOOM, N. G., 1975. Incompatibility and incongruity: two different mechanisms for the non-functioning of intimate partner relationships. *Proceedings of the Royal Society of London (B), 188:* 361–375.

KNOX, R. B., WILLING, R. R. & ASHFORD, A. E., 1972. Role of pollenwall proteins as recognition substances in interspecific incompatibility in poplars. *Nature, 237:* 381–383.

LAMM, R., 1950. Self incompatibility in *Lycopersicon peruvianum* Mill. *Hereditas, 36:* 509–510.

LAMM, R., 1953. Observations on *Lycopersicon peruvianum* crosses. *Tomato Genetics Coop. Report, 3:* 14–15.

LEWIS, D., 1958. Gene control of specificity and activity: loss by mutation and restoration by complementation. *Nature, 182:* 1620–1621.

LEWIS, D., 1960. Genetic control of specificity and activity of the S. antigen in plants. *Proceedings of the Royal Society of London (B), 151:* 468–477.

LEWIS, D., 1965. A protein dimer hypothesis on incompatibility. *Genetics Today, 3:* 657–663.

LEWIS, D. & CROWE, L. K., 1958. Unilateral interspecific incompatibility in flowering plants. *Heredity, 12:* 233–256.

LINSKENS, H. F., 1968. Egg-sperm interactions in higher plants. *Accademia Nazionale dei Lincei, Quaderno N. 104:* 47–56.

LUCKWILL, L. C., 1943. The genus *Lycopersicon*. An historical, biological and taxonomic survey of the wild and cultivated tomatoes. *Aberdeen University Studies, 120:* 44 pp.

LUNDQVIST, A., 1965. The genetics of incompatibility. *Genetics Today, 3:* 637–647.

MACARTHUR, J. W. & CHIASSON, L. P., 1947. Cytogenetic notes on tomato species and hybrids. *Genetics, 32:* 165–177.

MARTIN, F. W., 1961a. Complex unilateral hybridization in *Lycopersicon hirsutum*. *Proceedings of the National Academy of Sciences of the United States of America, 47:* 855–857.

MARTIN, F. W., 1961b. The inheritance of self-incompatibility in hybrids of *Lycopersicon esculentum* Mill. × *L. chilense* Dun. *Genetics, 46:* 1443–1454.

MARTIN, F. W., 1963. Distribution and interrelationships of incompatibility barriers in the *Lycopersicon hirsutum* Humb. and Bonpl. complex. *Evolution, 17:* 519–528.

MARTIN, F. W., 1966. Avoiding unilateral barriers in tomato species crosses. *Tomato Genetics Coop. Report, 16:* 19–20.

McGUIRE, D. C. & RICK, C. M., 1954. Self-incompatibility in species of *Lycopersicon* sect. Eriopersicon and hybrids with *L. esculentum*. *Hilgardia, 23:* 101–124.

MULLER, C. H., 1940. A revision of the genus *Lycopersicon*. *Miscellaneous Publications. United States Department of Agriculture, 382:* 29 pp.

PANDEY, K. K., 1967. Origin of genetic variability: Combinations of peroxidase isoenzymes determine multiple allelism of the S-gene. *Nature, 213:* 669–672.

PANDEY, K. K., 1968. Compatibility relationships in flowering plants: role of the S-gene complex. *American Naturalist, 102:* 475–489.

PANDEY, K. K., 1969. Elements of the S-gene complex. V. Interspecific cross-compatibility relationships and theory of the evolution of the S complex. *Genetica, 40:* 447–474.

RICK, C. M., 1953. Tests of compatibility of certain stocks of *L. peruvianum* var. *dentatum*. *Tomato Genetics Coop. Report, 3:* 21–22.

RICK, C. M., 1956. Genetic and systematic studies on accessions of *Lycopersicon* from the Galápagos Islands. *American Journal of Botany, 43:* 687–696.

RICK, C. M., 1961. Biosystematic studies on Galápagos tomatoes. *Occasional Papers of the California Academy of Sciences, 44:* 59–77.

RICK, C. M., 1963. Barriers to interbreeding in *Lycopersicon peruvianum*. *Evolution, 17:* 216–232.

RICK, C. M., 1967. Exploiting species hybrids for vegetable improvement. *Proceedings of the XVIIth International Horticultural Congress, 3:* 217–229.

RICK, C. M. & BUTLER, L., 1956. Cytogenetics of the tomato. *Advances in Genetics, 8:* 267–382.

RICK, C. M., KESICKI, E., FOBES, J. F. & HOLLE, M., 1976. Genetic and biosystematic studies on two new sibling species of *Lycopersicon* from interandean Peru. *Theoretical and Applied Genetics, 47:* 55–68.

RICK, C. M. & LAMM, R., 1955. Biosystematic studies on the status of *Lycopersicon chilense*. *American Journal of Botany 42:* 663–675.

SAMPSON, D. R., 1962. Intergeneric pollen-stigma incompatibility in the Cruciferae. *Canadian Journal of Genetics and Cytology, 4:* 38–49.

SMITH, P. G., 1944. Embryo culture of a tomato species hybrid. *Proceedings of the American Society for Horticultural Science, 44:* 413–416.

STOUT, A. B., 1952. Reproduction in *Petunia*. *Memoirs of the Torrey Botanical Club, 20:* 1–202.

SWAMINATHAN, M. S. & MURTY, B. R., 1957. One-way incompatibility in some species crosses in the genus *Nicotiana*. *Indian Journal of Genetics and Plant Breeding, 17:* 23–26.

ZERPA, D. M. DE, 1952. A case of incompatibility in tomato hybridization. *Tomato Genetics Coop. Report, 2:* 13–14.

35. Incompatibility and incongruity in tuber-bearing *Solanum* species

J. G. TH. HERMSEN AND EWA SAWICKA

Institute for Agricultural Plant Breeding, Agricultural University, Wageningen, TheNetherlands; and Potato Research Institute, Mlochów, Poland

Incompatibility is the inhibition of fusing of normal gametes owing to interaction of two mutually reactive identical proteins which are produced in pollen and pistil respectively. In tuber-bearing *Solanum* species incompatibility is gametophytically determined and based on either one or two loci.

Incongruity refers to all barriers to crossing which are independent of incompatibility. It is a general phenomenon in living organisms. Examples of incongruity in tuber-bearing *Solanum* are presented.

A special case of interspecific relations in *Solanum* is elaborated which may contribute to a better understanding of such relations.

CONTENTS

INTRODUCTION

There are many and various barriers to hybridization within and between species and genera. Attempts to find some leading thread running through this jungle of variation have been successful to a certain extent only. Many problems still have remained unsolved and different opinions exist on several phenomena. Two concepts are of fundamental significance in crosses between genotypes at different levels of relationship, viz. incompatibility and incongruity.

Within species and between closely related cross-fertilizing species incompatibility predominates and prevents undesirable inbreeding. Incompatibility reactions result from identity of alleles at one or more specified incompatibility loci in pollen and pistil.

In crosses between widely related species, genera and families, incompatibility can no longer be involved, because identity of incompatibility alleles must be lacking. Crossability barriers in such wide crosses are caused by so-called incongruity. According to Hogenboom (1973) incongruity is a pollen-pistil relation which is unsuccessful owing to incomplete (or rather non-matching) genetic information in the pollen partner relevant to physiological, biochemical or even mechanical barriers in the pistil partner.

There is no difference in opinion about incompatibility being the main mechan-

445

ism of non-crossability within most cross-fertilizing species. Neither is there a disagreement among scientists on incongruity (and not incompatibility) being involved in non-successful matings between widely related species, like tuber-bearing and non-tuber-bearing *Solanum* species, or between genera like *Nicotiana* and *Solanum*. However, between these extremes there is a large field of research on interspecific hybridization, where outstanding scientists have put forward very divergent hypotheses for explaining their results. Roughly three main theories can be found in the literature, all supported by experimental evidence and presented with great conviction.

The first theory, which in an excellent study was elaborated by Lewis and Crowe (1958) and expanded ingeniously by Pandey (1964, 1968, 1969a, b) assigns a dual function to the *S*-locus. Firstly a specific one, bringing about incompatibility, and secondly a non-specific one causing rejection of pollen from self-compatible species. Extensive research by Pandey, mainly on *Nicotiana* species, yielded new data on interspecific barriers which in the scope of the dual function theory could be explained only by assuming an extreme complexity of the *S*-locus.

The second theory rejects the hypothesis of a non-specific function of the *S*-locus in interspecific crosses. It even assumes that interspecific barriers, also unilateral ones, are completely independent of the *S*-locus. Although Grun and his collegues (Grun & Radlow, 1961; Grun & Aubertin, 1966) working with tuber-bearing *Solanum* species were the first to propose independent genes for rejection of pollen of the self-compatible species *S. verrucosum*, Hogenboom (1973) gave this theory a firm and broad basis in his model for incongruity in intimate partner relationships.

The third theory takes an intermediate position between "dual function" and "incongruity" theories. It was first launched by Abdalla (1970) as the "two powers competition" hypothesis and elaborated by Abdalla & Hermsen (1972). According to this theory self-incompatible and self-compatible populations act as two competing powers. Self-incompatible plants which are pollinated by self-compatible ones become self-compatible and then run the risk of inbreeding depression. The self-incompatible population by natural selection develops a system of dominant genes, not belonging to the *S*-locus, the so-called *UI*-genes by which *Sc*-pollen is rejected. Then the self-compatible population reacts by developing a new *Sc*-allele, etc. In this way a gene-for-gene relationship between specific *UI*-alleles and matching *Sc*-alleles is built up.

In this article incompatibility, incongruity and a special case of unilateral relations in tuber-bearing *Solanum* will be discussed.

INCOMPATIBILITY IN TUBER-BEARING *SOLANUM*

In the family *Solanaceae* the gametophytic system of incompatibility prevails. This is true also for tuber-bearing *Solanum* species. According to Hawkes (1963) 160 tuber-bearing *Solanum* species are known, of which nearly 100 are diploid. As far as they have been studied the diploids are self-incompatible with the exception of *S. verrucosum, S. etuberosum, S. brevidens, S. morelliforme, S. polyadenium* (Hawkes, 1956, 1963) and *S. fernandezianum* (Hermsen, unpubl.).

Some species are strictly self-incompatible. Other species besides self-incompatible genotypes also comprise partly or even completely self-compatible genotypes, e.g. *S. phureja* (Dodds, 1956; Cipar, 1964), *S. chacoense* (Hermsen, 1969; Sawicka, 1971), and dihaploid *S. tuberosum* (Olsder & Hermsen, 1976). Apart from some erratic results which were observed in almost all investigations, the one-locus gametophytic system was demonstrated in *S. chacoense* (Pushkarnath, 1953b), *S. kurtzianum*, *S. vernei*, *S. soukupii*, *S. sparsipilum*, *S. simplicifolium*, *S. megistacrolobum*, *S. sanctae-rosae*, *S. toralapanum* and *S. michoacanum* (Pandey, 1960b, 1962a, b), *S. phureja*, diploid *S. tuberosum* and *S. stenotomum* (Cipar, Peloquin & Hougas, 1964, 1967; Dodds & Paxman, 1962; Hermsen, 1972), whereas a two-loci gametophytic system was found in *S. pinnatisectum*, *S. ehrenbergii* and *S. bulbocastanum* (Pandey, 1962a, c). There are a few reports in which presumably erratic results were explained on the basis of a second locus (Pushkarnath, 1953a; Pandey, 1962b; Abdalla & Hermsen, 1971).

INCONGRUITY IN TUBER-BEARING *SOLANUM*

Using the term incongruity implies that incompatibility is not involved in the phenomena to be considered. In relation to incongruity, as far as it becomes manifest between pollination and fertilization, Hogenboom (1973) introduced the terms "barrier capacity" of the pistil and "penetration capacity" of the pollen. Barrier capacity (Bc) is the total of barriers of various kinds in the pistil against penetration of pollen tubes, whereas penetration capacity (Pc) is the total of genes (gene complexes) in the pollen needed for overcoming barriers to penetration in the pistil. Each species is characterized by a particular Bc and Pc.

Apart from incompatibility, plants from populations of a species as a rule can readily be intercrossed. This implies that within a species in general Bc and Pc are matching. Differences between the Bc/Pc levels in different species and genera become manifest by crossing them, preferably in a complete diallel. On the basis of the results, species or genera with equal Bc/Pc may be put into one group and different groups may be arranged in order of the level of Bc/Pc. In this way a stepwise crossability pattern is obtained as indicated schematically in Table 35.1. Such patterns have been found in several genera of the *Solanaceae*.

Table 35.1. Stepwise crossability pattern of five species groups on the basis of increasing levels of barrier/penetration capacity (Bc/Pc). A cross is successful (+), if Pc and Bc are matching or if there is a surplus of Pc in the male parent

♀	Pc−1 (Bc−1)	Pc−2 (Bc−2)	♂ Pc−3 (Bc−3)	Pc−4 (Bc−4)	Pc−5 (Bc−5)
(Pc−1) Bc−1	+	+	+	+	+
(Pc−2) Bc−2	−	+	+	+	+
(Pc−3) Bc−3	−	−	+	+	+
(Pc−4) Bc−4	−	−	−	+	+
(Pc−5) Bc−5	−	−	−	−	+

Results from intercrossing two groups of four tuber-bearing *Solanum* species are presented in Tables 35.2 and 35.3. The following conclusions could be drawn.

—Self-compatible species have a lower level of Pc/Bc than self-incompatible species. According to Hogenboom's theory (1973) on the evolution of self-compatible species this indicates that the self-compatible *Solanum* species used are ancient ones.

—Within the group of self-compatible species (Table 35.2) there is bilateral incongruity considering seed set, but slight differences in degree of pollen tube inhibition are observed in the reciprocal crosses.

—Absence of inhibition may or may not lead to seed set (cf. *ver × phu* with *etu × phu*). Apparently different barriers are involved.

—Table 35.3 shows that unilateral inhibition is not restricted to crosses between self-compatible and self-incompatible species, but also occurs in crosses between self-incompatible species (cf. *S. pinnatisectum × S. bulbocastanum* and its reciprocal).

In the small group of species tested, *S. verrucosum* had the lowest and *S. bulbocastanum* the highest level of Pc/Bc. In order to compare these species on a broader scale, *S. verrucosum* was reciprocally crossed with 19 species (diploids, tetraploids and one hexaploid). In all cases *S. verrucosum* was successful as a female. As a male it was only successful on diploid *S. gourlayi* and *S. vernei* and on certain genotypes of diploid *S. chacoense* and *S. tuberosum*. On the other hand *S. bulbocastanum* was not successful as a female in reciprocal crosses with eight

Table 35.2. Crossability pattern of three self-compatible (sc) and one self-incompatible (si) *Solanum*-species. Data from Abdalla (1970). +, Seed set; 1/5, stop of pollen tube growth at 1/5 of the style (1/1, no inhibition); *, very low pollen germination

	♂			
♀	sc *verrucosum*	sc *polyadenium*	sc *etuberosum*	si *phureja*
sc *verrucosum*	+	−1/5	−1/2	+1/1
sc *polyadenium*	−1/6*	+	−	−1/10
sc *etuberosum*	−1/3	−	+	−1/1
si *phureja*	−1/3	−1/10	−	+

Table 35.3. Crossability pattern of one self-compatible and three self-incompatible *Solanum* species

	♂			
♀	sc *verrucosum*	si *phureja*	si *pinnatisectum*	si *bulbocastanum*
sc *verrucosum*	+	+	+	+
si *phureja*	−	+		
si *pinnatisectum*	−	−	+	+
si *bulbocastanum*	−	−	−	+

species, but successful as a male on five of them. These results confirm the low Pc/Bc of *S. verrucosum* and the high Pc/Bc of *S. bulbocastanum*.

Differences in barrier and penetration capacity were found within tuber-bearing *Solanum* species as well as between species.

In an extensive crossing programme (nearly 5000 pollinated flowers) involving 22 genotypes of allotetraploid *S. acaule* ♀ and 28 genotypes of diploid *S. bulbo-castanum* ♂ it was found that *S. acaule* and *S. bulbocastanum* fell into three and four groups, respectively, on the basis of berry set (Hermsen, 1966). (See Table 35.4.). This indicates differences in Bc within *S. acaule* and differences in Pc within *S. bulbocastanum*. A genetic analysis of these differences was not carried out.

Table 35.4. Classification of genotypes of *S. acaule* and *S. bulbocastanum* on the basis of mutual crossability measured as percentage berry set. Bc, Barrier capacity (Bc 3 > Bc 2 > Bc 1); Pc, penetration capacity (Pc 4 > Pc 3 > Pc 2 > Pc 1)

Parent species	Groups	Number of clones	Average poll. fl. per clone	Percentage berry set (range)
S. acaule ♀	Bc–3	10	97	0
	Bc–2	4	160	7.2 (5.2–9.5)
	Bc–1	8	150	16.6 (13.3–18.8)
S. bulbocastanum ♂	Pc–1	3	97	0
	Pc–2	15	141	7.0 (2.4–11.4)
	Pc–3	7	130	16.0 (15.0–17.3)
	Pc–4	3	149	23.6 (22.9–24.3)

In this connection it is illuminating to refer to the clear-cut genetic analysis of differences of Bc in *Triticum aestivum* against pollen from *Secale cereale*. It was shown unambiguously that two active dominant barrier genes in wheat are responsible for inhibition of rye pollen (Lein, 1943). By means of intervarietal substitution lines these genes were localized on the chromosomes 5A and 5B (Riley & Chapman, 1967) and are believed to confer evolutionary advantage upon wheat by preventing the production of sterile wheat-rye hybrids.

To conclude the chapter on incongruity, it should be emphasized that investigations of incongruity between species may be hampered in different ways. When species are closely related, incompatibility may interfere with the results. When both self-compatible and self-incompatible species are involved the question may be raised whether the *S*-locus plays a role in unilateral inhibition. Finally, it is obvious, that incongruity barriers are not restricted to inhibition of pollen tube growth. Also embryo-abortion, male sterility, hybrid breakdown, etc. may be indications of incongruity. Such barriers may render a genetic analysis of incongruity unreliable or even impossible.

A SPECIAL CASE OF INTERSPECIFIC RELATIONS IN *SOLANUM*

As mentioned before, Grun & Aubertin (1966) found two types of plants in *S. chacoense*, one type rejecting *S. verrucosum* pollen (non-acceptor, NA), the other (recessive) one accepting *S. verrucosum* pollen (acceptor, A). Non-acceptance was based on 2–4 dominant genes, which are independent of the S-locus. Essentially the same results were recorded by Hermsen, Olsder, Jansen & Hoving (1974) for dihaploid *S. tuberosum*. Genetic analysis of 14 populations of the type A × A, NA × A and NA × NA led to two genetic models which could explain the ratios obtained. According to model 1 each of two dominant genes inhibits *S. verrucosum* pollen, only double recessives being acceptors. The second model assumes a dominant acceptor gene A and a dominant inhibitor I, only iiA.-plants being acceptors.

At our institute Sawicka analysed three F_1 populations from dihaploid *S. tuberosum* subsp. *andigena* × subsp. *tuberosum* for acceptance/non-acceptance of pollen from three individual I_1-plants of *S. verrucosum* (CPC 2247 × PI 195172) plant A3. Berry set (assessed on all F_1 plants) and pollen tube growth (assessed in a sample of plants) were used as a measure for acceptance. Berry set is the number of berries per 100 pollinated flowers. Pollen tube growth is expressed in four classes: 1 (strong inhibition); 2 (clear inhibition, few tubes reaching lower part of style); 3 (inhibition, a few tubes entering the ovary); 4 (no inhibition). The results are summarized in Table 35.5, in which the results from plants 45 and 51 are taken together, because of being equal.

It is clear that all three populations fall into four phenotypic classes: A (acceptor clones, berry set with all *S. verrucosum* pollen); NA (non-acceptor clones, no berry set); PA (pseudo-acceptor clones, low berry set) and finally a new group of plants, accepting pollen of plant numbers 45 and 51 of *S. verrucosum*, but rejecting pollen of plant 49. The indication DP derives from differentiating

Table 35.5. Average pollen growth type class and average berry set on acceptor (A), non-acceptor (NA), pseudo-acceptor (PA) and differentiating (DP) plants when pollinated with *S. verrucosum*. A3 selfed plant 45 (or 51) and 49. The Roman figures refer to the three *S. tuberosum* F_1s studied: I and II are of the type NA × NA, III is NA × A

| | | | *S. verrucosum* clone A3 selfed | | | |
| | | | I_1-plant 45 (or 51) ♂ | | I_1-plant 49 ♂ | |
F_1-plant group ♀		Number of ♀-plants pollinated	Average growth type class	Average berry set	Average growth type class	Average berry set
A	I	7	4.0 ⎤	43.0 ⎤	3.9 ⎤	32.0 ⎤
	II	8	4.0 ⎬3.9	55.8 ⎬46.5	3.5 ⎬3.6	45.6 ⎬40.3
	III	28	3.8 ⎦	40.7 ⎦	3.4 ⎦	43.3 ⎦
NA	I	60	1.4 ⎤	0	1.0 ⎤	0
	II	60	— ⎬1.2	0	— ⎬1.1	0
	III	40	1.0 ⎦	0	1.1 ⎦	0
PA	I	1	3.3 ⎤	7.7 ⎤	2.0 ⎤	7.1 ⎤
	II	5	2.6 ⎬2.1	7.3 ⎬6.6	2.5 ⎬2.2	3.8 ⎬5.9
	III	7	2.1 ⎦	4.8 ⎦	2.1 ⎦	6.7 ⎦
DP	I	2	3.6 ⎤	39.3 ⎤	1.9 ⎤	0
	II	1	4.0 ⎬3.8	43.8 ⎬42.8	1.7 ⎬1.7	0
	III	4	3.8 ⎦	45.3 ⎦	1.5 ⎦	0

penetration capacity of *S. verrucosum* plants. DP-plants enable us to distinguish *S. verrucosum* plants with different penetration capacity.

Two hypotheses may explain the occurrence of DP-plants.

—The first is based on the model for incongruity of Hogenboom (1973) and is summarized in Table 35.6 (PA class omitted).

The plants 45 and 51 are successful on A- and DP-plants because they have a surplus of Pc for A-styles and a matching Pc for DP-styles. Plant 49 is only successful on A-plants (Pc and Bc matching).

—The second hypothesis is based on the gene-for-gene relationship of *UI*- and *Sc*-alleles (Abdalla & Hermsen, 1971). If the *S. tuberosum* populations segregate for two *UI*-alleles and the original *S. verrucosum* plant A3 contains two *Sc*-alleles (one from each of both parent provenances), then the result of the pollinations could be as given in Table 35.7. Abdalla's hypothesis explains the occurrence of even two different DP-types of plants. That only one type was found, may be ascribed to the low number of *S. verrucosum* plants used. However, the low frequency of DP-plants (7 out of 223) does not fit the hypothesis.

Table 35.6. Hypothesis to explain the behaviour of A-, NA- and DP-plants in respect to pollen with different penetration capacity. Relative levels of Bc and Pc are indicated by number of asterisks. +, Acceptance (= no inhibition of pollen growth, normal seed set)

		Pc ♂	
Bc ♀		** 45 (51)	* 49
*	A	+	+
**	DP	+	−
***	NA	−	−

Table 35.7. Genotypes of *S. tuberosum* females and *S. verrucosum* males and phenotypes of F₁ plants. +/−, Acceptance/non-acceptance

Genotypes F₁ plants ♀	Genotypes I₁ plants ♂			Phenotypes F₁ plants
	Sc_1Sc_1	Sc_1Sc_2	Sc_2Sc_2	
$UI_1 ui_1 UI_2 ui_2$	−	−	−	NA
$UI_1 ui_1 ui_2 ui_2$	−	+	+	DP type 1
$ui_1 ui_1 UI_2 ui_2$	+	+	−	DP type 2
$ui_1 ui_1 ui_2 ui_2$	+	+	+	A

CONCLUDING REMARKS

If the Latin proverb "simplex sigillum veri" (simplicity is the mark of truth) holds good also in the case of wide crosses, the theory of Hogenboom (1973) is the right one, because until now all results can be explained in a simple, smooth way. Based on the same proverb Pandey's complicated theory (1964, 1968, 1969a) ought to be abandoned. The theory of Abdalla & Hermsen (1972) holds to the golden mean. However, scientific conclusions cannot be based on proverbs, which as a rule represent only part of the truth.

In this stage of research it is not yet possible either to accept or reject one theory. In tuber-bearing *Solanum*, experimental material is available which may throw new light upon the problems of incompatibility and incongruity and their possible interaction. Therefore the research is being continued.

REFERENCES

ABDALLA, M. M. F., 1970. *Inbreeding, heterosis, fertility and plasmon differentiation and* Phytophthora *resistance in* Solanum verrucosum *Schlechtd., and some interspecific crosses:* 213 pp. Doctor's Thesis, University of Wageningen.

ABDALLA, M. M. F. & HERMSEN, J. G. Th., 1971. A two-loci system of gametophytic incompatibility in *Solanum phureja* and *S. stenotomum. Euphytica, 21:* 345–350.

ABDALLA, M. M. F. & HERMSEN, J. G. Th., 1972. Unilateral incompatibility: hypotheses, debate and its implications for plant breeding. *Euphytica, 21:* 32–47.

CIPAR, M. S., 1964. Self-compatibility in hybrids between Phureja and haploid Andigena clones of *Solanum tuberosum. European Potato Journal, 7:* 152–160.

CIPAR, M. S., PELOQUIN, S. J. & HOUGAS, R. W., 1964. Inheritance of incompatibility in hybrids between *Solanum tuberosum* haploids and diploid species. *Euphytica, 13:* 163–172.

CIPAR, M. S., PELOQUIN, S. J. & HOUGAS, R. W., 1967. Haploidy and the identification of self-incompatibility alleles in cultivated potato groups. *Canadian Journal of Genetics and Cytology, 9:* 511–518.

DODDS, S. K., 1956. Sporadic self-fertility in *S. phureja. Report. John Innes (Horticultural) Institution, 47:* 19–20.

DODDS, S. K. & PAXMAN, G. J., 1962. The genetic system of cultivated diploid potatoes. *Evolution, 16:* 154–167.

GRUN, P. & RADLOW, A., 1961. Evolution of barriers to crossing of self-incompatible with self-compatible species of *Solanum. Heredity, 16:* 137–143.

GRUN, P. & AUBERTIN, M., 1966. The inheritance and expression of unilateral incompatibility in *Solanum. Heredity, 21:* 131–138.

HAWKES, J. G., 1956. A revision of the tuber-bearing Solanums. *Annual Report Scottish Plant Breeding Station:* 38–110.

HAWKES, J. G., 1963. A revision of the tuber-bearing Solanums, 2nd ed. *Scottish Plant Breeding Station Record:* 71–181.

HERMSEN, J. G. Th., 1966. Crossability, fertility and cytogenetic studies in *Solanum acaule* × *Solanum bulbocastanum. Euphytica, 15:* 149–155.

HERMSEN, J. G. Th., 1969. Induction of haploids and aneuhaploids in colchicine-induced tetraploid *Solanum chacoense* Bitt. *Euphytica, 18:* 183–189.

HERMSEN, J. G. Th., 1972. Self-compatibility among dihaploids from autotetraploid *Solanum tuberosum. Incompatibility Newsletter, 1:* 31–32.

HERMSEN, J. G. Th., OLSDER, J., JANSEN, P. & HOVING, E., 1974. Acceptance of self-compatible pollen from *Solanum verrucosum* in dihaploids from *S. tuberosum.* In H. F. Linskens (Ed.), *Fertilization in Higher Plants:* 37–40.

HOGENBOOM, N. G., 1973. A model for incongruity in intimate partner relationships. *Euphytica, 22:* 219–233.

LEIN, A., 1943. Die genetische Grundlage der Kreuzbarkeit zwischen Weizen und Roggen. *Zeitschrift für Induktive Abstammung und Vererbungslehre, 81:* 28–61.

LEWIS, D. & CROWE, L. K., 1958. Unilateral interspecific incompatibility in flowering plants. *Heredity, 12:* 233–256.

OLSDER, JANNY & HERMSEN, J. G. Th., 1976. Genetics of self-compatibility in dihaploids of *Solanum tuberosum* L. I. Breeding behaviour of two self-compatible dihaploids. *Euphytica, 25.* 597–607.

PANDEY, K. K., 1960a. Self-incompatibility system in two Mexican species of *Solanum. Nature, 185:* 483–484.

PANDEY, K. K., 1960b. Self-incompatibility in *Solanum megistacrolobum* Bitt. *Phyton, 14:* 13–19.

PANDEY, K. K., 1962a. Incompatibility in some *Solanum* crosses. *Genetica, 33:* 24–30.

PANDEY, K. K., 1962b. Interspecific incompatibility in *Solanum* species. *American Journal of Botany, 49:* 874–882.

PANDEY, K. K., 1962c. Genetics of incompatibility behaviour in the Mexican *Solanum* species *S. pinnatisectum. Zeitschrift für Vererbungslehre, 93:* 378–388.

PANDEY, K. K., 1964. Elements of the S-gene complex. The S_{FI} alleles in *Nicotiana*. *Genetical Research, 5:* 397–409.

PANDEY, K. K., 1968. Compatibility relationships in flowering plants: role of the S-gene complex. *American Naturalist, 102:* 475–489.

PANDEY, K. K., 1969a. Elements of the S-gene complex. IV. S-allele polymorphism in *Nicotiana* species. *Heredity, 24:* 601–619.

PANDEY, K. K., 1969b. Elements of the S-gene complex. V. Interspecific cross-compatibility relationships and theory of the evolution of the S-complex. *Genetica, 40:* 447–474.

PUSHKARNATH, 1953a. Studies on sterility in potatoes. IV. Genetics of incompatibility in *Solanum araccpapa*. *Euphytica, 2:* 49–58.

PUSHKARNATH, 1953b. Studies on sterility in potatoes. V. Genetics of self- and cross-incompatibility in *Solanum rybinii*. *Indian Journal of Genetics and Plant Breeding, 13:* 83–90.

RILEY, R. & CHAPMAN, V. ,1967. The inheritance in wheat of crossability with rye. *Genetical Research, 9:* 267–269.

SAWICKA, EWA, 1971. The self-compatibility in the series *Commersoniana* Buk. *Bulletin Instytut Ziemniaka, 8:* 5–12.

36. Haploid parthenogenesis in *Capsicum annuum* L.

E. POCHARD AND R. DUMAS DE VAULX

Plant Breeding Station, I.N.R.A. Montfavet-Avignon, France.

Spontaneous or induced haploids are useful for a better knowledge of *Capsicum* genetics and physiology. The unexpected phenotypic instability of some autodiploid progenies derived from haploids by colchicine treatment deserves further study.

CONTENTS

INTRODUCTION

Haploidy is not a novel subject. It is a rapidly expanding field of research and this is chiefly due to the discovery, in 1964, by Guha & Maheshwari, of the possibility to develop haploid androgenetic plants from microspores of *Datura innoxia*. The first case of haploid parthenogenesis on record was one observed in *Datura stramonium* (Blakeslee, Belling, Farnham & Bergner, 1922). Could it be that *Datura* and the Solanaceae, as a whole, have a prophetic part to play? The 1974 Guelph Symposium provided an occasion for an exhaustive review of both theoretical and practical consequences of haploidy. Presentation of fresh results is thus made easier. The aspects to be developed are those which are seldom taken up or those for which the results published to date are contradictory. This is what we are trying to do with research on *Capsicum*.

Christensen & Bamford (1943) reported the spontaneous development of haploids in *Capsicum annuum* in the form of twin embryos. They thus obtained haploid plants in four different varieties which they covered in a descriptive study. Toole & Bamford (1945) managed to produce diploid sectors through treatment with colchicine and then observed normal occurrence of meiosis (positive results in about 25 % of the cuttings or scions treated). Later on using other varieties of the same species, Morgan & Rappleye (1950), analysed the composition of the pairs or triplets (they made wrong use of the term 'Capsicum frutescens' because the genotypes employed: World Beater, Floral Gem, Goliath are typical Capsicum annuum according to Smith & Heiser (1951). They found, out of a total number of 78,000 germinated seeds, 291 twins and three triplets, that is 0.38 % polyembryony; among the surviving pairs or triplets they discovered 29 % n–2n and 1.4 % n–n. Their total finding was therefore a little over one spontaneous haploid per thousand plants. This figure seems very high and certainly ranks the pepper among those species that are the most fertile in haploid embryos, bearing comparison with maize before it was selectively bred (Chase, 1949: 0.39 to 1.49 per thousand). However, it is to be noted that haploids in maize are nearly always non-twins and therefore spring from a different mechanism. Using sw and C marker genes as well as other quantitative characters Morgan & Rappleye (1954) and then Campos & Morgan (1960) showed that in haplo-diploid pairs, the diploid partner is always of hybrid origin and the haploid partner always of the maternal type. In addition, two autodiploid lines sprung from diploidized haploids were compared: one of these, the G 168 A, issued from cv. Goliath, showed an extraordinarily high proportion of twin haploidy (24 per thousand), whereas P_1A (sprung from Perfection) produced no parthenogenetic haploids at all out of nearly ten thousand plants, but one androgenetic haploid plant.

We shall report below the results obtained from 1965 onwards, at the Montfavet-Avignon Plant Breeding Station. For the last ten years we have been able to observe haploid stock numbers of between two and three hundred each year; this represents a sample of over 2000 pepper haploids. This research work is being conducted with a mind to its being included in the general breeding program for Capsicum annuum. However the place actually taken up by haploidy in this program remains small. As will be seen, a lot of problems have yet to be solved before optimum use is possible.

ORIGIN OF HAPLOIDS

The use of haploidy in the breeding of the pepper is bound up with the possibility of getting haploid plants at sufficiently high rates in material of high agronomic value (Pochard & Dumas de Vaulx, 1971). This is also needed for direct investigation into the cytological origin of haploids. Four main methods can be put into effect in order to increase these rates: anther culture (Wang, Sun, Wang & Chien, 1973), selection from diploidized haploids (Morgan & Rappleye, 1950), treatment with chemicals (Montezuma de Carvalho, 1967, in Datura and Solanum phureja), or interspecific hybridization (Hougas, Peloquin & Gabert, 1964, in potato).

We show here a few results obtained through two different methods: selection of genotypes showing a high level of haploid parthenogenesis, and application of nitrous oxide on the floral organs. Until now the two other methods have been unsuccessful.

Breeding for high rates of parthenogenesis

The results are much less favourable than those already published (Morgan &

Rappleye, 1950). In the material currently handled in the collections or in the nurseries, the rate of haploidy is seldom above 0.5 per 1000 plants. Haplo-diploid twins are generally from five to ten times less numerous than purely diploid twins (conjoined twins are not included). Autodiploid lines derived from haploids only occasionally exceed the original population in the parthenogenetic trend. It is therefore necessary to complete a whole breeding program in order to improve this character, which shows low heritability; uncontrolled events of unfrequent occurrence play a major part in the induction of haploidy in the pepper. The most promising results (but yet far below those obtained by Campos & Morgan with the G 168A autodiploid) appear in the progeny of two 5 × 5 diallel crosses (A and B) performed in 1969 to combine the best autodiploid or inbred lines available at the end of the first breeding cycle. In the F2 generation, 125 haploids were recovered and 104 diploidized by treatment with colchicine. Their H1 and H2 selfed progeny were surveyed for the emergence of haploids in 1973, 1974 and 1975. We were able to bring into evidence significant differences between A and B, between crosses and between lines in the same cross. The results are summarized in Table 36.1.

For comparison of the parthenogenetic trends of the various genotypes it is necessary to produce the seeds under the same conditions. For example, conditions prevailing in the glasshouse are much less favourable than those encountered in the open field (about six times less haploids). On the other hand, the rate of purely diploid gemellity is not changed (Table 36.2).

The most surprising result comes from the comparison of sister H2–H2 autodiploid lines, sprung from the same haploid ancestor by two haplo-diploidizations followed by two selfing generations. In the four families studied, sister lines display

Table 36.1. Frequency of haploid plants in the initial material and after one or two breeding cycles

	Origin of the seeds	(3)	Number of plants	Twins (1) 2n-2n	Twins (1) n-2n	Solitary haploids (2)	Haploid gemellity per 1000	Haplo-diploid twins %	Diploid gemellity per 1000
Parents	Yolo Wonder		20,600	121	10	NR	0.49	8	5.9
	L 107		5120	6	0	NR	(0.10)	—	—
F8	(Y × 107) F8	10	26,090	77	18	NR	0.69	19	3.0
First breeding cycle	(Y × 107) H2	48	68,890	123	38	NR	0.61	24	2.0
	H2, Parents: (A)	5	21,860	51	10	NR	0.46	16	2.3
	H2, Parents: (B)	5	21,980	34	34	NR	1.55	50	1.6
Second breeding cycle	(A) F2	10	100,000	458	41	5	0.41	9	4.5
	(B) F2	10	14,200	38	29	2	2.04	43	2.7
	(A) H2	57	93,930	127	67	NR	0.71	35	1.4
	(B) H2	22	36,100	50	48	NR	1.32	49	1.4
	H2, best (A)	3	10,060	29	50	NR	4.97	63	2.9
	H2, best (B)	3	4940	19	23	NR	4.66	55	3.8
	H2-H2 best (A)	3	43,040	52	90	7	2.09	63	1.2
	H2-H2 best (B)	3	26,590	30	79	7	2.97	72	1.1

(1), Not conjoined twins and triplets; (2), haploids without partner; (3), number of progenies; NR, not recorded; F2, F8, second and eighth generation of selfing after a cross; H2, second generation of selfing after diploidization of a haploid (autodiploid line); H2-H2, sister autodiploid lines in H2, sprung from haploids collected in the H2 of an autodiploid line (two successive cycles of haplo-diploidization).

significantly different rate of haploidy in the selfed progeny (Fig. 36.1). New trials are undertaken in order to test the permanence of these properties in the H3 generation. We already know that the best line of each family gives again very high rates of haploidy (0.6 to 1 %). They are now being used in a third breeding course involving diallel reciprocal crosses.

Table 36.2. Influence of the growth conditions on the appearance of haploids in the seeds

Origin of the seeds	Number of plants	Twins 2n-2n	n-2n	Solitary haploids	Total rate of haploidy per 1000	Haplo- diploid	Diploid gemellity per 1000
Field production	101,900	414	188	21	2.05	45	4.1
Glasshouse production	258,200	1051	80	13	0.36	8	4.1

Field production, seeds harvested in open field from August to October; glasshouse production, seeds harvested underglass from September to December; same material from 'A' diallel cross, in H2.

Induction by means of nitrous oxide (N_2O)

Inhibition of the second nuclear division in the male gametophyte by mitoclasic agents has been demonstrated in *Datura* and *Solanum phureja* (Montezuma de Carvalho, 1967) and in *Capsicum annuum* (Dumas de Vaulx & Pochard, 1974). This frequently leads to the constitution of a single restitution sperm nucleus (Wangenheim *et al.*, 1960). By treating *S. phureja* pollen with colchicine, Monte-longo & Rowe (1969) increased the number of *Solanum tuberosum* haploids per

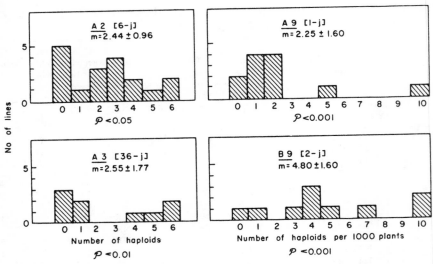

Figure 36.1. Frequency of haploid plants in the progeny of sister autodiploid lines issued from a single haploid ancestor by a second cycle of haplo-diploidization (H2-H2 lines). Test of the Poisson distribution for samples of 1000 plants (most of the actual samples fall between 1000 and 1500 plants; the number of haploids per 1000 is adjusted to the nearest integer); 4 families are studied, issued from 4 different crosses; 9 to 18 H₂-H₂ lines per family.

berry by a factor of two. One could hope that the applying of N_2O or colchicine to the growing pollen tube during second mitosis would increase the rate of emergence of solitary haploids in the pepper. Once more, the results are not exactly those we were expecting (Table 36.3). Treatment applications made from 4 to 8 hours after pollination greatly reduce fertility (in vitro, mitosis in the pollen tube occurs 6 to 8 hours after germination at 25° C). Solitary haploids appear at low rates (under 1 per thousand) but twin haploids are more frequent ($\times$ 2 à 5), especially in the auto-diploid line DL-1B issued from the Doux-des-Landes cultivar. This is also true for treatments affecting the female gametophyte only. In this case, fertility is restored to about 75 % of the control in DL-1B, 100 % in Yolo Wonder. Incidentally, a reduction in diploid gemellity can be observed, especially in Yolo Wonder. Thus N_2O effects are manifold and are dependent on plant genotypes.

Table 36.3. Effect of nitrous oxide (N_2O) on the occurrence of haploid plants in two pepper cultivars

Varieties	Treatments	Number of plants	Twins 2n-2n	Twins n-2n	Solitary haploids	Total rate of haploidy per 1000	Haplo-diploid twin %	Diploid gemellity per 1000	Number of seeds per fruit
Yolo Wonder	control	7800	40	3	0	0.38	7	5.12	130
	N_2O (T1)	4600	7	2	2	0.87	22	1.52***	62
	N_2O (T2)	2330	3	1	0	0.43	25	1.28**	120
Doux des Landes	control	5300	10	6	0	1.11	37	1.88	100
	N_2O (T1)	3200	3	19	0	5.90***	86***	0.93	19
	N_2O (T2)	8500	8	33	4	4.30***	79**	0.93	75

Control samples, not treated with N_2O, hand pollinated by cv. Nigrum; N_2O (T1), treated with N_2O for 4 h, from 4 to 8 h after pollinating by cv. Nigrum; N_2O (T2), treated with N_2O as in (T1) but before pollinating by cv. Nigrum.
*, **, ***, Significant at the 5, 1, 0.1 % level compared with control sample.

Hypothesis as to the origin of haploid embryos

For identifying haploids in experimental populations the pollen parent (cv. *Nigrum*) carries four semi-dominant marker genes A, MoA, C and up. Three androgenetic plants have so far been isolated among the solitary haploids of which two were in premeiotic N_2O treatments of the female parent. In the n-2n twins, the 2n plants are always hybrid, indicating a normal process of fecundation, as has been shown previously by Morgan & Rappleye (1954). The haploid sibling is always maternal. Parthenogenetic non-twin haploids appear at a rate which is about ten times lower than that of haploid twins, whatever the cause of the parthenogenetic trend (genotypic effect or N_2O treatment). One can therefore suppose that they are dependent on the same rare event. It can be seen, when studying the rate of haploidy in relation to the time of N_2O application, that the period of maximum sensitivity to treatment occurs two days before anthesis (Fig. 36.2A). At that time the female gametophyte has only one nucleus (meiosis occurs four days before anthesis). One can assume that normal differentiation of the macrosporocyte is modified by a change in polarity of the first divisions. A similar phenomenon can be seen in microspores following in vitro cultivation (Sunderland, 1974). The best candidates for embryogenesis seem to be the synergids. A synergid with two nuclei was observed several times in the A_2 (6–3) autodiploid line at the moment of gametic fusion. First division of the zygote appears only two or three days later.

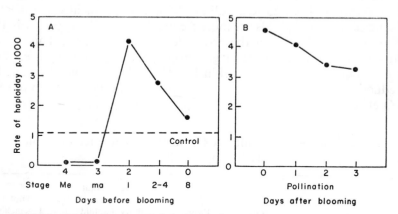

Figure 36.2. Study of the parthenogenetic induction (twins and non-twin haploids): effect of the stage of the female gametophyte on the response to N_2O treatment (A) and effect of delayed pollination (B). Autodiploid line DL-1B. A. N_2O treatment 0 to 4 days before blooming, pollination at blooming stage. B. Standard N_2O treatment 1 or 2 days before anthesis, pollination 0 to 3 days after blooming; emasculated flowers; male parent: cv. Nigrum. Me, Meiosis; ma, macrospores; 1, 2 . . . 8, number of nuclei in the female gametophyte.

The occasional n-n-2n triplets probably result from early cleavage of the haploid pro-embryo. It is interesting to note that one of the androgenetic haploids comprised two conjoined twins of the same genotype. The single haploids probably arise through very fast development of a parthenogenetic embryo which prevents growth of the zygotic embryo. The reverse is probably rather frequent. Very small torpedo-shaped embryos are often rejected during seed germination.

Further experimentation is necessary in order to detect any possible influence from the male gametophyte. Delayed pollination is not favourable to the production of haploid embryos (Fig. 36.2B).

PROPERTIES OF HAPLOID SPOROPHYTES

Haploid phenotypes

Any experienced observer can easily recognize haploid genotypes in *Capsicum annuum*, as early as the 4–5 leaf stage, without using marker genes. This property also helps towards recognition of diploidized shoots following treatment of haploid plants with colchicine.

The main characteristic is the formation of narrow, dull leaves. For example, in the Bastidon cultivar the length/width ratio is 1.93 in haploids, 1.65 in diploids and 1.30 in tetraploids (8th leaf on the main axis). Stomatal cell size is not a good differentiation criterion: haploid and diploid plant distributions overlap amply. On an average, the 2 n/n ratio, which theoretically borders on the cube root of 2, that is 1.26, in fact varies between 1.29 and 1.42. As regards the average surface of the epidermal cells, measured on adult leaves of a plant, the 2n/n ratio tends to increase from 2nd to 8th node above the first flower. The same is true when 4n and 2n are compared: cellular growth seems to be prolonged when the amount of DNA per cell is greater (Fig. 36.3).

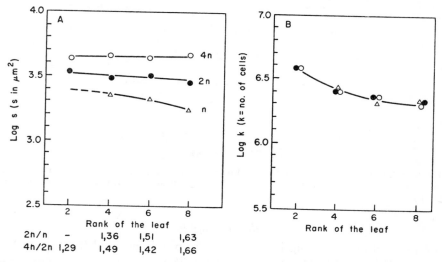

Figure 36.3. Comparison of n, 2n and 4n plants of the same cultivar: mean area and number of epidermal cells in the adaxial face of the leaves (cv. Bastidon). A. Log of the mean area (μm^2), stomata excluded. B. Log of the number of cells, stomata excluded; rank of the leaves: number of nodes above the first flower.

In n, 2n and 4n plants of the same genotype the average number of epidermal cells (adaxial face) is identical, at the same node, regardless of the level of ploidy. We do, however, know that varietal differences are very wide with regard to this character. The adult leaf surface, being essentially related to the number of cells, varies in the same manner as fruit size (Pochard, 1966). Thus, for a leaf at the second node (the one that usually carries the greatest number of cells), 0.7 and 3.4 million cells are to be observed in Enomi and Bastidon respectively (average weight of fruit 3 and 250 g). Whatever the rank of the leaf, this difference is consistent (Fig. 36.4). It can be inferred from this that the number of cells per leaf is controlled by specific genes and is not dependent on the amount of DNA in the cells.

Rapid ageing of the leaves of haploid plants can be noted: these turn yellow and drop off quickly. Susceptibility to powdery mildew and to red spider mite is extreme. On the contrary, tetraploids age slowly and have a better state of health than diploids. Shifriss (1974) noticed a similar phenomenon in *Ricinus communis* and claimed that the rate of senescence was slower when there was a greater amount of DNA per cell. It is probably the same trend which makes haploids completely parthenocarpic; about all the flowers set fruit and the plants covered in very small fruits. In an unfertilized diploid plant the parthenocarpic trend is weak until the plant is senescent.

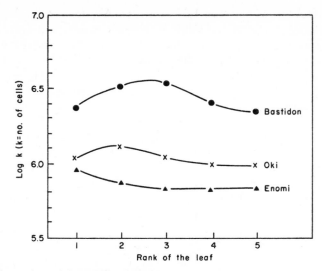

Figure 36.4. Comparison of three different cultivars for the number of epidermal cells in the adaxial face of the leaves. Bastidon, big fruited (250g); Oki, small fruited (40 g); Enomi, very small fruited (3 g).

Fertility of haploids

Seed production by selfing is unfrequent in purely haploid (monoploid) plants (Kimber & Riley, 1963; Magoon & Khanna, 1963). According to Jensen (1974), sterility could be related to the absence of dehiscence of the anther but dehiscence appears to be normal in haploid pepper. A small proportion of the pollen has normal appearance. It is actually functional: haploid plants completely isolated from all pollen can set seeds. In the original DL-1B haploid, maintained through vegetative propagation, a mean of 2.32 seeds per fruit was found in a sample of about 2000 fruits. Frequency of seeds does not follow the Poisson distribution: fruit with 0 or with from 5 to 10 seeds are in excess. (Fig. 36.5). As the branches are crowded with fruit, great quantities of seeds can be got from a single haploid

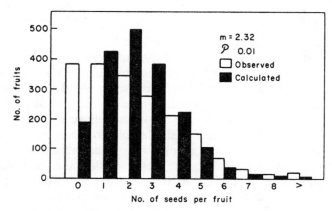

Figure 36.5. Distribution of the seeds in the fruits of the haploid plant DL-1B.

plant (80,000 from 300 cuttings of DL-1B: this is probably a fertility record for a monoploid plant). Morrison (1932) reported the occurrence of 10 seeds in the fruits of a haploid *Lycopersicon esculentum* cv. Marglobe. It is unlikely that spontaneous doubling of small sectors could account for that fertility. The haploid state appears very stable in the pepper, as well as in the tomato (Lindström, 1941). When doubling is achieved, diploid sectors soon become very large and stable. More probably, formation of a restitution nucleus during the first meiotic division enables the occurrence of normal gametes.

Meiosis in haploid pepper was studied by Christensen & Bamford (1943). They reported frequent associations between non-homologous chromosomes at the pachytene stage. In MI, chromosomes frequently appear in groups of 2, 3 or more. It was not possible to see if these associations were formed at random or if any particular chromosomes were affected. This type of pairing was observed by Sadasivaiah & Kasha (1973) in barley haploids. In the two reviews published in 1963, numerous examples of the same phenomenon are quoted. DNA exchanges between non-homologous chromosomes seem to be possible. We were able to confirm Christensen & Bamford's findings. Only a few pollen mother cells display 12 univalents during the first metaphase. However, no true bivalent associations are to be seen. Displacement of the chromosomes during first anaphase is most irregular. Chromatid bridges are clearly visible in more than 5% of the cells. Second division begins before completion of AI; the 12 chromosomes sometimes do not move and second mitosis seems to be normal, producing a dyad of the 12 : 12 or 11 : 13 type. A single giant restitution nucleus is sometimes formed, associated or not with micronuclei: the tetrad is then converted into a 'monad'.

Progeny of haploids

Chromosome countings show that aneuploids do, in fact, arise in big numbers (Table 36.4). In the DL-1B progeny 3.5% of the plants were found to contain from 25 to 28 chromosomes and 2.8% had from 35 to 37. It has therefore been possible to obtain 72 plants carrying 25 chromosomes. 65 of them have been sorted into phenotypic groups corresponding with primary trisomics. The properties of these trisomics have been published elsewhere (Pochard, 1970a). We have thus in hand a tool for making out the chromosome map of the pepper. The A, MoA, L^i, up, and C genes have already been located as well as two xantha mutants (Pochard, 1977). All these trisomics share the same basic genetic stock, as they spring from a single haploid plant (DL-1B). The analysis of certain quantitative characters

Table 36.4. Chromosome numbers in the selfed progeny of the haploid plant DL-1B

Chromosome number	12	24	25	26, 27, 28	35, 36, 37	Total	Twin haploids
Frequency of the plants	3	634	13	11	19	680	3
%	0.44	93.3	1.91	1.62	2.80	100	0.44

Cv. Doux des Landes, haploid plant propagated by cuttings, isolated in a greenhouse free from other pepper genotypes.

therefore seems possible. The fruit morphological type is comparable to that of the Cayenne pepper, but bigger. Through cross-breeding with a plant of the bell-pepper type, such as Yolo Wonder, very unequal distribution is to be observed in F2 offspring of the various trisomics as to the mean fruit weight. This hints at the existence of major genes or of gene groups which are accountable for the remarkable increase in fruit size in the bell-pepper type. The same holds true for the increase in fruit length. These investigations should be taken up again with the proper experimental equipment in order to minimize environmental influence.

Among the progeny of the DL-1B haploid are to be found approximately 0.4 % non-twin haploids and as many twins or triplets, that is 0.8 % to 1 % haploids in all. We therefore have a plentiful supply which could, for example, be used for trials in mutagenesis on dormant haploid embryos.

The phenotype of most of the 24 chromosome plants appears to be identical to that of the original variety and of the autodiploid lines obtained through colchicine doubling. However it is obvious that some plants are abnormal for many quantitative features and are more or less fertile.

Internal changes undergone by the genome following irregular exchanges probably account for this variation which it would be interesting to investigate in detail. The manifestation of haploidy is without doubt the most frequent mutation in *Capsicum annuum* (10^{-3} to 10^{-4} at every generation). We notice that these haploids are fertile and are accountable for a variation which is very unlikely to occur in diploids. It can be deduced that the part played by haploidy in the natural evolution of this species could be very important.

UTILIZATION IN PLANT BREEDING

Before using haploidy on a large scale in breeding programmes, it is necessary to gain information on the doubling ability of haploids and, above all, on the properties of the autodiploid lines which can be derived from them. On a partially autogamous plant like the pepper, it is not sure that the use of haploidy can afford better results than the conventional pedigree method. Walsh (1974) demonstrated the superiority of the traditional method as regards the probability of favourable gene associations, over the method based on isolation of haploids in the F2 generation. What remains is the time saving, which is most appreciable!

Comparing, by simulation studies, the mean value of the autodiploid lines arising from haploids gathered in the n generation, to pedigree lines in the n + 1 generation, Feyt & Pelletier (1976) drew similar conclusions. According to these authors, the main advantage of the use of haploidy in a breeding programme lies in the possibility to check progress made in the course of successive generations of the pedigree selection. This possibility is of special interest in cases of low heritability and when overdominance and epistasis are to be found.

Beyond these basic problems lie the hazards inherent in the achieving of a truly homozygous condition.

Chromosome doubling

Difficulties are mainly of a technical nature because of the toxicity of colchicine. Applications on cuttings (Pochard & Dumas de Vaulx, 1971) by means of soaking (0·3 % colchicine during 6 h) or through the action of N_2O applied under pressure

(24 h at 6 bars) show uneven results. Thus, in one particular trial, diploid sectors have been obtained in 17 % of the cuttings treated with colchicine, for a recovery rate of 47 % and a total number of 725 cuttings. Now utilized is another method which shows more stable results without causing mortality among the treated plants. At the end of a month 85 % of the plants show at least one diploidized fertile sector. It is possible to renew the treatment of the plants which have stayed haploid, so that the final yield borders on 100 %. In this method the plants used are in the 8–10 expanded leaves stage. The top is cut off just above the 8th leaf so as to lift apical dominance and a drop of colchicine solution is applied to the wounds left by excision of all axillary buds. Around these wounds appear new buds which develop from a very small number of cells. Thus the diploid sectors cover large areas and are easily recognized. When the L2 ontogenic layer is doubled, fertility is completely restored in female as well as in male organs. When the L3 alone is doubled flowers and leaves become nearly normal but fertility is not restored. Fertile tetraploid shoots are frequently found.

Characteristics of autodiploid lines

Authors generally acknowledge that autodiploid lines have qualities which compare with those of inbred lines in autogamous species such as barley (Park, Reinberg, Walsh & Kasha, 1974; Fedak, 1976) but also in allogamous species like maize (Thompson, 1954; Chase & Nanda, 1965; Chase, 1969) or cotton (Meredith, Bridge & Chism, 1970). The very survival of haploid embryos and seedlings from which the autodiploid lines spring is a sign of the elimination of recessive deleterious genes.

In tobacco, Legg & Collins (1968) appraised the autodiploid lines derived from parthenogenetic haploids and then those from androgenetic haploids (Collins & Legg, 1974). They observed several generations (H1 and H2: first and second inbred generations following diploidization), comparing these with traditional lines (C). On an average, only slight differences are visible between C and H as regards the six quantitative characters appraised. No change can be discerned between H1 and H2 when all lines are considered together. However, within one family, two H2 lines can be observed which differ in all six parameters. The authors believe that this is the result of pleiotropic mutations. Homogeneity is good in the other families.

Results in the pepper are much less promising. In fact, fertility of the H1 generation is generally below normal. In the cross Yolo Wonder × L 107, 20 to 30 % of the H1 derived from haploids (sorted out in the F3 generation) were semi-sterile (Fig. 30.6). This phenomenon has not been observed in any traditional selection. In the same cross, the only H2 lines studied were the 14 best which produced a mean of 0.91 g of seed per fruit (0.53 to 1.43), F9 lines and parents yielding an average of 1.11 g per fruit (0.80 to 1.55). It would be of interest to know if the differences between lines for the level of fertility remain constant through successive generations of selfing or if they are confined to the first generation derived from haploid plants.

In a particular autodiploid line (Y × 107)3–4 it has been possible to bring to light an important quantitative change between H1 and H5. As a matter of fact this line was well placed among candidates for commercial propagation. It actually combined, in H1, a semi-oblong shape of fruit (8×12 cm, average weight 250 g) with a short main axis (105 % of the shortest parent). In H5, it is obvious that the stem of $(Y \times 107)3$–4 has considerably increased in size; the main axis is now much longer than that of Yolo Wonder (122 %), which deprived it of all interest.

In H1, the family $(Y \times 107)3$–i was made up of 9 autodiploid lines, all with a short axis (on an average 29·4 cm ± 1.5) similar to Yolo Wonder (29 cm) while L 107 had a very long stem (40 cm). The F2 diploid mother plant was also very short. Perhaps transitory maternal influence could account for such facts.

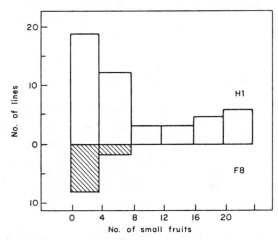

Figure 36.6. Comparison of H1 autodiploid lines to F8 inbred lines of the same cross: distribution of the different genotypes with regard to the small, few seeded fruits.

If it is true that some quantitative characters may change through generations of sexual reproduction after a haplo-diploidization, it will have to be taken into account in the appraisal of their agronomic worth.

Absence of fertilization and abnormal embryogenesis responsible for the emergence of haploid plants could favour the expression of certain maternal cytoplasmic characters. If, on the other hand, colchicine doubling restores the chromosome set to normal, it does not permit immediate achievement of nucleo-cytoplasmic balance.

In spite of this two-fold handicap (lack of fertility and instability) most of the autodiploid lines of *Capsicum annuum* are of great interest. On an average, their vigour is not weaker than that of inbred lines. In the cross observed by Pochard & Dumas de Vaulx (1971) some of them showed yield capacities close to those of the F1 hybrid derived from the two parents. It was moreover possible to establish a direct correlation between the average weight of fruit, estimated from the haploids themselves, and the same measurement taken from their H1 diploid progeny. The shape and pigmentation of fruit from haploid plants also allow for proper differentiation.

Several successive cycles of haplo-diploidization can be performed in order to test the genetical stability of the autodiploid lines. Theoretically, all progeny springing from the same truly homozygous plant should be identical unless mutations intervened during the successive generations. Do these mutations occur at a significant rate at each generation?

The few results already available show the occurrence of an unexpected variability through two successive haplo-diploidizations.

In 1975, significant quantitative differences were noticed between sister lines in the two studied families. Heterogeneity seems to be the rule rather than homogeneity which appears much below that of traditional inbred lines (Table 36.5).

Are the observed modifications stable? What is their origin? We cannot answer these questions but we believe these phenomena to be worth studying.

Mutagenesis in haploid plants

The genetic variability of *Capsicum annuum* is very remarkable. It can be further increased by crossbreeding with several related species (*C. chinense, C. frutescens, C. baccatum*). In spite of this, it could be of interest to bring about useful mutations in the breeding material. Besides, it is theoretically possible, owing to the haploid condition, to apply selective pressure to favour the occurrence of recessive mutations: for example, adjustment to climatic constraints, or lower susceptibility to a particular pathogenic agent. In practice, the difficulties in experimental mutagenesis are linked to the greatly increased susceptibility of the haploid plants towards physical or chemical agents and to the competition which arises between normal and mutated cells or tissues.

In order to obtain non-chimerical mutations, the gametophytes must be treated before DNA replication of the nucleus in the embryo forming cell. Devreux & Nettancourt (1974), treating tobacco uninucleate microspores at the G2 stage with X-rays obtained, among 150 haploid plantlets, 30 mutant phenotypes, part of them bearing chromosomal aberrations.

In another experiment in which radiation doses varied between 5 and 2000 rads, out of 84 haploid plants analysed, 12 were considered as mutants.

In cases of parthenogenesis like the one we are studying, it is rather difficult to apply the mutagenic treatment to the gametophyte or to the resting egg cell, as Cornu (1970) showed in diploid *Petunia hybrida*. We do not know the moment and the sequence of events leading to parthenogenesis. After a few preliminary trials, we studied the effect of γ rays on the cuttings (900 rads, at a flow rate of 4 rads a minute, enough to destroy half the DL-1B haploid cuttings) and that of EMS (0.2 %) applied by soaking the seeds. The latter method is most effective. Out of a sample comprising 210 haploid survivors, all direct offspring of DL-1B through self-fertilization, 198 sectorial mutations were discernible, with some plants showing two or three separate mutated sectors. Of these mutations, 168 were classified in four groups according to type: mutations in the photosynthetic apparatus, changes in stem and leaf structure, alteration of the floral organs and miscellaneous (Table 36.6). We only concerned ourselves with the third group and, within this, with cases in which the stamens were indehiscent and entirely wanting in pollen. Eight cases were identified and we proceeded to diploidize the sectors involved. Female fertility was then checked and 4 mutations: 403, 509, 705 and 712 were regarded as male-sterile and female-fertile. We were able to show that

Table 36.5. Differences between sister autodiploid lines after two haplo-diploidiz-ations and two generations of selfing (H2-H2 lines); comparison with the inbred lines of the original parents

5A. Families A2 (6-j) and B9 (2-j): four sister lines

| | H2 lines (i–j) | | | | family |
	(6–3)	(6–6)	(6–7)	(6–18)	A2 (6–j)
1	187	199*	184	188	L
2		0.02–0.05		> 0.50	
1	72	69	71	70	W
2		0.30–0.50		> 0.50	
1	2.60	2.88**	2.61	2.67	L/W
2		0.001–0.01		0.30–0.50	
1	0.49	0.57**	0.52	0.44**	Pe/Bl
2		0.001–0.01		0.001–0.01	

	(2–1)	(2–4)	(2–8)	(2–10)	B9 (2–j)
1	218**	248	234	243	L
2		0.01–0.001		0.10–0.20	
1	79**	87	87	79**	W
2		0.01		0.01–0.001	
1	2.76	2.85	2.71	3.09**	L/W
2		0.20–0.30		< 0.001	
1	0.56	0.61	0.56	0.63	Pe/Bl
2		0.20		0.05–0.10	

5B. Control: four inbred lines of the original parents

| | Inbred lines | | | | Varieties |
	3/75	4/75	6/75	7/75	Yolo Wonder
1	216	214	216	213	L
2		> 0.50		> 0.50	
1	91	91	92	90	W
2		> 0.50		0.50	
1	2.39	2.37	2.35	2.36	L/W
2		> 0.50		> 0.50	
1	0.53	0.55	0.59	0.58	Pe/Bl
2		0.50		> 0.50	

	5/75	6/75	8/75	9/75	L 107 ms
1	247	238	241	239	L
2		0.20		> 0.50	
1	80	77	74	77	W
2		0.10–0.20		0.20–0.30	
1	3.11	3.12	3.27	3.17	L/W
2		> 0.50		0.20–0.30	
1	0.70	0.73	0.77	0.67	Pe/Bl
2		0.30–0.50		0.02	

Measurement of adult leaves: L, total length of the leaf; W, width; Pe, length of the petiole; Bl, length of the leaf blade; (1), mean value (mm) (9 leaves, 1 leaf per plant); (2), probability for a random deviation (comparison of two adjacent lines).

Table 36.6. Classification of the mutations obtained on haploid plants selected in the progeny of the haploid DL-1B

I	37	*chlorophyll deficiencies* (albinos, xantha, viridis, virescent, dotted leaves).
(37%)	20	*chlorosis* (stem, margin of the leaves, old leaves . . .)
	5	*dull leaves* (dim, greyish colour)
II	51	*deformation of the leaves* 34 (curled, crinkled, folded)
(37%)		17 (broader and narrower leaves, changes in the bearing of the leaves)
	11	*changes in the length of internodes*
III	12	*modifications of the flower* (corolla, pedunculus . . .)
(17%)	8	*empty* or *non-dehiscent anthers* (4 male-sterile, female fertile)
	8	*change in the fruit shape or colour*
IV	16	*Miscellaneous:* hairiness, semi-lethality . . .
(9%)		

Seeds treated by EMS; 168 sectorial mutations on 190 haploid plants. I, Modification of the photosynthetic apparatus; II, structural modifications of the stems and leaves; III, modifications of flowers and fruit; IV, miscellaneous.

male-sterile mutations 509 and 705 are completely recessive and transmitted as mendelian characters. Male-sterile 509 has now been introduced into commercial hybrids and its use makes way for the extension of hybrid formulas (Pochard, 1970b, c; Breuils & Pochard, 1975). Male-sterile mutations thus seem to occur frequently: approximately one in 50 observable mutations. Twelve have been described to date (Daskaloff, 1973; Shifriss, 1973). However, the use of haploid material affords a considerable saving of time and means compared to traditional methods, which require inspection of at least 50 to 100 M_2 plants for just one M_1 plant, owing to a limitation of transfers to small sectors. Moreover a generation can be saved and there is no need for self-fertilization of all the flowers which is a tricky and rather ineffective process in the pepper because of the fragility of the flowers.

CONCLUSION

When induction methods for large scale androgenesis by in vitro culture become available, *Capsicum annuum* will afford a highly valuable material for research, probably on a level with tobacco. It will then be possible to compare the results of gynogenesis and androgenesis and, therefore, to investigate the effects which are peculiar to one source or to the other.

The few results already obtained on haploid peppers and on the derived auto-diploid lines afford insight into the lack of stability of the homozygous material during the successive generations of selfing and, mainly, by recurrent passing through the haploid state. This was unexpected and hence poorly studied in the preliminary experiments. It would be interesting to assess accurately the level of these effects and to know if they also appear through androgenesis performed in vitro. Nothing like this has been reported in tobacco arising from androgenesis, neither does this phenomenon seem to have been observed in barley (Park *et al.*, 1974). The origin of barley haploids is, however, quite different because they spring from interspecific hybridization: fertilization followed by nuclear fusion occurred normally.

The same is not true in the pepper as the haploid is derived from an unfertilized cell of the embryo sac; only the endosperm and the diploid partner were involved in fertilization and syngamy.

In any case we now know that induction and handling of a great number of haploids is no insuperable problem in the pepper. There are lines of agronomic worth available, which give rise to progenies comprising from 0.6 to 1 % parthenogenetic haploids. Research is being done to find out if some of these lines have, through their male gametophyte, the ability to induce development of haploid embryos in any material whatever. Owing to the high fertility of the pepper and to the possibility of vegetative reproduction, tens of thousands of seeds can easily be obtained from just one plant. An average rate of haploidy of 2 to 3 per 1000 would seem enough to give rise to at least one haploid among the direct progeny of any diploid plant.

Thus the production of homozygous lines can become just routine work requiring only two additional generations, if one considers H_2 as the first generation of use for the field experimentation. The breeder's main effort can then be directed towards improving the breeding system by maximizing the rate of recombination; rational use of haploidy could therefore lead to a complete change from traditional methods of breeding.

We have seen that haploidy affords numerous possibilities for discoveries in the field of physiology of growth and development and for induction and screening of mutations. Considerable effort remains to be made in research work so that all these promises may become reality.

ACKNOWLEDGEMENTS

We thank Mrs A. Florent and MM. G. Breuils and D. Chambonnet for their skilful technical assistance; we are grateful to Dr P. Pecaut for his valued advice and encouragement. We are indepted to Mrs E. Muratelle for the english translation and to Mrs M. Poitout for preparation of the manuscript.

REFERENCES

BLAKESLEE, A. F., BELLING, J., FARNHAM, M. E. & BERGNER, A. D., 1922. A haploid mutant in the Jimson weed, *Datura stramonium. Science, 55:* 646–647.

BREUILS, G. & POCHARD, E., 1975. Essai de fabrication de l'hybride de Piment Lamuyo-INRA avec utilisation d'une stérilité mâle génique (ms 509). *Annales de l'Amélioration des Plantes, 25:* 399–409.

CAMPOS, F. F. & MORGAN, Jr., D. T., 1960. Genetic control of haploidy in *Capsicum frutescens* L. following crosses with untreated and X-rayed pollen. *Cytologia, 25:* 362–372.

CHASE, S. S., 1949. Monoploid frequencies in a commercial doublecross hybrid maize and in its component singlecross hybrids and inbred lines. *Genetics, 34:* 328–332.

CHASE, S. S., 1969. Monoploids and monoploid-derivatives of maize (*Zea mays*). *Botanical Review 35:* 117–167.

CHASE, S. S. & NANDA, D. K., 1965. Comparison of variability in inbred lines and monoploid-derived lines of maize (*Zea mays* L.). *Crop Science, 5:* 275–276.

CHRISTENSEN, H. M. & BAMFORD, R., 1943. Haploids in twin seedlings of pepper, *Capsicum annuum* L. *Journal of Heredity, 34:* 99–104.

COLLINS, G. B. & LEGG, P. D., 1974. The use of haploids in breeding allopolyploid species. In K. J. Kasha (Ed.), *Haploids in Higher Plants.* University of Guelph, Canada.

CORNU, A., 1970. *Recherches sur l'induction et l'utilisation de mutations somatiques chez le* Petunia hybrida. Doctoral Thesis, University of Dijon, No. 4667 CNRS, 1970.

DASKALOFF, S., 1973. Three new male sterile mutants in pepper (*Capsicum annuum* L.). *Compte Rendu de l'Academie d'Agriculture de Bulgarie, 6:* 39–41.

DEVREUX, M. & de NETTANCOURT, D., 1974. Screening mutations in haploid plants. In K. J. Kasha (Ed.), *Haploids in Higher Plants*. University of Guelph, Canada.

DUMAS DE VAULX, R. & POCHARD, E., 1974. Essai d'induction de la parthénogénèse haploïde par action du protoxyde d'azote sur les fleurs de Piments (*Capsicum annuum* L.). *Annales de l'Amélioration des Plantes, 24:* 283–306.

FEDAK, G., 1976. Evaluation of doubled haploids in barley. *Zeitschrift für Pflanzenzüchtung, 76:* 147–151.

FEYT, H. & PELLETIER, G., 1976. Haploïdie et sélection. *Annales de l'Amélioration des Plantes, 26:* 365–386.

GUHA, S. & MAHESHWARI, S. C., 1964. *In vitro* production of embryos from anthers of *Datura. Nature, 204:* 497.

HOUGAS, R. W., PELOQUIN, S. J. & GABERT, A. C., 1964. Effect of seed parent and pollinator on frequency of haploids in *Solanum tuberosum. Crop Science, 4:* 593–595.

JENSEN, C. J., 1974. Chromosome doubling techniques in haploids. In K. J. Kasha (Ed.), *Haploids in Higher Plants*. University of Guelph, Canada.

KIMBER, G. & RILEY, R., 1963. Haploid Angiosperms. *Botanical Review, 29:* 480–531.

LEGG, P. D. & COLLINS, G. B., 1968. Variation in selfed progeny of doubled haploid stocks of *Nicotiana tabacum* L. *Crop Science, 8:* 620–621.

LINDSTRÖM, E. W., 1941. Genetic stability of haploid, diploid, and tetraploid genotypes in the tomato. *Genetics, 26:* 387–397.

MAGOON, M. L. & KHANNA, K. R., 1963. Haploids. *Caryologia, 16:* 191–234.

MEREDITH, N. R., Jr., BRIDGE, R. R. & CHISM, J. F., 1970. Relative performance of F1 and F2 hybrids from doubled haploids and their parent varieties in upland cotton, *Gossypium hirsutum* L. *Crop Science, 10:* 295–298.

MONTELONGO-ESCOBEDO, H. & ROWE, P. R., 1969. Haploid induction in potato: cytological basis for the pollinator effect. *Euphytica, 18:* 116–123.

MONTEZUMA de CARVALHO, J., 1967. The effect of N_2O on pollen tube mitosis in styles and its potential significance for inducing haploidy in potato. *Euphytica, 16:* 190–198.

MORGAN, D. T. & RAPPLEYE, R. D., 1950. Twin and triplet pepper seedlings. A study of polyembryony in *Capsicum frutescens. Journal of Heredity, 41:* 91–95.

MORGAN, D. T. & RAPPLEYE, R. D., 1954. A cytogenetic study of origin of multiple seedlings of *Capsicum frutescens. American Journal of Botany, 41:* 576–586.

MORRISON, G., 1932. The occurrence and use of haploid plants in the tomato with special reference to the variety Marglobe. *International Congress on Genetics, 1932* (2): 137.

PARK, S. J., REINBERGS, E., WALSH, E. J. & KASHA, K. J., 1974. Comparison of the haploid technique with pedigree and single seed descent methods in barley breeding. In K. J. Kasha (Ed.), *Haploids in Higher Plants*. University of Guelph, Canada.

POCHARD, E., 1966. Données expérimentales sur la sélection du Piment (*Capsicum annuum* L.). *Annales de l'Amélioration des Plantes, 16:* 185–197.

POCHARD, E., 1970a. Description des trisomiques du Piment (*Capsicum annuum* L.) obtenus dans la descendance d'une plantes haploïde. *Annales de l'Amélioration des Plantes, 20:* 233–256.

POCHARD, E., 1970b. Obtention de trois nouvelles mutations de stérilité mâle chez le Piment par traitements mutagènes appliqués à un matériel monoploïde. In B. Schweisguth (Ed.), *La Stérilité Mâle Chez les Plantes Horticoles*. Rept. EUCARPIA meeting, Centre National de la Recherche Agronomique, Versailles.

POCHARD, E., 1970c. Etude de la gamétogénèse chez les plantes normales, les plantes possédant la stérilité cytoplasmique Peterson et les mutants mâles-stériles mr 9, mc 509, mc 705 (*Capsicum annuum* L.). In B. Schweisguth (Ed.), *La Stérilité Mâle Chez les Plantes Horticoles*. Rept. EUCARPIA meeting, Centre National de la Recherche Agronomique, Versailles.

POCHARD, E., 1977. Localization of genes in *Capsicum annuum* L. by trisomic analysis. *Annales de l'Amélioration des Plantes, 27:* 255–266.

POCHARD, E. & DUMAS de VAULX, R., 1971. La monoploïdie chez le Piment (*Capsicum annuum* L.) *Zeitschrift für Pflanzenzüchtung, 65:* 23–46.

SADASIVAIAH, R. S. & KASHA, K. J., 1973. Non-homologous association of haploid barley chromosomes in the cytoplasm of *Hordeum bulbosum* L. *Canadian Journal of Genetics and Cytology, 15:* 45–52.

SHIFRISS, C., 1973. Additional spontaneous male-sterile mutant in *Capsicum annuum* L. *Euphytica, 22:* 527–529.

SHIFRISS, C., 1974. Regulation of plant growth in an autopolyploid series (an hypothesis). In K. J. Kasha (Ed.), *Haploids in Higher Plants*. University of Guelph, Canada.

SMITH, P. G. & HEISER, C. B., 1951. Taxonomic and genetic studies on the cultivated peppers *C. annuum* L. and *C. frutescens* L. *American Journal of Botany, 38:* 362–368.

SUNDERLAND, N., 1974. Anther culture as a means of haploid induction. In K. J. Kasha (Ed.), *Haploids in Higher Plants*. The University of Guelph, Canada.

THOMPSON, D. L., 1954. Combining ability of homozygous diploids of corn relative to lines derived by in-breeding. *Agronomy Journal, 46:* 133–136.

TOOLE, M. G. & BAMFORD, R., 1945. The formation of diploid plants from haploid peppers. *Journal of Heredity, 36:* 67–70.

WALSH, E. J., 1974. Efficiency of the haploid method of breeding autogamous diploid species: a computer simulation study. In K. J. Kasha (Ed.), *Haploids in Higher Plants.* The University of Guelph, Canada.

WANGENHEIM, K. V. von, PELOQUIN, S. J. & HOUGAS, R. W., 1960. Embryological investigations on the formation of haploids in the potato (*S. tuberosum*). *Zeitschrift für Vererbungslehre, 91:* 391–399.

WANG, Y. Y., SUN, C. S., WANG, C. C. & CHIEN, N. F., 1973. The induction of pollen plantlets of *Triticale* and *Capsicum annuum* from anther culture. *Scientia Sinica, 16:* 147–151.

VIII. Biosystematics of genera and sections

37. The genus *Brunfelsia* : a conspectus of the taxonomy and biogeography

T. PLOWMAN

Botanical Museum, Harvard University, Cambridge, Massachusetts, U.S.A.

The subgeneric classification and general evolutionary trends in the genus *Brunfelsia* are discussed. The ecology, distribution and past history of the species of *Brunfelsia* are also considered in detail, with special emphasis on the South American species.

CONTENTS

INTRODUCTION

Brunfelsia is a relatively advanced member of the Solanaceae and has been placed in the tribe Salpiglossideae, based on its zygomorphic corolla, didynamous stamens, capsular fruit, straight embryo, and imbricate corolla aestivation. The genus consists entirely of shrubs and small trees, which are confined to the New World tropics. Several species are cultivated as ornamentals. A number of medicinal plants are also known and widely employed in folk medicine. Certain species are valuable remedies for rheumatism and syphilis, and one, *B. uniflora*, still appears in the Brazilian pharmacopoeia. This and other species are thought to contain alkaloids, but the compounds have not been characterized chemically.

About 40 species of *Brunfelsia* are recognized, 21 of which are native to the West Indies, mostly as island endemics. Nineteen species are found in tropical South America and Panamá. One species, *B. nyctaginoides*, was described from Guatemala and southern Mexico but is an anomalous plant which has been excluded from *Brunfelsia* (Plowman, 1973a).

According to my studies, *Brunfelsia* consists of three distinct sections, one of them hitherto unrecognized. Section *Brunfelsia*, based on the type species *B. americana* (Fig. 37.1), is confined to the West Indies, and species occur on all the major islands. This group is characterized by a very long, narrow corolla tube, up to 24 times as long as the calyx, and by a clavate, bilobate stigma. The flowers are almost always white, fading to yellow with age, and are fragrant in the early evening—a typical Sphinx moth flower. The fruit of section *Brunfelsia* is a more

475

or less thick-walled capsule, usually orange or yellow in colour and dispersed by birds.

Section *Franciscea* was originally described as a separate genus based on *Brunfelsia uniflora* and three other Brazilian species. *Franciscea* differs from *Brunfelsia* in having a much shorter corolla tube in relation to the calyx, being only 1–5 times as long, and in having a distinctly bifurcate stigma. The flowers are always violet with a white 'eye' at the throat and fade to pale violet or pure white with age. They are rarely fragrant and are pollinated by butterflies. The fruits of section *Franciscea* are usually thin-walled and green. Section *Franciscea* includes some fifteen species, several of which are polymorphic and widely distributed in South America. Only one species, which is endemic in central Panamá, is found outside South America.

In uniting *Franciscea* with *Brunfelsia*, Bentham in 1846 astutely pointed out the intermediate nature of *B. guianensis* (Fig. 37.2), a species of the Guianas known to

BRUNFELSIA americana *L*.

Figure 37.1. *Brunfelsia americana*. Flowering branch with insert showing fruit.

have small, whitish flowers and a large, yellowish fruit. Miers, his contemporary and an avid student of the Solanaceae, opposed Bentham's view and again separated the two genera, transferring several species described as *Brunfelsia* back into *Franciscea* (1849, 1850). However, Miers conspicuously ignored Bentham's anomalous *B. guianensis* in presenting his arguments for maintaining *Franciscea*.

In the course of studying numerous herbarium specimens of *Brunfelsia* from northern South America, I discovered several plants which are closely related to *B. guianensis*, sharing with this species a series of characters which are similarly intermediate between *Brunfelsia* and *Franciscea*. These species comprise a new section which I have called *Guianenses* (Plowman, 1978), based on the type species *B. guianensis*. Section *Guianenses* is characterized by having relatively small, white to greenish flowers, with a corolla tube only 2–3 times as long as the calyx, and a bifurcate stigma. The fruits are either large, fleshy and yellowish, or small and

BRUNFELSIA guianensis *Benth.*

Figure 37.2. *Brunfelsia guianensis.* 1, Flowering branch; 2, fruiting branch; 3, corolla limb; 4, ovary and stigma; 5, longitudinal cut-away view of fruit; 6, seed with insert showing surface; 7, excised corolla tube showing stamens.

green. To date, four species have been collected which belong to this group, all occurring in northern South America and adjacent Panamá.

THE CARIBBEAN SPECIES

The three sections of *Brunfelsia* apparently were differentiated early in the history of the genus, probably in northern South America, which remains the geographical centre of *Brunfelsia*. At an early time, some populations were dispersed to and became established in the Lesser Antilles, an island archipelago formed by volcanic activity in the late Cretaceous or early Tertiary (Malfait & Dinkelman, 1972). These ancestral populations gave rise to section *Brunfelsia*, possibly as a result of ecological isolation caused by the development of extreme aridity in the southern-most members of this island chain (Fig. 37.4).

3 cm.

BRUNFELSIA
pauciflora (*Cham. & Schlecht.*) *Benth.*

Figure 37.3. *Brunfelsia pauciflora*. 1, Flowering branch; 2, fruiting branch; 3, capsule with excised calyx; 4, capsule showing seeds.

At the present time, only one species, *Brunfelsia americana*, is found naturally in the Lesser Antilles. This species reaches as far south as the island of St. Lucia and northward to Puerto Rico and the easternmost tip of Hispaniola. It is significant that *B. americana* is morphologically the least specialized species of section *Brunfelsia* and the only member of the section which has been dispersed to more than one island. This species exhibits relatively broad ecological tolerances and may represent a primitive form which has given rise to other species in the Caribbean area, such as *B. lactea*, a related species in Puerto Rico.

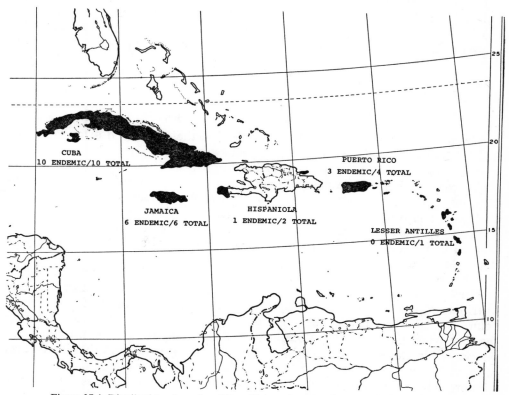

Figure 37.4. Distribution of species of *Brunfelsia* sect. *Brunfelsia* in the Caribbean region.

The remaining 20 species of section *Brunfelsia* are all island endemics, usually restricted to specialized habitats and local distributions. As a result, several of these species are very rare and in danger of extinction in the wild due to increasing pressures of human populations. In general, the size of the island determines the number of species encountered, following the model of island speciation and adaptive radiation.

In Puerto Rico, three endemics are found, in addition to *Brunfelsia americana*. *B. lactea* and *B. portoricensis* grow in wet, montane forests. *B. densifolia* occurs very locally on serpentine outcrops in western Puerto Rico. Only two species are known from Hispaniola: *B. americana* and *B. picardae*, the latter a very rare plant from southwestern Haiti.

Six species of *Brunfelsia* are native to Jamaica. Four of these grow in limestone woodlands at middle elevations: *B. maliformis*, *B. plicata*, *B. splendida*, and *B. membranacea*. Of these, only *B. maliformis* is relatively common throughout the island, the others being of very local distribution. *B. undulata* is found only in limestone woods on the north coast near sea level. Although widely cultivated, it is a rare plant in nature and should be considered an endangered species. Finally, *B. jamaicensis* is adapted to higher-altitude cloud forests in the Blue Mountains of eastern Jamaica.

In Cuba, about ten species of *Brunfelsia* occur, several of which are known from only a few collections. An especially rich area on this island is Oriente Province where seven species of *Brunfelsia* have been reported. Among these is *B. cestroides*, a species unique in the genus in having reddish flowers, which possibly are adapted for hummingbird pollination. Since it has not been possible to conduct field studies on the Cuban species, little is known of their ecology or actual distribution on the island.

THE SOUTH AMERICAN SPECIES

In comparison with the Caribbean brunfelsias, the origin and dispersal of the South American species is less readily understood. The present distribution of species suggests a complex pattern of evolution resulting from geographic and ecological isolation, wide ranging dispersal and long distant disjunctions. Particularly intriguing are those species of the circum-Amazonian region which have evolved in relatively uniform tropical forest habitats, where present ecological and geographical isolating mechanisms are minimal. These patterns of distribution can only be understood in the context of the past geological and climatic history of South America.

There is increasing evidence that major climatic changes have occurred in South America in the past, which are largely responsible for many plant and animal distribution patterns observed today (for reviews see Vuilleumier, 1971; Langenheim, 1973; Haffer, 1974). Beginning about the mid-Tertiary (Miocene) during the time when the Andes were being uplifted, a drying trend began, resulting in semi-arid climates which persisted into the Pliocene. It is thought that large areas of forest such as occur today in the Amazon basin were disrupted by these changes, with profound effects on the flora and fauna which inhabited the areas. Subsequent to the initial dry period, several major fluctuations of wet and dry periods apparently took place during the Pleistocene, which coincided with periods of glaciation.

In studies on animal speciation patterns in the Amazon, Haffer (1969, 1974) working with birds, and Vanzolini & Williams (1970) working with lizards, concluded independently that during recurrent dry periods, the presently immense Amazon forest was fragmented into smaller areas of wet forest habitat. These areas served as refugia for flora and fauna which were unable to adapt to newly formed, more arid and open habitats. Within these forest refugia, speciation continued under local selection pressures. When the climate once again became wetter, these restricted pockets of forest expanded over larger areas, bringing populations into contact once more. In such areas of secondary contact, Vanzolini & Williams (1970), using a statistical analysis of several characters, discovered complex patterns of variation. Similar patterns of variation and hybridization in proposed forest refugia have been observed in Heliconian butterflies by Brown & Mielke (1972).

Recently, botanists working on Amazonian groups have begun to perceive parallel patterns of distribution and variation in plants. Langenheim (1973) working with *Hymenaea* (Leguminosae), and Prance (1973) in comparing four families of Amazonian trees, cite areas of endemism and morphological differentiation which correspond strikingly to the refuge areas proposed by zoologists. It is significant, however, that the exact location and size of different refugia vary for each group of organisms under study. Prance (1973) and Haffer (1974) have summarized the location of refugia as proposed by various authors.

In contemplating the speciation patterns in *Brunfelsia* in South America, especially in the circum-Amazonian region, the concept of forest refugia offers an interesting model for interpreting the distribution of species and subspecies. Not a few taxa in *Brunfelsia* are located in areas which coincide closely with proposed refuge areas. Wide disjunctions in populations in certain species may also be explained by the effects of drastic climatic changes in the past.

Our knowledge of the actual distribution of many species of both plants and animals in tropical South America is still very incomplete owing to sparse collecting in certain areas and a total lack of collecting in others. Considerable caution is therefore necessary in interpreting plant distributions based upon extant collections. Broad and potentially erroneous conclusions may be based on what are merely artifacts of methodology. In order to determine the location and extent of endemic areas and refugia more correctly, additional data from many plant and animal groups must be collected and correlated with geological and paleoclimatic data. Until now, our knowledge of the geology and past climatic history of the Amazon is scanty, yet this information is necessary to corroborate theories derived from plant and animal distributions.

With these thoughts in mind, we can return to our discussion of the origin and evolution of *Brunfelsia* species in South America. There is in general very little overlap in the areas occupied by sections *Guianenses* and *Franciscea*, at least so far as present distributions are known. In areas where the two sections do overlap, such as in Bahia and Amapá in Brazil, the species may be ecologically distinct. Section *Guianenses* predominates in the Amazon basin and the Guianas; section *Franciscea* in the Andes and central and southern Brazil.

Section *Guianenses* apparently arose in the eastern Amazon and Guiana region, which is today the centre of distribution of the species. All the species of this group grow in wet, tropical lowland forests (Fig. 37.5). *Brunfelsia guianensis* occurs throughout Surinam and French Guiana, southward across the Amazon to just south of Belem do Pará and westward to the Rio Trombetas.

Brunfelsia martiana is known from a few scattered collections along the Amazon River from Belem westward to the Rio Japurá. Disjunct populations of *B. martiana*, known from two collections made over 100 years ago, have also been recorded from Bahia in the Atlantic rain forest of coastal Brazil. At the present time, the Amazon and Atlantic rain forests are effectively isolated by arid, open formations of northeastern Brazil. During a wetter period in the past, these two regions were apparently connected by more or less continuous wet forest environments which permitted the interchange and dispersal of species. Rizzini (1967) lists some 70

Amazonian species of plants which have been found in northeastern Bahia south
to Espiritu Santo in the Atlantic forest. Similar distributions have been pointed
out in Anolid lizards by Vanzolini & Williams (1970).

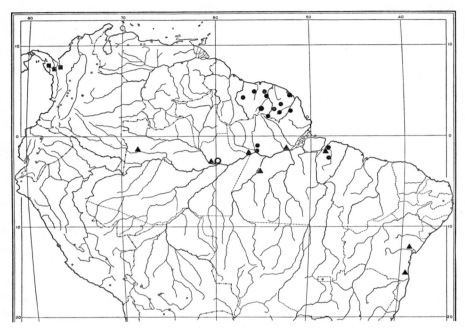

Figure 37.5. Distribution of species of *Brunfelsia* sect. *Guianenses* in northern South America.
■, *B. chocoensis*; ●, *B. guianensis*; ▲, *B. martiana*; ○, *B. amazonica*.

The third Amazonian species of section *Guianenses*, *Brunfelsia amazonica*,
grows only near Manaos in the central Amazon, in a very restricted area. Other
species and subspecies of plants in various families also show endemism at this
locality, e.g. *Hymenaea reticulata* (Leguminosae), *Virola multinervia* (Myristi-
caceae) and *Cariniana integrifolia* (Lecythidaceae).

Brunfelsia chocoensis is endemic in the basin of the Río Atrato in northwestern
Colombia and on Cerro Pirre in adjacent Panamá. This species presumably arose
after the final major uplift of the Andes in the late Pliocene-early Pleistocene, when
the lowland forests of the entire Chocó region were effectively cut off from the
Amazon basin. This area is well known for its high percentage of endemic species,
many of which show strong affinities with Amazonian species (Cuatrecasas, 1957).
In addition, the Amazon and Chocó regions have many species in common.

With a larger number of species and covering a much greater geographic and
ecological range, section *Franciscea* represents a complex history of speciation and
dispersal which occurred in different areas at different times. At present, two major
centres of speciation are recognized in this group—one in the eastern Andes
(Fig. 37.6), the other in southeastern Brazil (Figs 37.7–37.10). The greater number of species is found in southeastern Brazil, but the more primitive species

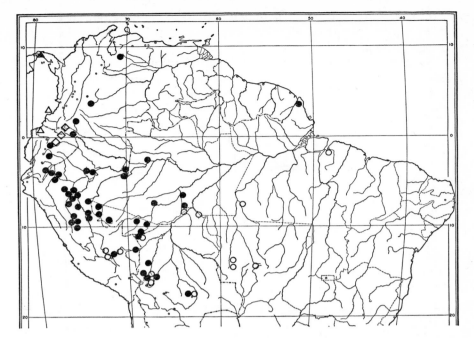

Figure 37.6. Distribution of species of *Brunfelsia* sect. *Franciscea* in northern South America. ▲, *B. dwyeri*; △, *B. macrocarpa*; ●, *B. grandiflora*; ◇, *B .chiricaspi*; ○, *B. mire.*

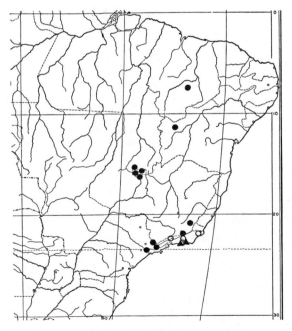

Figure 37.7. Distribution of species of *Brunfelsia* sect. *Franciscea* in eastern Brazil. ●, *B. obovata*; ▲, *B. latifolia*; ○, *B. bonodora.*

appear to be of Andean origin. These two regions today have one species and two pairs of vicarious species in common, indicating that gene flow existed between these areas in the past.

In view of the present widely scattered distribution of the species of section *Franciscea* and our general lack of data on the past history of the continent, it is difficult to pinpoint the area of origin of the group. Certainly some species occupied a much larger area in the past than at present, as a consequence of changing

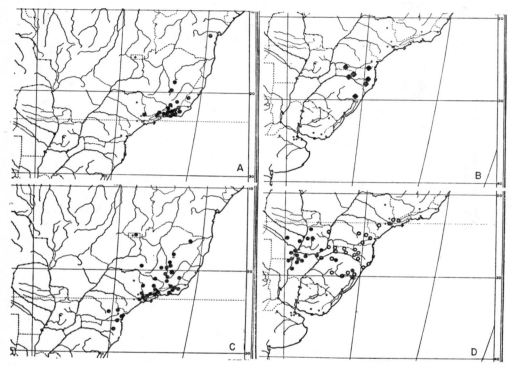

Figure 37.8. Distribution of species of *Brunfelsia* sect. *Franciscea* in southeastern Brazil. A, *B. hydrangeiformis*; B, *B. cuneifolia*; C, *B. brasiliensis*; D, ●, *B. australis*; ○, *B. pilosa*.

environmental conditions. It seems probable, however, that section *Franciscea* was derived from ancestral stock of section *Guianenses*, possibly through ecological isolation in adapting to upland habitats in the Andes or Brazilian highlands. More favourable climatic conditions in the past undoubtedly allowed for the dispersal of species to areas where they are now excluded.

Before the Andes became a significant barrier to gene flow, some ancestral stock of section *Franciscea* was dispersed to the Pacific coast of northern South America, where two isolated and now local species are found. Both of these are unique for section *Franciscea* in having relatively thick-walled fruits. *Brunfelsia macrocarpa* has a large, fleshy, yellow fruit recalling that of *B. guianensis* and some of the West Indian species. *B. macrocarpa* is known only from the wet coastal forests of northern Ecuador (1 collection) and from Gorgona Island off the coast of southern Colombia (3 collections). This area, which is the southern extremity of the Chocó phytogeographic province, has probably been little changed climatically since before the Andean orogeny. *B. macrocarpa* is comparable in its geographic origin to *B. chocoensis* of section *Guianenses*.

Brunfelsia dwyeri is allied to *B. macrocarpa* and restricted to Cerro Jefe in central Panamá, an area known to be rich in endemic species (Lewis, 1971). *B. dwyeri* is the only species of section *Franciscea* found outside South America and was probably isolated in the mountains of Panamá when the isthmus was repeatedly submerged and uplifted during the late Tertiary due to tectonic activity and rises in sea level (Bartlett & Barghoorn, 1973).

On the eastern slopes of the Andes, several species occur: *Brunfelsia grandiflora*, *B. chiricaspi*, *B. mire* and *B. uniflora*. Of these, *B. grandiflora* is the most widespread and is considered one of the most primitive species of section *Franciscea*, based on inflorescence, floral and indument characters. As in all remaining members of this section, the fruit is a thin-walled capsule. *B. grandiflora* is widely distributed in primary and secondary forests in western South America from Venezuela south to Bolivia and Brazil and is probably of Andean origin.

Two subspecies of *Brunfelsia grandiflora* have been recognized (Plowman, 1973b). *B. grandiflora* subsp. *grandiflora* has much larger flowers and fruits and grows in primary forests at elevations above 500 m in eastern Peru and Ecuador, although it has recently been found at lower elevations in western Brazil. *B. grandiflora* subsp. *schultesii* has a much broader ecological range and often survives in disturbed habitats, usually at elevations below 500 m. This taxon occurs over a larger geographic area in the foothills of the eastern Andes from Venezuela south to Bolivia and eastward to the Brazilian Amazon. In *B. grandiflora* subsp. *schultesii*, several morphologically distinct populations can be recognized in different parts of its range. These divergent populations are found in areas which closely correspond to proposed forest refugia, such as Villavicencio in Colombia, the Huallaga basin in Peru and the Yungas in Bolivia, apparently as a result of isolation in these areas during unfavourable climatic periods in the past.

Disjunct populations of *Brunfelsia grandiflora* subsp. *schultesii* have been found in the Oiapoque region of Amapá State in northern Brazil, some 2000 km from the nearest known populations in western Amazonia. Furthermore, a very similar but still poorly known species, *B. bonodora*, is reported from the State of Rio de Janeiro in southeastern Brazil. These disjunctions possibly represent relic populations of *B. grandiflora* which may have been more widely distributed in the past, with subsequent extinction of populations in the intervening areas during unfavourable climatic periods. Additional studies are needed to elucidate further the complex patterns of variation and geographical distribution observed in *B. grandiflora*.

Brunfelsia chiricaspi is closely related to *B. grandiflora* and probably arose from this species. *B. chiricaspi* is restricted to a narrow band east of the Andes from the Colombian Putumayo south to central Ecuador at 350–550 m elevation in primary forests. This region corresponds to the Napo—or more specifically the Putumayo—refuge area, which shows a high percentage of plant endemism. At present, *B. chiricaspi* is essentially sympatric with *B. grandiflora* in eastern Ecuador. Hybridization may occur, but additional studies are necessary to confirm this.

Brunfelsia mire is found in somewhat drier forests in the Andean foothills of southern Peru and Bolivia, and in Mato Grosso and Acre States of adjacent Brazil. Recently, collections have also been made in the Rio Tapajós and near Belem do Pará in the central and eastern Amazon respectively. A vicarious species, *B. hydrangeiformis*, occurs in humid forests of the Serra do Mar in southeastern Brazil.

Although it was first described less than 25 years ago, *Brunfelsia mire* is now known to be a wide ranging, polymorphic species of the circum-Amazonian area. Striking morphological differentiation is found in leaf and floral characters in each different isolated region where the species has been collected. Even in adjacent valleys in southeastern Peru (La Convención, Kosñipata) and northern Bolivia (Yungas), distinct forms of *B. mire* have evolved in partial isolation.

A fourth species of the eastern Andes is *Brunfelsia uniflora*, which here occurs only in fairly high elevation cloud forests in southern Bolivia and northernmost Argentina. *B. uniflora* is widely distributed in eastern Brazil and Venezuela. The Andean populations represent a wide disjunction and show appreciable ecological specialization and morphological differentiation from east Brazilian and Venezuelan populations.

It is evident from the distributions of *Brunfelsia uniflora* and other species mentioned above that gene flow existed in the past between the Andes of southern Peru and Bolivia and southeastern Brazil. The zone which intervenes between these areas is today a region of open vegetation formations with a high temperature and low rainfall, and includes parts of the northern Chaco and the interior mountains of the Brazilian shield. This area now serves as an effective barrier to the dispersal of many wet forest, montane species. Several workers (Goodspeed, 1954; Smith, 1962; Vuilleumier, 1971) have offered the opinion that this region may have been wetter and cooler during the past, probably in conjunction with periods of glaciation. A more favourable climate would have permitted the development of tropical forest and an avenue for dispersal and gene flow between the eastern Andes and southeastern Brazil. The cause of present aridity in the northern Chaco and adjacent areas is due in part to the rain shadow effect produced by the final uplift of the Andes (Smith, 1962), although climatic cycles may also have contributed to the general drying effect.

The coastal mountains of southeastern Brazil, the so-called Serra do Mar, are well known for their richness and endemism of plant and animal species, particularly in the ranges centered on Rio de Janeiro. The reasons for this great diversity are manifold. It is an area of rugged mountains and very high rainfall on the Atlantic-facing slopes; towards the interior, there is an increasingly drier climatic regime. The diverse topography and climate have created a wide range of habitats in a relatively small area, favouring ecological and geographic isolation of species. At least near the coast, conditions are nearly optimal for the development of the tropical forest ecosystem and apparently have remained so for a very long time (Tryon, 1944). Haffer (1974) suggests that the Serra do Mar was little affected by Pleistocene climatic cycles.

Once ancestral stocks of section *Franciscea* reached southeastern Brazil, they encountered optimal conditions for speciation. At least nine species originated in this general area, centered on Rio de Janeiro where six species occur; three others arose further south. It is clear that these species evolved *in situ* with the exception of two or three which are possibly of Andean origin.

Brunfelsia hydrangeiformis is a wet forest species ranging from Rio de Janeiro to São Paulo and adjacent Minas Gerais. As mentioned earlier, this species is closely related to *B. mire* and possibly arose from an early dispersal of *B. mire* into southeastern Brazil. Two distinct varieties of *B. hydrangeiformis* are recognized.

Another species of possible Andean origin is *Brunfelsia bonodora,* which as mentioned above is a close relative of *B. grandiflora. B. bonodora* is found only in primary forests near Rio de Janeiro but most nearly resembles populations of *B. grandiflora* subsp. *schultesii* from Bolivia. Another species which may have evolved from the same stock is *B. latifolia.* This species is very distinct in habit and is adapted to the maritime *restinga* vegetation near the city of Rio de Janeiro.

Brunfelsia brasiliensis is a polymorphic species which grows in a variety of habitats, including the wet, montane forests of the Serra do Mar and the drier, interior mountains of Minas Gerais, São Paulo and Distrito Federal. A distinct subspecies occurs at higher elevations in cloud forests of the Serra do Mar. *B. obovata* grows in the drier interior states but is ecologically specialized, usually growing in marshes (*brejos*), often in standing water. Another related species, *B. cuneifolia,* is found further south as an element of the *Araucaria* forests in the States of Santa Catarina and Paraná.

Brunfelsia pauciflora (= *B. calycina*) (Figs 37.3 and 37.9), one of the showiest

Figure 37.9. Distribution of *Brunfelsia pauciflora.*

members of the genus, grows naturally in very wet forests at low elevations near the coast from Espiritu Santo south to Santa Catarina. A widely disjunct sub-species has been collected in eastern Venezuela in the Imataca region, a forest refuge zone cited by Prance (1973). To date, no intervening populations of *B. pauciflora* have been discovered.

A somewhat similar situation is observed in *Brunfelsia uniflora* (Fig. 37.10). In

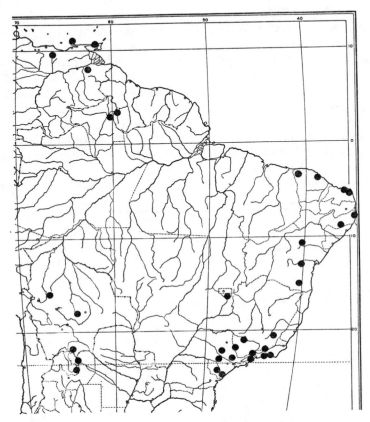

Figure 37.10. Distribution of *Brunfelsia uniflora*.

the Serra do Mar, this species grows in the understory of humid forests, recalling its habitat in the Bolivian Andes. However, it has also adapted to drier habitats northward along the coast of Brazil nearly to Belem do Pará, with consequent clinal variation in pubescence and calyx characters. *B. uniflora* has also been collected in sandstone areas in the Rio Branco-Roraima region of northern Brazil and Guyana, as well as along the north coast of Venezuela and Isla Margarita. The populations in the Roraima area are morphologically distinct and appear to have been isolated in this endemic area for some time.

Southwest from the Serra do Mar, ancestral populations of *Brunfelsia uniflora* gave rise to two distinct species, which are frequently confused with *B. uniflora* (Plowman, 1974). *B. pilosa* is adapted to the *Araucaria* forests of the planalto of

Santa Catarina and Paraná, but is also found in Rio Grande do Sul (Brazil) and Misiones Province (Argentina). *B. australis* evolved in the low lying basin of the Río Paraná in northern Argentina and Paraguay, an area which was inundated by the sea during interglacial periods (Vuilleumier, 1971). Both *B. pilosa* and *B. australis* are considered relatively recent species.

Several additional undescribed species of *Brunfelsia* are known to exist in central Brazil and Bolivia but their affinities and distributions are unknown due to lack of adequate collections. Future field work in these regions may shed new light on our knowledge of these species and on the phytogeography of *Brunfelsia* in general.

An interesting adjunct to these studies is the occurrence throughout the natural range of the South American brunfelsias of a genus of butterflies, *Methona* (Nymphalidae), the larvae of which feed exclusively on *Brunfelsia* leaves. *Methona* has recently been monographed by Lamas (1974). Although we have little data on host specificity, the various species and subspecies are found wherever *Brunfelsia* species occur in South America and Panamá, which the exception of the Chocó region. *Methona* is also absent from the Caribbean and apparently evolved as a genus after section *Brunfelsia* was isolated in the Caribbean area.

Based on distributional data, *Methona* species apparently feed on species of both sections *Guianenses* and *Franciscea*. Comparisons of the natural ranges of *Brunfelsia* and *Methona* reveal striking similarities, particularly in reference to the isolation of species and subspecies in forest refugia. Future collecting efforts which include the larvae and adults of *Methona* in conjunction with *Brunfelsia* species may reveal significant correlations in the evolution and distribution of these two groups.

SUMMARY

The genus *Brunfelsia* originated in northern South America and early differentiated into three distinct lines of evolution. Section *Brunfelsia* was isolated in the Caribbean region, entering this area most likely through the Lesser Antilles archipelago. Possibly due to subsequent arid conditions, this group was unable to re-invade South America and was dispersed to and formed additional species on the islands of the Greater Antilles.

In South America, the genus early split into two lines. One, ancestral to section *Guianenses*, developed in the Amazon lowlands, dispersing to the Pacific coast of Colombia and to the coastal forests of eastern Brazil. A second more successful group, giving rise to section *Franciscea*, adapted to upland forest habitats in the Andes and in southeastern Brazil. This group migrated to the western coast of northern South America where two species were formed in isolation. Several species of wet, montane forest habitats may have arisen on the eastern slopes of the Andes and dispersed to other parts of South America, especially southeastern Brazil. In the Serra do Mar region, adaptive radiation occurred under optimal conditions in a diversity of habitats, allowing for the development of several species in a relatively small area. From this core area, species were dispersed north- and southward, forming additional species and subspecies under conditions of ecological adaptation and geographical isolation.

The effects of past climatic changes played an important role in the present distribution of *Brunfelsia*, particularly in the geographic isolation of populations in endemic areas. These areas correspond to proposed forest refuges which were created during dry periods in the Pleistocene or earlier, when the extent of wet tropical forests was greatly restricted. This may also explain several puzzling disjunctions in certain species of *Brunfelsia* which are observed today.

ACKNOWLEDGEMENTS

I wish to thank Dr Richard Evans Schultes, Dr Bernice G. Schubert, Dr Elsa Zardini and Peggy Fiedler for reading the manuscript and offering useful suggestions. L. Teza Bates prepared the line drawings and her artwork is greatly appreciated. Research reported in this paper was supported in part by the National Institutes of Health Training Grant (T TO1 GM 00036-13) and by the National Science Foundation Evolutionary Biology Training Grant (GB 7346). The expense of illustrating the text was defrayed in part by a National Science Foundation Grant (GB 35047).

REFERENCES

BARTLETT, A. S. & BARGHOORN, E. S., 1973. Phytogeographic history of the Isthmus of Panamá during the past 12,000 years (A history of vegetation, climate, and sea-level change). In A. Graham (Ed.), *Vegetation and Vegetational History of Northern Latin America*. Amsterdam: Elsevier.

BENTHAM, G., 1846. Scrophulariaceae. In A. De Candolle, *Prodromus Systematis Naturalis Regni Vegetabilis*, *10:* 198–201.

BROWN, K. S., Jr. & MIEKLE, O. H. H., 1972. The Heliconians of Brazil (Lepidoptera: Nymphalidae). Pt. 2. Introduction and general comments, with a supplementary revision of the tribe. *Zoologica (N.Y.)*, *57:* 1–40.

CUATRECASAS, J., 1957. A sketch of the Vegetation of the North-Andean Province. *Proceedings of the Eighth Pacific Science Congress, 4:* 167–173.

GOODSPEED, T. H., 1954. *The Genus* Nicotiana. Waltham, Mass.: Chronica Botanica.

HAFFER, J., 1969. Speciation in Amazonian forest birds. *Science, 165:* 131–137.

HAFFER, J., 1974. Avian speciation in tropical South America. *Nuttall Ornithological Club Publication, 14.* Cambridge, Mass.

LAMAS, G., 1973. *Taxonomia e Evolução dos Gêneros* Ituna Doubleday *(Danainae) e* Paititia *gen. n.,* Thyridia Hübner *e* Methona Doubleday *(Ithomiinae) (Nymphalidae, Lepidoptera).* D.Sc. Thesis, Universidade de São Paulo, Brazil.

LANGENHEIM, J. H., 1973. Leguminous resin-producing trees in Africa and South America. In B. J. Meggers, E. S. Ayensu & W. D. Duckworth (Eds), *Tropical Forest Ecosystems in Africa and South America: A Comparative Review.* Washington D.C.: Smithsonian Institution Press.

LEWIS, W. H., 1971. High floristic endemism in low cloud forests of Panamá. *Biotropica, 3* (1): 78–80.

MALFAIT, B. T. & DINKELMAN, M. G., 1972. Circum-Caribbean tectonic and igneous activity and the evolution of the Caribbean Plate. *Bulletin of the Geological Society of America, 83:* 251–272.

MIERS, J., 1849. Observations upon several genera hitherto placed in the Solanaceae, and upon others intermediate between that family and the Scrophulariaceae. *Annals and Magazine of Natural History (Ser. 2), 3:* 161–179.

MIERS, J., 1850. XXIII: Contributions to the Botany of South America. *Annals and Magazine of Natural History (Ser. 2), 5:* 247–250.

PLOWMAN, T., 1973a. *The South American Species of* Brunfelsia *(Solanaceae).* Doctoral Dissertation, Harvard University, Cambridge, Mass. Harvard University.

PLOWMAN, T., 1973b. Four New Brunfelsias from Northwestern South America. *Botanical Museum Leaflets, 23* (6): 245–272.

PLOWMAN, T., 1974. Two New Brazilian Species of *Brunfelsia*. *Botanical Museum Leaflets, Harvard University, 24* (2): 37–48.

PLOWMAN, T., 1978. A new section of *Brunfelsia*: section *Guianenses*. In J. G. Hawkes (Ed.), Systematic notes on the Solanaceae. *Botanical Journal of the Linnean Society, 76:* 294–295.

PRANCE, G. T., 1973. Phytogeographic support for the theory of Pleistocene forest refuges in the Amazon Basin, based on evidence from distribution patterns in Caryocaraceae, Chrysobalanaceae, Dichapetalaceae and Lecythidaceae. *Acta Amazonica, 3* (3): 5–26.

RIZZINI, C. T., 1967. Delimitação, caracterização e relações da flora silvestre hileiana. In H. Lent (Ed.), *Atas do Simpósio sôbre a Biota Amazônica, 4 (Botânica):* 13–36. Rio de Janeiro: Conselho Nacional de Pesquisas.

SMITH, L. B., 1962. Origins of the flora of Southern Brazil. *Contributions from the United States National Herbarium, 35* (3): 215–249.

TRYON, R. M., Jr., 1944. Dynamic phytogeography of *Doryopteris. American Journal of Botany, 31:* 470–473.

VANZOLINI, P. E. & WILLIAMS, E. E., 1970. South American anoles: the geographic differentiation and evolution of the *Anolis chrysolepis* species group (Sauria, Iguanidae). *Arquivos de Zoologia, 19* (1–2): 1–12.

VUILLEUMIER, B. S., 1971. Pleistocene changes in the fauna and flora of South America. *Science, 173:* 771–779.

38. Biosystematics of the physaloid genera of the Solaneae in North America

JOHN E. AVERETT

Department of Biology, University of Missouri, St. Louis 63121, U.S.A.

A taxonomic survey of the following genera is presented: *Chamaesaracha, Leucophysalis, Quincula, Margaranthus*. Studies of the morphology, cytology, flavonoid patterns and other criteria are reported for each genus and the data are discussed with a view to elucidating intra- and inter-generic relationships. Short notes are added on *Physalis, Saracha, Jaltomata* and *Oryctes*.

A key to the identification of the above-mentioned eight genera in the tribe Solaneae is given at the end of the chapter, thus summarizing the main morphological differences by which the genera may be distinguished.

CONTENTS

INTRODUCTION

Chamaesaracha is included in the tribe Solaneae (Wettstein, 1895), subtribe Solaninae (Dunal, 1852). Because of their similar corolla forms, *Chamaesaracha* has generally been associated with *Physalis* and *Leucophysalis* (Averett, 1973). It is probably most closely related to *Leucophysalis*, differing in characters of the berry; whereas *Physalis* is separated by its inflated fruiting calyx, and to some extent by its flower. All of the species that now comprise *Leucophysalis* except three in China have been relegated, at one time or another, to *Chamaesaracha*.

The preferred treatment of the genera of the Solaninae, especially those surrounding *Physalis*, may be subject to some debate. For example, Waterfall (1958) included *Quincula*, a monotypic genus delimited by Rydberg (1896), within the confines of *Physalis*. He further suggested that *Physalis* and *Chamaesaracha* might be combined and treated as subgeneric categories, similar in concept to *Oenothera* as delimited by Munz (1965). Generic lines within the Solaninae are quite close in several instances, but if one begins merging genera it would be difficult to stop short of including all but a few of the more distinct genera of the subtribe (e.g., *Solanum* and *Capsicum*). For example, *Leucophysalis, Margaranthus* and *Withania* are all probably closer to *Physalis* than is *Chamaesaracha*. Hence, it seems preferable to afford generic status to these several smaller taxonomic entities of the

493

Solaninae, similar in concept to *Oenothera* as delimited by Raven (1964). In this chapter the distinguishing features and interrelationships of *Chamaesaracha* and genera closely allied with either *Chamaesaracha* or *Physalis* are discussed.

CHAMAESARACHA

Chamaesaracha was first recognized as a distinct genus by A. Gray in 1876, having previously been treated as a section of *Saracha*. A total of 18 species have been attributed to the genus, but in a revision of the genus (Averett, 1973) I have recognized 7 species, the remaining 11 being relegated to synonymy or transferred to other genera. Thus constituted, *Chamaesaracha* is a closely knit genus of perennial herbs largely restricted to the desert regions of the southwestern United States and northern Mexico.

Individual plants of *Chamaesaracha* are fairly common along roadsides in early and later summer, but populations become depauperate in the hot, dry, midsummer period, being found with some difficulty, except in areas of accumulated moisture. The plants often occur in closely clustered populations of 25–30 plants. Such populations probably originated from one or a few seeds, and are believed to be largely the result of vegetative reproduction from spreading rhizomes. In garden plots, individual plants form colonies 8–10 feet in diameter within a single growing season. Intrapopulational variation, both chemical and morphological, within these clusters is essentially absent, suggesting a single source for most of the individuals within a given population. Vegetative reproduction, along with perenniality of the rootstocks, must have much to do with the variation patterns, or lack of them, in the natural populations.

Since fruit with relatively high seed set are found in nature, the various insects that visit the flowers must be fairly effective pollinators. No consistent pollen vectors have been observed, but bee flies and several types of small beetles are commonly found on the flowers of *Chamaesaracha*. Evidence for self-sterility has not been obtained for any of the species, but in isolated plants in garden plots, fruit was observed only on *C. crenata*. Even in this species, however, seed set was reduced some 80 % from that found in nature. In the field, probably both self and out pollination occur, with outcrossing predominating.

Chromosome studies

Chromosome numbers have been determined from nearly 200 individuals representing the seven species of *Chamaesaracha* (Powell & Averett, 1967; Averett, 1970; Averett, 1973). Three of the species are typically diploid with $n = 12$ and four are typically polyploid. What are considered to be the more primitive, diploid species, *C. villosa*, *C. crenata* and *C. sordida*, have their centres of distribution in the Chihuahuan Desert of northern Mexico. Of these, *C. sordida* is the only species to have any appreciable distribution in the United States. Still, it has a distinctly southern distribution as compared to the polyploid species, *C. coronopus* and *C. coniodes*. *Chamaesaracha coronopus* is the most widespread species, extending essentially over the entire range of the genus; *C. coniodes* is also widespread, but does not extend so far west as *C. coronopus*. The remaining polyploid species are fairly restricted in distribution, with *C. edwardsiana* occurring primarily on the Edwards Plateau of south-central Texas and adjacent Mexico and *C. pallida* extending from the Edwards Plateau into western Texas and northern Mexico.

Flavonoid studies

Populations were sampled for flavonoids so as to represent as much of the total range, as well as intraspecific variability, of each of the species as possible. One to ten individuals from approximately 150 populations were analysed to determine the extent and significance of inter- and intrapopulational variation. Flavonoids were extracted overnight from leaf material with 85 % methanol and separated by two-dimensional paper chromatography. Solvent systems of t-butanol, glacial acetic acid, and water (3 : 1 : 1 v/v) and 15 % glacial acetic acid were employed. The compounds detected were then characterized structurally by ultraviolet spectroscopy and appropriate acid and enzyme hydrolyses. The techniques utilized for chromatographic and spectral analysis follow those presented by Mabry, Markham & Thomas (1971).

All of the compounds extracted from *Chamaesaracha* are based on either quercetin or kaempferol skeletons. Differences among the several compounds primarily reflect glycosidic substitutions, but a few methoxylated compounds also are present. Except for the methoxylated compounds, the flavonoids of *Chamaesaracha* are the same as those reported for *Leucophysalis* by Averett & Mabry (1971), but greater interspecific variability is found in *Chamaesaracha*.

Flavonoid data, in general, were found to substantiate conclusions based upon morphological data. Two-dimensional chromatographic patterns were, for the most part, species-specific, showing relatively little variation, either within or between populations of the same species. A notable exception was found in *Chamaesaracha coniodes* in which significant variations in the flavonoid patterns were observed. This species is also quite variable morphologically and much of the flavonoid variation probably reflects genetic heterogeneity from population to population. Within the more discrete populations, however, flavonoid patterns were essentially uniform, and this is also true of morphological variation.

Closely related taxa were found to have similar flavonoids; consequently, chemical data have been utilized for many of the taxonomic conclusions reached in the present study, especially in situations where morphological and chromosomal distinctions were not decisive.

Phylogeny

From the distributional data, it seems reasonable to suggest that *Chamaesaracha* had its origin in the highlands of northern Mexico, the ancestral species of which probably belonged to the more xeric members of the Madro-Tertiary geoflora (cf. Axelrod, 1958). With continued adaptation to xeric conditions, expansion from this area was probably concurrent with the spread of an arid environment to the north and east. In that the polyploid and derived species are marginal to the Chihuahuan Desert proper, it is difficult to imagine an opposing direction of migration and speciation.

The interpretation of evolutionary trends in *Chamaesaracha* is based for the most part on chromosomal and distributional data in combination with correlated morphological features. Some of the characters that I consider to be primitive and derived, at least within *Chamaesaracha*, are listed below:

	Primitive	*Derived*
1.	woody rootstock	subligenous root
2.	large, ovate, entire and thin leaves	small, lanceolate, lobed, and thickened or leathery leaves
3.	large flowers	small flowers
4.	O-methylation of flavonoids	absence of O-methylation of flavonoids
5.	increased glycosylation of flavonoids	decreased glycosylation of flavoids

The assessment of morphological characters generally follows classical consideration for those features (i.e., perenniality is generally considered primitive, as are large flowers). Polyploidy is considered to be derived in *Chamaesaracha*, quite apart from numerical considerations, because the polyploids possess fewer of the characters considered primitive in the listing above. Hence, the fairly widespread occurrence of polyploidy offers a sound basis upon which to establish evolutionary trends within the genus.

LEUCOPHYSALIS

Leucophysalis was established as a monotypic genus by Rydberg (1896) to accommodate *Physalis grandiflora* Hook., a species of the Great Lakes area of eastern North America. Fernald (1949: 82–83) recognized the similarity of *L. grandiflora* to certain species he thought to be *Chamaesaracha* and transferred the species to the latter genus. *Leucophysalis* is clearly distinct from *Chamaesaracha*, and is better treated as such (Averett, 1971, 1973, 1977). Two additional species, *L. nana*, a species of the Sierra Nevada of western North America, and *L. heterophylla*, a species from east-central China, were transferred to the genus (Averett, 1971). Averett (1977) considers *Physaliastrum* to be gongeneric with *Leucophysalis* and recognizes nine species, two North American and seven Asian.

Makino (1914) recognized two Japanese species then considered to be *Chamaesaracha* as comprising his new genus *Physaliastrum*. The genus was distinguished from *Leucophysalis* in having a campanulate corolla rather than the rotate, plicate corolla of *Leucophysalis*. In a discussion of his new genus, Makino suggested that the two Chinese species, which also had been relegated to *Chamaesaracha*, might also be within the confines of *Physaliastrum*. Kuang & Lu (1965), in a revisionary study of *Physaliastrum*, accepted one of the species *L. heterophylla*, and three other Chinese species to be within the genus. The Chinese species, however, were placed, in a section separate from the Japanese species. Seven species, four Chinese and three Japanese, were treated in the latter study.

There are two elements within Kuang and Lu's section *Physaliastrum*. I am treating two of the species, *L. savatieri* and *L. kimurai*, as a third section, section *Urceolateae*. With the removal of these species, the two sections recognized by Kuang & Lu appear to be natural groupings, with *L. nana* belonging with section *Physaliastrum* and *L. grandiflora* with the Chinese species (section *Leucophysalis*). The two sections would thereby contain the appropriate types of *Leucophysalis* and *Physaliastrum*. A case might be made for the recognition of both genera. However, flavonoid data indicate a close relationship between *L. grandiflora* and *L. nana*, the two being dissimilar only in the absence of a single compound in *L. nana* (Averett & Mabry, 1971). Further, although the two groups may be distinguished by floral characters and fruiting calyx length, they are more closely related than other genera of the tribe. That relationship would be obscured in the recognition of separate genera, and hence they are better treated as sections of the same genus as was done by Kuang and Lu. They should, however, be treated as *Leucophysalis* rather than *Physaliastrum*.

Generic relationships

Leucophysalis belongs within the tribe Solaneae and is probably most closely related to *Physalis*. The interrelationships of these and other genera of the tribe are more fully elaborated in a previous discussion (Averett, 1973). However, several points regarding origin and distribution deserve mention. The Solanaceae are best represented in the New World and, with the exception of *Solanum* and its segregates, are poorly represented in the flora of eastern Asia. Approximately seven genera and fewer than 25 species are found, and except for *Scopolia*, all of these are clearly associated with New World temperate and tropical taxa. *Leucophysalis*, has the greatest proliferation of species in northeast Asia with only one ranging into tropical areas, *L. yunnanense*; there are two species in North America. However, *Leucophysalis* is clearly related to *Physalis* which is American. The Asian species might represent a relatively old introduction with subsequent proliferation of species. The latter seems more likely than a secondary, or reintroduction of the taxon to the New World.

Another genus, *Archiphysalis* was recognized by Kuang (1966). The genus includes two species. Unaware of the transfer, I treated one of these as *Physalis* (*P. sinensis*; Averett, 1971). The other species, *A. chamaesarachoides*, has also been treated as *Physalis* (Makino, 1928; Ohwi, 1965). If the genus is distinct, it is quite close to *Physalis* and surely derived from physaloid stock.

Interspecific relationships

Striking Asian–North American disjunctions are found within *Leucophysalis*. Relationships between the floras of eastern Asia and North America, especially eastern North America, have long been recognized (Weber, 1965), but *Leucophysalis* is particularly notable in that closer affinities, in at least two instances, are found between species of Asia and North America than between species of the same continent.

Leucophysalis grandiflora and *L. nana* are North American. *L. savatieri* and *L. japonica* both occur in Japan and the latter is also found in adjacent eastern China. *Leucophysalis heterophylla* is found in the Hupeh and Anhwei provinces of China. *Leucophysalis japonica* and *L. nana* are ascending perennials of high elevations, and both have a fruiting calyx which does not exceed the berry. *L. grandiflora* and *L. heterophylla* are erect annuals inhabiting areas of low elevation, each possessing a fruiting calyx which exceeds the berry.

On the basis of the above characteristics, *L. grandiflora* and *L. heterophylla* appear to form a species pair, as do *L. nana* and *L. japonica*. The remaining species, *L. savatieri* and *L. kimurai* are spreading perennials of mountain woods of central

Japan, but the fruiting calyx is urceolate, exceeding the berry, and in *L. savatieri* the berry is ellipsoidal rather than globose. The latter species do not appear to be especially close to any of the other species of the genus, but are perhaps closer to *L. japonica* and *L. nana* than to *L. heterophylla* and *L. grandiflora*.

The relationships suggested above are largely based upon morphological considerations. A chromosome number of $n = 12$ has been reported for *L. nana* (Raven, 1959; reported as *Chamaesaracha nana*). This is the only species for which there is a chromosome count. Efforts to germinate seeds for chromosome counts and greenhouse material have been unsuccessful, probably because of the age of the available specimens. The flavonoid chemistry of the New World species has been examined (Averett & Mabry, 1971), but material for study of the Asian species was not available. It is notable, however, that the flavonoids of *L. grandiflora* and *L. nana* are identical except for the absence of a single compound from the complement in *L. nana*.

In spite of the lack of chemical and chromosomal data, the morphological features are distinctive and the relationships suggested above are made with considerable confidence. The presence of two sets of closely related disjunct species might add support to suggestions of an Arcto-Tertiary 'Bering' bridge or continuous distribution (cf. Sharp, 1971; Weber, 1965). Long distance dispersal would involve two separate introductions of two species of the same genus, events which do not seem likely.

QUINCULA

Quincula is represented by the single species *Q. lobata* (Torr.) Raf, which has generally been included in *Physalis*. Rafinesque (1832) first regarded the genus as distinct and cited *Physalis lobata* Torr. as the type species. Rydberg (1896) also considered the genus to be separate from *Physalis*, but Waterfall (1958) included it with *Physalis*. Having now examined several genera allied with *Physalis*, e.g. *Margaranthus*, *Leucophysalis*, and *Chamaesaracha*, I agree with the recognition of *Quincula* as a genus separate from *Physalis*. Certain aspects of the research on *Quincula* are in progress, but critical distinguishing features are given below.

Like *Chamaesaracha*, *Quincula* is restricted to the desert and adjacent dry plains of the southwestern United States and Northern Mexico. The centre of distribution is the Chihuahuan Desert. Plants are frequent in the spring and late summer, depending upon the rains. Plants occur in closely clustered populations, but more often are found singly or no more than a few individuals. Asexual reproduction does not seem to be as extensive in *Quincula* as it is in *Chamaesaracha*.

Chromosome data

A chromosome number of $x = 11$ has been found for *Quincula* (Menzel, 1950; Averett & Powell, 1972). Diploid populations are generally restricted to the Chihuahuan Desert while the tetraploid race extends into the dry grassland areas to the north. An autoploid origin is suspected for the tetraploid race (Averett & Powell, 1972).

With few exceptions, previously reported numbers of the Solaneae have a base chromosome number of $x = 12$, and a base number of $x = 11$ is uncommon in the whole of the Solanaceae. A chromosome number of $x = 11$ has been reported for 3 genera in Salpiglossidae. Menzel (1950) compared chromosomes of several $x = 11$ genera and concluded that the base number was independently derived in *Quincula*; a logical conclusion since there is no morphological basis for relating *Quincula* to the Salpiglossidae. From these data alone, Menzel supported the retention of *Quincula* as a distinct genus.

Morphological data

The berry of *Quincula* is enclosed in an inflated, accrescent calyx and the base of the flowering calyx is truncate—features which relate the genus to *Physalis*. However, the berry, unlike *Physalis*, is dry and placentation is basal. In *Physalis* placentation is axile and much enlarged; the latter contributing to the fleshiness of the berry. The berry of *Quincula* is more like that of *Chamaesaracha* and *Margaranthus*.

Additional characters which serve to distinguish *Quincula* from *Physalis* are found in the flower, seed and indument. In *Quincula*, the flower is rotate and entirely blue or purple. The rotate corolla is approached in a few species of *Physalis*, e.g. *P. acutifolia*, but it is generally campanulate. An entirely blue corolla is absent in *Physalis*. The seeds of *Quincula* are rugose-reticulate whereas in *Physalis* the seeds are punctate. *Quincula* is unique in having numerous, small crystalline vesicles scattered over the foliage.

Flavonoids

Flavonoids from approximately 30 populations of *Quincula* have been examined. The diploid populations have the 3-O-rutinosides of quercetin and kaempferol. The two flavonols plus 7-O-glucosides of the compounds are found in the tetraploids. These are flavonols that seem to be widespread and common in the Solaninae.

Origin and generic relationships

Quincula, like *Chamaesaracha*, appears to be derived from the Madro-Tertiary geoflora. With continued adaptation to arid conditions, the genus became established in what is now the Chihuahuan Desert. As drying continued in the grasslands to the north, the genus expanded into the adjacent region.

MARGARANTHUS

Margaranthus was established by Schlechtendal (1838) when he described the species *M. solanaceous*. Schlechtendal placed the genus in the Solaneae, noting its similarity in habit to *Physalis*. Subsequent to Schlechtendal's establishment of the genus, four additional species have been described; *M. tenuis* Miers (1850), *M. lemmoni* Gray (1883), *M. purpurascens* Rydberg (1896) and *M. sulphureus* Fernald (1900). The last of these species was recognized by Waterfall (1967) to be a species of *Physalis* and was transferred by him to that genus. Hence, four species remain in *Margaranthus*, but most workers have regarded it as monotypic (e.g. Correll & Johnston, 1970). Data now at hand (Averett & Judd, unpubl.) indicate that, indeed, there is only one species, *M. solanaceous*.

The distributional range of *Margaranthus* is entirely in the New World north of the equator. The genus ranges from eastern Arizona east into southern Texas and south through Mexico. Collections have also been made in Cuba and Curacao. Although the northern range of the genus is generally arid, *Margaranthus* occupies habitats of less extreme conditions such as higher elevations, shaded canyon floors, or cultivated sites.

Chromosome data

Little cytological data are available for *Margaranthus*. Menzel (1950) reported a count of $2n = 24$ from a single collection. Additional counts have not been made. From the one collection Menzel was able to do a karyotypic analysis and noted two pairs of chromosomes with median-submedian centromeres, and ten pairs with more or less subterminal centromeres. The total metaphase chromosome length in PDB treated material average *c.* 64 µm, or about midway of the *Physalis* size range. The ratio of total short arm length to total chromosome length fell between the range of the annual sections of *Physalis*, *Pubescentes* and *Angulatae*.

Flavonoids

The pigments extracted from the leaves of *Margaranthus* are based on the two aglycones quercetin and kaempferol (Averett & Judd, 1977). Two of each type were identified; the 3-O-glucoside and 3-O-rutinoside of quercetin and the 3-O glucoside and 3-O-rutinoside of kaempferol. The two quercetin-based compounds are present in readily detectable concentrations, but the kaempferol compounds are very weak and not detectable unless leaf extractions are heavily applied to the chromatography paper before developing. Material from 31 collections representing most of the distributional range of the species were examined, and the profile was uniform in each of the populations examined. For details of the experimental methods and voucher specimens, see Averett & Judd (1977).

Morphological and anatomical studies

Vegetatively and in features of the fruiting calyx, *Margaranthus* is very much like *Physalis*. The flower, however, is urceolate, constricted at the orifice, and purple. Because of this floral morphology, *Margaranthus* has enjoyed a variety of subtribal dispositions (see following section).

Recently, we have had the opportunity to examine the floral vasculature of *Margaranthus* and compare it with genera thought possibly to be related: *Physalis*, *Quincula*, *Chamaesaracha*, and *Salpichroa*. Floral vasculature of *Margaranthus* most nearly approaches that of *Physalis* (*Quincula* is more like *Chamaesaracha*).

Generic relationships

Margaranthus has been placed in the Solaninae (Dunal, 1852), Lyciinae (Wettstein, 1895) and Margaranthinae (Baehni, 1946). The latter two exclude *Physalis*. Menzel (1950) noted that the range of *Margaranthus* coincides in part with the area containing the largest number and diversity of *Physalis* species and that morphological and karyological affinities of *Margaranthus* with the *Pubescentes* and *Angulatae* suggest that the three have originated from a common stock, in particular the *Angulatae*. We have not examined *Physalis* in detail and can not evaluate the latter, but certainly, a relationship between *Margaranthus* and *Physalis* is supported by anatomical comparisons. Flavonoid data, at this point, are of limited value in such consideration because of the limited basis for comparison.

Origin

Less is known of *Margaranthus*, but the genus may be derived from the Madro-Tertiary geoflora. Unlike *Chamaesaracha* and *Quincula*, however, the genus never adapted to arid, desert environments and spread south rather than to the north.

MISCELLANEOUS GENERA

Physalis

Physalis is a genus of approximately 90 species, and although it has a world wide distribution, is principally New World. The North American species of *Physalis* have been treated in revisionary studies by Rydberg (1896) and Waterfall (1958, 1967, 1968), but subgeneric arrangements, intergeneric relationships, and phylogeny remain unclear. Cytological and genetic comparisons by Menzel (1950, 1951) give insight into certain of these questions, but additional data need to be accumulated.

Saracha and Jaltomata

Considerable taxonomic confusion has surrounded *Saracha* and *Jaltomata*. All of the species have been treated as *Saracha*, but essentially every worker dealing with the group has recognized two distinct taxonomic groups. The confusion involved the use of a correct name. Gentry (1973), by restoring the genus *Jaltomata*, has resolved this problem.

As separated, *Saracha* is a South American genus of perhaps five species restricted to high elevations in the Andes from Venezuela to Bolivia. *Jaltomata* occurs at lower elevations in South America and extends through Central America north to the southwestern United States. The species of *Jaltomata* and *Saracha* are poorly known taxonomically, but are being investigated by Tilton Davis at the University of Wyoming.

Oryctes

Oryctes is a monotypic genus of western Nevada and southeastern California. Rydberg (1896) included *Oryctes* in his treatment of *Physalis* and related genera, but its relationship to any of the physaloid genera, is obscure. It resembles *Chamaesaracha* in habit and fruit, but is easily distinguished by its small, tubular flowers and flat, orbicular seeds. Investigations into the intergeneric relationships of *Oryctes* are currently in progress.

KEY TO SELECTED NORTH AMERICAN GENERA OF THE TRIBE SOLANEAE

Flowers tubular, seeds orbicular . . . .*Oryctes*

Flowers campanulate, rotate, or urceolate; seeds reniform

    Fruiting calyx expanding under and not enclosing the berry

        Perennial shrubs, flowers campanulate . . .*Saracha*

        Herbaceous annuals or perennials, flowers rotate . . . .*Jaltomata*

    Fruiting calyx enclosing the berry

    Calyx inflated around the berry

        Flowers urceolate . . . .*Margaranthus*

        Flowers campanulate to rotate

            Flowers, purple; indument with crystalline vesicles; berry dry, with basal placentation.*Quincula*

            Flowers yellow, or with purple maculations in the throat; indument without crystalline vesicles; berry fleshy, with axile placentation *Physalis*

    Calyx tightly investing the berry

        Berry dry, placentation basal, seeds rugose-reticulate . . . .*Chamaesaracha*

        Berry fleshy, placentation axile, seeds punctate .*Leucophysalis*

REFERENCES

AXELROD, D. I., 1958. Evolution of the Madro-Tertiary Geoflora. *Botanical Review, 24:* 433–509.

AVERETT, J. E., 1970. *Systematics of* Chamaesaracha (*Solanaceae*): *a chemosystematic and cytotaxonomic study.* Ph.D. Dissertation, University of Texas at Austin.

AVERETT, J. E., 1971. New combinations in the Solaneae (Solanaceae) and comments regarding the taxonomic status of *Leucophysalis. Annals of Missouri Botanical Garden, 57:* 380–382.

AVERETT, J. E., 1973. Biosystematics of *Chamaesaracha* (Solanaceae). *Rhodora, 75:* 325–365.

AVERETT, J. E., 1977. New combinations and taxonomic notes in *Leucophysalis* (Solanaceae). *Annals of the Missouri Botanical Garden, 64:* 141–142.

AVERETT, J. E. & JUDD, J., 1977. The flavonoids of *Margaranthus* (Solanaceae). *Biochemical Systematics and Ecology, 5:* 279–280.

AVERETT, J. E. & MABRY, T. J., 1971. Flavonoids of the North American species of *Leucophysalis* (Solanaceae). *Phytochemistry, 10:* 2199–2200.

AVERETT, J. E. & POWELL, A. M., 1972. Chromosome numbers in *Physalis* and *Solanum* (Solanaceae). *Sida, 5:* 3–7.

BAEHNI, C., 1946. L'ouverture du bouton chez les fleurs de Solanées. *Candollea, 10:* 479.

CORRELL, D. & JOHNSTON, M., 1970. *Manual of the Vascular Plants of Texas.* Texas Research Foundation, Renner, Texas.

DUNAL, M. F., 1852. In A. de Candolle (Ed.), *Prodromus Systematis Naturalis Regni Vegetabilis, 13,* pt. I.

FERNALD, M. L., 1900. Some undescribed Mexican Phanerogams, chiefly Labiatae and Solanaceae. *Proceedings of the American Academy of Arts and Sciences, 35:* 566.

FERNALD, M. L., 1949. Contributions from the Gray Herbarium of Harvard University—No. CLXIX. Part II. Studies of eastern American plants. *Rhodora, 51:* 61–85.

GENTRY, J. L., 1973. Restoration of the genus *Jaltomata* (Solanaceae). *Phytologia, 27:* 286–288.

GRAY, A., 1876. Synopsis of the North American species of *Physalis*. *Proceedings of the American Academy of Arts and Sciences, 10:* 62.

GRAY, A., 1883. Contributions to North American Botany. *Proceedings of the American Academy of Arts and Sciences, 19:* 91.

KUANG, K., 1966. *Acta Phytotaxonomica Sinica, 11:* 59.

KUANG, K. & LU, A. 1965. Revisio Physaliastorum Makino. *Acta Phytotaxonomica Sinica, 10:* 347–359.

MABRY, T. J., MARKHAM, K. & THOMAS, M., 1971. *The Systematic Identification of Flavonoids.* New York: Springer-Verlag.

MAKINO, T., 1914. Observations on the flora of Japan. *The Botanical Magazine, 28:* 20–22.

MAKINO, T., 1928. A contribution to the knowledge of the flora of Japan. *Journal of Japanese Botany, 5:* 24.

MENZEL, M. Y., 1950. Cytotaxonomic observations on some of the genera of the Solaneae: *Margaranthus, Saracha,* and *Quincula. American Journal of Botany, 37:* 25–30.

MENZEL, M. Y., 1951. The cytotaxonomy and genetics of *Physalis*. *Proceedings of the American Philosophical Society, 95:* 132–183.

MIERS, J., 1850. Contributions to the Botany of South America. *Annals and Magazine of Natural History (Ser. 2), 5:* 74.

MUNZ, P., 1965. Onagraceae. *North American Flora (Ser. 2),* Part 5: 1–231.

OHWI, J., 1965. *Flora of Japan* (in English). Washington, D.C.: Smithsonian Inst.

POWELL, A. M. & AVERETT, J. E., 1967. Chromosome numbers of *Chamaesaracha* (Solanaceae) in Trans-Pecos Texas and adjacent regions. *Sida, 3:* 156–162.

RAFINESQUE, C. S., 1832. Twenty new genera of plants from the Oregon Mountains, &c. *Atlantic Journal, 149:* 236.

RAVEN, P. H., 1959. In, Documented chromosome numbers in plants. *Madroño, 15:* 49–50.

RAVEN, P. H., 1964. The generic subdivision of Onagraceae, tribe Onagreae. *Brittonia, 16:* 276–288.

RYDBERG, P. A., 1896. The North American species of *Physalis* and related genera. *Memoirs of the Torrey Botanical Club, 4:* 279–374.

SCHLECHTENDAL, D. F. L. von, 1838. *Index Seminum in Horto Academico. Halensi:* 8.

SHARP, A. J., 1971. Epilogue. In P. C. Holt (Ed.), *The Distributional History of the Biota of the Southern Applachians, Part II. Flora.* Research Division Monograph 2. Virginia Polytechnic Inst. and State University, Blacksburg, Va.

WATERFALL, U. T., 1958. A taxonomic study of the genus *Physalis* in North America north of Mexico. *Rhodora, 60:* 107–173.

WATERFALL, U. T., 1967. *Physalis* in Mexico, Central America, and the West Indies. *Rhodora, 60:* 224–225

WATERFALL, U. T., 1968. A new species of *Physalis* from the Galapagos Islands. *Rhodora, 70:* 408–409.

WEBER, W. A., 1965. Plant geography in the southern Rocky Mountains. In H. E. Wright, Jr. & D. G. Frey (Eds), *The Quaternary of the United States.* Princeton University Press.

WETTSTEIN, R. 1895. Solanaceae. In A. Engler & K. Prantl, *Die Natürlichen Pflanzenfamilien, IV* (36): 4–39.

39. *Mandragora*—taxonomy and chemistry of the European species

BETTY P. JACKSON

School of Pharmacy, Sunderland Polytechnic, Sunderland, U.K.

AND MICHAEL I. BERRY

School of Pharmacy, Liverpool Polytechnic, Liverpool, U.K.

Confusion in the literature concerning the nomenclature of the genus *Mandragora* has implied that a number of European species exist. These investigations suggest that only two species, *M. officinarum* L. and *M. autumnalis* Bertol. can be accepted and their morphology is described. The structure of the roots and rhizomes* of the two species has shown them to be indistinguishable. The chemistry of the roots and rhizomes of the two species is also similar; they contain tropane alkaloids and both tropic and tiglic esters, the significance of which in the chemotaxonomy of the Solanaceae is indicated. The roots and rhizomes also contain β-methylesculetin, sitosterol and free sugars.

CONTENTS

TAXONOMY

European species of *Mandragora* have a long history of usage in medicine, and many legends associated with the plants have arisen from the characteristic anthropomorphic appearance of the roots and rhizomes. There has been considerable confusion in the nomenclature of the genus and by choosing different and varying characters such as time of flowering, colour of the corolla and the size and shape of the fruits as taxonomic criteria, botanists in the past have given the impression that a number of species exist.

The Ancients distinguished two main European species which they called "female" and "male" Mandrakes. The female form is described by Dioscorides (see Gunther, 1933) as having "narrower and longer leaves than lettuce, of a poisonous and heavy scent . . . and amongst them ye apples like service berries pale, of a sweet scent, in which is ye seed as of a pear; ye roots two or three of a good bigness, wrapped within one another, black according to outward appearance, within white and of a thick bark". The male species he described as having leaves which are "greater, broad, smooth as of beet but ye apples twice as big—saffron in colour, sweet smelling . . . root greater and whiter . . .".

Parkinson (1629), also described Mandrake as being separated into two kinds,

* Technically, taproots (*Eds*)

the male and female, and his descriptions of the leaves and fruits of the two species are similar to those given by Dioscorides. In addition, he gives details of the flowers, those of the male Mandrake "rising up from the middle, among the leaves . . . every one upon a long slender stalk standing in a whitish green husk consisting of five pretty large round pointed leaves, of a greenish white colour"—while those of the female Mandrake are similar but "of a bluish-purple colour". It is interesting to note that he did not distinguish between the two species on the basis of the colour of the roots, but stated that both kinds were "black without and white within". Parkinson also refers to a second form of the male Mandrake which he describes as having leaves of a more greyish green colour and somewhat folded together. He also writes "The male flowers in March and the fruit is ripe in July. The female if it is well preserved flowers not until August or September, so that without extraordinary care we never see the fruits thereof in our garden".

Parkinson first drew attention to one character which was the cause of considerable confusion in the identification of the two types. It had previously been suggested that the male Mandrake had globose fruits and the female pear-shaped or oblong fruits. However, Parkinson states that the female plant has "small round fruits or apples and not long like a pear".

The first attempt at a botanical classification of *Mandragora* was that of Tournefort (1719). A fuller description of the genus was published by Jussieu (1789) in which it was placed with *Atropa* in the order Solaneae. This link with the genus *Atropa* is reflected in some of the names given to *Mandragora* at that time. In *Species Plantarum* (1753) Linnaeus had described a single species of *Mandragora*, *M. officinarum*, but in successive editions of the work (1762, 1764) this was changed to *Atropa acaulis*. Woodville (1794), applied the name *Atropa Mandragora* to the same plant.

Sprengel (1825), who edited the sixteenth edition of Linnaeus' *Systema Vegetabilium*, named two species of *Mandragora*, *M. vernalis* Bertol. and *M. autumnalis* Bertol. He thus referred to the work of Bertoloni (1824) who described two species, *M. vernalis* and *M. officinarum*, and claimed that the latter species was the same as that earlier described by Linnaeus. Bertoloni stated that *M. vernalis* Bertol. was equivalent to the species which he had previously named *M. officinalis* Bertol., and *M. officinarum* Bertol. was the same as *M. autumnalis* Bertol. Sprengel preferred to use the specific name "*autumnalis*" which Bertoloni had originally given to the second species rather than the later name "*officinarum*".

In a later work, Bertoloni (1835) described a third species of *Mandragora* which he called *M. microcarpa* Bertol. This he characterized as having smaller, more globose fruits than *M. officinarum* Bertol. and was the form known as "*Mandragora minore*" in Italy and "*petite Mandragora*" in France. The existence of this additional species was supported by De Candolle (1852) who distinguished *M. microcarpa* Bertol. from the other species by the smaller size of the underground parts. Cesati, Passerini & Gibelli (1867) also described two autumn flowering species, *M. officinarum* L., which had an ovoid-oblong fruit equal in length to the calyx and flowered from September to October in Calabria and Sicilia, and *M. microcarpa* Bertol., which had a smaller, globose fruit, shorter than the calyx and flowered from October to November in Sardinia and Napoletano. Von Heldreich (1886) again recognized two autumn flowering species which he called *M. autumnalis*

Spreng. and *M. microcarpa* Bertol. He maintained that, whilst in *M. autumnalis* the leaves and flowers develop simultaneously, the first flowers of *M. microcarpa* Bertol. "arise and develop from the middle of September, in dense clusters; the plant has no leaves or at the most incompletely mature, small leaves. These subsequently develop little by little, though they never become very large—they are broader and shorter even than in *M. autumnalis* Spreng. . . . The flowers are the largest and most beautiful of the genus, a very beautiful violet, 1½ inches long". His description of the fruit is similar to that of previous workers.

More recently, Palhina (1939) also distinguished two autumn flowering species growing in Portugal, *M. autumnalis* Spreng. and *M.* (β) *microcarpa* Bertol., each with characteristics of the fruits similar to those given by Cesati *et al.* (1867) for the Italian species.

Other workers have suggested that *M. microcarpa* Bertol. should be considered as a variety of *M. autumnalis* and not as a distinct species (Arcangeli, 1894; Fiori & Paoletti, 1908; Fiori, 1929; Maugini, 1959).

It is now generally considered that *M. microcarpa* Bertol. is not sufficiently different from *M. autumnalis* Bertol. to be regarded as a distinct species or variety. This view was held by Moris (1858) who named *M. officinarum* L. as being synonymous with *M. autumnalis* Bertol., *M. microcarpa* Bertol. and *Atropa mandragora* L. Parlatore (1883) also proposed only two species, uniting *M. microcarpa* Bertol. with *M. autumnalis* Bertol. In addition, he chose not to use the name *M. officinarum* L. for the spring flowering species because he maintained it was not clear to which species Linneaus had given this name and its use had consequently led to considerable confusion.

Von Heldreich (1886), in addition to accepting *M. microcarpa* Bertol. as a true species, also recognized a fourth European species which he called *M. haussknechtii* Heldr. This species, he claimed, was more clearly related to the autumn flowering species although it flowered in winter and spring. It had large, violet flowers and could be distinguished from *M. autumnalis* Bertol. by having longer leaves and fruits which were larger, more ovoid and waxy yellow in colour. He further claimed that Professor Haussknecht, after whom he had named the species, had found a hybrid between *M. vernalis* Bertol. and *M. haussknechtii* Heldr. (*M. hybrida* Hausskn. et Heldr.) which showed characters intermediate between the two parent species. He pointed out that such a hybrid could easily arise as *M. vernalis* Bertol. and *M. haussknechtii* Heldr. flower and fruit simultaneously over a long period.

The validity of von Heldreich's claims concerning *M. haussknechtii* was examined by Vierhapper (1915) and he was unable to confirm the hybrid nature of *M. hybrida* Hausskn. et Heldr. on examining the original specimen. Rather, he considered that *M. haussknechtii* and *M. hybrida* were more likely to be varieties of *M. officinarum* L. and not true species. Vierhapper did, in fact, suggest that four distinct taxa could be distinguished within the species *M. officinarum* L. as follows:

(1) an autumn flowering form found over the whole area of distribution of the species; this variety was equivalent to *M. autumnalis* Bertol.

(2) and (3) late spring flowering forms, of which one occurs in the western and the other in the eastern part of the areas in which the species is found; these are distinctly different from (1) although clearly closely related. The western form he

named *M. hispanica* Vierhapper and the eastern form he said was equivalent to *M. haussknechtii* Heldr.

(4) a form which flowers in autumn but occurs only in the centre of the area of distribution and is morphologically distinct; this form was equivalent to *M. femina* Gersault.

Vierhapper described the characters of his new species *M. hispanica* in detail, and claimed that it could be distinguished from the other varieties by the characters of the leaves. He also emphasized the differences in the flowering times amongst the four varieties. *M. femina*, he stated, flowered in September and October while *M. autumnalis* produced flowers and fruits from September through to February. Both *M. hispanica* and *M. haussknechtii* commenced their flowering periods in January and Vierhapper claimed to have collected samples of flowers and fruits of *M. autumnalis* at the same time as flowers of *M. hispanica*, in the same localities.

In addition to the four varieties of *M. officinarum* L., Vierhapper also recognized a true spring flowering species under the name *M. mas* Gersault (equivalent to *M. vernalis* Bertol.). He thus added considerably to the confusion in the nomenclature of the genus by maintaining that the plant described by Linnaeus as *M. officinarum* referred to the autumn flowering and not the spring flowering species as had been assumed by earlier workers.

Tercinet (1950) carried out an extensive investigation into the taxonomy of *Mandragora* and was unable to support the work of von Heldreich and Vierhapper. He decided that there were only two Mediterranean species and these he defined as follows:

> "*Mandragora autumnalis* Bertol.—Female Mandragora (syn. *M. microcarpa* Bertol., *M. officinalis* Moris*, *M. officinarum* Bertol. but not Linnaeus, *M. femina* Gersault) which grows principally in S. Italy (where it has the name *M. femmina*), in Spain, Greece, Crete and Asia Minor. It is also found in North Africa, particularly Tunisia and Morocco."

> "*Mandragora officinarum* L.—Male Mandragora (syn. *M. acaulis* Gaertn., *M. neglecta* G. Don, *M. officinalis* Miller but not Moris*, *M. praecox* Sweet, *M. vernalis* Bertol. *M. mas* Gersault), which grows in the same countries as the previous species but is more resistant to cold and can appear further north."

Tercinet thus agreed basically with Engler & Prantl (1897) and Lazaro & Ibiza (1921) who also had recognized only two Mediterranean species. Engler & Prantl, however, had assigned the authorship of *M. autumnalis* to Sprengel and not to Bertoloni because they considered *M. autumnalis* Bertol. to be synonymous with *M. microcarpa* Bertol. and, as they wished to stress that there was only one autumn species, by using the name *M. autumnalis* Spreng. they hoped to avoid confusion which might be caused by reference to the work of Bertoloni. Maugini (1959) supported Tercinet's use of the name *M. autumnalis* Bertol. but, like Parlatore, she advocated the name *M. vernalis* Bertol. for the spring flowering species in preference to *M. officinarum* L.

---

* The name *M. officinalis* was used by Miller (1768) for the species which he regarded as synonymous with *M. officinarum* L. His description is that of the spring flowering species. Moris (1827) recognized the existing name *M. officinalis* but the specimens to which he gave this name flowered in the autumn. Bertoloni (1824) on the other hand, specified that *M. officinalis* Bertol. is synonymous with *M. vernalis* Bertol. Hence Moris applied to the autumn flowering species the name that Bertoloni and Miller had given to the spring flowering species.

In the *Notulae Systematicae*, No. 13 for the *Flora Europaea*, edited by Heywood (1972), Hawkes recommends that the name *M. officinarum* L. should be retained as representing the type species of the genus.

Tercinet's classification includes reference to some of the variations noted by other workers, and he outlines the main characters of the two species as follows:

"*M. autumnalis* Bertol. thrives in shaded places, on river banks and it also sprouts up in stony habitats where conditions are sufficiently sheltered and irrigated. It is occasionally found around stone tombs in cemeteries. It flowers in autumn during October and November. Its root is long, fusiform, whitish within, blackish at the surface, entire or bifurcated and, in some, more or less similar to the lower part of the human body. The leaves are rather large, the margin obtuse at the apex, the base acute, a pale sea-green colour, shiny above, paler beneath, more or less bristling with hairs, and ciliate at the margin, with an elongated petiole. The flowers succeed each other over a long period in the plant and have reddish-green peduncles, enlarged and pentagonal in their upper region. The corollas are large, about 3 times as long as the calyx, pale violet in colour. The fruit is ovoid, obtuse, with a small elongated calyx which surrounds it. The colour of the fruit is yellowish-fawn at maturity and it has a very foetid odour.

"*M. officinarum* L. produces its leaves and flowers at the beginning of spring (March and April). Its root is similar to that of the female Mandragora, but larger and pale on the outside (ordinarily of a dirty white). The glabrous leaves are more uniformly large and diffuse a strong nauseous odour. The flowers are numerous, compact, on pale green peduncles, shorter than the leaves and hairy. The corolla is a whitish-green or slightly yellow. The outer side of the corolla is covered with hairs which, viewed under the microscope, present the appearance of masses having a swollen head, comprising about 15 cells and supported by a slender, 2 or 3 celled, uniseriate stalk. The fruits are much larger than those of the previous species, being as large as a small apple (pomme d'api—a particular species of small red apple called *malum Appiarum*), globose, glossy, yellow and extending well beyond the calyx. Certain authors in antiquity called *M. officinarum* L., because of this latter character, *Mandragora fructu rotundo* or *fructu majore*; in this case *M. autumnalis* Bertol. was known as *Mandragora fructu pyri*. Flahaut called attention to the fact that *M. officinarum* L. has an embryo which is much larger than that of other species of Solanaceae he had examined."

In *Flowers of Europe* (Polunin, 1969) *Mandragora* is represented by a single species, *M. officinarum* L., which is described as having violet flowers which appear in the spring and autumn. The accompanying illustration shows the characters of *M. autumnalis* Bertol. In a later work (*Flowers of S.W. Europe*, Polunin & Smythies, 1973) a similar illustration is included but the plant is here called *M. autumnalis* Bertol. (*M. officinarum*) and the reference to flowers appearing in the spring is omitted. It would seem that, if Tercinet's simplified classification were to be generally accepted, it would greatly assist in the identification of the Mediterranean species, as it is in *Flora Europaea*, Vol. 3.

ANATOMY

Apart from the brief reference to the microscopy of the hairs on the corolla included in Tercinet's description of *M. officinarum* L. given above, very little detailed work on the anatomical characters of the two species has been carried out. Maugini (1959) described the structure of the roots, rhizomes* and leaves of *M. autumnalis* Bertol. but her account did not include drawings to illustrate the general histological features. In view of the confusion which exists in the identification of the European species, it seemed of interest to investigate in detail the microscopy of the two species in order to determine whether or not there were any histological features which could be used to distinguish between them. Work on the roots and rhizomes has already been completed (Berry & Jackson, 1976) and this has shown that, both morphologically and anatomically, there are no marked differences between those obtained from the two species. Roots and rhizomes of *M. officinarum* L. attain a larger size than those of *M. autumnalis* Bertol. but both frequently show the characteristic anthropomorphic development. Contrary to the descriptions given in the literature, the colour of the cork layers cannot be used to distinguish between the two species as all the specimens examined were pale greyish or reddish brown externally on the roots with darker brown on the rhizomes. A preliminary comparison of the leaves has indicated that *M. autumnalis* Bertol. has considerably more trichomes, especially on the upper surface, than *M. officinarum* L. and it is anticipated that similar differences may be detected when comparison of the histology of the flowers and fruits has been completed. It is hoped that these anatomical investigations will provide useful information which can be applied to the differentiation of the species and serve as additional criteria in the evaluation of any so-called varieties.

CHEMISTRY

In spite of the considerable interest shown in the Mandrake plant throughout the ages and the traditional use of the roots in herbal medicine, surprisingly little work has previously been published on the chemical constituents. Any information which was available could not always be evaluated satisfactorily because it was not clear on which species the work had been carried out. Ahrens (1889), for example, examined samples which he called Sicily, Venice and Trieste mandrake, the first of which he thought was equivalent to *M. vernalis*, but he did not attempt fully to authenticate his material. Hesse (1901) also analysed samples from Trieste which he stated were *M. officinalis* Miller, adding the note that "this is differentiated, to be sure, by Bertoloni into *M. vernalis*, *M. autumnalis* and *M. officinarum* but this distinction has not generally been remembered and considerable contradiction is often found". Staub (1942), investigated samples which he called *M. autumnalis* Spr. for non-alkaloid constituents, but for his work on the alkaloids he used a commercial sample of *Radix Mandragorae offic.* which he considered to be a mixture of *M. autumnalis* Spr. and *M. vernalis* Bertol. (Staub, 1962).

It therefore seemed appropriate to carry out a full investigation of the constituents of authenticated samples of *M. autumnalis* Bertol. and *M. officinarum* L. to determine whether or not any differences exist which could be used as an additional means of distinguishing between the two species. Because of the established use of the underground parts of the plants in medicine the investigations were carried out on roots and rhizomes.

* Technically taproots (*Eds*)

The results of an extensive investigation into the alkaloid constituents has shown the presence, in both species, of hyoscyamine, hyoscine, cuscohygrine, apoatropine, 3α-tigloyloxytropane and 3,6-ditigloyloxytropane. Belladonnine was also detected in the dried roots (Jackson & Berry, 1973). Work on the non-alkaloid constituents has confirmed the presence of sitosterol and β-methylesculetin and has demonstrated that four free sugars, namely rhamnose, glucose, fructose and sucrose are also present; similar results were obtained for both species. The occurrence of β-methylesculetin in the fruits of *Mandragora* species has been reported by Tercinet who suggested that this probably explained the legends about the 'glowing' of the fruits in the evening.

Although this work did not indicate any differences in the constituents between the roots of the two *Mandragora* species, it did yield information of considerable interest in the wider chemotaxonomic field within the family Solanaceae. *Mandragora* is placed by Wettstein (1897) in the tribe Solaneae which includes a number of genera in which the occurrence of tropane alkaloids has been established. Amongst these genera, however, there is only one other, namely *Scopolia*, in which the presence of both tropic and tiglic acid esters has been reported. On the other hand, within the tribes Daturae and Salpiglossideae, are included several other genera in which both types of esters occur. It is possible, therefore, that *Mandragora* and *Scopolia* can be considered as a chemotaxonomic subgroup linking the Solaneae with similar subgroups within the Daturae and Salpiglossideae.

REFERENCES

AHRENS, F. B., 1889. Über das Mandragorin. *Berichte der Deutschen Chemischen Gesellschaft, 22*: 2159.

ARCANGELI, G., 1894. *Compendio della Flora Italiano*, 2nd ed: 390. Torin—Rome.

BERRY, M. I. & JACKSON, B. P., 1976. European Mandrake (*Mandragora officinarum* L. and *M. autumnalis* Bertol.); the structure of the rhizome and root. *Planta Medica, 30*: 281–290.

BERTOLONI, A., 1824. *Plante dell' orto botanico di Bologna. Giorn. Arcad. 21* (10): 191 (ex *Virid Bon: 6*).

BERTOLONI, A., 1835. *Commentarius de Mandragoris*. Bononiae.

CANDOLLE, A. L., P. De, 1852. *Prodromus Systematis Naturalis Regni Vegetabilis, 13*: 465. Paris.

CESATI, V., PASSERINI, G. & GIBELLI, G., 1867. *Compendio della Flora Italiana*: 367. Milan.

ENGLER, A. & PRANTL, K., 1897. *Die Naturlichen Pflanzenfamilien*, Part IV, Eb. 27.

FIORI, A. & PAOLETTI, G., 1908. *Flora Analitica d'Italia*. Padova.

FIORI, A., 1929. *Nuova Flora Analitica d'Italia, 2*: 316. Florence.

GUNTHER, R. T., 1933. *The Greek Herbal of Discorides, illustrated by a Byzantine, A.D. 512, translated by J. Goodyear 1655*. Oxford: OUP.

HELDREICH, Th. VON., 1886. Bemerkungen uber die Gattung *Mandragora. Mitteilungen der Geographischen Gesellschaft in Jena, 4*: 75–80.

HESSE, O., 1901. Über die Alkaloidie der Mandragorawurzel. *Journal für Praktische Chemie, 172*: 274–286.

HEYWOOD, V. H. (Ed.), 1972. Flora Europaea. Notulae Systematicae, No. 13 (Solanaceae). *Botanical Journal of the Linnean Society, 65*: 356–357.

JACKSON, B. P. & BERRY, M. I., 1973. Hydroxytropane tiglates in the roots of *Mandragora* species. *Phytochemistry, 12*: 1165–1166.

JUSSIEU, A-L. de, 1789. *Genera Plantarum*: 125. Reprinted 1964 by J. Cramer, Wheldon & Wesley Ltd.

LAZARO, B. & IBIZA, B. L., 1921. *Compendio de la Flora Espanola*, 3rd ed., *3*: 231. Madrid.

LINNAEUS, C., 1753. *Species Plantarum*: 181.

LINNAEUS, C., 1762. *Species Plantarum*: 252.

LINNAEUS, C., 1764. *Species Plantarum*: 259.

MAUGINI, E., 1959. Ricerche sul Genere *Mandragora, Nuovo Giornale Botanico Italiano e Bolletino della Societa Botanica Italiana* (n.s.), *66* (1–2): 34–60.

MILLER, P., 1768. *The Gardener's Dictionary*, 8th ed. London: printed by the author.

MORIS, Y. H., 1827. *Stirpium Sardoarum Elenchus*, Fasc. 1, 33, Carali ex typis Regiis.

MORIS, Y. H., 1858. *Flora Sardoa, 3*: 159–161. Taurinii.

PALHINA, R. T., 1939. *Flora de Portugal*, 2nd ed.: 637. Lisbon: Bertrand.

PARKINSON, J., 1629. *Paradisi in Sole Paridisus Terrestris:* 377–379. Reprinted 1904, Methuen & Co., London.

PARLATORE, F., 1883. *Flora Italiana, 4:* 697–700. Florence.

POLUNIN, O., 1969. *Flowers of Europe:* 372. Oxford: OUP.

POLUNIN, O. & SMYTHIES, B. E., 1973. *Flowers of S.W. Europe* p. 327. Oxford: OUP.

SPRENGEL, C. K. (Ed.), 1825. *Systema Vegetabilium, 1:* 699. Göttingen.

STAUB, H., 1942. Non-alkaloid constituents of Mandrake root. *Helvetica Chimica Acta, 25:* 649–683.

STAUB, H., 1962. The alkaloid constituents of Mandragora root. *Helvetica Chimica Acta, 45:* 2297.

TERCINET, L., 1950. *Mandragore Qui-es-tu?:* 117. Paris: published by the author.

TOURNEFORT, J. P. de, 1719. In A. de. Jussieu (Ed.), *Institutiones rei Herbariae.*

VIERHAPPER, F., 1915. Betreige zur Kenntniss der Flora Creta. *Österreichische Botanische Zeitschrift, 65:* 124–138.

WETTSTEIN, R., 1897. In A. Engler & K. Prantl (Eds), *Die Naturlichen Pflanzenfamilien*, Part IV, Vol. *3b:* 4.

WOODVILLE, W., 1794. *Medical Botany*, Part II: 35–36. London: James Phillips.

40. Biosystematic and taxometric studies of the *Solanum nigrum* complex in eastern North America

CHARLES B. HEISER, JR., DONALD L. BURTON AND EDWARD E. SCHILLING, JR.

Department of Biology, Indiana University, Bloomington, U.S.A.

The difficulties of the taxonomic recognition and treatment of the members of the *Solanum nigrum* complex are well exemplified by a study of the group in eastern North America, a region well known botanically. Most previous authors have recognized only one or two species (excluding the casually adventive species) in this area. On the basis of morphological studies using taxometric analyses, and artificial hybridization, we are able to distinguish four elements, all diploids; three of these are recognized taxonomically. (1) *S. pseudogracile* is the new name for the species that has been treated as *S. gracile* by most American authors and as *S. nigrescens* by D'Arcy. It is primarily a coastal species, extending from North Carolina to Mississippi, but found throughout most of Florida. It is quite distinct from the other eastern species, and appears to be most similar to *S. douglasii*. (2) *S. nodiflorum* Jacq. is a tropical species that is well represented in Florida. Edmonds has recently concluded that *S. americanum* Miller is an earlier name for *S. nodiflorum*. We have examined the lectotype and have some doubts whether it represents the species that Miller described. (3) *S. americanum* Miller is the most widespread species in the eastern United States. Hybrids between *S. nodiflorum* and *S. americanum* show reduced fertilities. Moreover, when *S. americanum* is used as the female parent the hybrids show various degrees of abnormal development; whereas the reciprocal combination gives normal appearing plants. (4) There is still another entity that we have designated as S-211 which is represented by several collections from Florida. S-211 gives hybrids with reduced fertility when crossed with other species. It is quite distinct from *S. americanum* in greenhouse cultures, but we have found it nearly impossible to distinguish from that species on the basis of herbarium specimens.

CONTENTS

INTRODUCTION

In a study of the *Solanum nigrum* complex some years ago (Heiser, Soria & Burton, 1965), two species from Florida had to be assigned numbers rather than names because of uncertainty regarding the correct names to be employed. Since that time three studies (Edmonds, 1972; Henderson, 1974; D'Arcy, 1974) have appeared that bear directly or indirectly upon the species of eastern North America. There is, however, still some disagreement as to the number of species in this area, as well as to the names that should be used for them. In the present paper we shall attempt to justify the presence of four species, all diploids, in the area east of 89° longitude in North America (excluding casually adventive species such as *S. nigrum*).

Although the taxonomic difficulties in section *Solanum* are generally well known, a brief review seems in order. The breeding system is primarily autogamous with occasional outcrossing. Barriers to gene exchange are often incomplete between closely related "species", so that occasional hybridization between "species" is not uncommon. When followed by inbreeding this may result in the formation of new populations different from either parent. The opportunity for such hybridization is enhanced by the weedy nature of most "species". Thus there are a large number of "microspecies" or "semispecies" (Grant, 1971) in this section. The problem comes in deciding how many of these deserve taxonomic recognition. Lumpers have tended to call all members of this complex *S. nigrum*—certainly an extreme view, for the complex includes diploids, tetraploids, and hexaploids which are morphologically distinguishable and which apparently seldom if ever interbreed in nature. At the other extreme are the splitters, who have recognized almost any variant as a species. Georg Bitter, who published a great number of new species in *Fedde's Repertorium* in the second decade of this century, belongs to this category.

A second difficulty stems from the nature of herbarium material in this group. Both flowers and mature fruit are usually necessary for identification, and frequently one or the other is lacking. Many taxonomically important characters, such as colour of the fruit and leaves and position of the calyx lobes in fruit, commonly become obscured in herbarium material. Moreover, vegetative parts of the plants may vary depending on environmental conditions, so that summer and winter collections of the same plant may look strikingly different.

A third difficulty arises in trying to associate the names used by earlier taxonomists with the plants we find today. Not only are the early descriptions brief and often vague, but they frequently fail to include characters we consider diagnostic. Many of Dunal's (1852) descriptions fall into this category. Even Bitter's very lengthy descriptions often fail to point out diagnostic characters of his species, and rarely does he mention the species to which his new one is related. Type material is often fragmentary and difficult to locate.

Since Dunal's time no one has attempted a world-wide revision of this section. There have been several regional studies but unfortunately these often leave a large number of problems unsolved. The most ambitious one to date is Edmond's (1972; Gray, 1968) treatment of the species of South America, but as she points out, further knowledge of the Central American and Mexican species will be necessary for a better understanding of the South American species.

Unfortunately, there are still disagreements among the recent workers not only as to circumscription of species but also in the application of names. This study will attempt to clarify some of these problems. Edmonds (1972) states that in places her treatment is tentative. Indeed, all taxonomic treatments must be tentative, especially when done on a regional basis—a restriction that applies here. Although the present study employs both biosystematic and taxometric data, many questions remain. In a group such as this, there is still a subjective element involved in delimiting the species.

KEY TO *SOLANUM* SECTION *SOLANUM* OF EASTERN NORTH AMERICA

Anthers exceeding 2·0 mm in length, usually 2·5–3·0 mm; style usually conspicuously exserted, surpassing anthers by 1 mm or more; sclerotic granules absent...............................1. *S. pseudogracile*
Anthers less than 2·0 mm; style usually not exserted or at most to 1 mm beyond anthers; sclerotic granules present or absent

    Calyx lobes strongly reflexed in mature fruit; sclerotic granules absent*; flowers 2–14 per inflorescence, usually more than 7 in largest inflorescences*; fruiting pedicels erect* (occasionally deflexed with age or in winter), to 8 mm long; fruit shiny, black
2. S. nodiflorum

    Calyx lobes adherent or spreading in mature fruit, occasionally somewhat reflexed; sclerotic granules present; flowers usually less than 6 per inflorescence; fruiting pedicels deflexed, to 13 mm long; fruit dull or shiny black, or rarely green in *S. americanum*

        Sclerotic granules usually more than 4, sometimes less; seedling leaves red on undersurface; undersurface of leaves glabrous or puberulent; anthers 1·4–2·0 mm long.........3. *S. americanum*
        Sclerotic granules at most 5, usually 4 or less; all leaves green on undersurface; undersurface of leaves puberulent; anthers 1·4–1·6 mm long 4. S-211

TAXONOMY

1. *S. pseudogracile* (S-210 group). In 1886 Gray applied the name *S. gracile* Link to a species of the eastern United States, a treatment that was adopted by Small (1903) and that has been rather widely followed. Heiser *et al.* (1965) realized that this name applied to another species and used an accession number (S-210) to refer to that species. That this species is distinct from other species in the eastern United States is clear, but there has been some question as to its relation to certain other New World species.

D'Arcy in his study of the Florida species of *Solanum* annotated material of this species from several herbaria as *S. douglasii* Dunal. After Edmonds (1972) reduced *S. douglasii* to synonymy under *S. nigrescens* Mart. & Gal., D'Arcy (1974) adopted the latter name for the species in Florida. Henderson (1974) has raised some question as to whether Edmond's inclusion of *S. douglasii* within *S. nigrescens* is justified. We shall use the name *S. douglasii* here for the widespread species of western North America whose range extends from California southward into Mexico and eastward into western Texas. Neither it nor the S-210 group has been found in eastern Texas so that the two apparently are discontinuous in their distributions. Although D'Arcy (1974) states that many plants of Florida and California "can be matched with complete confidence", we believe that there are actually two distinct elements represented. Heiser *et al.* (1965) found that plants from the two regions linked at a low level in their phenogram; however, only one sample of each was used. In order to obtain a larger sample for taxometric analysis we had to resort to herbarium material, although we feel that living material is

---

* Character not always applicable to specimens from outside of eastern North America.

more suitable for such studies. Thirty characters were scored from 47 herbarium specimens. In the graph (Fig. 40.1) resulting from principal component analysis the specimens from the eastern United States form a group that is largely separate from specimens from the western United States and Mexico. In the phenogram (Fig. 40.2) obtained by using cluster analysis, specimens from the two geographical regions were divided into different major clusters. It is of note that the characters that are most heavily weighted for the first axis of Fig. 40.1 include those by which we would distinguish S-210 from *S. douglasii*. In general, the S-210 group has shorter peduncles, fewer flowers per inflorescence, smaller flowers, larger pollen, and lacks sclerotic granules in the fruits. These results, of course, do not necessarily indicate whether the groups shown represent species, subspecies, or lesser categories.

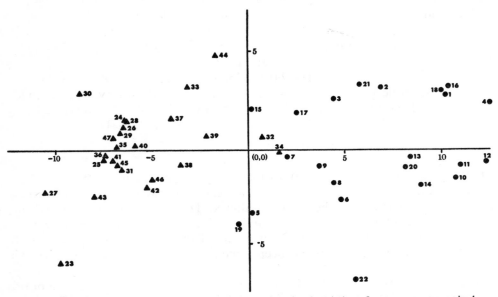

Figure 40.1. Graph of *Solanum pseudogracile* (▲) and *S. douglasii* (●) on first two axes recognized by principal component analysis based on 30 characters. Each entry represents a single herbarium specimen.

Crossing data support distinction of the S-210 group from *S. douglasii* at the specific level. Four accessions of the S-210 group and three of *S. douglasii* from different geographical sources were used. Berries were produced in only a few of the attempts in crosses between the two species, and these contained aborted seeds (Verhoek 1966; Burton, unpubl.).

D'Arcy (1974) also reports *S. nigrescens* from the West Indies, and has called attention to differences between the plants there and material from Florida. We have not yet studied living material from the West Indies, but we have seen a number of herbarium specimens in the collections of the New York Botanical Garden and the United States National Herbarium. While not all material can presently be assigned to species, we have not seen any specimens which could be identified as belonging to the S-210 group from the West Indies.

Although using the name *S. nigrescens* for the species in Florida, D'Arcy (1974) stated that he thought that an earlier name might be found. This species may have reached Europe in early collections, perhaps from the Carolinas. Possibly Dillenius' (1732) *S. nigrum vulgari simili, caulibus exasperatus* is this species, although it has generally been interpreted as representing *S. americanum* (see species number 3). The excellent drawing in Dillenius' work is a good match for some specimens of the S-210 group except that it does not show the characteristic exserted style, and it is said to come from Virginia. Linnaeus based his *S. nigrum* var. *virginicum* on Dillenius' plate, and Miller (1768) cites Dillenius in his description of *S. scabrum*. Henderson (1974), however, points out that Miller in his description states that his species has entire leaves and thus excludes Dillenius' drawing from consideration as the type for it shows dentate leaves. Our species may have either entire or dentate leaves, so it is not excluded from *S. scabrum*, *sensu* Miller, but the specimen that Henderson has chosen as the lectotype clearly belongs to another species, for which the name *S. melanocerasum* All. has previously been used. As long as this specimen is accepted as the lectotype, the name *S. scabrum* cannot be used for S-210.

One other name that might possibly refer to S-210 is *S. besseri* Weinm. ex Roemer and Schultes. The source is stated to be America, and the brief description might

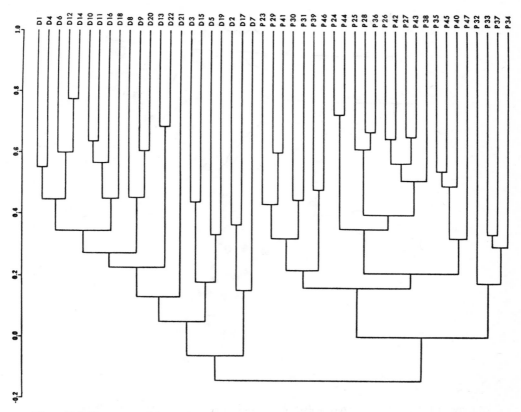

Figure 40.2. Phenogram of *S. pseudogracile* (P) and *S. douglasii* (D) based on same data as for Fig. 40.1.

apply to S-210 (or, for that matter, to a number of other species). Gray (1886) was of the opinion that *S. besseri* was not from North America but does not give the reasons for his decision. We have been unable to locate a type specimen of *S. besseri*.

Under the circumstances a new name must be proposed. This might be done at the subspecific rank under *S. douglasii*. However, we feel that our data justify ranking it as a taxonomic species. Therefore, we have proposed the name *Solanum pseudogracile* Heiser (Heiser, 1978) for this taxon.

S. pseudogracile extends along the coast from North Carolina to Florida, where it is found throughout much of the state, thence along the coast to Mississippi (Fig. 40.3). This species often appears to occur in relatively undisturbed sites along the seacoast, and its distribution parallels that of many native plants. While weeds may often attain distributions very similar to that of native plants (Shinners, 1948), we feel that *S. pseudogracile* may represent a primary element in the United States, rather than an introduced weed. Like most species of this group it shows variation in a number of characters. Particularly striking is the variation in the leaves, which may vary from large and dentate to small and entire. This variation, however, probably represents an age difference rather than stemming from genetic differences, for we find that a given plant grown in the greenhouse may exhibit both leaf forms,

Figure 40.3. *Solanum pseudogracile*. Records shown in triangles are reports from Radford *et al.* (1964) as *S. gracile*.

the large dentate leaves being found on young plants, whereas older plants may have only small, entire, leaves (Verhoek, 1966).

2, 3. *S. nodiflorum* and *S. americanum*. The next two species must be discussed together. In 1949 Stebbins & Paddock demonstrated the existence of two diploid species in the United States that had hitherto been confused with the hexaploid *S. nigrum*. They used the names *S. nodiflorum* Jacq. and *S. americanum* Miller for these species. The former is a tropical species that extends into Pacific North America as far as Washington and which we find also well established in Florida (Fig. 40.4).

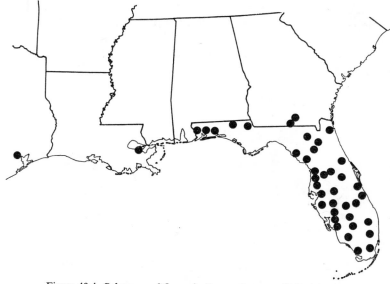

Figure 40.4. *Solanum nodiflorum* in the southeastern United States.

The latter is primarily a species of the eastern United States extending to Canada and the Great Plains (Fig. 40.5). They found that these species showed clear morphological differences and when intercrossed gave hybrids with highly reduced fertility. Their treatment of these species was subsequently accepted in a number of Floras in the United States.

There has apparently been some confusion recently regarding the distinction of *S. americanum* and *S. nodiflorum*. D'Arcy (1974), for example, places all material of these two species from Florida under *S. americanum*. We have grown accessions of both species from Florida, as well as from other geographical sources, and have also examined specimens from a number of herbaria. Our taxometric studies (Heiser *et al.*, 1965, Schilling & Heiser, 1976, and Fig. 40.6) indicate that two discrete groups are present. A reliable distinguishing character in the eastern United States is that *S. nodiflorum* lacks sclerotic granules whereas *S. americanum* usually has at least four per fruit. The two species also differ in a number of other characters (Table 40.1), although there is overlap in some of them. It may be noted here that although some previous authors have also used life duration to distinguish the two, this appears to be controlled by seasonal conditions, and both of these species may perennate under favourable conditions.

Crossing data for these two species also tend to support recognizing them as species. Although some of our earlier crosses gave hybrids with moderate fertility (Heiser *et al.*, 1965), a number of crosses made more recently (Burton, unpubl.) give hybrids generally showing 10% or less pollen stainability. We have also observed phenotypic differences in the reciprocal hybrids. With *S. nodiflorum* as the female parent, the hybrids are normal in appearance, whereas the reciprocal cross produces plants showing various degrees of abnormality which we have called a virus-like syndrome. The degree of abnormality varies, apparently depending on the strain employed in the crosses and the daylength under which the hybrids are grown. In some hybrids there is a slight distortion of the leaves, whereas at the other extreme only dwarf, twisted plants that fail to flower are produced. From various F_2's and backcrosses we have grown, it appears that the inheritance of the abnormality is not Mendelian, and apparently is controlled by the cytoplasm. Whatever the basis of this virus-like syndrome it would seem that it might strengthen the barrier to gene flow between the two species.

In his treatment of the Florida Solanums, D'Arcy (1974) described *S. americanum*

Figure 40.5. *Solanum americanum* (●), herbarium specimens thought to belong to the S-211 group (○), and material assigned to the S-211 group based on living plants (♦). *S. americanum* is much more common than indicated on the map and may well be found as a garden weed in every county of some of the states.

var. *baylisii*. This is the same taxon that Baylis (1958) had treated as *S. gracile* Otto. Although we have not seen the type (based on a plant cultivated in New Zealand), it seems clear from his description that D'Arcy's variety is the same as the material that we (Heiser *et al.*, 1965) had also previously called *S. gracile*, seeds of which (S-144) we had received from Dr Baylis. On the basis of morphology and crossing bahaviour (Heiser *et al.*, 1965) we feel that this is a species quite distinct from *S. americanum*. Henderson (1974) used the name *S. gracilius* Herter for this species. D'Arcy cites only one specimen of his variety from Florida (*Cooley and Eaton 6666*, USF), and from our examination of this sheet we are uncertain as to what species it should be referred.

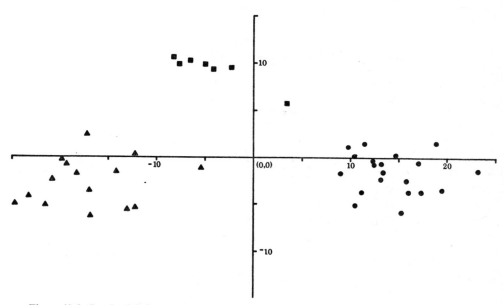

Figure 40.6. Graph of *Solanum americanum* (●), *S. nodiflorum* (▲), and S-211 group (■) on first two axes recognized by principal component analysis based on 44 characters. Each entry represents a greenhouse population of five plants.

Solanum nodiflorum exhibits considerable variability, and Gray (1968) and Henderson (1974) recognized infraspecific taxa within this species. Moreover, crosses between different strains of *S. nodiflorum* may yield very low F_1 pollen stainability (Henderson, 1974; Burton, unpubl.). Thus it may be that more than one taxon is represented by *S. nodiflorum*. Delimitation of these taxa must await a study of material from throughout the range of *S. nodiflorum*.

Recently Edmonds (1971) has concluded that *S. americanum* is an earlier name for *S. nodiflorum*, which, if correct, means that the names we have used here will have to be changed. Her decision was based on the examination of a specimen in the Miller Herbarium of the British Museum which she has selected as the lectotype. We have examined the Miller specimen, and while we feel that it is nearer to *S. nodiflorum* than to any other species known to us, it nevertheless raises several questions. In his discussion of *S. americanum*, Miller (1768) states that it 'grows

Table 40.1. Comparison of selected characters of the S-211 group, *S. americanum* and *S. nodiflorum* based on greenhouse cultures

Character	S-211 group			S. americanum			S. nodiflorum		
	Mean	S.D.	range	Mean	S.D.	Range	Mean	S.D.	Range
Days to flowering	47.0	2.6	44–51	34.9	4.7	28–41	39.4	2.8	35–44
Height of stem bifurcation	25.0 cm	3.0	20–29	11.0 cm	3.4	6–17	21.0 cm	4.3	15–31
Number of flowers per infl.	5.7	0.5	5–7	5.1	0.4	4–6	9.7	2.1	7–12
Corolla length	5.2 mm	0.4	4.7–5.8	5.5 mm	0.3	5.0–6.0	4.6 mm	0.5	3.8–5.4
Calyx length	2.0 mm	0.1	1.8–2.2	2.3 mm	0.5	1.9–3.9	1.6 mm	0.2	1.3–1.9
Style length	2.9 mm	0.2	2.6–3.2	2.8 mm	0.2	2.5–3.1	2.2 mm	0.3	1.8–3.9
Stigma length	0.20 mm	0.02	0.15–0.25	0.28 mm	0.02	0.25–0.40	0.17 mm	0.04	0.10–0.20
Fruiting pedicel length	6.8 mm	0.6	6.0–8.0	9.4 mm	1.4	7.2–12.8	6.8 mm	0.9	5.0–8.2
Sclerotic granule number	3.7	0.6	2–5	6.3	2.0	1–11	0.0	0	—
Fruiting calyx position	adherent			adherent/spreading			reflexed		
Fruiting pedicel position	deflexed			deflexed			erect		
Colour of young leaf undersurface	green			red			green		
Pubescence of lower leaf surface	dense			usually sparse			usually dense		
Stem colour	purple			usually green			green or purple		
Number of accessions (each of 5 plants)	7			20			15		

naturally in Virginia'. An examination of material from Virginia revealed nothing approaching *S. nodiflorum*. Moreover, sclerotic granules are clearly evident in several fruits on the specimen, and on this basis we think it is unlikely that Miller's source could have been from anywhere in eastern North America, although it could be from tropical America where plants of *S. nodiflorum* may contain sclerotic granules.

Although the lectotype matches *S. nodiflorum* in characters of the calyx, in many other characters it falls in the zone of overlap between *S. nodiflorum* and *S. americanum*; and in the somewhat deflexed pedicels on some of the infructescences it is nearer to *S. americanum*. The specimen shows good fruit set and two of the fruits have what appear to be filled seeds. Pollen stained in aniline blue in lactophenol was examined, and it was found that only 15% of the grains took on a deep blue stain and appeared to be normal sized. We do not know how reliable pollen counts are on specimens over 200 years old, but we have found that specimens 90 years old give counts of over 90% stainable pollen. We are aware, of course, that causes other than hybridity can result in pollen abortion. Nevertheless, the possibility exists that this specimen is a hybrid or a hybrid derivative of *S. americanum* and *S. nodiflorum*.

If the plant is of hybrid derivation how might it have originated? It seems very unlikely that seeds coming from Virginia could have been of hybrid origin unless they were sent by a collector who was growing both species. The only place where the two species grow together naturally is Florida. A more likely source of origin would have been Miller's garden. If Miller had been growing both species there (although he includes no other species in his Dictionary that might be interpreted as *S. nodiflorum*), and if his herbarium specimen was made after the first year in which the plants were grown, a hybrid could well have been present.

The foregoing is all highly speculative, although perhaps of some interest, but the pertinent question is whether this particular specimen can be used to typify Miller's species. Miller's description could have been based (1) upon this specimen, which could be either *S. nodiflorum*, or a hybrid involving *S. americanum*, (2) upon plants of *S. nodiflorum*, seeds of which most likely would have come from tropical America, (3) upon plants of *S. americanum*, as we are interpreting this species or, (4) upon plants which included both *S. nodiflorum* and *S. americanum*. A case might be made for any one of these alternatives, but we have selected the third for the following reasons: (i) In his discussion Miller states that the flowers of this species are white on the inside and purplish without. The corolla of *S. nodiflorum* is nearly always white, although at times it may have a violet or bluish tinge. On the other hand, a rather deep purple colour is frequently observed in *S. americanum* and its hybrids. It is impossible to make out any purplish colour in the lectotype, although it may have faded with age. (ii) Miller states that there are but a few flowers in each umbel. We, of course, have no way of knowing what Miller meant by "few", but the flowers of *S. americanum* are generally only four to six in each inflorescence whereas they are usually more than six in *S. nodiflorum*. The lectotype shows as many as nine flowers to an umbel. (iii) Miller states that the leaves may have few indentures on their edges. The leaves of both *S. americanum* and *S. nodiflorum* may be either toothed or entire. The leaves on the lectotype are nearly all entire or slightly sinuate. (iv) Although the accuracy of some of Miller's comments as to his geographical sources may be questioned, nevertheless we feel that consideration should be given to his statement that this plant grows naturally in Virginia.

Although we must admit that some uncertainty may exist as to the correct application of Miller's *S. americanum*, we shall continue to use this name for the widespread eastern North American plant, as has been done in most recent Floras. If this name is employed in this sense it means that another lectotype must be selected, or if none is available a neotype will have to be designated. This course of action also means that the well established name, *S. nodiflorum*, can continue to be used for the widespread tropical plant.

So far we have located only two early names other than *S. nigrum* var. *virginicum* that likely refer to *S. americanum*. Following his description of *S. crenato-dentatum*, Dunal cites an *Asa Gray* collection from New York. From our examination of the International Documentation Centre microfiche of this specimen we are fairly certain that it represents *S. americanum*. However, Dunal also cites several other collections under his species that do not belong to *S. americanum*. One of these, *Gaudichaud 112* from Chile, Edmonds (1972) places in synonymy with *S. furcatum* Dunal; two others, *Gay 2* from Chile and *Berlandier 118* from Mexico, she places with *S. nigrescens*. D'Arcy (1973) also gives *S. crenato-dentatum* as a synonym of *S. nigrescens* and designates *Gay s.n.* as the lectotype, although this specimen is not cited with Dunal's original description. While a case might be made for selecting another lectotype, we feel that the name *S. crenato-dentatum* must be rejected as it is based on several entirely discordant elements. A second name that apparently applies, however, is *S. ptycanthum* Dunal (1852), the type locality of which, Savannah, Georgia, falls within the range of our species, and the brief description could well apply to *S. americanum*.

4. S-211. In 1960 a *Solanum* was collected in Florida (S-211) that appeared to be different from our other accessions. It was found that crosses with other species failed or gave hybrids with reduced fertility (Heiser *et al.*, 1965). Since that time we have acquired several additional collections that are similar to S-211. Several new hybrids have been secured involving these collections and the results of the crosses with the other species treated here are summarized in Table 40.2. The source of the accessions used for the crosses is given in Table 40.3. A taxometric study (Fig. 40.6), moreover, reveals that the S-211 group is well set off from both of these species morphologically. We find that it is readily recognized in the greenhouse, but herbarium specimens are difficult or impossible to distinguish from *S. americanum*. Many of the characters by which we distinguish the two in the greenhouse are not apparent in herbarium collections, and other characters show some overlap (Table 40.1). From our tentative determination of herbarium material, however, we feel that the S-211 group may prove to be much more common than *S. americanum* in Florida (Fig. 40.5).

Whether taxonomic recognition should be given to the S-211 group poses something of a dilemma. Its morphological distinctness and the results of the hybridizations suggest that it deserves taxonomic recognition, although the moderate fertility in crosses with *S. nodiflorum* perhaps implies that it is not a "full-fledged" species. On the other hand, the difficulty of distinguishing it from *S. americanum* in herbarium collections indicates that little purpose is served at present in giving it any taxonomic ranking.

Table 40.2. Pollen stainability of hybrids involving S-211 group

A\B	S-211				S. pseudo-gracile			S. americanum								S. nodiflorum					
	211	551	557	558	210	342	565	118	264	317	485	459	460	4769G	4769P	122	251	561	6684	CRAN	988
211		97 / 96	96 / 97	96 / 96	ab / 69*	no / 3	ab / ng	9 / 24	7 / 8	11 / —	7	13	7	10 / 16	8 / 12	1 / 40	6 / —	6	41 / 60	35 / 47	23 / 22
551	96 / 97		96 / 95	95 / 95	ab / ab	— / ng	ab / —	13	8 / 8	10 / 15	6	9	6	13	10	0 / 55	0 / 0	0	83 / 17	p / L	— / —
557	97 / 96	95 / 96		94 / 96	ng / ab	— / —	— / ab	8		13		7				4 / 42	L		68	L / L	— / —
558	96 / 96	95 / 95	96 / 94		p / ng	— / —	— / —			18		—				3 / 44	8	8	66	p	p

Upper figure, A × B; lower figure, B × A; blank, cross not attempted.

—, Cross failed; L, fruit lost; no, fruit harvested, no seeds; ab, fruit harvested, aborted seeds; ng, fruit harvested, seeds failed to germinate; p, fruit harvested, seeds not planted.

The figure indicates the percentage of stainable pollen grains, based on at least 5 hybrid plants except for exception noted.

See Table 2 for source of accessions.

*Pollen stainability for the hybrid 210 × 211, reported as 69% by Heiser et al. (1965), was apparently based on a single plant. Attempts to repeat this cross have been unsuccessful.

Table 40.3. Sources of parental strains of hybrids listed in Table 1

Accession number	Collector	State or county	County or province
211	C. B. Heiser	Florida	Palm Beach
551	D. L. Burton	Florida	Monroe
557	D. L. Burton	Florida	Brevard
558	D. L. Burton	Florida	Brevard
210	C. B. Heiser	Florida	Broward
342	C. R. Bell	North Carolina	Onslow
565	D. L. Burton	Georgia	Glynn
118	W. P. Stoutamire	Florida	Highlands
264	G. E. Morley	Kansas	Republic
317	C. B. Heiser	Massachusetts	Hampden
458	C. B. Heiser	Indiana	Tippecanoe
459	C. B. Heiser	Indiana	Tippecanoe
460	D. L. Burton	Iowa	Dubuque
4769G	C. B. Heiser	Missouri	St. Louis (City)
4769P	C. B. Heiser	Missouri	St. Louis (City)
122	C. Rick	Peru	Lima
251	J. McCaskill	California	Solano
561	D. L. Burton	Florida	Indian River
6684	C. B. Heiser	Costa Rica	Heredia
CRAN	M. P. Coons	Costa Rica	Puntarenas
988	G. J. Gastony	Mexico	Vera Cruz

ACKNOWLEDGEMENTS

This study was carried out with the aid of a grant from the National Science Foundation. The authors would also like to thank Dr W. Greuter for his examination of the type of *S. ptycanthum* Dunal, and the curators of the following herbaria for the loan of specimens: BM, FLAS, GA, LAF, MISS, NO, NY, TEX, UNA, US, USF, VPI and WILLI. A list of the specimens examined for this study may be obtained by writing to the Curator of the Herbarium, Department of Plant Sciences, Indiana University, Bloomington, Ind., 47401, U.S.A.

REFERENCES

BAYLIS, G. T. S., 1958. A cytogenetical study of New Zealand forms of *Solanum nigrum* L., *S. nodiflorum* Jacq., and *S. gracile* Otto. *Transactions of the Royal Society of New Zealand, 85:* 379–385.
D'ARCY, W. G., 1973. Solanaceae. In E. Woodson, Jr., R. W. Schery & Collaborators (Eds), Flora of Panama. *Annals of Missouri Botanical Garden, 60:* 573–780.
D'ARCY, W. G., 1974. *Solanum* and its close relatives in Florida. *Annals of Missouri Botanical Garden, 61:* 819–867.
DILLENIUS, J. J., 1732. *Hortus Elthamensis.* London.
DUNAL, M. F., 1852. Solanaceae. In A. de Candolle (Ed.), *Prodromus Systematis Naturalis Regni Vegetabilis, 13,* 1.
EDMONDS, J. M., 1971. *Solanum.* In W. T. Stearn, Taxonomic and nomenclatural notes on Jamaican gamopetalous plants. *Journal of the Arnold Arboretum, Harvard University, 52:* 634–635.
EDMONDS, J. M., 1972. A synopsis of the taxonomy of *Solanum* Sect. *Solanum* (Maurella) in South America. *Kew Bulletin, 27:* 95–114.
GRANT, V., 1971. *Plant Speciation.* New York: Columbia University Press.
GRAY, A., 1886. *Synoptical Flora of North America.* Washington: Smithsonian Institution.
GRAY, J. M., 1968. *The Taxonomy of the* Morella *Section of the Genus* Solanum L. *within South America.* Ph.D. thesis. University of Birmingham.

HENDERSON, R. J. F., 1974. *Solanum nigrum* L. (Solanaceae) and related species in Australia. *Contributions from the Queensland Herbarium*, No. 16.

HEISER, C. B., SORIA, J. & BURTON, D. L., 1965. A numerical taxonomic study of *Solanum* species and hybrids. *American Naturalist, 99:* 471–488.

HEISER, C. B., 1978. *S. pseudogracile*—a new species of *Solanum*, section *Solanum* (*Maurella*). In J. G. Hawkes (Ed.), Systematic notes on the Solanaceae. *Botanical Journal of the Linnean Society, 76:* 294.

MILLER, P., 1768. *The Gardeners Dictionary*, 8th ed. London.

RADFORD, A. E., AHLES, H. E. & BELL, C. R. ,1964. *Manual of the Vascular Flora of the Carolinas*. Univ. of North Carolina Press, Chapel Hill.

SHINNERS, L. H., 1948. Geographic limits of some alien weeds in Texas. *Texas Geographical Magazine, 12* (Spring No.): 16–25.

SMALL, J. K., 1903. *Flora of the Southeastern United States*. New York.

SCHILLING, E. E. & HEISER, C. B., 1976. Re-examination of a numerical taxonomic study of *Solanum* species and hybrids. *Taxon, 25:* 451–462.

STEBBINS, G. L. & PADDOCK, E. F., 1949. The *Solanum nigrum* complex in Pacific North America. *Madroño, 10:* 70–81.

VERHOEK, S. E., 1966. *Identification of One Species of the* Solanum nigrum *Complex by Genetic and Numerical Taxonomic Procedures*. Master's thesis, Indiana University.

41. Biosystematics of *Solanum* L., section *Solanum* (*Maurella*)

JENNIFER M. EDMONDS

Botany School, University of Cambridge, U.K.

The taxonomic complexity of the widespread and variable species associated with the section *Solanum* is now generally accepted. Experimental and numerical taxonomic studies have shown that the section is taxonomically critical for several reasons. Many species are very variable genetically and can exhibit considerable phenotypic plasticity; they form a polyploid series; infra- and interspecific hybridization can occur leading to complex population variation, and the species also show inherent discordant variation. Allowances must be made for these complications in any classification of the section, and they are all difficult to assess from herbarium material. In addition, there is considerable nomenclatural confusion, largely resulting from the publication of over 300 specific and infraspecific post-Linnaean names.

A provisional taxonomic framework of the section, based on herbarium studies, and supported by cytological data, hybridisation experiments, information from seed protein electrophoresis and numerical clustering behaviour, has been established. This largely deals with the South American and European taxa, where respectively 18 and 6 species are so far recognized. Different kinds of experimental and numerical data are also proving useful as indicators of the genome relationships of some polyploid *Solanum* species. Attempts are currently being made to resynthesise some of these polyploids. Particular emphasis is being placed on *S. nigrum* L., the black nightshade, where an increasing amount of evidence is pointing to a trigenomic origin for this Old World hexaploid species.

CONTENTS

INTRODUCTION

The section *Solanum* is one of the largest, most widespread and most variable species groups of the genus *Solanum* L. It is usually known as the section *Morella*

(*Maurella* (Dun.) Dumort.; *Morella* (Dun.) Bitt.), and centres around the species known as the black, common or garden nightshade, *Solanum nigrum* L. This weed of disturbed and ruderal habitats is the type species of both the section and the genus. The species group is, in fact, often referred to as the "*Solanum nigrum complex*". Although the section takes its name from this predominantly Eurasian species, it is composed of a number of distinct taxa, with the species showing their greatest concentration and diversity in the tropics, particularly in South America.

The taxonomic complexity of the species associated with the section *Solanum* is now generally accepted (see Stebbins & Paddock, 1949; Heiser, 1963; Symon, 1970; Venkateswarlu & Rao, 1972; D'Arcy, 1973, 1974, for example). Recent classical, experimental and numerical studies have shown that this complexity can be attributed to a number of different causes. Allowances must be made for these complications in any classification of the section.

SOURCES OF TAXONOMIC COMPLEXITY

Historical

Certain taxa belonging to the section *Solanum* have been the subject of extensive taxonomic study. Dillenius (1732) seems to have been the first to consider the species group, delimiting taxa with four polynomials. Linnaeus (1753) later modified Dillenius' work, describing six varieties under the binomial *S. nigrum*. Since then, the *S. nigrum* group has been reclassified innumerable times. Characters used by later taxonomists to separate additional taxa often differed only slightly from those given for species by earlier workers, and synonymy is extensive within the section. Although over 300 post-Linnaean specific and infraspecific names have now been published, some authors continue to treat different members of the section *Solanum* as belonging to one highly variable species, namely *S. nigrum* (e.g. Tandon & Rao, 1964, 1966a, b; Rao, 1971; Rao & Tandon, 1969a; Rao, Khan & Khan, 1971; Venkateswarlu & Rao, 1971, 1972).

A bibliography of the more important taxonomic treatments of species belonging to the section *Solanum* is given in Edmonds (1977a). Two of the most industrious workers on the species group must be specially mentioned, however, as they are responsible for a great deal of the current taxonomic confusion. Dunal (1813, 1852) described or recognized 60 species belonging to this section, plus many infraspecific taxa. Most of these taxa are difficult to distinguish from his rather brief diagnoses, which usually fail to provide any quantitative measurements. Moreover, he not only cited numerous syntypes, but in many cases, cited the same specimens under different species, which has led to considerable nomenclatural confusion. Bitter (1912–1922), later described approximately 100 new species belonging to the section *Solanum*. Most of these were slight morphological variants of species already described, and the taxonomic confusion here has been increased by the destruction of many of his types in the Berlin-Dahlem herbarium, during the last war.

Such nomenclatural and taxonomic problems can only be solved by studying the older literature, and by searching various European herbaria for authentic specimens which, though undocumented, may still exist. Indeed, many duplicates of the specimens cited by Bitter have now been located.

Phenotypic plasticity

Species belonging to the section *Solanum* display varying amounts of phenotypic plasticity, particularly in their vegetative features. Senescence may be accompanied by smaller and fewer flowers and fruits than usual, while the gene for anthocyanin pigmentation in flowers seems to be dependent on light intensity and temperature for its expression in some species. In describing this variation, some authors have emphasised the difficulties of defining the limits within which features such as plant habit, leaf size and form, stem winging and flower colouration, for example, are genetically fixed (e.g. Baylis, 1958; Henderson, 1974; Edmonds, 1977a).

Genetic variation

In addition, certain species exhibit considerable genetic variation, both florally and vegetatively. Leaf margins, for example, may vary from entire to sinuate-dentate in different populations of the same species, e.g. *S. americanum* Mill., *S. furcatum* Dun. and *S. nigrum*. The subspecies of both *S. nigrum* and *S. villosum* Mill. are characterized by different indumentum types, while populations of these two species also exhibit different berry colours (see Edmonds, 1977a). Though flower colour is phenotypically plastic in some species, in others it seems to be under genetic control, as in the occasional purple striping on the corolla of *S. opacum* A.Br. & Bouché and *S. retroflexum* Dun., for example (Henderson, 1974).

The limits of both phenotypic plasticity and genetic variation that are acceptable within taxa can only be assessed from observations of living populations, preferably in their natural habitats, coupled with observations of genetic segregation, following artificial hybridisation between the variants concerned.

Polyploidy

The species of the section constitute a polyploid series, with $2n=2x=24$, $2n=4x=48$ and $2n=6x=72$. Octoploid plants ($2n=8x=96$) have been reported on two occasions (Heiser, 1963; Edmonds, 1977a).

Though living material is necessary to determine the chromosome numbers of taxa, herbarium material can sometimes provide clues of the ploidy levels concerned. Both pollen diameter and stomatal length tend to increase with ploidy level. Pollen sizes are not always directly proportional to chromosome number in this group (see Heiser, 1963; Henderson, 1974), but, if used in conjunction with stomatal sizes, a good estimate of the ploidy level may usually be obtained.

The histograms in Figure 41.1 illustrate the mean measurements of these characters for a number of accessions of known chromosome number, of which 56 were diploid, 20 were tetraploid and 19 were hexaploid. The pollen was measured in acetocarmine in the form of a glycerol jelly (Marks, 1954), and the stomatal aperture lengths from polystyrene epidermal peels taken from comparable sites on comparable leaves of herbarium specimens. The significance of difference between the three major ploidy levels was investigated by means of a *t*-test. For the stomata, the differences between the diploid and tetraploid measurements ($t=8·77$, $n=74$), between the diploid and hexaploid measurements ($t=17·43$, $n=73$) and between the tetraploid and hexaploid measurements ($t=5·16$, $n=37$) were all highly significant ($P>0·01$). Similarly, for the pollen, the differences between the diploid and tetraploid measurements ($t = 10 \cdot 23$, $n = 74$) and between the diploid and hexaploid measurements ($t = 7 \cdot 42$, $n = 73$) were also highly significant ($P > 0 \cdot 01$), though the difference between the tetraploid and hexaploid measurements was not significant ($t = 1 \cdot 5$, $n = 37$, $P = 0 \cdot 1$–$0 \cdot 2$). These stomatal measurements refer to aperture lengths, and may be smaller than those given in the literature for complete stomatal lengths (e.g. Ahmad, 1964; Sharma & Sen, 1969).

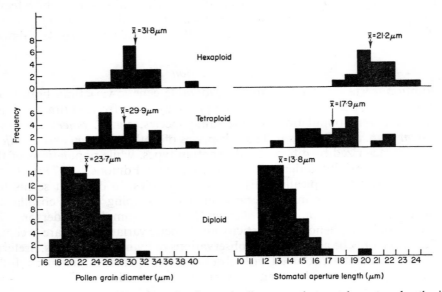

Figure 41.1. Frequency histograms of pollen grain diameter and stomatal aperture lengths in various *Solanum* accessions of known chromosome number. The frequency plotted represents the mean of ten measurements for each accession.

Natural hybridisation

Artificial hybridisation studies have demonstrated that morphological divergence within the section is usually accompanied by genetical isolation, but that nearly all species can be induced to cross with at least one related species, even if differing ploidy levels are involved. Although the crossability behaviour observed or induced in cultivation may be very different from that operative in natural habitats, the possibility remains that most species in the section are capable of hybridisation, at least initially. The species are predominantly self-pollinating, but out- and cross-breeding can occur, and natural infra- and interspecific hybridisations, especially among the smaller-flowered diploids, have now been reported by several workers (Stebbins & Paddock, 1949; D'Arcy, 1974; Henderson, 1974). Indeed, Stebbins & Paddock considered some of the diploid entities to be "species in the making, which have developed partial but not complete isolating mechanisms". Such taxa are usually separated eco-geographically, but in intermediate habitats F_1 generation plants may become established, which may back-cross with the parental populations, resulting in introgression. D'Arcy (1974) even referred to the diploids found in Florida as a syngameon (Grant, 1963), consisting of "partially interbreeding, partially isolated, sympatric population systems".

Natural hybrids have also been reported at higher ploidy levels, e.g. infraspecific hybrids of the hexaploid *S. nigrum* (Venkateswarlu & Rao, 1972). Similarly, hybrids between taxa of different ploidy levels have been recorded, e.g. between the hexaploid *S. scabrum* Mill. and the diploid *S. nodiflorum* Jacq., and probably between the hexaploid *S. nigrum* and the tetraploid *S. villosum* (both Henderson, 1974), and between hexaploid *S. nigrum* and the diploid *S. sarrachoides* Sendtn. (Leslie, 1978).

Natural hybridisation is probably more widespread in this section than generally supposed. Though this may be accompanied by genetic breakdown in the F_1 or F_2 generations, it may also be followed by backcrossing to the parental species. Such phenomena would result in morphologically and genetically complex population variation, and the collection of specimens from such populations would account for much of the morphological difficulty experienced in the herbarium.

When hybridisation is suspected, herbarium material can sometimes provide clues of its occurrence, through intermediate morphology, through empty but mature peduncles (where F_1 sterility is involved) and through pollen stainability. Large numbers of collapsed, unstained pollen grains are usually indicative of a hybrid origin (Gray, 1968; Henderson, 1974), as all species in this section usually exhibit very high pollen stainabilities (Edmonds, 1977a).

Discordant variation

During a population differentiation study*, the degree of discordance in 32 *Solanum* populations, representing 20 different species, was investigated using a simple randomisation technique (see Jardine & Sibson, 1971). Discordance values were calculated between pairs of dissimilarity matrices derived from pairs of random subsets of 41 morphological attributes. As the size of the random subsets increased, so the mean of the discordances decreased, approaching an asymptotic value, which can be interpreted as a measure of the amount of inherent discordance which must be tolerated however large the set of characters considered. With a character sample size of 15, a relatively large amount of inherent discordance was still evident in the differentiation between these *Solanum* populations. Indeed, such a mosaic pattern of variation, where the relative extents to which populations are differentiated do not coincide for different sets of attributes, might be expected where populations are largely autogamous, but are occasionally out- or cross-pollinated, resulting in some recombinants, as in many of the Solanums used in this analysis.

The discordance study also indicated that a large number of characters must be used for the differentiation of *Solanum* populations. The relatively slow decrease in discordance values with an increase in character sample size, demonstrated that around 15 characters are necessary to provide a stable estimate of the relative dissimilarities in these populations of *Solanum*.

PROVISIONAL TAXONOMIC FRAMEWORK

In spite of these complications, the section *Solanum* is clearly composed of a number of distinct taxa, which can be thought of as good species. Standard

---

* Full details of the numerical analyses mentioned in this paper are given in Jardine & Edmonds (1974), or Edmonds (1977b).

orthodox procedures have been used to construct a provisional taxonomic framework, which can be modified as experimental data become available. This framework is based on South American and European representatives, though a few species from other areas have been included. Currently, 18 species are recognised in South America (Edmonds, 1972, 1977a, 1978), and six in Europe (Hawkes & Edmonds, 1972), with some of the species being subdivided infraspecifically. There are some additional species in Central and North America, Australasia and Africa which are also known to be good. Though neither the total number of taxa recognised nor the names used for them can be considered final, the section *Solanum* is probably composed of approximately 30 species.

Schematic keys are used to summarise the South American and European species in Edmonds (1977a). The taxonomic value of the different classes of morphological characters used to differentiate the various species, and selected intuitively during the classical work, derived considerable support from later numerical analyses. Inflorescence and infructescence characters were extensively used for the major dichotomies in my identificatory keys, particularly those which could be numerically measured or objectively assessed. Such characters proved to be the best predictors of the overall pattern of dissimilarities in populations of *Solanum*, and dissimilarity matrices based on these sets of attributes gave the most meaningful hierarchic clusters.

South American glandular species

The South American species can be conveniently divided into species with predominantly glandular or with predominantly eglandular hairs. Of these, the glandular species are relatively easy to distinguish morphologically, being readily separated on inflorescence and infructescence characters. They include the diploid species *S. sinuatiexcisum* Bitt., *S. fiebrigii* Bitt., *S. glandulosipilosum* Bitt., *S. tweedianum* Hook., *S. insulae-paschalis* Bitt. and *S. sarrachoides* Sendtn., and two tetraploids *S. excisirhombeum* Bitt. and *S. fragile* Wedd. These species are all fairly localised in their distribution in South America, apart from *S. sarrachoides* which is found throughout most of the subcontinent.

The morphological diversity of these glandular species is accompanied by genetic isolation. Although I have not yet managed to obtain living material of *S. sinuatiexcisum* or *S. fiebrigii*, among the other diploids, the isolation of *S. glandulosipilosum*, *S. tweedianum* and *S. insulae-paschalis* was found to be maintained by failure of initial crosses, by the development of seedless berries or empty or inviable seeds, by F_1 sterility or by F_2 breakdown. Similar stages of genetical breakdown were encountered in hybridisations involving *S. sarrachoides*, though any "successful" crosses terminated in F_1 sterility. At the tetraploid level, the isolation of the two species *S. fragile* and *S. excisirhombeum* has so far proved to be maintained by failure to cross, or by the development of abortive, empty or inviable seeds. These isolatory mechanisms apply to hybridisations attempted both between the glandular species, and between the glandular and the eglandular taxa. When used in hybridisation attempts between ploidy levels, any F_1 progeny raised were sterile. Details of the hybridisation studies mentioned in this paper, and of reports of similar crosses by other workers, are given in Edmonds (1977a).

Seed protein electrophoretic data* supported the morphological distinction and genetical isolation of these glandular taxa. Acrylamide gels of the five species *S. glandulosipilosum*, *S. tweedianum*, *S. sarrachoides*, *S. insulae-paschalis* and *S. excisirhombeum* all showed species-specific band patterns. Calculation of the percentage similarities of these patterns (after the method used by Vaughan & Denford, 1968) resulted in low values for these glandular species, particularly when compared with each other, but also when compared with the other taxa analysed electrophoretically.

Similarly, some of the hierarchic clusters delimited by the highest dissimilarity levels during the population study using the single-link cluster method (see Jardine & Sibson, 1971), and plotted from the dissimilarity matrix of all 41 attributes, contained these glandular South American species. In particular, the accessions of *S. tweedianum*, *S. glandulosipilosum*, *S. insulae-paschalis* and *S. fragile* were conspicuously separated at very high dissimilarity levels. Though not so isolated, *S. sarrachoides* also separated into its own cluster at a relatively high level. The tetraploid *S. excisirhombeum* was the only glandular taxon showing an exception to this behaviour, tending to cluster with members of the *S. nigrum* complex.

South American eglandular species

In contrast, the eglandular South American species are less easy to distinguish morphologically, especially in the herbarium, and some are isolated by incomplete sterility barriers. Although many exhibited characteristic protein band patterns, these gave variable and relatively high percentage similarity values when compared with one another. These similarity values did, in fact, reflect the morphological variability of the taxa concerned; those exhibiting the greatest infraspecific variation gave the most variable percentages, e.g. *S. americanum*, while those showing little infraspecific variation gave more stable percentages, e.g. *S. interandinum* Bitt. In addition, these eglandular taxa were found to cluster more heterogeneously in the single-link cluster analysis, and were generally differentiated at lower dissimilarity levels than their glandular relatives.

Among the most morphologically distinct eglandular taxa are the large-flowered diploids *S. itatiaiae* Glaz. *ex* Edmonds and *S. aloysiifolium* Dun. (in which I have provisionally recognized two varieties, *aloysiifolium* and *polytrichostylum* (Bitt.) Edmonds). At the other extreme, is the tiny-flowered diploid which is usually known as *S. nodiflorum* Jacq., but which I have been calling *S. americanum* (Edmonds, 1971). This is probably the most widespread and variable member of the section *Solanum*, and is the species to which the word "complex" could be most aptly attached. It is distributed throughout the world, and its vegetative divergence is often accompanied by variable genetic behaviour. I have provisionally recognised two varieties of *S. americanum* in South America (i.e. var. *americanum* and var. *patulum* (L.) Edmonds). The reasons for this, and some of the different taxonomic treatments accorded the variability found in this species are given in Edmonds (1977a). The correct name for the taxon that I am calling *S. americanum* may well prove to be *S. nodiflorum*; Heiser, Burton & Schilling (this Volume,

* Details of the seed protein electrophoresis mentioned in this paper are given in Edmonds & Glidewell (1977).

p. 513) argue the case for retaining Miller's epithet *americanum* for the closely related but distinct species found in eastern North America.

Other eglandular species include the smaller-flowered diploids *S. pentlandii* Dun., *S. zahlbruckneri* Bitt., *S. nigrescens* Mart. & Gal. and *S. sublobatum* Willd. (which is variously referred to as *S. ottonis* Hyl., *S. gracile* Dun. or Otto, or *S. gracilius* Hert. by other workers); the tetraploid *S. interandinum* (from which two auto-octoploid plants were reported by Heiser, 1963), and the two hexaploids *S. furcatum* and *S. arequipense* Bitt.

Despite difficulties in their initial morphological delimitation, these eglandular taxa are considered to be good, though variable, species. They were found to maintain their isolation, both from one another, and from other members of the section, through mechanisms analogous to those operative in the glandular species, with the number of mechanisms actually observed in each taxon generally reflecting the relative use of the species concerned in the hybridisation programme. Nevertheless, most of these taxa could be induced to cross with at least one other in cultivation. The isolation of the majority of these eglandular species is undoubtedly assisted by their fairly restricted eco-geographic distributions in South America, but interspecific hybridisation is possible where their distributions overlap. When different ploidy levels are involved, any resultant progeny would be sterile but natural hybridisation among the diploids, particularly among those of similar floral sizes, could lead to backcrossing and introgression.

This problem is particularly relevant to the smaller-flowered species *S. americanum* (both in the sense used here (i.e. cf. *S. nodiflorum*) and in the sense used by Heiser *et al.* (1978) for the different taxon found in eastern North America), *S. sublobatum*, and to an additional taxon found in southern North America and Mexico, *S. douglasii* Dun. These are among the diploids which Stebbins & Paddock (1949) described as "borderline species", and which D'Arcy (1974) described as a syngameon. Though they are again usually separated eco-geographically, they are the species in which introgression is most likely to occur, if they grow together for any length of time (see Stebbins & Paddock, 1949, for example). Similar intergradation might also occur between other eglandular taxa, particularly in artificially disturbed habitats, since the degree of genetical isolation so far observed in some of these species could also be described as incomplete.

European species

Three of the species most widely distributed in South America are also found as casuals, or are locally naturalized, in Europe. These are the glandular species *S. sarrachoides* (where it is often referred to as *S. nitidibaccatum* Bitt., and which might prove to be a distinct infraspecific variant), and the two eglandular taxa *S. sublobatum* and *S. americanum*.

Another species found as a casual in Europe, usually as an escape from cultivation, is the hexaploid *S. scabrum*. Commonly known as the garden huckleberry, it has been widely cultivated for its fruits, but is a species of unknown origin. It is morphologically distinct and genetically isolated from other hexaploid Solanums. The seed proteins of this species gave a characteristic band pattern and low percentage similarity values when compared with the patterns of other species in the section.

The remaining European taxa are native; they are the tetraploid *S. villosum* in which the subspecies *villosum* and *alatum* (Moench) Edmonds are currently recognized, and the hexaploid *S. nigrum*, in which two subspecies *nigrum* and *schultesii* (Opiz) Wess. are also recognised. Both species are widely distributed in Eurasia, are highly variable, and are often very difficult to distinguish in the herbarium. Their subspecies are based on eglandular or glandular-haired indumenta, and they exhibit infraspecific reciprocal compatibility, with regular Mendelian segregation in the F_2 (Edmonds, 1977a and ined.). Their crossability behaviour is therefore very different from that observed between the glandular and eglandular South American taxa. Both *S. villosum* and *S. nigrum* appear to be genetically isolated from other members of the section, again exhibiting a range of isolating mechanisms.

The morphological variability of these species was found to be reflected in the hierarchic clusters obtained from the numerical analysis. Though accessions of both species were included in the same major clusters, they were differentiated at comparatively low levels, with those of *S. villosum* becoming separated in the dendrogram derived from the dissimilarity matrix of all attributes. The protein band patterns of *S. nigrum* were, however, acceptably species-specific. A good degree of band matching was possible between the five morphological variants used, and the similarity values calculated from these accessions were high. The patterns obtained from the variants of *S. villosum* were more variable, and this was reflected in their lower similarity values. Comparisons between the band patterns of these two species and those of the other *Solanum* taxa examined electrophoretically yielded comparatively high values. These two European species are therefore analogous to the eglandular South American taxa in their variability, and in the difficulties associated with their delimitation.

GENOME RELATIONSHIPS

The relative usefulness of the different kinds of taxonomic data so far considered could thus be said to reflect the relative morphological distinction of the various taxa. Data from these sources are, however, proving more useful as indicators of genome relationships in the section, particularly of the polyploid species.

With the exception of *S. nigrum*, the polyploid members of the section *Solanum* have generated little interest. The origin of only two other species has been the subject of speculation, namely the hexaploid *S. furcatum* (Paddock, 1941), and Luther Burbank's so-called Wonderberry (see Heiser, 1969), which is now thought to be the South African tetraploid *S. retroflexum*. Attempts are currently being made to resynthesise some of these polyploid Solanums, using information derived from experimental and numerical sources to indicate their possible progenitors.

Progenitors of S. nigrum

Synthesis from S. villosum *and* S. americanum (S. nodiflorum)

This work was started with the resynthesis of *S. nigrum* itself, since the derivation of this hexaploid from the tetraploid *S. villosum* and the diploid *S. americanum* (i.e. *S. nodiflorum*) through the amphiploidy of a sterile triploid is now fully established. Using conspecific taxa, Tandon & Rao (1964, 1966a, 1974), Venkateswarlu & Rao (1969, 1971, 1972) and Khan, Rao & Khan (1972), produced such hexaploids artificially, while Soria & Heiser (1959) reported spontaneous doubling of analogous sterile triploid hybrids.

I readily obtained sterile triploids by crossing accessions of *S. americanum* with *S. villosum*, and fertile branches arose on these plants following the application of 0.25% colchicine to axillary buds for 24 hours. Seeds collected from these branches germinated well, giving rise to vigorous hexaploid F₂ and F₃ generation plants, which flowered profusely, and set seed spontaneously (Plate 41.1).

These hexaploids (M1165) were morphologically uniform, and "gigas" mutants such as those reported by Rao & Khan (1970) were not encountered. Their pollen fertilities rapidly reached the percentages displayed by natural hexaploid accessions, though the numbers of seeds/berry remained much lower (Table 41.1). Both the

Table 41.1. Fertility data of progeny derived from artificially-induced polyploids

Cross ♀	Cross ♂	Generation	Accession no.	Ploidy level (x = 12)	% Germination	% Pollen stainability	Seed nos./berry Range	Seed nos./berry Average
S. sublobatum × *S. americanum*		F₁	M467	2x	c. 80.0	4.8	0	0
				↓ colch.				
T355	T198	F₁	M467	4x	—	—	33–36	34
(2x)	(2x)	F₂	M467/1	4x	73.3	80.8	14–50	25
		F₂	M467/2	4x	91.7	83.8	13–29	20
		F₃	M467/1	4x	100.0	82.2	7–35	26
S. americanum × *S. villosum*		F₁	M1165	3x	37.5	1.0	0	0
				↓ colch.				
T310	T95	F₁	M1165	6x	—	—	2–55	22
(2x)	(4x)	F₂	M1165/1	6x	50.0	94.2	3–35	23
		F₂	M1165/3	6x	100.0	93.8	24–48	37
		F₂	M1165/6	6x	96.7	92.2	6–29	16
		F₃	M1165/1	6x	100.0	91.8	3–20	10
		F₃	M1165/3	6x	33.3	93.8	6–58	32
		F₃	M1165/6	6x	64.0	89.6	8–45	31
S. nigrum × *S. sarrachoides*		F₁	M1155	4x	c. 80.0	4.2	0	0
				↓ colch.				
C60	C62	F₁	M1155	8x	—	—	6–27	17
(6x)	(2x)	F₂	M1155/1	8x	60.0	40.2	23–55	39
		F₂	M1155/5	8x	50.0	23.0	6–20	9
		F₂	M1155/12	8x	80.0	30.8	2–14	7
		F₃	M1155/1	8x	13.3	47.1	0–6	2
		F₃	M1155/12	8x	6.3	50.0	1	1

sterile triploids and their amphiploid derivatives generally resembled plants raised from similar parentage by other authors, with respect to their morphology and, when reported, their seed germination, pollen stainability and seed set values (e.g. Tandon & Rao, 1964, 1966a; Chennaveeraiah & Patil, 1968; Venkateswarlu & Rao, 1969, 1971, 1972, and by Rao *et al.* 1971). No male or female sterility was apparent such as that reported by Venkateswarlu & Rao (1971, 1972) in similarly derived hexaploids.

The hexaploid derivatives were morphologically similar to naturally-occurring *S. nigrum*, and closely resembled the glandular subspecies *schultesii* (cf. Plate 41.1).

This was expected in that the tetraploid used for the artificial synthesis was the glandular subspecies *villosum* of *S. villosum*. Future synthesis of a similar amphiploid using the eglandular tetraploid subspecies *alatum* should, theoretically, result in hexaploid plants resembling the eglandular *S. nigrum* subspecies *nigrum*.

Reciprocal backcrossing of the F₃ generation plants to various accessions of natural *S. nigrum* resulted in a high percentage of fruit set, with some seed numbers/ berry approaching those of the natural parental accessions, especially when the subspecies *schultesii* (T309) was involved (Table 41.2). Seed harvested from such crosses appeared healthy and viable, but has not yet been grown. Tandon & Rao (1966a, 1974) and Venkateswarlu & Rao (1969, 1972) too, found reciprocal crosses between such natural and artificial hexaploids easy.

Table 41.2. Berry and seed set data recorded from crosses between natural and artificially- induced *S. nigrum*

	Cross	% Berry set	Seed nos./berry Range	Average
Using *S. nigrum* subsp. *nigrum*	M1165/1 × T72	60	1–23	10
	C60 × M1165/1	75	14–31	24
	M1165/1 × C60	100	3–32	16
Using *S. nigrum* subsp. *schultesii*	M1165/1 × T309	80	3–23	14
	M1165/6 × T309	75	42–62	51
	T309 × M1165/1	92	6–19	11
	T309 × M1165/6	75	30–37	34

Protein band patterns obtained from the electrophoresis of natural *nigrum*, synthesized *nigrum* and from a raw mixture of the protein extracts of the parental diploid and tetraploid accessions, were all very similar, with the high number of homologous bands resulting in very high percentage similarity values. In addition, more or less identical chromatographic profiles were obtained from anthocyanins extracted from the synthesised hexaploid and natural *S. nigrum* subsp. *schultesii*.

Accessions of *S. nigrum* were closely associated with those of *S. americanum* and *S. villosum* in the hierarchic dendrogram based on all characters, as well as in the dendrograms plotted from dissimilarity matrices of various selections of attributes. Similar association was also apparent from the use of a non-hierarchic cluster method (B₂ in Jardine & Sibson, 1971). Heiser, Soria & Burton (1965), and Schilling & Heiser (1976), using different numerical analyses, also found that samples of *S. nigrum*, *S. villosum* and various hybrids between *S. nodiflorum* and either *S. villosum* or *S. nigrum* clustered together or were closely associated. Alloploids derived from *S. villosum* × *S. nodiflorum* showed a high level of correlation with *S. nigrum*, again reflecting its origin.

Derivation of S. villosum

Though *S. americanum* and *S. villosum* (or taxa conspecific with them) are relatively indisputable as progenitors of *S. nigrum*, opinions still vary as to its precise origin. From meiotic chromosomal behaviour, several earlier workers considered that *S. nigrum* was either an autohexaploid (Jørgensen, 1928; Ellison,

1936; Nakamura, 1935, 1937), or an autoallopolyploid, containing two sets of identical genomes (Günther, 1959; Stebbins, 1950; Magoon, Ramanujam & Cooper, 1962; Rao, 1971). Others, on similar evidence, favoured allopolyploidy and three different sets of genomes (Sharma & Bal, 1961; Tandon & Rao, 1964, 1966a, 1974; Chennaveeraiah & Patil, 1968; Khan et al., 1972; Henderson, 1974). Oinuma (1949) proposed that S. nigrum was originally an autopolyploid, whose genomes later evolved into a homoeologous state.

Clarification of these conflicting reports clearly hinges upon the artificial synthesis of the tetraploid involved. It has already been established that S. villosum is not an autotetraploid form of the diploid S. americanum (Bhaduri, 1945; Tandon & Rao, 1966a, b, 1974; Rao & Tandon, 1969b; Venkateswarlu & Rao, 1972), and S. nigrum cannot therefore be an autohexaploid. Stebbins (1950), however, suggested that S. nigrum might contain "four genomes from S. nodiflorum or some other species closely related to it, and two from some diploid species as yet not identified". This would necessitate the derivation of S. nigrum through hybridisation between an unknown diploid and S. americanum, followed by amphiploidy, backcrossing to americanum and more amphiploidy. If Stebbins' proposal is correct, we are looking for one other diploid progenitor of S. nigrum, but if this species is an allohexaploid, two other diploid entities are involved.

Possible contribution of S. sarrachoides

During my crossability programme (Edmonds, 1977a), only 15.9% of the inter-specific crosses attempted between ploidy levels resulted in true hybrid progeny. Where crosses were successful, there must have been sufficient genetic harmony between the parental accessions for the development of viable F_1 embryos, and perhaps such parental accessions possessed a common genome. The sterile F_1 progeny derived from crosses involving S. nigrum as one parent were thus re-examined, to determine whether this might be so. In fact, S. nigrum was successfully crossed twice with its proven diploid progenitor S. americanum. It was therefore interesting to find that of the 31 successful interpolyploid hybridisations, 14 involved S. nigrum as one parent, and S. sarrachoides as the other. On this basis, could S. sarrachoides be a second diploid progenitor of S. nigrum, via its partici-pation in the origin of the tetraploid S. villosum?

The tetraploid plants raised from these crosses were all vigorous, morphologi-cally intermediate and sterile. Following colchicine treatment, fertile branches arose on plants of one accession (M1155) from which vigorous octoploid F_2 and F_3 generations were raised (Plates 41.2 and 41.3). The F_2 plants were moderately fertile, spontaneously setting 2–55 seeds/berry, though their pollen stainability values remained low (<40%). Seeds of the F_3 generations germinated poorly, and, though their pollen stainability improved over the values recorded for the F_2, seed development decreased to 0–6 seeds/berry (Table 41.1). This was not, however, accompanied by any visible reduction in vegetative or floral vigour.

The low seed viability and development could have been due to genic imbalance between the embryo and endosperm, possibly caused by the duplication of similar genomes in the hybrid, if S. nigrum does indeed contain a set of genomes from S. sarrachoides. This would also result in irregular meiosis in the pollen mother cells, leading to some pollen sterility, thereby accounting for the low stainability values recorded. Similar meiotic disturbances in the embryo-sac would lead to some ovule abortion. The drastic increase in this ovule abortion in the F_3 generation could have been due to the effects of such an imbalance being accentuated by successive inbreeding.

Having raised these plants several years ago, I was particularly interested when some large populations of *S. nigrum* and *S. sarrachoides* were recently found growing together in a Cambridgeshire market garden field. Among them were some very vigorous and conspicuous plants (Leslie, 1978), which looked morphologically similar to my artificial hybrids between *S. nigrum* and *S. sarrachoides*. These plants proved to be sterile tetraploids; their pollen stainability values averaged 0.96%, and they developed only small purple, parthenocarpic berries. These two species therefore seem to be sufficiently related to hybridise naturally where their distributions overlap, despite their different ploidy levels.

Additional support for the participation of *S. sarrachoides* in the origin of *S. nigrum* was derived from seed protein electrophoresis, where the band patterns obtained from raw mixtures of their protein extracts matched those of *S. nigrum* alone remarkably well, and resulted in a very high percentage similarity value. Moreover, the higher R_f band pattern of the octoploid derivative was also very similar to that of the raw mixture. The clustering behaviour of *S. sarrachoides* during the numerical analyses also demonstrated its close affinity with *S. nigrum*. This diploid was closely associated with accessions of *S. nigrum*, *S. villosum* and *S. americanum* in the hierarchic dendrograms, and in the non-hierarchic clustering.

Some support for the contribution of *S. sarrachoides* to the chromosome complement of *S. nigrum* can be derived from reports by earlier workers. Jørgensen (1928) raised highly fertile decaploid plants from a sterile pentaploid hybrid between *S. nigrum* and *S. villosum* (using the conspecific *S. luteum* Mill.). Meiosis was regular in these plants, they bred true, and produced fruits freely. Tandon & Rao (1964, 1966a, 1974) raised similar sterile pentaploid hybrids, and Westergaard (1948) also raised an amphiploid of a comparable cross. If the known participation of *S. villosum* in the ancestry of *S. nigrum* contributed to the success of these hybridisations and subsequent amphiploid derivatives, then the successful hybridisation and amphiploidy of *S. nigrum* and *S. sarrachoides* could be analogous.

Ellison (1936) was probably the first to hybridise *S. nigrum* and *S. sarrachoides* (using *S. nitidibaccatum*). Meiosis in these sterile tetraploids usually revealed 24 bivalents on the first metaphase plates, with the number of univalents in other instances varying from two to six. The second division nearly always resulted in 24 regularly arranged chromosomes at metaphase, and the majority of pollen grains appeared normal. Ellison concluded that the 24 bivalents formed in the first division of meiosis resulted from the pairing of 12 *nigrum* and 12 *nitidibaccatum* chromosomes, and the pairing *inter se* of the remaining 24 *nigrum* chromosomes. Ellison cited the latter as evidence for *S. nigrum* being an autohexaploid. While this has been disproved by later work, his results do suggest that there must have been sufficient homology between the chromosome sets of *S. nitidibaccatum* and *S. nigrum* to allow regular bivalent formation. Larsen (1943) and Westergaard (1948) produced amphiploids of *S. nigrum* and *S. nitidibaccatum*, but gave no details of the plants.

The most obvious anomaly in the suggestion that *S. nigrum* contains a set of genomes from *S. sarrachoides* is that this species seems to be reproductively isolated from all other diploids (Edmonds, 1977a). Attempts to hybridise this species might be more successful if material from the same geographical source were used, as, so far, I have been crossing *S. sarrachoides* collected in Europe with various diploid species collected in the Americas. If it is completely isolated genetically, however, it could be argued that this isolation, arising gradually, became complete after the evolution of *S. nigrum* in the Old World. There seems to be some morphological divergence of the more common variant of *S. sarrachoides* found in Europe from the form prevalent in South America, and this could be accompanied by some genetical divergence.

Autoallo- or allo-polyploid origin

If *S. nigrum* does contain a set of genomes from *S. sarrachoides*, two of the three sets of genomes involved in the lineage of this hexaploid would be known. Is, however, the black nightshade an autoallohexaploid or an allopolyploid? The seed protein electrophoresis provided some evidence that the hexaploid does not contain four genomes of the taxon that I am calling *S. americanum*, if it does indeed contain *S. sarrachoides*. The band patterns obtained from raw mixtures of *S. sarrachoides* with either *S. americanum* var. *americanum* or var. *patulum* did not match those of *S. villosum*, giving low percentage similarity values. These results therefore suggest that *S. nigrum* does not contain four genomes of *S. nodiflorum*, as proposed by Stebbins (1950).

Meiotic regularity is characteristic of natural *S. nigrum*. Multivalent and univalent formation at meiosis have rarely been reported (i.e. only by Nakamura, 1935, 1937; and Günther, 1959), and might be expected where autopolyploidy was involved. Rao (1971) considered *S. nigrum* to be an autoallopolyploid, explaining the absence of quadrivalent formation in the natural hexaploids by its genetic suppression. Genetic control of bivalent formation is, of course, possible, but its operation would be extremely difficult to demonstrate. Many authors, however, by studying meiosis in triploid hybrids raised from crossing various taxa conspecific with *S. americanum* and *S. villosum*, have reached a different conclusion.

Tandon & Rao (1964, 1974) found that most of the chromosomes were present as univalents (24–36) in pollen mother cells of progeny derived from crossing the Indian diploid and tetraploid "*S. nigrum*". They attributed this to the dissimilarity of the three genomes in this triploid, with respect to the majority of chromosomes. Chennaveeraiah & Patil (1968) also reported univalents at meiosis in similar sterile triploids, and concluded that the very low frequency of trivalent formation, the presence of as many as 36 univalents at diakinesis, large numbers of univalents at metaphase I, and complete sterility, all revealed the non-homologous nature of the three genomes in the hybrid. Similarly, Rao *et al.* (1971) attributed the occurrence of univalents in similar sterile hybrids to lack of homology between the genomes, with respect to some of their chromosomes. Khan *et al.* (1972) too, concluded that the wide range of meiotic abnormalities, the high percentage of pollen sterility, and the failure to set fruit, observed in hybrids between *S. villosum* × *S. nodiflorum*, *S. luteum* × *S. nodiflorum*, and tetraploid "*S. nigrum*" × *S. nodiflorum*, indicated that the three genomes in these triploid hybrids were not homologous. Finally, Venkateswarlu & Rao (1972) considered that the poor germination, survival capacity and sterility encountered in some of the progeny raised from selfing synthesised *S. nigrum* was partly due to disharmonious gene combination in the sterile triploid hybrids, which was passed on to the hexaploid progeny on chromosome doubling.

Other possible diploid progenitors

Any possible third diploid progenitor of *S. nigrum* must not only comply with morphological criteria, but also with features of geographical distribution. Both *S. nigrum* and *S. villosum* are Old World species, which have not yet been found in South America, though they have been sparingly introduced into parts of North America. The diploid *S. americanum* is found more or less throughout the world, whereas the second possible diploid progeniton, *S. sarrachoides*, is native to South America, but is found as a casual in Europe.

Another diploid native to South America and also found as a casual in Europe is *S. sublobatum*. It is characterized by strongly deflexed peduncles, usually subtended by small obovate leaves, features which were, in fact, noted in the octoploids derived from *S. nigrum* and *S. sarrachoides*. *S. sublobatum* also develops broadly ovate, purple, dull berries, which would comply with the morphological requirements of *S. nigrum*. Indeed, morphological extrapolation of numerous characters, both of this diploid and of *S. sarrachoides*, can provide many features found in *S. nigrum*.

In addition, just as natural *S. nigrum* could be successfully crossed with *S. sarrachoides* and with *S. americanum*, so could *S. sublobatum* (Edmonds, 1977a). This species was also implicated as a possible diploid contender for the ancestry of *S. nigrum* by the non-hierarchic clustering pattern, and it was included in the same major clusters as *S. americanum*, *S. villosum* and *S. nigrum* in many of the hierarchic dendrograms, especially in that based on the dissimilarity matrix of all attributes.

Recently, however, my attention has been drawn to some populations of *S. sarrachoides* which also exhibit deflexed penduncles, so the chief morphological evidence for the inclusion of *S. sublobatum* in *S. nigrum* is diminishing. The precise mode of origin of *S. nigrum* is therefore far from being completely understood, though the species almost certainly originated polytopically. The morphological variation found in *S. nigrum* is extensive, and is identical to that exhibited by its proven tetraploid progenitor *S. villosum*. The undisputed diploid ancestor *S. americanum* is probably the most morphologically variable species in the section *Solanum*. The combination of such morphologically different biotypes during the formation of the triploid hybrids would account for the variability found in *S. nigrum* itself. Depending upon the origin of the different biotypes contributing to the genome complement of *S. nigrum*, the hexaploid species could be said to have several different modes of origin.

If *S. nigrum* is indeed an allopolyploid, it is quite possible that the third progenitor will prove to be a diploid not yet investigated experimentally, and which may not even be found in Europe. The hexaploid could have evolved in Africa, or, as is more probable, in Asia, where *S. nigrum* and its two known progenitors are particularly variable. The Asian representatives of the section, especially the diploid taxa, are taxonomically complex, and have not yet been revised. It is also possible that Stebbins is right, and that *S. nigrum* might still prove to contain four genomes from the taxon which I have referred to as *S. americanum*, in the form of two infraspecific variants. The lack of homologous bands between the seed proteins of *S. villosum*, and mixtures of extracts from *S. sarrachoides* with the two varieties of *S. americanum* recognised in this work, may have been indicative of (a) the absence of *S. sarrachoides* from *S. nigrum* rather than that of four *americanum* genomes, or (b) that different infraspecific variant(s) of *S. americanum* were concerned in the ancestry of this hexaploid. If *S. nigrum* did contain genomes from two infraspecific variants of *S. americanum*, it could be regarded as a segmental allopolyploid. Clearly, a considerable amount of further experimental work, especially incorporating the meiotic pairing behaviour of any relevant hybrids, is necessary, before the precise origin of this Old World hexaploid can be resolved.

Progenitors of S. interandinum

Another polyploid whose re-synthesis has been attempted is the tetraploid *S. interandinum*. Amphiploid derivatives (M467) of the sterile progeny derived from crossing the diploids *S. sublobatum* and *S. americanum* closely resembled this Ecuadorian species. Seeds of all generations germinated well, with the pollen stainability stabilising around 80% by the F_2 generation and berries setting an average of 23 seeds/berry spontaneously (Table 41.1.) These values were, however, considerably lower than the corresponding parental values, and of those found in the natural tetraploid. Moreover, variation in height and vegetative vigour were still evident in the F_3 generation, which might be indicative of a different ancestry, either through one, or even both, of these diploid species.

In the numerical analysis, *S. sublobatum* clustered next to, or close to, *S. interandinum* in many of the hierarchic dendrograms, while *S. americanum* was also found in the same major cluster as these two species in the dendrogram based on all attributes. From the non-hierarchic clustering pattern, *S. sublobatum* was the most strongly implicated diploid for participation in the ancestry of *S. interandinum*. Moreover, Heiser *et al.* (1965) and Schilling & Heiser (1976) reported an amphiploid of a similar cross between *S. gracile* and *S. nodiflorum* which, in their numerical analyses, was closely associated with various strains of *S. interandinum*.

As far as the geographical distribution of these species goes, *S. americanum* is widespread throughout South America, whereas *S. sublobatum* is found only in Uruguay, parts of Argentina and southern Brazil. *S. interandinum*, on the other hand, is largely confined to Ecuador. While the restriction of *S. sublobatum* to south-western South America makes its participation in the natural tetraploid unlikely, it is not impossible, since this species is found as a casual in Europe, and therefore probably elsewhere.

If further experimental work, incorporating meiotic analyses and the crossability behaviour of this tetraploid hybrid with natural *S. interandinum* precludes such an origin, the non-hierarchic clustering suggests the Peruvian-Bolivian-Ecuadorian species *S. zahlbruckneri* as a likely alternative diploid progenitor. This would correlate well morphologically and geographically. Moreover, *S. zahlbruckneri* was readily crossed with both *S. sublobatum* and *S. americanum*, resulting in vigorous, partially fertile F_1 progeny (Edmonds, 1977a).

Progenitors of other polyploid species

The non-hierarchic clustering pattern proved particularly useful, in that it not only confirmed previously established experimental data on genome relationships in the section *Solanum*, but also indicated possible progenitors of polyploids whose artificial synthesis has not yet been attempted. Thus, it strongly suggested that the glandular tetraploid *S. excisirhombeum* contains a set of genomes from the diploid *S. sarrachoides*. These two species also clustered closely together in many of the hierarchic dendrograms. Another diploid closely associated with this tetraploid in most of the dendrograms was *S. americanum*. On comparative morphological grounds, such an allopolyploid origin for *S. excisirhombeum* through the initial hybridisation of *S. americanum* and *S. sarrachoides* would be quite acceptable.

At the hexaploid level, the participation of the tetraploid *S. interandinum* in the ancestry of *S. furcatum* was strongly implicated by the non-hierarchic clustering pattern, and this correlates well with the morphology and distribution of these two species. Even if *S. sublobatum* does not prove to be ancestral to *S. interandinum*, it has almost certainly been involved in the derivation of *S. furcatum*, since these are the only two *Solanum* species characterized by *consistently* strongly deflexed peduncles. Close association between *S. furcatum* and *S. sublobatum*, and between *S. furcatum* and *S. interandinum*, was also apparent in many of the hierarchic dendrograms. Another diploid probably involved in the origin of *S. furcatum* is *S. douglasii*. Such a possibility was first proposed by Paddock (1941), and these two species were again closely associated in most of the hierarchic dendrograms, and in the non-hierarchic clustering pattern.

Similarly, among the possible diploid contributors to the genome complement of the hexaploid *S. arequipense*, the non-hierarchic clustering suggested *S. pentlandii*, *S. aloysiifolium* and *S. douglasii*. Again, these diploids were closely associated with the hexaploid in many of the hierarchic dendrograms.

Such possible participants in the origin of these polyploids, and of the others found in the section *Solanum*, are currently being investigated.

CONCLUSIONS

The polyploid members of the section *Solanum* are probably mostly allopolyploids, with all species so far investigated cytologically, showing regular bivalent formation at meiosis. From the work now completed, it is becoming clear that these polyploids have probably arisen from comparatively few diploid species contributing genomes in different combinations. This is a further factor complicating the taxonomy of this section. The possession of common genetic material would, however, explain the similar ranges of morphological variation encountered in species belonging to the section *Solanum*.

The supplementation of conventional classical work with data from a variety of taxonomic methods is gradually clarifying the genome relationships in the section, but a great deal of work remains to be completed. It is hoped, however, that continued research on this species group, incorporating a variety of taxonomic approaches, will eventually lead to a stable classification of the section *Solanum* in the not too distant future.

ACKNOWLEDGEMENTS

I am indebted to Dr S. M. Glidewell who was responsible for the seed protein electrophoresis, and to Dr N. Jardine, who initiated and gave invaluable advice on the population differentiation studies. Grateful thanks are also due to Dr H. L. K. Whitehouse and Dr S. M. Walters for reading the manuscript, and to Professor P. W. Brian for the provision of facilities in the Botany School. Grateful acknowledgement is made to the Science Research Council who supported some of this work by Research Grants. Finally, sincere thanks are due to my former supervisor, Professor J. G. Hawkes, for his continual advice and encouragement.

REFERENCES

AHMAD, K. J., 1964. Epidermal studies in *Solanum*. *Lloydia, 27:* 243–249.

BAYLIS, G. T. S., 1958. A cytogenetical study of New Zealand forms of *Solanum nigrum* L., *S. nodiflorum* Jacq. and *S. gracile* Otto. *Transactions of the Royal Society of New Zealand, 85* (3): 379–385.

BHADURI, P. N., 1945. Artificially raised autotetraploid *S. nigrum* L. and the species problem in the genus *Solanum*. *Proceedings of the Indian Science Congress Association*, Sect. 8, Abstr. 39.

BITTER, G., 1912–1922. *Solana* nova vel minus cognita I–XXI. *Feddes Repertorium Novarum Specierum Regni Vegetabilis, 10*; 529–565 (1912): *11*; 1–18, 202–237, 241–260, 349–394, 431–473 (1912): *11*; 481–491, 562–566 (1913): *12*; 1–10, 49–90, 136–162, 433–467, 542–555 (1913): 13; 88–103, 169–173 (1914): *15*; 93–98 (1917): *16*; 10–15, 79–103 (1919): *16*; 389–409 (1920): *18*; 49–71, 301–309 (1922).

Solana africana I. *Botanische Jahrbücher, 49*; 560–569 (1913): II. *Botanische Jahrbücher, 54*; 416–506 (1917). Solanaceae andinae. *Botanische Jahrbücher, 54*; Beibl. 119: 5–17 (1916).

Die papuasischen Arten von *Solanum*. *Botanische Jahrbücher, 55*: 59–113 (1917).

CHENNAVEERAIAH, M. S. & PATIL, S. R., 1968. Some studies in the *Solanum nigrum* L. complex. *Genetica Iberica, 20:* 23–36.

D'ARCY, W. G., 1973. *Solanaceae*. In Flora of Panamá by Woodson, R. E., Schery, R. W. and Collaborators, Part 9, *Annals of Missouri Botanical Garden, 60:* 573–780.

D'ARCY, W. G., 1974. *Solanum* and its close relatives in Florida. *Annals of Missouri Botanical Garden, 61*, No. 3: 819–867.

DILLENIUS, J. J., 1732. *Hortus Elthamensis, II:* 366–368.

DUNAL, M. F., 1813. *Histoire Naturelle, Médicale et Economique des* Solanum. Paris, Strasbourg & Montpellier.

DUNAL, M. F., 1852. *Solanaceae*. In A. P. De Candolle (Ed.), *Prodromus Systematis Naturalis Regni Vegetabilis, 13* (1): 1–690. Paris.

EDMONDS, J. M., 1971. *Solanum* L. In W. T. Stearn, Taxonomic and nomenclatural notes on Jamaican gamopetalous plants. *Journal of the Arnold Arboretum, 52:* 634–635.

EDMONDS, J. M., 1972. A synopsis of the taxonomy of *Solanum* Sect. *Solanum* (*Maurella*) in South America. *Kew Bulletin, 27* (1): 95–114.

EDMONDS, J. M., 1977a. Taxonomic studies on *Solanum* L. section *Solanum* (*Maurella*). *Botanical Journal of the Linnean Society, 75:* 141–178.

EDMONDS, J. M., 1977b. Numerical taxonomic studies on *Solanum* L. section *Solanum* (*Maurella*). *Botanical Journal of the Linnean Society, 76:* 27–51.

EDMONDS, J. M., 1978. A supplement to the taxonomy of *Solanum* Sect. *Solanum* (*Maurella*) in South America. In preparation.

EDMONDS, J. M. & GLIDEWELL, S. M., 1977. Acrylamide gel electrophoresis of seed proteins from some *Solanum* (section *Solanum*) species. *Plant Systematics and Evolution, 127:* 277–291.

ELLISON, W., 1936. Synapsis and sterility in a *Solanum* hybrid. *Journal of Genetics, 32:* 473–477.

GRANT, V., 1963. *The Origins of Adaptations*. New York: Colombia University Press.

GRAY, J. M., 1968. *Taxonomy of the Morella Section of the Genus* Solanum *L. within South America*. Unpublished Ph.D. thesis, University of Birmingham, England.

GÜNTHER, E., 1959. Cytologische Untersuchungen an *Solanum nigrum* L. *Bericht der Deutschen Botanischen Gesellschaft, 72:* 14.

HAWKES, J. G. & EMONDS, J. M., 1972. *Solanum* L. In T. G. Tutin *et al.* (Eds), *Flora Europaea, III:* 197–199. Cambridge: Cambridge University Press.

HEISER, C. B. Jr., 1963. Estudio biosistematico de *Solanum* (*Morella*), en el Ecuador. *Ciencia y Naturaleza, 6* (2): 50–58.

HEISER, C. B. Jr., 1969. *Nightshades, The Paradoxical Plants*. San Francisco: W. H. Freeman.

HEISER, C. B. Jr., BURTON, D. L. & SCHILLING, E. E., Jr., 1978. Biosystematic and taxometric studies of the *Solanum nigrum* complex in eastern North America. In J. G. Hawkes *et al.* (Eds), *The Biology and Taxonomy of the Solanaceae:* 513–527. London: Academic Press.

HEISER, C. B., SORIA, J. & BURTON, D. L., 1965. A numerical taxonomic study of *Solanum* species and hybrids. *American Naturalist, 99*, No. 909: 471–488.

HENDERSON, R. J. F., 1974. *Solanum nigrum* L. (*Solanaceae*) and related species in Australia. *Contributions from the Queensland Herbarium*, No. 16: 1–78.

JARDINE, N. & SIBSON, R., 1971. *Mathematical Taxonomy.* New York: Wiley.

JARDINE, N. & EDMONDS, J. M., 1974. The use of numerical methods to describe population differentiation. *New Phytologist, 73:* 1259–1277.

JØRGENSEN, C. A., 1928. The experimental formation of heteroploid plants in the genus *Solanum. Journal of Genetics, 19* (2): 133–211.

KHAN, A. H., RAO, G. R. & KHAN, R., 1972. Interrelationship within the *Solanum nigrum* Complex *Proceedings of the Indian Science Congress, 59* (3): 364.

LARSEN, P., 1943. The aspects of polyploidy in the genus *Solanum*. II. Production of dry matter, rate of photosynthesis and respiration, and development of leaf area in some diploid, autotetraploid and amphidiploid Solanums. *Kongelige Danske Videnskabernes Selskabs Skrifter, 18* (2): 1–52.

LESLIE, A. C., 1978. The occurrence of *Solanum nigrum* L. × *S. sarrachoides* Sendtn. in Britain. *Watsonia, 12* (1): 29–32.

LINNAEUS, C., 1753. *Species Plantarum,* 1st ed.: 186.

MAGOON, M. L., RAMANUJAM, S. & COOPER, D. C., 1962. Cytogenetical studies in relation to the origin and differentiation of species in the genus *Solanum* L. *Caryologia, 15* (1): 151–252.

MARKS, G. E., 1954. An aceto-carmine glycerol jelly for use in pollen-fertility counts. *Stain Technology, 29* (5): 277.

NAKAMURA, M., 1935. Preliminary note on the polyploidy in *Solanum nigrum* Linn. *Journal of the Society of Tropical Agriculture, 7:* 255–256.

NAKAMURA, M., 1937. Cytogenetical studies in the genus *Solanum*. I. Autopolyploidy of *Solanum nigrum* Linn. *Cytologia, Fujii Jubilee, 1:* 57–68.

OINUMA, T., 1949. Cytological studies on the genus *Solanum* with special reference to the karyotypes. *Japanese Journal of Genetics, 24:* 182–189.

PADDOCK, E. F., 1941. Natural and experimental polyploidy in *Solanum douglasii* Dunal and its relatives. *American Journal of Botany, 28:* 727.

RAO, M. K., 1971. Cytology of a pentaploid hybrid and genome analysis in *Solanum nigrum* L. *Genetica, 42* (1): 157–164.

RAO, G. R. & TANDON, S. L., 1969a. Relationship between tetraploid *Solanum nigrum* and *Solanum luteum. Science and Culture, 35* (12): 688–689.

RAO, G. R. & TANDON, S. L., 1969b. Cytogenetical studies in relation to origin of the Indian tetraploid *Solanum nigrum* L. and its taxonomic relationship with *Solanum luteum* Mill. *Journal of Cytology and Genetics, 4:* 19–24.

RAO, G. R. & KHAN, A. H., 1970. Genetic improvement of *Solanum nigrum. Science and Culture, 36* (11): 614–615.

RAO, G. R., KHAN, R. & KHAN, A. H., 1971. Structural hybridity and genetic system of *Solanum nigrum* complex. *Botanical Magazine Tokyo, 84:* 335–338.

SCHILLING, E. E., Jr. & HEISER, C. B., Jr., 1976. Re-examination of a numerical taxonomic study of *Solanum* species and hybrids. *Taxon, 25* (4): 451–462.

SHARMA, A. K. & BAL, A. K., 1961. Cytological studies in several species of non-tuberiferous *Solanum* with special reference to polyploid types of *S. nigrum* L. *Proceedings 48th Indian Science Congress Pt. III Abstr.:* 295–296.

SHARMA, K. D. & SEN, D. N., 1969. Polymorphism in the stomata of *Solanum nigrum* Linn. *Current Science, 38:* 394–395.

SORIA, J. & HEISER, C. B., Jr., 1959. The garden huckleberry and the sunberry. *Baileya, 7:* 33–35.

STEBBINS, G. L., Jr., 1950. *Variation and Evolution in Plants.* New York: Colombia University Press.

STEBBINS, G. L., Jr. & PADDOCK, E. F., 1949. The *Solanum nigrum* complex in Pacific North America. *Madroño, 10:* 70–81.

SYMON, D. E., 1970. Dioecious Solanums. *Taxon, 19* (6): 909–910.

TANDON, S. L. & RAO, G. R., 1964. Cytogenetical investigations in relation to the mechanism of evolution in hexaploid *Solanum nigrum* L. *Nature, 201:* 1348–1349.

TANDON, S. L. & RAO, G. R., 1966a. Interrelationship within the *Solanum nigrum* complex. *Indian Journal of Genetics and Plant Breeding, 26* (2): 130–141.

TANDON, S. L. & RAO, G. R., 1966b. Biosystematics and origin of the Indian tetraploid *Solanum nigrum* L. *Current Science, 35* (20): 524–525.

TANDON, S. L. & RAO, G. R., 1974. *Solanum nigrum* L. In J. Hutchinson (Ed.), *Evolutionary Studies in World Crops:* 109–117. Cambridge: Cambridge University Press.

VAUGHAN, J. G. & DENFORD, K. E., 1968. An acrylamide gel electrophoretic study of the seed proteins of *Brassica* and *Sinapis* species, with special reference to their taxonomic value. *Journal of Experimental Botany, 19:* 724–732.

VENKATESWARLU, J. & RAO, M. K., 1969. Chromosome numerical mosaicism in some hybrids of the *Solanum nigrum* complex. *Genetica, 40:* 400–406.

VENKATESWARLU, J. & RAO, M. K., 1971. Inheritance of fruit colour in the *Solanum nigrum* complex. *Proceedings of the Indian Academy of Science, 74* (3), Sect. B: 137–141.

VENKATESWARLU, J. & RAO, M. K., 1972. Breeding system, crossability relationships and isolating mechanisms in the *Solanum nigrum* complex. *Cytologia, 37:* 317–326.

WESTERGAARD, M., 1948. The aspects of polyploidy in the genus *Solanum*. III. Seed production in auto-polyploid and allopolyploid Solanums. *Kongelige Danske Videnskabernes Selskabs Skrifter, 18* (3): 1–18.

EXPLANATION OF PLATES

PLATE 41.1.

Artificial synthesis of *S. nigrum* from the diploid *S. americanum* and the tetraploid *S. villosum*.

A. *S. americanum* (T310) ♀.
B. *S. villosum* subsp. *villosum* (T95) ♂.
C. Sterile triploid hybrid (M1165).
D. Fertile hexaploid derivative (M1165/1), F_2 generation.
E. Natural *S. nigrum* subsp. *schultesii* (T309).

PLATE 41.2.

Artificial synthesis of an octoploid derivative of *S. nigrum* ($6x$) and *S. sarrachoides* ($2x$).

A. *S. nigrum* (C60) ♀.
B. *S. sarrachoides* (C62) ♂.
C. Sterile tetraploid hybrid (M1155), showing colchicine application to the axillary buds.
D. Fertile octoploid derivative (M1155/1), F_2 generation.

PLATE 41.3.

F_2 generation field plants of the octoploid derivative M1155/12.

A. Flowers, × 0.7.
B. Berries, × 0.6. Note that the peduncles are usually subtended by small leaves, and many peduncles are strongly deflexed.

Plate 41.1

(*Facing p. 548*)

42. Systematic and evolutionary consideration of species of *Solanum*, section *Basarthrum*

GREGORY J. ANDERSON

Biological Sciences Group, University of Connecticut, Storrs, USA

The group most closely related to the potato and allied wild species (section *Petota*) is *Solanum*, section *Basarthrum*. The section comprises one cultivated and 22 wild species. The cultigen, *S. muricatum*, has a long history of use in the Andes of southern Colombia, Ecuador and Peru. It is proposed that the Peruvian *S. basendopogon* could be involved in the ancestry of *S. muricatum*. Some individuals of *S. muricatum* are self-compatible, and most plants produce some fruits with seeds. The wild species are usually found at altitudes between 2000–3000 m from central Mexico south to southern Peru. The centre of species diversity in the section is in northwest South America (Colombia, Ecuador, Peru). Nearly one-half of the species in the section are known from only one to three localities. All taxa and artificial hybrids assayed have a chromosome number of $n = 12$ and no meiotic abnormalities. Only four of the 12 species studied so far are self-compatible; all others are self-incompatible. Self-compatible species have styles which end in the region of the terminal anther pores; the styles of self-incompatible species exceed the staminal column. Autogamous species are shown to be morphologically as variable as allogamous species. In contrast with the easily-made interspecific hybrids and relatively high F_1 fertility among the tuberous species, crosses in section *Basarthrum* fail most of the time, seedless fruits are produced and/or F_1 hybrids are often sterile. There are a number of examples of non-reciprocal crossability, seed set and F_1 fertility. It is concluded that, in addition to ecogeographic factors, *S*-gene interactions and possibly cryptic structural chromosome differences are the prominent barriers to free gene flow among the species. Crossing studies generally support the current taxonomic system, but they, together with morphological data, indicate a need for reconsideration of the placement of certain taxa. The systematic value of pollen and seed morphology is described.

CONTENTS

INTRODUCTION

As a result of the excellent works of Correll (1962) and Hawkes & Hjerting (1969) we have a good understanding of the taxonomy of the tuberous solanums in tropical regions. This knowledge facilitates more detailed studies including experimental analyses. Although "biosystematic" studies in tropical areas suffer because species in the tropics are not as well known as those in temperate zones, such studies are important in that they give a clearer idea not only of the biological limits of the taxa, but also of potentially important evolutionary mechanisms.

My interest in section *Basarthrum* (Bitt.) Bitt. began a number of years ago with work on the origin of the cultigen, *S. muricatum* (the "pepino"). In this effort, a number of allied wild species were assembled and grown and my interest in the group was extended to their intriguing and complex patterns of variation. What follows is a synthesis of what is known of the biosystematics of the cultivated and wild taxa of section *Basarthrum*. In an effort to put this material into perspective, comparisons will be made with the closely related tuberous section, *Petota* Dumort. The most extensive and thorough biosystematic work treating this latter section is that by Hawkes & Hjerting (1969); this work has been frequently consulted in the preparation of the present paper.

The species of section *Basarthrum* are of interest from a biosystematic view for several reasons, not the least of which is their degree of complexity which prompted Correll (1962) to consider them even more difficult to define than some of the tuberous species. Furthermore, Correll (1962) points out that section *Basarthrum* is the group most closely related to the tuberous species (section *Petota*).

Systematic and evolutionary interpretations of section *Basarthrum* may thus have some general application to the tuberous species as well. In treating the Venezuelan species in section *Basarthrum*, Brücher (1968b) adds other practical reasons for studying the section. He considers these taxa to be immune to late blight and to some viruses that infect the tuberous species.

To facilitate understanding of the discussion that follows, a listing of the 29 taxa in section *Basarthrum* is given below. The series are as given by Correll (1962) but with two additional species (*S. heiseri*, *S. trachycarpum*). The (*) symbol indicates that those species have been treated biosystematically previously (Anderson, 1975a, 1976a) and the (†) symbol marks those included in a study of pollen morphology (Anderson & Gensel, 1976).

Series *Muricata* Corr.—*S. muricatum* Ait. (†), *S. muricatum* f. *glaberrimum* Corr., *S. muricatum* var. *protogenum* Bitt.

Series *Caripensa* Corr.—*S. basendopogon* Bitt. (*, †), *S. basendopogon* f. *obtusum* Corr. (†), *S. caripense* H. et B. ex Dun. (*, †), *S. caripense* var. *stellatum* Corr. (*, †), *S. filiforme* Ruiz et Pavon, *S. heiseri* Anderson (*, †), *S. trachycarpum* Bitt. et Sod. (*, †).

Series *Appendiculata* Rydb.—*S. appendiculatum* H. et B. ex Dun. (†), *S. connatum* Corr. (†), *S. inscendens* Rybd. (†), *S. skutchii* Corr., *S. subvelutinum* Rydb., *S. tacanense* Lundell, *S. brevifolium* H. et B. ex Dun. (†), *S. carchiense* Corr., *S. chimborazense* Bitt. et Sod., *S. sodiroi* Bitt., *S. sodiroi* var. *dimorphophyllum* (Bitt.) Corr., *S. tabanoense* Corr. (*, †), *S. tetrapetalum* Rusby (†).

Series *Articulata* Corr.—*S. sanctae-marthae* Bitt. (*, †).

Series *Suaveolentia* Rydb.—*S. fraxinifolium* Dun. in DC, (†), *S. fraxinifolium* var. *protoxanthum* (Bitt.) Bitt., *S. suaveolens* Kunth et Bouché (*, †), *S. taeniotrichum* Corr.

Series *Canensa* Corr.—*S. canense* Rybd. (*, †).

DISTRIBUTION

The distributions of the wild species in section *Basarthrum* range from very restricted to very broad. Forty-two per cent of the species are rare, i.e., found in only one to three localities. At the other extreme, *S. suaveolens* has a broader north-south distribution (more than 3000 km) than any other species in section *Petota* or *Basarthrum*. The most broadly distributed species in Hawkes & Hjerting's (1969) treatment of section *Petota* for southern South America is *S. chacoense* Bitt. with an east-west range of 1300 km and a north-south distribution of some 1800 km. Both *S. canense* and *S. caripense* have a narrower east-west distribution than this, but have at least as great a north-south distribution. It is of some interest to note that two (*S. canense*, *S. suaveolens*) of these three widely distributed species are self-compatible and are quite weedy (Correll, 1962). This may in a large part account for their wide distribution.

Sixty-two per cent of the 26 wild taxa in the section are found exclusively or primarily in South America and 30% exclusively or primarily in Mexico and Central America; 19% of the taxa are found in both areas. Thus, for section *Basarthrum*, the greatest species diversity is in South America. The only series which show any marked diversity in Central America and Mexico are *Appendiculata* and *Suaveolentia*. Although six of the species in the former are found only in Central America and Mexico, only one of the latter is confined to this area. The series *Appendiculata* is also the least homogeneous of all those in the section, and perhaps some of the species (primarily the Central American species) should be recognized in a separate series or section (see Correll, 1962).

In a discussion of the origin of Argentine potatoes, Hawkes & Hjerting (1969) supported the hypothesis that the place of origin of the tuber-bearing species of *Solanum* was in Mexico. The origin of section *Basarthrum* remains a problem, though species diversity would suggest South America to be a more likely site than either Central America or Mexico. If both of these hypotheses regarding origins are accepted, then Correll's (1962) suggestion of parallel rather than lineal relationship between sections *Basarthrum* and *Petota* is supported.

In a recent review Raven & Axelrod (1974) indicate that *Solanum* probably was present in both North and South America prior to the Eocene, and that most movement of species between these continents, other than that by long-distance dispersal over water, took place sometime within the last six million years (Raven & Axelrod, 1974; Savage, 1974). If section *Basarthrum* is native to South America this would then be the time period in which at least the green-fruited members (virtually all the members of the series *Caripensa*, *Suaveolentia*, *Canensa* and some of the South American species of series *Appendiculata*) extended their ranges north. Some of the Central American species in series *Appendiculata* climb high up tree trunks and bear small red fruits; these are probably bird dispersed in part and thus could have been moved greater distances. It is interesting however, that virtually all of these red-fruited species are confined to central and southern Mexico and Guatemala. In any event, the ancestral taxa appear to have reached nuclear North America sufficiently long ago for considerable speciation to have occurred.

The altitudinal distribution pattern is similar to the latitudinal pattern; some species are very restricted, and others occupy a wide altitudinal range. For example, based on the herbarium specimens cited by Correll (1962), *S. caripense* has a remarkable altitudinal distribution of 3000 m (from 800 to 3800 m). This far exceeds *S.*

chacoense (2300 m) cited by Hawkes & Hjerting (1969) as having the widest altitudinal distribution of any of the tuber-bearing species they studied.

There is little correlation between altitude and placement in a given series. Very generally, there are three altitudinal groups, all of which cross series lines (subspecific categories and species with unknown altitudes excluded): (a) usually less than 1000 m—*S. canense* and *S. suaveolens*; (b) 1500–2495 m—*S. appendiculatum*, *S. connatum*, *S. fraxinifolium*, *S. inscendens*, *S. sanctae-marthae*, *S. skutchii*, *S. sodiroi*, *S. subvelutinum*, *S. tacanense*, *S. tetrapetalum*, *S. trachycarpum;* (c) greater than 2500 m—*S. basendopogon*, *S. brevifolium*, *S. caripense*, *S. chimborazense*, *S. heiseri*, *S. tabanoense*, *S. taeniotrichum*. There is some suggestion (Correll, 1962; Anderson, 1971) that closely related species (such as *S. caripense* and *S. fraxinifolium*) may hybridize thus potentially broadening each species' altitudinal tolerance. Hawkes & Hjerting (1969) make a similar proposal to account for the broad altitudinal range of *S. chacoense*.

BREEDING SYSTEMS

Before one can fully understand the systematics of a group, it is necessary to have an appreciation of the breeding system. Thirteen of the 23 species in the section have been assayed (Heiser, 1964; Brücher, 1968a; Anderson, 1975a, 1976a). Of these the eight following do not set fruit on self-pollination or in pollinator-free greenhouses and are thus considered self-incompatible: *S. appendiculatum*, *S. basendopogon*, *S. caripense*, *S. connatum*, *S. heiseri*, *S. inscendens*, *S. sanctae-marthae*, *S. tabanoense*. Three others produce fruit abundantly in pollinator-free greenhouses and on self-pollination: *S. canense*, *S. suaveolens*, and *S. trachycarpum*. Preliminary studies indicate that at least some of the populations of *S. fraxinifolium* are also self-compatible. *Solanum muricatum* is variable; Heiser (1964) and Brücher (1968a) report collections that are self-incompatible; some of those I have grown do set fruit in pollinator-free greenhouses. Whether *S. muricatum*, or for that matter, the other species listed above are self-compatible or apomictic is problematic. Correll (1962:26) reports no knowledge of apomixis in sections *Basarthrum* and *Petota*, and Hawkes & Hjerting (1969) give no examples of it in their work. However, the interspecific hybrid triploids produced with *S. trachycarpum* as the female parent (Anderson, 1975a) indicate that further study of the breeding system of the self-compatible species (including *S. muricatum*) is warranted. Most of the species in section *Petota* are also self-incompatible (Grun, 1961; Pandey, 1962; Hawkes & Hjerting, 1969); only two of the 19 wild species studied by Hawkes & Hjerting (1969) are self-compatible.

There is an interesting correlation in section *Basarthrum* between self-compatibility and the length by which the style exceeds the staminal column (Anderson, 1976a). With the exception of *S. fraxinifolium* which has not been studied in detail, the self-compatible species in this section all have styles which do not exceed or exceed by less than 1.9 mm the staminal column. This has the obvious advantage of placing the style very near the anther pores. In section *Petota*, the two self-compatible wild species listed by Hawkes & Hjerting (1969) differ: *S. acaule* Phil. has a system like that described above, whereas the styles of *S. brevidens* Bitt. apparently exceed the staminal column by 1–2 mm.

It has been assumed that the variation of allogamous species is greater than that of autogamous species. Hawkes & Hjerting (1969) suggest that the high degree of intraspecific variation of the tuberous species they studied is due to the pre-dominant outbreeding system. However, Allard (1965) showed in a comparison of two California annual grasses that though there is less variability within families (populations grown from individual collections) for the autogamous species, there is at least as much variability between families (between individuals from different collections) as there is in the allogamous species. A similar comparison of five species from section *Basarthrum* supports Allard's conclusions (Table 42.1). Thus,

Table 42.1. Morphological variability of self-compatible and self-incompatible *Basarthrum* species (Values are coefficients of variation)

Species	Number of leaflets	Length of terminal leaf(let) blade	Number of flowers	Style length
Self-compatible				
S. canense				
within family	6.3	10.1	23.5	10.9
between family	17.3	31.7	29.6	21.2
S. suaveolens				
within family	10.7	13.4	11.1	20.1
between family	18.3	24.5	35.1	23.4
S. trachycarpum				
within family	—[a]	13.0	15.8	11.3
Self-incompatible				
S. basendopogon				
between family	—[a]	29.8	29.6	13.9
S. caripense				
within family	24.8	21.9	24.6	12.3
between family	28.3	23.7	35.2	15.5

a, Data not included because leaves primarily simple.

one may expect, as Hawkes & Hjerting (1969) indicate, that autogamous species may exist as a series of more or less homogeneous lines, but not that autogamous species are any less variable overall. The variability of the autogamous species is easily explained, as Allard (1965) proposed, by a low level of outcrossing.

There is virtually nothing known of another important component of the breeding system, the anthecology. The pollination of the tuberous species is also apparently poorly known; Hawkes & Hjerting (1969) report very few observations of floral visitors. In a preliminary survey, I have found no ultraviolet patterns on the flowers of any of the following (larger flowered) species: *S. caripense, S. fraxinifolium, S. heiseri, S. muricatum, S. trachycarpum, S. tabanoense.*

CHROMOSOME NUMBER AND BEHAVIOUR

Correll (1962) indicates that there is little known cytologically of section *Basarthrum*. Since his publication, a number of counts (14 of the 23 species) have been made all of which indicate $n = 12$ or $2n = 24$ (Heiser, 1963, 1964, 1969; Brücher, 1966, 1968a, b, 1970; Anderson, 1975a, 1976b; Brücher, 1970, records a count for *S. muricatum* made by Vilmorin & Simonet in 1928). In most of the species, meiosis takes place at around sunrise and it is very difficult to find meiosis I stages after 08.00 hrs.

Studies of chromosome behaviour at meiosis by Heiser (1964) and Anderson (1975a, 1976a) of some of the species and of interspecific hybrids show relatively good chromosome pairing and no evidence of large interchanges. The possibility of pericentric inversions has not yet been eliminated; the chromosomes are small and karyotype studies remain to be done. The only notable variance from normal chromosome behaviour occurred in two hybrids between *S. trachycarpum* (♀) and *S. caripense* (Anderson, 1975a). Both of these proved to be relatively sterile and triploid. The only other hybrid secured with *S. trachycarpum* as the female was with *S. canense*; the F_1 in this case was sterile but diploid (Anderson, 1976a). Thus, as speculated elsewhere (Anderson, 1975a, 1976a), differences among species do not seem to involve gross chromosomal changes, but instead genetic differences and/or cryptic structural changes in chromosomes.

Hawkes & Hjerting (1969) report that the same may be true of tuberous species. Seventeen of the 19 wild species they studied are also diploid and the interspecific hybrids examined by them and others (see references in Hawkes & Hjerting, 1969) generally show good pairing. Polyploidy is however apparently more common among the Mexican species in section *Petota*. Hawkes & Hjerting (1969) present evidence from the work of Swaminathan that cryptic structural differences may be involved in interspecific hybrids in the tuberous species; he found a lower percentage of quadrivalent formation in artificial amphidiploids than in artificial autotetraploids.

Hawkes & Hjerting (1969) point out that for the South American tuberous species, our "knowledge of the process of genome differentiation . . . is in its infancy"; this statement applies as well to section *Basarthrum*.

CROSSABILITY AND BREEDING BARRIERS

The success and hybrid products of crossing attempts are useful indicators in biosystematic studies. One gains an idea of species boundaries and perhaps relationships, but more importantly, some insight into evolutionary mechanisms. As pointed out by many others (Grun, 1961; Hawkes & Hjerting, 1969; Levin, 1971; Pickersgill, 1971) the data from such studies need to be treated carefully because they are generated under artificial conditions and thus both success and failure in crossing attempts must be interpreted with caution. Nevertheless, this technique remains valuable because of the unique information generated.

The questions to be considered below are: (a) what, if any, are the barriers to interspecific hybridization in section *Basarthrum*; (b) if barriers exist, how do they compare with those described for section *Petota*; and (c) do crossing studies generally support the current taxonomic scheme? In section *Basarthrum*, nine of the 23 species have been tested in an extensive crossing programme (Anderson, 1975a, 1976a). Over 80% of the nearly 5000 intra- and interspecific crosses attempted failed to yield any fruits. Ninety-two per cent of the more than 1550 interspecific crosses failed (Table 42.2). About 50% of the interspecific combinations attempted (not including results from *S. muricatum*) yielded fruit, but less than half of the fruits had seeds which germinated to give F_1 hybrids (Table 42.2). Only two had high F_1 fertility (Table 42.2). The F_1 hybrids from the other five combinations all showed various degrees of sterility ranging from no pollen production at all to around 50% pollen stainability. One of the fertile combinations (between *S.*

basendopogon and *S. caripense*) was grown through an F_3 generation without any loss of fertility (Anderson, 1976a).

Thus, nearly all of the species tested so far seem to be isolated by pre-fertilization or seed development barriers. Most of the species are separated further by low F_1 hybrid fertility. Since there have been no cytological phenomena observed to account for the latter, the lower fertility in hybrids seems, as concluded above, to be due to genic or cryptic structural differences.

Table 42.2. Interspecific crossing success of *Basarthrum* species

	% Crosses successful	% Combinations[a] successful	% Combinations[b] with seed	% F_1s[c] fertile	% F_2s[d] fertile
Anderson (1975a, 1976a)	8	50	22	29	—
Grun (1961)	45	99	High	High	—
Hawkes & Hjerting (1969)	—	79	High	High	< 20
Pandey (1962)					
Both Mexican and South American species	—	34	31	—	—
South American species only	—	66	45	—	—

a, The number of crossing combinations between species yielding fruit, divided by the total number of combinations attempted. b, Includes only those combinations attempted. c, The proportion of the F_1s secured which were fertile. d, The proportion of the F_2s secured which were fertile.

All of this is in rather direct contrast to the situation found in most of the tuberous species (Table 42.2). Grun (1961), Hawkes & Hjerting (1969) and Ugent (1970) have all found interspecific crossing to be readily accomplished, and in general, the F_1s have a fertility approaching that of the parental species. Grun (1961) in fact stated that one of the problems of the systematics of *Solanum* has been that results from crossability indicate a closer relationship for the species than does the morphology.

In working with only South American tuberous species in two series (eight species), Grun (1961) found good crossing success, seed set and high F_1 hybrid fertility (Table 42.2). Hawkes & Hjerting (1969) indicate that about 80% of the 33 different interspecific combinations attempted among the 19 wild tuberous species from southern South America were successful. Most of the F_1 hybrids secured were fertile. However, they found a "breakdown" (primarily in fertility) in more than 80% of the F_2 combinations. Thus the barriers between some of the South American tuberous species operate only at the F_2 level. This relative ease of hybridization and fertility of F_1 hybrids should allow for more hybridization and gene flow among the wild South American species in section *Petota*, than in section *Basarthrum*. However, as Hawkes & Hjerting iterate, the success of interspecific hybridization in the field or greenhouse is not necessarily a valid indicator of the situation to be expected in nature; they considered that only about 9.5% of the specimens from nature that they examined were hybrids.

Pandey (1962) carried out crosses within and between both Mexican (five wild species) and South American (eight wild, one cultivated species) species and concentrated study on pollen tube growth and seed set to gain an idea of inter-specific compatability. The results are different from those described above (Table 42.2), in that only about 34% of the combinations were successful, but most of these showed good pollen tube growth with at least partial seed set (i.e., not a full complement of seeds). This discrepancy with the results described previously seems to be almost entirely due to the inclusion of the Mexican species. Combinations among South American species only were successful almost 65% of the time, with at least partial seed set in about 45% of the combinations. Pandey concluded that the interspecific barriers were primarily inhibition to pollen tube growth, and secondarily failure to set fruit even after normal fertiliz-ation. Doubtless such a system is also operative with the species in section *Basarthrum*.

Based on the results from crossing studies, Hawkes & Hjerting (1969) and others have postulated that the most important barrier to free hybridization among the tuberous species in nature is ecogeographic. In addition to ecogeographic isolation, the taxa in section *Basarthrum*, and at least some of the Mexican tuberous species, also have ensured their specific integrity by imposition of the rather complete barriers described above.

It is of interest to note that in section *Basarthrum* crossing is considerably less successful with narrow endemics such as *S. heiseri* and *S. sanctae-marthae* than with some of the more widely distributed species. This could be taken to suggest that these species arose not by ecogeographical isolation, since they would then not necessarily have developed the apparently strong barriers to interspecific hybridization, but by some change in the *S*-gene complex.

Lewis & Crowe (1958) and Grun & Radlow (1961) have postulated that in general crosses would not be as successful in combinations between self-incom-patible (SI) females × self-compatible (SC) males as in: SC♀ × SI♂. It is hypothe-sized that this unilateral incompatibility serves to prevent the incorporation of self-compatibility genes into self-incompatible systems. Data from the tuberous section (Grun, 1961; Pandey, 1962) support this hypothesis as do the results of most crosses in section *Basarthrum* (Anderson, 1975a, 1976a).

There are in section *Basarthrum* (Anderson, 1975a, 1976a) as in section *Petota* (Grun, 1961; Pandey, 1962; Hawkes & Hjerting, 1969) reciprocal differences in crossing success and seed set, and differential success in crossing depending upon which individual plants or collections of a species are involved.

Although there are a number of exceptions (see following section), the results of the crossing tests generally support the present taxonomic scheme. A higher proportion of crosses between members of different series fail than do those between members of the same series. The highest F_1 fertility has been achieved in an intra-series hybrid (*S. caripense* and *S. basendopogon*). The only other com-bination with relatively high fertility was between *S. caripense* and *S. tabanoense*, species from different series, but as suggested below and elsewhere (Anderson, 1975a, Anderson & Gensel, 1976), the latter should be placed in the same series as *S. caripense*. Grun (1961) found very strong support in crossing studies for the taxonomic scheme in series *Tuberosa* Rydb. and *Commersoniana* Buk. of section *Petota*. Interspecific crossing and seed set were significantly better within than between the two series. However, fertility of the hybrids was the same whether the crosses were made within or between the series. In view of Hawkes & Hjerting's (1969) experience with F_2s, it would be of some interest to compare the "breakdown" of intra- with inter-series F_2s. The virtually complete lack of compatibility between the South American and Mexican species reported by Pandey (1962) offers further support for the taxonomic system in section *Petota*.

MORPHOLOGICAL FEATURES AND NOTES ON THE SERIES

Morphological features

Pollen and seed morphology have not previously been used in the taxonomy of section *Basarthrum*. Though the differences are slight in comparison to some other groups, there are features of the pollen useful in the systematic treatment of the group. Specifically, the density of exine granules, the presence or absence of apertures, and pollen diameter have been used to gain further insight into the relationships of the taxa (Anderson & Gensel, 1976).

Seed size and the degree of development of a wing encircling the seed are heretofore unexplored characters which also seem taxonomically useful (Fig. 42.1; and

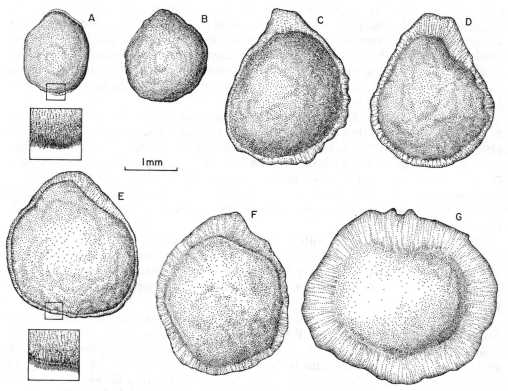

Figure 42.1. Seeds of selected species in *Solanum*, section *Basarthrum*. These seven examples represent the sampled extremes of variation in section *Basarthrum*. A, *S. canense* (series *Canensa*); B, *S. caripense* (series *Caripensa*); C, *S. appendiculatum* (series *Appendiculata*); D, *S. tetrapetalum* (series *Appendiculata*); E, *S. tabanoense* (series *Appendiculata*); F, *S. taeniotrichum* (series *Suaveolentia*); G, *S. sanctae-marthae* (series *Articulata*).

Anderson, 1976a). Furthermore, a preliminary scanning electron microscope analysis of seeds of several species in section *Basarthrum* revealed that they are covered with a vesture that shows variation among species. The full extent of distribution of these features and their systematic reliability are currently under investigation.

Notes on the series

Series Muricata

This series is comprised of but one species, *S. muricatum*. Though the fruits of other species in section *Basarthrum* are occasionally collected and eaten, *S. muricatum*, the pepino, is the only truly cultivated species in the section.

There has not been unanimity regarding the place of origin or the progenitor, but it is generally agreed that the pepino is an old cultivated plant. Towle (1961) records evidence of early pottery representations of the fruit and indicates that "the early chroniclers" report the growing of pepinos along the coast of Peru in pre-historic times.

We know little of the origin of the pepino. This may be because it has been in cultivation for so long that it shows little resemblance to its progenitor(s); or it may be due to the fact that it is grown on such a limited scale that there is little material available, and little economic interest in it. Although there is disagreement as to the ancestral home of the pepino (Colombia—Bukasov, 1930; Peru—Towle, 1961; Ecuador—Schultes and Romero-Castañeda, 1962) it seems clear that the Andean region from southern Colombia to southern Peru is the site of original domestication. Today, the pepino can be found in markets as far north as Mexico (Correll, 1962).

Only Brücher (1970) has stated that there are wild species yet to be discovered from series *Muricata*, thus implying an as yet undiscovered wild ancestor. Correll (1962) and Heiser (1964, 1969) have proposed consideration of *S. tabanoense* and *S. caripense* respectively as being involved in the ancestry of the pepino. Heiser (1969) proposed several reasons, including relatively fertile F_1s, in support of *S. caripense* as a more likely wild ancestor than *S. tabanoense*. Brücher (1966, 1968a, 1970), without documentation, considers *S. tabanoense* as the most likely progenitor.

The only hybrids secured with *S. muricatum* to date are with *S. caripense* and *S. tabanoense* (Heiser, 1964, 1969). As stated above, the former was rather fertile; the latter gave plants with lower fertility. In crosses with *S. basendopogon* and *S. sanctae-marthae*, Anderson (1976a) secured fruit set, but no viable seeds were produced.

Another species not previously proposed which deserves consideration as the progenitor of the pepino is the Peruvian *S. basendopogon*. As mentioned above, some limited degree of success in hybridization with the pepino was achieved with this species, but more important is the similarity in habit and leaf morphology of some robust plants of *S. basendopogon* to *S. muricatum*. In addition, *S. basendopogon*, although it has many more flowers, has a branched inflorescence as do some individuals of *S. muricatum*. If the predominantly simple-leaved (to 3-foliate) *S. basendopogon* were involved in the origin of the pepino, the proposal made by Bitter (1913; also supported by Brücher, 1968a) that the simple-leaved forms of

S. muricatum represent the end point of an extreme reduction series, would have to be modified. Even if *S. basendopogon* is not involved, it does not seem entirely correct to postulate an extreme reduction series to explain the simple-leaved forms. The pepino is known only from cultivation, and as a result, the leaf, fruit and flower variability may as well represent a mosaic of human selection for attractive plants, sweeter fruits, brightly coloured flowers, etc., having little to do with selection in any particular direction.

As described above, some plants of *S. muricatum* are apparently self-incompatible and others self-compatible. Another variable feature is seed set. Popenoe (1924) and Correll (1962) describe pepino fruits as usually seedless; Heiser (1964), and Brücher (1968a) cite collections of fruits with seeds. Although there is variability from plant to plant, virtually all the collections of the pepino that I have grown (from Colombia, Ecuador, and Peru) that have fruited produce on the average half of the fruits with seeds (mean number of seeds ranges from 13–120 seeds per fruit). However, Heiser (pers. commun.) reports growing cuttings from Peru that gave only seedless fruits.

Series Caripensa *and* Suaveolentia

All species in series *Caripensa* except *S. filiforme* have been treated in detail elsewhere (Anderson, 1975a, 1976a; Anderson & Gensel, 1976). This series is comparatively homogeneous in gross morphology and pollen morphology (Anderson & Gensel, 1976), *S. trachycarpum* being the most distinct on both counts. A preliminary study of seed morphology also shows homogeneity; none of the species has winged seeds (Fig. 42.1). In terms of crossability, fertile F_1 hybrids between *S. basendopogon* and *S. caripense* have been made while those produced by *S. caripense* and *S. heiseri* and between *S. caripense* and *S. trachycarpum* showed low fertility. A number of seedless fruits have been produced in other interspecific crosses. Inter-population variability in crossing is high in some of the species, e.g., in a study of more than 20 population samples of *S. caripense* (Anderson, 1975a) the variability in crossing tests was shown to be as high as the most extreme morphological variation.

Some species in other series are closely allied to those in series *Caripensa*. *Solanum tabanoense* is similar in overall morphology (Anderson, 1975a) and pollen morphology (Anderson & Gunsel, 1976). In addition, crosses between *S. caripense* and *S. tabanoense* yielded F_1s some of which were highly fertile (Anderson, 1975a). The seeds (Fig. 42.1) of *S. tabanoense* are somewhat larger than average for series *Caripensa*.

Brücher (1970) questioned the need for establishing the series *Caripensa*. He felt that *S. caripense* and the species *S. fraxinifolium* and *S. suaveolens* from series *Suaveolentia* were very similar. In 1966, although admitting the lack of a detailed study, Brücher proposed these three as synonymous; later (1968b, 1970) he treated only *S. caripense* and *S. fraxinifolium* as synonyms. Although Brücher presents little evidence for his claims regarding the treatment of the species, he may be correct in proposing that at least some should be treated together in the same series. *Solanum fraxinifolium* is morphologically very similar (including seed morphology) to some of the more highly compound-leaved races of *S. caripense*. An F_1 hybrid with moderate fertility has been produced between these two species.

Although the rare *S. taeniotrichum* has a highly branched inflorescence and pre-liminary studies indicate that it has somewhat larger seeds with a rather pro-nounced wing (Fig. 42.1), it is similar in general morphology to *S. fraxinifolium* (Correll, 1962). *Solanum suaveolens* is similar to some individuals of *S. fraxini-folium*; however, it shows a stronger relationship with *S. canense* (see below).

Series Appendiculata

There has been little additional study of this heterogenous series since Correll's work (1962). Six of the 12 taxa in the series are rare. It seems to be made up of two rather distinct units, one Mexican and Central American, and the other South American. The latter group is quite heterogenous. The only species studied in detail, *S. tabanoense*, seems at this time better placed with another series (*Cari-pensa*). *Solanum brevifolium* and *S. tetrapetalum* are vegetatively somewhat similar to the Central American element of the series. They also have larger pollen than *S. tabanoense*, although not as large as that of the Central American species, and an exine granule density close to that of the Central American species (Ander-son & Gensel, 1976). Seeds of *S. brevifolium* are still unknown, those of *S. tetra-petalum* (Fig. 42.1) have a very dense vesture, and a moderately broad wing.

The primary dichotomy used by Correll (1962) to separate the morphologically similar Mexican and Central American species is the length by which the style exceeds the staminal column. We have grown from seed two collections each of *S. appendiculatum* and *S. connatum*. Each collection was grown from seeds of a single fruit. In all four collections there is segregation for style length; of the 22 plants that have flowered to date, seven have long styles and 15 short styles. This striking situation is being investigated further. Preliminary observations show a correlation between styles that exceed the staminal column and the following: greater absolute style length (anther length and stamen length are not significantly different between the two groups); higher pollen fertility (80% versus 40%); inaperturate pollen grains (vs. the typical tricolporate condition). Whether these phenomena indicate a heterostylous-type breeding system, evolution toward a dioecious condition, hybrid segregation or simply aberrant plants, is at this time unknown. However, since the collections are from two distinct localities, the latter two explanations seem unlikely.

The pollen of some of the species in this part of the series is significantly larger and has a much more densely granulated exine than other species in section *Basar-thrum* (Anderson & Gensel, 1976). The complete distribution among the species of the unique inaperturate pollen will be very helpful in characterizing some of these closely related taxa. From a preliminary examination it appears that the seeds of two species (*S. appendiculatum*, Fig. 42.1; *S. inscendens*) from this part of the series are somwhat larger than those of species in other series (except *S. sanctae-marthae*) and are encircled by a narrow wing.

Series Articulata

The single species in this series, *S. sanctae-marthae*, restricted to the Santa Marta mountains in northern Colombia, has been discussed in detail elsewhere (Anderson, 1976a). It is similar in leaf and floral morphology (although the inflorescence is branched) to *S. fraxinifolium* (and to some races of *S. caripense*) but has the unique feature of individual articulation of the leaflet petioles, has purple unstriped fruits, and has a very distinctive large seed with a large wing (Fig. 42.1). It also bears primarily 3–4 celled hairs, and thus perhaps does not belong in section *Basarthrum* (Anderson, 1976a).

Series Canensa

Correll (1962) recognized a distinct series for this species and considered it to be only superficially similar to *S. suaveolens* with which he had placed it earlier (Correll, 1952). However, these two species show general overall similarity in gross morphology (Anderson, 1976a), pollen (Anderson & Gensel, 1976) and seed morphology, and breeding behaviour (Anderson, 1976a). On these grounds, I consider these two taxa to be more closely associated with each other than with species of other series, and have postulated (Anderson, 1976a) a phylogenetic relationship between them.

Correll (1962) and Hawkes & Hjerting (1969) point out a possible connection between sections *Basarthrum* and *Petota* in the morphological similarity of the autogamous non-tuberous *S. canense* with the autogamous non-tuberous *S. brevidens* in series *Etuberosa* Juz. This proposed relationship deserves further study.

ACKNOWLEDGEMENTS

I am grateful to: C. B. Heiser, Jr., who originally suggested the problem of the origin of the pepino, made large amounts of seed available and made critical comments on the manuscript; M. J. Anderson who contributed to all phases of this work; M. J. Gallo for the drawing; P. Lilliquist for aid in growing the plants; J. A. Slater for critical comments on the manuscript. Portions of the work have been supported by grants from the Society of the Sigma Xi and from the University of Connecticut and University of Nebraska Research Foundations.

REFERENCES

ALLARD, R. W., 1965. Genetic systems associated with colonizing ability in predominantly self-pollinated species. In H. G. Baker & G. L. Stebbins (Eds), *The Genetics of Colonizing Species: 49–75.* New York: Academic Press.

ANDERSON, G. J., 1971. *The Variability and relationships of Solanum caripense.* Ph.D. Thesis, Indiana Univ., 161 pp. Univ. Microfilms. Ann Arbor, Mich. (Diss. Abstr. 32 (8): 4442).

ANDERSON, G. J., 1975a. The variation and evolution of selected species of *Solanum* section *Basarthrum.* *Brittonia, 27:* 209–222.

ANDERSON, G. J., 1975b. Chromosome numbers in *Solanum.* In A. Löve (Ed.), IOPB chromosome number reports XLVIII. *Taxon, 24:* 370.

ANDERSON, G. J., 1976a. The variation and evolution of selected species of *Solanum* section *Basarthrum* II. *Brittonia, 29:* 116–128 (1977).

ANDERSON, G. J., 1976b. Chromosome numbers in *Solanum.* In A. Löve (Ed.), IOPB chromosome number reports LIII. *Taxon, 25:* 497.

ANDERSON, G. J. & GENSEL, P. G., 1976. Pollen morphology and the systematics of *Solanum* section *Basarthrum. Pollen et Spores, 18:* 533–552.

BITTER, G., 1913. Solana nova vel minus cognita. *Repertorium Specierum Novarum Regni Vegetabilis, 12:* 441–444.

BRÜCHER, H., 1966. *Solanum caripense* HBK (subsect. *Basarthrum*) in Venezuela. *Feddes Repertorium, 73:* 216–221.

BRÜCHER, H., 1968a. Die genetischen Reserven Südamerikas für die Kulturpflanzenzüchtung. *Theoretical and Applied Genetics, 38:* 9–22.

BRÜCHER, H., 1968b. Wildkartoffeln (subsect. *Basarthrum* und *Hyperbasarthrum*) der kordilleren Venezuelas. *Feddes Repertorium, 77:* 31–46.

BRÜCHER, H., 1970. Chromosomenzahlen argentinischer, chilenischer und venezolanischer Wildkartoffeln (*Solanum*, sect. *Tuberarium*). *Cytologia, 35:* 153–170.

BUKASOV, S. M., 1930. The cultivated plants of Mexico, Guatemala, and Colombia. *Trudy po Prikladnoĭ Botanike, Genetike i Selektsii*, Suppl. 47.

CORRELL, D. S., 1952. Section *Tuberarium* of the genus *Solanum* of North America and Central America. *Agricultural Monograph, No. 11.* United States Department of Agriculture.

CORRELL, D. S., 1962. *The Potato and Its Wild Relatives.* Renner, Texas: Texas Research Foundation.

GRUN, P., 1961. Early stages in the formation of internal barriers to gene exchange between diploid species of *Solanum. American Journal of Botany, 48:* 79–89.

GRUN, P. & RADLOW, A., 1961. Evolution of barriers to crossing of self-incompatible with self-compatible species of *Solanum. Heredity, 16:* 137–143.

HAWKES, J. G. & HJERTING, J. P., 1969. The potatoes of Argentina, Brazil, Paraguay, and Uruguay, a biosystematic study. Annals of Botany Memoirs, No. 3. Oxford: Clarendon Press.

HEISER, C. B., Jr., 1963. Numeración chromósomica de plantas ecuatorianas. *Ciencia y Naturaleza (Ecuador) 6:* 2–6.

HEISER, C. B., 1964. Origin and variability of the Pepino (*Solanum muricatum*): a preliminary report. *Baileya, 12:* 151–158.

HEISER, C. B., 1969. *Solanum caripense* y el origen de *Solanum muricatum. Revista Politécnica (Quito Ec.), 1* (3): 1–7.

LEVIN, D. A., 1971. The origin of reproductive isolating mechanisms in flowering plants. *Taxon, 20:* 91–113.

LEWIS, D. & CROWE, L. K., 1958. Unilateral interspecific incompatibility in flowering plants. *Heredity, 12:* 233–256.

PANDEY, K. K., 1962. Interspecific incompatibility in *Solanum* species. *American Journal of Botany, 49:* 874–882.

PICKERSGILL, B., 1971. Relationships between weedy and cultivated forms in some species of chili peppers (genus *Capsicum*). *Evolution, 25:* 683–691.

POPENOE, W., 1924. Economic fruit-bearing plants of Ecuador. *Contributions of the United States National Herbarium, 24* (5): 101–134.

RAVEN, P. H. & AXELROD, D. I., 1974. Angiosperm biogeography and past continental movements. *Annals of the Missouri Botanical Garden, 61:* 539–673.

SAVAGE, J. M., 1974. The Isthmian link and the evolution of Neotropical mammals. *Contributions in Science, 260:* 1–51.

SCHULTES, R. E. & ROMERO-CASTAÑEDA, R., 1962. Edible fruits of *Solanum* in Colombia. *Botanical Museum Leaflets, 19:* 235–286.

TOWLE, M. A., 1961. *The Ethnobotany of pre-Columbian Peru.* New York: Wenner-Gren Foundation for Anthropological Research, Inc.

UGENT, D., 1970. *Solanum raphanifolium*, a Peruvian wild potato species of hybrid origin. *Botanical Gazette, 131:* 225–233.

43. Dispersal and speciation in *Solanum*, section *Brevantherum*

KEITH E. ROE

Life Sciences Library, The Pennsylvania State University, U.S.A.

A brief account is given of the taxonomic position of *Solanum*, section *Brevantherum*. The section is distributed widely in the Americas and shows two areas of species diversity, Mesoamerica and the northern parts of South America. The species seem to be mostly found in open habitats and include several successful weeds.

The spread from the Americas into Africa and Asia of three of these weeds, *S. erianthum*, *S. mauritianum* and *S. umbellatum* is described and their distribution is related to 16th century trade routes.

Isolating mechanisms, species hybridization and the general reproductive biology of species in section *Brevantherum* are briefly discussed.

Section *Brevantherum* is a member of subgenus *Solanum* which includes plants lacking prickles and having short, blunt anthers. The section is characterized by its sympodial growth habit, terminal inflorescences and stellate pubescence. There are 27 known species in the section, all native to the American tropics and subtropics.

Section *Brevantherum* is linked to other groups by intermediate species which are excluded from the section on the basis of few or even one character, such as hair type. For example, both *S. bullatum* and *S. oblongifolium* have dendritic pubescence, yet the former is placed in section *Brevantherum*, the latter in section *Anthoresis*. Hence, although it is an easily recognized group, section *Brevantherum* has probably not had a long history of evolutionary independence. It might be said that intermediate species form a continuum between this section and others.

The species of section *Brevantherum* are distinguished morphologically from one another by a variety of characters, many of which are cryptic, for example hair types and features of the bud (Roe, 1972). The pattern of a diversity of unique characters which have evolved in the various species fits in with their believed mode of speciation by isolation of populations following disruption of ranges or long-distance dispersal. Under such conditions one might expect to find fixation of an assortment of seemingly non-adaptive characters.

Phenetic relationships among the 27 species suggest a basic subdivision within the section (Roe, 1974). The 17 species with complex hair types are distinguished from the ten species with ordinary stellate hairs.

Geographic distribution of the section indicates two main centres of speciation, one in southern Mexico and Central America, the other in western South America (Fig. 43.1). These regions are associated with geologically recent tectonic and volcanic activities, resulting in both geographic isolation of populations and creation of open habitats to which the species are adapted. Related endemic and localized species are often geographically replacing or are isolated by altitude.

Two especially successful weeds, *S. erianthum* and *S. mauritianum*, are widely naturalized in the Old World (Fig. 43.2). A third weedy species, *S. umbellatum*, has been collected in western Africa. The distributions of these species are inter-

Figure 43.1. Distribution of *Solanum* section *Brevantherum* in the Americas (shaded area). Ranges of localized species are outlined (after Roe, 1969).

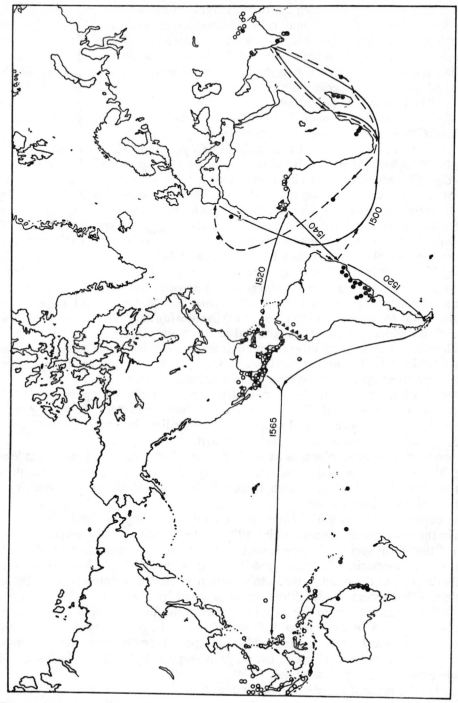

Figure 43.2. Distributions of *Solanum erianthum* (O), *S. mauritianum* (●), and *S. umbellatum* (▲). Arrows show some routes of exploration, slave ships and trade fleets and dates by which these activities had commenced (after Schurz, 1939; Marchant, 1941; Merrill, 1954; Clark, 1964; Boxer, 1969).

esting in that they coincide, in part, with the old Spanish and Portuguese trade routes which began in the 16th century (Schurz, 1939; Marchant, 1941; Clark, 1964; Boxer, 1969) and which may have initiated a means of their dispersal, either by accident or by design (cf. Merrill, 1954: 238, 340). The fruits of *S. erianthum* are reportedly edible (Tanaka, 1976) while an observation regarding the local Brazilian species, *S. bullatum*, may hold significance for the above, more weedy taxa. The leaves of this species make attractive forage for horses (Hunnicutt, 1919), a feature likely viewed as potentially useful and possibly attributed to other species if noticed by the early colonists. Further work on the ethnobotany of *S. erianthum* and *S. mauritianum* could prove worthwhile.

The species of this section are colonizers of open ground, including volcanic talus, forest openings, stream borders, and areas of human disturbance such as roadsides. The widespread species seem adapted to diverse habitats, for example, *S. erianthum* is found in the wet lowlands of southern Mexico while it grows on dry coral limestone in southern Florida. Other species have less ecological tolerance, e.g. *S. hazenii* is restricted to coastal lowlands, and local species, such as *S. brevipedicellatum*, appear quite restricted ecologically. This species is apparently disjunct between the Guatemala/Chiapas area and Jalisco in western Mexico. In the latter region, it is found in subtropical microhabitats in moist ravines associated with Musaceae and other tropical plants within an otherwise Pine-Oak forest. The western Mexican disjunct is typical of recently established, small populations (Wright, 1943; Mayr, 1963; Grant, 1971: 13) in showing as much diversity in flower colour (both white and violet) as is found in the main range.

Ecological isolation is effective between at least one pair of species, *S. erianthum* being found on lower, wet ground, with *S. axillifolium* found on higher, drier areas where these species are sympatric in southern Mexico.

Hybridization appears to be rare in the section. Only one putative hybrid between *S. erianthum* and *S. axillifolium* has been collected. The only other suspected crosses involve several specimens intermediate between *S. mauritianum* and *S. granuloso-leprosum* from southern Brazil.

Reproduction in these plants is by seed and, seemingly more common in at least some, like *S. erianthum* and *S. mauritianum*, by adventitious shoots from shallow roots to form large colonies. Both these species are self-compatible, one more criterion of a successful weed.

The colourful berries and frequently isolated plants suggest bird dispersal of seeds in the widespread species (White, 1929) and even some local ones (Hunnicutt, 1919). Other local species, however, such as *S. atitlanum*, appear to have diverged in rather the opposite direction, producing dull, greenish-black berries which readily drop from the plant immediately upon maturity or, seemingly, even before ripening. This precocious deciduous habit is aided by abscission of pedicels and even entire inflorescences in the case of *S. brevipedicellatum*, unlike the persistent inflorescences of *S. erianthum* and *S. mauritianum*. In *S. brevipedicellatum*, and perhaps other local species, the resulting dispersal on-the-spot, or topochory (Zohary, 1962), may be of selective advantage in keeping the seeds within favourable microhabitats (cf. Harper & White, 1974).

Clearly, the present knowledge of section *Brevantherum* raises more questions than it answers. Features of heterostyly in local species, genetic isolation in sympatric species, and the rarity of seedlings are apparent facts but their underlying mechanisms remain to be discovered. One can hope, however, that with the existing taxonomic diversity available for study in this section, future work on it and related groups will help us cope with the phylogenetic questions which remain unanswered, or even yet to be asked.

REFERENCES

BOXER, C. R., 1969. *The Portuguese Seaborne Empire*. New York: Knopf.

CLARK, W. R., 1964. *Explorers of the World*. New York: Natural History Press

GRANT, V., 1971. *Plant Speciation*. New York: Columbia University Press.

HARPER, J. L. & WHITE, J. W., 1974. The demography of plants. *Annual Review of Ecology and Systematics,* 5: 419–463.

HUNNICUTT, B. H., 1919. A forage plant from the Solanaceae family. *Journal of Heredity, 10:* 184–187.

MARCHANT, A., 1941. Colonial Brazil as a way station for the Portuguese India fleets. *Geographical Review, 31:* 454–465.

MAYR, E., 1963. *Animal Species and Evolution*. Cambridge: Belknap Press.

MERRILL, E. D., 1954. *The Botany of Cook's Voyages and Its Unexpected Significance in Relation to Anthropology, Biogeography, and History*. Waltham, Mass.: Chronica Botanica.

ROE, K. E., 1969. *A revision of* Solanum *section* Brevantherum. Ph.D. thesis, University of Wisconsin, U.S.A.

ROE, K. E., 1972. A revision of *Solanum* section *Brevantherum. Brittonia, 24:* 239–278.

ROE, K. E., 1974. A simple technique for measuring phenetic similarity in *Solanum* using edge-punched cards. *Taxon, 23:* 707–713.

SCHURZ, W. L., 1939. *The Manila Galleon*. New York: Dutton.

TANAKA, T., 1976. *Tanaka's Cyclopedia of Edible Plants of the World*. Tokyo: Keigaku.

WHITE, C. T., 1929. *Solanum auriculatum*—A "wild tobacco". *Queensland Agricultural Journal, 32:* 194–197.

WRIGHT, S., 1943. Isolation by distance. *Genetics, 28:* 114–138.

ZOHARY, M., 1962. *Plant Life of Palestine: Israel and Jordan*. New York: Ronald Press.

44. Patterns in biogeography in *Solanum*, section *Acanthophora*

M. NEE

Botany Department, University of Wisconsin, Madison, Wisconsin, U.S.A.

This section consists of shrubs with weedy tendencies; the centre of origin is in south-eastern South America. Four species are mapped and their ranges discussed; similar patterns can be expected in other species of *Solanum*. Dates of earliest collection, knowledge of relationships and variability within a species are helpful in interpreting ranges and histories.

CONTENTS

INTRODUCTION

Solanum is apparently in the process of explosive speciation. Subgeneric divisions are usually based on loose correlations of characters such as spine type, hair types and distribution, calyx shape, leaf shape and general aspect. Few sections are as clearly distinct as one would like. Roe (1972, and this volume, p. 563), for example, mentions several species excluded from section *Brevantherum* Seithe on the basis of only one character. There are certain relatively homogeneous groups in subgenus *Leptostemonum* such as section *Micracantha* Dun. and section *Aculeigerum* Seithe, but I confess to being unable to see clearcut divisions among the majority of species in the subgenus.

Section *Acanthophora* Dun. is a member of the prickly subgenus *Leptostemonum* and well merits its name; all the species are heavily armed with sharp slender spines and some species have stouter broad-based spines as well. Among the myriad species of subgenus *Leptostemonum* (approximately half of all the *c.* 1500 species of *Solanum*), this section of about 15 species can be fairly unambiguously delimited by the combination of: (1) the upper leaf surface bearing only simple hairs, (2) glabrous fruit, and (3) non- to only slightly accrescent calyx.

Subgenus *Leptostemonum* is generally characterized (as is the evidently closely related subgenus *Brevantherum* (Seithe) D'Arcy) by stellate hairs on most parts of the plants. Simple hairs are by no means uncommon, but to have the upper surface of the leaf bearing only simple hairs (or in a few cases no hairs at all) is unusual. In section *Acanthophora* these simple hairs are several celled, uniseriate, shiny, hyaline and translucent. A comparison of these simple hairs with the central ray of the stellate hairs on the same plant will show that they are identical. During evolution the basal side rays of the stellate hairs have disappeared. Single celled simple hairs, often gland tipped, may also occur on the upper surface of the leaf as well as elsewhere on the plant. A. Seithe (1962, and this Volume, p. 307) has provided a thorough overview of hair types and their distribution in *Solanum*, and Roe (1971) has provided a terminology for the hair types. In a few other species of subgenus *Leptostemonum*, there are only simple hairs (or no hairs at all) on the upper leaf surface, but the hair structure and a combination of other characters clearly indicate that their affinities lie outside section *Acanthophora*.

Solanum pectinatum Dun. (= *S. hirsutissimum* Standl.) and a few other close relatives of *S. quitoense* Lam. also have only simple hairs on the upper leaf surface, but these species have a densely hirsute ovary and fruit and thus belong in a related section. Very few species of section *Acanthophora* have a calyx which is appreciably accrescent. The elongate herbaceous calyx lobes of *S. incarceratum* R. & P. are certainly an isolated evolutionary line, but the slightly accrescent, stoutly prickly calyx of *S. capsicoides* All. might indicate a relationship to *S. sisymbriifolium* Lam. of section *Cryptocarpum* Dun.

Section *Acanthophora* itself may be separable into two subsections based on seed morphology. One subsection would include *S. mammosum* L., *S. aculeatissimum* Jacq. of Africa, *S. palinacanthum* Dun. (= *S. claviceps* Griseb.) of South America and the *S. globiferum* (*sensu* Gentry and Standley, 1974, probably not Dunal)—*S. viarum* Dun. complex of Mexico, Guatemala and south-eastern South America. These species have seeds which do not differ greatly from those of almost all other *Solanum* species.

The other subsection would include those species whose seeds have a flattened margin forming a wing around the seed. Since this winged seed is unique in *Solanum*, this group is probably monophyletic, but until mature fruits and seeds are known from all species, I prefer not to divide the section; few other characters correlate with seed morphology. *Solanum atropurpureum* Schrank, *S. incarceratum* R. & P., *S. acerifolium* H. & B. ex Dun., and *S. capsicoides* All., among others, belong in this group. The evolutionary significance of the wing is not clear. Perhaps it would aid in local dispersal by wind or water for the species such as *S. capsicoides* which has dry fruits at maturity which split irregularly to release the seeds. However, the smaller fruits of others turn yellow and juicy at maturity, obviously an adaptation to bird dispersal.

Solanum atropurpureum Schrank, *Syll. Ratisb.*, *1*:200 (1824)

(Figure 44.1)

This fast growing, prolifically fruiting shrub of secondary growth, cut-over forest and pastures is common in south-eastern South America. It has the narrowest range of the four species to be discussed—south-eastern Brasil, Uruguay, Paraguay and north-eastern Argentina. This is also the centre of diversity of section *Acanthophora*; the greatest concentration of species is found here and species growing elsewhere can be shown to have their ultimate origin here.

Figure 44.1. Distribution of *Solanum atropurpureum*.

The disjunct population in Colombia calls for an explanation. All of the specimens for which there are label data are from pastures, a habitat common enough now in the highly disturbed northern Andes. The collections seem quite restricted, mostly in the vicinity of the large city of Medellín. Finally, all the collections are relatively recent, the earliest being from 1922 (Depto. Antioquia: Medellín, *F. W. Pennell 10971* (GH, PH)). All the evidence indicates a recent introduction; either of two methods of introduction by man from south-eastern South America seems possible. The seed could have been inadvertently carried in agricultural produce, for example. The other good possibility is that this rather striking plant with deeply divided purple leaves has been planted in Colombia as a horticultural curiosity and escaped locally; it has been cultivated in many European botanical gardens for over 150 years.

An additional piece of evidence suggests that the species was recently introduced through a limited number of seeds. There is much variation in the leaf lobing of *S. atropurpureum*. Some of the difference has to do with age and vigour of the individuals, position of the leaves on the plant and habitat, but some appears to be genetic. In southern South America there is the whole range of leaf shapes from only slightly lobed to the characteristic highly pinnatifid undulate leaves; the Colombian collections are uniformly of this latter shape.

Solanum acerifolium H. & B. ex Dunal, *Solan. Syn.*: 41 (1816)

(Figure 44.2)

S. quinquangulare Willd. ex Roem. & Schult. *Syst.*, *4*: 669 (1819).
S. acerosum sensu D'Arcy, 1973, non Sendtner.

This species is a common weedy shrub of secondary growth, coffee fincas and pastures in the mountains from southern Mexico to the northern Andes at altitudes from 1000 to 2400 m.

S. acerifolium shows a pattern that is apparently the result of a natural introduction in the distant past similar to the recent human introduction of *S. atropurpureum* into a part of the same region. There are many very early collections from all parts of its range indicating that it is pre-Columbian. None of its nearest relatives occur within its range; its affinities are clearly with a complex of species in the South American centre of the section. *Solanum acerifolium* has been established in its range long enough to diverge into two very weak segregates. The Central American populations tend to have sparsely pilose pedicels and leaves shaped like those of *M. Nee et al. 14152* (Panamá: Prov. Chiriquí: Cerro Pando (WIS)) from the mountains of western Panamá. South American collections tend to have nearly glabrous pedicels and to have few-lobed to occasionally entire leaves, e.g. Mexico: Edo. Vera Cruz: Jalapa, 25 Aug 1866, *M. Hahn s.n.* (P); Guatemala: Depto. Alta Verapaz: Cobán, *P. C. Standley* 69501 (F). Some, in fact, are very reminiscent of leaves of certain species of *Acer*.

The leaves in Fig. 44.2 were deliberately chosen to show that variation of leaf shape is only a tendency, since all shapes can be found in Central America. Similarly, all the shapes can be found in South American collections. *Solanum acerifolium* and *S. quinquangulare* were named for their very different leaf shapes before the range of variation in this one species was known.

There is a great need for careful collections of leaf variation on individual plants, within populations and between populations. Many species of subgenus *Leptostemonum* show great variation in leaf shapes; young vigorous branches tend to have large highly lobed leaves while flowering branches on the same plants may

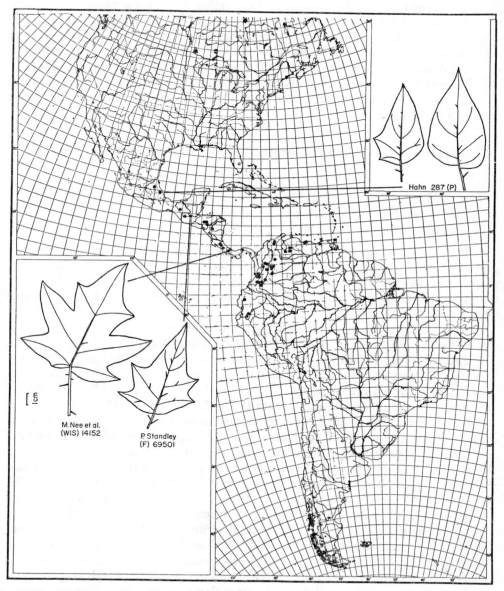

Figure 44.2. Distribution of *Solanum acerifolium* with representative leaf shapes. Mexico, *Hahn 287* (P); Guatemala, *Standley 69501* (F); Panamá, *M. Nee et al. 14152* (WIS).

have small entire leaves, often also devoid of spines (see Roe, 1966, for a well documented example).

There is no fossil evidence of this group of *Solanum* and no real data on evolutionary rates so that the time of introduction of *S. acerifolium* from south-eastern South America is little more than a matter of speculation. The rise of the Andean system, mostly in the Pliocene and Pleistocene would seem to set an upper limit. With the undoubtedly drastic, but as yet little understood, climatic fluctuations of the Pleistocene (see van der Hammen & González, 1960, for a pioneering study), there may have been times when direct migration of the ancestor of *S. acerifolium* would have been possible; there surely were times when possibilities of long distance dispersal were more likely than today. It should be pointed out, however, that the distance involved of some 3000 km between Ecuador and south-eastern Brasil is less than the distance between Hawaii and the nearest landmasses, and Hawaii has received a significant number of introductions of seeds by long distance dispesral by birds.

Solanum capsicoides Allioni, *Melanges philos.-math. soc. roy. Turin*, 5: 64 (1774)

(Figure 44.3)

S. ciliatum Lam. *Illust.*, 2: 21 (1799).
S. ciliare Willd. *Enum. Hort. Berol.*: 237 (1809).
S. aculeatissimum auct., non Jacq.

This is a weedy lowland species of drier, more open habitats than the two preceding species; like most widespread *Solanum* species it has a wide ecological amplitude. *Solanum capsicoides* is certainly a very successful introduction to a number of tropical and subtropical areas where now there are an abundance of open weedy habitats.

It is quite common in southern China and some neighbouring islands and is obviously an introduced weed. For some reason it has spread only slowly in Africa. The three collections from Liberia span a range of 118 years: 1834, *Dr Skinner s.n.* (NY); 1896, *M. Dinklage 1661* (B); and 1952, *P. M. Daniel 420* (MO). I have seen only one specimen from southern Africa: Durban, 1895, *J. Medley Wood 5718* (K). The Sierra Leone collection is from 1955, *H. D. Jordan 2115* (K).

It has often been assumed that this species is Caribbean in origin as it is abundant on the islands and surrounding mainland and a large number of specimens have accumulated in herbaria. It is adventive here, however; the native range is eastern Brasil not far from the coast, where it was found by most of the very earliest collectors. The first collections of the Caribbean region are: Jamaica, 1849–50, *R. C. Alexander s.n.* (F); Cuba, 1860–64, *C. Wright 3029* (GH, MO, NY, US); and Florida, Clay Co., Hibernia, 1869, *Wm. M. Canby s.n.* (GH, NY, US).

The precise native range may never be determined with any greater accuracy than "coastal Brasil south of the Amazon". Section *Acanthophora*, in common with almost all groups of *Solanum*, has great weedy tendencies, the species being found in areas of soil disturbance and high light levels. In nature, landslides, openings in forests from fallen trees and eroded areas along streams provided the proper habitat. The agricultural practices of pre-Columbian Indians certainly must have

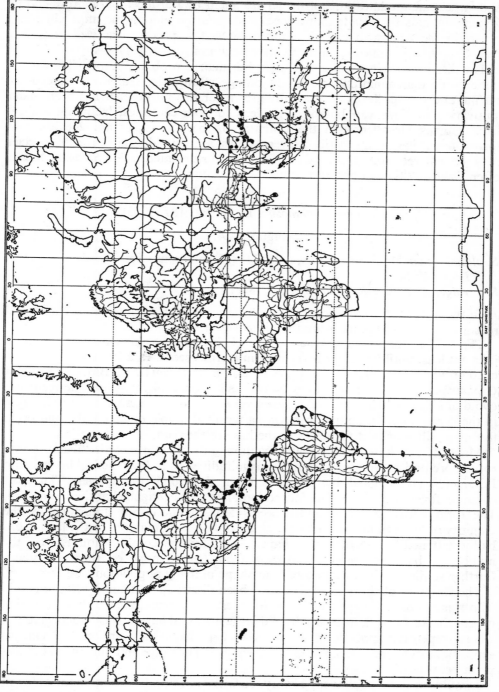

Figure 44.3. Distribution of *Solanum capsicoides*.

expanded the ranges of many *Solanum* species. It was the beginning of pervasive and large scale modifications of the environment by European agriculture and its grazing animals that made many species of *Solanum* among the most abundant plants in the American tropics. Unfortunately this process was well underway before the first botanical collections were begun in the early 1700's.

The dispersal mechanism is uncertain but man is probably responsible for the initial wide dissemination; the bright orange mature dry fruits are quite ornamental and the species is occasionally grown for this purpose. It has been under cultivation in European botanical gardens for two centuries, including a sheet from Hortus Upsaliensis in the Linnaean herbarium (sheet no. 248.30). Unfortunately it has been described several times, and an older name for this species may yet emerge from the residue of poorly documented older *Solanum* names often listed as "Solana non satis nota".

Solanum mammosum L. *Sp. Pl.*: 187 (1753)

(Figure 44.4)

S. platanifolium Hook. *Bot. Mag.*, t. 2618 (1826), non sensu Sendtner *in* Martius *Fl. Bras.*, *X*: 58.
S. cornigerum Andre *Rev. Hortic.*, *33*: 1868.

Of the four species considered, *S. mammosum* is the most interesting for several reasons, the most striking being the bizarre fruit.

The fruit in the genus *Solanum* is remarkably uniform in structure, a globose juicy (at least until maturity) berry with numerous seeds. In a few species in unrelated groups, e.g. *S. muricatum* Ait. of section *Basarthrum* (Bitt.) Bitt. or species of *Cyphomandra* Sendtn. (sometimes treated as a section of *Solanum*), the fruits are ovoid or ellipsoid. The fruits of *S. mammosum*, c. 7 by 5 cm, are among the largest in the genus, and are unique in having a terminal nipple or mammilla and often five mammillae or protuberances at the base of the fruit.

Many descriptions of this plant mention only the form with "fructu aureo rotundiore pyri parvi inversi forma & magnitudine" as in Plukenet's Almagestrum Botanicum (1696) cited by Linnaeus when describing the species. The most common form of the fruit does indeed look like a pear attached at the wrong end. The form with basal mammillae seems less common, but unfortunately the majority of herbarium specimens are either of flowering material or the fruit is too badly flattened to ascertain the shape. An additional complication is that the number of basal mammillae may vary from five to none on the same plant. Further studies of the variation of fruit shape in living populations are necessary.

The structure of this curious fruit has been discussed in detail by Miller (1969) who was unable to suggest a reason for these basal mammillae although he dismissed the idea that they were galls, a disease or casual teratological abnormalities. "Their repetitious prevalence suggests that perhaps they are merely curious expressions of an inheritable genomic anomaly." The range and human uses of this plant provide some clues.

I would suggest the following history for *S. mammosum*. It belongs to the group of section *Acanthophora* which does not have winged seeds, but like the three species discussed previously, it seems to be of south-eastern South American affinities. The closest relatives are from this region, but *S. mammosum* is definitely native to the Caribbean region where the earliest botanists recorded it; the Brasilian collections are all recent as are the scattered introductions in the Old World.

By long distance dispersal or more direct migration, the ancestors became established in the seasonally dry areas of Venezuela or neighbouring regions. Prance (1973) defends the view that during dry periods in the Pleistocene, the Amazonian forest was restricted to small refugia, leaving broad corridors of open savanna-like vegetation between northern South America and southern Brasil. This view of Amazonian forest fragmentation is probably too extreme; it has not yet been corroborated by paleobotanic or geologic evidence.

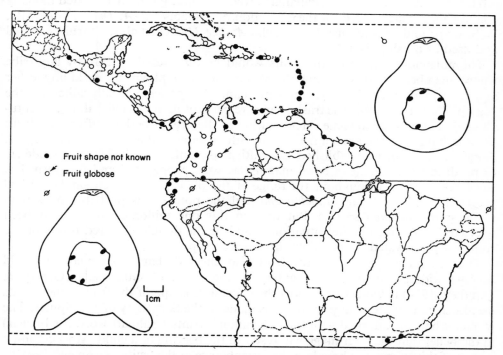

Figure 44.4. Distribution of fruit types of *Solanum mammosum*; scattered introductions into the Old World not shown.

Four collections may be of especial significance: Venezuela, Edo. Monagas, Maturín, 1971, *Nee & Mori 4179* (WIS); Edo. Guárico, Hatovecerra, 1968, *T. Plowman 1913* (GH); Colombia, Depto. Meta, Los Micos, *T. Plowman 4236* (seen as living material at University of Birmingham, England, 1976; these plants are a perfect match for the illustration of *S. platanifolium* Hook.); and Panamá, Prov. Panamá, Piria, 1967, *J. A. Duke 14408* (MO, US, WIS). All four collections (*Nee & Mori 4179* and *Plowman 1913* are accompanied by photographs) produce definitely globose fruit as would be expected from the near universal occurrence of this shape in *Solanum* fruits. The Venezuelan and Colombian collections are from the savanna areas where primitive *S. mammosum* very likely grew in naturally disturbed habitats. The Panamanian collection was reported as cultivated by the inland Cuna Indian tribe as a medicine and as a cockroach poison (Duke, 1970). The Panamanian area is a lowland tropical rain forest region where the plant would not be expected to survive in the wild.

Herbarium specimens and floristic and cultural references indicate a quite widespread belief in its ability to kill cockroaches (Venezuela: Edo. Táchira: San Cristóbal, "fruits mixed with sugar to poison cockroaches", *W. A. Archer 3195* (US)); kill rats (Costa Rica: Prov. Guanacaste: El Arenal, *P. C. Standley &* *J. Valerio 45216* (US); Peru: Huánuco: Tingo María, *J. F. Macbride 5066* (F, US)); as medicine to cure colds (Honduras: Depto. Atlántida: Tela, *P. C. Standley 53596* (F, US); El Salvador: Depto. Ahuachapán: Ahuachapán, *P. C. Standley 20318* (GH, NY, US)); as medicine (Panamá: Prov. Panamá: Paria, "Cultivated in many inland Cuna villages, used for the extraction of maggots", *J. A. Duke 14408* (MO, WIS, US)); and to catch fish (Venezuela, *Archer 3195* cited above, "fruits crushed and used as a fish poison".).

The poisonous nature of the fruits would be discovered quickly by any tribe whose members tried sampling them; utilitarian employment in killing other animals would ensure its travelling with man. The species is ideally suited for the "campfollower" existence as trash heaps are perhaps its preferred habitat. Eventually it would become to some extent cultivated. Before, or perhaps after the advent of Amerindians in the area, a fruit form with an elongated apical nipple appeared. Some time after being established as a cultivated plant, individuals with a tendency to produce the anomalous basal mammillae were propagated and this form has since been spread throughout the range of the plant. I have seen similar "basal protuberances" on *Solanum melongena* and the potential is probably present in many spiny *Solanum* species, but only in this one minor cultivar has it been selected for. A nice parallel situation would be the double flowered forms of the commonly cultivated floripondios, *Brugmansia* spp.

The ornamental nature of the fruit also ensured that *S. mammosum* would become widespread. The commentary in the Flora of Guatemala (Gentry & Standley, 1974) is indicative of how the suggestive shape of the fruit influences its use. Each January a great pilgrimage is made to the Sanctuary of Esquipulas. The women's hats are adorned with the "chichihua" fruits (Nahuatl word for breasts) in the belief that the pilgrimage will result in the bearing of a child. Most common names of this species are based on the resemblance of the fruits to women's breasts.

Another mutation has occurred which would be unviable in the wild but has been perpetuated because the species is cultivated. There is a population of totally spineless plants in Guatemala: Depto. Alta Verapaz: Cubilquitz, alt. 350 m, April 1913, *H. von Türckheim 4132* (US—2 sheets); Finca Sepacuite, 18 March 1902, *O. F. Cook & R. F. Griggs 32* (US); Secanquim, alt. 250–600 m, April 1904, *O. F. Cook & C. B. Doyle 80* (US); Secanquim, alt. 550 m, April 1905, *H. Pittier 177* (NY)). It is common in some sections of subgenus *Leptostemonum* (notably in section *Torva* Nees) for the flowering branches to lack spines while the lower stem or young shoots are quite prickly, but in section *Acanthophora* a lack of spines is as conspicuously "abnormal".

P. Madhavadian (1968) reported the first count of $n=11$ in *Solanum* from a collection of *S. mammosum* from India. Heiser (1971) has confirmed this count and gives references to previous reports of $n = 12$, the number found in almost all of the species of subgenus *Leptostemonum* that have been counted. Heiser (1971) also reports on a crossing experiment between a plant of *S. mammosum* with basal mammillae and one without them; the hybrid had an intermediate fruit shape. Further work on the genetics of *S. mammosum* should be highly rewarding.

CONCLUSION

Many *Solanum* species are adapted to the scattered temporary spots of disturbance in the native vegetation and have succulent berries much sought after by birds to ensure their dispersal. This combination has meant the spread at different ages for different groups to most other parts of the world from the centre of the genus (and the family Solanaceae) in South America. The genus has proliferated in Central America, profiting both by geographic contiguity and the sharing of migration routes of hundreds of species of birds. Africa is the third richest area with Asia and Australia receiving the odds and ends. A discussion of the times of dispersal and the groups involved must await a comprehensive review of the genus *Solanum* as a whole.

At the present time dealing with a section or other well defined small group is more profitable. Most sections will probably show some of the patterns discussed above. Almost all species will have expanded their ranges in the last few centuries unless they have very restricted habitat requirements, especially if this depends on large areas of undisturbed vegetation. A few hardy species in each section may have the potential to become common, widespread weeds.

In this last case, the problem of determining the geographical origin of the species may be difficult, especially if they became widespread before 1700. The typification of many of the obscure early names in the genus must be based on widespread species and taxonomic confusion is bound to result until these names are either adopted or relegated to synonymy. A good case is *Solanum capsicoides* All., an almost totally ignored but quite validly published name which must now be taken up. There are, however, a large number of very old herbarium specimens in European herbaria of this species. It has obviously been known and cultivated for a long time and an earlier name may still appear.

ACKNOWLEDGEMENTS

I wish to thank the Davis Fund of the University of Wisconsin Botany Department for funds which have enabled me to visit many European and American herbaria as well as for several field trips to the American tropics.

REFERENCES

D'ARCY, W. G., 1973. Solanaceae. In, Flora of Panamá. *Annals of Missouri Botanical Garden, 60:* 573–780.
DUKE, J. A., 1970. Ethnobotanical observations on the Choco Indians. *Economic Botany, 24:* 344–366.
GENTRY, J. L., Jr. & STANDLEY, P. C., 1974. Solanaceae. In, Flora of Guatemala. *Fieldiana: Botany, 24* (X).
HEISER, C. B., Jr., 1971. Notes on some species of *Solanum* (Section *Leptostemonum*) in Latin America. *Baileya, 18:* 59–65.
MADHAVADIAN, P., 1968. Chromosome numbers in south Indian Solanaceae. *Caryologia, 21:* 343–347
MILLER, R. H., 1969. A morphological study of *Solanum mammosum* and its mammiform fruit. *Botanical Gazette, 130:* 230–237.

PRANCE, G. T., 1973. Phytogeographic support for the theory of Pleistocene forest refuges in the Amazon basin. *Acta Amazonica, 3:* 5–28.

ROE, K. E., 1966. Juvenile forms in *Solanum mitlense* and *S. blodgettii* and their importance in taxonomy. *Sida, 2:* 381–385.

ROE, K. E., 1971. Terminology of hairs in the genus *Solanum. Taxon, 20:* 501–508.

ROE, K. E., 1972. A revision of *Solanum* section *Brevantherum* (Solanaceae). *Brittonia, 24:* 239–278.

SEITHE, A., 1962. Die Haararten der Gattung *Solanum* L. und ihre taxonomische Verwertung. *Botanisches Jahrbücher, 81:* 261–336.

VAN DER HAMMEN, T. & GONZALEZ, E., 1960. Upper Pleistocene and Holocene climate and vegetation of the "Sabana de Bogota". *Leidsche Geologische Mededelingen, 25:* 261–315.

45. Speciation in *Solanum*, section *Androceras*

MICHAEL D. WHALEN

Department of Botany, University of Texas, Austin, Texas, U.S.A.

Most species of section *Androceras* have arisen initially by divergence of geographically isolated populations. Effective geographical isolation has been achieved in two ways: by fragmentation of broad ranges into smaller disjunct areas and by establishment of isolated peripheral populations. When disjunctions exist, whatever their mode of origin, divergence may proceed on its own or may be hastened by occasional hybridization with other species. The latter possibility appears to be left open by general interspecific chromosomal uniformity. Two kinds of reproductive barriers operate between sympatric species of section *Androceras:* mechanical isolation, due to floral size differences, and the failure of hybrid seed to develop properly following interspecific cross-pollinations.

CONTENTS

INTRODUCTION

Although the total number of species of *Solanum* can hardly be guessed at until taxonomic understanding of the genus has advanced considerably from its current level, Willis's (1973) estimate of 1700 cannot be far above the mark. Such species-rich genera as *Solanum* pose a fundamental question. Is their extreme diversity attributable to great age alone or have unusual modes of speciation led to higher rates of evolutionary cladogenesis in these genera than in others? Perhaps for very large genera like *Solanum* both explanations are important. Systematic study of such taxa may provide us with initial clues regarding any unique features of their evolution. Some results of a taxonomic investigation of *Solanum* section *Androceras* are presented here and examined for what they reveal about speciation in that group.

Two sections of *Solanum*, section *Petota* (=*Tuberarium*) (the potatoes) and section *Solanum* (the *S. nigrum* group), have long been objects of active systematic research. The common occurrence of polyploidy in these two sections has led to the prevalent misconception that polyploid series characterize the genus *Solanum* as a whole. However, outside these two well-studied groups, polyploidy appears to be relatively infrequent (Moore, 1973). Speciation in section *Androceras* has been entirely on the diploid level, all of its taxa consistently possessing twelve pairs of chromosomes (Whalen, in prep.). In this respect it is a more typical group of *Solanum*. In addition, the section exhibits a full spectrum of speciational stages and may yield valuable insights into the nature of diploid speciation in the genus. Regrettably, much of the story told here must be a condensation of data that need publication at greater length elsewhere.

SECTIONAL DELIMITATION AND AFFINITIES

Section *Androceras* consists of 12 species of prickly, mostly annual herbs. It is a morphologically well-marked group, having enantiostylous, heterandrous, weakly zygomorphic flowers and prickly, accrescent calyx tubes, which tightly and completely invest the berries. The leaves are prickly and pinnatifid, the inflorescences raceme-like with all but the terminal few flowers fertile, and the seeds brown to black and lenticular. Most of the species are confined to Mexico and the southwestern United States, but two extend beyond this region, one, *S. angustifolium*, reaching Honduras and the other, *S. rostratum*, extending to the north-central United States. Relationships of section *Androceras* within *Solanum* are obscure. It resembles section *Nycterium* (Canary Islands and India), with which it was combined as a genus by several early authors (e.g. Ventenat, 1805). *S. tridynamum* of Mexico is an interesting species that appears to be intermediate between the two sections. Plants of section *Androceras* also look much like those of section *Cryptocarpum*, as recently reconstituted in a strict sense (D'Arcy, 1972) and typified by *S. sisymbriifolium*. The sections *Nycterium* and *Cryptocarpum* do not seem to be closely related however, and they cannot both be accepted as near relatives of section *Androceras*.

DISPERSAL AND COLONIZATION

Since isolation of populations is an important first step in plant speciation, any dispersal or life-history strategies of section *Androceras* species which significantly affect it need to be understood. Although two species of the section are rhizomatous or woody-based perennial herbs, most of them are annuals which colonize open, sunny places, often disturbed areas like roadsides, abandoned fields and dry streambeds. Many of them are restricted in range, but they all exhibit attributes that are characteristic of weedy species (Baker, 1973): (1) rapid vegetative growth with continuous production of numerous flowers and seeds as long as conditions permit, (2) self-compatibility, with a mechanism to encourage some degree of outcrossing, and (3) pollination by many genera of polylectic bees (i.e. lack of specialized pollinator requirements). A feature on Baker's list which is conspicuously absent in most species of section *Androceras* is any special adaptation for long-distance dispersal. Only *S. rostratum*, by far the weediest of the species, appears to be effectively dispersed by natural means. In the windy, open Great Plains of the United States, where it assumes a tight, congested growth habit, it is a common tumbleweed (Barrell, 1975). The other species are more "stemmy", as are most Mexican forms of *S. rostratum*, and do not lend themselves well to this type of dispersal, neither does the rough Mexican topography which they inhabit. The popular interpretation of the fruits of section *Androceras* as burrs (Ridley, 1930;

Van der Pijl, 1972) is in error. Although the berries, tightly enclosed by the prickly, accrescent calyx tubes, closely resemble burrs, they do not function as such. No abscission layer develops in the tough and fibrous pedicels. Instead, the dry calyx tube splits open gradually between the main veins, beginning at the apex, and simultaneously ruptures the papery remains of the berry wall, which adhere to it. The seeds are then released individually.

The single dispersal strategy upon which most of the species of section *Androceras* have come to depend is the production of large numbers of seeds per plant. The probability of the haphazard transport of a *few* seeds is thus improved as are the chances that one or a few plants will establish a new population upon reaching a suitable habitat. With this type of "hit and miss" dispersal strategy, the arrival of seeds from one population to the confines of another is doubtless a rare event. The resulting restriction of gene flow is enhanced in the pollination phase by polylectic bees which are likely to visit other species between stops at *Solanum* stands.

Preliminary electrophoretic study of allozyme variation in section *Androceras* has been suggestive concerning colonization and interpopulational gene flow in the group. Populations of the two perennial species are polymorphic at a substantial proportion of the loci examined, while those of the annual taxa are usually monomorphic at all loci. The easiest explanation is that most of the populations of annuals studied were founded by one or a few individuals (thus passing through genetic "bottlenecks") and that acquisition of new alleles through mutation or gene flow is not rapid enough to keep up with their loss through the fluctuations in population size (including population extinctions) that characterize colonizing annuals. Several examples are known in section *Androceras* of nearby populations of the same annual species that are fixed or nearly fixed for different alleles at one enzyme locus. It is of interest that annual populations in which polymorphisms *were* observed were very large and quite likely relatively old ones, which (like the presumably long-lived populations of perennials) may have had a chance to acquire new alleles by mutation or infrequent dispersal events from neighbouring populations.

POLLINATION

As divergent populations expand their ranges and become sympatric, selection may operate to encourage development of parental generation barriers to hybridization between them (Grant, 1971). The mechanism by which flowers of section *Androceras* are pollinated is described by Bowers (1975) for *S. rostratum*. My own field observations indicate that it lends itself well to the establishment of mechanical barriers, which prevent or substantially reduce the frequency of interspecific matings. In flowers of section *Androceras*, one anther is longer than the others and is declined to one side but is incurved at the tip, where pollen is released from apical pores. The style is about equal to it in length and is similarly declined but to the opposite side. It also is incurved at the tip. The sides to which the long anther and style project are reversed in alternate flowers of an inflorescence, and the open flowers of a given plant are at any one time roughly half left-handed and half right-handed. As a pollen collecting bee works the four shorter anthers in the centre of a flower, its body is in such a position that the abdomen is in contact on one side with the tip of the long anther and on the other side with the stigma. Left-handed flowers thus pollinate right-handed ones and vice-versa. The successful operation of this mechanism requires, however, that the visiting bees be sufficiently large to contact the stigma and the tip of the long anther while working the short, central anthers. In section *Androceras* two modes of floral size are evident in the various taxa. Most species possess flowers *c.* 2·5–3·5 cm in diameter, while those of some are only about 1 cm across. The latter are visited only by small bees, such as *Nomia* and *Exomalopsis*, while the former are visited by large bees like *Bombus*, *Xylocopa*, and *Protoxaea*, as well as by small ones. The large bees seem to ignore the small-flowered species. Probably the flowers do not provide sufficient pollen to make their visits profitable. Small bees visit the large forms freely, where they function only as pollen thieves, too small to touch the stigma or the tip of the long anther while working the shorter anthers. When one of them returns to a sympatric, small-flowered species, cross-pollination is not effected, since pollen from the large flowers has not been deposited on the sides of the abdomen but only on the venter. Virtually complete reproductive isolation of sympatric species is possible if they are sufficiently different in floral size. This potentiality appears to have been important in the evolutionary history of several species of section *Androceras*.

A TAXONOMIC OVERVIEW

A species diversity map of section *Androceras* (Fig. 45.1A) reveals that it has three major centres of distribution: one in the highlands of south-central Mexico near Mexico City, a second in the rugged northern portion of the Chihuahuan Desert, and a third near the Pacific Coast of Mexico. Based on morphology, three groups of closely related species can be discerned in the section, and each has, as its centre of development, one of the three regions of species richness shown in Fig. 45.1A. The groups have recently been named as series (Whalen, 1976).

The type series, *Androceras*, is the most widespread of the three (Fig. 45.1B), but most of its distribution is accounted for by the native range of *S. rostratum*. The series is otherwise restricted to south-central and southern Mexico and includes all but one of the five species present there. Plants of series *Androceras* usually have yellow, pentagonal corollas, stellate pubescence on the stems and weakly ridged or unridged seeds.

The species of series *Violaciflorum* are concentrated in the northern Chihuahuan Desert. The corollas of all four of them are stellate-pentagonal and violet or dark blue with light green, star-shaped central markings. The stems have simple, mostly glandular hairs and the seeds are usually reticulately ridged.

Series *Pacificum* is found along the pacific slope of the Sierra Madre Occidental. All three of the species located there are its members. They have white, deeply stellate corollas with black central star-like markings, simple and often glandular hairs on the stems and radially ridged seeds.

Chemical as well as morphological characters mark the three series (Whalen, 1978). The flavonoid chemistry of series *Androceras* is quite simple, dominated by kaempferol-, quercetin- and isorhamnetin-3-0-glycosides. Series *Violaciflorum* and *Pacificum* produce, in addition, flavones, C-glycosylflavones, 8-hydroxy-

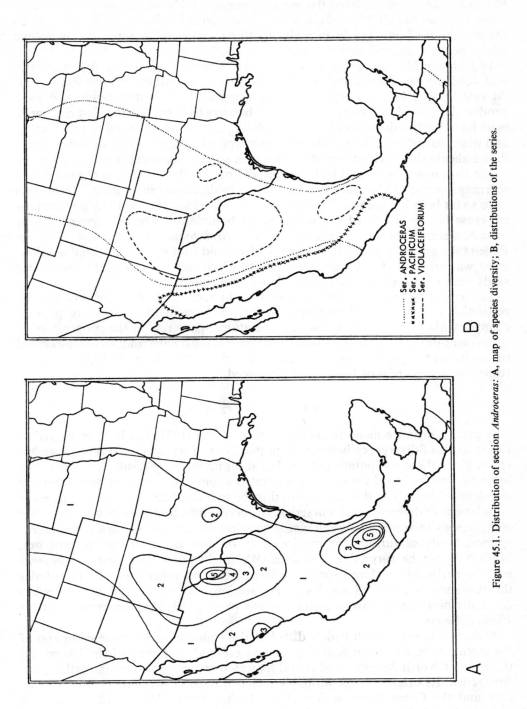

Figure 45.1. Distribution of section *Androceras*: A, map of species diversity; B, distributions of the series.

flavonoids and methoxylated flavonoid aglycones. 8-Hydroxyflavonoids of series *Violaceiflorum* are all flavonols, while those of series *Pacificum* are flavones. With the exception of one species in each, the three series are also marked by unique alleles at the esterase-3 enzyme locus.

In Hidalgo, Mexico, where *S. heterodoxum* of series *Violaceiflorum* is sympatric with *S. rostratum* of series *Androceras*, hybrid plants are ordinarily not encountered. At one locality however, the two species appear to hybridize freely and have produced a hybrid swarm, suggesting that barriers to hybridization that ordinarily exist have broken down locally. Intermediate plants were collected at this locality and were characterized by meiotic regularity and 93–99% pollen stainability. These data indicate that early speciational events in section *Androceras*, which gave rise to the lines now represented by the series, did not involve any major chromosomal rearrangements. Although experimental hybridizations in section *Androceras* have so far been limited to a few species grown in the summer of 1975, one reciprocal cross has been carried out across series boundaries. The two parent species were *S. rostratum* of series *Androceras* and *S. citrullifolium* of series *Violaceiflorum*. Pollen tube growth and fertilization took place, and fruit was set, but seed development was arrested at any early stage. From 25 crosses to each parent, no viable seeds were obtained, and no hybrid plants were raised.

The emerging picture of early evolution in section *Androceras* involves the restriction of an ancestral prototype to three refugia in Mexico, where phyletic change, without major chromosomal alterations, gave rise to the progenitors of the present-day series. The locations of the refugia must have roughly corresponded to the current regions of species richness for the section (Fig. 45.1A), where the respective series are now found well developed.

SERIES *ANDROCERAS*

A genetic distance measure similar to that of Nei (1971) has been employed to assess levels of divergence between 21 populations of ten species in section *Androceras*. Since these computations were based on preliminary results involving only seven enzyme loci, and since we are interested in only one aspect of them, they are not detailed here. The distances proved to be two to three times greater between populations of different species in series *Androceras* than for interspecific population comparisons in series *Violaceiflorum* suggesting greater age of series *Androceras* species. Study of further enzyme loci may or may not bear this pattern out, but it will not be surprising if it does. We might guess that series *Androceras* was the earliest to undergo species differentiation on other grounds. It includes the two most widespread species of the section, and all five of its species are decidedly more distinct from each other morphologically than are those of series *Violaceiflorum*.

S. rostratum is the most widely distributed member of series *Androceras* and of the section (Fig. 45.2A). It is an aggressive weed and, with man's help, has spread throughout North America and found its way to every continent on earth except Antarctica. Its original range probably included most of Mexico north of Mexico City and the Great Plains region of the United States. When U.S. material is compared with Mexican material of this species, some degree of geographical

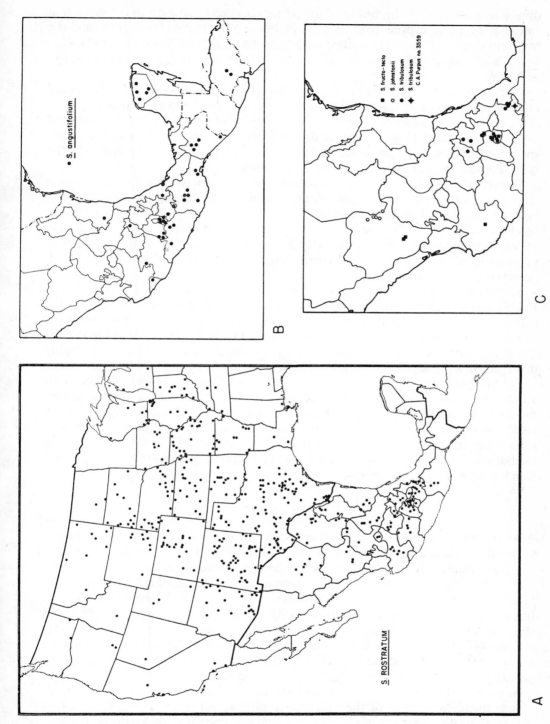

Figure 45.2. Distributions of the species of series *Androceras*: A, *S. rostratum*; B, *S. angustifolium*; C, *S. fructo-tecto*, *S. johnstonii* and *S. tribulosum*.

differentiation is apparent, but local differences between populations are often of a similar magnitude, especially in Mexico. Plants of *S. rostratum* from Texas and the Mexican state of Hidalgo hybridize readily in the experimental garden, indicating that widely separated populations do not exhibit any degree of reproductive isolation.

While *S. rostratum* extends from central Mexico northwards, another widespread and weedy taxon of series *Androceras*, *S. angustifolium* (often known as *S. cornutum*), is more tropical in its preferences and ranges from central Mexico south-east as far as Honduras (Fig. 45.2B). It differs from *S. rostratum* in its large, widely spaced prickles and its glaucescent stems, which often lack stellate hairs. In the latter feature, it approaches series *Violaceiflorum*.

The remaining three taxa of series *Androceras* are more restricted in distribution (Fig. 45.2C). *S. fructo-tecto* is an edaphic specialist, preferring but not requiring volcanic ash soils. It is native to volcanic areas in the region of Mexico City, but has been collected in the vicinity of the city of Durango (where volcanic soils also occur) as early as the 1880's. It was doubtless brought there by man. A distinctive plant, it is taller, more erect and hairier than *S. rostratum* and has broad-based, recurved prickles on the stem.

S. tribulosum is a beautiful plant of high elevations, ranging from eastern Puebla north-west to Querétaro. Its pale, sky-blue flowers are unusual in the otherwise yellow-flowered series *Androceras*, with which its stellate-pubescent stems align it. One peripheral population (marked C. A. Purpus 3559 in Fig. 45.2C) is cut off from the rest of the species range by the Tehuacán Desert. It is highly divergent morphologically and may represent an initial stage in species formation.

A fifth species of series *Androceras*, *S. johnstonii* (Whalen, 1976), is a native of stony, calcareous soils of a restricted area in eastern Durango. It is a unique member of the series in being a long-lived perennial. The shoots emerge annually from woody bases that eventually become large and gnarled. *S. johnstonii* was among the species which were cultivated in the summer of 1975. Early failure in seed development was exhibited in both of the reciprocal crosses with *S. rostratum*, which it most closely resembles.

SERIES *PACIFICUM*

As the name denotes, the taxa of series *Pacificum* are found in the westernmost parts of Mexico, adjacent to the Pacific Ocean (Fig. 45.3). A northern species, *S. lumholtzianum*, is an inhabitant of southern Arizona, Sonora, and the northern half of Sinaloa. Its relative, *S. grayi*, overlaps with it in southern Sonora and in Sinaloa, from there extending south-east to Guerrero. *S. lumholtzianum* differs from *S. grayi* in its unique androecium with anthers of three lengths, in the ovoid shape and whitish colouring of its fruiting structures, in its larger and differently shaped seeds, and in its more intricately dissected leaves. Although the two species are partially sympatric now, their distributions hint that they originated by geographical subdivision of a single, ancestral species.

S. grayi and *S. lumholtzianum* both have medium-large flowers that are similar in size throughout most of their respective ranges, but in the zone of overlap, a conspicuous displacement of floral characters in *S. grayi* occurs. Figure 45.4 shows variation in style and long anther length for both species along a transect from Arizona to Guerrero. The transect includes the zone of overlap and is based on herbarium specimens. For reasons already discussed, the small flowers that characterize *S. grayi*, where it is sympatric with *S. lumholtzianum*, would be expected to bestow on it a considerable degree of mating isolation from the latter species. The precise geographic correspondence of interspecific sympatry with the small-flowered condition in *S. grayi* is an indication that selection has been at work. Interestingly, the small-flowered race of *S. grayi* must be as effectively isolated

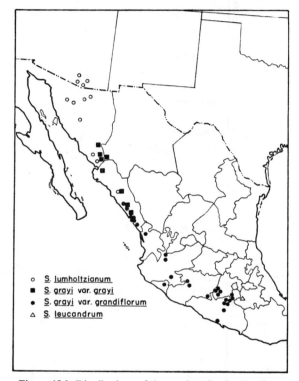

Figure 45.3. Distributions of the species of series *Pacificum*.

from its larger flowered parent race as it is from *S. lumholtzianum*. The transition between the two races is more abrupt than would be expected if they were interbreeding freely (Fig. 45.4). We may be seeing an initial stage in a unique process of plant species formation. It is curious that reciprocal floral displacement is not observed. *S. lumholtzianum* does not respond to the presence of *S. grayi* by making its flowers larger. Quite likely, it is already utilizing the largest pollen collecting bees available, and there is simply no "bigger bee" niche for it to fill.

Gene flow between populations of the annual species, *S. grayi*, is sufficiently restricted that it permits two populations nearly fixed for different alleles at the ADH–2 enzyme locus to exist only 10 km apart near Mazatlán, Sinaloa. With this in mind, it is no surprise that a population established in western Puebla, nearly 100 km east of the main body of *S. grayi*'s distribution, has become so radically divergent

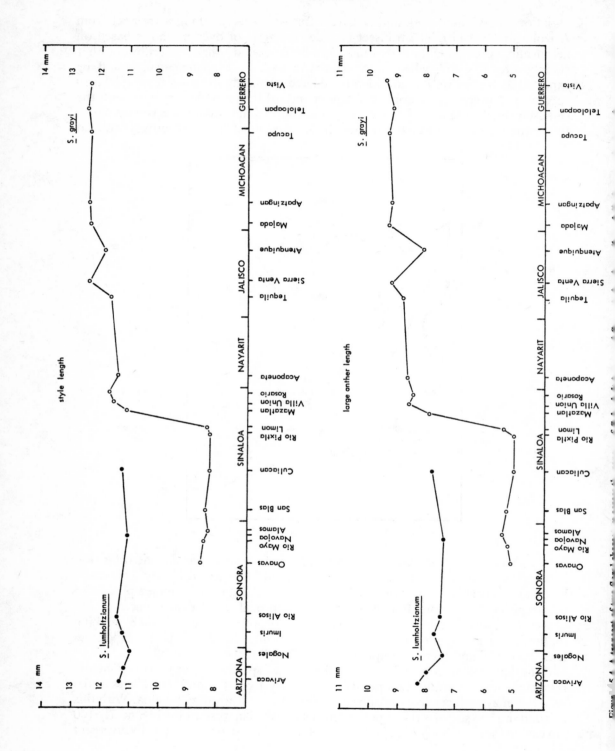

Figure 5.1. A transect of two Sierra Lobatae, an example of [...]

that it has been deemed to deserve description as a distinct species, *S. leucandrum* (Whalen, 1976), (Fig. 45.3). This species differs from *S. grayi* in its much smaller flowers, in the shape of its anthers, in the presence of scattered stellate hairs on its stems and in its less dissected leaves. In addition, it produces flavonoid compounds that are in general more highly glycosylated than those of *S. grayi*. *S. leucandrum* provides an excellent example of the origin of a species by divergence of an isolated, peripheral population. If it expands its range and becomes sympatric with *S. grayi*, it will already be reproductively isolated from that species by floral size differences, in addition to any internal isolation mechanisms that may have evolved incidentally during the period of separation.

SERIES *VIOLACEIFLORUM*

Of the three species groups in section *Androceras*, series *Violaceiflorum* is clearly the most evolutionarily exuberant. Each of its four species shows some degree of geographical differentiation. Like series *Androceras*, this group possesses one perennial species, *S. tenuipes*, which is a denizen of calcareous soils in the northern Chihuahuan Desert and forms rather large clones by means of corky, perennial, underground rhizomes. It is sympatric through much of its range with the annual, *S. citrullifolium*, also a member of series *Violaceiflorum*, which it resembles. But it seems to be effectively isolated from that species. No apparent hybrids between the two have been observed. *S. tenuipes* consists of two morphologically distinct and essentially disjunct geographical races (Fig. 45.5A): an eastern one with intricately dissected leaves and a western one with broad-lobed leaves. Their geographic separation makes it clear that they have evolved in isolation. *S. tenuipes* is distinguished from all the other species of series *Violaceiflorum* by its extremely rich flavonoid chemistry. Methanol extracts of its foliage yield an array of almost 30 flavonoid compounds, nearly half of which are unique to the species.

The annual *S. citrullifolium*, characterized by its large, blue to violet flowers, is more extensively divided geographically and occurs in several disjunct areas (Fig. 45.5B). One in central Texas is widely separated from the nearest part of the range to the west in the mountains of Trans-Pecos Texas and northern Coahuila. The plants of both regions are very similar morphologically, but there are some consistent differences between them, probably reflecting a very early stage of genetic differentiation under geographical isolation. For taxonomic purposes, the two forms are classed together as variety *citrullifolium*. A third disjunct in southeastern Chihuahua and western Coahuila (variety *setigerum*) is conspicuously divergent and appears to have arisen through past hybridization with *S. lumholtzianum* of series *Pacificum*, a matter which will be discussed further in connection with a race of *S. heterodoxum*, which has probably had a related history. A fourth disjunct of *S. citrullifolium*, variety *knoblochii* in western Chihuahua, is a poorly understood taxon and may represent an additional species.

Small flowers, about 1 cm in diameter, make *S. heterodoxum* unique in series *Violaceiflorum*. Since this species is now widespread and geographically partitioned, it is impossible to ascertain what the mode of its origin may have been. But it is enticing to think that floral displacement, as observed in *S. grayi*, may have had a role in its evolution. Whether selected or not, its small flowers affect its potential for hybridization with any large-flowered species of section *Androceras* that it encounters. As in *S. citrullifolium*, two widely spaced, disjunct elements of *S. heterodoxum* (Fig. 45.5C) have experienced only a minor degree of morphological differentiation but for taxonomic purposes are maintained as distinct varieties. They produce virtually identical flavonoid compounds. One of them, variety *novomexicanum*, is in northern New Mexico and the other, variety *heterodoxum*, far to the south in Mexico, extending from San Luis Potosí to Puebla. Between these two regions stretches the full length of the Chihuahuan Desert. A third variety of *S. heterodoxum*, which is morphologically and chemically distinct, lies just to the south of the range of variety *novomexicanum*. It occupies some of the northern reaches of the Chihuahuan Desert, as well as the semi-arid lands peripheral to it.

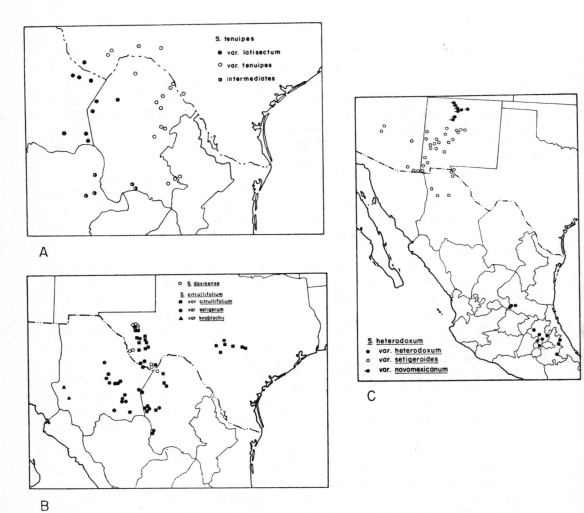

Figure 45.5. Distributions of the species of series *Violaceiflorum*: A, *S. tenuipes*; B, *S. davisense* and *S. citrullifolium*; C, *S. heterodoxum*.

This race has been named variety *setigeroides* (Whalen, 1976), because of its resemblance to *S. citrullifolium* variety *setigerum*, which is slightly further south still, in the Mexican state of Chihuahua.

Differentiation of the last two named varieties from their parent species has likely been the result of past introgressive hybridization with *S. lumholtzianum* of

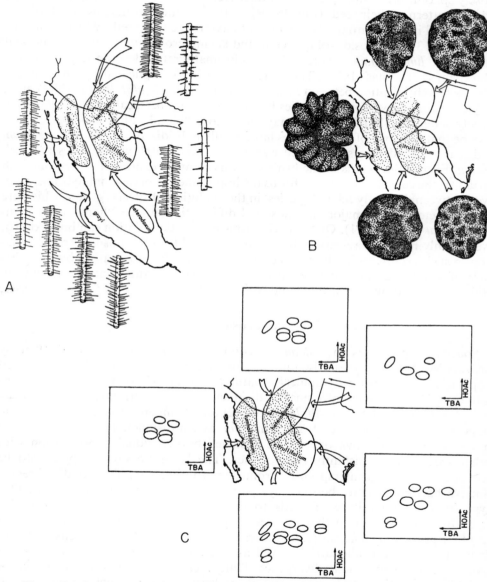

Figure 45.6. Summary of evidence that the geographical differentiation of *S. citrullifolium* var. *setigerum* and *S. heterodoxum* var. *setigeroides* from their parent species involved an episode of hybridization with *S. lumholtzianum*. (The two vars. are shown as the stippled portions of their respective species ranges). A, Armament; B, Seeds; C, flavonoid profiles. Further explanation in text.

series *Pacificum*, which now lies just across the Sierra Madre Occidental. In
Fig. 45.6, some of the evidence that suggests this hypothesis is summarized.
S. citrullifolium variety *setigerum* and *S. heterodoxum* variety *setigeroides* are
shown as the stippled regions of their respective species ranges. They approach
S. lumholtzianum (to which they are geographically adjacent) in possessing unique,
closely spaced, bristle-like prickles on their stems (Fig. 45.6A), in the weak radial
ridging pattern of their seeds (Fig. 45.6B) (both anomalous traits in series *Violaceiflorum*), and in chromatographic profiles of flavonoid compounds, which show two
unidentified 3–O–glycosides of quercetin and kaempferol that are otherwise present
only in *S. lumholtzianum* (Fig. 45.6C) (chromatograms developed per Mabry,
Markham & Thomas, 1970). The picture is further complicated by an apparent
history of hybridization between *S. heterodoxum* and *S. citrullifolium* in this region.
In fact, preliminary allozyme studies hint that what has been called *S. citrullifolium*
variety *setigerum* (Bartlett, 1909) may have more "blood" of *S. heterodoxum*. It
may be that only one of these species has actually hybridized with *S. lumholtzianum*
and that the syndrome so acquired was simply "passed on" within series *Violaceiflorum*. Even if the details of the history of this complex cannot be sorted out with
certainty, the case clearly stresses for us the importance of occasional hybridization,
even between distinctly related species, in the evolution of species groups which are
not characterized by major chromosomal differentiation (the "homogamic complex" *sensu* Grant, 1971). Other mechanisms of genetic isolation can break down,
even if only rarely, to re-expose the latent potential for chromosomal recombination in such groups, a fact that was clearly impressed upon us earlier by the existence of a hybrid swarm, composed of fully fertile individuals, at one locality in
Hidalgo, formed by *S. rostratum* and *S. heterodoxum*, members of different series.

CONCLUSIONS

   Speciation processes in *Solanum* section *Androceras* are not fundamentally
unique. Most species have arisen initially by divergence of geographically isolated
populations. Effective geographical isolation has been of two types: (1) fragmentation of broad ranges into smaller disjunct areas, as currently exemplified by
S. heterodoxum and *S. citrullifolium*, and (2) establishment of isolated, peripheral
populations, as in the case of *S. leucandrum*. When disjunctions exist, divergence
may proceed on its own or may be hastened by occasional hybridization with
other species, even distantly related ones. The latter possibility appears to be
left open by general interspecific chromosomal uniformity. Two kinds of reproductive barriers operate between sympatric species of section *Androceras*. One
type involves the failure of seeds to develop properly following interspecific
hybridization, a situation that has been reported in innumerable species crosses
in other groups of *Solanum* (e.g. Anderson, 1975; Grun, 1961; Jørgensen, 1928).
It is found between allopatric as well as sympatric species and probably arises
as a by-product of genetic divergence. Differences in floral size result in mating
isolation in section *Androceras*, and in at least one case, such a difference is
clearly the product of selection. This type of barrier is unique in *Solanum*, because
it depends on the specialized pollination syndrome exhibited by section *Androceras*
flowers. It becomes increasingly apparent that, although geographic divergence

and hybrid seed failure are common in the genus, unique modes of species isolation often operate in the various species groups.

If speciation in section *Androceras* has been particularly rapid, it is likely that two factors have been among those responsible. First, isolated populations appear to be established frequently and with great facility, and second, briefer periods of geographic isolation may be required than for other plant groups to allow for divergence sufficient to result in reproductive barriers. For example, enough differences in seed characters to cause hybrid seed breakdown may accumulate quite rapidly in some isolated populations. My own field studies have shown that seed characters exhibit considerable variation in many populations of section *Androceras*, and abortive seeds are frequently found in berries collected at localities where only one species is present and there is no hint of hybridization. Thus genetic factors capable of causing varying degrees of seed failure may even segregate within ordinary populations.

Thorough studies of genetic distances between species, based on allozyme data, and comparisons of such studies for many large and small genera of plants will eventually provide definitive answers concerning relative rates of species formation in different groups (Avise, 1975). Allozyme investigations now in progress will form a basis for comparison of speciation rates in section *Androceras* with other plant genera, large and small, that have been similarly studied.

NOMENCLATURAL NOTES

The botanical names used in this paper are here listed, followed by their authors and, in parentheses, synonyms which are commonly encountered in the literature and in herbaria. Sections: *Androceras* (Nutt.) Bitt. ex Marzell, *Cryptocarpum* Dun., *Nycterium* (Vent.) Dun., *Petota* Dumort (= *Tuberarium* (Dun.) Bitt.). Species: *S. angustifolium* Mill. *non* Lam. (*S. cornutum* Lam., *S. macroscolum* Fern.), *S. citrullifolium* A. Br., *S. davisense* M. D. Whalen, *S. fructo-tecto* Cav. (*S. fontanesianum* Dun.), *S. grayi* Rose, *S. heterodoxum* Dun., *S. johnstonii* M. D. Whalen, *S. leucandrum* M. D. Whalen, *S. lumholtzianum* Bartl., *S. rostratum* Dun., *S. sisymbriifolium* Lam., *S. tenuipes* Bartl., *S. tribulosum* Schauer, *S. tridynamum* Dun. (*S. amazonium* Ker.).

ACKNOWLEDGEMENTS

This research has been supported by a grant-in-aid from Sigma Xi and by a predoctoral grant (DEB 76–02189) from the National Science Foundation of the United States. I would like to express my gratitude to both of these organizations.

REFERENCES

ANDERSON, G. J., 1975. The variation and evolution of selected species of *Solanum* section *Basarthrum*. *Brittonia*, 27: 209–222.

AVISE, J. C., 1976. Genetic differentiation during speciation. In F. J. Ayala (Ed.), *Molecular Evolution*. Sunderland, Mass.: Sinauer.

BAKER, H. G., 1973. Characteristics and modes of origin of weeds. In H. G. Baker & G. L. Stebbins (Eds), *The Genetics of Colonizing Species*. New York & London: Academic Press.

BARRELL, J., 1975. *Red Hills of Kansas, Crossroads of Plant Migrations*. Rockford, Ill.: Natural Land Institute.

BARTLETT, H. H., 1909. The purple-flowered Androcerae of Mexico and the southern United States. *Contributions of the Gray Herbarium, 36: 627–629.*

BOWERS, K. A. W., 1975. The pollination ecology of *Solanum rostratum* (Solanaceae). *American Journal of Botany, 62:* 633–638.

D'ARCY, W. G., 1972. Solanaceae studies 2: Typification of subdivisions of *Solanum. Annals of Missouri Botanical Garden, 59:* 262–278.

DUNAL, M. F., 1852. Solanaceae. In de Candolle, *Prodromus, 13* (1): 1–690.

GRANT, V., 1971. *Plant Speciation.* New York & London: Columbia.

GRUN, P., 1961. Early stages in the formation of internal barriers to gene exchange between diploid species of *Solanum. American Journal of Botany, 48:* 79–89.

JØRGENSEN, C. A., 1928. The experimental formation of heteroploid plants in the genus *Solanum. Journal of Genetics, 19:* 133–211.

MABRY, T. J., MARKHAM, K. R. & THOMAS, M. B., 1970. *The Systematic Identification of Flavonoids.* New York, Heidelberg & Berlin: Springer-Verlag.

MOORE, R. J. (Ed.), 1973. *Index to Plant Chromosome Numbers 1967–1971.* Utrecht.

NEI, M., 1971. Interspecific gene differences and evolutionary time estimated from electrophoretic data on protein identity. *American Naturalist, 105:* 385–398.

RIDLEY, H. N., 1930. *The Dispersal of Plants Throughout the World.* Ashford: Reeve.

VAN DER PIJL, L., 1972. *Principles of Dispersal in Higher Plants,* 2nd ed. New York, Heidelberg & Berlin: Springer-Verlag.

VENTENAT, E. P., 1805. *Jardin de la Malmaison.* Paris.

WHALEN, M. D., 1976. New taxa of *Solanum* section *Androceras* from Mexico and adjacent United States. *Wrightia, 5:* 228–239.

WHALEN, M. D., 1978. Foliar flavonoids of *Solanum* section *Androceras:* A systematic survey. *Systematic Botany,* (in press).

WILLIS, J. C., 1973. *A Dictionary of Flowering Plants and Ferns,* 8th ed. London: Cambridge University Press.

IX. Biosystematics of domesticates

46. Crossability relationships between some species of *Solanum, Lycopersicon* and *Capsicum* cultivated in Nigeria

Institute of Agricultural Research and Training, Ibadan, Nigeria

Five diploid *Solanum* species, namely *S. aethiopicum, S. dasyphyllum, S. gilo, S. macrocarpon,* and *S. melongena,* hexaploid *S. nigrum, Lycopersicon esculentum* and two *Capsicum* species were used in making crosses.

Fully fertile hybrids were obtained from the interspecific cross *S. aethiopicum* × *S. gilo,* while crosses of *S. aethiopicum* with *S. dasyphyllum, S. macrocarpon* and *S. melongena* gave partially fertile hybrids. Some crosses of the local scarlet egg plant, *S. gilo,* with the exotic purple egg plant, *S. melongena,* produced partially fertile hybrids, but others did not. Crosses between diploids and hexaploids did not produce viable seeds.

Nearly all pollinations of *Lycopersicon esculentum* with *Solanum* species and with *Capsicum* species resulted in normal-size parthenocarpic fruits; the stimulus of pollination was essential for fruit development.

CONTENTS

INTRODUCTION

The family Solanaceae includes a large number of economic plants that are grown in Nigeria for their edible leaves, and/or fruits. Such cultivars include many *Solanum* species, *Lycopersicon esculentum* and *Capsicum* species. There are also some wild *Solanum* species which are valued for their therapeutic properties (Watt & Breyer-Brandwijk, 1962). The cultivated vegetables are mainly used to supplement the starchy foods, and also to provide essential nutrients, like vitamins, minerals and to some extent protein (Oke, 1965; Oyenuga & Fetuga, 1975; Tindal, 1968).

Studies involving interspecific crosses with *Solanum* species have been widely reported (reviewed by Magoon, Ramanujani & Cooper, 1962), but crosses with related genera such as *Lycopersicon* and *Capsicum* are few. Rick (1951, 1960) successfully hybridized *L. esculentum* with *Solanum lycopersicoides* and *S. pennellii,* respectively. Viable hybrids were obtained from both crosses, the former being possible only through embryo culture. He further reported that the F$_1$ hybrids from

L. esculentum × *S. lycopersicoides* were sterile, while those from *L. esculentum* × *S. pennellii* showed a high degree of fertility. Wann & Johnson (1963) carried out comprehensive crosses between five *Lycopersicon* species and 15 *Solanum* species. They reported that other than *S. lycopersicoides* and *S. pennellii*, only *S. ochranthum* and *S. etuberosum* were cross-compatible with *Lycopersicon* species. All other *Solanum* species did not give ovary stimulation when crossed with *Lycopersicon* species. The failure to obtain stimulation was reportedly due to non-germination of pollen grains. Fruit development when *S. ochranthum* was pollinated with either *L. chilense* or *L. peruvianum* was not preceded by fertilization, while in the *L. esculentum* × *S. etuberosum* cross, post-fertilization development was incomplete, and this resulted in fruits with inviable seeds.

In Nigeria, a large number of *Solanum* species, possibly indigenous and easily recognized by the different sizes and shapes of their fruits (Plate 46.1), and also various cultivars of *Capsicum* and *L. esculentum*, are widely grown. It is however unfortunate, that in spite of their popularity and economic importance, there have been very few attempts to investigate the relationships between these and related species and genera for the possibility of gene exchange. Nsowah (1969), compared the yields of local and exotic egg plants and reported that one of the factors which limit productivity of the local egg plant is the poor fruit yield. There are however, related *Solanum* species which produce heavily, but of which the taste of the fruits is unacceptable due to their extreme bitterness. Similarly, some of the major constrains which limit tomato production in the hot, humid tropics are the incidence of diseases (Simons, 1972) and high production costs largely due to staking (Adelana, 1975; Simons, 1975). The related genus *Solanum* is relatively free from most of the diseases known to affect tomatoes, and also bears fruits heavily without staking. It is therefore clear that there is a need to explore the possibilities of interspecific gene transfer by hybridization.

This paper is a report on the crossability relationships between several local *Solanum* species, and also between *L. esculentum* and *Solanum* and *Capsicum* species.

MATERIALS AND METHODS

The following plant materials were used*:—five diploid *Solanum* species ($2x = 24$), namely *S. melongena* (cv. Long Purple), wild *S. dasyphyllum* (cv. Igbagba elegun), *S. macrocarpon* (cv. Igbagba), *S. aethiopicum* (cv. Osun), *S. gilo*† (cv. Ikan) and two other breeding lines of *S. gilo*, H9 derived from a varietal cross between mottled white and mottled green fruited *S. aethiopicum*, and H10 from an interspecific cross *S. aethiopicum* × *S. dasyphyllum*; two varieties of hexaploid *S. nigrum* ($6x = 72$) (Odu and Ogumo); six cultivars of *Lycopersicon esculentum* and two of *Capsicum frutescens* and *C. annuum*.

*In a previous paper (Omidiji, 1975) these and other plants were given different names, but have now been re-identified by the editor (R.N.L.) and the author (M.O.O.), viz. *S. gilo* cv. Igba was called *S. melongena*; *S. aethiopicum* cv. Osun was *S. macrocarpon*; *S. macrocarpon* cv. Igbagba was *S. incanum*; *S. nigrum* var. Odu was *S. nodiflorum*; and *S. nigrum* var. Ogumo, which is probably *S. scabrum*, was *S. nigrum*.

†There is considerable confusion as to the correct identification and nomenclature of *S. gilo* cvs. Igba and Ikan, which are synonyms, and of the breeding lines H9 and H10 which are derived from other species. Furthermore *S. aethiopicum* cv. Osun is probably a variety of *S. indicum*—Ed. (R.N.L.)

Seeds of all *Solanum* species were soaked overnight in 2000 p.p.m. solution of gibberellic acid (GA) to aid emergence (Spicer & Dionne, 1961). They were grown in seed boxes for about three weeks then transplanted into the field. Seeds of *Capsicum* and *Lycopersicon* were sown without GA treatment.

All pollinations were carried out in the field. Flowers were emasculated at late bud stage, and pollinated the following day. An estimate of pollen fertility in both parental species and hybrids was carried out on pollen grains stained in aceto-carmine. All deeply stained grains were counted as functional, while the poorly stained ones with irregular outline were classified as degenerate.

Pollen tube growth in some incompatible crosses was studied according to the techniques of Sanyal (1958). Chromosome numbers in the *Solanum* species were determined from root tips prefixed overnight in para-dichlorobenzene and stained in Feulgen after hydrolysis in N.HCl for 10 minutes.

RESULTS

The results of all the crosses of *Solanum* species are shown in Table 46.1.

Table 46.1. Results of crosses between *Solanum* species

♀	2x	S. macro-carpon (Igbagba)	S. dasy-phyllum (Igbagba elegun)	S. aethio-picum (Osun)	S. gilo (Igba)	S. melon-gena	S. nigrum (Odu)	S. nigrum (Ogumo)
S. macrocarpon	24	A	A	D	—	E	D	D
S. dasyphyllum	24	A	A	—	—	—	—	—
S. aethiopicum	24	B	B	A	A	B, E	C	C
S. gilo H9	24	—	—	—	A	B, E	—	—
„ H10	24	—	—	—	A	C	—	—
S. melongena	24	D	—	D	D	A	—	—
S. nigrum (Odu)	72	C	—	C	—	—	A	A
S. nigrum (Ogumo)	72	C	—	C	—	—	A	A

A, Cross produced fully fertile hybrids; B, cross produced partially fertile hybrids; C, cross produced fruits but no viable seeds; D, cross produced no fruits; E, cross produced parthenocarpic fruits.

Crosses producing fully fertile hybrids

The crosses between *S. aethiopicum* and *S. gilo* (both diploid), and also those between the two hexaploid varieties of *S. nigrum*, produced fully fertile hybrids. These crosses were fairly easy to make.

Crosses producing partially fertile hybrids

Three *Solanum* species, *S. macrocarpon*, *S. dasyphyllum* and *S. melongena*, when crossed to *S. aethiopicum*, all produced partially fertile hybrids, as did also *S. melongena* with a breeding line of *S. gilo* (H9). The percentage of stainable pollen grains when *S. aethiopicum* was crossed with *S. macrocarpon* and *S. dasyphyllum* were 17% and 18% respectively, while that of H9 × *S. melongena* was 13%. All the parental *Solanum* species gave high pollen stainability ranging between 94%

and 98%. Some of the plants from the crosses *S. aethiopicum* × *S. melongena* and H9 × *S. melongena* produced undersized parthenocarpic fruits. All the fruits set on the F₁ plants of the *S. macrocarpon* × *S. melongena* cross were undersized and parthenocarpic (Fig. 46.1. 1), although the F₁ hybrid when produced had about 2% stainable pollen.

Crosses producing fruits but no viable seeds

Most crosses between diploid and hexaploid *Solanum* species, and also the cross *S. gilo* H10 × *S. melongena*, resulted in ovary stimulation, but the berries were incompletely developed and had inviable seeds.

Crosses producing no fruits

Some crosses using *S. macrocarpon* or *S. melongena* as the female parents did not produce ovary stimulation. Two of these crosses namely *S. macrocarpon* × *S. aethiopicum* and *S. melongena* × *S. gilo*, were studied for pollen tube growth. In both, 24 hours after pollination, only a few pollen grains had germinated, and the pollen tubes in the style were short and their growth was abnormal. In the parental species, *S. aethiopicum* or *S. macrocarpon*, however, pollen grains germinated profusely and many pollen tubes were traced to the base of the style.

Crosses producing parthenocarpic fruits

Fruits were easily set when *Lycopersicon esculentum* was pollinated with various *Solanum* and *Capsicum* species, with a success rate of (10–) 60 (–70)%. However, all the fruits were parthenocarpic. Crosses were made with *S. aethiopicum*, *S. gilo*, *S. dasyphyllum*, *S. macrocarpon*, *S. nigrum* and *Capsicum* species.

The crosses of *L. esculentum* × *S. aethiopicum* and *L. esculentum* × *C. annuum* were studied for pollen tube growth. In both cases, eight hours after pollination there was profuse germination of pollen grains and many tubes were seen within the style. The germination of pollen grains and growth of the tubes were similar to those of *L. esculentum* after controlled pollination which produced normal fruits with seeds.

Morphology of F₁ progeny

All F₁ plants were vigorous, and were usually intermediate morphologically, but in some crosses the characters of one parent were predominant. F₁ plants from *S. aethiopicum* × *S. gilo* were hairy like the pollen parent, but set clustered fruits as the pistillate parent. All the plants from the crosses *S. aethiopicum* × *S. melongena* and *S. gilo* (H9) × *S. melongena* showed the dominant characters of the pollen parent, but those from *S. macrocarpon* × *S. melongena* showed the characters of the pistillate parent, except for the excessive spininess. Those from *S. aethiopicum* × *S. macrocarpon* resembled the pollen parent, but unlike both parents they were spiny and hairy. The plants from the crosses between *S. aethiopicum* and *S. dasyphyllum* were like the pollen parent, being hairy and heavily spined.

DISCUSSION

Solanum aethiopicum and *S. gilo* produced fully fertile hybrids when crossed, so it can be inferred that they are closely related and the differences between them are only of varietal nature. In fact Omidiji (1974) has shown from genetic studies that the morphological differences between them are due to few gene differences.

The hybrids of *S. aethiopicum* with *S. dasyphyllum*, *S. macrocarpon* and *S. melongena* were only partially fertile indicating that these three *Solanum* species are isolated from *S. aethiopicum* and have very limited possibilities for gene exchange.

The incongruity between the local cultivated egg plant *S. gilo* and the exotic *S. melongena* indicates the effect of domestication on the development of genetic barriers. Thus whilst crosses of the cultivar of *S. gilo* with *S. melongena* failed to set fruits, those of two breeding lines of *S. gilo*, H9 and H10, succeeded. Fruits from H10 × *S. melongena* were sterile, but those of H9 × *S. melongena* contained viable seeds and the F_1 produced 13% stainable pollen. The barrier to crossability between the cultivated *S. gilo* and *S. melongena* was probably acquired during domestication of the local egg plant. The ease of crossability of the as yet uncultivated breeding lines with *S. melongena* supports this hypothesis. Both breeding lines are morphologically similar to *S. gilo*, which has been shown to be of hybrid origin (Omidiji, 1976).

The lack of ovary development in all crosses in which *S. macrocarpon* and *S. melongena* were pistillate parents was probably due to slow and abnormal growth of pollen tubes resulting from crosses in which the style of the maternal parent is longer and stouter than that of the staminate parent. Under such conditions, pollen tubes which are normally adapted to short slender styles might not grow fast enough in such massive stylar tissues to effect any fertilization. Such unidirectional pattern of success in crossability has been widely reported in crosses involving *S. melongena* (Sarvayya, 1936; Rajasekaran, 1971; Rangasamy & Kadambavanasundaram, 1973). Magoon *et al.* (1962) have postulated that failure of fruit development in crosses with *S. melongena* as the female parent may be due to development of an isolation mechanism which tends to inhibit growth of pollen from other *Solanum* species on its own style.

Although cross-fertilization between *L. esculentum* and closely related *Solanum* species has been reported (Rick, 1951, 1960), in this study only parthenocarpic fruits were developed when *L. esculentum* was hybridized with *Capsicum* and *Solanum* spp.

Parthenocarpy is a natural phenomenon, but can also be induced by artificial chemicals or by pollen extracts from unrelated plant genera. The parthenocarpic fruit development reported here in intergeneric crosses was probably a response to hormones released through the stimulus of pollination, and cannot be taken as evidence of close relationship between these genera. However, perhaps the new techniques of somatic hybridization by protoplasmic fusion will help geneticists and taxonomists to asses the affinities between the genera *Capsicum*, *Lycopersicon* and *Solanum*.

ACKNOWLEDGEMENTS

The author is thankful to Dr Ojomo, the Acting Director of his Institute, for help with his manuscript, and to the Vice Chancellor of the University of Ife for giving him financial support to attend the Solanaceae Conference.

REFERENCES

ADELANA, B. O., 1975. Effect of staking on tomato yields in Western State of Nigeria. *I.A.R. & T. Annual Research Review.* (Mimeographed).

MAGOON, M. L., RAMANUJANI, S. & COOPER, D. C., 1962. Cytogenetical studies in relation to the origin and differentiation of species in the genus *Solanum. Caryologia, 15:* 151–252.

NSOWAH, G. F., 1969. Genetic variation in local and exotic varieties of garden eggs. Variation in morphological and physiological characteristics. *Ghana Journal of Science, 9:* 61–78.

OKE, O. L., 1965. Chemical studies of some Nigerian vegetables. *Experimental Agriculture, 1:* 125–129.

OMIDIJI, M. O., 1974. Genetic studies with fertile interspecific hybrids of three *Solanum* L. species. In, *Proceedings of Second Annual Conference, Genetics Society of Nigeria:* 72–76.

OMIDIJI, M. O., 1975. Interspecific hybridization in the cultivated, non-tuberous *Solanum* species. *Euphytica, 24:* 341–353.

OMIDIJI, M. O., 1976. Evidence concerning the hybrid origin of the local garden egg plant (*Solanum gilo*). *Nigerian Journal of Science, 10:* 123–135.

OYENUGA, V. O. & FETUGA, B. L., 1975. Dietary importance of fruits and vegetables. In, *Proceedings of First National Seminar on Fruits and Vegetables, Ibadan:* 122–131.

RAJASEKARAN, S. S., 1971. Cytological studies on the F_1 hybrid *Solanum xanthocarpum* × *S. melongena* and its amphidiploid. *Caryologia, 24:* 261–267.

RANGASAMY, P. & KADAMBAVANASUNDARAM, M., 1973. Interspecific hybridization in *Solanum—Solanum melongena* L. × *S. indicum* L. *Madras Agricultural Journal, 60:* 1687–1694.

RICK, C. M., 1951. Hybrids between *Lycopersicon esculentum* and *Solanum lycopersicoides*. *Proceedings of the National Academy of Sciences of the United States of America, 37:* 741–744.

RICK, C. M., 1960. Hybridization between *Lycopersicon esculentum* and *Solanum pennellii*. *Proceedings of the National Academy of Sciences of the United States of America, 46:* 78–82.

SANYAL, P., 1958. Studies on the pollen tube growth in six species of *Hibiscus* and their crosses in vivo. *Cytologia, 23:* 460–467.

SARVAYYA, C. V., 1936. The first generation of an interspecific cross in *Solanum* between *S. melongena* and *S. xanthocarpum. Madras Agricultural Journal, 24:* 139–142.

SIMONS, J. H., 1972. Screening of tomatoes for resistance to diseases. *I.A.R. & T. Annual Report, 1972/73:* 35–36.

SIMONS, J. H., 1975. The economics of tomato cultivation by improved methods. *I.A.R. & T. Annual Research Review.* (Mimeographed).

SPICER, P. B. & DIONNE, L. A., 1961. Use of gibberellin to hasten germination of *Solanum* seed. *Nature, 189:* 327–328.

TINDAL, H. D., 1968. *Commercial Vegetable Growing.* Oxford University Press.

WANN, E. V. & JOHNSON, K. W., 1963. Intergeneric hybridization involving species of *Solanum* and *Lycopersicon. Botanical Gazette, 124:* 451–455.

WATT, J. M. & BREYER-BRANDWIJK, M. G., 1962. *The Medicinal and Poisonous Plants of Southern and Eastern Africa.* 2nd ed. Edinburgh & London: Livingstone.

EXPLANATION OF PLATE

PLATE 46.1

a–m. Fruits of different *Solanum* species and hybrids.

a, b. *S. macrocarpon*, edible (a) and inedible (b) varieties.

c, d, e, f. Four different cultivars of *S. gilo* (and related species—Ed.).

g. Clustered fruits of *S. nigrum.*

h. Clustered fruits of *S. aethiopicum.*

i. *S. melongena.*

j. *S. macrocarpon.*

k. *S. aethiopicum.*

l. Parthenocarpic undersized fruit of *S. macrocarpon* × *S. melongena.*

m. Intermediate sized fruit of *S. aethiopicum* × *S. melongena.*

Plate 46.1

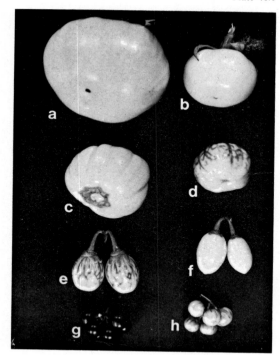

(*Facing p. 604*)

47. The barriers to hybridization between *Solanum melongena* and some other species of *Solanum*

N. NARASIMHA RAO

Ithanagar, Tenali, A.P., India

Barriers to crossability have developed to various degrees at different levels between *Solanum melongena* and many other species of *Solanum*. Complete failure of fruit-set, formation of parthenocarpic fruits, production of shrunken seeds or well-developed but non-germinable seeds, and seedling mortality are different phenomena indicative of crossing failure. Partial incompatibility barriers permit crosses of many combinations to be made only in one direction. The failure of crosses is generally a post-fertilization phenomenon. Embryo-endosperm incompatibility, or reaction of a gene or gene complexes of the male parent with the cytoplasm of the female parent, or complementary action of a gene or gene complexes derived from different genomes, are possible causes of failure of crosses.

Morphological differences are not correlated with inability to cross, nor does an inability to cross imply lack of genetic relationships. The failure of crosses is not only due to general genotypic incompatibilities but to specific gene controlled reactions also. Crossability studies between these species do not necessarily indicate the cytogenetic relationships of the taxa or their systematic positions.

For successful crosses, a measure of the affinity between the constituent parents, based on their ability to cross and produce hybrids, is devised and is presented in the form of a 'Crossability Index'.

CONTENTS

INTRODUCTION

The genus *Solanum* Section *Leptostemonum* is characterized by certain peculiarities with regard to crossability among its species. Although there are good prospects of transferring yield attributes, hardiness, and resistance to pests and diseases from wild germ plasm to *Solanum melongena* and other species cultivated for food or alkaloids, little has yet been achieved. This is mainly because no one has continued breeding programmes beyond the F_2 generation or through backcrosses. However, the hybrids obtained among the taxa of this group have thrown much light on the cytogenetic relationships of different species and on the nature of species differentiation, besides providing clues as to the potential of interspecific hybridization in the improvement of cultivated species. Natural hybridization, even between similar and sympatric species, is impossible or rare in this Section of *Solanum*.

SURVEY OF HYBRIDS IN *SOLANUM*

Artificial hybridization at generic, specific, sub-specific and varietal levels in the Section *Leptostemonum* has been achieved by many different investigators as shown in Table 47.1. As will be seen from the table, the number of successful crosses so far obtained is limited. Inter-generic crosses were effected with difficulty between a few species of *Solanum* and other genera such as *Petunia*, *Lycopersicon* and *Capsicum*. Successful crosses at the inter-specific level have been reported between *S. melongena*, on the one hand, and *S. xanthocarpum*. *S. integrifolium*, *S. tamago*, *S. incanum*, *S. grandiflorum*, *S. indicum*, *S. aethiopicum*, *S. gilo*, *S. aculeatissimum*, *S. cumingii*, and *S. zuccagnianum* on the other, besides some successful attempts between various wild species. Intra-specific hybrids were reported between the cultivated forms of *S. melongena* and its close wild relatives such as *S. melongena* var. *insanum*, *S. melongena* var. *potangi* and "*S. melongena* var. *bulsarensis*". It has been reported that barriers to crossability have developed even between strains or varieties of the same species, indicating progressive differentiation of the taxa with incipient isolation. It is a matter of interest that crosses between species with the same chromosome number or even between those with different ploidy levels have

Table 47.1. Hybrids of *Solanum melongena* and other taxa reported in the literature, arranged according to their fertility

---

A. Fertile intra-specific hybrids of *Solanum melongena*

S. melongena × *S. m.* var. *insanum*	Rai, 1959; Narasimha Rao, 1966, 1968
S. m. var. *insanum* × *S. melongena*	Swaminathan, 1949; Mittal, 1950; Babu Rao, 1965; Narasimha Rao, 1966, 1968; Rajasekaran, 1968
S. m. var. *potangi* × *S. melongena*	Swaminathan, 1949; Mittal, 1950
S. m. var. *pumilo* × *S. melongena*	Fukumotoh, 1962
S. m. ssp. *occidentale* var. *bulgaricum* × *S. m.* ssp. *subspontaneum* var. *leucoum*	Rajki, Cicer & Pal, 1964

B. Fertile inter-specific hybrids with *Solanum melongena*, *S. incanum*, and allied species

---

S. incanum × *S. melongena*	{ Swaminathan, 1949 Mittal, 1950; Babu Rao, 1965
All combinations of *S. incanum*, *S. melongena* and *S. m.* var. *insanum*	Narasimha Rao, 1968
S. melongena × *S. cumingii*	Capinpin *et al.*, 1963; Fukusawa, 1964
S. melongena × *S. aethiopicum*	Ignatova, 1971

C. Fertile intra- and inter-specific hybrids with *S. gilo*, *S. aethiopicum*, *S. indicum* and allied species

---

S. gilo × *S. indicum*	Nasrallah & Hopp, 1963
S. gilo × *S. zuccagnianum*	Narasimha Rao, 1968
S. aethiopicum × *S. gilo*	Omidiji, 1975
(called *S. macrocarpon* × *S. melongena*; See footnote 2, this Volume, Chapter 46)	
S. integrifolium var. *inerme* × *S. integrifolium*	Fukumotoh, 1962
S. indicum × *S. indicum*	Krishnappa & Chennaveeraiah, 1965

Table 47.1. (continued)

D. Inter-specific hybrids of *S. melongena*, *S. incanum*, *S. gilo*, *S. indicum* etc. reported to be partially sterile, semi-sterile or partially fertile (p) or sterile (s).

S. aethiopicum × *S. macrocarpon*	p Omidiji, 1975
(called *S. macrocarpon* × *S. incanum*; see footnote 1, this Volume, Chapter 46)	
S. gilo × *S. integrifolium*	p Hagiwara *et al.*, 1963
S. incanum × *S. gilo*	p Narasimha Rao, 1968
S. incanum × *S. indicum*	p Narasimha Rao, 1968
S. incanum × *S. integrifolium*	p Narasimha Rao, 1968
S. integrifolium × *S. incanum*	p Narasimha Rao, 1968
S. indicum × *S. gilo*	p Narasimha Rao, 1968
S. indicum × *S. integrifolium*	p Narasimha Rao, 1968
S. indicum × *S. melongena*	p Krishnappa & Chennaveeraiah, 1965; Rajasekaran, 1968; Narasimha Rao, 1968; Rangasamy & Kandambavanasundaram, 1973a, b, 1974a, b
S. melongena × *S. indicum*	s Nasrallah & Hopp, 1963
S. indicum × *S. m.* var. *insanum*	p Rajasekaran, 1968; Narasimha Rao, 1968
S. integrifolium × *S. melongena*	s Tatebe, 1936; Fukumotoh, 1962
S. integrifolium × *S. melongena*	p Hagiwara & Iida, 1938, 1939; Tatebe, 1941; Miwa *et al.*, 1958; Narasimha Rao, 1968
S. melongena × *S. integrifolium*	s Berry, 1953
S. melongena × *S. integrifolium*	p Katarzin, 1965; Narasimha Rao, 1968; Ludilov, 1974
S. integrifolium × *S. m.* var. *insanum*	p Narasimha Rao, 1968
S. m. var. *insanum* × *S. integrifolium*	p Narasimha Rao, 1968
S. integrifolium × *S. tamago*	p Tatebe, 1941
S. melongena × *S. gilo*	s Nasrallah & Hopp, 1963
S. melongena × *S. gilo*	p Narasimha Rao, 1968
S. melongena × *S. grandiflorum*	p Ramirez, 1959
S. melongena × *S. macrocarpon*	s Rajasekaran, 1961
(erroneously called *S. melongena* × *S. m. bulsarensis—Ed.*)	
S. melongena × *S. tamago*	p Tatebe, 1941
S. m. var. *insanum* × *S. gilo*	p Narasimha Rao, 1968
S. m. var. *pumilo* × *S. integrifolium*	s Fukumotoh, 1962
S. zuccagnianum × *S. m.* var. *insanum*	—Narasimha Rao, 1968
S. incanum × *S. xanthocarpum*	s Swaminathan, 1949; Mittal, 1950
S. xanthocarpum × *S. indicum*	s Mittal, 1950, Rajasekaran, 1968; Narasimha Rao, 1968
S. xanthocarpum × *S. melongena*	s Sarvayya, 1936; Hiremath, 1952; Rajasekaran, 1968
S. xanthocarpum × *S. melongena*	p Swaminathan, 1949
S. xanthocarpum × *S. m.* var. *insanum*	s Babu Rao, 1965; Rajasekaran, 1968
S. xanthocarpum × *S. trilobatum*	s Rajasekaran, 1968

E. Inter-generic hybrids, reported to be partially sterile (p) or sterile (s)

Lycopersicon esculentum × *S. pennellii*	p Rick, 1963
Lycopersicon esculentum × *S. lycopersicoides*	p Wann & Johnson, 1963
Lycopersicon esculentum × *S. melongena*	s ⎫
Capsicum annuum × *S. integrifolium*	s ⎬ Miwa *et al.*, 1958
S. integrifolium × *Petunia violacea*	s ⎭

been successful in the subgroup *Pachystemonum*, while many unsuccessful attempts at hybridization have been reported between the species belonging to the subgroup *Leptostemonum*, even though all species of the latter group are diploids.

While the tuberiferous group of the genus *Solanum* has been extensively studied, little attention has been paid to the non tuber-bearing group and hence only limited information is available on the latter group with regard to hybridization and cytogenetical aspects.

CROSSABILITY OF TEN *SOLANUM* SPECIES

Some of the results of 12 years study of *Solanum* species in Section *Leptostemonum*, by the author, may serve to illustrate his work on crossability relationships in this group and bring out some salient facts. Ten taxa belonging to the subsection *Asterotrichotum* of the Section *Leptostemonum* in the genus *Solanum* (Dunal, 1852) have been chosen: *viz. S. melongena* cultivar Pusa Purple Long, *S. melongena* var. *insanum*, *S. incanum*, *S. integrifolium*, *S. gilo*, and *S. zuccagnianum* from the series *Melongena*; *S. xanthocarpum* and *S. indicum* from the series *Oliganthes*; *S. sisymbrifolium* from the series *Cryptocarpum*, and *S. khasianum* which was not dealt with by Dunal. (*S. khasianum* C.B.Clarke var. *chatterjeeanum* Sengupta is *S. viarum* Dun.-*Ed.*)

Crosses were attempted in all possible combinations between these ten taxa. The development of fruits, seed-set and the germination of hybrid seeds were examined to find out the nature of barriers at different levels.

Out of 90 possible cross combinations, including reciprocals, there was no fruit set at all in 47, and only parthenocarpic fruits in four. Of the remaining 39 combinations where there was seed-set, one produced shrunken seeds and 12 produced seeds which appeared normal but did not germinate. Hybrid seedlings were obtained from the remaining 26 crosses, but the seedlings from two crosses died before transplantation. Thus mature hybrids grown successfully to flowering stage were derived from only 24 crosses. Full details of the results of these experiments are presented in Fig. 47.1.

CROSSABILITY INDEX

For successful crosses a Crossability Index was used to measure the crossing affinity between each pair of parents. This Index was derived from the commonly used seed set crossability index (% seed set in crosses / % seed set in selfs $\times 100\%$) and the Plants per Pollination formula of Marks (1965).

For each cross the following data were recorded:

A—percentage of fruits set,
B—average number of seeds per berry,
C—percentage germination of the seeds (and time taken),
D—percentage survival of the germinated seedlings.

A formula was then devised to standardize the values in the crosses (A^c, B^c, C^c and D^c), by the values obtained when the mother plant was selfed (A^s, B^s, C^s and D^s). Thus:-

$$\text{Crossability Index} = \frac{\text{Crossing efficiency of the cross}}{\text{Selfing efficiency of the female parent}} \times 100,$$

$$= \frac{A^c \times B^c \times C^c \times D^c}{A^s \times B^s \times C^s \times D^s} \times 100.$$

This is only a tentative formula, but it appears to work efficiently in assessing the crossability affinities between any two species which can cross and produce a hybrid.

BARRIERS TO CROSSABILITY

Barriers to crossability appear to have developed to various degrees at different levels. Failure of the crosses occurred by complete failure of fruit-set, parthenocarpic fruit-set, production of shrunken seeds, production of well developed but non-germinable seeds, and by seedling mortality.

♀ \ ♂	S. melongena cultivar	S. melongena var. insanum	S. incanum	S. integrifolium	S. gilo	S. indicum	S. zuccagnianum	S. xanthocarpum	S. khasianum	S. sisymbrifolium
S. melongena cultivar	▪	●	●	●	●	▲	▲	▲	⊖	▲
S. melongena var. insanum	●	▪	●	●	●	▲	▲	▲	▲	▲
S. incanum	●	●	▪	●	●	●	▲	▲	▲	▲
S. integrifolium	●	●	●	▪	⊕	⊕	▲	▲	▲	▲
S. gilo	⊕	⊕	⊕	⊕	▪	▲	●	▲	⊕	⊕
S. indicum	●	●	◖	●	●	▪	◖	⊕	▲	▲
S. zuccagnianum	▲	●	▲	▲	▲	▲	▪	⊕	▲	▲
S. xanthocarpum	⊕	●	⊕	▲	▲	●	▲	▪	▲	▲
S. khasianum	▲	▲	○	▲	▲	▲	▲	▲	▪	▲
S. sisymbrifolium	▲	▲	○	▲	○	▲	▲	▲	○	▪

▪ Normal seeds set on selfing
● Normal seeds which grew well
◖ Normal seeds, but seedlings died
⊕ Normal seeds, but did not grow
⊖ Fruits set, but seeds shrunken
○ Fruits set, but parthenocarpic
▲ Cross unsuccessful

Figure 47.1. Diagrammatic representation of the results of the crosses attempted between *Solanum* species.

Pollen germination and stylar incompatibility

Pollen germination and growth was observed *in vivo* in self-matings, and in compatible and incompatible cross matings. No evidence was found of inhibition of pollen germination on the stigmas, which took about 30 minutes in all cases. Likewise pollen tubes reached the middle of the style in 50–55 minutes and the ovarian cavity in 75–80 minutes. There was no stylar incompatibility in any of the combinations, and any differences in length of the styles did not matter.

The ovules were observed to increase in size for three to four days. Although it is possible that this growth was merely stimulated by entry of the pollen tubes into the ovarian cavity, it is here assumed that it was the result of fertilization, and thus subsequent degeneration, and the failure of crosses in cases where there was no fruit set at all, or only parthenocarpic fruits, or fruits with shrunken seeds, were presumed to be post-fertilization phenomena.

Collapse of the young embryos

When shrunken seeds were produced it was assumed that there was genetic incompatibility between the embryo and endosperm due to interspecific lethal genes. The collapse of the embryo in early stages of development, called "somato-plastic sterility" has been reported by many workers in many taxa, e.g. by Brink & Cooper (1947) in crosses of *Nicotiana rustica* with the pollen of other species of *Nicotiana, Petunia* and *Lycopersicon*, and by Wann & Johnson (1963) in crosses between *Lycopersicon* and *Solanum*.

Failure to germinate

The tetrazolium chloride test indicated that viable embryos were present in the normal sized but non-germinable seeds. Dissection showed that the embryos were of normal size. Failure of germination was shown not to be due to mechanical obstruction by the seed coat. It is possible that it is due to embryo and endosperm incompatibility. Further studies are needed, especially by culture of excised embryos, a technique which has seldom been used successfully in this Section of *Solanum*.

Death of seedlings

The mortality of seedlings from the crosses of *S. indicum* with *S. incanum* and with *S. zuccagnianum* was not accidental. Repeated sowings of seeds in three consecutive seasons produced similar mortality, which may have been due to lethal genes.

CYTOPLASMIC FACTORS CAUSING ONE-WAY INCOMPATIBILITY

Partial incompatibility barriers permitted crosses of many combinations to be made in only one direction, i.e., reciprocal differences were observed between the species with regard to crossability.

To obtain an insight into possible causes for these differences in reciprocal crossability, the hybrid between *Solanum indicum* and *S. melongena*, which was only achieved with *S. indicum* as the female parent, was backcrossed to both parents, and F_2 and backcross progenies were again crossed to both parents. When *S. indicum* was used as the female parent, all crosses were successful, but when *S. melongena* was the female parent, incompatibility was shown by the pollen from *S. indicum*, from all F_1 plants, from eight of the ten F_2 plants, from seven of the ten progeny from the backcross to *S. melongena*, and from all ten of the progeny from the backcross to *S. indicum*. Thus it was observed that cytoplasmic influence persisted even up to the F_2 and BC_1 generations without any modification. There is also a clear case of breakdown of incompatibility in F_2 and BC_1 progenies indicating the existence of cytoplasmic gene action bringing about incompatibility between *S. indicum* and *S. melongena* in the cytoplasm of the latter species. The cytoplasm of *S. melongena* is unaccommodative to particular genes or gene complexes of *S. indicum*, while the cytoplasm of *S. indicum* can tolerate the genotype of *S. melongena*. The number of genes involved in bringing about this incompatibility could not be determined in these limited studies.

Similar observations were made by Rangasamy and Kadambavanasundaram (1973b) on the nature of one-way incompatibility in *S. indicum* and *S. melongena*, although they had less data.

An instance of reaction of a nuclear factor with cytoplasm to induce differential sterility in reciprocal crosses was deduced by Buck (1960) in his crosses among tuberiferous solanums. He noticed that crosses of a number of diploid species as males, with some clones of *S. verrucosum* ($2n = 24$) as female, gave only male-sterile hybrids, whereas the few reciprocal hybrids obtained proved to be completly male-fertile. On backcrossing the male-sterile hybrid in the *S. verrucosum* cytoplasm to the non-*verrucosum* parent, only male-sterile progeny resulted, but backcrossing to *S. verrucosum* produced equal numbers of male-sterile and male-fertile plants, indicating the reaction of a nuclear factor with cytoplasm of *S. verrucosum* to produce male-sterility in the hybrids.

The influence of ploidy on the cytoplasmic effect was also studied by crossing tetraploids, produced by colchicine treatment of the same cultivar of *S. melongena*, with diploid *S. indicum*. Crosses were easily obtained in both directions producing triploid hybrids. Evidently the expression of the cytoplasm is modified with the change in the ploidy level. The effect of polyploidization on the cytoplasm has also been reported by Magoon, Hougars & Cooper, (1958) in tuberiferous solanums. They observed a large difference in the pollen fertility between the cross *S. pinnatisectum* × *S. jamesii* and its reciprocal when the species are at tetraploid level, and almost identical values for the two crosses at the diploid level.

The factor making certain crosses successful in one direction only appears to be operating throughout the family Solanaceae. Such reciprocal differences with regard to crossability have been reported in *Solanum melongena* and other species by Sarvayya (1936). Tatebe (1936), Nasrallah & Hopp (1963), Krishnappa & Chennaveeraiah (1965), Babu Rao (1965), and Rajasekaran (1968). Such a type of breeding system appears to be prevalent in related genera also, as reported for *Datura* by Blakeslee, Murry & Satina (1935) and Satina (1959), in *Nicotiana* by Ramanujam & Joshi (1942), in *Lycopersicon* by MacArthur & Chiasson (1947) and McGuire & Rick (1954), and in *Capsicum* by Ohta (1961). No satisfactory explanation has been given so far for this limitation. McGuire & Rick (1954) suggested, based on their studies on the crosses between *Lycopersicon esculentum* and *L. peruvianum* where the cross was possible only with the former species as female parent, that two independent genes from *L. esculentum* render its pollen incapable of fertilizing *L. peruvianum* but concluded by saying that additional cytogenetic information was needed for a thorough understanding of this limitation.

CONCLUSIONS

The present studies have shown that isolation barriers are not always accompanied by morphological differentiation in this group of species: two varieties of *Solanum melongena* differed in their ability to cross with other species. Similar observations have been made by Babu Rao (1965). Krishnappa & Chennaveeraiah (1965), and Rangasamy & Kadambavanasundaram (1973a, b, 1974a, b), reported interstrainal differences in cultivated *S. melogena* in their ability to cross with *S. indicum*. Even crosses between different strains of *S. melongena* and of *S. indicum* were reported to be unsuccessful by Krishnappa & Chennaveeraiah (1965).

Racial differences in crossability between species of *Datura* were also encountered by Blakeslee & Satina (1949). This is evidence of the influence of genes or gene complexes, distributed in different genomes, acting as a barrier against successful hybridization, possibly in a complementary way.

Because of the peculiarities exhibited by *Solanum* species with regard to crossability, the information obtained by crossability studies alone can not offer any clues as to the cytogenetic relationships of the species involved and their systematic positions, even though the ability of species to cross with one another to give rise to viable offspring is normally accepted as one of the principles used in assessing relationships. In *Solanum* species, where sometimes even different strains of the same species would not cross with one another, where the reciprocal differences with regard to crossability were modified with a change in ploidy level, where morphological similarities are not necessarily associated with an ability to cross, and morphological differences are not always associated with an inability to cross, crossability studies could only complement the evidence obtained from other sources. A similar caution has been given by Davis & Heywood (1965) for the use of crossability relationships in arriving at the true affinities of the taxa in different genera where crossability is influenced by minor genetic factors instead of total genotypic differences.

ACKNOWLEDGEMENTS

Grateful thanks are due to Dr B. W. X. Ponnaiya and Dr P. Madhava Menon formerly of the Department of Cytogenetics and Plant Breeding, Coimbatore Agricultural College and Research Institute, for helping me with this work during the period 1964–1971. I am grateful also to the University of Birmingham and the Linnean Society of London for having invited me to read this paper.

REFERENCES

BABU RAO, L., 1965. *Cytomorphological Studies on Certain Interspecific Hybrids of Non-tuberiferous* Solanum *species*. M.S.(Ag.) Dissertation, University of Madras, India.

BERRY, S., 1953. *Some Aspects of Breeding in* S. melongena L. M.Sc. Dissertation, University of New Hampshire, Durham, U.S.A.

BLAKESLEE, A. F., MURRY, M. J. & SATINA, S., 1935. Crossability in relation to taxonomic classification in the genus *Datura. American Naturalist, 69:* 57.

BLAKESLEE, A. F. & SATINA, S., 1949. Differences in crossability between species of *Datura* due to individual races used in the cross. *American Journal of Botany, 36:* 795.

BRINK, R. A. & COOPER, D. C., 1947. The endosperm in seed development. *Botanical Review, 13:* 423–541.

BUCK, R. W., 1960. Male sterility in interspecific hybrids of *Solanum. Journal of Heredity, 51:* 13–14.

CAPINPIN, J. M., LUNDE, M. & PANCHO, J. V., 1963. Cytogenetics of interspecific hybrid between *Solanum melongena* Linn. and *S. cumingii* Dunal. *Philippines Journal of Science, 92:* 169–178.

DAVIS, P. H. & HEYWOOD, V. H., 1965. *Principles of Angiosperm Taxonomy*. Edinburgh & London: Oliver & Boyd.

DUNAL, M. F., 1852. "Solanaceae". *DC Prodr., 13:* 1–690.

FUKUMOTOH, K., 1962. Nuclear instability and chromosomal mosaicism in high polyploids of *Solanum* species and Hybrids. *Japanese Journal of Botany, 18:* 19–53.

FUKUSAWA, C. A., 1964. Genetics of clustered and solitary fruit segregants from the interspecific cross between *Solanum melongena* L. and *S. cumingii* Dunal. *Araneta Journal of Agriculture, 11:* 55–75.

HAGIWARA, T. & IIDA, H., 1938. On the species hybrid between *Solanum integrifolium* Poir. and *Solanum melongena* L. *Botany and Zoology, 6:* 858–864.

HAGIWARA, T. & IIDA, H., 1939. Inter-specific crosses between *Solanum integrifolium* and the egg plant and the abnormal individuals which appeared in F$_2$. *Botany and Zoology, 7:* 1520–1528.

HAGIWARA, T., SASAKI, H., INABA, T. & YANASE, Y., 1963. Genetical studies on habit in the progeny of spontaneous hybrids of *Solanum gilo*. *Bull. Coll. Agric. vet. Med. Nihon. Univ.*, No. 16: 47–58 (Japanese).

HIREMATH, K. G., 1952. *A Study of Inter-specific Hybridization in Two Species of Genus* Solanum, *i.e.,* Solanum melongena *L. and* S. xanthocarpum *Schrad. and Wendl. M.Sc.(Ag.)* Dissertation, Karnatak University, Dharwar, India.

IGNATOVA, S. I., 1971. (The morphological and biological characteristics of F_1 hybrids of egg-plant from *Solanum melongena* × *S. aethiopicum.*) Sb. statei moloody-dykh uchenykh i aspirantov. N II ovoshch. kh-va, No. 4, 200–303 (Ru). *Referativnyi Zhurnal, 8:* 55.30.

KATARZIN, M. S., 1965. (Distant hybridization of the egg plant.) *Trudȳ po Prikladnoĭ Botanike, Genetike Selektsii, 37:* 122–129. (Russian).

KRISHNAPPA, D. G. & CHENNAVEERAIAH, M. S., 1965. Breeding behaviour in non-tuber bearing *Solanum* species. *Journal of the Karnatak University of Science, 9:* 82–96.

LUDILOV, V. A., 1974. (Ways of increasing fertility in inter-specific egg-plant hybrids.) *Sel' skokhozyaistve-nnaya Biologia, 9* (6): 933–936 (Pu: 5 ref) N-i institut ovoschchnogo Khozyaistva, Novocherkassk, U.S.S.R.

MacARTHUR, J. W. & CHIASSON, L. P., 1947. Cytogenetic notes on tomato species and hybrids. *Genetics, 32:* 165–177.

MAGOON, M. L., COOPER, D. C. & HOUGAS, R. W., 1958. Induced polyploids of *Solanums* and their crossability with *S. tuberosum. Botanical Gazette, 119:* 224–233.

MARKS, G. E., 1965. Cytogenetic studies in tuberous *Solanum* species. II. Species relationships in some South and Central American species. *New Phytologist, 64:* 293–306.

McGUIRE, D. C. & RICK, C. M., 1954. Self-incompatibility in species of *Lycopersicon* sect. *Eriopersicon* and hybrids with *L. esculentum. Hilgardia, 23:* 101–124.

MITTAL, S. P., 1950. *Studies in Non-tuberiferous Species and Hybrids of* Solanum. M.Sc. Dissertation I.A.R.I., New Delhi.

MIWA, T., SAITO, Y. & YAMAMOTO, S., 1958. The effect of plant hormones on interspecific and intergenic hybridization in the Solanaceae. *Bulletin of the Faculty of Agriculture, University of Miyazaki, 4:* 153–165.

NARASIMHA RAO, N., 1966. *Studies on the breeding behaviour of some inter- and intra-specific hybrids of non-tuberiferous species of* Solanum. M.Sc.(Ag.) Dissertation, University of Madras, India.

NARASIMHA RAO, N., 1968. A note on the crossability relationships between some non-tuberiferous species of *Solanum. Madras Agricultural Journal, 55:* 146–149.

NASRALLAH, M. E. & HOPP, R. J., 1963. Inter-specific crosses between *Solanum melongena* L. (egg plant) and related *Solanum* species. *Proceedings of the American Society for Horticultural Science, 83:* 571–574.

OHTA, Y., 1961. Cytogenetical studies in the genus *Capsicum*. I. *C. frutescens* × *C. annuum. Japanese Journal of Genetics, 36:* 105–111.

OMIDIJI, M. O., 1975. Interspecific hybridization in the cultivated non-tuberous *Solanum* species. *Euphytica, 24:* 341–353.

RAI, U. K., 1959. Cytogenetic studies in *S. melongena* L. I. Chromosome morphology. *Caryologia, 12:* 299–316.

RAJASEKARAN, S., 1961. *Cytogenetic Studies in* Solanum melongena *L.* S. melongena *var. bulsarensis Argikar and their Hybrid and Study of Colchicine Induced Polyploidy in* S. melongena *L.* M.Sc.(Ag.) Dissertation, University of Poona, India.

RAJASEKARAN, S., 1968. *Cytogenetic Studies on Sterility in Certain Inter-specific Hybrids of* Solanum. Ph.D. Thesis, Annamalai University, India.

RAJKI CICER, E. & PAL, G., 1964. Study of the phenomenon of heterosis in the first and following generations of sexual hybrids and grafts of egg plant varieties. *Biológiai Közlemények, 11:* 131–143.

RAMANUJAM, S. & JOSHI, A. B., 1942. Inter-specific hybridization in *Nicotiana*. A cytogenetical study of the hybrid *N. glauca* Grah. × *N. plumbaginifolia* Viv. *Indian Journal of Genetics and Plant Breeding, 2:* 80–97.

RAMIREZ, D. A., 1959. Cytology of Philippine plants. II. *Solanum grandiflorum* Ruiz and Pav. *Philippine Agriculturist, 43:* 375.

RANGASAMY, P. & KADAMBAVANASUNDARAM, M., 1973a. A study on the inheritance of certain qualitative characters in the cross between *Solanum indicum* L. and *Solanum melongena* L. *South Indian Horticulture, 21:* 1–16.

RANGASAMY, P. & KADAMBAVANASUNDARAM, M., 1973b. Incompatibility in interspecific hybrid *Solanum indicum* L. × *S. melongena* L. *Madras Agricultural Journal, 60:* 1617–1621.

RANGASAMY, P. & KADAMBAVANASUNDARAM, M., 1974a. Variation pattern of quantitative traits in the second generation and back-cross progenies of the cross between *S. indicum* L. × *S. melongena* L. *South Indian Horticulture, 22:* 158–163.

RANGASAMY, P. & KADAMBAVANASUNDARAM, M., 1974b. A Cytogenetic analysis of sterility in interspecific hybrid *Solanum indicum* Linn × *Solanum melongena* Linn. *Cytologia, 39:* 645–654.

RICK, C. M., 1960. Hybridization between *Lycopersicon esculentum* and *Solanum pennellii*: phylogenetic and cytogenetic significance. *Proceedings of the National Academy of Sciences of the United States of America, 46:* 78–82.

SARVAYYA, Ch. V., 1936. The first generation of an inter-specific cross in *Solanums*, between *Solanum melongena* and *S. xanthocarpum. Madras Agricultural Journal, 24:* 139–142.

SATINA, S., 1959. Growth processes in the embryo and seed. In Blakeslee, *The Genus* Datura. New York: Ronald Press.

SWAMINATHAN, M. S., 1949. *Cytotaxonomic Studies in the Genus* Solanum. M.Sc. Dissertation, I.A.R.I., New Delhi.

TATEBE, T., 1936. Genetic and Cytological studies on the F_1 hybrid of scarlet or tomato egg plant (*S. integrifolium* Poir.) × *S. melongena* L. *Botanical Magazine, Tokyo, 50:* 457–462.

TATEBE, T., 1941. (On inter-specific hybrids involving the egg plant and related species.) *Botany and Zoology, 9:* 36–37 (Japanese).

WANN, E. V. & JOHNSON, K. W., 1963. Inter-generic hybridization involving species of *Solanum* and *Lycopersicon. Botanical Gazette, 124:* 451–455.

48. Chemotaxonomy of the cultivated eggplant—a new look at the taxonomic relationships of *Solanum melongena* L.

KIT PEARCE AND RICHARD N. LESTER

Department of Plant Biology, University of Birmingham, U.K.

Analysis of seed proteins by electrophoresis and immunology, and of phenolics by chromatography, was used to compare 27 species of *Solanum*. Electrophoresis produced identical patterns of proteins from seeds of both modern and primitive cultivars of *Solanum melongena*, similar patterns for *S. incanum*, but less similar for the other species of the section *Melongena*. *S. aethiopicum*, *S. indicum*, and several other species in section *Oliganthes*, and also *S. anomalum* of section *Torva* formed another distinct group. A basic similarity between these two groups was shown by immunological techniques. These results are in agreement with the species relationships indicated by hybridization experiments.

CONTENTS

INTRODUCTION

The cultivated eggplant, *Solanum melongena* L., is a species of considerable economic importance in many tropical and subtropical parts of the world. Together with many other *Solanum* species, its taxonomic relationships are not well understood, and most investigations have been limited to studies of morphological characters (Bitter, 1923; Seithe, 1962) and the evaluation of crossability relationships (Rao, 1969; Rajasekaran, 1969 a, b, 1970 a, b, 1971 a, b, c; Krishnappa & Chennaveeraiah, 1964; Capinpin, Lunde & Pancho, 1963 *etc.*).

The cultivated eggplant is morphologically similar to a number of taxa (*vide* Bhaduri, 1951). In addition to the primitive cultivars and the wild species *S. incanum*, a number of other putative wild varieties and species have been described at one time or another. Names that have been used include *S. insanum* L. (and other authors) and *S. melongena* L. var. *insanum* Prain, *S. sanctum* L., *S. album* Lour. and *S. cumingii* Dun., and incorrectly, *S. coagulans* Forssk. The natural group to which all these taxa belong has here been termed the "eggplant complex".

Cultivated varieties of eggplant can usually be easily distinguished from the remainder of the taxa comprising the eggplant complex by their enlarged and often purple fruit and almost complete lack of prickles on the stem, leaves and calyx. It

615

is also easy to distinguish members of the eggplant complex from other *Solanum* species. There has, however, been much difficulty in establishing the taxonomic limits, status and nomenclature of the taxa belonging to the eggplant complex, due in part to the anthropogenic factors involved with the introduction of the wild taxon into cultivation and the existence of weedy forms, and the genotypic variability and phenotypic plasticity characteristic of *Solanum* species. Furthermore, the truly wild species *S. incanum* is widespread throughout the Old World tropics, and appears to comprise several geographically or ecologically localized infraspecific taxa.

In recent years numerous hybridization experiments involving the cultivated eggplant, its closest relatives, and a number of other species, have been made. Much of this work has been carried out in India, using indigenous and naturalized species mostly from sections *Melongena* and *Oliganthes*. Some of the crossing results, together with other kinds of data were examined by Bhaduri (1951), who summarized the relationships of the cultivated eggplant apparent at that time.

The taxa used for the present investigation were chosen from a wide geographical range, including the New World and Australasia, as well as the Old World. They represent several sections and series of the subgenus *Leptostemonum*, the majority being from sections *Oliganthes* (Dun.) Bitt., *Torva* Nees, and *Melongena* Dun. Bitter's "Solana Africana" (1913, 1917, 1921, 1923), an exhaustive treatment of many Old World species in which morphologically similar taxa have been placed in the same series or subseries, is the basis for the arrangement of species in Table 48.4. The names of the series are according to D'Arcy (1972). Many samples of *S. melongena*, including primitive cultivars and apparently wild varieties, and also of the closely related wild species *S. incanum*, were obtained from as many different geographical locations as possible.

The chemotaxonomic techniques of immunoelectrophoresis, immunodiffusion, polyacrylamide gel electrophoresis and thin layer chromatography were applied to a wide range of taxa. New taxonomic evidence was provided by these methods, which, in conjunction with that from the results of numerous crossability studies reported and evaluated in detail elsewhere (Pearce, 1975), has allowed a reconsideration of some of the taxonomic relationships of the cultivated eggplant.

MATERIALS

Many seed samples of a wide range of taxa were obtained from diverse sources (Pearce, 1975) and grown to provide leaf and corolla material for chromatographic analysis, seeds for protein analyses and voucher specimens for the confirmation of identifications. The names used are those generally accepted and have not been examined for nomenclatural accuracy.

METHODS

Protein analysis

Seeds were ground finely, defatted, and the proteins extracted in phosphate buffered saline, pH 7.0, for immunological assay (Crowle, 1961; Lester, 1969), or in Tris-glycine buffer, pH 8.3, for polyacrylamide gel electrophoresis (PAGE). Antisera to crude protein extracts of *S. melongena* (S.46) and to *S. gilo* (S.395) were raised in rabbits. The methods of immunoelectrophoresis (IEP), immunodiffusion with absorption (ID) and PAGE were those used by Lester (this Volume, p. 285). For ID the number of precipitin arcs produced was counted, and for IEP the arcs were further characterized and identified. The patterns of bands obtained by PAGE were examined carefully and individuals with similar patterns, in terms of the position, width and density of the bands, subjectively grouped together.

Thin layer chromatography (TLC)

Healthy mature air-dried leaves and fresh corollas were extracted in 1% HC1 in methanol. Leaf components were separated in two dimensions on a thin layer of cellulose powder, using first 2% ethanoic acid, then butan-l-ol: ethanoic acid: water (6 : 2 : 1). The latter solvent was used for the one-dimensional separation of corolla pigments. The chromatograms were examined for fluorescent spots under ultraviolet light, and re-examined with ammonia vapour.

Analysis of data

The results from PAGE and ID were interpreted directly. The much larger amounts of data obtained in the IEP and TLC studies were analysed by computer. Data used were the presence or absence of individual characterized arcs (IEP) and the presence and absence of individual spots (TLC of leaf and corolla phenolics). The results of IEP were subjected to principal components analysis, and those from the TLC studies to cluster analysis, using Jaccard's similarity coefficient.

RESULTS AND DISCUSSION

It is appropriate to consider the affinities of the cultivated eggplant at two distinct levels in the taxonomic hierarchy, namely its relationships (1) with wild eggplant varieties and species, and (2) within the whole genus *Solanum*.

(1) *The taxonomic relationships of the cultivated eggplant with wild eggplant varieties and species*

A close similarity between the cultivated eggplant and its wild relatives was demonstrated by each of the chemotaxonomic techniques employed in the present investigation.

The principal components analysis of the IEP data for a wide range of taxa indicated a high level of serological similarity between *S. melongena* and *S. incanum*, and indeed between these species and a number of other species. In projections of the first and second components of variation of data obtained using antisera to both *S. melongena* (Fig. 48.1) and *S. gilo* (Fig. 48.2) virtually all the individuals of *S. melongena* and *S. incanum* and those of species belonging to the series *Aethiopica* of section *Oliganthes*, together with a few individuals of other taxa fall into a single cluster. These individuals showed in IEP a similar complement of arcs to those of the reference reactions. Similar clusters were produced when the first and third components of variation were plotted (Pearce, 1975). A further principal components analysis for only those individuals in the clusters resulted in no further subdivision (Fig. 48.3). On close examination, the reference reactions in IEP for accessions of *S. melongena* and *S. gilo* were found to be very similar, and the test reactions for *S. incanum* very similar to the reference reaction for *S. melongena*.

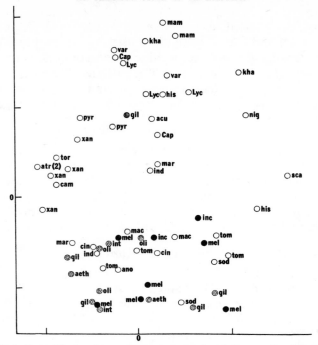

Figure 48.1. Projection of first and second components of variation produced by Principal Components Analysis of data from IEP studies using antiserum to *S. melongena*. (For explanation of code of abbreviations, see Table 48.4.).

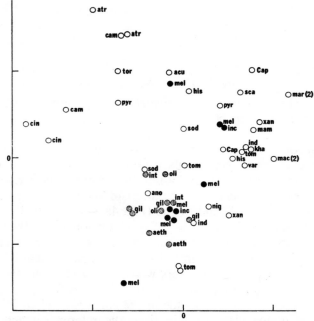

Figure 48.2. Projection of first and second components of variation produced by Principal Components Analysis of data from IEP studies using antiserum to *S. gilo*. (For code of abbreviations see Table 48.4.).

Similar indications of the close relationship between the cultivated eggplant and *S. incanum* were provided by the results of the ID studies. In tests using antiserum to *S. melongena* absorbed with seed protein extracts of *S. nigrum*, or of *S. sodomaeum*, accessions of *S. incanum*, as well as accessions of a number of other species, gave reactions which were closely similar to those of the reference reaction in terms of the number of arcs produced (Tables 48.2 and 48.3), whereas about ten other species produced fewer arcs or none.

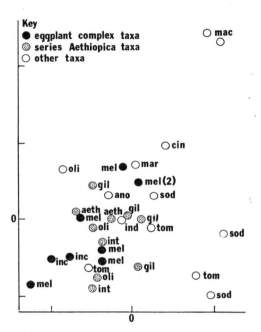

Figure 48.3. Projection of first and second components of variation produced by Principal Components Analysis of data from IEP studies using antiserum to *S. melongena*, for only those individuals in the main cluster in Fig. 48.1. (For code of abbreviations, see Table 48.4.)

PAGE studies were made of seed protein extracts of many modern cultivated varieties of eggplant, a semi-cultivated variety of eggplant from Indonesia (S.1355 and S.1360), and several accessions of *S. incanum* from a wide geographic range. Virtually identical patterns of protein bands were obtained for all the accessions of the cultivated eggplant, and the semi-cultivated and wild varieties gave very similar patterns (Plate 48.1). All the accessions of *S. incanum* gave a considerably weaker reaction, with differences in the slower-moving bands near the beginning of the gels (Plate 48.1), indicating, as expected, a less close relationship between *S. incanum* and the cultivated eggplant than between the cultivated eggplant, the primitive cultivars, and the putative wild variety.

None of the other species tested by PAGE gave protein spectra resembling those obtained for *S. melongena* and *S. incanum* (Plate 48.2). Thus PAGE demonstrated the close relationships of the cultivated eggplant and its nearest relatives, and the separateness of this group of taxa from all other species of *Solanum*.

A chromatographic study of the leaf phenolics of six species showed that *S.*

melongena and *S. incanum* were generally similar to each other and distinct from species in series *Aethiopica*. However, both *S. indicum* and *S. marginatum* were included in the main cluster of the dendrogram (Fig. 48.4).

Further chromatographic studies using a number of morphologically distinct accessions of *S. melongena* (cultivated, semi-cultivated and wild) and accessions of *S. incanum* from several different localities showed a diversity of patterns with relatively little similarity (50 % similarity or less). All nine accessions of non-prickly cultivated eggplant were clustered together, but this cluster was hardly distinct from the remaining assemblages of prickly and wild eggplants and various accessions of *S. incanum* (Fig. 48.5). Similar conclusions were drawn from the results of one-dimensional chromatography of flower pigments (Pearce, 1975).

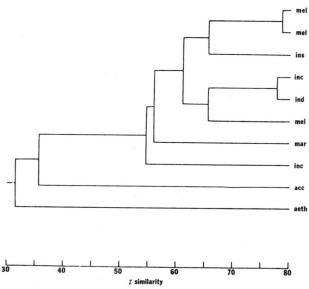

Figure 48.4. Phenogram resulting from cluster analysis of chromatographic data from *S. melongena* and other *Solanum* species.

The resolution of the chromatographic studies was lowered to some extent by technical difficulties in obtaining physiologically equivalent leaves and in recognizing some of the spots. Nevertheless the results of chromatography tended to support those from PAGE in indicating the close similarities of various cultivated eggplant accessions with *S. incanum* and the relative distinctness of these taxa from the other species of *Solanum*.

(2) *The relationships of the cultivated eggplant with other species in the genus* Solanum

The most valuable evidence as to the taxonomic relationships of the cultivated eggplant and its close relatives with other *Solanum* species was provided by the extensive immunoelectrophoretic survey. Tests involving a wide range of taxa and using antisera to both *S. melongena* and *S. gilo* indicated high levels of serological similarity among and between the members of the eggplant complex, and also taxa belonging to the series *Aethiopica* (*S. aethiopicum*, *S. gilo*, *S. integrifolium* and *S. olivare*). Indeed members of these two groups of species clustered together whichever antiserum was employed. The techniques of IEP did not discriminate between members of the eggplant complex, the series *Aethiopica*, and some other taxa which became included in the same cluster (Figs. 48.1 and 48.2). These indi-

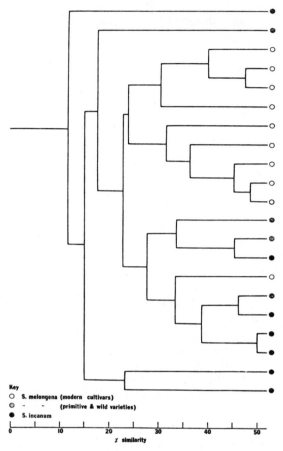

Figure 48.5. Phenogram resulting from cluster analysis of chromatographic data from the cultivated eggplant and its nearest wild relatives.

viduals (Table 48.1) appear to be closely related to *S. melongena* and *S. incanum*. All the accessions tested of the species in series *Aethiopica*, *S. indicum*, *S. sodomaeum*, *S. anomalum*, and *S. tomentosum*, are included in the main clusters indicated by one or both antisera (Table 48.1). Of all the taxa tested by IEP these have the closest serological affinity with, and can be considered most closely related to *S. melongena*. Accessions of *S. marginatum*, *S. macrocarpon* and *S. cinereum* were also included in the clusters indicated by antiserum to *S. melongena*, and therefore these taxa may be considered to be more closely related to this species. With antiserum to *S. gilo*, *S. xanthocarpum* and *S. nigrum* were included in the main cluster. *S. nigrum* is morphologically distinct from the eggplant and belongs not to subgenus *Leptostemonum* but to subgenus *Solanum*. Unexpected and unexplained strong serological results involving this species have, however, been encountered previously (Lester, this Volume, p. 285).

Table 48.1. Species and accessions included with the eggplant complex taxa in clusters obtained by analysis of IEP data

Species		Antiserum	
		S. melongena	*S. gilo*
Section *Oliganthes*			
Series *Aethiopica*	*S. aethiopicum*	S.225, S.1267	S.225, S.1267
	S. gilo	S.395, S.1236	S.395, S.1236
	S. integrifolium	S.240	S.240, S.344
	S. olivare	S.156, S.279	S.156, S.279
Series *Austroafricana*	*S. tomentosum*	S.927, S.1155, S.1278, S.1320	S.927, S.1155, S.1320
Series *Afroindica*	*S. indicum*	S.857	S.871
Series ?	*S. cinereum*	S.202, S.390	
Section *Torva*	*S. anomalum*	S.964	S.964
Section *Melongena*			
Series *Incaniformia*	*S. marginatum*	S.256	
Series *Sodomela*	*S. sodomaeum*	S.101, S.210	S.101
	S. xanthocarpum		S.1264
Series *Macrocarpon*	*S. macrocarpon*	S.133, S.1269	
Section *Solanum*	*S. nigrum*		S.275

When a further analysis was made of IEP data for those individuals which were clustered together in tests using antiserum to *S. melongena*, most of the individuals still formed a single cluster, (Fig. 48.3) but *S. macrocarpon* was excluded and *S. cinereum*, *S. sodomaeum*, and one accession of *S. tomentosum* were located on the periphery of the cluster, and can thus be considered to have less serological affinity with the taxa comprising the eggplant complex and series *Aethiopica* than the accessions of *S. anomalum*, *S. tomentosum* and *S. indicum* located within the main cluster.

Immunodiffusion tests were made using a similar range of taxa. When antiserum to *S. melongena* was absorbed with an extract of *S. nigrum*, relatively little absorption occurred, and, as expected, a large number of arcs was obtained for most of the tests reactions. Those taxa which produced nine or more arcs (an arbitary figure) included all accessions belonging to the eggplant complex and series *Aethiopica*, and also some accessions of *S. anomalum*, *S. indicum*, *S. sodomaeum*, *S. xanthocarpum*, *S. tomentosum*, *S. marginatum*, *S. hispidum* and *S. pyracanthum* (Table 48.2). With the exception of the last two species all these taxa were included in the clusters obtained from the IEP studies.

An extract of *S. sodomaeum* effected almost complete absorption of the antiserum to *S. melongena*, indicating its closer relationship. Those taxa whose extracts still produced one or more arcs included all the eggplant complex and members of series *Aethiopica* and also *S. indicum*, *S. macrocarpon*, *S. mammosum*, *S. atropurpureum*, *S. hispidum* and *S. aculeatissimum* (Table 48.3). In the IEP studies the last four of these species had shown very little serological relationship with the taxa of the eggplant complex.

In summary it may be stated that a close serological affinity to *S. melongena* is shown by the species in series *Aethiopica* together with *S. indicum* and *S. anomalum*, slightly less by *S. sodomaeum* and *S. tomentosum*, and yet less by *S. macrocarpon* and *S. marginatum*.

Table 48.2. Species with nine or more arcs after immunodiffusion with absorption, using antiserum to *S. melongena* absorbed with an extract of *S. nigrum*

Species	Accession Number(s)
S. melongena	S.448, S.241, S.46, S.657
S. incanum	S.859
S. aethiopicum	S.1267, S.225
S. gilo	S.1236, S.395
S. integrifolium	S.344, S.240
S. olivare	S.279, S.156
S. indicum	S.871
S. anomalum	S.964
S. hispidum	S.17
S. marginatum	S.128
S. pyracanthum	S.145
S. sodomaeum	S.210, S.101
S. tomentosum	S.1155
S. xanthocarpum	S.1417

Table 48.3. Species with one or more arcs after immunodiffusion with absorption, using antiserum to *S. melongena* absorbed with an extract of *S. sodomaeum*.

Species	Accession number(s)
S. melongena	S.657, S.448, S.46
S. incanum	S.859
S. aethiopicum	S.1267
S. gilo	S.395, S.1236
S. integrifolium	S.344, S.240
S. olivare	S.279, S.156
S. indicum	S.871
S. aculeatissimum	S.1307
S. atropurpureum	S.668
S. macrocarpon	S.133, S.1269, S.1319
S. mammosum	S.587

The intensive PAGE study of the *S. melongena* complex was extended to include species from the same and other subseries, series and even other sections of the genus *Solanum* (Table 48.4). Widely different protein band patterns were obtained, of which none was found to be similar to the patterns obtained for members of the *S. melongena* complex (Plate 48.2). The lack of similarity of the patterns of other members of section *Melongena* either to each other or to *S. melongena* itself contrasts strongly with the high degree of homogeneity within the eggplant complex.

The PAGE results suggest that the placing of *S. marginatum* and *S. panduraeforme* within section *Melongena* should be re-examined.

The members of the series *Aethiopica*, which constitute a morphologically homogeneous group, produced electrophoretic patterns which have a similar series of very deeply staining, slow-moving protein bands, (Plate 48.3). The slowest band was common to all these individuals. Patterns from *S. scalare*, *S. anomalum* and *S. indicum* all included slow-moving protein bands similar in every respect to those characteristic of members of series *Aethiopica*. One accession of *S. tomentosum* had similar, though weaker bands. The close affinity of *S. scalare*, *S. indicum* and *S. anomalum* with members of the series *Aethiopica* is clearly indicated by these results. No other taxa produced similar patterns. These results

Table 48.4. Species examined by chemotaxonomic methods and/or hybridization experiments

		Code
Subgenus *Solanum*		
Section *Solanum*	*S. nigrum* L.	nig
Subgenus *Leptostemonum* (Dun.) Bitt.		
Section *Acanthophora* Dun.	*S. aculeatissimum* Jacq.	acu
	S. atropurpureum Schrank	atr
	S. ciliatum Lam.	cil
	S. khasianum C. B. Clarke*	kha
	S. mammosum L.	mam
Section *Cryptocarpum* Dun.	*S. campanulatum* R.Br.	cam
Section *Melongena* Dun.		
Series *Incaniformia* Bitt.		
Subseries *Campylacantha* Bitt.	*S. panduraeforme* E. Meyer	pan
Subseries *Euincana* Bitt.	*S. melongena* L.	mel
	S. incanum L.	inc
	S. marginatum L. fil.	mar
Series *Macrocarpon* Dun.	*S. macrocarpon* L.	mac
Series *Sodomela* (Lowe) Bitt.	*S. sodomaeum* L.	sod
	S. xanthocarpum Schrad. et Wendl.	xan
Series ?	*S. variabile* Mart.	var
Section *Oliganthes* (Dun.) Bitt.		
Series ?	*S. cinereum* R.Br.	cin
Series *Aethiopica* Bitt.	*S. aethiopicum* L.	aeth
	(= *S. zuccagnianum* Dun.)	
	S. gilo Raddi	gil
	S. integrifolium Poir.	int
	S. olivare Pail. et Bois	oli
	S. sp. "Accra" (S.1482)	acc
Series *Afroindica* Bitt.	*S. indicum* (L.p.pte) Nees	ind
	(= *S. scalare* C.H. Wr.)	sca
Series *Austroafricana* Bitt.	*S. tomentosum* L.	tom
Series *Pyracanthum* Bitt.	*S. pyracanthum* Jacq.	pyr
Section *Torva* Nees		
Series ?	*S. hispidum* Pers.	his
Series *Anomalum* Bitt.	*S. anomalum* Thonn.	ano
Series *Eutorvum* Bitt.	*S. torvum* Sw.	tor
Other genera		
	Capsicum annuum	Cap
	Lycopersicon esculentum	Lyc

*This is var. *chatterjeeanum* (correct name *S. viarum* Dun., see p. 231 and Solanaceae Newsletter, 5: 5, 1978).

lead one to consider the transfer of *S. anomalum* from section *Torva* and *S. scalare* and *S. indicum* from series *Afroindica*, into series *Aethiopica*.

Only a few species were compared with the *S. melongena* complex by chromatography, namely *S. marginatum* from the same section, series and subseries as *S. melongena*, and three species from section *Oliganthes* (Fig. 48.4). The inclusion of *S. indicum* within the main cluster of *S. melongena* allies is anomalous, and needs further investigation. *S. marginatum* was at the edge of the cluster. *S. aethiopicum* and *S. sp.* "Accra" showed little relationship to the main cluster or even to each other, although they are similar morphologically. The chromatographic methods used here provided little information on the relationships of the *S. melongena* complex to other species of *Solanum*.

CONCLUSIONS

The taxonomic homogeneity of the eggplant complex, and the separateness of this group of taxa from all of the other *Solanum* species tested has been demonstrated clearly by the present PAGE investigations, and also, though less definitely, by the chromatographic studies. The internal homogeneity of this group has been confirmed by a number of crossability studies (Pearce, 1975), which are considered in detail elsewhere but which may be summarized as follows: the production of fertile F_1 hybrids, with abundant stainable pollen and normal F_2 generation plants has been obtained *only* from crosses between taxa belonging to the eggplant complex, and not from crosses between members of this complex and other *Solanum* species.

Although some discrimination between the cultivated varieties and other members of the complex was achieved with the technique of polyacrylamide gel electrophoresis, much more material, representing a range of primitive cultivars and putative wild varieties from all of tropical Asia needs to be examined. The traditional confusion as to the taxonomic limits of the various taxa belonging to the complex also needs to be clarified. Only then can an attempt be made to answer the dependent questions as to the rank of the "wild" taxa, and the relationship of these to the cultivated eggplant, both in phylogenetic terms and with regard to the ancestry of the cultivated species.

Evidence of the relationships of the members of the eggplant complex with other species in the genus *Solanum* is provided not only by the serological studies reported here, but also by the results of the extensive crossability studies of Pearce (1975) and others. Pearce's results can be summarized as follows. Members of the eggplant complex such as *S. incanum* and *S. melongena* are easily crossed and produce healthy F_1 progeny with 60–100% stainable pollen. Some other species (*viz. S. indicum, S. torvum, S. anomalum, S. aethiopicum, S. sp.* "Accra", *S. gilo., S integrifolium, S. olivare,* and *S. tomentosum*), when crossed with the eggplant complex taxa, gave healthy F_1 progeny which had a low proportion of stainable pollen (10–30%). Yet other species (*S. sodomaeum, S. marginatum, S. scalare, S. macrocarpon, S. cinereum,* and *S. integrifolium* again) were capable of producing F_1 hybrids with members of the eggplant complex, but the hybrids showed abnormalities, or failed to survive, or had very little stainable pollen. Crosses with some other species such as *S. khasianum, S. campanulatum, S. aculeatissimum, S. xanthocarpum* and *S. pyracanthum* produced seed which did not germinate. Similar results have been reported in the literature, *S. indicum* and members of the series *Aethiopica* having the closest relationship to members of the eggplant complex, and *S. xanthocarpum* and *S. macrocarpon* showing less relationship.

The results obtained from the crossability studies agree well with those obtained from the serological studies, a similar range of taxa with close affinities with the *S. melongena* group being obtained in each case. The present experimental investigations have indicated that *S. indicum* (section *Oliganthes*, series *Afroindica*), *S. anomalum* (section *Torva*, series *Anomalum*), *S. tomentosum* (Section *Oliganthes*, series *Austroafricana*), and *S. aethiopicum*, *S. gilo*, *S. integrifolium*, *S. olivare*, and *S.* sp. "Accra" (section *Oliganthes*, series *Aethiopica*) are the taxa most closely related to *S. melongena*. Slightly less closely related are *S. macrocarpon* (section *Andromonoecum*, series *Macrocarpon*), *S. sodomaeum*, *S. xanthocarpum* (section *Andromonoecum*, series *Sodomela*), *S. marginatum* (section *Andromonoecum*, series *Incaniformia*) and *S. cinereum* (section *Oliganthes*). Three different sections and seven different series are represented by these taxa and it is interesting to note that some species placed by Bitter in sections *Oliganthes* and *Torva* appear to be more closely related to the *S. melongena* group than some species which he placed in the same section (*S. sodomaeum*, *S. macrocarpon* and *S. xanthocarpum*) and even the same subseries (*S. marginatum*).

On the basis of the evidence from serology and crossability tests it seems necessary to reconsider the taxonomic position of *S. melongena*.

ACKNOWLEDGEMENTS

The authors are particularly grateful to Miss Sarah Marsh for growing the plants, to Mrs Marion Hood for technical assistance, and to Miss D. A. Cadbury whose financial support made this project possible.

REFERENCES

BHADURI, P. N., 1951. Inter-relationship of non-tuberiferous species of *Solanum* with some consideration on the origin of brinjal (*S. melongena* L.). *Indian Journal of Genetics, 11:* 75–82.

BITTER, G., 1913. Solana Africana, Part I. *Botanische Jahrbücher für Systematik, Pflanzengeschichte und Pflanzengeographie, 49:* 560–569.

BITTER, G., 1917. Solana Africana, Part II. *Botanische Jahrbücher für Systematik, Pflanzengeschichte und Pflanzengeographie, 54:* 416–506.

BITTER, G., 1921. Solana Africana, Part III. *Botanische Jahrbücher für Systematik, Pflanzengeschichte und Pflanzengeographie, 57:* 248–286.

BITTER, G. 1923. Solana Africana, Part IV. *Repertorium Novarum Specierum Regni Vegetabilis, 16:* 1–320.

CAPINPIN, J. M., LUNDE, M. & PANCHO, J. V., 1963. Cytogenetics of interspecific hybrid between *Solanum melongena* Linn. and *S. cumingii* Dunal. *Philippine Journal of Science, 92:* 169–178.

CROWLE, A. J., 1961. *Immunodiffusion.* New York: Academic Press.

D'ARCY, W. G., 1972. Solanaceae studies II: Typification of subdivisions of *Solanum*. *Annals of Missouri Botanical Garden, 59:* 262–278.

KRISHNAPPA, D. G. & CHENNAVEERAIAH, M. S., 1964. Breeding behaviour in non-tuber bearing *Solanum* species. *Journal of the Karnatak University of Science, 9–10:* 87–96.

LESTER, R. N., 1969. An apparently asystematic reaction of Vicilin from *Pisum* with an antiserum to *Lotus* seed proteins. *Archives of Biochemistry and Biophysics, 133:* 305–312.

PEARCE, K. G., 1975. *Solanum melongena* L. and Related Species. Ph.D. Thesis, University of Birmingham, U.K.

RAJASEKARAN, S., 1969a. Cytogenetic studies on the inter-relationship of some common *Solanum* species occurring in South India. *Annamalai University Agricultural Research Annals, 1:* 49–61.

RAJASEKARAN, S., 1969b. Cytology of the hybrid *Solanum indicum* L. × *S. melongena* var. *insanum* Prain *Current Science, 39:* 22.

RAJASEKARAN, S., 1970a. Cytogenetic studies of the F₁ hybrid *Solanum indicum* L. × *S. melongena* L. and its amphidiploid. *Euphytica, 19:* 217–224.

RAJASEKARAN, S., 1970b. Cytogenetic studies on the F₁ hybrid of *Solanum macrocarpon* L. (*S. melongena* var. *bulsarensis* Argikar) × *S. melongena* L. *Annamalai University Agricultural Research Annals, 2:* 21–29.

RAJASEKARAN, S., 1971a. A cytomorphological study of the F₁ hybrid, *Solanum xanthocarpum* Schrad. and Wendl. × *S. melongena* var. *insanum* Prain. *Madras Agricultural Journal 58:* 308–309.

RAJASEKARAN, S., 1971b. Cytological studies on F₁ hybrid (*Solanum xanthocarpum* Schrad. and Wendl. × *S. melongena* L.) and its amphidiploid. *Caryologia, 23:* 261–267.

RAJASEKARAN, S. & SIVASUBRAMANIAN, V., 1971c. Cytology of the F₁ hybrid of *Solanum zuccagnianum* Dun. × *S. melongena* L. *Theoretical and Applied Genetics, 41:* 85–86.

RAO, N. N., 1969. *Cytogenetic Investigations on Certain Non-tuberiferous Species in the Genus* Solanum *L.* Ph.D. Thesis, University of Madras.

SEITHE, A., 1962. Die Haararten der Gattung *Solanum* L. und ihre taxonomische Verwertung. *Botanische Jahrbücher für Systematik, Pflanzengeschichte und Pflanzengeographie, 81:* 261–336.

EXPLANATION OF PLATES

PLATE 48.1

PAGE patterns obtained for *S. melongena* (cultivated, primitive and wild varieties) and *S. incanum.*

PLATE 48.2

PAGE patterns obtained for species from a wide taxonomic range. (For code of abbreviations see Table 48.4.)

PLATE 48.3

PAGE patterns obtained for members of series *Aethiopica* and *Afroindica* and *S. anomalum.* (For code of abbreviations see Table 48.4.)

Plate 48.1

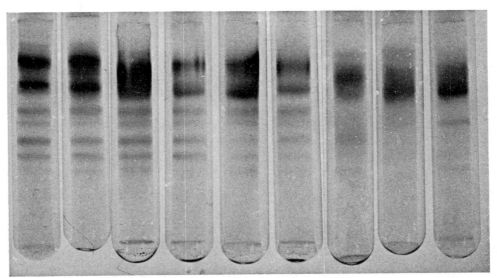

S.448   S.1243   S.1239   S.1384   S.1360   S.1490   S.1398   S.1270   S.1301
modern   primitive cultivars   wild   S. incanum

Plate 48.2

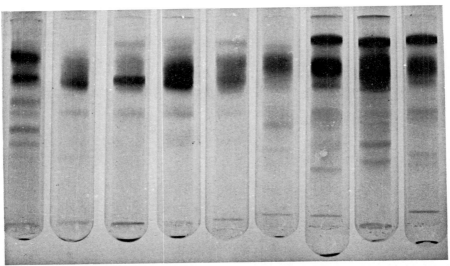

S.448   S.1301   S.1320   S.128   S.101   S.1417   S.964   S.1267   S.1380
mel   inc   tom   mar   sod   xan   ano   aeth   ind

(*Facing p. 628*)

Plate 48.3

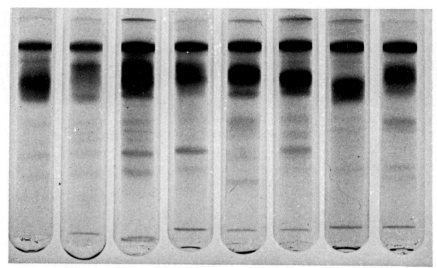

S.1306   S.964   S.1267   S.1482   S.240   S.225   S.1335   S.1380
sca   ano   aeth   acc   gil   aeth   ind   ind

49. *Solanum melongena* and its ancestral forms

REAYAT KHAN

Department of Botany, Aligarh Muslim University, Aligarh, India

Solanum melongena, the brinjal, is widely grown in India as a vegetable with nutritive and medicinal value. The morphological characters are distinctive, yet so variable as to cause taxonomic confusion, which is reviewed here. A few hybridization experiments have been made with *S. incanum*, *S. xanthocarpum*, *S. zuccagnianum*, *S. indicum*, *S. integrifolium*, *S. hispidum*, *S. anomalum*, *S. sodomaeum*, *S. macrocarpon* and *S. gilo*, with some success, indicating that *S. melongena* has some genetic relationship with all ten of these species, but further studies are needed.

A survey of literature, both modern and ancient indicates that *Solanum melongena* originated in Asia, probably in the Indo-Burma region, but possibly in more than one centre. Ancient Sanskrit texts contain several different names for the brinjal.

CONTENTS

INTRODUCTION

The problem of the phylogenetic affinities of *Solanum melongena*, the determination of its ancestral forms, and the place of origin of the various cultivated varieties, have engaged the attention of biosystematists and students of the origin of cultivated plants for a long time. Definite and precise answers to the various questions are, however, not yet available. The present paper is an attempt to assess the results of the efforts that are being made in this direction, and, to point out some of the difficulties that have to be faced. It may be made clear at the very outset that, as far as the biosystematic aspects are concerned, only a beginning has been made and not much work has been done so far in a planned manner, nor on as extensive a scale as is considered desirable. The problem will have to be tackled from various points of view before even tentative conclusions can be arrived at. One fundamental obstacle is the difficulty of finding among extant organisms the ancestors of long established species. Although it is always a very fascinating exercise to attempt to determine the evolutionary origin and development of organisms, one has to be very cautious in drawing definite and final conclusions.

SOLANUM MELONGENA, ITS IMPORTANCE

S. melongena, the eggplant, also known as aubergine, brinjal, or Guinea Squash, is widely cultivated in India and many other countries. Its unripe fruit is a common and popular vegetable. It grows throughout the year in India and is available in all seasons. It has a large number of cultivated varieties adapted to the conditions in different geographical regions. The fruit varies a great deal in colour, shape and size. It is cooked in a variety of ways. It may be roasted, fried, stuffed, cooked as curry, pickled, or prepared in some other manner. From the point of view of nutritive value it is comparable with many other common vegetables. Analysis has shown the fresh weight composition of the fruit to be:- moisture 92.7%, protein 1.4%, fat 0.3%, minerals 0.3%, fibre 1.3%, other carbohydrates 4.0%. The mineral constituents present are (mg/100 gm edible matter): calcium 18, magnesium 16, phosphorus 47, iron 0.9, sodium 3, potassium 200, copper 0.17, sulphur 44, chlorine 52; small quantities of manganese and iodine are also reported to be present. The vitamins present are: vitamin A 124 I.U., thiamine 0.04 mg, riboflavin 0.11 mg, nicotinic acid 0.9 mg, vitamin C 12 mg, and choline 52 mg per 100 gm of edible matter (Aykroyd, 1966; Chadha, 1972).

The eggplant is widely used in medicine. According to the ayurvedic system, the white varieties are said to be good for patients suffering from diabetes. Roots of eggplant are credited with antiasthmatic properties. In Guiana their juice is employed to cure otitis and toothache. Leaves are said to possess narcotic properties and are used in cholera, bronchitis, dysuria and asthma. Extracts of the plant inhibit several types of bacteria; the pulp of the fruit is more effective than the juice. The eggplant is reported to promote intrahepatic metabolism of cholesterol. Both leaf and fruit, fresh or dry, produce a marked drop in blood cholesterol level. The decholesterolizing effect is attributed to the presence of magnesium and potassium salts in the tissues of the plant. Experimental results, however, have not been confirmed by clinical trials. All this suggests the desirability of further investigations of the medicinal properties of the plant (Chadha, 1972).

SOME DISTINCTIVE FEATURES OF S. MELONGENA

S. melongena belongs to a non-tuberiferous group of species of Solanum. It is an erect or prostrate, branched herb or subshrub, about 1 m tall, woolly or scurfy and spiny, but the spines are inconspicuous in some cultivated varieties. The leaves are large, ovate or oblong-ovate, shallowly sinuate-lobed, becoming nearly glabrous above but remaining densely tomentose beneath. The flowers are large, mostly in clusters. The calyx is woolly and often spiny, persistent and accrescent. The spreading, lobed, purplish corolla is 2.5 cm or more in diameter. Heterostyly is a common feature (or rather andromonoecy by brachystyly—Ed.). The fruit is a large berry, showing great variety in size, shape and colour. It may be oblong, obovoid, ellipsoid, egg-shaped, more or less spherical, cylindrical, elongate, etc. The colour may be almost black, various shades of purple from dark to very light, blue, green, yellowish, striped variously, or white, ripening to brown or orange.

TAXONOMIC CONSIDERATIONS

In order to be able to discuss the affinities and origin of a taxon, it is essential that its taxonomic status, its delimitation from other taxa, and its classification into subcategories should be clearly understood. Unfortunately the limits of the species *S. melongena* are not well-defined, and, its classification into subspecies or varieties is in a state of confusion.

Some authorities include only the cultivated, edible forms in the species and hold the opinion that wild forms of eggplant are not known. Others include, within the species *S. melongena*, taxa that have been regarded as separate species. *S. esculentum* Dunal is generally regarded as synonymous with *S. melongena* Linn. either at the level of species or variety. While some taxonomists think that *S. melongena* has not been found wild, others feel that *S. insanum* Roxb. and *S. incanum* Linn., which are wild taxa and considered to be distinct species, are really varieties of *S. melongena* (see De Candolle, 1886; Prain, 1903).

Filov (1940) classified the various forms of *S. melongena*, both cultivated and wild, on an agro-ecological basis. He distinguished five subspecies: (i) *S. melongena* Linn. subsp. *agrestis* Fil.; (ii) *S. m.* subsp. *occidentale* Gaz.; (iii) *S. m.* subsp. *orientali-asiaticus* Fil.; (iv) *S. m.* subsp. *palestinicum* Fil. and (v) *S. m.* subsp. *arabico-italicum* Fil. (syn. *S. ovigerum* Dun.). The first subspecies, namely, *agrestis*, includes wild plants with "extremely bitter and inedible fruits" which, according to Filov, are found only in India. Subspecies *occidentale*, according to him, arose in western Asia Minor under conditions of adequate humidity and high temperature. Subspecies *orientali-asiaticus* originated in Japan and to some extent in China according to him. Subspecies *palestinicum* is a hydrophytic type and adapted to withstand high temperature. Subspecies *arabico-italicum* Fil. (syn. *S. ovigerum* Dunal) has bitter fruits and is of ornamental value only (see also Bhaduri, 1951).

Bailey (1947), in Volume I of his *The Standard Cyclopedia of Horticulture*, recognizes three varieties of equal rank. He writes: "There are three main types of eggplants, as follows: The commoner garden varieties, *Solanum Melongena* var. *esculentum*, Bailey . . .; the long fruited or "serpent" varieties, *S. Melongena* var. *serpentinum*, Bailey; the Early Dwarf Purple type var. *depressum*, Bailey. . .". In Volume III of the *Cyclopedia*, however, he recognizes *S. m.* var. *esculentum* Nees (*S. esculentum* and *S. ovigerum* Dun.) which is cultivated for its fruit and distinguishes two "well-marked subvarieties", *viz.*, var. *serpentinum* Bailey, the Snake Eggplant in which fruits are "greatly elongated and curled at the end" and, var. *depressum* Bailey, the Dwarf Purple Eggplant which is low and diffuse, many of the branches resting on the ground.

Bhaduri (1951) mentions collection of two varieties from Orissa and Travancore. The former was provisionally named "potangi" after the name of the place of collection. Dr D. B. Deb (pers. comm.) informs me that this variety does not exist. Agrikar (1952) erected a new variety *"bulsarensis"* which later proved to be nothing but *Solanum macrocarpon* Linn.

According to "The Wealth of India" (Chadha, 1972) "There are four main botanical varieties: (i) *S. m.* var. *incanum* (Linn.) Kuntze (syn. *S. incanum* Linn., *S. coagulans* Forsk); (ii) *S. m.* var. *melongena* (syn. *S. melongena* var. *esculenta* Nees); (iii) *S. m.* var. *depressum* Bailey; and (iv) *S. m.* var. *serpentinum* (Desf.) Bailey (syn. *S. serpentinum* Desf.)".

At least eight names have been considered as varieties or synonyms of *S. melongena* by one authority or another: *S. esculentum* Dun., *S. insanum* Roxb., *S. incanum* Linn., *S. ovigerum* Dun., *S. serpentinum* Desf., *S. macrocarpon* Linn., *S. coagulans* Forsk., and *S. longum* Roxb.

The brief account given above reveals the confusion prevailing with reference to the delimitation and subclassification of the species *S. melongena*. This confusion will have to be cleared up before the problem of the origin and evolution of this species can be solved.

AFFINITIES AND ORIGIN OF *S. MELONGENA*

When two taxa are hybridized, the results may conform to one of the following possibilities: (i) fertile hybrids may be produced, (ii) the hybrids obtained may be only partially fertile, (iii) the hybrids may be sterile, and (iv) no hybrid may be obtained at all. The first alternative, that is, production of fully fertile hybrids, is undoubtedly an indication of the closest genetic relationship. But even when the hybrids are partially or completely sterile, the fact that hybrids can be produced, does indicate some degree of genetic relationship. If no hybrid is obtained at all, the two taxa may or may not have any genetic affinities: evidence from other sources such as cytology, comparative morphology, geographical distribution, etc., will have to be considered to come to a decision.

Hybridization experiments between *S. melongena* and other species of *Solanum* have been performed only in a few cases. The results obtained so far suggest that hybridization will have to be carried out on an extensive scale before we can hope to unravel the mystery of the origin of *S. melongena*.

Bhaduri (1951) has reported that *S. melongena*, as female parent, produced fertile hybrids with two of its own varieties, viz., *S. m.* var. *insanum* and *S. m.* var. *potangi*; and, as male parent, with *S. incanum*. According to him, it also produced partially fertile hybrids as male parent with *S. xanthocarpum*. Later, Rajasekaran reported more or less similar results. He also reported sterile hybrids from the crosses between *S. macrocarpon* and *S. melongena*, between *S. indicum* and *S. melongena* and between *S. zuccagnianum* and *S. melongena*. (Rajasekaran, 1969, 1970a, b, Rajasekaran & Sivasubramanian, 1971).

Dr E. Pochard (pers. comm.) has succeeded in obtaining hybrids from crosses of four different species of *Solanum* with *S. melongena*, either by embryo culture or by selfing and back-crossing with *S. melongena*. The four species are *S. integrifolium*, *S. macrocarpon*, *S. anomalum* and *S. sodomaeum*. Dr. M. O. Omidiji (pers. comm.) has hybridized *S. melongena* with *S. macrocarpon*, *S. incanum* and *S. gilo*.

In the Department of Botany, Aligarh Muslim University, experiments on interspecific hybridization have been conducted involving 15 cultivated varieties of *S. melongena* and six other species of *Solanum*, namely, *S. xanthocarpum*, *S. sysimbrifolium*, *S. indicum*, *S. integrifolium*, *S. hispidum* and *S. incanum*.

S. indicum has produced hybrids as the male parent with three varieties of *S. melongena*, and both as male and as female parent with one variety. *S. integrifolium* succeeded in producing hybrids as the female parent with five varieties of *S. melongena*, and, as the male parent with two varieties. *S. hispidum* has been successful as male parent in producing hybrids with nine varieties of *S. melongena*. The greatest success, so far, has been obtained with *S. incanum*: as female parent, it succeeded with fourteen varieties of *S. melongena* and with one variety as both male and female parent (Khan & Rao, 1976).

Most of the hybrids obtained were fertile although in a few cases the hybrid produced seedless fruits. Sometimes the first fruits formed were seedless but those developing later set seeds. F_3 progeny has been obtained from interspecific crosses between various varieties of *S. melongena* on the one hand, and *S. indicum* and *S. incanum* on the other. The hybrids obtained from crosses of *S. melongena* with *S. integrifolium* and with *S. hispidum* are at present in the F_1 stage. Most of them exhibit vigorous growth. Fruits have developed and most of them have seeds. Whether the seeds will germinate or not is yet to be ascertained. Data obtained from tests of pollen fertility and chromosome configuration at metaphase I of meiosis are significant. In hybrids from crosses with *S. integrifolium*, the pollen fertility varies from 18% to 34%. The chromosome configuration at MI in seven out of eight crosses is 11 bivalents and 2 univalents. The meiosis is fairly regular. It is, therefore, expected that it will not be difficult to obtain the F_2 generation. The F_1 hybrids obtained from crosses with *S. hispidum* are also quite vigorous although the pollen fertility is very low, i.e., 4% or less. Meiosis also is highly irregular. The number of bivalents is six or less; univalents range in number from 12 to 24.

Thus it is seen that, in interspecific crosses, *S. melongena* has been reported to produce hybrids, which may be sterile, partially fertile or fully fertile, with several species, namely, *S. incanum*, *S. xanthocarpum*, *S. zuccagnianum*, *S. integrifolium*, *S. anomalum*, *S. sodomaeum*, *S. macrocarpon*, *S. gilo*, *S. indicum* and *S. hispidum*. It is, therefore, obvious that *S. melongena* has some degree of genetic relationship with all these ten species. Similar results may be obtained if hybridization of this taxon is extended, involving more diploid species of *Solanum*. Discussion of the phylogenetic affinities of *S. melongena* and its origin from ancestral forms will have to wait till more extensive and comprehensive biosystematic data become available.

Bhaduri (1951) has suggested that the nearest ancestors of the cultivated forms of *S. melongena* may be (i) wild varieties of *S. melongena* such as *S. m.* var. *insanum* and *S. m.* var. *potangi*; (ii) *Solanum incanum* L.,(iii) hybrids of some of the varieties under (i) and *S. incanum*. This seems to be an oversimplification of the problem, especially in view of the results obtained recently which show that *S. melongena* can produce hybrids with at least ten different species, and we believe that this number will increase with more work on interspecific hybridization.

THE PLACE OF ORIGIN

Although the genus *Solanum* is predominantly Central and South American, the eggplant is probably a native of Asia.

De Candolle, in his *Origin of Cultivated Plants* published in 1886, states that the species *S. melongena* has been known in India from a very remote epoch. He further writes, "Thunberg does not mention it in Japan though several varieties are now cultivated in that country. The Greeks and Romans did not know the species, and no botanist mentions it in Europe before the beginning of the 17th century, but its cultivation must have spread towards Africa before the Middle Ages. The Arab physician, Ebn Baithar, who wrote in the 13th century, speaks of it, and he quotes Rhasis who lived in the 9th century. . . . . A sign of antiquity in Northern Africa is the existence of a name, tabendjalts, among the Berbers or Kabyles of the province of Algiers. . ."

Hooker (1885), in his *Flora of British India*, states as follows: "De Candolle says it is a native of Asia, not America, and Sendtner fixes its origin in Arabia; all this appears uncertain.". Watt (1908) does not consider the eggplant to be native of India; he writes: "Introduced into India and now extensively cultivated.".

Vavilov (1928, 1931, 1951) mentions the occurrence of *S. melongena* L. in the Indian Centre of origin of cultivated plants which excludes North West India, Punjab and North West Frontier but includes Assam and Burma. He also mentions the occurrence of a special form of *S. melongena* with small fruits in the Chinese Centre of origin of cultivated plants. According to him the place of origin of *S. melongena* is the Indo-Burma region.

Bailey (1947), in his *The Standard Cyclopedia of Horticulture*, (Vol. I: 1101) states that "The first reports of its use as vegetable come from India" but later, in the same great work of reference, (Vol. III: 3182) he writes: "Original habitat probably S. W. Asia."

Thus most of the authorities seem to agree that Asia is the original home of *S. melongena*. Asia, however, is a vast continent. Different regions of Asia can claim the credit of being the original home of *S. melongena*. According to Filov (1940), his *S. melongena* subsp. *agrestis* is native to India, his subspecies *occidentale* arose in western Asia Minor and his subspecies *orientali-asiaticus* originated in Japan and to some extent in China. This is an interesting and attractive hypothesis because it reconciles several separate opinions. In China the cultivation of the eggplant has been known for the last 1500 years. However, since several authorities have stated that the first reports of its cultivation have come from India and since Vavilov considers the Indo-Burma region as its original home, it is considered desirable to seek references to this plant in the Sanskrit language, which was the language of ancient India.

"Namalinganusasana" popularly known as "Amara-Kosha", believed to be the oldest Sanskrit dictionary, was compiled before A.D. 700, probably during the 4th or 5th century (Banerji, 1971; Swamy, 1976). It contains five different names for the eggplant. The dictionary entitled "Shabdakalpadrumah" (Deb Bahadur, 1891) lists 33 Sanskrit names for the eggplant, which is very remarkable. Another dictionary, by Monier-Williams (1899), refers to three ancient sources in connection with one of the Sanskrit words for the eggplant, namely, "varttaka". These are (i) "Sushruta-Sanhita", (ii) "Markandeya Purana" and (iii) "Harivansa". It is estimated that while the first of these was compiled in the 6th century A.D. and the second in the 4th century A.D., "Harivansa" was written not later than 300 A.D., probably between the 3rd century B.C. and the 3rd century A.D.

The Persian and Arabic words for the eggplant are significant in this connection. Two of the 33 Sanskrit names for the eggplant are "bhantaki" and "vaatin-gan". The Persian word for eggplant is "baadangaan" or "baatangaan". The Arabic word is "baadanjaan". Steingass (1892) has suggested the derivation of Persian "baatangaan" from Sanskrit "bhantaki". The Arabic "baadanjaan" seems to have been derived from Sanskrit "vaatingan" or "bhantaki" either directly or indirectly through Persian. In Arabic there is no alphabet for the sound represented in the Sanskrit alphabet as "ga". In transliteration of foreign words into Arabic the sound "ga" is substituted by "ja". Thus the Sanskrit "bhantaki" or "vaatingan" become the Persian "baatangaan" or "baadangaan" which in Arabic is written as "baadanjaan". Yule & Burnell (1968) have also considered the words for eggplant in different languages. They conclude that the Sanskrit word was probably the original.

The extraordinarily large number of words for the eggplant in Sanskrit and the fact that some of them are not only descriptive but also highly complimentary, suggest that the eggplant was not only common, but was also quite popular in ancient days with the people who spoke Sanskrit. For example, "kantavrintaki", "kantalu", and "kantapatrika" refer to the spiny character of the plant; "nidralu" refers to the narcotic or hypnotic properties of parts of the plant (Nadkarni, 1927); "nilphala" means the blue fruit; "shakasreshta" means an excellent vegetable; "rajakushmand" means the royal pumpkin. These records, which are among the most ancient available so far, undoubtedly suggest that the eggplant was first cultivated by the Sanskrit speaking people. The occurrence of a large number of varieties distributed all over the world may suggest parallel evolution of the various types of cultivated eggplant (Bhaduri, 1951; Chadha, 1972).

If it is finally established that *Solanum melongena* originated in the Indo-Burma region, it will be of special interest from the point of view of origin of cultivated plants in general. *Solanum* is a very large genus. Asia is a vast continent. India is so large in area and exhibits such a great variety of geographical and climatic regions that it is described as a subcontinent. Yet, out of the nearly 2000 species of *Solanum*, only 37 are native to Asia. Hooker's (1885) *Flora of British India* mentions 27 species, six of which have been transferred to the genus *Lycianthes*. Only 17 of the remaining 21 are considered native to India, that is less than 1% of the total number of species of *Solanum*. In view of this, it is really interesting that *Solanum melongena* should have originated in the Indo-Burma region and not in Central and South America, the home of the genus *Solanum*.

CONCLUSION

Solanum melongena has been reported to produce hybrids, which may be sterile, partially fertile or fully fertile, with ten different species of *Solanum*. It has, therefore, some degree of genetic relationship with all these species. Similar results may be obtained if hybridization is extended to involve more species of *Solanum*. Conclusions regarding the phylogenetic affinities of the eggplant, and its origin from ancestral forms, will have to wait till more extensive and comprehensive biosystematic data become available.

Although the problem of origin of *S. melongena* is yet to be solved, most of the evidence seems to indicate that it originated in Asia. S. W. Asia including Arabia, the Indo-Burma region, Japan and China have been suggested as probable places of origin by different authors. There seems to be a strong probability that the Sanskrit speaking people were the first to cultivate it. The wide distribution of the cultivated varieties throughout the world indicates the possibility of parallel evolution.

ACKNOWLEDGEMENTS

I am grateful to Professor R. S. Tripathi, Department of Sanskrit, Aligarh Muslim University, for suggesting references to Sanskrit literature, and, to Dr G. R. Rao, Reader in Botany, for assistance. I also thank the authorities of the Department of Agriculture of the United States of America for financing a research project under their PL 480 programme.

REFERENCES

AGRIKAR, G. P., 1952. *Solanum melongena* var. *bulsarensis* var. novo Agrikar. *Current Science, 15:* 226.

AYKROYD, W. R., 1966. In C. Gopalan & S. C. Balasubramanian, *The Nutritive Value of Indian Foods and the Planning of Satisfactory Diets,* 6th ed. I.C.M.R. Special Report Series, New Delhi.

BAILEY, L. H., 1947. *The Standard Cyclopedia of Horticulture.* New York.

BANERJI, S. C., 1971. *A Companion to Sanskrit Literature.* Delhi.

BHADURI, P. N., 1951. Inter-relationship of non-tuberiferous species of *Solanum* with some consideration on the origin of brinjal (*S. melongena* L.). *Indian Journal of Genetics and Plant Breeding, 11:* 75–82.

CHADHA, Y. R. (Ed.), 1972. *The Wealth of India, Raw Materials, IX.* New Delhi.

DEB BAHADUR, R., 1891 (Skakabd, 1813). *Shabdakalpadrumah.* Calcutta.

DE CANDOLLE, A., 1886. *Origin of Cultivated Plants.* 2nd ed., reprinted 1959. New York.

FILOV, A. I., 1940. An agro-ecological classification of eggplants and a study of their characters. *Compte Rendu de l'Académie des Sciences de l'U.R.S.S.* (*N.S.*), *26:* 815–818.

HOOKER, J. D., 1885. *The Flora of British India.* London.

KHAN, R. & RAO, G. R., 1976. Some results of interspecific pollination with *Solanum melongena*—a popular vegetable crop. *Indo-Soviet Symposium on Embryology of Crop Plants* (*Abstracts*). New Delhi.

MONIER-WILLIAMS, M., 1899. *A Sanskrit-English Dictionary.* 1951 ed. Oxford.

NADKARNI, K. M., 1927. *Indian Materia Medica.* Bombay.

PRAIN, D., 1903. *Bengal Plants.* Calcutta.

RAJASEKARAN, S., 1969. A note on the crossability of *Solanum xanthocarpum* Schrad. and Wendl. with the eggplant (*S. melongena* L.) *Annamalai University Agricultural Magazine, 10, 11:* 70–71.

RAJASEKARAN, S., 1970a. Cytogenetic studies on the F_1 hybrid of *Solanum macrocarpon* L. (*S. melongena* var. *bulsarensis* Agrikar) × *S. melongena* L. *AUARA, 2:* 21–28.

RAJASEKARAN, S., 1970b. Cytogenetic studies of the F_1 hybrid *Solanum indicum* L. × *S. melongena* L. and its amphidiploid. *Euphytica, 19:* 217–224.

RAJASEKARAN, S. & SIVASUBRAMANIAN, V., 1971. Cytology of the F_1 hybrid of *Solanum zuccagnianum* Dun. × *S. melongena* L. *Theoretical and Applied Genetics, 41:* 85–86.

STEINGASS, F., 1892. *A Comprehensive Persian-English Dictionary.* London.

SWAMY, B. G. L., 1976. Sources for a history of plant sciences in India. II. The Rgvedic soma plant. *Indian Journal of the History of Science, 11:* 11–32.

VAVILOV, N. I., 1928. Geographical centres of our cultivated plants. *Proceedings of the 5th International Congress on Genetics, N.Y.:* 342–369.

VAVILOV, N. I., 1931. Mexico and Central America as the principal centre of origin of cultivated plants of the New World. *Bulletin of Applied Botany, Genetics and Plant Breeding, 26.*

VAVILOV, N. I., 1951. The origin, variation, immunity and breeding of cultivated plants. *Chronica Botanica, 13:* 1–364.

WATT, G. 1908. *The Commercial Products of India.* Reprinted 1966. New Delhi.

YULE, H. & BURNELL, A. C., 1968. *Hobson-Jobson,* 2nd ed. Delhi.

50. Evolution and polyploidy in potato species

J. G. HAWKES

Department of Plant Biology, University of Birmingham, U.K.

A discussion of the nature of polyploidy in potatoes centres on the mode of origin of polyploid series in different groups of species. The main evolutionary development in the tuber-bearing solanums has taken place at the diploid level and polyploids do not form a very high proportion of the total number of species. Furthermore, diploids seem to be just as successful as polyploids amongst the wild species.

Amongst cultivated species, whose polyploid relationships are better known than the wild ones, the tetraploid *S. tuberosum* is more successful in terms of yield and cultivation area than the diploids, triploids and pentaploids.

CONTENTS

INTRODUCTION

The tuber-bearing species of the genus *Solanum* form a relatively small group in comparison with the genus as a whole. Conservative estimates indicate that there are about 160 wild and seven cultivated species in this group, whilst figures of up to 2000 species have been calculated for the genus. Potatoes belong to section *Petota* (formerly Tuberarium), subsection *Potatoe* (formerly Hyperbasarthrum) (D'Arcy, 1972). They are further subdivided into some eighteen series.

Most species in *Potatoe* can be crossed together with relative ease, forming highly fertile offspring. Many natural hybrids have also been reported (see Hawkes & Hjerting, 1969). This contrasts markedly with the difficulties encountered with other sections of *Solanum* (see Anderson's and Roe's papers, this Volume, pp. 549 & 563).

This ease of crossing has been taken by some workers to indicate that potatoes are a relatively young group that has evolved rapidly in a short period of time. It has also been suggested that the true tuber-bearing species (excluding the related but non-tuberiferous series *Juglandifolia* and *Etuberosa*) arose in Mexico, where genome differentiation is clearly exhibited (see Ramanna & Hermsen, this Volume,

p. 647). From Mexico a small and probably uniform group of species migrated to South America in Eocene times before the bridge between North and South America was severed. Here it evolved rapidly into a large range of species and series, though with very little genomic evolution. After the north-south land bridge was restored in Pliocene times it is postulated that two waves of return migration to Mexico resulted in the formation of groups of tetraploid and hexaploid species respectively when they hybridized with some of the original species in that area.

This hypothesis is not altogether proven, but as Ramanna & Hermsen have shown (this Volume, p. 647), and as was pointed out earlier by Hawkes (1958), there is a certain amount of cytogenetical evidence to support it, thus enabling a system of genome formulae to be postulated.

Returning now to a general discussion of polyploidy in potatoes it must be stressed that although polyploidy is frequent in this group its primary evolutionary development has taken place at the diploid level. The base number for the whole group, as for most of the rest of the genus, is $x = 12$. In the wild species, ploidy levels extend to hexaploid ($2n = 72$), whilst in the cultivated ones the highest ploidy level discovered is pentaploid ($2n = 60$). Odd-numbered polyploids (triploids and pentaploids) can be maintained by means of vegetative reproduction, and aneuploids are rare, apparently always resulting from some past history of hybridization between different ploidy levels.

For the species whose chromosome numbers have been determined, most are diploid (74%), very few are triploid (4.5%), rather more are tetraploid (11.5%), 2.5% are pentaploid and 5% are hexaploid. Some 2.5% contain a mixture of cytotypes (Hawkes, 1978). This survey does not include triploid cytotypes of diploid species, which we shall discuss below.

It has been customary to speak of potatoes as containing a polyploid series, but in fact there are several. These seem to have arisen independently in distinct taxonomic series, such as Conicibaccata ($2x, 4x, 6x$), Acaulia ($4x, 5x\ 6x$), Tuberosa (wild—$2x, 4x, 6x$; cultivated—$2x, 3x, 4x, 5x$), and Longipedicellata/Demissa ($3x, 4x, 5x, 6x$) (see Table 50.1).

WILD SPECIES POLYPLOIDS

Triploids

These occur both as auto and allotriploids, the latter derived from spontaneous species hybridization. Since they contain three sets of chromosomes, multivalents and univalents are frequent at meiosis, and thus they are sterile or almost so. Vegetative reproduction enables triploids to persist for very long time periods, however.

What appear to be autotriploid cytotypes of diploid species have been discovered in some ten species and subspecies (*S. bulbocastanum*, *S. cardiophyllum* subsp. *cardiophyllum* and subsp. *ehrenbergii*, *S. jamesii*, *S. commersonii* subsp. *commersonii* and subsp. *malmeanum*, *S. maglia*, *S. medians*, *S. microdontum* and *S. venturii*). It seems very probable that more will be discovered in due course if tuber, rather than seed collections are made, since the triploids do not normally set seed and

Table 50.1 Classification and chromosome number of the more important wild and cultivated potato species

Series	Species arranged according to chromosome number ($x = 12$)				
	2x	3x	4x	5x	6x
I. Juglandifolia	*S. juglandifolium* *S. lycopersicoides*				
II. Etuberosa	*S. brevidens* *S. etuberosum*				
III. Morelliformia	*S. morelliforme*				
IV. Bulbocastana	*S. bulbocastanum* *S. clarum*	*S. bulbocastanum*			
V. Pinnatisecta	*S. cardiophyllum* *S. jamesii* *S. pinnatisectum* *S. trifidum*	*S. cardiophyllum* *S. jamesii*			
VI. Commersoniana	*S. chacoense* *S. commersonii* *S. tarijense*	*S. calvescens* *S. commersonii*			
VII. Circaeifolia	*S. capsicibaccatum*				
VIII. Conicibaccata	*S. chomatophilum* *S. violaceimar-moratum*		*S. colombianum* *S. oxycarpum*		*S. moscopanum*
IX. Piurana	*S. piurae*		*S. tuquerrense*		
X. Acaulia			*S. acaule*	(*S. acaule*)	*S. acaule*
XI. Demissa				*S. × edinense* *S. × semidemissum*	*S. brachycarpum* *S. demissum*
XII. Longipedicellata		*S. × vallis-mexici*	*S. fendleri* *S. polytrichon* *S. stoloniferum*		
XIII. Polyadenia	*S. polyadenium* *S. lesteri*				
XIV. Cuneoalata	*S. infundibuliforme*				
XV. Megistacro-loba	*S. boliviense* *S. megistacrolobum* *S. raphanifolium* *S. sanctae-rosae* *S. toralapanum*	*S. × bruecheri*			
XVI. Ingaefolia	*S. rachialatum*				
XVII. Olmosiana	*S. olmosense*				
XVIIIa. Tuberosa (wild)	*S. × berthaultii* *S. bukasovii* *S. canasense* *S. gourlayi* *S. kurtzianum* *S. leptophyes* *S. maglia* *S. microdontum* *S. oplocense* *S. sparsipilum* *S. spegazzinii* *S. vernei* *S. verrucosum*	*S. maglia* *S. microdontum*	*S. gourlayi* *S. oplocense* *S. sucrense*		*S. oplocense*
XVIIIb. Tuberosa (cultivated)	*S. × ajanhuiri* *S. goniocalyx* *S. phureja* *S. stenotomum*	*S. × chaucha* *S. × juzepczukii*	*S. tuberosum* subsp. *tuberosum* *S. tuberosum* subsp. *andigena*	*S. × curtilobum*	

thus can be discovered only if tubers are collected. They often occur in mixed populations with diploid cytotypes of the same species and in the complete absence of tetraploids. It is thus assumed that they are formed from the union of normal haploid with 'unreduced' diploid gametes. Experimental studies show that diploid gametes are not uncommon in this group of *Solanum* (see for instance, Propach, 1938; von Wangenheim, 1954). Tarn (1967) synthesized the autotriploids of *S. commersonii* by means of diploid × synthetic autotetraploid crosses and showed that the synthetics were identical morphologically, cytologically and biochemically with the naturally occurring forms. Autotriploids of the other species mentioned above have not been studied in such detail but they certainly appear to have had a similar origin. No autotriploids have yet arisen experimentally from intraspecific diploid × diploid crosses, but this may well be due to the fact that no-one has looked for them to a very great extent.

Some autotriploids are extremely vigorous, as for example *S. maglia* from Chile and Argentina, where only two diploids have been discovered amongst hundreds of thousands of triploids (Hawkes & Hjerting, 1969). An even more extreme case is that of *S. calvescens* from Brazil (Minas Gerais), where all plants examined from two distinct collections were triploid (Tarn & Hawkes, unpubl.). Probably if more collections were examined one or two diploids might be discovered.

Autotriploids seem to be very vigorous and can grow at the expense of the diploids providing that conditions are appropriate for continued vegetative reproduction. Where conditions are changing rapidly or potatoes need to re-colonize frequently in appropriate habitats because of their inability to compete with climax vegetation, then triploids are eliminated, and re-colonization of diploids from seed takes place.

Naturally occurring triploid hybrids have been reported also, such as the wild species *S. × vallis-mexici* and *S. × bruecheri*. The former was synthesized by Marks (1958) from the tetraploid *S. stoloniferum* (4x) and the diploid *S. verrucosum* (2x). Marks also produced a synthetic hexaploid from his triploid hybrid and was later able to match it with a naturally occurring (but unnamed) hexaploid form. The triploid *S. × bruecheri* seems to be a hybrid of *S. acaule × S. megistacrolobum*, and this hybrid was synthesized by Hawkes & Hjerting (1969), comparing well with the naturally occurring hybrid.

Latin names were given to these triploid hybrids before their true nature was discovered, but the question should be raised as to whether this is appropriate. The present writer takes the pragmatic approach that it is useful to have a name for a naturally occurring entity, especially if it is found to be quite frequent, and to fit into a particular ecological niche. On the other hand, the ranges of cytotypes seen in such genera as *Dioscorea* and *Saccharum*, as well as the autotriploid cytotypes discussed above, hardly merit distinct names since they are almost impossible to distinguish morphologically. This pragmatic approach may not appeal to the purist but it is certainly useful and does not seem to contravene the rules of botanical or horticultural nomenclature.

Tetraploids

Tetraploids are fairly well known in potatoes because the cultivated *S. tuberosum* is a tetraploid. However, this will be dealt with later. The wild tetraploids *S.*

acaule and *S. stoloniferum* behave as allotetraploids, and there is reason to believe that all the other wild tetraploids behave similarly.

Thus, in Series Conicibaccata there are four tetraploids known: *S. oxycarpum*, *S. agrimonifolium* and *S. longiconicum* from Mesoamerica, and *S. colombianum* from South America. The origin of none of these is known, and the same must be said for *S. tuquerrense* (4x) in Series Piurana.

The whole Series Longipedicellata is tetraploid (*S. stoloniferum*, *S. fendleri*, *S. hjertingii*, *S. papita* and *S. polytrichon*), apart from the triploid *S.* × *vallis-mexici*, which has already been discussed. We have no idea as to their origin, though it is believed that they all had a common ancestry since no genetic break-down occurs in F$_2$ hybrids between them (Hawkes, 1966). It would seem possible that they are formed as hybrids between Mexican B-genome forms and A-genome forms newly arrived from South America in Pliocene times, but the actual parental species have not yet been identified (Hawkes, 1958). The work of Matsubayashi (1955) with *S. stoloniferum* (4x) × *S. chacoense* (2x) hybrids lends strength to this hypothesis.

We have very little idea as to the origin of the tetraploid *S. acaule* but it has been suggested that one of its diploid parents may have been a species in Series Megistacroloba. The parent species must have been closely related since when crosses between *S. acaule* (4x) and *S. chacoense* (2x) are made the resulting hybrids behave very much like autotriploids (Propach, 1937). One might assume that homologues pair preferentially in *S. acaule* as a tetraploid but in triploid hybrids where exact homologues are absent, then homoeologous pairing takes place, thus making them appear to be autotriploids.

The Bolivian weed tetraploid potato, *S. sucrense*, has been studied by Astley (1975) and is considered to be derived from natural crosses between a tetraploid cytotype of the wild *S. oplocense* and the cultivated tetraploid *S. tuberosum* subsp. *andigena*.

S. oplocense itself needs further study since it occurs in Argentina and Bolivia as 2x, 4x and 6x cytotypes, the latter showing occasional 5x sectors. We assume, but with really no evidence, that these are autopolyploids. Obviously more work is needed.

More study is needed also with *S. gourlayi* from Argentina which occurs as diploid and tetraploid cytotypes. Is the tetraploid an autopolyploid?

Pentaploids

Two pentaploid wild species are known, *S.* × *edinense* and *S.* × *semidemissum*. They are hybridogenic and the same remarks on their nature apply as were made on the hybridogenic triploids. *S.* × *edinense* subsp. *salamanii* occurs as a weed of high altitude potato fields and was evidently derived from *S. demissum* × *S. tuberosum* subsp. *andigena* crosses (Ugent, 1967). The typical subspecies was derived from *S. tuberosum* subsp. *tuberosum* and was formed in Europe.

S. × *semidemissum* is very similar to the hexaploid *S. demissum* and is apparently a hybrid of *S. demissum* (6x) × *S. verrucosum* (2x), the latter having contributed an unreduced gamete. Experimental proof of the origin of both these pentaploids is lacking as yet.

Pentaploid cytotypes occur occasionally in *S. acaule* (northern Peruvian forms) and *S. oplocense*. Their exact nature is unknown.

Hexaploids

The Mexican Series Demissa consists entirely of hexaploid species, apart from the pentaploid *S.* × *semidemissum*, which has already been discussed. These hexaploids comprise the well-known Phytophthora-resistant *S. demissum* as well as *S. hougasii*, *S. brachycarpum*, *S. iopetalum* and *S. guerreroense*. They are well-differentiated from each other genomically, apart from the common A-genome derived from the diploid *S. verrucosum* (Marks, 1955). It is possible that they have all been derived by hybridization of *S. verrucosum* with a range of tetraploid species after the second return-migration from South America (Hawkes, 1958) and might possess the genome formula $A_1A_4B_{(1-4)}$, where B_{1-4} are derived from a range of species, possibly including *S. oxycarpum*, *S. agrimonifolium*, *S. papita*, *S. hjertingii* and other members of Series Longipedicellata. Experimentation is now needed to check this hypothesis.

It may be asked whether any data exist on the climate of the Central American isthmus during the times postulated for migrations of potato species from North to South America. At the present day the climate is too warm and the mountains too low for the cool-temperate potato species to use them as "stepping stones". However, Bartlett & Barghoorn (1973) have reviewed the vegetational history of the area, quoting their own work and that of others to the effect that vegetation characteristic of colder climates existed in Costa Rica at least 600–800 m below its present level (Martin, 1960) and that at 12,000 B.P. in the Canal Zone of Panamá temperatures at least 2.5°C lower than those of the present day were indicated, indicating a shift of climatic zones downwards by about 500 m.

Finally, we must mention the South American hexaploid, *S. moscopanum*, in Series Conicibaccata, the hexaploid subspecies *albicans* of *S. acaule* in northern Peru and the hexaploid cytotype of *S. oplocense* in northern Argentina. We have no idea whatever as to how any of these evolved from ancestors of lower ploidy levels.

CULTIVATED SPECIES POLYPLOIDS

Triploids

Work on the nature of the cultivated triploid species, *S.* × *chaucha*, has been undertaken by Jackson (1975) who was able to verify it as a hybrid between the cultivated species *S. tuberosum* subsp. *andigena* (4x) × *S. stenotomum* (2x) through the synthesis of artificial hybrids. Very few seeds are produced from this cross, and tetraploids as well as triploids are produced, the former being derived from normal gametes of the tetraploid parent and unreduced ones of the diploid parent (see also Jackson, Rowe & Hawkes, 1976).

The other cultivated triploid species, *S.* × *juzepczukii*, has been investigated by Hawkes (1962) and later by Schmiediche (1977). Synthetic hybrids were produced between *S. acaule* (wild, 4x) and *S. stenotomum* (cultivated, 2x), which agreed morphologically, cytologically and biochemically with the naturally occurring forms. This cultivated triploid inherits the frost resistance of its wild ancestor and is thus very valuable economically in the high Andes of Peru and Bolivia.

A similar case is that of *S.* × *ajanhuiri* (Huamán, 1975) which inherits its frost resistance from the wild species *S. megistacrolobum*. This has hybridized, again with *S. stenotomum*, to provide a valuable series of frost-resistant hybrids—but this time at the diploid level only.

Tetraploids

Solanum tuberosum has been considered by earlier workers as an autotetraploid, since meiosis is irregular, with quadrivalent + trivalent frequencies ranging from 1.70 to 5.24 (reported by Hawkes, 1958). These do not differ very much from the figures obtained in autotetraploid *S. phureja* (Lamm, 1945) of 3.7 and 4.0, whilst other autotetraploids reported in the literature are rather similar. Perhaps significantly, the quadrivalent + trivalent frequencies reported for subspecies *andigena* range from 2.1 (Lamm, 1945) to 1.36 (Swaminathan & Howard, 1953). Swaminathan's work (1953) is of interest here, since in autotetraploids the quadrivalent frequencies ranged from 3.45 to 4.56, whilst in amphidiploids of species which were taxonomically distinct the quadrivalent frequencies dropped to 1.19–0. In crosses of "species" thought to be taxonomically identical the quadrivalent frequencies approached those of the autotetraploids. These data therefore give some support to the hypothesis that *S. tuberosum* may be an amphiploid hybrid of two distinct though perhaps somewhat related diploid species.

We have now established with some degree of confidence (Cribb, 1972) that *S. tuberosum* originated as a hybrid, subsequent to domestication, between the diploid cultivated *S. stenotomum* and the weed diploid *S. sparsipilum*, with subsequent chromosome doubling.

Since this event would have taken place in the Andes it was evidently the Andean subspecies, *andigena*, which was formed first. Morphological and biochemical comparisons have been made between the natural and synthetic *andigena* lines, with extremely good matching in every case. Further evidence comes from studies at the dihaploid level (see Woodcock & Howard, 1975) which show segregation of characters from the two original diploid parents, especially the calyx of *S. stenotomum* which is supressed in the tetraploids but shows up again in dihaploids. The auto-allotetraploid nature of *S. tuberosum* seems now to be reasonably certain, where partial genome evolution of the two diploid parents does not allow us to classify it as a true auto or a true allotetraploid.

Pentaploids

There is only one cultivated pentaploid potato species, *S. curtilobum*. This shows a marked degree of frost resistance and some morphological similarity to *S. juzepczukii* and *S. acaule*. Hawkes (1962) established, on the basis of synthetic hybrids, that *S. curtilobum* was derived from the cross: *S. juzepczukii* (unreduced 36-chromosome gamete) × *S. tuberosum* subsp. *andigena* (24-chromosome gamete). The hybrids thus formed showed an extremely good match, both cytologically and morphologically with the naturally occurring species.

CONCLUSIONS

To sum up, although it appears at first glance that we know a great deal about polyploidy in the tuber-bearing *Solanum* species, the fact is that we know something about the triploids and pentaploids but very little about the tetraploids and hexaploids. The cultivated species are much better known than the wild ones. This is because it was agreed with the International Potato Center (CIP) in Peru that the cultigens should take higher priority, and in consequence collaborative research between CIP and Birmingham has up to now concentrated on the cultivated species only.

We now need to return to the wild species and to arrive at some idea of the origin and nature of the tetraploids and hexaploids. It has been stated many times that apparent allopolyploids may have been formed as autopolyploids in the first place and have been selected over long time periods for fertility through genetic control which promotes bivalent pairing. This may have been so with the tetraploid and hexaploid cytotypes of *S. acaule* and *S. oplocense* and the tetraploid cytotype of *S. gourlayi*. However, it seems much less likely in Mexico with the tetraploid Series Longipedicellata and the hexaploid Series Demissa, since in both of these there is clear evidence of genome differentiation.

We have thus arrived at an exciting stage in understanding cultivated species origins and evolution, and it would seem that a concentration of effort on the wild species could throw a great deal of light on their relationships, especially within the various polyploid series that we have been describing.

How successful are the higher polyploids, in terms of evolutionary adaptation? Some tetraploids, such as *S. tuberosum*, *S. acaule* and *S. stoloniferum* are very successful, when measured by their wide distribution areas and their abundance. There are large numbers of tetraploids in Series Conicibaccata, whilst Series Longi-pedicellata is wholly tetraploid. On the other hand, the very successful Series Tuberosa and other related ones such as Commersoniana, Cuneoalata, Megistacro-loba and others are almost wholly diploid. Furthermore, although the tetraploid species *S. acaule* is so successful, the diploid wild *S. chacoense* from Argentina and other countries is as successful as the best of the wild tetraploids. Only the culti-vated *S. tuberosum* could be said to surpass them all, and here of course, factors other than polyploidy must be taken into account. Again, only *S. demissum* amongst the hexaploids is really successful when measured by the parameters given above.

So we can conclude that polyploidy as such is not the only road to success, in terms of wide distribution and large numbers of individuals. However, amongst the cultivated species the tetraploids are much more successful than diploids, triploids and pentaploids. Where sexual reproduction is at a premium, diploids possess a selective advantage; where stable habitats are available, polyploids can then com-pete advantageously.

Probably tetraploidy is the optimum ploidy level for the group. Artificial octoploids of *S. tuberosum* are unthrifty and slow growing, and hexaploids, although found in wild species, do not seem on average to be better adapted or more successful than diploids, and are not found amongst the cultivated group of species.

REFERENCES

ASTLEY, D., 1975. *Studies on the Taxonomic Relationships of* Solanum sucrense—*a Bolivian Weed Potato Species*. Ph.D. Thesis, University of Birmingham.

BARTLETT, A. S. & BARGHOORN, E. S., 1973. Phytogeographic history of the Isthmus of Panama during the past 12,000 years. In A. Graham (Ed.), *Vegetational History of Northern Latin America*, Ch. 7: 203–299. Amsterdam: Elsevier.

CRIBB, P., 1972. *Studies on the Origin of* Solanum tuberosum *L. subsp.* andigena (*Juz. et Buk.*) *Hawkes—the Cultivated Tetraploid Potato.* Ph.D. Thesis, University of Birmingham.

D'ARCY, W. G., 1972. Solanaceae Studies II: Typification of subdivisions of *Solanum. Annals of Missouri Botanical Garden, 59:* 262–278.

HAWKES, J. G., 1958. Potatoes: Taxonomy, cytology and crossability. In H. Kappert & W. Rudorf (Eds), *Handbuch der Pflanzenzüchtung, Züchtung der Knollen und Wurzelfruchtarten, 3:* 1–43. Berlin & Hamburg: Paul Parey.

HAWKES, J. G., 1962. The origin of *Solanum juzepczukii* Buk. and *S. curtilobum* Juz. et Buk. *Zeitschrift für Pflanzenzüchtung, 47:* 1–14.

HAWKES, J. G., 1966. Modern taxonomic work on the *Solanum* species of Mexico and adjacent countries. *American Potato Journal, 43:* 81–103.

HAWKES, J. G., 1978. Biosystematics of the potato. In P. M. Harris (Ed.), *The Potato Crop: The Scientific Basis for Improvement:* 15–69. London: Chapman & Hall.

HAWKES, J. G. & HJERTING, J. P., 1969. *The Potatoes of Argentina, Brazil, Paraguay and Uruguay— a Biosystematic Study:* pp. 525. Oxford: Oxford University Press.

HUAMÁN, Z., 1975. *The Origin and Nature of* Solanum ajanhuiri *Juz. et Buk.—a South American Cultivated Diploid Potato.* Ph.D. Thesis, University of Birmingham.

JACKSON, M. T., 1975. *The Evolutionary Significance of the Triploid Cultivated Potato,* Solanum × chaucha *Juz. et Buk.* Ph.D. Thesis, University of Birmingham.

JACKSON, M. T., ROWE, P. R. & HAWKES, J. G., 1976. The enigma of triploid potatoes: a reappraisal. *American Potato Journal, 53:* 395.

LAMM, R., 1945. Cytogenetic studies in *Solanum,* Sect. Tuberarium. *Hereditas, Lund, 31:* 1–128.

MARKS, G. E., 1955. Cytogenetic studies in tuberous *Solanum* species. I. Genomic differentiation in the group Demissa. *Journal of Genetics, 53:* 262–269.

MARKS, G. E., 1958. Cytogenetic studies in tuberous *Solanum* species. II. A synthesis of *Solanum × vallis-mexici* Juz. *New Phytologist, 57:* 300–310.

MARTIN, P. S., 1960. Effect of Pleistocene climate change on biotic zones near the equator. *Year Book of the American Philosophical Society, 1960:* 265–267.

MATSUBAYASHI, M., 1955. Studies on the species differentiation in the section *Tuberarium* of *Solanum.* III. Behaviour of meiotic chromosomes in F_1 hybrid between *S. longipedicellatum* and *S. schickii* (= *S. stoloniferum × S. chacoense*—Ed.) in relation to its parent species. *Science Reports of the Hyogo University of Agriculture, 2:* 25–31.

PROPACH, H., 1937. Cytogenetische Untersuchungen in der Gattung *Solanum,* Sect. *Tuberarium.* II. Triploide und tetraploide Artbastarde. *Zeitschrift für Induktive Abstammungs- u. Vererbungslehre, 73:* 143–154.

PROPACH, H., 1938. Cytogenetische Untersuchungen in der Gattung *Solanum,* Sect. *Tuberarium.* IV. Tetraploide und sesquidiploide Artbastarde. *Zeitschrift für Induktive Abstammungs- u. Vererbungslehre, 74:* 376–387.

SCHMIEDICHE, P. E., 1977. *Biosystematic Studies on the Cultivated Frost-resistant Potato Species* Solanum juzepczukii *and* Solanum curtilobum. Ph.D. Thesis, University of Birmingham.

SWAMINATHAN, M. S., 1953. Studies on the inter-relationships between taxonomic series in the section *Tuberarium,* genus *Solanum.* I. *Commersoniana* and *Tuberosa. American Potato Journal, 30:* 271–281.

SWAMINATHAN, M. S. & HOWARD, H. W., 1953. The cytology and genetics of the potato (*Solanum tuberosum*) and related species. *Bibliographia Genetica, 16:* 1–192.

TARN, T. R., 1967. *The Origin of Two Polyploid Species of* Solanum, *Sect.* Tuberarium. Ph.D. Thesis, University of Birmingham.

UGENT, D., 1967. Morphological variation in *Solanum × edinense,* a hybrid of the common potato. *Evolution, 21:* 696–712.

WANGENHEIM, K. H. FRH. VON, 1954. Zur Ursache der Kreuzungsschwerigkeiten zwishcen *Solanum tuberosum* L. und *S. acaule* Bitt. bzw. *S. stoloniferum* Schlechtd. et Bouché. *Zeitschrift für Pflanzenzüchtung, 34:* 7–48.

WOODCOCK, K. M. & HOWARD, H. W., 1975. Calyx types in *Solanum tuberosum* dihaploids, *S. stenotomum, S. sparsipilum* and their hybrids. *Potato Research, 18:* 460–465.

51. Genome relationships in tuber-bearing Solanums

M. S. RAMANNA AND J. G. Th. HERMSEN

Institute of Plant Breeding (I.v.P.), Agricultural University, Wageningen, The Netherlands

Chromosomes in the tetraploid hybrids of the diploid wild potato, *Solanum verrucosum*, and the hexaploid wild potato, *S. demissum*, show regular bivalent pairing. This also occurs in dihaploid *S. tuberosum* × *S. demissum* crosses, agreeing in both cases with a genome formula of A_1 for the diploids and A_1A_4B for the hexaploids, using the symbols proposed earlier by Hawkes. Evidently homoeologous pairing of the A_4B sets takes place. Pentaploid hybrids are also formed, due to the functioning of unreduced gametes from the diploid parents. These show high trivalent formation and lower bivalent formation, compared with the tetraploids.

Examination of pachytene chromosomes gave no evidence for preferential pairing of homoeologues.

In hybrids between the diploid Mexican species, *S. pinnatisectum* and *S. bulbocastanum*, pachytene analysis reveals evidence of structural differences resulting in loops at certain points on the chromosomes.

Some evidence for genome differentiation in South American potato species is also presented; this runs counter to general beliefs but appears to be indisputable.

CONTENTS

INTRODUCTION

The study of chromosome pairing in interspecific hybrids is a convenient way to distinguish the constituent genomes of the parental species. There is a fairly extensive literature on chromosome pairing in the interspecific hybrids of tuber-bearing Solanums (Swaminathan & Howard, 1953; Marks, 1955, 1965; Kawakami & Matsubayashi, 1957; Matsubayashi, 1962). From the available information Marks (1965) has made a few generalizations for Solanums: (1) meiosis in species hybrids is characterized by complete chromosome pairing; (2) there is no reduction of chiasmata in the hybrids as compared with parents; (3) there are no gross chromosomal structural differences, with one exception (Marks, 1968), between species. These three characteristics of hybrid meiosis apply not only to interspecific hybrids of tuber-bearing Solanums but also to intergeneric hybrids such as *Lycopersicon esculentum* × *S. pennellii* (Khush & Rick, 1963), and *Lycopersicon esculentum* × *S. lycopersicoides* (Menzel, 1962). It appears difficult to make a clear distinction between different genomes in Solanums owing to the absence of cyto-logically detectable meiotic abnormalities in species hybrids.

However, based on the available cytological data, and other evidence, Hawkes (1958) has proposed tentatively the following genome constitutions for different groups of Solanums: ". . . the genome A_1 for diploid South American species; A_2A_3 for the tetraploid South American species *S. acaule*; A_4B for the tetraploid Mexican species *S. stoloniferum* and its relatives; A_1A_4B for the hexaploid Mexican species *S. demissum* and its relatives (see also Hawkes, 1972). Furthermore, it is concluded that the South American species have evolved by "cryptic genome evolution" whereas Mexican wild potatoes, which include allotetraploid and allohexaploid groups of species, show considerable genome differences.

In the present paper some of our observations on chromosome pairing in interspecific hybrids of Mexican and South American species will be described, and the nature of genome differentiation in tuber-bearing solanums discussed.

CHROMOSOME PAIRING IN THE F_1-HYBRIDS OF *S. VERRUCOSUM* × *S. DEMISSUM* AND *S. TUBEROSUM*-DIHAPLOID × *S. DEMISSUM*

S. demissum is an allohexaploid ($2n = 6x = 72$) Mexican species which shows considerable genome differentiation. The evidence for genome differentiation has been derived from the study of chromosome pairing in the species, and its triha-ploid ($2n = 3x = 36$). In the first place the 72 chromosomes of *S. demissum* form regularly 36 bivalents at metaphase I of meiosis showing the typical behaviour of an allohexaploid. The regular bivalent formation in the hexaploid gives the impres-sion that there is preferential pairing between the homologous chromosomes of three distinct genomes. The distinctness of genomes is also evident from the chromosome pairing in the trihaploids in which a range of bivalents (mean 9.95 bivalents) and univalents have been observed (Bains & Howard, 1950; Dodds, 1950; Howard & Swaminathan, 1953; Marks, 1955). The formation of bivalents in trihaploid *S. demissum* may be assumed to follow from the pairing between two genomes which are more homologous to each other, and the univalents may repre-sent the genome which is more differentiated than the other two. Using the genome symbols proposed by Hawkes (1958), the two closely related genomes are A_4 and B, while the more differentiated genome is A_1. The genomes A_1A_4B of Hawkes (1958) are equivalent to ABB_1 of Marks (1955) and VD_1D_2 of Kawakami & Matsubayashi (1957). The genome A_1 of *S. demissum* has been envisaged by Hawkes (1972) to be identical with the A_1 genome of *S. verrucosum*, and this is also the same genome of the diploid South American species.

In view of comparing the homologies between the A_1 genome of *S. verrucosum*, as well as the diploid South American species with the A_1 genome of *S. demissum*, the interspecific hybrids mentioned in Table 51.1 are of interest. These hybrids were obtained by crossing *S. verrucosum* × *S. demissum*, and *S. tuberosum*-dihaploid × *S. demissum*.

In the diploid by hexaploid crosses, surprisingly, both pentaploids and tetraploids were obtained (Table 51.1). The pentaploids must originate from functional un-reduced egg cells of the diploid female parents. The typical chromosome associa-tions observed in tetraploid hybrids of both F_1s, which are predominantly bivalents, are illustrated in Plate 51.1 A,B, and the average chromosome associations/cell are indicated in Table 51.2. The high frequency of bivalent formation observed in tetraploid F_1s is in agreement with the previous observations (Kawakami & Matsubayashi, 1957; Matsubayashi, 1962).

A reasonable explanation for the bivalent formation in these hybrids appears at first sight to be the result of preferential pairing between the genomes which are more homologous to each other, that is, A_1 of *S. verrucosum*, or *S. tuberosum*-dihaploid, may pair preferentially with A_1 of *S. demissum* and A_4B of *S. demissum* may pair with each other.

Table 51.1. List of hybrids of *S. verrucosum* × *S. demissum* and *S. tuberosum*-dihaploid × *S. demissum*, their chromosome numbers and the genome constitution

Hybrids	Chromosome number 2n	Numbers of F_1s	Genome constitution*
S. verrucosum × *S. demissum*	48	2	$A_1A_1A_4B$
S. verrucosum × *S. demissum*	60	1	$A_1A_1A_1A_4B$
S. tuberosum-dihaploid × *S. demissum*	48	7	$A_1A_1A_4B$
S. tuberosum-dihaploid × *S. demissum*	60	5	$A_1A_1A_1A_4B$

* Genome symbols according to Hawkes (1958).

Table 51.2. Chromosome associations in *S. verrucosum* × *S. demissum*, and *S. tuberosum*-dihaploid × *S. demissum*

Hybrid	Chromosome number 2n	No. of cells studied	Average chromosome associations/cell				
			V	IV	III	II	I
Metaphase I associations							
S. verrucosum × *S. demissum*							
(i) (ver. 29 × dms WAC 3158) pl. 3	48	62	—	0.74	0.48	21.0	0.90
(ii) (ver. 27 × dms WAC 3195) pl. 1	48	37	—	0.81	0.38	21.43	0.95
(iii) (ver. 27 × dms WAC 3158) pl. 2	60	32	0.19	0.97	7.03	14.03	6.65
S. tuberosum-dihaploid × *S. demissum*							
(i) (GB39 × dms WAC 3158) pl. 1	48	29	—	0.93	0.66	20.31	1.27
(ii) (GB39 × dms WAC 3158) pl. 11	60	22	0.32	0.86	7.46	13.54	6.78
Pachytene association							
S. verrucosum × *S. demissum*							
(ver. 29 × dms WAC 3158) pl. 3	48	31	—	3.42	—	—	—

In order to establish whether there is preferential pairing in the tetraploid hybrids, chromosome associations were studied at pachytene stage. Chromosome 4 in the tetraploid *S. verrucosum* × *S. demissum* hybrid is especially suitable for the purpose since it is one of the easily identifiable chromosomes. Out of the four homoeologous chromosomes, two have clearly shorter short arms than the corresponding arms of the remaining pair (Plate 51.1 C,D). If preferential pairing is the rule, or if it predominates, in a large number of cases the chromosomes with shorter short arms should pair together as will be the case with longer short arms. Out of the 24 associations studied, only 11 cases produced evidence of preferential pairing of chromosomes with shorter short arms, whereas in 9 cases there was no preferential pairing, i.e. longer-shorter arms were associated. In four cases they could not be classified. From these results it is clear that, at least in the case of one set of chromosomes, preferential pairing is not the rule.

However, a more convincing evidence for the absence of preferential pairing is derived from the observation of quadrivalent associations at pachytene stage (Plate 51.2A-F). At this stage the quadrivalents can be detected since all the four homoeologues are found together, showing exchange of pairing partners (Plate 51.2A-F). The frequencies of different types of chromosome associations could not be analysed at pachytene stages, but it was possible to estimate approximately the number of quadrivalent associations in exceptionally well spread out cells (but this is in any case an under-estimate), and the data are presented in Table 51.2. From the table it is apparent that quadrivalent frequency is higher at pachytene stage than at metaphase I by a factor of nearly 5. This means the high frequency of bivalent formation at metaphase I (Plate 51.1A) may not be due to preferential pairing, but may be due to the lack of a sufficient number of chiasmata to hold all the homoeologous chromosomes as a quadrivalent, even though they were paired together at pachytene stage. In this connection it may be pointed out that for the formation of a quadrivalent at least 3 chiasmata are necessary to hold all the four homoeologous chromosomes together at metaphase, that is, 0.75 chiasma/chromosome.

According to Marks (1965), with one exception, less than 0.50 chiasma/chromosome has been recorded in some comparable tetraploid *Solanum* hybrids. In view of this it might be argued that the bivalent formation in some of the tetraploid hybrids is due to the restriction of chiasma formation rather than to preferential pairing.

If there is indeed a restriction in chiasma formation between the homoeologous chromosomes of different genomes in the tetraploid hybrids reported here, it may be expected that A_1A_1 and A_4B are more likely to be connected as bivalents. The occurrence of a high frequency of trivalents in pentaploid hybrids can also be explained accordingly, since there may be no restriction in chiasma formation between three similar A_1 genomes and the more closely related genomes A_4 and B of *S. demissum*.

The formation of a high frequency of bivalents in tetraploid hybrids of *S. verrucosum* × *S. demissum* and *S. tuberosum*-dihaploid × *S. demissum*, confirms the contention of Hawkes (1972) that the A_1 genome of *S. verrucosum* and *S. tuberosum* is similar to the A_1 genome of *S. demissum*.

NATURE OF GENOME DIFFERENTIATION IN MEXICAN SPECIES

There is evidence for genome differentiation in *S. demissum* (Bains & Howard, 1950; Dodds, 1950; Howard & Swaminathan, 1953; Marks, 1955; Kawakami & Matsubayashi, 1957; Matsubayashi, 1962) and in other Mexican species (Marks, 1955, 1965), but the exact nature of genome differentiation has not been explained. The failure of one of the genomes (A_1) of *S. demissum* to pair, or to form chiasmata with the other two genomes (A_4B) may be due to either genetic causes or chromosome structural differentiation. Since nothing is known about the genetic control of chromosome pairing in *Solanum* species, it is futile to speculate on this point at this stage. As far as the chromosome structural differentiation of the A_1 genome is concerned, no direct observation can be made either in the hexaploid or in the trihaploid-*S. demissum*. This is because chromosome pairing is too complicated for a proper analysis at the pachytene stage. However, the diploid Mexican species *S. verrucosum* has the genome A_1, which is the same as that of the A_1 of *S. demissum*, and the former diploid species can be crossed with other Mexican and South American diploid species for comparisons. Such diploid species hybrids are especially suitable for genome comparisons because at pachytene stage a direct comparison can be made for morphological or structural differences that may exist between the two species.

In the F_1 hybrids of *S. verrucosum* × *S. bulbocastanum*, the latter of which is also a diploid Mexican species, pachytene studies have revealed the existence of small structural differences between some of the chromosomes (Hermsen & Ramanna, 1976). In spite of the presence of small chromosomal structural differences between chromosomes, there is normal bivalent formation at metaphase I. The colchicine-doubled F_1 of *S. verrucosum* × *S. bulbocastanum* shows a high frequency of bivalents (Hermsen & Ramanna, 1976) as is observed in the case of *S. verrucosum* × *S. demissum* and *S. tuberosum* × *S. demissum*.

The existence of small structural differences is also observed in another F_1 hybrid between two diploid Mexican species, viz. *S. pinnatisectum* × *S. bulbocastanum*. At the pachytene stage the bivalents show small loop-like structures. For example, the chromosome 11 of *S. bulbocastanum* has a thick block of heterochromatin, adjacent to the centromere on the long arm (Plate 51.3A), whereas such a block of heterochromatin is absent in the corresponding chromosome of *S. pinnatisectum* (Plate 51.3B). In the F_1 a loop-like structure can be very clearly recognized (Plate 51.3 C, D) in 18–20% of bivalents of chromosome 11. It is noteworthy that the loop is always localized in correspondence with the heterochromatic block. Such loops occur in other bivalents in both darkly stained (chromatic or heterochromatic) parts as well as in the lightly stained (achromatic or euchromatic) parts of the bivalents, but they can be revealed only by excellent staining of bivalents and careful observations. The loops can be interpreted in several ways. They may be artefacts, or be caused by spontaneous chromosome aberrations. These interpretations are improbable because the frequency of loops is too high. Differences in chromosome contraction or length might be another explanation. However, with this hypothesis it is hard to explain why the loops occur at distinct places on the chromosomes. Because the loops are constantly occurring and localized on chromosomes, they appear to result from structural differentiation brought about by duplications, deletions, transpositions or in some cases small inversions or re-inversions.

Such loop formations representing chromosome structural changes were previously reported in *Solanum* species hybrids (Gottschalk & Peters, 1956; Gottschalk, 1972). Preliminary studies on *S. tuberosum*-dihaploids × *S. verrucosum* have indicated the presence of a low frequency of loop-like structures in pachytene bivalents (Ramanna, unpubl.).

GENOME DIFFERENTIATION IN SOUTH AMERICAN SPECIES

As pointed out earlier, South American *Solanum* species have been supposed to have evolved by "cryptic structural genome differentiation"—meaning that the changes in the chromosomes are too small to be detected by usual observations on meiosis in interspecific hybrids. While this explanation is a reasonable one in view of other considerations (Swaminathan, 1953; Hawkes, 1972), the presence of cytologically detectable structural differences may not be entirely lacking in inter-specific hybrids of diploid South American species (Ramanna, unpubl.). In addition, our extensive studies on the dihaploids of *S. tuberosum* var. Gineke, have indicated the presence of structural differences within the genomes that constitute the so-called autotetraploid *S. tuberosum* (Ramanna & Wagenvoort, unpubl.).

Additional evidence for the existence of chromosome structural differences within the genomes of *S. tuberosum*, var. Gineke, has been obtained from the study of meiotic chromosome pairing in the monohaploids of *S. tuberosum* $(2n = x = 12)$ (Breukelen, Ramanna & Hermsen, 1976). In some of the monohaploids bivalent-, trivalent- and quadrivalent-like structures are formed. This indicates that some of the non-homologous chromosomes of the monohaploids may possess homologous segments (duplications or transpositions). Pairing and chiasma formation within these segments may lead to bivalent and multivalent formations.

CONCLUSIONS

It may be concluded that the chromosomes of different genomes of tuber-bearing Solanums are not structurally differentiated to such an extent that it results in the failure of chromosome pairing at meiosis. Nevertheless, the existence of small chromosomal structural differences between the genomes can be demonstrated. The presence of such small structural changes as well as "cryptic structural changes" may be responsible for the suppression of crossing over between the homoeologous chromosomes of different genomes. The diploid-like behaviour of some of the polyploids, such as *S. demissum* and the colchicine-doubled tetraploids of diploid F_1s may be due to the presence of a considerable amount of chromosomal structural differentiation.

REFERENCES

BAINS, G. S. & HOWARD, H. W., 1950. Haploid plants of *Solanum demissum. Nature, 166:* 795.
BREUKELEN, E. W. M. VAN, RAMANNA, M. S. & HERMSEN, J. G. TH., 1976. Monohaploids (2n = x = 12) from autotetraploid *Solanum tuberosum* (2n = 4x = 48) through two successive cycles of female parthenogenesis. *Euphytica, 24:* 567–574.
DODDS, K. S., 1950. Polyhaploids of *Solanum demissum. Nature, 166:* 795.
GOTTSCHALK, W. & PETERS, N., 1956. Das Konjugationsverhalten partiell homologer Chromosomen. *Chromosoma, 7:* 708–725.
GOTTSCHALK, W., 1972. The study of evolutionary problems by means of cytological methods. *Egyptian Journal of Genetics and Cytology, 1:* 73–84.
HAWKES, J. G., 1958. Kartoffel I. Taxonomy, cytology and crossability. In Rudorf & Kappert (Eds), *Handbuch der Pflanzenzüchtung, 3:* 1–43.
HAWKES, J. G., 1972. Evolutionary relationships in wild tuber-bearing *Solanum* species. *Symp. Biol. Hung., 12:* 65–69.
HERMSEN, J. G. TH. & RAMANNA, M. S., 1976. Barriers to hybridization of *Solanum bulbocastanum* Dun. and *S. verrucosum* Schlechtd. and structural hybridity in their F_1 plants. *Euphytica, 25:* 1–10.
HOWARD, H. W. & SWAMINATHAN, M. S., 1953. The cytology of haploid plants of *Solanum demissum. Genetica, 26:* 381–391.
KAWAKAMI, K. & MATSUBAYASHI, M., 1957. Studies on the species differentiation in the Section Tuberarium of *Solanum.* V. Genomic affinity between *Solanum verrucosum* and *S. demissum. Science Reports of the Hyogo University of Agriculture, 3:* 17–21.
KHUSH, G. & RICK, C. M., 1963. Meiosis in hybrids between *Lycopersicon esculentum* and *Solanum pennellii. Genetica, 33:* 167–183.

MARKS, G. E., 1955. Cytogenetic studies in tuberous *Solanum* species. I. Genomic differentiation in the group Demissa. *Journal of Genetics, 53:* 262–269.

MARKS, G. E., 1965. Cytogenetic studies in tuberous *Solanum* species. III. Species relationships in some South and Central American species. *New Phytologist, 64:* 293–306.

MARKS, G. E., 1968. Structural hybridity in a tuberous *Solanum* hybrid. *Canadian Journal of Genetics and Cytology, 10:* 18–23.

MATSUBAYASHI, M., 1962. Studies on the species differentiation in *Solanum*, Sect. Tuberarium VIII. Genome relationships between *S. demissum* and certain diploid *Solanum* species. *Seiken Zihô, 13:* 57–68.

MENZEL, M. Y., 1962. Pachytene chromosomes of intergeneric hybrid *Lycopersicon esculentum* × *Solanum lycopersicoides. American Journal of Botany, 49:* 605–615.

SWAMINATHAN, M. S., 1953. Studies on the inter-relationships between the taxonomic series in the section Tuberarium, Genus *Solanum*. I. *Commersoniana* and *Tuberosa. American Potato Journal, 30:* 206–222.

SWAMINATHAN, M. S. & HOWARD, H. W., 1953. The cytology and genetics of the potato (*Solanum tuberosum*) and related species. *Bibliographia Genetica, 16:* 1–192.

EXPLANATION OF PLATES

PLATE 51.1

A. Metaphase I chromosome association showing 24 bivalents in tetraploid F_1 of *S. verrucosum* × *S. demissum* (ver 29 × dms WAC 3158) pl. 3. B. Metaphase I chromosome association showing 1 III + 20 II + 5 I in tetraploid F_1 of *S. tuberosum*-dihaploid × *S. demissum* (GB39 × dms WAC 3158) pl. 1. C, D. Association of chromosome 4 at pachytene in tetraploid F_1 of *S. verrucosum* × *S. demissum* (ver 29 × dms 3158) pl. 3, showing a pair of shorter short arms, and a pair of longer short arms preferentially paired.

PLATE 51.2

A–F. Quadrivalent associations in *S. verrucosum* × *S. demissum* tetraploid F_1 showing exchange of pairing partners.

PLATE 51.3

A. Chromosome 11 of *S. bulbocastanum* showing a thick block of heterochromatin adjacent to the centromere in the long arm at pachytene stage. B. Pachytene bivalent of chromosome 11 of *S. pinnatisectum* without a thick block of heterochromatin in the region corresponding to that of *S. bulbocastanum* in A. C, D. Chromosome 11 in the F_1 of *S. pinnatisectum* × *S. bulbocastanum* showing a loop in the region corresponding to the heterochromatic block shown in A.

Plate 51.1

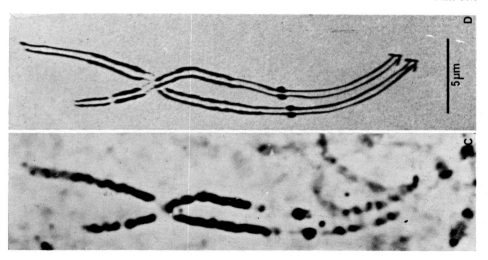

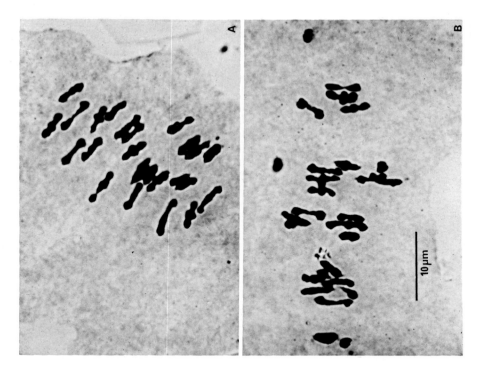

Plate 51.2

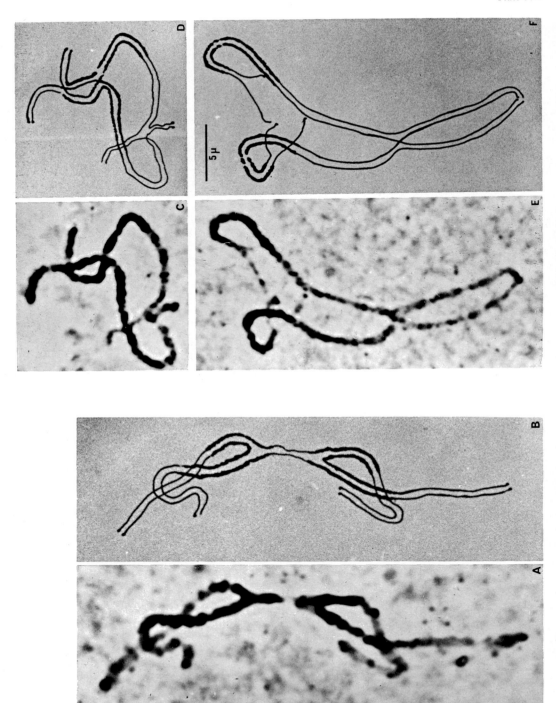

Plate 51.3

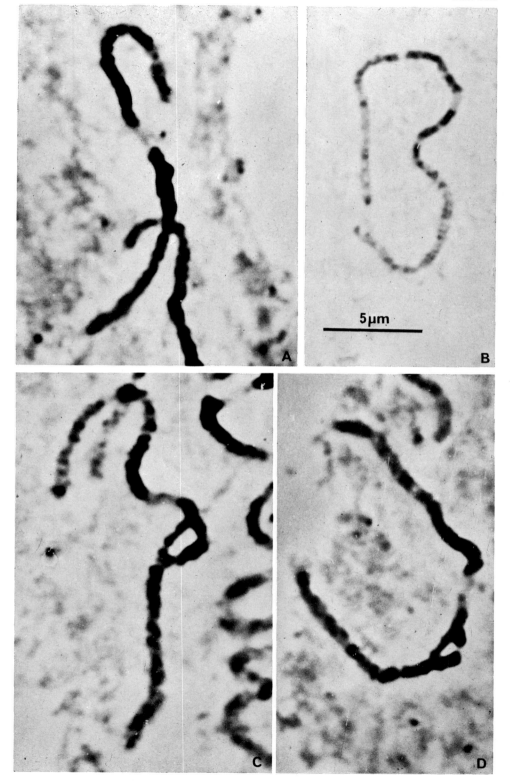

52. Evolution of the cultivated potato: a cytoplasmic analysis*

PAUL GRUN

Departments of Horticulture and Biology, The Pennsylvania State University, Pennsylvania, U.S.A.

The potato raised in the northern hemisphere, *Solanum tuberosum* subspecies *tuberosum*, is similar to Chilean biotypes of the same species in morphology and in the nature of its cytoplasmic factors. The cytoplasmic factors of both strongly contrast with those of *S. tuberosum* subsp. *andigena*, the Andean potato from which subsp. *tuberosum* is generally felt to have evolved. Subspecies *tuberosum* contains cytoplasmic factors that are sensitive to chromosomal genes of subsp. *andigena* and interact with them to produce male and female sterilities. Direct evolution of subsp. *tuberosum* by simple mass selection from subsp. *andigena* is difficult to visualize since a change to the sensitive form of the cytoplasmic factors would be a necessary part of the process. This change, in the presence of the subsp. *andigena* genes, would sterilize the evolving subsp. *tuberosum* forms in which it occurred and inhibit their role in the evolutionary sequence.

Historical evidence is presented that subsp. *tuberosum* was imported from Chile to the Northern Hemisphere after the late blight epiphytotics of the 1840s and it became founder material for our present day cultivated potato.

It is suggested that Chilean subsp. *tuberosum* may trace its ancestry to the complex of wild, ruderal and cultivated forms that occur in northwestern Argentina. Evidence is presented of a similarity between the cytoplasm of one wild taxon of the area, *S. chacoense* f. *gibberulosum*, and that of *S. tuberosum* subsp. *tuberosum*. A form of subsp. *tuberosum*, the variety *chubutense*, which is native to the eastern slope of the Andes at southern latitudes, could have evolved from the complex of forms in northwestern Argentina. Variety *chubutense* could, in turn, be a direct ancestor of Chilean subsp. *tuberosum*, a relationship suggested by their relative geographic proximities and their morphological intergradation.

CONTENTS

ORIGIN OF THE CULTIVATED POTATO

Morphological evidence

The first stages of the evolution of the cultivated potato could have involved either *S. stenotomum* Juz. & Buk. or *S. vernei* Bitt. & Wittm., two diploid species similar in morphology to *S. tuberosum* L. subsp. *andigena* (Juz. & Buk.) Hawkes, the tetraploid potato commonly raised in the Andes mountains. Either *S. stenoto-*

* Authorized for publication July 8, 1976 as paper No. 5119 in the Journal Series of The Pennsylvania Agricultural Experiment Station. Supported by a Grant from the National Science Foundation (GB-17838).

mum, after hybridization with *S. sparsipilum* (Bitt.) Juz. & Buk. (Hawkes, 1956, 1967), or *S. vernei* (Brucher, 1964) could have undergone a chromosome doubling in prehistoric times and subsequently given rise to the tetraploid subsp. *andigena*.

Potatoes of subsp. *andigena* were imported in to Europe some time during the 1570s and 1580s. The historical evidence of possible importation routes favours this view (Salaman, 1949; Hawkes, 1956, 1967) and the morphology of preserved herbarium specimens processed between the late 1500s and the early 1800s confirms it (Salaman, 1937, 1946; Hawkes, 1944, 1956; Simmonds, 1964). Some change in the subsp. *andigena* clones probably occurred during this period, especially in respect to its adaptation to northern latitudes. Nevertheless, most of the plants were still morphologically subsp. *andigena* (Simmonds, 1964). However, the plant from the mid-1700s in Linnaeus' herbarium (No. 248–12) is the type for *S. tuberosum* (Hawkes, 1956) and consequently must be considered as subsp. *tuberosum*.

In the 1840s late blight was inadvertently imported from Central America and almost totally decimated the cultivated potatoes then being raised in Europe and North America. As a result, catastrophic selection (Lewis, 1966) occurred, leading to very rapid changes in the potatoes raised there. By one hundred years later the end product, *S. tuberosum* subsp. *tuberosum*, differed from typical subsp. *andigena* in a number of features, particularly including the dimensions of its leaflets (Simmonds, 1964).

This change has been visualized as occurring as a result of mass selection for late blight resistance coupled with continued selection for desirable agronomic characters. In an effort to determine whether this was the mechanism that produced subsp. *tuberosum* Simmonds (1966) and Glendinning (1975a,b) have undertaken to repeat it by means of mass selection experiments utilizing an originally diverse collection of subsp. *andigena* clones. The potato produced, named Neo-tuberosum, resembles subsp. *tuberosum* in some agronomic characteristics, including quantity of haulm, amount of flowering, stolon shortness, yield, time of initiation of tubering and late blight and virus resistance. Neo-tuberosum supplies a welcome addition of adapted genotypes that should be extremely useful in broadening the genetic base of potato breeding germ plasm.

There is still some uncertainty, however, whether this artificial selection of Neo-tuberosum was a recreation of the actual evolution of subsp. *tuberosum*. The morphological characters which have been most consistently useful in discriminating between subsp. *andigena* and subsp. *tuberosum* have been those concerned with leaflet dimensions. It is these that were used in developing a discriminant function, X, which has been used to identify clones as belonging to one or the other subspecies (Simmonds, 1964). While Neo-tuberosum resembles subsp. *tuberosum* in many agronomic properties on which selection was practised, apparently the leaflet form did not change concomitantly, with the result that Neo-tuberosum still has the leaf of subsp. *andigena*. Possibly this could imply that subsp. *andigena* biotypes that were ancestral stock for Neo-tuberosum were not as predominantly from Colombia as were the original potatoes imported to Europe in the 1500s. The Colombian clones are said to be more *tuberosum*-like in their leaflet character (Hawkes, 1956). Alternatively, it is possible that selection for agronomic characters of potatoes did not result in change in leaf form because genes for leaf form and agronomic characters are not functionally related or linked to one another. If this is the case one would still have to account for the form of the subsp. *tuberosum* leaf that eventually did evolve. It would be necessary to assume that farmers selected this leaf shape because, for some unknown reason, they liked it, even though it was not related to crop yield or quality.

One source of skepticism concerning the hypothesis that post late blight mass selection in Europe explains the origin of subsp. *tuberosum* is the fact that such a hypothesis requires acceptance of the assumption that a strong coincidence occurred. Under the influence of selection following the late blight epiphytotic in Europe the potato that appeared was morphologically closely similar to forms of subsp. *tuberosum* that have been known to be present on islands off the coast of Chile since prehistoric times. Existence of the Chilean potatoes was reported by Drake in 1577, by Cavendish in 1587, by Darwin in 1834, and they have recently been seen growing in uninhabited beach areas there (Correll, 1962; Sykin, 1971). Since late blight was not present in this area until recently, it must be assumed that subsp. *tuberosum* evolved in Chile under the force of selection pressures very different from those that operated on subsp. *andigena* of the 1840s in Europe. One could argue that there was a similarity in evolution of these potatoes in northern Europe and southern Chile because in both places the pressure was for adaptation to low altitude and similar latitudes far from the equator. In northern Europe, however, subsp. *andigena* was subject to altitude and latitude selection from 1570 until 1840 without any change in leaf morphology to the subsp. *tuberosum* form; the change occurred only after the devastating late blight attack (Simmonds, 1964).

Although such parallel evolution may have occurred, such an interpretation is suspect at the start because the simpler explanation exists that the evolution occurred but once, in Chile, and the product raised today in the Northern Hemisphere was simply imported from that region.

Cytoplasmic evidence

The origins of the cultivated potato can be analyzed by a technique entirely separate from the morphology of the plants which formed the basis of the discussion above. Since the 1950s, when the studies of Koopmans (1952, 1954, 1955, 1959) were published, an increasing body of information has been developed concerning the cytoplasmic components of the heredity of these wild and cultivated potatoes. These cytoplasmic components are non-chromosomal genes, presumably genes of chloroplasts, of mitochondria, or possibly of a regularly inherited virus. They are the sorts of factors commonly referred to as plasmon factors or determinants of cytoplasmic sterilities.

The analysis of cytoplasmic factors involves genetic analysis of reciprocal hybrids, for plants of parental species usually express no sterilities, being typically normal and fertile. One detects the fact that these species have different cytoplasmic factors by the differences among their progeny in response to insertion of tester chromosomal genes that are cytoplasm-sensitive. *Solanum stenotomum*, for example, has a cytoplasmic entity, the deformed flowers sensitive factor, symbolized $[df^s]$ (Grun, Aubertin & Radlow, 1962). This factor is expressed only when the chromosomal allele *df* is inserted, following crossing, into plants having this $[df^s]$ factor. Combination of the deformed sensitive cytoplasmic factor together with the homozygous alleles *dfdf*, produces plants having either short anthers (Fig. 52.1) or no anthers. *Solanum chacoense* f. *gibberulosum* (Juz. & Buk.) Corr., the taxon which houses the *df* chromosomal allele, does not have the deformed sensitive factor and so even though *dfdf* plants occur in this species all its flowers have normal appearing anthers.

The genus *Solanum* is somewhat unusual in that a rather large number of contrasting sorts of cytoplasmic factors have been identified in plants belonging to its

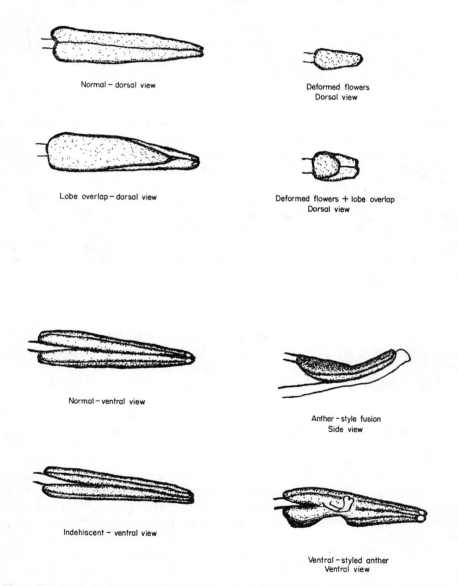

Figure 52.1. Typical appearance of anthers of normal plants and of anthers of plants expressing various sorts of cytoplasmic male sterilities.

section *Tuberarium*. To date, nine different cytoplasmic factors have been found within the group of plants thought to be involved in *S. tuberosum* evolution alone. These factors can be recognized by their responses to different chromosomal genes and each has a different morphological expression. Two of these factors produce abnormalities due to an interaction between recessive alleles and cytoplasm-sensitive factors while the other seven result from an interaction between dominant alleles and various cytoplasm sensitive factors (Table 52.1).

Table 52.1. Sorts of cytoplasmic sterilities that operate in the species involved in the evolutionary pathways of cultivated potatoes*

| Cytoplasmic factor | Expression | Cytoplasm sensitive genes | | |
		Dominance	Number	Penetrance
[df^s]	Deformed flowers having short anthers or none	Recessive	1	High
[lo^s]	Donsal anther lobes under-developed	Recessive	1	Variable
[Sp^s]	Sporads in meiotic stages in anthers of open flowers	Dominant	1	High
[SM^s]	Predominantly or only shrivelled empty microspores in anthers of open flowers	Dominant	?	High, but not expressed in sporo-cytes developmentally stopped at sporad stage
[In^s]	Anther pores do not open normally	Dominant	≥ 5	High, but a few terminal anther pores may open in an otherwise indehiscent plant
[TA^s]	Anther width reduced	Dominant	?	Consistently expressed. Difficult to evaluate since genic basis uncertain
[ASF^s]	Anthers fused to style	Dominant	?	Variable. Often associated with ventral styled anthers
[VSA^s]	Small styles on ventral surface of anthers	Dominant	?	Variable. Often associated with anther-style fusion
[Fm^s]	Cytoplasmic female sterility	Dominant	?	Low. Modified by male fertility flowering time and insect activity

* Based on descriptions in Grun, Aubertin & Radlow (1962), Grun & Aubertin (1965, 1966), Grun (1970a, b 1974) and Grun, Ochoa & Capage (1977).

There is no reason at present for assuming that *Solanum* is unique in having this many cytoplasmic factors, for it is possible that other genera have as many or more. What has been unusual in *Solanum* is that the course of evolution in this genus has produced great evolutionary divergence among forms that have not yet evolved biologically-based isolating mechanisms. The result is that one can cross many distinct species to produce not only F_1s but also advanced generation derivatives, and so unusually wide tests for cytoplasmic differentiation can be run. Other genera may also have diverse cytoplasmic factors, but they have not been as easy to recognize because not as many different sorts of interspecific progenies have been obtained to test for cytoplasmic diversity. There is evidence, however, for extensive cytoplasmic differentiation in such genera as *Epilobium*, *Streptocarpus*, and *Aegilops-Triticum* (summarized in Grun, 1976).

Characteristic expressions of sterilities produced by nine different cytoplasmic factors present in the cultivated potato and its close relatives are listed in Table 52.1. Typical appearances of some that affect the external morphology of the anther are illustrated in Fig. 52.1. Distribution of these factors among the taxa thought to be involved in evolution of the cultivated potato is shown in Table 52.2.

Table 52.2. Cytoplasmic sensitivites of cultivated potato species and some of their relatives[1]

Taxon	$[df^s]$	$[lo^s]$	$[Sp^s]$	$[SM^s]$	$[In^s]$	$[TA^s]$	$[ASF^s]$	$[VSA^s]$	$[Fm^s]$
S. tuberosum									
ssp. *tuberosum*									
Northern Hemisphere	−	−[3]	+	+	+	+	+	+	+
Coastal Chile	−	−	+	+	+	+	+	+	?
ssp. *andigena*	−	+	−	−	−	−	−	−	−
S. stenotomum-phureja	+	(+)	−	−	−	−	−	−	−
S. vernei	−	?	−	−	−	−	−	−	−
S. maglia[2]	(−)	?	(−)	(−)	(−)	?	(−)	(−)	(−)
S. chacoense f.									
gibberulosum	−	(−)	+	+	+	?	?	?	?

[1] Based on data in reference cited in Table 52.1, (*S. maglia* results form Grun, unpubl.). Values in parentheses are results of tests of cytoplasm of single clones.
+, Presence of indicated cytoplasmic factor; −, absence of indicated cytoplasmic factor.

[2] Based on test of the cytoplasm of a triploid clone from Valparaiso Province, Chile.

[3] One exception exists, the clone Hindenburg apparently having $[lo^s]$.

Do the distributions of the various cytoplasmic factors within these species of *Solanum* correspond with what would be expected based on the proposed pathways of evolution of the cultivated potato described at the beginning of this discussion? The pathway postulated an evolution of *S. tuberosum* subsp. *andigena* from *S. stenotomum* or *S. vernei*. The known cytoplasmic factors of subsp. *andigena* are identical to those so far identified in *S. vernei*. They differ from those of *S. stenotomum* only in the presence of the deformed flowers factor in *S. stenotomum*, which is absent from subsp. *andigena*. The overall pattern is one of close similarity between *S. vernei* and *S. stenotomum* and between both these species and their presumed derivative, *S. tuberosum* subsp. *andigena*.

The same cannot be said concerning the evolution of *S. tuberosum* subsp. *tuberosum* from its commonly assumed antecedent, *S. tuberosum* subsp. *andigena*. Subspecies *tuberosum* differs from subsp. *andigena* in eight of the nine cytoplasmic sensitivities listed in Table 52.2. In view of this cytoplasmic dissimilarity the evolution of subsp. *tuberosum* from subsp. *andigena* by simple mass selection is difficult to visualize. For subsp. *tuberosum* to have arisen from subsp. *andigena* would have required not only selection for genes basic to changes in leaf morphology and late blight resistance, but also changes in the cytoplasm from the resistant to the sensitive form of seven of the eight factors. These seven factors ($[Sp^s]$, $[SM^s]$, $[In^s]$, $[TA^s]$, $[ASF^s]$, $[VSA^s]$, and $[Fm^s]$) condition partial or complete male or female sterility in the presence of specific dominant chromosomal alleles, and the genotype of subsp. *andigena* is a warehouse of just these dominant alleles. If the cytoplasm itself changed, the clones in which the change occurred would be expected to show increased sterility.

The nature of this sterility would control the subsequent course of events. Mutations of cytoplasmic factors to female sterility would remove themselves from the evolutionary line of descent because such cytoplasmic factors are only transmitted through the eggs. Male sterility might not immediately be disadvantageous to the plant bearing it because potatoes are usually propagated asexually. If a farmer found a male sterile clone with improved tuber or disease resistance qualities, and propagated it he would end up with a field composed exclusively of such male sterile derivatives which would set berries to a very limited extent or not at all. In any case, if a berry did mature in such a field it would likely be courtesy of pollen from a subsp. *andigena* in the area, and would produce seed that, in turn, would develop into male sterile plants. This course of events would not necessarily bother the farmer since he was concerned principally with tuber yield, but it would be fatal to this as a probable pathway for the evolution of *S. tuberosum* subsp. *tuberosum*.

The brevity of the time span during which subsp. *tuberosum* evolved in the Northern Hemisphere, a process restricted to the one hundred years between the 1840s and the 1950s (Simmonds, 1964), adds to the difficulty in visualizing this pathway. Rapid evolution by way of mass selection does not seem likely to have occurred if the species was simultaneously undergoing sterilization.

As was mentioned in an earlier section, the mass selection hypothesis requires the assumption that very different selection pressures in coastal Chile produced, by coincidence, the same morphological entity, *S. tuberosum* subsp. *tuberosum*. The data in Table 52.2 require that the coincidence be even more striking, for the Northern Hemisphere clones of subsp. *tuberosum* and those from coastal Chile have been found to have identical cytoplasmic factors.

What historical evidence is there for a direct importation of subsp. *tuberosum* from Chile to become the basic stock of our cultivated potato? Following the late blight attack, the Reverend C. E. Goodrich of Utica, New York, developed the concept that the potato had somehow become "devitalized" by continual asexual propagation. His solution to this problem involved "regeneration of the race by procurement of seed in its highest condition of vitality. This will probably be best done by importation from some country of which the potato is a native . . ." (Goodrich, 1848). Through the good offices of his brother-in-law, and the not insignificant sum, at that time, of two hundred dollars, he imported several potatoes bought in the market place in Panama, as well as several lots imported from the Andes. The Andean material did not tuber well in the northern latitudes of Utica. Of the eight clones which he believed had come up the Pacific coast from Chile to the market in Panama, seven were too late for good tuber production in Utica, but one did well and he named it Rough Purple Chili. An open pollinated berry of this clone produced progeny that included his clone Garnet Chili, and Garnet Chili in turn produced Early Rose. Early Rose and its offspring are known to have been extremely important parts of the lineage of many subsequent clones (Hawkes, 1956; Grun, 1970b).

The identity of Rough Purple Chili has been in question, however. Salaman

(1949) believed it was a form of subsp. *andigena* from the Andes, while Hawkes (1956) thought it more likely that it came from Chile. A few years ago a sample of Garnet Chili was obtained for experimental study (Grun, Ochoa & Capage, 1977). The clone was found to have the leaf geometry, as evaluated by Simmonds' (1964) discriminant function X, associated with subsp. *tuberosum*. The clone also expressed cytoplasmic sensitivity to dominant genes for sporads, shrivelled microspores, indehiscence, thin anthers, anther-style fusion and ventral-styled anthers, as do other clones of subsp. *tuberosum*. Rough Purple Chili represents, therefore, a known example of direct importation of subsp. *tuberosum* from Chile to the Northern Hemisphere.

Aside from Rough Purple Chili it is not directly known where other founder stocks of our cultivated potato came from. Siebeneick (1948) presented a pedigree chart which suggests that about 12 clones were important original contributors. One of these, the Goodrich import from Cuzco, has been found (Grun, 1974) probably to be a form of subsp. *andigena*, but the background of the others is not clear. The accumulated evidence certainly suggests that there was a genetic input from both subsp. *andigena* and subsp. *tuberosum*, but the leaf morphology and the cytoplasmic factors point to the predominant role of direct imports from Chile.

ORIGIN OF *S. TUBEROSUM* SUBSPECIES *TUBEROSUM* IN CHILE

If the cultivated potato evolved in the Andes and later spread to Chile, an understanding of its line of descent requires an explanation of the course of events that led to its establishment there. Subspecies *tuberosum* in the Chonos Archipelago exists in at least two forms. Natives in this region grow this potato in their gardens and have been doing so since prehistoric times. The Araucanian Indians cultivate many forms of potatoes which they call "Poñi" and have native names for each clone, names that are not of Spanish origin. Many terms and expressions concerning potato culture exist in their language, and all are different from those of the Andean highlanders (Bukasov, 1966; Sykin, 1971). In addition to these cultivated potatoes there are in the Chonos Archipelago populations of subsp. *tuberosum* that apparently are growing in the wild. Such "wild" forms occur in uninhabited areas along the beaches, and such may have been growing there for a long time, for the Araucanians have a separate term in their language, "Malla", for wild potatoes. At present there is no objective evidence that these wild forms are the original ones from which Araucanians derived their cultivated clones by selection. It is possible that the wild forms were derived instead from clones which escaped from cultivation. A number of collectors including Brücher, Sykin and Ochoa have visited the sites of such wild potatoes in recent times and all concur that the area is highly inaccessible by land. The possibility still exists, however, that fishermen or hunters came to the beach areas by boat, camped there, and advertently or inadvertently planted some tubers.

Samples of these "wild" forms were collected by Ochoa in 1969, one at Esmeralda Beach on the Island of Chiloé and the other at Low Bay, Guaytecas Island. These were included in the analysis, results of which are listed in Table 52.2, and were found to have the same cytoplasmic factors as the cultivated subsp. *tuberosum* of the area (Grun, Ochoa & Capage, 1977).

The only other wild tuber-bearing *Solanum* species native to the Chilean coast is *S. maglia*. This species occurs principally in beach areas, but far north of Chiloé Island. The fact that most clones of *S. maglia* are triploid has made it difficult to evaluate its role, but a start has been made (Grun, unpubl.). *Solanum maglia* flowers prolifically when grown in the environment in central Pennsylvania and a few berries set spontaneously by open pollination. Seed obtained from these berries produced several fertile diploid plants. These were tested by reciprocal crossings, and found to have cytoplasmic characteristics quite different from those of subsp. *tuberosum* (Table 52.2). Because, in addition to its general sterility, *S. maglia* differs from *S. tuberosum* subsp. *tuberosum* in morphology, ploidy level, geographic distribution, and cytoplasmic factors, it seems unlikely that this species played a direct role in the evolution of subsp. *tuberosum*.

A possible pathway to the potato plants of Chiloé could begin with *S. tuberosum* subsp. *andigena* and one or more of the variable complex of wild, weedy, and primitive cultivated species of northwestern Argentina with which it could have crossed. This process could have involved ruderal forms described by Brücher (1966). The hybrid products and their derivatives may have been maintained by Indians of the region or dug, upon need, by these Indians in a primitive gathering agriculture of the type described by Ugent (1970). Some of these forms could have been traded with Indians who lived farther south, where they could have crossed with the indigenous *S. chacoense*, or other indigenous species that had sensitive cytoplasmic factors to give rise, following further selection on the eastern slope of the Andes, to forms of subsp. *tuberosum* such as the variety *chubutense* (Hawkes & Hjerting, 1969). This variety occurs in southern Argentina along the mountain chains in a north-south range about 500 miles long including the area directly across the mountains from the Chonos Archipelago. Its morphology grades into that of the forms found on Chiloé Island (Correll, 1962). It is not difficult to visualize an importation of *S. tuberosum* var. *chubutense* and the subsequent further evolution of subsp. *tuberosum* from it on Chiloé Island.

Solanum tuberosum var. *chubutense* could also, of course, be a direct evolutionary descendent of *S. tuberosum* subsp. *andigena*. The principal basis for the suggestion that subsp. *tuberosum* evolved from certain interspecific hybridizations is the distinctive cytoplasmic factors which subsp. *tuberosum* contains. Thus far the survey of cytoplasmic factors of wild and cultivated species of *Solanum* has uncovered only one other taxon that contains at least some of the cytoplasmic factors of subsp. *tuberosum* (see Table 52.2). It is *S. chacoense* f. *gibberulosum*, a wild species of Series Commersoniana from the Province of Córdoba in Argentina (Table 52.2). There is, of course, a great morphological difference between *S. chacoense* and *S. tuberosum*, for the genes governing morphology of *S. tuberosum* are those of species of the series Tuberosa. *Solanum chacoense* can easily be crossed with species of the series Tuberosa, however, (Choudhuri, 1944; Swaminathan, 1953; Grun, 1961) and hybrids are often fertile. There is, in fact, evidence of introgression between *S. chacoense* and *S. simplicifolium*, another species of the series Tuberosa (Hawkes, 1962).

It would not be difficult to conceive of backcross progeny of a hybridization between a primitive cultivated potato of series Tuberosa and *S. chacoense* maintaining the overall morphology of series Tuberosa and the cytoplasmic factors of

S. chacoense. Furthermore, hybrid vigour and range of adaptation to growth at lower elevations could have been acquired from the *S. chacoense* gentoype. Such a plant could have entered the primitive agriculture of southern Andean foothill Indians, a location that would favour subsequent introduction into Chile.

REFERENCES

BRÜCHER, H., 1964. El origen de la papa (*Solanum tuberosum*). *Physis, 24:* 439–452.

BRÜCHER, H., 1966. Eine polyploide Serie von "Ruderalkartoffeln" (*Solanum* sect. *Tuberarium*) aus de argentinischen Kortillere. *Der Züchter, 36:* 189–196.

BUKASOV, S. M., 1966. Die Kulturarten der Kartoffeln und ihre wildwachsenden Vorfahren. *Zeitschrift für Pflanzenzüchtung, 55:* 139–164.

CHOUDHURI, H. C., 1944. Cytological and genetical studies in the genus *Solanum*. II. Wild and native cultivated 'diploid' potatoes. *Transactions of the Royal Society of Edinburgh, 61:* 199–219.

CORRELL, D. S., 1962. *The Potato and Its Wild Relatives.* Renner, Texas: Texas Research Foundation.

GLENDINNING, D. R., 1975a. Neo-Tuberosum: new potato breeding material. I. The origin, composition, and development of the Tuberosum and Neo-Tuberosum gene pools. *Potato Research, 18:* 256–261.

GLENDINNING, D. R., 1975b. Neo-Tuberosum: new potato breeding material. II. A comparison of Neo-Tuberosum with unselected Andigena and with Tuberosum. *Potato Research, 18:* 343–350.

GOODRICH, C. E., 1848. The potato disease. *Transactions of the New York Agricultural Society, 8:* 403–426.

GOODRICH, C. E., 1863. The potato. Its diseases—with incidental remarks on its soils and culture. *Transactions of the New York Agricultural Society, 23:* 103–134.

GRUN, P., 1961. Early stages in the formation of internal barriers to gene exchange between diploid species of *Solanum*. *American Journal of Botany, 48:* 79–89.

GRUN, P., 1970a. Changes of cytoplasmic factors during the evolution of the cultivated potato. *Evolution, 24:* 188–198.

GRUN, P., 1970b Cytoplasmic sterilities that separate the cultivated potato from its putative diploid ancestors. *Evolution, 24:* 750–758.

GRUN, P., 1974. Cytoplasmic sterilities that separate the Group Tuberosum cultivated potato from its putative. tetraploid ancestor. *Evolution, 27:* 633–643.

GRUN, P., 1976. *Cytoplasmic Genetics and Evolution.* New York: Columbia Univ. Press.

GRUN, P., AUBERTIN, M. & RADLOW, A., 1962. Multiple differentiation of plasmons of diploid species of *Solanum*. *Genetics, 47:* 1321–1333.

GRUN, P. & AUBERTIN, M., 1965. Evolutionary pathways to cytoplasmic male sterility in *Solanum*. *Genetics, 51:* 399–409.

GRUN, P. & AUBERTIN, M., 1966. Cytological expressions of a cytoplasmic male sterility in *Solanum*. *American Journal of Botany, 53:* 295.301.

GRUN, P., OCHOA, C. & CAPAGE, D., 1977. Evolution of cytoplasmic factors in tetraploid cultivated potatoes (Solanaceae). *American Journal of Botany, 64:* 412–420.

HAWKES, J. G., 1944. Potato collecting expedition in Mexico and South America. II. Systematic classification of the collections. *Bulletin. Imperial Bureau of Plant Breeding and Genetics,* Cambridge, pp. 142.

HAWKES, J. G., 1956. Taxonomic studies on the tuber-bearing Solanums. I. *Solanum tuberosum* and the tetraploid species complex. *Proceedings of the Linnean Society of London, 166:* 97–144.

HAWKES, J. G., 1962. Introgression in certain wild potato species. *Euphytica, 11:* 26–35.

HAWKES, J. G., 1967. The history of the potato. *Journal of the Royal Horticultural Society, 92:* 207–365.

HAWKES, J. G. & HJERTING, J. P., 1969. *The Potatoes of Argentina, Brazil, Paraguay, and Uruguay. A Biosystematic Study.* Oxford at the Clarendon Press.

KOOPMANS, A., 1952. Changes in sex in the flowers of the hybrid *Solanum rybinii* × *S. chacoense*. *Genetica, 26:* 359–380.

KOOPMANS, A., 1954. Changes in sex in the flowers of the hybrid *Solanum rybinii* × *S. chacoense*. *Genetica, 27:* 273–285.

KOOPMANS, A., 1955. Changes in sex in the flowers of the hybrid *Solanum rybinii* × *S. chacoense*. III. Data about the reciprocal cross *Solanum chacoense* × *S. rybinii*. *Genetica, 27:* 465–471.

KOOPMANS, A., 1959. Changes in sex in the flowers of the hybrid *Solanum rybinii* × *S. chacoense*. IV. Further data from the reciprocal cross *S. chacoense* × *S. rybinii*. *Genetica, 30:* 384–390.

LEWIS, H., 1966. Speciation in flowering plants. *Science, 152:* 167–172.

SALAMAN, R. N., 1937. The potato in its early home and its introduction into Europe. *Journal of the Royal Horticultural Society, 62:* 61–77; 112–123; 153–162; 233–266.

SALAMAN, R. N., 1946. The early European potato: its character and place of origin. *Journal of the Linnean Society of London (Botany), 53:* 1–27.

SALAMAN, R. N., 1949. *The History and Social Influence of the Potato.* Cambridge: Cambridge Univ. Press.

SIEBENEICK, H., 1948. Die deutschen und ausländischen Kartoffelsorten 1947/1948. *Schriftenreihe für Kartoffelwirtschaft, 2 & 3*, 1948.

SIMMONDS, N. W., 1964. Studies of the tetraploid potatoes. II. Factors in the evolution of the Tuberosum Group. *Journal of the Linnean Society of London (Botany), 59:* 43–56.

SIMMONDS, N. W., 1966. Studies of the tetraploid potatoes. III. Progress in the experimental re-creation of the Tuberosum Group. *Journal of the Linnean Society of London (Botany), 59:* 279–288.

SWAMINATHAN, M. S., 1953. Studies on the interrelationships between taxonomic series in the section Tuberarium genus *Solanum*. I. *Commersoniana* and *Tuberosa*. *American Potato Journal, 30:* 271–281.

SYKIN, A. G., 1971. Zur Frage der Abstammung und wildwachsenden Vorfahren chilinischer Kulturkartoffeln. *Zeitschrift für Pflanzenzüchtung, 65:* 1–14.

UGENT, D., 1970. The potato. *Science, 170:* 1161–1166.

53. Biosystematic studies in *Lycopersicon* and closely related species of *Solanum*

CHARLES M. RICK

Department of Vegetable Crops, University of California, Davis

The species of *Lycopersicon* (including *Solanum pennellii*) are closely allied and well isolated from other genera, according to all tested criteria—genetic variability at all taxonomic levels, geographic distribution, comparative chromosome cytology, viability and fertility of interspecific F_1 hybrids, and the fate of later generations. Speciation in the genus is illustrated by two examples—divergence within *L. peruvianum* and differentiation between the sibling species, *L. chmielewskii* and *L. parviflorum*.

Solanum section *Petota*, subsection *Potatoe*, [formerly section *Tuberarium*, subsection *Hyperbasarthrum*] series *Juglandifolia* is the group most closely related and probably ancestral to *Lycopersicon*. All have 12 chromosomes morphologically similar to those of *Lycopersicon*. Close relationships exist between *S. lycopersicoides* and *S. rickii* and between *S. juglandifolium* and *S. ochranthum*, but a huge morphological and genetic hiatus exists between these two groups and between them and *Lycopersicon* and other *Solanum* species. *S. lycopersicoides* can be hybridized with certain tomato species, but the F_1 hybrids suffer complete genic and chromosomal sterility; it does not cross with other *Solanum* species except *S. rickii*; it thus constitutes a very imperfect link between the two genera.

Species of *Lycopersicon* have therefore evolved entirely via gene substitution, whereas chromosomal differentiation has also occurred in their divergence from *Solanum*. The evidence does not justify merging the two genera.

CONTENTS

INTRODUCTION

Solanum-Lycopersicon interrelationships have generated much interest amongst systematists, geneticists, and plant breeders. These taxa present great diversity of mating systems, ecological adaptations, plant-animal interactions, patterns of speciation, and other intriguing biological features. Abundant material is available, and much of it is amenable to biosystematic investigation. I shall attempt to summarize the results of various tests applied to the natural relationships in *Lycopersicon* and between that genus and *Solanum*. Criteria for these tests included: genetic variability for morphological, physiological, and electrophoretic characters (at the individual, population, race, and species levels) in relation to geographic distribution, comparative chromosome cytology, interspecific crossability, viability and fertility of the hybrids, and the cytogenetic fate of natural and experimental introgression in subsequent generations. As a summary of the present status of such relations, this paper necessarily blends new results with previously published data.

For the taxonomic history of this group and descriptions of the species, the reader is referred to Correll (1962), Luckwill (1943), Muller (1940), Rick (1963b, 1971), Rick & Lamm (1955), and Rick, Kesicki, Fobes & Holle (1976). Full descriptions of each species obviously cannot be presented here; instead, attention will be limited to items that are pertinent to phylogenetic relationships. This treatment is divided into two main sections, the first dealing with the biosystematics of *Lycopersicon*, the second with inter-relationships between the two genera.

LYCOPERSICON

Gross morphology

The species of *Lycopersicon*, as defined here, form a cohesive group in respect to their mutually possessing the following characters: (1) herbaceous perennial growth, (2) sprawling or prostrate habit, (3) pinnately segmented leaves, (4) stem organization in sequences of 2– or 3–leaved sympodia, (5) cymose inflorescences, (6) yellow corolla and anthers, (7) anthers connate or connivent, and (8) fruit a soft berry. Whereas certain of these characters appear in the closely related genera, the entire combination is not found in any species thereof.

Mating systems

A complete range is found from the almost completely autogamous *L. cheesmanii* and *parviflorum* to the obligately outcrossed, self-incompatible biotypes of *L. chilense, hirsutum, peruvianum*, and *S. pennellii*. Self-fertility with various degrees of facultative outcrossing is found in *L. chmielewskii, esculentum, pimpinellifolium*, and the self-compatible biotypes of *L. hirsutum* and *S. pennellii*. Domestication of *L. esculentum* was accompanied by a transition from exserted to inserted stigmas and consequent change from facultative outcrossing to enforced autogamy (Rick, 1976). Derived autogamy is considered to have played an important role in the evolution of *L. cheesmanii* (Rick, 1963b; Rick & Fobes, 1975) and *L. parviflorum* (Rick *et al.*, 1976). Altogether, the tomato species therefore provide useful material for studies on the evolution of mating systems and their bearing on the nature of speciation.

Crossability relations

Accessions of each species were hybridized in all possible combinations, direct and reciprocal. Each combination was attempted on a minimum of six flowers on each of numerous plants of each accession. All such hybridizations were made in air-conditioned greenhouses, under which standard conditions of self- or sib-crosses of all accessions yielded full fruit sets and full seed complements. Results of this survey are summarized in terms of degree of success in yielding F_1 progeny in the conventional polygon style in Fig. 53.1.

These hybridization tests cleave the genus into two major intra-crossable, inter-incrossable groups One ("*peruvianum* group") consists of *L. chilense* and *peruvianum*, the other ("*esculentum* group"), the remaining seven species. The barrier between these two groups can be broken only by the application of embryo culture, which succeeds only when the member of the *esculentum* complex is employed as the female parent. Hogenboom (1972a, b) has ingeniously demonstrated that the reciprocal cross can be made if highly selected inbred variants of *L. peruvianum* are used, but the occurrence of such hybridizations in nature is highly improbable.

Within the *peruvianum* complex, a chain relationship links the two component species, but only via several bridging accessions. *L. peruvianum* var. *humifusum* is differentiated from the bulk of the species by a similar barrier (Rick, 1963a and below). Within the *esculentum* group, the taxa are more intercrossable. Although unilateral relationships are common, hybrids can be obtained in nearly all combinations without the need of special techniques. It is especially fortunate for tomato (*L. esculentum*) improvement that this species can be thus hybridized with every other one in the genus. The transmission of many valuable genes has thereby been facilitated to numerous superior cultivars already in cultivation (Stevenson & Jones, 1953).

F_1 hybrids

F_1 hybrids of all successful combinations within *Lycopersicon* display normal or near-normal chromosome behaviour. The chromosomes are generally completely paired throughout their lengths at pachytene, and the resultant bivalents form chiasmata and segregate normally at anaphase of the two meiotic divisions (Afify, 1933; Khush & Rick, 1963; McGuire & Rick, 1954; Lesley & Lesley, 1943; Rick, 1960; Sawant, 1958, and others). It is therefore evident that speciation in this genus has taken place almost entirely by gene mutation and to a very minor extent by chromosomal differentiation. Thus, in respect to crossability, comparative chromosomal morphology and pairing, a remarkable degree of coherence is displayed by the tomato species. These indices of close relationship are matched by the aforementioned concordance in taxonomic traits and contrast, as the following section reveals, with their interrelationships with their nearest relatives in the parental genus *Solanum*.

It is difficult to generalize in respect to fertility of hybrids and comportment of later generations. The situation varies from complete fertility with no cytogenetic irregularities in later generations (as in *L. esculentum* × *L. pimpinellifolium*) to combinations with appreciable F_1 (genic) sterility and of inviability, reduced recombination, modified segregation ratios, and other problems in F_2 and BC generations (as in *L. esculentum* × *S. pennellii*), as well as intermediate situations in other interspecific combinations. Thus, reproductive and geographic barriers serve to prevent gene flow between nearly all combinations of the tomato species. The features of two combinations will be presented to exemplify the nature of speciation within *Lycopersicon*.

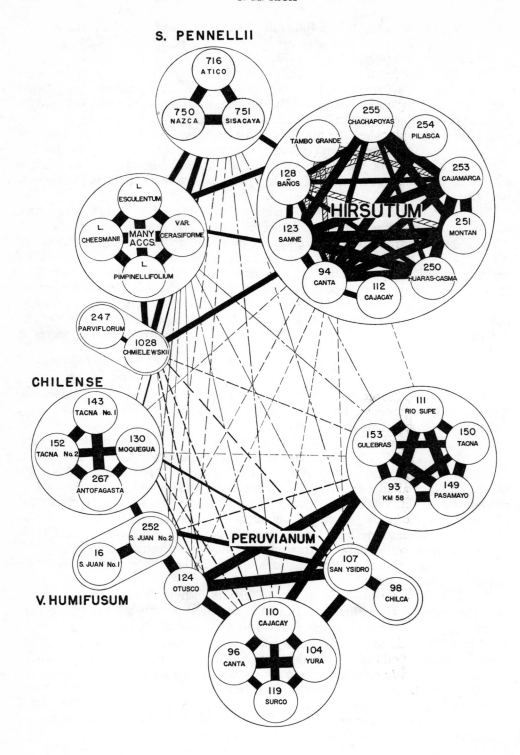

Speciation in L. peruvianum (*Rick, 1963a*)

L.peruvianum is a wondrously polymorphic species. A high level of variability is evident between individuals of the same population, between populations of a given race, and between races. This variation, manifest in multitudinous morphological as well as electrophoretic characters, is so extreme that one seldom encounters two plants of identical genotype. Differentiation is abetted by the strict self-incompatibility and geographic isolation of some races in the many narrow valleys of the western Andes.

Hybridization experiments revealed a complete barrier between var. *humifusum* at higher elevations in the Jequetepeque drainage in the NE extreme of the *peruvianum* distribution and most of the other forms of the species. Isolation is not complete, however, because intermediate forms either in the lower reaches of the valley or at approximately the same elevation in the contiguous valleys to the south are freely compatible with var. *humifusum* as well as with all other tested accessions of *L. peruvianum*.

The existence of these "bridging" races permits a genetic analysis of the control of this incompatibility. Thus, the hybrids between var. *humifusum* acc. San Juan (S) and the bridging race Otusco (O) can be tested against a representative of the coastal races, Culebras (C). In the incompatible combinations (S × C, C × S), fruits are set, but embryos cease development at early stages, the pattern of breakdown differing in the reciprocal crosses. F_1 × C yields only a few viable seeds in reciprocal crosses with S. The F_2 and BC generations of O × C display a continuous curve of degree of incompatibility with S, but no combinations approximated complete compatibility. Although the populations were too small to permit satisfactory estimation of the number of responsible genes, the genetic control is clearly multigenic. Various considerations lead to the conclusion that such barriers were acquired secondarily to geographic isolation and other kinds of genetic differentiation.

Although less extensively investigated, the barriers between the self-incompatible species are of the same nature in respect to the stage of breakdown and other aspects, but they tend to be more severe.

Differentiation between a pair of sibling species

The second example is selected from the *esculentum* group to illustrate the nature of differentiation amongst self-compatible forms (Rick *et al.*, 1976). Chmielewski (1962, 1968b) described elements of the provisionally labelled "*L. minutum*" and justified separation of it from *L. esculentum*. Although the two can be readily crossed and the F_1 hybrids are fertile, the segregating generations are plagued by low viability, partly attributable to interactions with the defoliator (*Df*) gene (Chmielewski, 1968a) and by very low fruitfulness.

Figure 53.1. Crossability polygon for *Lycopersicon* species. All completely intercrossable combinations are enclosed in second-order circles. Solid lines represent intercompatible crosses; the width of the lines, the degree of compatibility. Dashed lines designate cross failures; dot-dashed lines, combinations that yield F_1 progeny only via embryo culture. The crossability relations of the sibling species *L. chmielewskii* and *L. parviflorum* are similar except that the former is incompatible with *L. hirsutum* and *S. pennellii*. Chart does not indicate degree of fertility of F_1 hybrids or fate of later generations.

Travels to the interandean habitat of this group permitted us to observe natural populations and to triple the number of living collections. It soon became evident from field observations that the complex consists of two distinct forms—one with tiny flowers and barely exserted stigmas, essentially completely self-pollinated; the other with larger flowers, well-exserted stigmas, and mating by mixed self- and cross-pollination. No intermediate forms were found in the wild, but F_1 hybrids can be produced experimentally between the two types. Although highly fertile, such F_1s produce seeds that germinate poorly.

All available accessions were characterized by electrophoretic tests of allozymes at 14 loci, of which eight were polymorphic. Alternative alleles of *Got-3* distinguished the two taxa in all accessions; the same held for alleles of *Prx-3* in the region of sympatry; and variation at other polymorphic loci corresponded to a varying extent with the morphological separation of the two types. The smaller flowered taxon, *L. parviflorum*, proved to be exceedingly uniform morphologically and electrophoretically; except for minor variation at two loci, every tested individual of every accession had the same genotype. The larger flowered *L. chmielewskii* displays much more genetic variability, as detected in intra- and inter-populational polymorphy, heterozygosity, and direct evidence of outcrossing. According to all available evidence, the latter is subject to considerable outcrossing, whereas the former breeds by strict autogamy. Thus, the morphological and enzymatic divergence between the two is reinforced by the isolation engendered by the autogamy of *L. parviflorum* and poor reproductivity of the hybrids. Evidently the latter species evolved from *L. chmielewskii* by acquiring the self-pollinating habit, subsequently developing other differentiation traits. Since the two are very closely related but are reproductively isolated, they fit the category of sibling species.

The status of S. pennellii

In the foregoing summary of genetic relationships, it is noteworthy that *Solanum pennellii* Corr. is most closely allied to species of the *esculentum* group. Possessing the key anther traits of *Solanum*, this species was described and placed in section *Tuberarium* [now *Petota*], series *Juglandifolia* (Correll, 1958), but subsequently it was reclassified as the monotypic representative of the new section *Neolycopersicon* (Correll, 1962).

S. pennellii is closely affiliated with the tomato species in respect to not only chromosomal similarity, crossability, hybrid fertility, and experimental introgression, but also gross morphology. Aside from anther dehiscence, it is concordant with *Lycopersicon* in general features of plant habit, branching pattern, leaf shape, colour and form of corolla, and shape, form, and consistency of fruits.

Limited chemosystematic evidence also links this species with *Lycopersicon*, not *Solanum*. In polyacrylamide gel electrophoresis of leaf proteins, West (1973) ascertained that *S. pennellii* resembles *Lycopersicon* (actually closest to *L. esculentum*) more than closely affiliated *Solanum* species. Dr R. Durbin (unpubl.) has discovered that, like species of *Lycopersicon*, *S. pennellii* is insensitive to tentotoxin, whereas all tested spp. of *Solanum* section *Petota* are sensitive. In our isozyme survey, although of a preliminary nature, *S. pennellii* likewise shows closer relationships with *Lycopersicon*.

Attention should also be drawn to some bizarre characteristics common to all of our accessions of *S. pennellii*. In the basal articulation of its pedicels, it differs from *Lycopersicon* and all closely related species of *Solanum*. Also, unique to *pennellii* is a water-conserving stomatal regulation characteristic, which enables it to withstand the extremely xerophytic conditions of its habitat (Rick, 1973). In view of these anomalous, specialized characters, *S. pennellii* could scarcely be considered as a model, or even as an approximation, of a surviving link between the two genera.

From these and other examples of interspecific barriers, it is clear that a varied assortment of isolation mechanisms tends to impede gene exchange between species of *Lycopersicon*. Despite the feasibility of producing hybrids in almost every combination under experimental conditions, surprisingly few instances of natural introgression are known (hybridization and introgression between *L. esculentum* and *L. pimpinellifolium* were documented by Rick, 1950, 1958, and others). The various mechanisms therefore seem to be effective in maintaining the integrity of the tomato species. Speciation occurred via gene substitution, not by chromosomal differentiation, abetted by various isolating mechanisms. Nevertheless, as explained above, in spite of the development of these barriers, the species of *Lycopersicon* are a relatively coherent group in respect to all tested criteria.

LYCOPERSICON-SOLANUM INTERRELATIONSHIPS

General

Crosses between the two genera have been attempted periodically by many workers. Such species as the following cultigens or conveniently available weeds belonging to several sections of *Solanum* have been tested unsuccessfully as parents: *S. dulcamara, gilo, hyporhodium, incanum, integrifolium, melongena, muricatum, nigrum, pseudocapsicum, quitoense, tuberosum*, and undoubtedly others. Cultivars of *Capsicum, Cyphomandra*, and *Physalis* species have also been assayed. Omidiji (this Volume, p. 599), for example, reports failure of many such combinations. Wann & Johnson (1963) tested 18 *Solanum* species section *Tuberarium*; no reaction was obtained in any cross except that in which pollen of *S. etuberosum* stimulated ovary enlargement and fertilization in *L. esculentum*, but endosperm development lagged and no viable seeds were obtained. Chmielewski (1968b) tested crossabilities between *L. parviflorum* and the tomato-like *Solanum* species without success.

Our tests were limited to those Solanums with greatest morphological affinities to *Lycopersicon*—namely, section *Tuberarium* (Dun.) Bitt. [now *Petota* Dumort], subsection *Hyperbasarthrum* Bitt. [now *Potatoe* G. Don], series *Juglandifolia* Rydb. Only four species are known for this group: *S. juglandifolium* Dun., *S. ochranthum* Dun., *S. lycopersicoides* Dun., and *S. rickii* Corr. In common, they have pinnately divided leaves, yellow-pigmented corollas, and lack tubers—a combination of tomato-like characters that immediately differentiates them from all other *Solanum* species. They exhibit other *Lycopersicon* characters, but these are not common to all of them. All have 12 chromosomes similar to those of *Lycopersicon*.

Morphology

A feature of special interest in this group is their great morphological diversity. Although often overlooked, this heterogeneity was recognized by Correll (1962) and led to his partitioning *S. pennellii* into a new, monotypic section *Neolycopersi-con*. But the remaining four species are so discordant that they seem to have been classified together in series *Juglandifolia* only as a matter of taxonomic expediency. They fall into two groups—*juglandifolium-ochranthum* and *lycopersicoides-rickii*, within which considerable morphological and cytogenetic affinity exists, but between which the differences greatly exceed those between other species in Sect. *Tuberarium*. These characters are briefly enumerated as follows, the unique features in italics:

S. ochranthum (Plate 53.1A, B). A robust *woody liana* with large, showy inflorescences, prominently displayed from the tops of shrubs or low trees to a height of *15 m*; leaves soft; fruits with *three or more locules* (otherwise known only in cultivars of *L. esculentum*), *5–6 cm diameter, pericarp woody* even to maturity; seeds very large and *winged; sympodia of 9–12 leaf nodes;* moist habitats, 1800–3500 m.

S. juglandifolium (Plate 53.1C). Generally with the same features as *S. ochranthum*, but all parts smaller in size; *leaves scabrous; sympodia 8–10 noded;* liking wet habitats, 1500–3000 m.

S. lycopersicoides (Plate 53.1E, F). *Dense, woody shrub, 1.5–2 m tall; leaf segments blunt*, strongly serrate; large, multiflowered inflorescence; *anthers white; upper style bent abruptly by 90°;* fruit a soft berry, the outer pericarp ripening to a *black* colour; *sympodia 4–5 noded;* habitat: dry quebradas, 2800–3200 m.

S. rickii (Plate 53.1D). *Low, erect, perennial herb;* anther and style as in *lycopersicoides;* fruit a small berry *ripening to a papery texture and a pale yellow colour;* sympodia 3–5 noded; rare endemic in dry open situations, 3000 m.

When these deviating characters are taken into account, this group seems like a very strange conglomeration of species to be classified in the same series of a subsection of a section, of a subgenus of *Solanum!*

Crossability tests

We have attempted to acquire, cultivate, and subject these species to various biosystematic tests. These efforts have been frustrated by several problems, one of which is dearth of accessions: some are known from only a single collection (*S. rickii*), the rest from very few. Under our conditions some are reluctant to flower, whilst *S. juglandifolium* has never flowered in our cultures. Another problem is seasonability of flowering, which limits hybridizations with tomato species that do not bloom at the same time. The limitation imposed by restricted flowering and accessions could lead to erroneous conclusions. That all biotypes of a species may not behave in the same fashion in crossability tests is well known and is exemplified by the example mentioned above in the differentiation within *L. peruvianum*.

Mindful of these limitations, we made nearly all possible crosses between the flowering species (*lycopersicoides, ochranthum, pennellii*, and *rickii*). The outcome of these tests is summarized in Fig. 53.2. Despite the aforementioned limitations, the results were consistent in the following respects.

(1) *S. pennellii* hybridizes unilaterally with *L. esculentum, cheesmanii, pimpinellifolium, parviflorum, hirsutum*, and *S. lycopersicoides* whereas it does not cross directly or reciprocally with the other species of *Lycopersicon* or *Solanum*. In this respect it behaves as if a member of the "*esculentum* group". The *esculentum-pennellii* hybrids are heterotic, display almost normal chromosome behaviour (Khush & Rick, 1963), and have about 25% normal fertility.

(2) *S. lycopersicoides* hybridizes unilaterally with *L. esculentum, pimpinellifolium, cheesmanii,* and *S. pennellii* to yield small but viable hybrid seeds. All other crosses failed except with *S. rickii.* Hybrids of the first aforementioned combination are exceedingly heterotic and completely sterile (Rick, 1951). Detailed cytological studies by Menzel (1962) and Menzel & Price (1966) revealed nearly complete pairing at pachytene and extensive formation of synaptonemal complexes, but diminished chiasma formation and distribution, resulting in reduced pairing at metaphase. When the number is doubled, parental chromosomes pair preferentially, the bivalent pairing being highly regular; pollen stainability is much increased; and viable seeds are produced, albeit at a very low fertility level (Menzel, 1964; Rick, 1951).

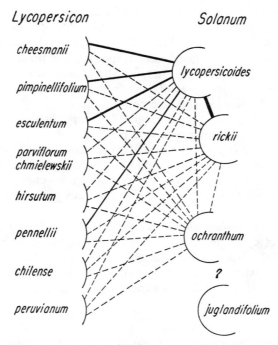

Figure 53.2. Crossability relations between *Lycopersicon* and *Solanum*. Solid lines designate inter-compatible combinations; the width of the lines, the degree of compatibility. No combinations were more than half as compatible as intraspecific matings. Dashed lines designate cross failures.

(3) *S. rickii* showed incompatibility in every cross except with *S. lycopersicoides.* The hybrids have not been studied cytologically, but their moderately high pollen and seed fertility imply regular chromosome behaviour.

(4) *S. ochranthum* failed to cross in every combination. The failure of flowering in *S. juglandifolium* prevented crossability tests. Its remarkable morphological similarity to *S. ochranthum* suggests, however, that its phylogenetic relations are stronger with the latter than with any other species considered.

The results with *S. pennellii* support the previous conclusions that its affinities lie with *Lycopersicon,* not *Solanum.* The crossability between *S. lycopersicoides* and

S. rickii is also consistent with morphological similarities. Further, the complete reproductive isolation of *S. ochranthum* matches its extreme morphological divergence from the group and thereby provides another example of the agreement between crossability and morphological resemblance.

Conclusions

Finally, we must face the question: what are the natural relationships between *Solanum* and *Lycopersicon*? Any conclusions must take the following facts into consideration:

(1) According to all available information, *S. pennellii* bears a much stronger affinity to *Lycopersicon* than to *Solanum*.

(2) The only genetic tie between the two genera is provided by *S. lycopersicoides*, but, though the closest living approximation of an intermediate, it provides only a weak link at best. Although crossable with *Lycopersicon*, the hybrids suffer profound genic and chromosomal sterility. Also, except for the closely allied *S. rickii*, it is genetically isolated from other *Solanum* species. The production of such sterile hybrids is scarcely an argument for merging the genera; otherwise, *Lolium* should not be separated from *Festuca*, *Raphanus* from *Brassica*, *Cattleya* from *Schomburgkia*, or scores of other intergeneric combinations.

(3) All species of *Lycopersicon* (including *S. pennellii*) are intercrossable and yield hybrids with complete chromosome pairing and of complete to somewhat impaired fertility. They are concordant, furthermore, in the following series of morphological characters, which, though sometimes seen in *Solanum* species, are not all associated in the same species: sprawling perennial herbaceous habit, indeterminate sympodial growth with 2–3 leaf nodes per sympodium, and fusion of anthers to form a tube.

The boundary between *Solanum* and *Lycopersicon* is profound—a fact evident not only in terms of cytogenetic evolution but also in morphological and physiological differentiation. These considerations lead to the conclusion that separation of the two genera is justified and that they should be recharacterized taxonomically to accommodate the aforementioned experimental evidence.

ACKNOWLEDGEMENTS

Support from BMS75–03024 and other grants from the National Science Foundation is gratefully acknowledged.

REFERENCES

AFIFY, A., 1933. The cytology of the hybrid between *Lycopersicon esculentum* and *L. racemigerum* in relation to its parents. *Genetica, 15:* 225–240.

CHMIELEWSKI, T., 1962. Cytogenetical and taxonomical studies on a new tomato form. Part I. *Genetica Polonica, 3:* 253–264.

CHMIELEWSKI, T., 1968a. New dominant factor with recessive lethal effect in tomato. *Genetica Polonica, 9:* 39–48.

CHMIELEWSKI, T., 1968b. Cytogenetical and taxonomical studies on a new tomato form. Part II. *Genetica Polonica, 9:* 97–124.

CORRELL, D. S., 1958. A new species and some nomenclatural changes in *Solanum*, section *Tuberarium*. *Madroño, 14:* 232–236.

CORRELL, D. S., 1962. *The Potato and Its Wild Relatives.* Renner, Texas: Texas Research Foundation.

HOGENBOOM, N. G., 1972a. Breaking breeding barriers in *Lycopersicon*. 4. Breakdown of unilateral incompatibility between *L. peruvianum* (L.) Mill. and *L. esculentum* Mill. *Euphytica, 21:* 397–404.

HOGENBOOM, N. G., 1972b. Breaking breeding barriers in *Lycopersicon*. 5. The inheritance of the unilateral incompatibility between *L. peruvianum* (L.) Mill. and *L. esculentum* Mill. and the genetics of its breakdown. *Euphytica, 21:* 405–414.

KHUSH, G. S. & RICK, C. M., 1963. Meiosis in hybrids between *Lycopersicon esculentum* and *Solanum pennellii. Genetica, 33:* 167–183.

LESLEY, M. M. & LESLEY, J. W., 1943. Hybrids of the Chilean tomato. *Journal of Heredity, 34:* 199–205.

LUCKWILL, L. C., 1943. The genus *Lycopersicon*, an historical, biological, and taxonomic survey of the wild and cultivated tomatoes. *Aberdeen University Study:* 120.

MCGUIRE, D. C. & RICK, C. M., 1954. Self-incompatibility in species of *Lycopersicon* sect. *Eriopersicon* and hybrids with *L. esculentum. Hilgardia, 23:* 101–124.

MENZEL, M. Y., 1962. Pachytene chromosomes of the intergeneric hybrid *Lycopersicon esculentum* × *Solanum lycopersicoides. American Journal of Botany, 49:* 605–615.

MENZEL, M. Y., 1964. Preferential chromosome pairing in allotetraploid *Lycopersicon esculentum-Solanum lycopersicoides. Genetics, 50:* 855–862.

MENZEL, M. Y. & PRICE, J. M., 1966. Fine structure of synapsed chromosomes in F_1 *Lycopersicon esculentum-Solanum lycopersicoides* and its parents. *American Journal of Botany, 53:* 1079–1086.

MULLER, C. H., 1940. A revision of the genus *Lycopersicon. United States Department of Agriculture, Miscellaneous Publication,* 382.

RICK, C. M., 1950. Pollination relations of *Lycopersicon esculentum* in native and foreign regions. *Evolution, 4:* 110–122.

RICK, C. M., 1951. Hybrids between *Lycopersicon esculentum* Mill. and *Solanum lycopersicoides* Dun. *Proceedings of the National Academy of Sciences of the United States of America, 37:* 741–744.

RICK, C. M., 1958. The role of natural hybridization in the derivation of cultivated tomatoes of western South America. *Economic Botany, 12:* 346–367.

RICK, C. M., 1960. Hybridization between *Lycopersicon esculentum* and *Solanum pennellii:* phylogenetic and cytogenetic significance. *Proceedings of the National Academy of Sciences of the United States of America, 46:* 78–82.

RICK, C. M., 1963a. Barriers to interbreeding in *Lycopersicon peruvianum. Evolution, 17:* 216–232.

RICK, C. M., 1963b. Biosystematic studies on Galápagos tomatoes. *Occasional Papers of the California Academy of Sciences, 44:* 59–77.

RICK, C. M., 1971. The genus *Lycopersicon*, In I. L. Wiggins & D. M. Porter (Eds) *Flora of the Galápagos Islands*. Stanford, Calif.: Stanford Univ. Press.

RICK, C. M., 1973. Potential genetic resources in tomato species: clues from observations in native habitats, In A. Hollaender & A. Srb (Eds) *Genes, Enzymes, and Populations*. New York: Plenum.

RICK, C. M., 1976. Tomato (Family *Solanaceae*), In N. W. Simmonds (Ed.) *Crop Plant Evolution*. London: Longman.

RICK, C. M. & FOBES, J. F., 1975. Allozymes of Galápagos tomatoes: polymorphism, geographic distribution and affinities. *Evolution, 29:* 443–457.

RICK, C. M., KESICKI, E., FOBES, J. F. & HOLLE, M., 1976. Genetic and biosystematic studies on two new sibling species of *Lycopersicon* from interandine Perú. *Theoretical and Applied Genetics, 47:* 55–68.

RICK, C. M. & LAMM, R., 1955. Biosystematic studies on the status of *Lycopersicon chilense. American Journal of Botany, 42:* 663–675.

SAWANT, A. C., 1958. Cytogenetics of interspecific hybrids, *Lycopersicon esculentum* Mill. × *L. hirsutum* Humb. and Bonpl., *Genetics, 43:* 502–514.

STEVENSON, F. J. & JONES, H. A., 1953. Some sources of resistance in crop plants. *In USDA Yearbook, Plant Diseases.*

WANN, E. V. & JOHNSON, K. W., 1963. Intergeneric hybridization involving species of *Solanum* and *Lycopersicon. Botanical Gazette, 124:* 451–455.

WEST, H. R., 1973. *A Chemotaxonomic Study of the Genus* Lycopersicon *(Tourn.) Mill.*, M.Sc. Thesis, Univ. Birmingham.

EXPLANATION OF PLATE

PLATE 53.1

A–F. Species of *Solanum*, series *Juglandifolia:* photographs taken in native habitats.

A. *S. ochranthum* LA113. Río Urubamba, below Machupicchu (Cuzco) Perú. Note trilocular nature of transected fruit. Fruits 5–6 cm diameter.

B. *S. ochranthum* LA129. Guaca (Carchi) Ecuador. Vines ascending trees in background. Note inflorescences emerging at top of picture.

C. *S. juglandifolium* LA1235. Alluriquin (Pichincha) Ecuador. Vines climbing in shrubs. Detail of leaf, inflorescence, and fruit. Scale is 3 cm wide.

D. *S. rickii* LA469. Chuquicamata (Antofagasta) Chile. Upper half of plant, approximately 25 cm high.

E. *S. lycopersicoides* LA461. Palca (Tacna) Perú. Detail of foliage, inflorescence, and fruit. Fruits 7–9 mm diameter.

F. *S. lycopersicoides* LA461. Note plant habit and associated vegetation. Plant 1.5 m high.

Plate 53.1

(*Facing p. 678*)

54. Numerical taxonomic studies on variation and domestication in some species of *Capsicum*

BARBARA PICKERSGILL

Department of Agricultural Botany, University of Reading, Reading, England

C. B. HEISER

Department of Plant Sciences, Indiana University, Bloomington, Indiana, U.S.A.

AND

J. McNEILL

Biosystematics Research Institute, Agriculture Canada, Ottawa, Ontario, Canada

Variation patterns in the white-flowered domesticated species of *Capsicum* and their immediate wild relatives were studied using some techniques of numerical taxonomy. Group average and nearest-neighbour clustering and principal co-ordinate analyses all show that *C. baccatum* is clearly distinct from the *C. annuum*–*C. chinense*–*C. frutescens* complex and that domesticated forms of *C. baccatum* are derived from wild forms of the same species. Domesticated forms of *C. annuum*, *C. chinense* and *C. frutescens* also prove clearly distinguishable from one another and from domesticated *C. baccatum* despite considerable parallel evolution in both qualitative and quantitative characters. However, wild forms of *C. annuum*, *C. chinense* and *C. frutescens* form an intergrading and poorly-differentiated complex which cannot readily be divided into distinct taxa.

These analyses suggest independent allopatric domestication of the four chili peppers studied, followed by further divergence of the cultigens after domestication. As in many crop plants, the postulated evolutionary relationships within and between the various wild and domesticated taxa cannot easily be depicted by the formal hierarchy of taxonomic categories.

CONTENTS

INTRODUCTION

The taxonomic history of *Capsicum*, in which the number of species recognized has varied from one (Bailey, 1923) to over 60 (Dunal, 1852) is such as to suggest that this is a difficult genus. In fact, much of the proliferation of binomials in *Capsicum* is due to the not uncommon phenomenon of taxonomic inflation in an economically important group. Like most cultivated plants, chili peppers show striking intraspecific variability, particularly in that part of the plant used by man, in this instance the fruit. Furthermore, since most domesticated peppers are predominantly inbreeding, forms which differ in fruit characters will usually breed true. It is thus not surprising that many "species" recognised in the last century and based solely on fruit characters are now considered to represent simply horticultural variants within a much more limited number of "good" species. Most recent workers (Hazenbus, 1958; Heiser & Pickersgill, 1969; Hunziker, 1958; Terpó, 1966) are in reasonable agreement on the number and distinguishing characteristics of these domesticated species. They are *C. pubescens* Ruiz & Pav., *C. annuum* L., *C. baccatum* L., and the complex of forms at present included in *C. chinense* Jacq. and *C. frutescens* L. Of these, *C. pubescens* is morphologically, cytogenetically and to some extent ecologically distinct and has not been studied further by us. The remaining species are distinguished mainly by floral characters and characters of the calyx at fruit maturity. Other characters, notably size, shape, position, colour and pungency of the fruit, show parallel variation within each of the domesticated species and provide a striking example of Vavilov's Law of Homologous Series in variation (Vavilov, 1922). They are consequently unreliable for species delimitation.

As well as being morphologically distinct, the domesticated species also have rather distinct geographic distributions (Fig. 54.1). Domesticated *C. annuum* seems to have been confined to Mesoamerica in pre-Spanish times (Smith & Heiser, 1957), though it is now much more widespread. Domesticated *C. baccatum* occurs mainly in western and southern tropical South America. Domesticated *C. chinense* is widespread in the West Indies and northern and eastern South America, and overlaps with *C. baccatum* in the Andes and in Brazil.

It has been argued elsewhere (Pickersgill, 1971) that these three cultigens were domesticated independently, from wild forms which were already specifically distinct, in different places and probably at different times. Suitable wild ancestors are known in each group, and the distribution of these is also shown in Fig. 54.1. Wild forms of *C. annuum* occur virtually throughout the pre-Columbian range of domesticated *C. annuum*, whereas wild forms of *C. baccatum* have a more limited distribution than their domesticated counterparts. There are two possible candidates for the wild ancestor of domesticated *C. chinense*. On the one hand, there appear to be some genuinely wild forms within *C. chinense*. These have the small erect deciduous red fruits characteristic of all wild peppers and the calyx constriction diagnostic of *C. chinense* (Smith & Heiser, 1957). They occur in eastern lowland South America, particularly Amazonia. On the other hand, there are the wild forms which have hitherto been treated as a distinct species, *C. frutescens*. These forms are widespread, weedy, and undoubtedly closely related to *C. chinense*. Nevertheless, *C. chinense* and *C. frutescens* have been maintained as distinct species, while the wild and domesticated forms within *C. annuum* and *C. baccatum* have been distinguished only at the varietal level and no taxonomic recognition at all has been accorded to the wild forms of *C. chinense*. This lack of a uniform

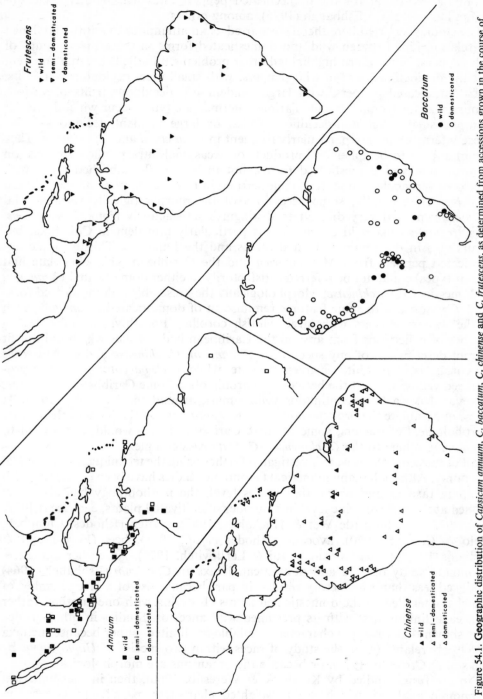

Figure 54.1. Geographic distribution of *Capsicum annuum, C. baccatum, C. chinense* and *C. frutescens,* as determined from accessions grown in the course of these studies.

taxonomic treatment for the domesticated peppers and their putative wild ancestors has perturbed Eshbaugh (1975), among others.

The studies reported here therefore started as an attempt to quantify and depict
the relationships between wild and domesticated forms in these three groups of
chili peppers. This in turn highlighted other problems. Firstly, it became apparent
that the distinction between wild peppers, with small erect deciduous red fruits,
and domesticated peppers, with large pendent non-deciduous fruits of various
colours, was far from clear-cut. Various intermediate types occur which have, for
example, small erect non-deciduous fruits, or large pendent deciduous fruits.
These intermediates are particularly frequent in *C. annuum* and *C. chinense*. They
do not appear to be ephemeral products of occasional intraspecific hybridisation
between wild and domesticated forms, but include well-established types with
their own vernacular names (e.g. chile serrano in *C. annuum*, pimenta de cheiro in
C. chinense). Secondly, some accessions of the morphologically variable wild
C. annuum proved very difficult to distinguish satisfactorily from *C. frutescens*.
This "*frutescens*-like wild *C. annuum*" is particularly prevalent in Colombia, but
occurs also in the southern United States and the Bahamas. Thirdly, some domesticated peppers from Mesoamerica and the Caribbean (chili chocolate and
similar types) could not be referred satisfactorily to either domesticated *C. annuum*
or domesticated *C. chinense*. Morphologically they resemble a domesticated form
of *C. frutescens*, with some features reminiscent of domesticated *C. annuum*, such
as the solitary flowers with large whitish corollas. Fourthly, an embarrassing
number of collections from around the Caribbean had corollas which did not fit
current descriptions of any species. In *C. annuum*, *C. chinense* and *C. frutescens*
the corolla is plain white or greenish-white, while in *C. baccatum* there are pronounced yellow spots at the base of the corolla lobes. Some Caribbean collections,
however, have a faint continuous yellow mark around the base of the corolla
lobes, quite unlike the broken spotting of *C. baccatum*. On the basis of their other
morphological characters, some of these Caribbean forms would be assigned to
C. annuum, others to the *C. chinense*—*C. frutescens* complex.

All of these problems were investigated further using the techniques of numerical
taxonomy. Although many numerical taxonomic studies have been concerned with
grouping taxa at and above the species level, the methodology has also been
applied at the infra-specific level in a number of cultivated plants, e.g. rice (Chu &
Oka, 1972), sorghum (de Wet & Huckabay, 1967), mango (Rhodes, Campbell,
Malo & Carmer, 1970), avocado (Rhodes *et al.*, 1971), beet (Ford-Lloyd &
Williams, 1975), *Cannabis* (Small, Jui & Lefkovitch, 1976). An earlier numerical
taxonomic study of wild and domesticated taxa in *Capsicum* (Eshbaugh, 1964)
produced less than satisfactory results, in part because use of a large number of
size characters led to the domesticated forms clustering with one another, rather
than each linking first with its presumed wild ancestor. Studies in other groups
have shown that when the characters available are limited, either because the taxa
are closely related (as in the study of races within two species of *Haplopappus* by
Jackson & Crovello, 1971), or because the organisms are morphologically simple
(as in the fungi studied by Kendrick & Weresub, 1966), then in the numerical
taxonomic analyses, which use unweighted characters, parallel or convergent
evolution in several of these characters may obscure the basic divisions perceived

by the taxonomist, who uses his own subjective character weighting. Since the species of *Capsicum* included in our analyses show extensive homologous variation, it was of some methodological interest to determine whether satisfactory discrimination between them could be achieved.

MATERIALS AND METHODS

Data were collected over a period of about 10 years from plants grown in the experimental field at Indiana University. Five to eight plants (occasionally fewer) were grown for each accession and the individual accessions treated as OTUs in the subsequent analyses. Voucher specimens of each accession were deposited in the herbarium of Indiana University and a list of accessions included in these analyses can be provided on request. The characters used are listed in Table 54.1. Of the

Table 54.1. Characters used in the numerical analyses

1. Leaf area (cm²)
2. Leaf pubescence (square root of no. hairs per cm²)
3. Leaf texture (smooth; rugose; very rugose)
4. Leaf colour (green; purple)
5. Peduncle no. per node
6. Peduncle length (mm)
7. Peduncle width (mm)
8. Peduncle length: peduncle width
9. Calyx teeth (absent; slight; present)
10. Calyx constriction (absent; present)
11. Corolla colour (greenish; greenish-white; white)
12. Corolla anthocyanin (absent; present)
13. Corolla spot (absent; present)
14. Other corolla markings (yellow mark absent; yellow mark present)
15. Corolla length (mm)
16. Corolla tube length (mm)
17. Depth of corolla lobing ((corolla length—corolla tube length)/corolla length)
18. Anther colour (yellow; purple)
19. Filament constriction (absent; present)
20. Stamen length (mm)
21. Pistil length (mm)
22. Style exsertion ((pistil length—stamen length) + 5)
23. Filament anthocyanin (absent; present)
24. Style anthocyanin (absent; present)
25. Immature fruit anthocyanin (absent; present)
26. Immature fruit colour (cream; pale green; green; dark green)
27. Red pigments in mature fruit (absent; present)
28. Chlorophyll retention in mature fruit (retained; not retained)
29. Inhibition of pigment synthesis in mature fruit (uninhibited; partially inhibited; strongly inhibited)
30. Fruit position (erect; pendent)
31. Fruit pungency (pungent; non-pungent)
32. Fruit dispersal (deciduous; non-deciduous)
33. Fruit length (cm)
34. Fruit width (cm)
35. Fruit size (fruit length × fruit width)
36. Fruit shape (fruit width/fruit length)
37. Inner wall of fruit (rough; smooth)
38. No. seeds per fruit
39. Seed width (mm)

total of 39 characters, 12 were considered useful primarily in separating species, while 17 were useful primarily in separating wild from domesticated peppers. The remaining 10 characters included attributes such as pubescence, which varied in the accessions under study but whose variation was not obviously correlated either with differences between species or with changes associated with domestication. There were 21 qualitative and 18 quantitative characters. Since chili peppers are predominantly inbreeding and since obviously hybrid accessions were omitted from these analyses, there was no intra-accession variation in qualitative characters and hence no need to use the character state frequency procedure of McNeill (1974). Instead, qualitative characters were treated simply as binary or ordered multistate characters. For quantitative characters, the values used were means obtained from single measurements on every plant of each accession, transformed or otherwise adapted where necessary (see Table 54.1) to ensure that a unit difference was of approximately equal significance throughout the range. Several accessions were grown in each of two different years. These paired sets of data were treated as distinct OTUs and provided some estimate of the degree of dissimilarity which could be produced by different growing conditions in different years.

The OTU by OTU similarity assessment was made using Gower's general coefficient of similarity (Gower, 1971), which assesses qualitative and quantitative characters simultaneously. Negative matches (e.g. absence of calyx teeth in each of a pair of OTUs) were treated as similarities. Where data were missing for a particular character, the procedure used in effect eliminates that character from calculation of the similarity coefficients for the particular OTU concerned. Clustering was performed using group average and nearest-neighbour methods. The matrix of OTU—by—OTU similarities was used to derive the original co-ordinates of the OTUs, by the method of Gower (1966). Principal co-ordinate analyses were then carried out. Further details of these methods are given by Sneath & Sokal (1973) and McNeill (1975).

Data were available on more accessions than could be handled without requiring unreasonably long run times. The accessions used were therefore selected to represent most of the geographic range and all the major morphological variants within each species (including examples of homologous variation among the domesticated peppers) and to include all the anomalous accessions whose positions were in doubt. The first run involved 194 OTUs, made up of 22 wild *C. annuum*, 41 domesticated *C. annuum*, 13 wild *C. baccatum*, 32 domesticated *C. baccatum*, 62 *C. chinense* and 24 *C. frutescens*.

In the diagrammatic representations of the results, individual accessions are indicated by symbols rather than accession numbers. These symbols represent the opinion of one of us (B.P., usually in agreement with C.B.H.) as to the identity of the accessions prior to the start of the numerical analyses. These subjective determinations are shown simply to indicate the degree of agreement between the computer groupings and the taxa previously recognised; they were not taken into account in any way in arriving at the computer groupings. The cytological data available (Pickersgill, 1971, and unpubl.) have also been summarised on the diagrams but were likewise not used in the computations of dissimilarities.

RESULTS

First run

Group average dendrogram (Fig. 54.2)

With only a few exceptions, OTUs representing the same accession grown in different years cluster at the first levels of fusion. In general, therefore, dissimilarities between accessions are much greater than dissimilarities within accessions produced by variation in environmental conditions.

The dendrogram also shows that domesticated peppers are somewhat more heterogeneous (the within-group dissimilarities are greater) than wild peppers, as one would expect from a comparison of variable cultigens with their less variable wild relatives. Despite this heterogeneity, domesticated accessions of *C. annuum*,

C. baccatum and *C. chinense* form their own distinctive groups before fusing with groups representing any of the other taxa. The basic distinction between domesticated *C. annuum*, domesticated *C. baccatum* and domesticated *C. chinense* has thus not been obscured even though many examples of homologous variation were included and the analysis used more characters which separate wild from domesticated peppers than which separate species.

The major division in the dendrogram distinguishes *C. baccatum* from the other species. Domesticated *C. annuum* and domesticated *C. chinense* fuse more or less simultaneously with each other and with a group containing predominantly wild and semi-domesticated accessions of *C. annuum*, *C. chinense* and *C. frutescens*. Within this latter group, wild *C. frutescens* forms one subgroup and wild and semi-domesticated accessions of *C. annuum* comprise another subgroup. Between these two is a subgroup consisting mainly of *frutescens*—like wild *C. annuum*. Accessions with the yellow corolla markings are scattered throughout this section of the dendrogram.

In the *C. baccatum* group there is a major division into domesticated versus wild forms, but the subgroup representing wild *C. baccatum* does fuse with the subgroup representing domesticated *C. baccatum* rather than with the other wild peppers. This suggests that the clustering of wild *C. annuum* and *C. frutescens* with each other rather than with domesticated *C. annuum* and *C. chinense* respectively cannot be attributed to use of too many characters separating wild from domesticated peppers, but is an accurate reflection of the biological situation.

Minimum spanning tree (Fig. 54.3)

This is derived from the nearest-neighbour analysis and shows accurately the distance of each accession from its nearest neighbour or neighbours. The only defined distances are those along the lines in the diagram: the angles of branching, spatial separation of branches and hence all other distances are arbitrary.

In almost every case where one accession was represented by two OTUs, these two OTUs are nearest neighbours and are separated by some of the shortest distances in the whole tree. Again, therefore, environmentally induced differences are small compared to genetic dissimilarities between accessions.

The accessions representing *C. baccatum* once more form a distinct group, with a clear separation between wild and domesticated accessions. They join the other taxa through a link between a domesticated *C. baccatum* and a domesticated *C. annuum*. The minimum spanning tree, like the group average dendrogram, thus shows wild *C. baccatum* to be distinct from the other wild peppers in this study, although some convergence has occurred between domesticated *C. baccatum* and domesticated *C. annuum*. Domesticated *C. annuum* itself makes another discrete group, although some domesticated forms with erect fruits are placed among semi-domesticated and wild *C. annuum* rather than with the main group of domesticated *C. annuum*.

Most of the "backbone" of the tree consists of various semi-domesticated and wild accessions of *C. annuum*. The single wild accession of *C. chinense*, together with most of the semi-domesticated accessions of *C. chinense*, are rather incongruously placed, with their near neighbours being wild *C. annuum*, not other accessions of *C. chinense*. Conversely, two Costa Rican accessions of wild *C. annuum* find their

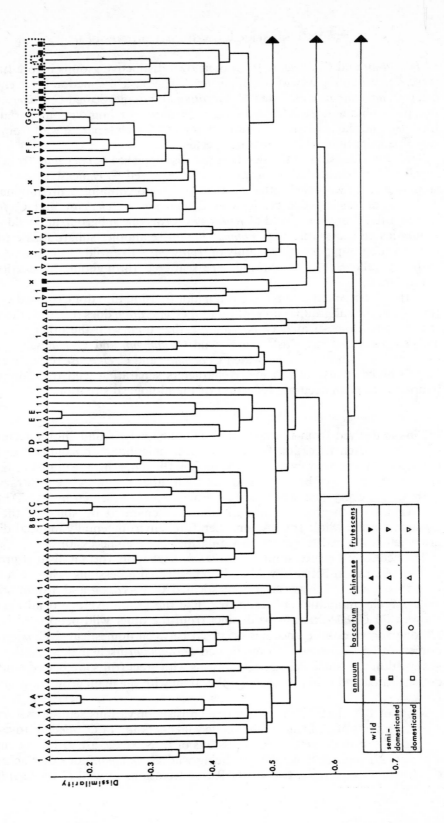

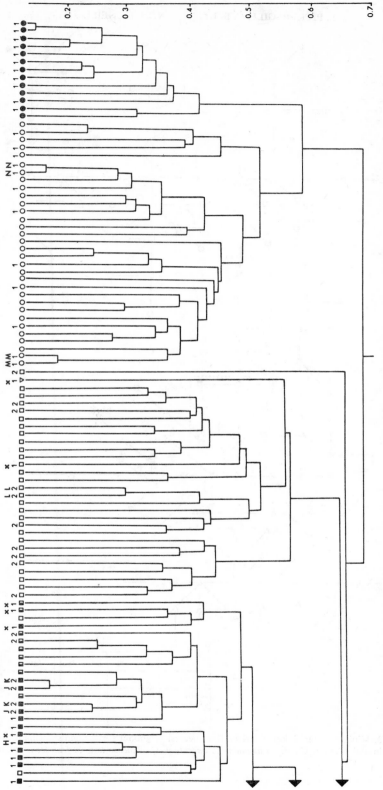

Figure 54.2. Dendrogram derived by group average sorting and representing relationships within and between *Capsicum annuum*, *C. baccatum*, *C. chinense* and *C. frutescens* (A,A; B,B; etc.: same accessions grown in different years, X: accessions with faint yellow corolla markings, 1: accessions with one pair of acrocentric chromosomes, 2: accessions with two pairs of acrocentric chromosomes, dotted line encloses accessions of "*frutescens*-like wild *C. annuum*").

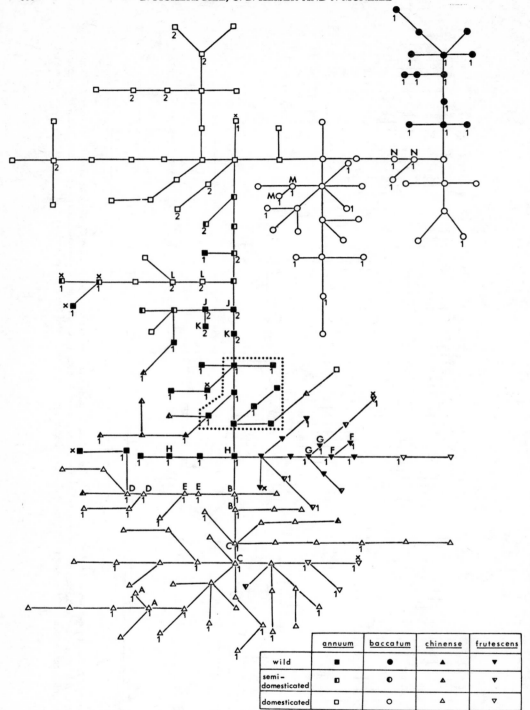

Figure 54.3. Minimum spanning tree derived from nearest neighbour sorting and showing relationships within and between *Capsicum annuum*, *C. baccatum*, *C. chinense* and *C. frutescens* (symbols etc. as in Fig. 54.2).

near neighbours in a group of small-fruited forms of *C. chinense*, not among the other wild accessions of *C. annuum*. Wild forms of *C. frutescens* again make a coherent group, but this group has a wild *C. annuum* (H) as its nearest neighbour, not *C. chinense*. However, *frutescens*-like wild *C. annuum* does find its near neighbours among other accessions of wild *C. annuum* rather than among *C. frutescens*. Accessions with the yellow corolla markings are again scattered throughout the diagram.

Principal co-ordinates analysis

The positions of the OTUs on the first and second principal axes (or co-ordinates) derived from the original pairwise distances among the OTUs are shown in Fig. 54.4. The first axis accounts for approximately 22% of the total variation and separates the species. The second axis accounts for about 16% of the total variation and separates wild from domesticated peppers. The similarities between OTUs representing the same accession grown in different years are less apparent in this analysis than in the group average dendrogram or the minimum spanning tree. However, *C. baccatum* again emerges as clearly distinct from the other taxa.

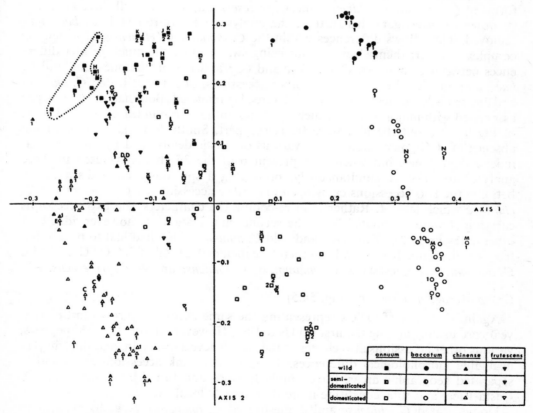

Figure 54.4. Plot of accessions representing *Capsicum annuum*, *C. baccatum*, *C. chinense* and *C. frutescens* on the first two axes recognized by a principal co-ordinates analysis (symbols etc. as in Fig. 54.2).

Domesticated *C. annuum* and domesticated *C. chinense* are also distinct, with domesticated *C. frutescens* tending to occupy a position between them. There is no sharp demarcation between wild and domesticated peppers, especially in the *C. annuum–C. chinense–C. frutescens* group, where the intergradation is virtually continuous.

As in the group average dendrogram and the minimum spanning tree, although wild *C. annuum*, wild *C. frutescens* and *frutescens*–like wild *C. annuum* tend to form their own groups, these are by no means distinct. In an attempt to resolve them further, the positions of the OTUs on the first and third principal axes were plotted. The third axis accounts for about 8 % of the total variation but apparently serves mainly to separate domesticated *C. annuum* from domesticated *C. chinense* and does not contribute usefully to further separation of the wild forms.

Accessions with the yellow corolla markings occur in the groups representing *C. frutescens*, wild *C. annuum*, and domesticated *C. annuum* and among the wild-domesticated intermediates.

Second run

An interesting outcome of the first run was the lack of distinction between wild forms of *C. annuum*, *C. chinense* and *C. frutescens*, apparent in all three analyses. In order to investigate this further, the analyses were re-run with *C. baccatum* removed. This allows differences within the *C. annuum–C. chinense–C. frutescens* complex to assert themselves without being swamped by the much greater differences between *C. annuum*, *C. chinense* and *C. frutescens* on the one hand and *C. baccatum* on the other hand. Two other accessions, one domesticated *C. annuum* and one semi-domesticated *C. chinense*, were also removed. Both had the phenotype associated with major gene *A*, which produces anthocyanin in leaves, several parts of the flower, and the immature fruits (Lippert, Smith & Bergh, 1966). In the absence of *A*, pigmentation of the various organs is determined by different and independent genes, but when *A* is present 6 of the 39 characters used in these analyses are affected simultaneously, producing a spuriously large dissimilarity between the two accessions carrying *A* and other accessions of *C. annuum* and *C. chinense* which lack *A*. Rather than re-coding the characters, the two accessions carrying *A* were eliminated from the second run. Data were however added on other accessions of *C. frutescens* and wild *C. annuum* which had had to be omitted from the first run. The second run therefore included a total of 200 OTUs: 56 wild *C. annuum*, 39 domesticated *C. annuum*, 61 *C. chinense* and 44 *C. frutescens*.

Group average dendrogram (Fig. 54.5)

Again, the pairs of OTUs representing the same accessions grown in different years are usually among the first OTUs to fuse. However, among the wild peppers, various other accessions also cluster at these early levels of fusion, and thus display negligible morphological differences. These clusters link accessions from widely separated geographic areas, for example Jamaica and Brazil, or Guatemala and Costa Rica, and thus constitute more than merely local races.

Domesticated *C. chinense* and domesticated *C. annuum* emerge as even more distinct than they were in the first run, though some erect-fruited forms of each have been included in the central complex of wild and semi-domesticated peppers.

Two large groups can be distinguished in this central complex: one representing "good" *C. annuum* and the other representing "good" *C. frutescens*. Between these is a group containing mainly accessions of *frutescens*-like wild *C. annuum*. This group fuses with *C. frutescens* rather than with wild *C. annuum*. Wild *C. frutescens* and the group containing most accessions of wild *C. annuum* once more fuse together rather than fusing with their presumed derivatives, domesticated *C. chinense* and domesticated *C. annuum* respectively. A further group of wild *C. annuum* joins the main group of wild peppers only after the groups representing wild *C. annuum* and wild *C. frutescens* have fused. This apparently rather distinct group of wild *C. annuum* consists mainly of accessions from Costa Rica and includes those which in the minimum spanning tree from the first run found their nearest neighbours among *C. chinense* rather than *C. annuum*. Domesticated *C. frutescens* forms a distinct, although internally somewhat heterogeneous, group which joins the main *C. annuum*–*C. frutescens* complex at a still greater level of dissimilarity.

Accessions with yellow markings on the corolla can now be seen to be particularly common in wild *C. annuum*, and to be characteristic of particular groups within this morphologically variable taxon. However comparable forms also occur, although less commonly, in domesticated *C. annuum* and in *C. frutescens*.

Minimum spanning tree (Fig. 54.6)

Domesticated *C. annuum* and domesticated *C. chinense* again form cohesive but separated groups. Most of the accessions of domesticated *C. frutescens* appear closer to domesticated *C. chinense* than to wild *C. frutescens*, though a few do occur with their putative wild relatives. Amongst the wild peppers, wild and semi-domesticated *C. annuum* constitute one group, which is somewhat distinct from *C. frutescens* and *frutescens*-like *C. annuum*. The latter now has *C. frutescens* and *C. chinense* as near neighbours instead of other accessions of wild *C. annuum*, which were its near neighbours in the first run. However, wild and semi-domesticated accessions of *C. chinense* still have wild *C. annuum*, not *C. frutescens*, as their near neighbours. The group of Costa Rican wild *C. annuum*, which in the first run joined *C. chinense* rather than *C. annuum*, again failed to join the other accessions of wild *C. annuum*, but this time is linked with *C. frutescens* (F). The distinction between wild forms of *C. frutescens* and *C. annuum* is thus again far from complete, and again the *C. frutescens* group joins domesticated *C. chinense* via a wild form of *C. annuum* rather than directly.

As in the group average dendrogram, yellow corolla markings are characteristic of particular lines within wild *C. annuum*, but otherwise occur sporadically in various parts of the diagram.

Principal co-ordinates analysis

In this run, the first axis accounts for about 18% of the total variation and the second axis about 11%. The first axis separates wild from domesticated peppers, while the second axis separates species. Thus, when *C. baccatum* is removed from the analysis, the largest proportion of the variation in the characters studied is associated with change under domestication, not with differentiation into species.

A plot of the positions of the OTUs on axis 1 against axis 2 (Fig. 54.7) again produces a good separation between domesticated *C. annuum* and domesticated

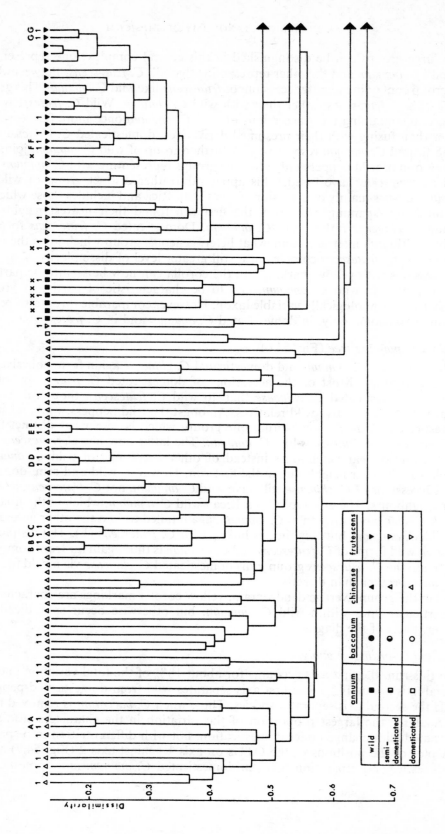

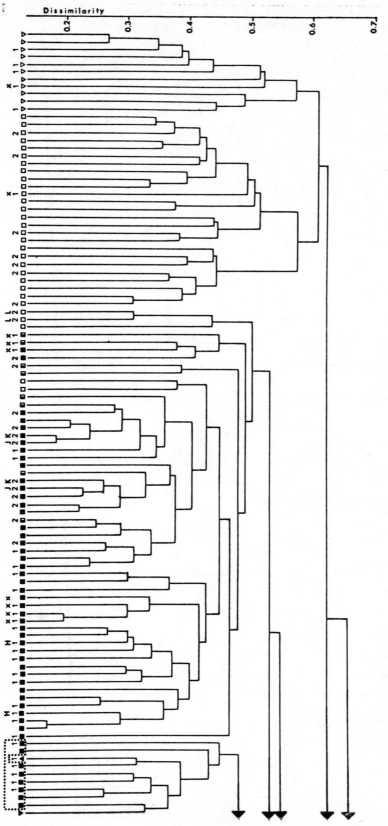

Figure 54.5. Dendrogram derived by group average sorting and representing relationships within and between *Capsicum annuum*, *C. chinense* and *C. frutescens* (symbols etc. as in Fig. 54.2).

693

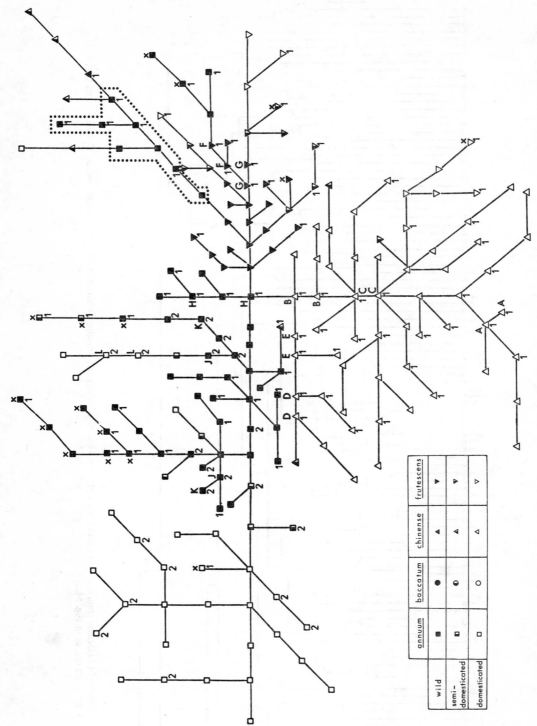

Figure 54.6. Minimum spanning tree derived from nearest neighbour sorting and showing relationships within and between *Capsicum annuum*, *C. chinense* and *C. frutescens* (symbols etc. as in Fig. 54.2).

	annuum	baccatum	chinense	frutescens
wild	▣	●	▲	▶
semi-domesticated	▢	◉	▲	▽
domesticated	☐	○	△	▷

C. chinense and again shows that any division between wild and domesticated forms within these two groups is distinctly arbitrary. Most of the accessions of domesticated *C. frutescens* occupy an intermediate position between domesticated *C. annuum* and domesticated *C. chinense*. This was expected on their morphological characters, but was obscured in both the group average dendrogram and the minimum spanning tree, which allow intermediate forms to show similarity to only one, not both, of the groups between which they lie.

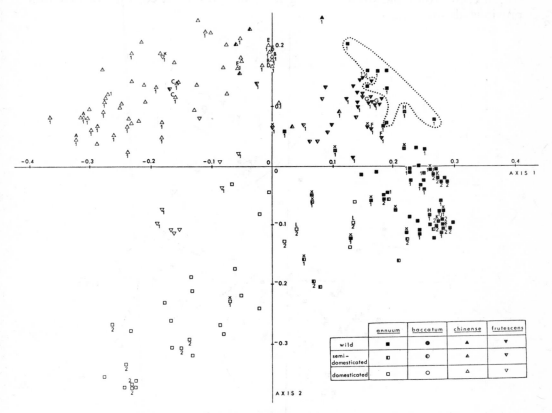

Figure 54.7. Plot of accessions representing *Capsicum annuum*, *C. chinense* and *C. frutescens* on the first two axes recognized by a principal co-ordinates analysis (symbols etc. as in Fig. 54.2).

In both this and the earlier principal co-ordinates analysis (Fig. 54.4) the domesticated accessions of each species tend to scatter about a straight line, of approximately comparable slope for each group. In Fig. 54.4 virtually all the wild forms of *C. baccatum* occur along the same line as the domesticated forms but in Figs 54.4 and 54.7 only "typical" wild *C. annuum* (not Costa Rican wild *C. annuum*, nor *frutescens*-like wild *C. annuum*) occurs along the same line as domesticated *C. annuum*. When the line from domesticated *C. chinense* is projected back it encounters wild *C. chinense* not wild *C. frutescens*, while wild *C. frutescens* lies on the same line as domesticated *C. frutescens*. If these projections do indeed indicate the probable wild forms which gave rise to the various domesticates, then they support suggestions of independent domestication and also provide some reason for retaining *C. frutescens* and *C. chinense* as distinct taxa.

Although two, possibly three, domesticated taxa can be recognized in Fig. 54.7, the corresponding wild forms still do not form clear-cut groups. There is some separation between "typical" wild *C. annuum* (predominantly below axis 1) and *C. frutescens* and *frutescens*-like wild *C. annuum* (predominantly above axis 1), but these distinctions are no sharper than those between wild and domesticated forms within each group and division along these lines does not correspond well with previous identifications.

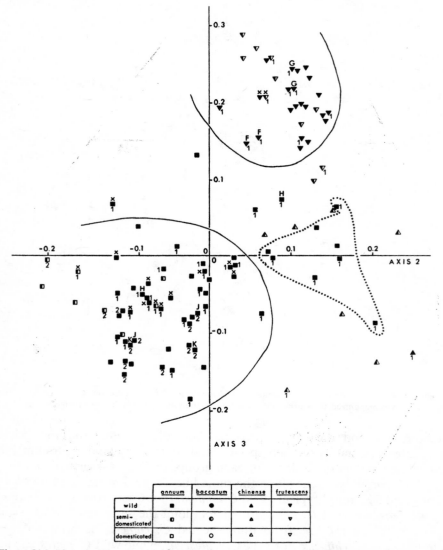

	annuum	baccatum	chinense	frutescens
wild	■	●	▲	▼
semi-domesticated	◧	◑	◮	▾
domesticated	▢	○	△	▽

Figure 54.8. Plot of wild and semi-domesticated accessions of *Capsicum annuum*, *C. chinense* and *C. frutescens* on the second and third axes recognized by a principal co-ordinates analysis (symbols etc. as in Fig. 54.2; solid lines enclose "typical" *C. annuum* and "typical" *C. frutescens*).

The third axis accounts for about 7% of the total variation. In an attempt to discriminate further between the wild taxa, the positions of the OTUs were plotted on axis 2 (already shown to separate species) against axis 3. The results are shown in Fig. 54.8 in which positions of the wild and semi-domesticated accessions only are shown. The domesticated peppers have already been adequately distinguished by their positions on axes 1 and 2 and, since neither axis 2 nor axis 3 discriminates between wild versus domesticated peppers, adding the domesticates to Fig. 54.8 merely obscures the picture. In Fig. 54.8, for the first time in these analyses, a satisfactory separation between wild *C. annuum* and wild *C. frutescens* is achieved. When boundary lines are drawn around "typical" wild *C. annuum* and "typical" wild *C. frutescens*, as indicated in Fig. 54.8, then the two clusters of points can be seen to be appreciably separated. The points outside these clusters represent *frutescens*-like wild *C. annuum*, together with most of the other Colombian accessions of wild *C. annuum*, some of the Costa Rican accessions of wild *C. annuum* which tended to form anomalous groupings in other analyses, and some semi-domesticated accessions of *C. frutescens*. As in the minimum spanning tree, wild *C. chinense* appears quite distinct from wild *C. frutescens*.

In both Figs 54.7 and 54.8 accessions with yellow corolla markings are well scattered throughout the diagrams.

DISCUSSION AND CONCLUSIONS

Of the problems raised at the start of these studies, the most easily dismissed concerns the status of the accessions with faint yellow markings on the corolla. In all the analyses which have been carried out, these accessions occur among most of the major groupings recognized: domesticated *C. annuum*, the domesticated *C. chinense—C. frutescens* complex, wild *C. frutescens* and, most abundantly, in wild *C. annuum*. This suggests that these accessions have no major features in common other than the markings on the corolla and their circum-Caribbean distribution. Some of these accessions showed segregation for presence versus absence of the corolla mark and this, together with preliminary results from inheritance studies, indicates that presence of the corolla mark is probably under fairly simple genetic control. Occurrence of similar variants in both *C. annuum* and the *C. chinense— C. frutescens* group could result from hybridisation but, since there is no indication of convergence in other morphological characters, it seems more likely to constitute yet another example of independent but homologous variation in these closely-related species. The role, if any, of these corolla markings in pollination biology remains to be investigated, but the morphological analyses reported here provide no basis for suggesting that these forms deserve any taxonomic recognition.

At the start of these studies there were thought to be three wild/domesticated pairs, belonging respectively to *C. annuum*, *C. baccatum* and the *C. chinense– C. frutescens* complex. The numerical analyses have confirmed that *C. baccatum* is distinct and that domesticated forms of *C. baccatum* are most probably derived from wild forms of the same species, as Smith and Heiser (1957) and Eshbaugh (1970) have indicated. In the *C. annuum–C. chinense–C. frutescens* group, however, our studies suggest three rather than two distinct groups of domesticates and a poorly differentiated complex of wild forms. The numerical taxonomic analyses used only morphological data, but when combined with data on karyotype morphology (Pickersgill, 1971 and unpubl.) and geographic distribution, they suggest a possible course of evolution in this group.

All the species of *Capsicum* discussed here have the same chromosome number, $2n = 24$. The basic karyotype in this group consists of one pair of acrocentric chromosomes plus 11 pairs of meta- and submetacentric chromosomes. The basic karyotype is found in *C. chinense*, *C. frutescens*, *frutescens*-like wild *C. annuum*, Costa Rican wild *C. annuum* and several other wild forms of *C. annuum*. Domesticated *C. annuum* has a distinctive karyotype, with two pairs of acrocentric chromosomes. This karyotype is found also in wild *C. annuum* in central Mexico. The second pair of acrocentric chromosomes appears to have arisen as a result of an unequal reciprocal translocation between two of the pairs of meta- or submetacentrics in the basic complement (Koompai, 1976; Pickersgill, unpubl.). Translocation heterozygotes produced by appropriate crosses between wild and domesticated, or within wild, forms of *C. annuum* have their pollen stainability reduced to 30–50%. Thus there are sterility barriers within *C. annuum* as marked as those between *C. chinense* and *C. frutescens* or between wild *C. frutescens* and wild *C. annuum*.

It seems probable that the immediate ancestor of the present-day *C. annuum*–*C. chinense*–*C. frutescens* group was a variable complex of wild forms carrying the basic karyotype and perhaps similar in appearance to the *frutescens*-like wild *C. annuum* found today. This ancestral complex probably occurred throughout South America, Mesoamerica, southern North America and the West Indies. It presumably resembled similar modern wild peppers in being self-compatible and occurring in small populations separated by distances too great to permit much inter-populational gene exchange. These factors would favour morphological differentiation in different parts of the range: into "typical" wild *C. annuum* in Mesoamerica, wild *C. chinense* in Amazonia, "typical" wild *C. frutescens* in South America and the West Indies, and so on. From this complex of partially differentiated wild forms the domesticated peppers developed, by independent domestication of local wild peppers in highland Mexico (*C. annuum*), lowland South America (*C. chinense*), and possibly also lowland Mesoamerica (*C. frutescens*).

Multiple domestication has not always been favourably regarded, but it is more probable for crops such as chili peppers than for man's staple crops. In staples such as the cereals and legumes, yield is a primary consideration, and indigenous wild forms are at a competitive disadvantage compared with domesticates possessed by neighbouring cultures. In this situation a prehistoric group is likely to acquire its domesticates from its agricultural neighbours, rather than to repeat the domestication process itself. In chili peppers, however, yield is unlikely to be a primary consideration for primitive cultivators. Wild peppers produce numerous small fruits which are satisfactory in quality and other characteristics, witness the continued exploitation of wild peppers today. In chili peppers, then, the sequence of changes resulting in domestication could well have occurred independently several times among different prehistoric cultures.

It is easier to suggest the probable course of evolution in the *C. annuum*–*C. chinense*–*C. frutescens* group than to suggest a suitable taxonomic treatment. Domesticated *C. chinense* and domesticated *C. annuum* are sufficiently distinct morphologically and geographically, and hybrids between them are sufficiently sterile (see Pickersgill, 1971), to provide a case for treating them as different species. Domesticated *C. frutescens* has been less studied but may prove to deserve equal rank with domesticated *C. annuum* and domesticated *C. chinense*. However, as has been shown, these three domesticates are not sharply separated from their respective wild forms (hence the current treatment of wild versus domesticated peppers as distinct only at the varietal level) and the wild forms intergrade among themselves to an extent that makes it difficult to propose that they should be correspondingly divided into three species. The situation is similar to that mentioned by Smartt (1976) in *Vigna*, where one wild species, *V. radiata* var. *sublobata*, appears to have given rise to two specifically distinct cultigens, *V. radiata* var. *radiata* and *V. mungo*. Such situations are hard to represent adequately within the constraints of the formal taxonomic hierarchy. For the moment, it seems best to retain the three specific names *C. annuum*, *C. chinense* and *C. frutescens*, in the senses in which they are now customarily used, particularly since the numerical taxonomic analyses suggest that wild *C. chinense* and wild *C. frutescens* do after all differ significantly and that wild *C. chinense* is a more plausible ancestor than wild *C. frutescens* for domesticated *C. chinense*. The specific names can then be qualified where necessary by descriptive phrases, without any implications of taxonomic rank, as we have done here, pending more detailed investigations of the wild peppers.

ACKNOWLEDGEMENTS

Experimental work carried out at Indiana University was made possible by a National Science Foundation grant to C. B. Heiser. We are also extremely grateful to all those, too many to mention individually, who have collected chili peppers for us so assiduously, in so many parts of Latin America, over so many years.

REFERENCES

BAILEY, L. H., 1923. *Capsicum. Gentes Herbarum, 1:* 128–129.

CHU, YAW-EN & OKA, HIKO-ICHI, 1972. The distribution and significance of genes causing F_1 weakness in *Oryza breviligulata* and *O. glaberrima. Genetics, 70:* 163–173.

DE WET, J. M. J. & HUCKABAY, J. P., 1967. The origin of *Sorghum bicolor*. II. Distribution and domestication. *Evolution, 21:* 787–802.

DUNAL, F., 1852. *Capsicum*. In Alphonse de Candolle, *Prodromus Systematis Naturalis Regni Vegetabilis, 13* (1): 411–429. Paris: Masson.

ESHBAUGH, W. H., 1964. *A Numerical Taxonomic and Cytogenetic Study of Certain Species of the Genus Capsicum.* Ph.D. thesis, Indiana University Library.

ESHBAUGH, W. H., 1970. A biosystematic and evolutionary study of *Capsicum baccatum* (Solanaceae). *Brittonia, 22:* 31–43.

ESHBAUGH, W. H., 1975. Genetic and biochemical systematic studies of chili peppers (*Capsicum*—Solanaceae). *Bulletin of the Torrey Botanical Club, 102:* 396–403.

FORD-LLOYD, B. V. & WILLIAMS, J. T., 1975. A revision of *Beta* section *Vulgares* (Chenopodiaceae), with new light on the origin of cultivated beets. *Botanical Journal of the Linnean Society, 71:* 89–102.

GOWER, J. C., 1966. Some distance properties of latent root and vector methods used in multivariate analysis. *Biometrika, 53:* 325–338.

GOWER, J. C., 1971. A general coefficient of similarity and some of its properties. *Biometrics, 27:* 857–874.

HAZENBUS, V. L., 1958. Pepper—*Capsicum* Tourn. In P. M. Zhukovsky (Ed.), *Flora of Cultivated Plants of the USSR, XX. Vegetable Plants, Family Solanaceae:* 394–487. Moscow & Leningrad: State Agricultural Publishing Office.

HEISER, C. B. & PICKERSGILL, B., 1969. Names for the cultivated *Capsicum* species (Solanaceae). *Taxon, 18:* 277–283.

HUNZIKER, A. T., 1958. Synopsis of the genus *Capsicum*. *Comptes Rendus 8ieme Congrès International de Botanique, Paris, Sections 3–6*, 73–74.

JACKSON, R. C. & CROVELLO, T. J., 1971. A comparison of numerical and biosystematic studies in *Haplopappus*. *Brittonia, 23:* 54–70.

KENDRICK, W. B. & WERESUB, L. K., 1966. Attempting neo-Adansonian computer taxonomy at the ordinal level in the Basidiomycetes. *Systematic Zoology, 15:* 307–29.

KOOMPAI, P., 1976. *Some Barriers to Interspecific Crossing and Gene Exchange in Five Species of* Capsicum. M.Phil. thesis, Reading University Library.

LIPPERT, L. F., SMITH, P. G. & BERGH, B. O., 1966. Cytogenetics of the garden crops. Garden pepper, *Capsicum* sp. *Botanical Review, 32:* 24–55.

MCNEILL, J., 1974. The handling of character variation in numerical taxonomy. *Taxon, 23:* 699–705.

MCNEILL, J., 1975. A generic revision of Portulacaceae tribe Montieae using techniques of numerical taxonomy. *Canadian Journal of Botany, 53:* 789–809.

PICKERSGILL, B., 1971. Relationships between weedy and cultivated forms in some species of chili peppers (genus *Capsicum*). *Evolution, 25:* 683–691.

RHODES, A. M., CAMPBELL, C., MALO, S. E. & CARMER, S. G., 1970. A numerical taxonomic study of the mango, *Mangifera indica* L. *Journal of the American Society of Horticultural Science, 95:* 252–256.

RHODES, A. M., MALO, S. E., CAMPBELL, C. W. & CARMER, S. G., 1971. A numerical taxonomic study of the avocado (*Persea americana* Mill.). *Journal of the American Society of Horticultural Science, 96:* 391–395.

SMALL, E., JUI, P. Y. & LEFKOVITCH, L. P., 1976. A numerical taxonomic analysis of *Cannabis* with special reference to species delimitation. *Systematic Botany, 1:* 67–84.

SMARTT, J., 1976. Comparative evolution of pulse crops. *Euphytica, 25:* 139–143.

SMITH, P. G. & HEISER, C. B., 1957. Taxonomy of *Capsicum sinense* Jacq. and the geographic distribution of the cultivated *Capsicum* species. *Bulletin of the Torrey Botanical Club, 84:* 413–420.

SNEATH, P. H. A. & SOKAL, R. R., 1973. *Numerical Taxonomy.* San Francisco: Freeman.

TERPÓ, A., 1966. Kritische Revision der wildwachsenden Arten und der kultivierten Sorten der Gattung *Capsicum* L. *Feddes Repertorium Specierum Novarum Regni Vegetabilis, 72:* 155–191.

VAVILOV, N. I., 1922. The law of homologous series in variation. *Journal of Genetics, 12:* 47–89.

55. A preliminary biochemical systematic study of the genus *Capsicum* — Solanaceae

MICHAEL J. MCLEOD, W. HARDY ESHBAUGH AND SHELDON I. GUTTMAN

Departments of Botany and Zoology, Miami University, Oxford, Ohio, U.S.A.

Using isoenzyme analysis the authors have examined the relationships of the domesticated chili peppers and their hypothetical progenitor species as well as several wild taxa. The data indicate a high correlation with previously reported results from genetic and chemotaxonomic research.

CONTENTS

INTRODUCTION

Although chili peppers cannot be considered as one of the world's major economic crops they are one of the major spice crops and have significant economic importance for individual countries or localized geographic regions. The genus *Capsicum* includes approximately 20–30 species/taxa of New World origin several of which have been spread throughout the world since the time of Columbus. Recent authors recognize three or four domesticated species. However, many of the wild taxa are in a state of semi-cultivation in which they are found not only in the wild but also in and on the edge of the crop field where they are maintained, harvested, and later sold in the marketplace alongside the several domesticated species. Some workers have suggested that the domesticated Capsicums have had but a single origin (Davenport, 1970; Jett, 1973) while other investigators believe in multiple independent origins of these taxa in quite different geographical regions and perhaps even at different time intervals (Pickersgill, 1972; Heiser, Eshbaugh & Pickersgill, 1971). It is also possible that an individual domesticated species might have arisen more than once from the ancestral wild gene pool, especially when one considers the wide geographic range of a wild progenitor type and the number of civilizations which might have had contact with the potential ancestral stock.

Systematic treatments of the genus *Capsicum* have recognized as few as a single species (Bailey, 1923) to as many as 63 (Dunal, 1852) species. Our understanding of the taxonomy of the genus will not be complete until we have better collections and field work throughout South America and especially in Paraguay and Brazil. Within the past ten years several new species have been discovered and described. Nonetheless, various investigations have given us a much clearer understanding of the domesticated species. The studies of Pickersgill (1966, 1971), Smith & Heiser (1957), Hirose, Nishi & Takashima (1960), Novak & Betlach (1971), and Eshbaugh (1964, 1970, 1978 in press), just to mention a few, have contributed in determining something of the breeding behaviour and relationships of several wild and domesticated species. The cytological investigations of Ohta (1962), Chenna-veeraiah & Habib (1966), Shopova (1966), and Pickersgill (1971) have been useful for grouping several taxa on the basis of specific karyotypes. Pickersgill's research has suggested the potential use of the karyotype in establishing the centre of origin for the domesticated taxon of *C. annuum* from the extensive geographic range of the hypothetical ancestral taxon. Numerical taxonomic analysis has been attempted on a limited basis (Eshbaugh, 1964, 1970) (see also Pickersgill *et al.*, this Volume, p. 679). Systematic biochemical research has been confined to a single study by Ballard, McClure, Eshbaugh & Wilson (1970) using flavonoid pattern analysis. However, all the aforementioned investigations have had one phenomenon in common. They tended to separate the domesticated species into distinct groupings.

In the belief that still another approach, that of horizontal starch gel electro-phoresis, might be useful in gaining further insights into the evolutionary relation-ships of the domesticated chili peppers, we initiated such an investigation in 1975. The precedent and usefulness of this technique in investigating the evolutionary systematics of domesticated plant species within the Solanaceae has been estab-lished by such studies as Sheen's (1972) research on *Nicotiana* species including various varieties and hybrids, Natarella & Sink's (1975) investigation of several species of *Petunia* and their cultivars, and Rick & Fobes' (1975) study of culti-vated *Lycopersicon* and its related species.

MATERIALS AND METHODS

Our investigation of chili peppers included the following domesticated taxa and associated wild species: *C. annuum* var. *annuum* and *C. annuum* var. *aviculare*; *C. baccatum* var. *pendulum*, *C. baccatum* var. *baccatum*, and *C. praetermissum*; *C. chinense* and *C. frutescens*; and *C. pubescens*, *C. eximium*, and *C. cardenasii*. We also included wild *C. chacoense* and an unnamed taxon from the region of Ayacucho, Peru, in our study material. This unnamed species is currently under investigation by Pickersgill. The species was first collected and brought to the author's (W.H.E.) attention by Paul G. Smith in 1959. Smith later suggested the name *C. tovari* for the taxon. Heiser (1976) recently used this name for this species but no formal Latin description or binomial has ever been published for the taxon. Therefore, we have chosen to refer to this material as an unnamed species.

To date, our investigation has included 807 individuals from approximately 274 different collections. One of the difficulties encountered in working with domesti-cated species is the definition of a population. In this investigation the domesti-cated population was defined as the individuals that came from a geographic zone where the potential for gene exchange existed. In some cases a population might be far removed from its place of growth and origin. For instance, the three different collections of *C. pubescens* from the coast of Peru all had their origin from the Arequipa area. These collections attained their geographic distribution pattern through marketing practices and are thus removed from the true centre of growth. Defining populations in these terms is risky especially when one is dependent on an interview process to determine the place of origin of a domesticated taxon. In the case of the wild related species true populations could be obtained by sampling from the wild. However, even some of the wild material came from markets and we were again dependent on the interview process to corroborate the place of origin of these samples (see Figs 55.1 and 55.2 for population locations, the actual geographic range of the species can be seen in a recent paper by Eshbaugh, 1975).

Seeds of the various taxa of *Capsicum* were germinated and grown to maturity in the greenhouse and experimental field at Miami University. New growth (leaf tissue) on each plant was harvested and homogenized in a 0.5 M phosphate buffer (pH 7.0) with 0.5 ml of mercaptoethanol added per 100 ml of buffer (Levin, 1975). Homogenates were centrifuged for one hour at 25,000 **g** and 0° C. The supernatant was decanted and stored at −60° C until subjected to electrophoresis.

Standard methods of horizontal starch gel electrophoresis utilizing the techniques of Selander *et al.* (1971) were performed, with the following exceptions: The modified lithium hydroxide buffer was adapted from Ashton & Braden (1961) and Almgard & Clapham (1975). The enzymes catalase and peroxidase were stained using the techniques of Almgard & Norman (1970). Gels made with buffers 1 and 2 were at a 12.5% (w/v) concentration and those made with buffer 3 at a 15% concentration. The following 15 enzymes were resolved: esterase (EST), catalase (CAT), peroxidase (PRX), general protein (TP), 6-phosphogluconate dehydrogenase (6-PGDH), lactate dehydrogenase (LDH), phosphoglucomutase (PGM), glutamate dehydrogenase (GDH), malate dehydrogenase (MDH), indophenol oxidase (IPO), glutamate oxaloacetate transaminase (GOT), isocitrate dehydrogenase (IDH), malic enzyme (ME), phosphoglucoisomerase (PGI), and xanthine dehydrogenase (XDH). The enzymes stained, buffer systems used, and number of loci scored per enzyme are listed in Table 55.1.

Genetic distance was determined using Nei's (1972) measure of genetic distance.

RESULTS

Zones of activity which varied independently of other zones of activity on a gel were considered to be controlled by single gene loci. Either one, two, or three banded patterns within each zone of activity (locus) were found. A single banded pattern which was of identical mobility for the entire genus (for example TP) was considered to be encoded by a monomorphic allele. Phenotypes of single bands or double bands, each of equal intensity (for example PGM) are consistent with what one would expect if the protein is a monomer. The single band would represent a homozygote and the double band would represent a heterozygote. Patterns consisting of a single band or three bands, with the middle band being most intense, are expected if the protein is active as a dimer. Again, the single band represents a homozygote and the three banded pattern would indicate a heterozygote.

In those enzymes in which several loci were present each locus was assigned a different number. The locus with the greatest anodal mobility from the origin was

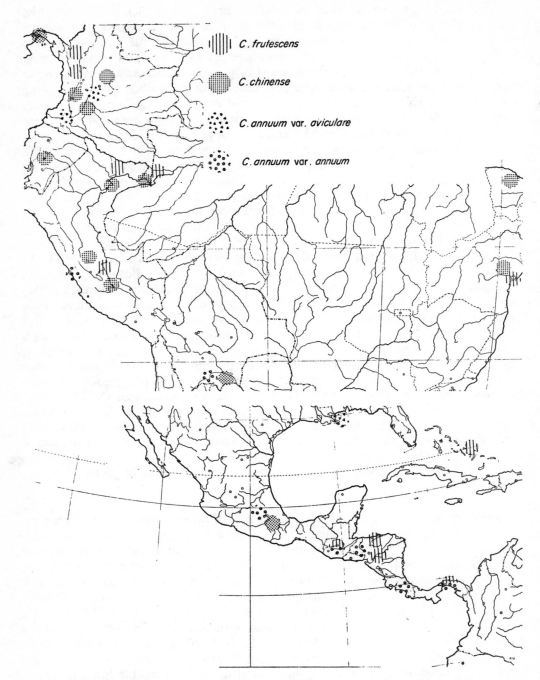

Figure 55.1. Distribution of populations included in this investigation of *C. annuum* var. *annuum*, *C. annuum* var. *aviculare*, *C. chinense*, and *C. frutescens* in South America (upper) and Central America, Mexico, and the West Indies (lower).

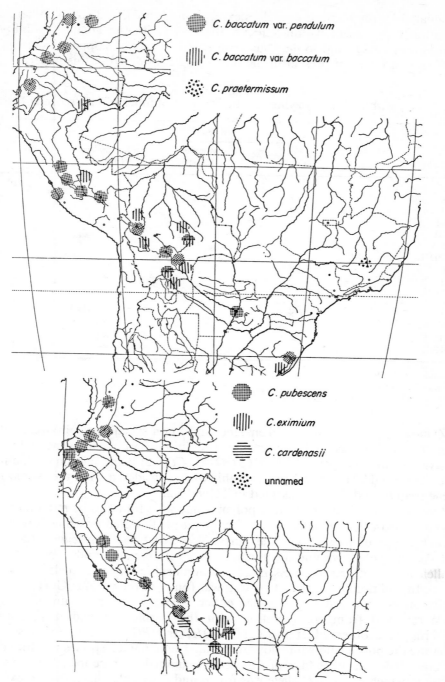

Figure 55.2. Distribution of populations included in this investigation of *C. baccatum* var. *pendulum*, *C. baccatum* var. *baccatum*, and *C. praetermissum* in South America (upper); *C. pubescens*, *C. eximium*, *C. cardenasii*, and an unnamed species in South America (Lower).

designated by the number one, the next as number two, and so on. Where allelic variation was present within a locus, the allele with the greatest mobility was called *a*, the next *b*, and so on. One or more letters may be skipped, indicating that those alleles were not present in the taxa being considered in this paper.

Table 55.1. Enzyme, gel buffer, and number of loci scored

	Buffer			
Enzyme[a]	LiOH[1]	Modified LiOH[2]	Tris maleic EDTA[1]	No. Loci Scored
TP	×			1
PRX	×			1
CAT	×			1
GDH		×		1
MDH		×		2
LDH		×		1
6-GPDH		×		1
PGI			×	1
IDH			×	3
PGM		×		2
IPO		×		1
EST	×			4
ME			×	1
GOT			×	5
XDH			×	1

[1] Selander *et al.* (1971).
[2] Ashton & Braden (1961).
[a] For explanation see text p. 703.

Zymograms of 15 proteins (14 enzymes and 1 non-specific protein) were obtained after electrophoresis and histochemical staining. Twenty-six genetic loci were analyzed with four loci (*TP*, *EST*-6, *GOT*-3, *GOT*-5) being monomorphic for the same allele and 22 loci being polymorphic when considered over the whole genus. Gene frequency data are presented in Table 55.2.

Many of those loci which were polymorphic were found to be monomorphic or essentially so when considered only for the purple flowered or white flowered taxa or subgroups of those taxa. The purple flowered taxa (*Capsicum cardenasii*, *C. eximium*, *C. pubescens*, and the unnamed species) were all found to be fixed for the *a* allele at *GDH*, and to be essentially monomorphic for the *c* allele at the *EST-7* locus. In addition, a consistent pattern was found at the various GOT loci for the entire purple flowered group. When either *C. pubescens* or the unnamed species were removed from consideration with the purple flowered taxa, consistent patterns were observed at the *CAT*, *LDH*, and *PGM*-2 loci, which served to separate the purple and white flowered taxa. The unnamed species was found to be different from all other taxa of *Capsicum* examined at three loci. It was the only taxon in which the XDH^a allele was found and it was also fixed for the extremely rare MDH-1^a and PRX-2^b alleles.

The white flowered taxa were found to be not only electrophoretically distinct from the purple flowered taxa, but also to contain two subgroups readily dis-

Table 55.2. Allele frequencies at each of the Polymorphic loci[1]

Enzyme	Allele	Taxon											
		1 card	2 exim	3 pube	4 unn	5 bacb	6 bacp	7 prae	8 chac	9 anav	10 anan	11 frut	12 chin
PRX-2	a	0.89	0.99	1.00		1.00	1.00	1.00	1.00	1.00	1.00	1.00	1.00
	b	0.11	0.01		1.00								
CAT	a	1.00	1.00										
	b			1.00	1.00	1.00	1.00	1.00	1.00	1.00	1.00	1.00	1.00
GDH	a	1.00	1.00	1.00	1.00								
	c					1.00	1.00	1.00	1.00	1.00	1.00	1.00	1.00
MDH-1	a	0.05	0.02	0.04	1.00					0.03	0.04		
	b	0.95	0.92	0.96		1.00	1.00	1.00	1.00	0.97	0.96	1.00	1.00
	c		0.06										
MDH-2	a	0.03	0.05	0.03									
	b	0.97	0.95	0.97	1.00	1.00	1.00	1.00	1.00	1.00	1.00	1.00	1.00
LDH	a	1.00	1.00	1.00									
	b				1.00	1.00	1.00			1.00	1.00	1.00	1.00
6-PGDH	a	0.54	0.83	0.95	1.00	0.99	0.99	1.00	1.00	1.00	1.00	1.00	0.88
	b	0.46	0.17	0.05		0.01	0.01						0.12
PGI-2	a			1.00						0.28	0.11		
	b									0.42	0.74	0.90	0.04
	c					1.00	1.00	1.00		0.30	0.14	0.10	0.96
IDH-1	a	1.00										0.21	
	b		1.00	1.00	1.00	1.00	1.00	1.00	1.00	0.95	0.35	0.79	0.93
	c									0.05	0.65		0.07
IDH-2	a	1.00								0.03	0.31		0.93
	b			1.00		1.00	1.00	1.00	1.00	0.93	0.69	1.00	0.07
	c									0.04			
IDH-3	a									0.07	0.21		
	b			1.00		1.00	1.00	1.00		0.93	0.79	1.00	1.00
PGM-1	a	0.07	0.05										
	b	0.93	0.95	1.00		0.99	1.00	1.00	1.00	0.23	0.69	1.00	0.89
	c				1.00	0.01				0.77	0.31		0.11
PGM-2	a	0.02		0.12									
	b	0.81	0.95	0.50		0.04		1.00			0.37		
	c	0.17	0.05	0.13	1.00	0.96	1.00		1.00	1.00	0.73	1.00	1.00
	d			0.24									
IPO-1	a							0.22		1.00	1.00	0.77	0.65
	b			0.41		1.00	1.00	0.78	1.00			0.23	0.35
	c		1.00	0.59	1.00								
EST-3	a				0.5	0.11	0.04			0.14	0.01	0.96	0.34
	b	1.00	1.00	1.00	0.5	0.78	0.92	0.07	1.00	0.82	0.93	0.04	0.66
	c					0.11	0.03	0.62		0.04	0.06		
	d							0.31					
EST-7	a			0.01			0.02						
	b			0.06		1.00	0.97	1.00		1.00	1.00	1.00	1.00
	c	1.00	1.00	0.93	1.00				1.00				
EST-8	a	1.00	0.40	0.01	0.94	1.00	1.00	1.00		0.61	1.00	1.00	1.00
	b		0.10	0.99	0.06					0.39			
	c		0.50										
ME	a	1.00	0.90	1.00	1.00	0.92	1.00		1.00			0.05	
	b		0.10			0.08				1.00	0.94	0.95	1.00
	c							1.00			0.06		
GOT-1	a		0.31	1.00									
	n	1.00	0.69		1.00	1.00	1.00	1.00	1.00	1.00	1.00	1.00	1.00
GOT-2	a		0.22			0.98	1.00	1.00					
	n	1.00	0.78	1.00	1.00	0.02			1.00	1.00	1.00	1.00	1.00
GOT-4	a	1.00	1.00	1.00	1.00	0.03			1.00	1.00	1.00	1.00	1.00
	n					0.97	1.00	1.00					
XDH	a				1.00								
	b	1.00	1.00	1.00		1.00	1.00	1.00	1.00	1.00	1.00	1.00	1.00

[1] The taxon number corresponds to the taxa in Table 55.3. No data have been included in this table where patterns were not consistently resolved. The *n* allele at the GOT loci signifies a null allele.

tinguishable at two GOT loci. *Capsicum baccatum* var. *baccatum*, *C. baccatum* var. *pendulum*, and *C. praetermissum* were characterized by being essentially mono-morphic (major allele at a frequency of 0.95 or higher) for $GOT-2^a$ and $GOT-4^n$. *Capsicum annuum* var. *annuum*, *C. annuum* var. *aviculare*, *C. chinense*, and *C. frutescens* were fixed for the null allele at $GOT-2$ and the *a* allele at $GOT-4$.

DISCUSSION

When discussing the systematics, genetics, and evolution of a group that contains both wild and cultivated forms, there are several facts which must be considered. The first is that domestication can, and in the case of the peppers does, mean travel. The original ranges of the various taxa have been extended and forms which are supposed to have been at one time allopatric are now found in sympatry. These migrations were at first achieved by the Indians, and later by the Spanish and modern man. As a result the opportunity for interbreeding now exists where it possibly did not originally. It has, however, been postulated that genetic separation between wild and domesticated forms is maintained by a strong tendency towards self-pollination in the domesticates, as well as geographical and agricultural isolation (Eshbaugh, 1970, 1975).

The second fact that must be considered is that under domestication certain characters have been selected for which lead to a convergence of morphological characteristics and make taxonomy more difficult (D'Arcy & Eshbaugh, 1974). Wild peppers typically have small, red, erect, deciduous fruits. Under cultivation the domesticated plants become characterized by large fruits, with fewer fruits per plant, and a persistent, pendent fruit with a variety of colours (Pickersgill, 1969).

Interspecific breeding data resulting from many independent investigations have been at best somewhat contradictory. Smith and Heiser (1957), Hirose *et al.* (1960), and Eshbaugh (1964; and in press) have suggested that *C. pubescens* is genetically isolated from the other domesticated taxa. However, Chennaveeraiah and Habib (1966) have reported obtaining F_1's in crosses between *C. pubescens* and the other three domesticates. In each case only a small number of viable seeds in a single fruit was obtained. With the exception of the above report, the purple flowered taxa are generally considered genetically isolated from the white flowered taxa. However, Eshbaugh (1964) did obtain a single successful cross between *C. eximium* and *C. baccatum*.

Experimental hybrids, which exhibited considerable sterility, have been reported between *C. baccatum* and *C. frutescens* (Pickersgill, 1971; Heiser *et al.*, 1971), but no natural hybrids between the two taxa have ever been reported. Likewise, hybrids have been experimentally produced between *C. frutescens* and *C. praetermissum* and *C. frutescens* and *C. annuum*, with appreciable sterility in the latter cross (Smith & Heiser, 1951; Heiser & Smith, 1958; Heiser *et al.*, 1971). The white flowered taxa form a group of six cross-fertile species (*C. baccatum*, *C. praetermissum*, *C. annuum*, *C. frutescens*, *C. chinense*, and rarely *C. chacoense*) which usually form sterile F_1 progeny.

Pickersgill (1971) has noted that interspecific crosses between wild species are more likely to result in fertile hybrids than crosses between the various domesticated taxa. The crossing relationships between some of the domesticated and wild *Capsicum* are summarized in Fig. 55.3.

As mentioned in the results section, electrophoretically, three main groups were found in the genus. These three groups will be discussed independently. Genetic distance (D) between all taxa of *Capsicum* is presented in Table 55.3.

The purple flowered Capsicum

There are four species which can be considered in this group. These include *Capsicum cardenasii*, *C. eximium*, *C. pubescens*, and an unnamed species. All are wild except *C. pubescens*. *Capsicum cardenasii* and the unnamed species both have a very narrow geographic range. The other two species, *C. eximium* and *C. pubescens*, have much wider ranges.

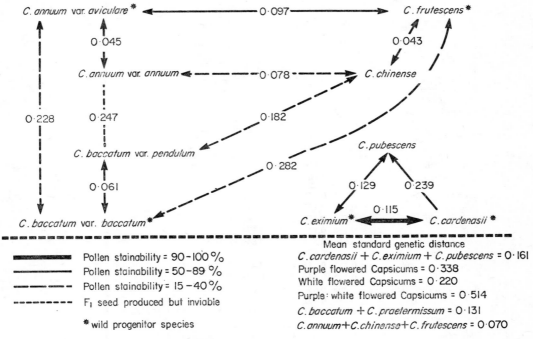

Figure 55.3. Summary crossing diagram from Eshbaugh (1975), with Nei's standard genetic distance coefficients, based on starch gel electrophoresis data, superimposed to show the correlation of these two independent analyses.

Breeding data have led to the suggestion that *C. cardenasii* and *C. eximium* are the wild taxa from which *C. pubescens* was derived (Eshbaugh, 1978, in press). *Capsicum cardenasii*, which is self-incompatible, can be easily crossed to *C. eximium* with the resulting hybrids exhibiting 90–100% pollen viability. Crosses between *C. cardenasii* or *C. eximium* and *C. pubescens* are also relatively easy to make although when *C. cardenasii* is used as the female parent the cross becomes much more difficult (Eshbaugh, 1964, 1975, in press).

Genetic data (Table 55.3) yield much the same results as the breeding data. Based on the standard genetic distance (D), *C. cardenasii* and *C. eximium*, and *C. eximium* and *C. pubescens* are relatively similar, while *C. cardenasii* and *C.*

pubescens are somewhat more distantly related. These data are consistent with the suggestion by Eshbaugh (in press) that *C. cardenasii*, *C. eximium*, and *C. pubescens* compose a single species complex. If so, then *C. cardenasii* may well represent a morphological extreme of *C. eximium*.

The unnamed species has been the subject of limited research. Preliminary crossing data (unpublished) indicate that it does not readily cross with any other species of *Capsicum*. Electrophoretically, this species is not closely related to any other taxon in the genus. It is possible that this species is highly representative of the ancestral genome based on genetic distance, although much more work is necessary before this can be verified.

Table 55.3. Nei's (1972) standard genetic distance (and standard error[1]) between taxa of *Capsicum*

Taxon	1	2	3	4	5	6	7	8	9	10	11
1. *C. cardenasii*											
2. *C. eximium*	0.115										
	0.058										
3. *C. pubescens*	0.239	0.129									
	0.106	0.065									
4. Unnamed species	0.506	0.514	0.542								
	0.161	0.162	0.176								
5. *C. baccatum* var *baccatum*	0.641	0.525	0.584	0.538							
	0.188	0.157	0.179	0.172							
6. *C. baccatum* var *pendulum*	0.593	0.480	0.529	0.560	0.061						
	0.178	0.147	0.166	0.174	0.046						
7. *C. praetermissum*	0.655	0.515	0.609	0.846	0.197	0.134					
	0.192	0.155	0.183	0.237	0.087	0.071					
8. *C. chacoense*	0.311	0.355	0.383	0.404	0.275	0.194	0.320				
	0.115	0.120	0.134	0.138	0.110	0.091	0.124				
9. *C. annuum* var *aviculare*	0.568	0.488	0.457	0.487	0.228	0.232	0.331	0.261			
	0.170	0.152	0.147	0.159	0.098	0.970	0.116	0.103			
10. *C. annuum* var *annuum*	0.451	0.446	0.464	0.508	0.261	0.247	0.307	0.245	0.045		
	0.147	0.139	0.152	0.162	0.102	0.099	0.106	0.098	0.020		
11. *C. frutescens*	0.506	0.480	0.448	0.570	0.282	0.262	0.282	0.225	0.097	0.091	
	0.153	0.148	0.156	0.175	0.105	0.102	0.112	0.094	0.047	0.048	
i12. *C. chinense*	0.501	0.435	0.451	0.562	0.243	0.182	0.224	0.160	0.066	0.078	0.043
	0.158	0.137	0.147	0.172	0.098	0.083	0.099	0.079	0.033	0.034	0.027

[1] Standard error (Nei & Roychoudhury, 1974) appears immediately below standard genetic distance.

The white flowered taxa

The white flowered taxa as a group are genetically distinct from the purple flowered group. At the same time, the white flowered taxa form a somewhat cohesive group with a mean standard genetic distance (0.220) which is less than the D between *Capsicum cardenasii* and *C. pubescens*. However, based on standard genetic distance, there appear to be two subgroups within the broader white flowered group.

Subgroup 1

The first white flowered subgroup is composed of the *Capsicum baccatum* complex and *C. praetermissum*. *Capsicum baccatum* var. *baccatum* is a wild taxon and *C. baccatum* var. *pendulum* is the cultivar. The *Capsicum baccatum* complex had long been considered to consist of two separate species, *C. microcarpum* (the wild form) and *C. pendulum* (the domesticate). However, Eshbaugh (1964, 1968, 1970) combined the two taxa under the name *C. baccatum*, based on crossing data and numerical analysis.

Capsicum praetermissum is a wild taxon from Brazil (Heiser & Smith, 1958). Hunziker (1971) reduced this species to a variety of *C. baccatum*. Eshbaugh (1975) mentioned unpublished data which suggested that *C. baccatum* and *C. praetermissum* possess identical flavonoids. Electrophoretic data, though preliminary, tend to support the idea of a close relationship between these three taxa.

Subgroup 2

This subgroup consists of the *Capsicum annuum* complex and both *C. chinense* and *C. frutescens*. Emboden (1961) and Pickersgill (1966, 1971) have demonstrated a very close relationship between *Capsicum annuum* var. *annuum* (the domesticated form) and *C. annuum* var. *aviculare* (the wild taxon) based on cytogenetic and breeding data.

Capsicum chinense is a cultivated species which is considered most closely related to the wild *C. frutescens*. Pickersgill (1966, 1971) has suggested that these two species are so close as to represent a single wild-domesticated species pair much the same as the *C. baccatum* or *C. annuum* complexes. Electrophoretic data tend to suggest the close relationship of *C. chinense* and *C. frutescens*, and also a very close relationship between the *C. chinense/C. frutescens* complex and *C. annuum*. This relationship will be further explored in a manuscript now in preparation.

There remains one species to be discussed, *Capsicum chacoense*. This is a wild, white flowered species which is found in southern Paraguay, northern Argentina, and Bolivia (Hunziker, 1950). It is not a well studied species. Data included in this paper are highly preliminary because of a small sample size. However, the data suggest that *C. chacoense* is the white flowered taxon which is most closely related to the purple flowered taxa, and is almost equally distant from the other white flowered taxa.

ACKNOWLEDGEMENTS

We would like to thank the National Geographic Society (Grant Nos. 901 and 1025), The American Philosophical Society (Penrose Fund 5479), Society of the Sigma Xi, and The National Science Foundation (GB-31932 and DEB76-11478) for their financial assistance of this investigation. We are grateful to the National Science Foundation, The Miami University Alumni Association, and Miami University for travel support to attend the International Symposium on the Biology and Taxonomy of the Solanaceae.

Finally, we are indebted to Diana K. McLeod for her assistance in collecting field material and data.

REFERENCES

ALMGARD, C. & CLAPHAM, D., 1975. Isozyme variation distinguishing 18 *Avena* cultivars grown in Sweden. *Swedish Journal of Agricultural Research, 5:* 61–67.

ALMGARD, C. & NORMAN, T., 1970. Biochemical techniques as an aid to distinguish some cultivars of barley and oats. *Agriculture, Horticulture and Genetics, 28:* 117–123.

ASHTON, G. C. & BRADEN, A. W. H., 1961. Serum beta-globulin polymorphism in mice. *Australian Journal of Biological Science, 14:* 248–253.

BAILEY, L. H., 1923. *Capsicum. Gentes Herbarum, 1:* 128–129.

BALLARD, R. E., MCCLURE, J. W., ESHBAUGH, W. H. & WILSON, K. G., 1970. A chemosystematic study of selected taxa of *Capsicum. American Journal of Botany, 57:* 225–233.

CHENNAVEERAIAH, M. S. & HABIB, A. F., 1966. Recent advances in the cytogenetics of *Capsicums. Proceedings of the Autumn School Botany, 1966:* 69–90.

D'ARCY, W. G. & ESHBAUGH, W. H., 1974. New world peppers (*Capsicum*—Solanaceae) north of Colombia: A resumé. *Baileya, 19:* 93–105.

DAVENPORT, W. A., 1970. Progress report on the domestication of *Capsicum* (chili peppers). *Proceedings, Association of American Geographers, 2:* 46–47.

DUNAL, F. M., 1852. *Capsicum.* In A. P. de Candolle (Ed.), *Prodromus Systematis Naturalis Regni Vegetabilis, 13:* 411–429. Paris: Masson.

EMBODEN, W. A., JR., 1961. A preliminary study of the crossing relationships of *Capsicum baccatum. Butler University Botanical Studies, 14:* 1–5.

ESHBAUGH, W. H., 1964. *A Numerical, Taxonomic and Cytogenetic Study of Certain Species of the Genus* Capsicum. Ph.D. dissertation, Indiana University, Bloomington.

ESHBAUGH, W. H., 1968. A nomenclatural note on the genus *Capsicum. Taxon, 17:* 51–52.

ESHBAUGH, W. H., 1970. A biosystematic and evolutionary study of *Capsicum baccatum* (Solanaceae). *Brittonia, 22:* 31–43.

ESHBAUGH, W. H., 1975. Genetic and biochemical systematic studies of chili peppers (*Capsicum*—Solanaceae). *Bulletin of the Torrey Botanical Club, 102:* 396–403.

ESHBAUGH, W. H., 1978. A biosystematic and evolutionary study of the *Capsicum pubescens* complex. *National Geographical Society Research Reports, 1970* (in press).

HEISER, C. B., JR., 1976. Peppers—*Capsicum* (Solanaceae). In N. W. Simmonds (Ed.), *Evolution of Crop Plants:* 265–268. London: Longmans.

HEISER, C. B., JR., ESHBAUGH, W. H. & PICKERSGILL, BARBARA, 1971. The domestication of *Capsicum*—A reply to Davenport. *Professional Geographer, 23:* 169–170.

HEISER, C. B., JR. & SMITH, P. G., 1958. New species of *Capsicum* from South America. *Brittonia, 10:* 194–201.

HIROSE, T., NISHI, S. & TAKASHIMA, S., 1960. Studies on the inter-species crossing in *Capsicum.* I. Crossability. *Scientific Reports of Kyoto Prefectural University of Agriculture, 12:* 40–46.

HUNZIKER, A. T., 1950. Estudios sobre Solanaceae. I. Sinopsis de las especies silvestres de *Capsicum* de Argentina y Paraguay. *Darwiniana, 9:* 225–247.

HUNZIKER, A. T., 1971. Estudios sobre Solanaceae. VII. Contribución al conocimiento de *Capsicum* y generos afines (*Witheringia, Acnistus, Athenaea,* etc.) Tercera Parte. *Kurtziana, 6:* 241–259.

JETT, S. C., 1973. Comment on Pickersgill's "Cultivated plants as evidence for cultural contacts." *American Antiquity, 38:* 223–335.

LEVIN, D. A., 1975. Interspecific hybridization, heterozygosity, and gene exchange in *Phlox. Evolution, 29:* 37–51.

NATARELLA, N. J. & SINK, K. C., JR., 1975. Electrophoretic analysis of proteins and peroxidases of selected *Petunia* species and cultivars. *Botanical Gazette, 136:* 20–26.

NEI, M., 1972. Genetic distance between populations. *American Naturalist, 106:* 283–292.

NEI, M. & ROYCHOUDHURY, A. K., 1974. Sampling variances of heterozygosity and genetic distance. *Genetics, 76:* 379–390.

NOVÁK, F. & BETLACH, J., 1971. Systematika a nomenklatorické poznámky k rodu *Capsicum* L. (Solanaceae). *Genetika a Šlechtěni, 7:* 141–150.

OHTA, Y., 1962. Karyotype analysis of *Capsicum* species. *Seiken Zihô, 13:* 93–99.

PICKERSGILL, BARBARA, 1966. *The Variability and Relationships of* Capsicum chinense Jacq. Ph.D. dissertation, Indiana University, Bloomington.

PICKERSGILL, BARBARA, 1969. The domestication of chili peppers. In P. J. Ucko & G. W. Dimbleby (Eds), *The Domestication and Exploitation of Plants and Animals:* 443–450. London: Duckworth.

PICKERSGILL, BARBARA, 1971. Relationships between weedy and cultivated forms in some species of chili peppers (genus *Capsicum*). *Evolution, 25:* 683–691.

PICKERSGILL, BARBARA, 1972. Cultivated plants as evidence for cultural contacts. *American Antiquity, 37:* 97–104.

RICK, C. M. & FOBES, J. F., 1975. Allozyme variation in the cultivated tomato and closely related species. *Bulletin of the Torrey Botanical Club, 102:* 376–384.

SELANDER, R. K., SMITH, M. H., YANG, SUH Y., JOHNSON, W. E. & GENTRY, J. B., 1971. Biochemical polymorphism and systematics in the genus *Peromyscus*. I. Variation in the old-field mouse (*Peromyscus polinotus*). *University of Texas Publication, 7103:* 49–90.

SHEEN, S. J., 1972. Isozymic evidence bearing on the origin of *Nicotiana tabacum* L. *Evolution, 26:* 143–154.

SHOPOVA, M., 1966. Studies in the genus *Capsicum*. I. Species differentiation. *Chromosoma, 19:* 340–348.

SMITH, P. G. & HEISER, JR., C. B., 1951. Taxonomic and genetic studies on the cultivated peppers, *Capsicum annuum* L. and *C. frutescens* L. *American Journal of Botany, 38:* 362–368.

SMITH, P. G. & HEISER, JR., C. B., 1957. Breeding behavior of cultivated peppers. *Proceedings of the American Society of Horticultural Science, 70:* 286–290.

X. Resolutions

56. Resolutions

The following resolutions were passed during the last session of the Symposium on 17 July (see also: J. G. Hawkes, R. N. Lester, & D. C. D. Langley, 1977. *Solanaceae Newsletter*, No. 4, May, 1977. Department of Plant Biology, University of Birmingham):

This Conference recommends:

(1) That botanists be encouraged to collect Solanaceae material in the form of adequate seed samples from natural populations and deposit suitable material in a national or international seed bank for long-term storage.

(2) That such material should be made freely available to all *bona fide* research workers.

(3) That care should be taken when collecting material under (1) above to avoid over-collection of rare or endangered species unless by so doing they can be preserved in seed banks rather than dying out completely in the wild.

(4) That collections of woody material be sampled wherever possible and that a wood collection with voucher herbarium material be established in a suitable institute.

(5) That flower and fruit collections in spirit or pickling solution be made wherever possible on collecting expeditions and be deposited in a suitable institute, together with voucher herbarium specimens.

(6) That Solanaceae workers, when making herbarium or living collections, be encouraged to offer duplicate material to colleagues and/or institutions in the countries where the collections were made, according to the present international agreement.

(7) That, in view of the value of the present Symposium to workers with this family, a future Solanaceae conference be held, possibly in 1981, in a country and city to be determined in due course.

(8) That, in connection with (7) above, an international committee be established to consider the organization of the next meeting.

(9) That publication of the Solanaceae Newsletter be carried on, and that the Department of Plant Biology at Birmingham, U.K. be asked to continue editing it for the next three years.

(10) That all workers in the Solanaceae be encouraged to support the News-letter by supplying information requested and occasional notes on obituaries, new findings and other material of interest to the scientific community and by paying subscription costs upon request.

(11) That co-operative research be initiated in various taxonomic groups between workers in a number of different disciplines and that the Solanaceae Newsletter should report on such arrangements and the results of these investi-gations from time to time.

(12) That when name changes are considered these be reported first in a pre-liminary form in the Solanaceae Newsletter, inviting comments from workers with the taxa concerned. Comments received might then be discussed by the proposer of such name changes before valid publication is made.

XI. Indexes

Taxonomic index

General index